W. Greiner · J. Reinhardt

QUANTUM ELECTRODYNAMICS

Greiner
Quantum Mechanics
An Introduction 3rd Edition

Greiner
Quantum Theory
Special Chapters
(in preparation)

Greiner · Müller
Quantum Mechanics
Symmetries 2nd Edition

Greiner
Relativistic Quantum Mechanics
Wave Equations

Greiner · Reinhardt
Field Quantization
(in preparation)

Greiner · Reinhardt
Quantum Electrodynamics
2nd Edition

Greiner · Schäfer
Quantum Chromodynamics

Greiner · Maruhn
Nuclear Models
(in preparation)

Greiner · Müller
Gauge Theory of Weak Interactions

Greiner
Mechanics I
(in preparation)

Greiner
Mechanics II
(in preparation)

Greiner
Electrodynamics
(in preparation)

Greiner · Neise · Stöcker
Thermodynamics and Statistical Mechanics

Walter Greiner · Joachim Reinhardt

QUANTUM ELECTRODYNAMICS

With a Foreword by
D. A. Bromley

Second Corrected Edition
With 155 Figures,
and 52 Worked Examples and Exercises

Springer

Professor Dr. Walter Greiner
Dr. Joachim Reinhardt

Institut für Theoretische Physik der
Johann Wolfgang Goethe-Universität Frankfurt
Postfach 11 19 32
D-60054 Frankfurt am Main
Germany

Street address:

Robert-Mayer-Strasse 8–10
D-60325 Frankfurt am Main
Germany

Title of the original German edition: *Theoretische Physik,* Band 7: Quantenelektrodynamik, 2., überarbeitete und erweiterte Auflage 1994 © Verlag Harri Deutsch, Thun 1984, 1994

ISBN 3-540-58092-1 2. Auflage Springer-Verlag Berlin Heidelberg New York
ISBN 0-387-58092-1 2nd edition Springer-Verlag New York Berlin Heidelberg

ISBN 3-540-55802-0 1. Auflage (broschiert) Springer-Verlag Berlin Heidelberg New York
ISBN 0-387-55802-0 1st edition (Softcover) Springer-Verlag New York Berlin Heidelberg

ISBN 3-540-52078-3 1. Auflage (gebunden) Springer-Verlag Berlin Heidelberg New York
ISBN 0-387-52078-3 1st edition (Hardcover) Springer-Verlag New York Berlin Heidelberg

This volume was originally published with the title: *Theoretical Physics* – Text and Exercise Books Vol. 4.

CIP data applied for

Printed and bound by Hamilton Printing Co., Rensselaer, NY., U.S.A.

9 8 7 6 5 4 3

Typesetting: Camera ready copy from the authors using a Springer $\mathrm{T_EX}$ macro package
Production Editor: P. Treiber
SPIN 10123143 56/3140 - 5 4 3 2 - Printed on acid-free paper

Foreword to Earlier Series Editions

More than a generation of German-speaking students around the world have worked their way to an understanding and appreciation of the power and beauty of modern theoretical physics – with mathematics, the most fundamental of sciences – using Walter Greiner's textbooks as their guide.

The idea of developing a coherent, complete presentation of an entire field of science in a series of closely related textbooks is not a new one. Many older physicists remember with real pleasure their sense of adventure and discovery as they worked their ways through the classic series by Sommerfeld, by Planck and by Landau and Lifshitz. From the students' viewpoint, there are a great many obvious advantages to be gained through use of consistent notation, logical ordering of topics and coherence of presentation; beyond this, the complete coverage of the science provides a unique opportunity for the author to convey his personal enthusiasm and love for his subject.

The present five-volume set, *Theoretical Physics*, is in fact only that part of the complete set of textbooks developed by Greiner and his students that presents the quantum theory. I have long urged him to make the remaining volumes on classical mechanics and dynamics, on electromagnetism, on nuclear and particle physics, and on special topics available to an English-speaking audience as well, and we can hope for these companion volumes covering all of theoretical physics some time in the future.

What makes Greiner's volumes of particular value to the student and professor alike is their completeness. Greiner avoids the all too common "it follows that . . . " which conceals several pages of mathematical manipulation and confounds the student. He does not hesitate to include experimental data to illuminate or illustrate a theoretical point and these data, like the theoretical content, have been kept up to date and topical through frequent revision and expansion of the lecture notes upon which these volumes are based.

Moreover, Greiner greatly increases the value of his presentation by including something like one hundred completely worked examples in each volume. Nothing is of greater importance to the student than seeing, in detail, how the theoretical concepts and tools under study are applied to actual problems of interest to a working physicist. And, finally, Greiner adds brief biographical sketches to each chapter covering the people responsible for the development of the theoretical ideas and/or the experimental data presented. It was Auguste Comte (1798–1857) in his *Positive Philosophy* who noted, "To understand a science it is necessary to know its history". This is all too often forgotten in modern physics teaching and the bridges that Greiner builds to the pioneering figures of our science upon whose work we build are welcome ones.

Greiner's lectures, which underlie these volumes, are internationally noted for their clarity, their completeness and for the effort that he has devoted to making physics an integral whole; his enthusiasm for his science is contagious and shines through almost every page.

These volumes represent only a part of a unique and Herculean effort to make all of theoretical physics accessible to the interested student. Beyond that, they are of enormous value to the professional physicist and to all others working with quantum phenomena. Again and again the reader will find that, after dipping into a particular volume to review a specific topic, he will end up browsing, caught up by often fascinating new insights and developments with which he had not previously been familiar.

Having used a number of Greiner's volumes in their original German in my teaching and research at Yale, I welcome these new and revised English translations and would recommend them enthusiastically to anyone searching for a coherent overview of physics.

Yale University
New Haven, CT, USA
1989

D. Allan Bromley
Henry Ford II Professor of Physics

Preface to the Second Edition

The need for a second edition of our text on Quantum Electrodynamics has given us the opportunity to implement some corrections and amendments. We have corrected a number of misprints and minor errors and have supplied additional explanatory remarks at various places. Furthermore some new material has been included on the magnetic moment of the muon (in Example 5.6) and on the Lamb shift (in Example 5.8). Finally, we have added the new Example 3.17 which explains the equivalent photon method.

We thank several colleagues for helpful comments and also are grateful to Dr. R. Mattiello who has supervised the preparation of the second edition of the book. Furthermore we acknowledge the agreeable collaboration with Dr. H. J. Kölsch and his team at Springer-Verlag, Heidelberg.

Frankfurt am Main,
July 1994

Walter Greiner
Joachim Reinhardt

Preface to the First Edition

Theoretical physics has become a many-faceted science. For the young student it is difficult enough to cope with the overwhelming amount of new scientific material that has to be learned, let alone obtain an overview of the entire field, which ranges from mechanics through electrodynamics, quantum mechanics, field theory, nuclear and heavy-ion science, statistical mechanics, thermodynamics, and solid-state theory to elementary-particle physics. And this knowledge should be acquired in just 8–10 semesters, during which, in addition, a Diploma or Master's thesis has to be worked on or examinations prepared for. All this can be achieved only if the university teachers help to introduce the student to the new disciplines as early on as possible, in order to create interest and excitement that in turn set free essential new energy. Naturally, all inessential material must simply be eliminated.

At the Johann Wolfgang Goethe University in Frankfurt we therefore confront the student with theoretical physics immediately, in the first semester. Theoretical Mechanics I and II, Electrodynamics, and Quantum Mechanics I – An Introduction are the basic courses during the first two years. These lectures are supplemented with many mathematical explanations and much support material. After the fourth semester of studies, graduate work begins, and Quantum Mechanics II – Symmetries, Statistical Mechanics and Thermodynamics, Relativistic Quantum Mechanics, Quantum Electrodynamics, the Gauge Theory of Weak Interactions, and Quantum Chromodynamics are obligatory. Apart from these a number of supplementary courses on special topics are offered, such as Hydrodynamics, Classical Field Theory, Special and General Relativity, Many-Body Theories, Nuclear Models, Models of Elementary Particles, and Solid-State Theory. Some of them, for example the two-semester courses Theoretical Nuclear Physics or Theoretical Solid-State Physics, are also obligatory.

This volume of lectures deals with the subject of *Quantum Electrodynamics*. We have tried to present the subject in a manner which is both interesting to the student and easily accessible. The main text is therefore accompanied by many exercises and examples which have been worked out in great detail. This should make the book useful also for students wishing to study the subject on their own.

When lecturing on the topic of quantum electrodynamics, one has to choose between two approaches which are quite distinct. The first is based on the general methods of quantum field theory. Using classical Lagrangian field theory as a starting point one introduces noncommuting field operators, builds up the Fock space to describe systems of particles, and introduces techniques to construct and evaluate the scattering matrix and other physical observables. This program can be realized either by the method of canonical quantization or by the use of path integrals. The theory of quantum electrodynamics in this context emerges just as a particular example of the general formalism. In the present volume, however, we do not follow this general but lengthy path; rather we use a "short cut" which arrives at the same results with less effort, and which has the advantage of great intuitive appeal. This is the propagator formalism, which was introduced by R.P. Feynman (and, less well known, by E.C.G. Stückelberg) and makes heavy use of Green's functions to describe the propagation of electrons and photons in space-time.

It is clear that the student of physics has to be familiar with both approaches to quantum electrodynamics. (In the German edition of these lectures a special volume is dedicated to the subject of field quantization.) However, to gain quick access to the fascinating properties and processes of quantum electrodynamics and to its calculational techniques the use of the propagator formalism is ideal.

The first chapter of this volume contains an introduction to nonrelativistic propagator theory and the use of Green's functions in physics. In the second chapter this is generalized to the relativistic case, introducing the Stückelberg–Feynman propagator for electrons and positrons. This is the basic tool used to develop perturbative QED. The third chapter, which constitutes the largest part of the book, contains applications of the relativistic propagator formalism. These range from simple Coulomb scattering of electrons, scattering off extended nuclei (Rosenbluth's formula) to electron–electron (Møller) and electron–positron (Bhabha) scattering. Also, processes involving the emission or absorption of photons are treated, for instance, Compton scattering, bremsstrahlung, and electron–positron pair annihilation. The brief fourth chapter gives a summary of the Feynman rules, together with some notes on units of measurement in electrodynamics and the choice of gauges.

Chapter 5 contains an elementary discussion of renormalization, exemplified by the calculation of the lowest-order loop graphs of vacuum polarization, self-energy, and the vertex correction. This leads to a calculation of the anomalous magnetic moment of the electron and of the Lamb shift. In Chap. 6 the Bethe–Salpeter equation is introduced, which describes the relativistic two-particle system.

Chapter 7 should make the reader familiar with the subject of quantum electrodynamics of strong fields, which has received much interest in the last two decades. The subject of supercritical electron states and the decay of the neutral vacuum is treated in some detail, addressing both the mathematical description and the physical implications. Finally, in the last chapter, the theory of perturbative quantum electrodynamics is extended to the treatment of spinless charged bosons.

An appendix contains some guides to the literature, giving references both to books which contain more details on quantum electrodynamics and to modern treatises on quantum field theory which supplement our presentation. We should mention that in preparing the first chapters of our lectures we have relied heavily on the textbook *Relativistic Quantum Mechanics* by J.D. Bjorken and S.D. Drell (McGraw-Hill, New York 1964).

We enjoyed the help of several students and collaborators, in particular Jürgen Augustin, Volker Blum, Christian Borchert, Snježana Butorac, Christian Derreth, Bruno Ehrnsperger, Klaus Geiger, Mathias Grabiak, Oliver Graf, Carsten Greiner, Kordt Griepenkerl, Christoph Hartnack, Cesar Ionescu, André Jahns, Jens Konopka, Georg Peilert, Jochen Rau, Wolfgang Renner, Dirk-Hermann Rischke,

Jürgen Schaffner, Alexander Scherdin, Dietmar Schnabel, Thomas Schönfeld, Stefan Schramm, Eckart Stein, Mario Vidovic, and Luke Winckelmann.

We are also grateful to Prof. A. Schäfer for his advice. The preparation of the manuscript was supervised by Dr. Béla Waldhauser and Dipl. Phys. Raffaele Mattiello, to whom we owe special thanks. The figures were drawn by Mrs. A. Steidl.

The English manuscript was copy-edited by Mark Seymour of Springer-Verlag.

Frankfurt am Main,
March 1992

Walter Greiner
Joachim Reinhardt

Contents

Contents of Examples and Exercises

1. Propagators and Scattering Theory

1.1 Introduction

In this course we will deal with quantum electrodynamics (QED), which is one of the most successful and most accurate theories known in physics. QED is the quantum field theory of electrons and positrons (the electron–positron field) and photons (the electromagnetic or radiation field). The theory also applies to the known heavy leptons (μ and τ) and, in general, can be used to describe the electromagnetic interaction of other charged elementary particles. However, these particles are also subject to nonelectromagnetic forces, i.e. the strong and the weak interactions. Strongly interacting particles (hadrons) are found to be composed of other particles, the quarks, so that new degrees of freedom become important (colour, flavour). It is believed that on this level the strong and weak interactions can be described by "non-Abelian" gauge theories modelled on QED, which is the prototype of an "Abelian" gauge theory. These are the theories of quantum chromodynamics (QCD) for the strong interaction and quantum flavourdynamics for the weak interaction. In this course we will concentrate purely on the theory of QED in its original form. Quantum electrodynamics not only is the archetype for all modern field theories, but it also is of great importance in its own right since it provides the theoretical foundation for atomic physics.

There are two approaches to QED. The more formal one relies on a general apparatus for the quantization of wave fields; the other, more illustrative, way originates from **Stückelberg** and **Feynman**, and uses the propagator formalism. Nowadays a student of physics has to know both, but it is better, both in terms of the physics and teaching, if it is obvious at an early stage why a formalism was developed and to what it can be applied. Almost everyone is keen to see as early as possible how different processes are actually calculated. Feynman's propagator formalism is the best way to achieve this. Consequently, it will be central to these lectures. References to the less intuitive but more systematic treatment of QED based on the formalism of quantum field theory are given in the appendix.

For the moment we turn to a more general discussion of scattering processes. The aim here is to calculate transition probabilities and scattering cross sections in the framework of Dirac's theory of electrons and positrons. These calculations will be exact in principle; practically, however, they will be carried out using perturbation theory, that is an expansion in terms of small interaction parameters. Because we have to describe the creation and annihilation processes of electron–positron pairs, the formalism has to be relativistic from the beginning.

In Feynman's propagator method, scattering processes are described by means of integral equations. The guiding idea is, that positrons are to be interpreted as electrons with negative energy which move in the reverse time direction. This idea was first formulated by E.C.G. Stueckelberg and was used extensively by R.Feynman.[1] Feynman was rewarded with the Nobel price for his formulation of quantum electrodynamics, together with **J. Schwinger** and **S. Tomonaga** in 1965. The latter gave alternative formulations of QED, that are mutually equivalent. In the following we want to convince ourselves of the power of Feynman's formulation of the theory. The more or less heuristic rules obtained in this way fully agree with the results that can be obtained with much more effort using the method of quantum field theory.

1.2 The Nonrelativistic Propagator

First it is useful to remember the definition of Green's functions in nonrelativistic quantum mechanics. The concepts and methods to be acquired here are then easily transferred to relativistic quantum mechanics.

We shall mainly consider quantum-mechanical scattering processes in three dimensions, where one particle collides with a fixed force field or with another particle. A scattering process develops according to the scheme outlined in Fig. 1.1. In practice, one arranges by means of collimators D that the incoming particles are focussed in a well-defined beam. Such a collimated beam is in general not a wave, which extends to infinity, e.g. of the form $\exp(ikz)$, but a superposition of many plane waves with adjacent wave vectors $\boldsymbol{k}$, i.e. a wave packet. Nevertheless, in the stationary formulation of scattering theory for simplicity one often represents the incoming wave packet by a plane wave. Then one has only to ensure that interference between the incoming wave packet and the scattered wave is impossible at the position of the detector which is far removed from the scattering centre. If plane waves are used in calculations, therefore one has to exclude this interference explicitly.[2]

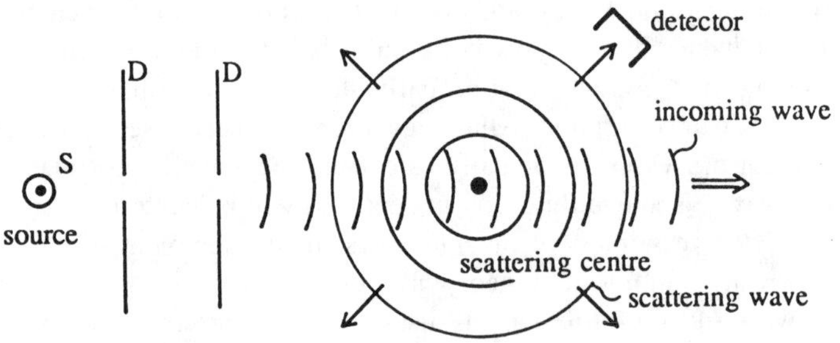

Fig. 1.1. Schematic representation of an experimental arrangement to measure a scattering process. Collimators D ensure that, at the position of the detector no interference occurs between incoming and scattered waves

[1] See for example R.P. Feynman: Phys. Rev. **76**, 749 (1949).

[2] For a more detailed discussion of the wave-packet description see for example M.L. Goldberger and K.M. Watson: *Collision Theory* (Wiley, New York 1964), Chap. 3, or R.G. Newton: *Scattering Theory of Waves and Particles* (McGraw-Hill, New York 1966), Chap 6.

In scattering processes we consider wave packets, which develop in time from initial conditions, which were fixed in the distant past. So in general, one does not consider stationary eigenstates of energy (i.e. stationary waves). A typical question for a scattering problem is then: What happens to a wave packet that represents a particle in the distant past and approaches a centre of scattering (a potential or another particle)? What does this wave look like in the distant future?

Here the generalized Huygens principle helps us to answer these questions. If a wave function $\psi(\boldsymbol{x},t)$ is known at a certain time t, then its shape at a later time t' can be deduced by regarding every spatial point $\boldsymbol{x}$ at time t as a source of a spherical wave that emerges from $\boldsymbol{x}$. It is plausible to assume that the intensity of the wave, which emerges from $\boldsymbol{x}$ and arrives at $\boldsymbol{x}'$ at time t', is proportional to the initial exciting wave amplitude $\psi(\boldsymbol{x},t)$. Let us call the constant of proportionality

$$\mathrm{i}G(\boldsymbol{x}',t';\boldsymbol{x},t) \quad . \tag{1.1}$$

The generalized Huygens principle can thus be expressed in the following terms:

$$\psi(\boldsymbol{x}',t') = \mathrm{i}\int d^3x\, G(\boldsymbol{x}',t';\boldsymbol{x},t)\psi(\boldsymbol{x},t) \quad , \quad t' > t \quad . \tag{1.2}$$

Here $\psi(\boldsymbol{x}',t')$ is the wave that arrives at $\boldsymbol{x}'$ at time t'. The quantity $G(\boldsymbol{x}',t';\boldsymbol{x},t)$ is known as the *Green's function* or *propagator*. It describes the effect of the wave $\psi(\boldsymbol{x},t)$, which was at point $\boldsymbol{x}$ in the past (at time $t < t'$), on the wave $\psi(\boldsymbol{x}',t')$, which is at point $\boldsymbol{x}'$ at the later time t'. If the Green's function $G(\boldsymbol{x}',t';\boldsymbol{x},t)$ is known, the final physical state $\psi(\boldsymbol{x}',t')$, which develops from a given initial state $\psi(\boldsymbol{x},t)$, can be calculated using (1.2). Knowing G therefore solves the complete scattering problem. Or, in other words: Knowing G is equivalent to the complete solution of Schrödinger's equation. First, however, we want to gain some mathematical insight and discuss the various ways of defining Green's functions.

1.3 Green's Function and Propagator

To explain the mathematical concepts it is best to start with Schrödinger's equation,

$$\begin{aligned} \mathrm{i}\hbar\frac{\partial\psi(\boldsymbol{x},t)}{\partial t} &= \hat{H}\psi(\boldsymbol{x},t) = \left(\hat{H}_0 + V(\boldsymbol{x},t)\right)\psi(\boldsymbol{x},t) \quad , \\ \hat{H}_0 &= -\frac{\hbar^2}{2m}\nabla^2 \quad , \end{aligned} \tag{1.3}$$

which describes the interaction of a particle of mass m with a potential source fixed in space. If we replace m by the reduced mass $\mu = m_1m_2/(m_1+m_2)$, (1.3) remains valid for the (nonrelativistic) two-body problem. The differential equation (1.3) is of first order in time, i.e., there are no higher-order time derivatives. Therefore, the first derivative with respect to time, $\partial\psi(\boldsymbol{x},t)/\partial t$, can always be expressed by $\psi(\boldsymbol{x},t)$, which is obviously the meaning of (1.3). From this, in turn, it follows that, if the value of $\psi(\boldsymbol{x},t)$ is known at one certain time (e.g. t_0) and at all spatial points $\boldsymbol{x}$, i.e. if $\psi(\boldsymbol{x},t_0)$ is known, one can calculate the wave function $\psi(\boldsymbol{x},t)$ at any point and any times (at earlier times ($t < t_0$) as well as at later times ($t > t_0$)).

Furthermore, since Schrödinger's equation is linear in ψ, the superposition principle is valid, i.e. solutions can be linearly superposed and the relation between wave functions at different times ($\psi(\boldsymbol{x}, t)$ and $\psi(\boldsymbol{x}, t_0)$) has to be linear. This means that $\psi(\boldsymbol{x}, t)$ has to satisfy a linear homogenous integral equation of the form

$$\psi(\boldsymbol{x}', t') = \mathrm{i} \int d^3x \, G(\boldsymbol{x}', t'; \boldsymbol{x}, t)\psi(\boldsymbol{x}, t) \quad , \tag{1.4}$$

where the integration extends over the whole space. This relation also defines the function G, which is called the Green's function corresponding to the Hamiltonian $\hat{H}$. It is important to note that relation (1.4) – in contrast to (1.2) – makes no difference between a propagation of ψ forward in time ($t' > t$) or backward in time ($t' < t$). However, in most cases it is desirable to distinguish clearly between these two cases. For forward propagation one therefore defines the *retarded Green's function* or propagator by

$$G^+(\boldsymbol{x}', t'; \boldsymbol{x}, t) = \begin{cases} G(\boldsymbol{x}', t'; \boldsymbol{x}, t) & \text{for} \quad t' > t \\ 0 & \text{for} \quad t' < t \end{cases} \quad . \tag{1.5}$$

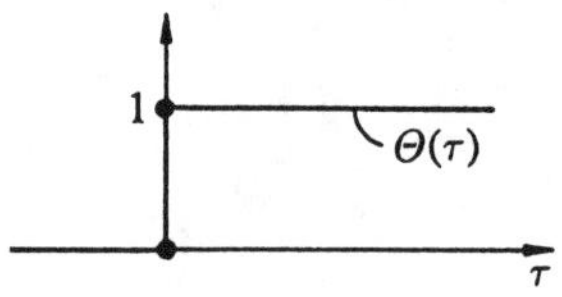

Fig. 1.2. The unit step function

It is now useful to introduce the step function $\Theta(\tau)$ (Fig. 1.2):

$$\Theta(\tau) = \begin{cases} 1 & \text{for} \quad \tau > 0 \\ 0 & \text{for} \quad \tau < 0 \end{cases} \quad . \tag{1.6}$$

With this the causal evolution of $\psi(\boldsymbol{x}', t')$ from $\psi(\boldsymbol{x}, t)$, with $t' > t$, can be formulated as follows:

$$\Theta(t' - t)\psi(\boldsymbol{x}', t') = \mathrm{i} \int d^3x \, G^+(\boldsymbol{x}', t'; \boldsymbol{x}, t)\psi(\boldsymbol{x}, t) \quad . \tag{1.7}$$

For $t' < t$ this relation is trivial because of (1.5) and (1.6), which together give $0 = 0$, and for $t' > t$ it is identical with (1.4). Equation (1.7) ensures that the original wave packet $\psi(\boldsymbol{x}, t)$ develops into a later $\psi(\boldsymbol{x}', t')$ with $t' > t$. Hence there exists a causal connection between $\psi(\boldsymbol{x}', t')$ and $\psi(\boldsymbol{x}, t)$. We will return to this question in Sect. 1.6 and Exercise 1.1. If one wants to describe the evolution backwards in time, it is useful to introduce the *advanced Green's function* G^-:

$$G^-(\boldsymbol{x}', t'; \boldsymbol{x}, t) = \begin{cases} -G(\boldsymbol{x}', t'; \boldsymbol{x}, t) & \text{for} \quad t' < t \\ 0 & \text{for} \quad t' > t \end{cases} \quad . \tag{1.8}$$

Then the determination of the former wave packet $\psi(\boldsymbol{x}', t')$ from the present one $\psi(\boldsymbol{x}, t)$, with $t' < t$, proceeds according to the relation

$$\Theta(t - t')\psi(\boldsymbol{x}', t') = -\mathrm{i} \int d^3x \, G^-(\boldsymbol{x}', t'; \boldsymbol{x}, t)\psi(\boldsymbol{x}, t) \quad , \tag{1.9}$$

which is again trivial for $t' > t$ because of (1.6) and (1.8) and is identical with (1.4) for $t' < t$.

EXERCISE

1.1 Properties of G

Problem. Show the validity of the following relations:
a) if $t' > t_1 > t$:

$$G^+(\boldsymbol{x}', t'; \boldsymbol{x}, t) = \mathrm{i} \int d^3x_1 G^+(\boldsymbol{x}', t'; \boldsymbol{x}_1, t_1) G^+(\boldsymbol{x}_1, t_1; \boldsymbol{x}, t) \quad ,$$

b) if $t' < t_1 < t$:

$$G^-(\boldsymbol{x}', t'; \boldsymbol{x}, t) = -\mathrm{i} \int d^3x_1 G^-(\boldsymbol{x}', t'; \boldsymbol{x}_1, t_1) G^-(\boldsymbol{x}_1, t_1; \boldsymbol{x}, t) \quad ,$$

c) if $t > t_1$:

$$\delta^3(\boldsymbol{x} - \boldsymbol{x}') = \int d^3x_1 G^+(\boldsymbol{x}', t; \boldsymbol{x}_1, t_1) G^-(\boldsymbol{x}_1, t_1; \boldsymbol{x}, t) \quad ,$$

d) if $t < t_1$:

$$\delta^3(\boldsymbol{x} - \boldsymbol{x}') = \int d^3x_1 G^-(\boldsymbol{x}', t; \boldsymbol{x}_1, t_1) G^+(\boldsymbol{x}_1, t_1; \boldsymbol{x}, t) \quad .$$

Solution. a) The first two assertions (a) and (b) are readily understood because of relations (1.7) and (1.9), respectively. If we consider the propagation of an arbitrary wave packet $\psi(\boldsymbol{x}, t)$ into the future, we are able to conclude that

$$\psi(\boldsymbol{x}', t') = \mathrm{i} \int d^3x \, G^+(\boldsymbol{x}', t'; \boldsymbol{x}, t) \psi(\boldsymbol{x}, t) \tag{1}$$

if $t' > t$. $\psi(\boldsymbol{x}, t)$ can be chosen at any arbitrary time t. Thus we can also insert an intermediate step:

$$\begin{aligned}
\psi(\boldsymbol{x}', t') &= \mathrm{i} \int d^3x_1 G^+(\boldsymbol{x}', t'; \boldsymbol{x}_1, t_1) \psi(\boldsymbol{x}_1, t_1) \\
&= \mathrm{i} \int d^3x_1 G^+(\boldsymbol{x}', t'; \boldsymbol{x}_1, t_1) \mathrm{i} \int d^3x \, G^+(\boldsymbol{x}_1, t_1; \boldsymbol{x}, t) \psi(\boldsymbol{x}, t) \\
&= \mathrm{i} \int d^3x \, \mathrm{i} \int d^3x_1 G^+(\boldsymbol{x}', t'; \boldsymbol{x}_1, t_1) G^+(\boldsymbol{x}_1, t_1; \boldsymbol{x}, t) \, \psi(\boldsymbol{x}, t)
\end{aligned} \tag{2}$$

If we compare relations (1) and (2), assertion (a) follows.

b) The proof of case (b) proceeds along similar lines:

$$\psi(\boldsymbol{x}', t') = -\mathrm{i} \int d^3x \, G^-(\boldsymbol{x}', t'; \boldsymbol{x}, t) \psi(\boldsymbol{x}, t) \tag{3}$$

if $t' < t$. Again we insert an intermediate step:

Exercise 1.1.

$$\begin{aligned}\psi(\boldsymbol{x}',t') &= -\mathrm{i}\int d^3x_1 G^-(\boldsymbol{x}',t';\boldsymbol{x}_1,t_1)\psi(\boldsymbol{x}_1,t_1)\\ &= -\mathrm{i}\int d^3x_1 G^-(\boldsymbol{x}',t';\boldsymbol{x}_1,t_1)(-\mathrm{i})\int d^3x\, G^-(\boldsymbol{x}_1,t_1;\boldsymbol{x},t)\psi(\boldsymbol{x},t)\\ &= -\mathrm{i}\int d^3x\,(-\mathrm{i})\int d^3x_1 G^-(\boldsymbol{x}',t';\boldsymbol{x}_1,t_1)G^-(\boldsymbol{x}_1,t_1;\boldsymbol{x},t)\psi(\boldsymbol{x},t) \qquad (4)\end{aligned}$$

if $t' < t_1 < t$. Comparing relations (3) and (4) assertion (b) follows.

c) The proof of relations (c) and (d) proceeds similarly. We first write

$$\begin{aligned}\psi(\boldsymbol{x}',t) &= \mathrm{i}\int d^3x_1 G^+(\boldsymbol{x}',t;\boldsymbol{x}_1,t_1)\psi(\boldsymbol{x}_1,t_1)\\ &= \mathrm{i}\int d^3x_1 G^+(\boldsymbol{x}',t;\boldsymbol{x}_1,t_1)(-\mathrm{i})\int d^3x\, G^-(\boldsymbol{x}_1,t_1;\boldsymbol{x},t)\psi(\boldsymbol{x},t)\\ &= \int d^3x\int d^3x_1 G^+(\boldsymbol{x}',t;\boldsymbol{x}_1,t_1)G^-(\boldsymbol{x}_1,t_1;\boldsymbol{x},t)\psi(\boldsymbol{x},t) \qquad (5)\end{aligned}$$

if $t > t_1$. For a constant time t, $\psi(\boldsymbol{x},t)$ can be expressed with the help of the δ function as

$$\psi(\boldsymbol{x}',t) = \int d^3x\,\delta(\boldsymbol{x}-\boldsymbol{x}')\psi(\boldsymbol{x},t) \quad . \qquad (6)$$

The comparison of relations (5) and (6) yields assertion (c).

d) The proof of (c) can be exactly copied

$$\begin{aligned}\psi(\boldsymbol{x}',t) &= -\mathrm{i}\int d^3x_1 G^-(\boldsymbol{x}',t;\boldsymbol{x}_1,t_1)\psi(\boldsymbol{x}_1,t_1)\\ &= \int d^3x\int d^3x_1 G^-(\boldsymbol{x}',t;\boldsymbol{x}_1,t_1)G^+(\boldsymbol{x}_1,t_1;\boldsymbol{x},t)\psi(\boldsymbol{x},t) \qquad (7)\end{aligned}$$

if $t < t_1$. Comparing (7) with the integral representation (6) proves (d).

1.4 An Integral Equation for ψ

Now we aim for a formal definition of the Green's function. To this end we still want to proceed in a physical, illustrative manner to ensure that the propagator method is understood. Since the motion of a free particle is completely known, the *free Green's function* $G_0(\boldsymbol{x}',t';\boldsymbol{x},t)$ can be explicitly constructed (see Example 1.3). However, if we switch on a potential $V(\boldsymbol{x},t)$, then G_0 is modified to $G(\boldsymbol{x}',t';\boldsymbol{x},t)$ and the question arises how the Green's function G (including the interaction) is calculated from the free Green's function G_0.

To answer this we assume that the interaction potential $V(\boldsymbol{x},t)$ acts at time t_1 for a short time interval Δt_1. The potential during this interval is then $V(\boldsymbol{x}_1,t_1)$. For times preceeding t_1 the wave function is that of a free particle, i.e. for $t < t_1$ the particle propagates according to the free propagator G_0. At $t = t_1$, $V(\boldsymbol{x}_1,t_1)$ acts, and a scattered wave is created, which can be calculated from Schrödinger's equation

$$\left(\mathrm{i}\hbar\frac{\partial}{\partial t_1} - \hat{H}_0\right)\psi(\boldsymbol{x}_1,t_1) = V(\boldsymbol{x}_1,t_1)\psi(\boldsymbol{x}_1,t_1) \quad . \tag{1.10}$$

As already mentioned, $V(\boldsymbol{x}_1, t_1)$ acts only during the time interval Δt_1. We denote the resulting wave with the help of the free wave ϕ as

$$\psi(\boldsymbol{x}_1,t_1) = \phi(\boldsymbol{x}_1,t_1) + \Delta\psi(\boldsymbol{x}_1,t_1) \quad , \tag{1.11}$$

where ϕ solves the free Schrödinger equation

$$\left(\mathrm{i}\hbar\frac{\partial}{\partial t_1} - \hat{H}_0\right)\phi(\boldsymbol{x}_1,t_1) = 0 \tag{1.12}$$

and where the scattered wave $\Delta\psi(\boldsymbol{x}_1, t_1)$ is zero for $t < t_1$. Inserting (1.11) into (1.10) and taking into account (1.12), we find

$$\left(\mathrm{i}\hbar\frac{\partial}{\partial t_1} - \hat{H}_0\right)\Delta\psi(\boldsymbol{x}_1,t_1) = V(\boldsymbol{x}_1,t_1)(\phi(\boldsymbol{x}_1,t_1) + \Delta\psi(\boldsymbol{x}_1,t_1)) \tag{1.13}$$

and, neglecting the small term $V\Delta\psi$ on the right-hand side,

$$\left(\mathrm{i}\hbar\frac{\partial}{\partial t_1} - \hat{H}_0\right)\Delta\psi(\boldsymbol{x}_1,t_1) = V(\boldsymbol{x}_1,t_1)\phi(\boldsymbol{x}_1,t_1) \quad . \tag{1.14}$$

This differential equation can be integrated in the time interval t_1 to $t_1 + \Delta t_1$. Taking into account that $\Delta\psi(\boldsymbol{x}_1, t_1) = 0$ we get

$$\mathrm{i}\hbar\Delta\psi(\boldsymbol{x}_1,t_1+\Delta t_1) = \int\limits_{t_1}^{t_1+\Delta t_1} dt' \left(\hat{H}_0\Delta\psi(\boldsymbol{x}_1,t') + V(\boldsymbol{x}_1,t')\phi(\boldsymbol{x}_1,t')\right) \quad . \tag{1.15}$$

The first term on the right-hand side is of second order with respect to the small quantities $\Delta\psi$ and Δt_1. Then in first-order accuracy the scattered wave is given by

$$\Delta\psi(\boldsymbol{x}_1,t_1+\Delta t_1) = \frac{-\mathrm{i}}{\hbar}V(\boldsymbol{x}_1,t_1)\phi(\boldsymbol{x}_1,t_1)\Delta t_1 \quad . \tag{1.16}$$

Since the potential $V(\boldsymbol{x}_1, t_1)$ is assumed to vanish after the time interval Δt_1, the scattered wave propagates according to the free propagator G_0 too, and we obtain at the later time $t' > t_1$

$$\begin{aligned}\Delta\psi(\boldsymbol{x}',t') &= \mathrm{i}\int d^3x_1\, G_0(\boldsymbol{x}',t';\boldsymbol{x}_1,t_1)\Delta\psi(\boldsymbol{x}_1,t_1)\\ &= \int d^3x_1\, G_0(\boldsymbol{x}',t';\boldsymbol{x}_1,t_1)\frac{1}{\hbar}V(\boldsymbol{x}_1,t_1)\phi(\boldsymbol{x}_1,t_1)\Delta t_1 \quad .\end{aligned} \tag{1.17}$$

Here we have replaced $t_1 + \Delta t_1$ by t_1 which is justified in the limit of infinitesimal time intervals. Note that $\phi(\boldsymbol{x}_1, t_1)$ is the wave that arrives at space-time point $(\boldsymbol{x}_1, t_1)$ before it is scattered at the potential $V(\boldsymbol{x}_1, t_1)$. Then the potential $V(\boldsymbol{x}_1, t_1)$ acts for a short time period Δt_1. It modifies the incoming wave to $1/\hbar V(\boldsymbol{x}_1, t_1)\phi(\boldsymbol{x}_1, t_1)\Delta t_1$ and this "perturbed" wave propagates freely, described by the propagator $G_0(\boldsymbol{x}', t'; \boldsymbol{x}, t)$ from $(\boldsymbol{x}_1, t_1)$ to $(\boldsymbol{x}', t')$. The total wave $\psi(\boldsymbol{x}', t')$,

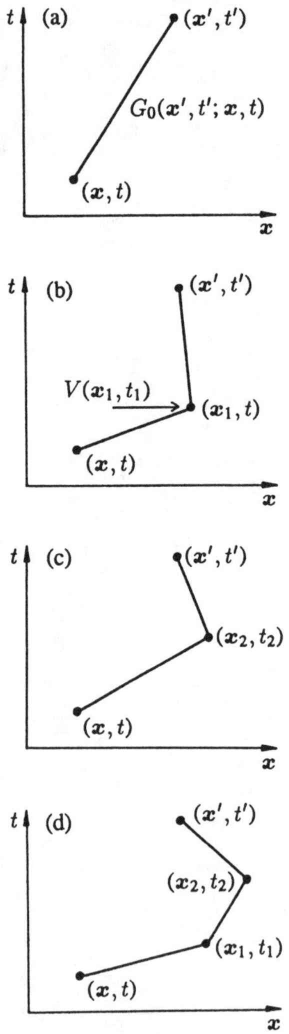

Fig. 1.3a–d. Graphs illustrating scattering processes. **(a)** describes the free motion (propagation) of a particle from space–time point $(\boldsymbol{x}, t)$ to $(\boldsymbol{x}', t')$. In **(b)** the particle moves from $(\boldsymbol{x}, t)$ to $(\boldsymbol{x}', t')$ too, but is scattered once at the intermediate point by potential $V(\boldsymbol{x}_1, t_1)$. Graph **(c)** shows the same as graph **(b)**, but the scattering now takes place at $(\boldsymbol{x}_2, t_2)$ instead of $(\boldsymbol{x}_1, t_1)$. Finally, **(d)** represents a double scattering event at $(\boldsymbol{x}_1, t_1)$ and $(\boldsymbol{x}_2, t_2)$

which originates from an arbitrary wave packet ϕ in the distant past by scattering once at the potential $V(\boldsymbol{x}_1, t_1)$ during the time period Δt_1, is then given by

$$\begin{aligned} \psi(\boldsymbol{x}', t') &= \phi(\boldsymbol{x}', t') + \Delta\psi(\boldsymbol{x}', t') \\ &= \phi(\boldsymbol{x}', t') + \int d^3x_1\, G_0(\boldsymbol{x}', t'; \boldsymbol{x}_1, t_1)\frac{1}{\hbar}V(\boldsymbol{x}_1, t_1)\phi(\boldsymbol{x}_1, t_1)\Delta t_1 \\ &= \mathrm{i}\int d^3x\, \Big(G_0(\boldsymbol{x}', t'; \boldsymbol{x}, t) \\ &\quad + \int d^3x_1 \Delta t_1\, G_0(\boldsymbol{x}', t'; \boldsymbol{x}_1, t_1)\frac{1}{\hbar}V(\boldsymbol{x}_1, t_1)G_0(\boldsymbol{x}_1, t_1; \boldsymbol{x}, t)\Big)\phi(\boldsymbol{x}, t) \quad . \end{aligned} \tag{1.18}$$

Comparing this with (1.2) or (1.4) we can identify the expression in brackets as the propagator $G(\boldsymbol{x}', t'; \boldsymbol{x}, t)$:

$$\begin{aligned} G(\boldsymbol{x}', t'; \boldsymbol{x}, t) &= G_0(\boldsymbol{x}', t'; \boldsymbol{x}, t) \\ &\quad + \int d^3x_1\, \Delta t_1\, G_0(\boldsymbol{x}', t'; \boldsymbol{x}_1, t_1)\frac{1}{\hbar}V(\boldsymbol{x}_1, t_1)G_0(\boldsymbol{x}_1, t_1; \boldsymbol{x}, t) \,. \end{aligned} \tag{1.19}$$

Now we have achieved our goal of calculating the propagator G from the free propagator G_0 – at least for the simple case of an interaction $V(\boldsymbol{x}_1, t_1)$, which is turned on only during the short time interval Δt_1. The various terms in (1.19) can be illustrated in space–time diagrams, as in Fig. 1.3. The first term of (1.19) corresponds to the free propagation of the wave packet from space–time point $(\boldsymbol{x}, t)$ to $(\boldsymbol{x}', t')$. This is represented Fig. 1.3a. The second term in (1.19) describes the free propagation from space–time point $(\boldsymbol{x}, t)$ to $(\boldsymbol{x}_1, t_1)$. Here the particle is scattered at the potential $V(\boldsymbol{x}_1, t_1)$ during the time interval Δt_1. Afterwards it propagates again freely to space–time point $(\boldsymbol{x}', t')$. This process is illustrated in Fig. 1.3b.

If we switch on a second potential $(\boldsymbol{x}_2, t_2)$ at time $t_2 > t_1$ for a time interval Δt_2, then an additional scattering wave is created, whose contribution $\Delta\psi(\boldsymbol{x}', t')$ to the total wave $\psi(\boldsymbol{x}', t')$ at time $t' > t_2$ can immediately be written down according to (1.17):

$$\begin{aligned} \Delta\psi(x') &= \int d^3x_2\, G_0(x'; x_2)V(x_2)\psi(x_2)\Delta t_2 \\ &= \mathrm{i}\int d^3x\, d^3x_2\, \Delta t_2 G_0(x'; x_2)V(x_2) \\ &\quad \times \left(G_0(x_2; x) + \int d^3x_1\, \Delta t_1\, G_0(x_2; x_1)V(x_1)G_0(x_1; x)\right)\phi(x) \,. \end{aligned} \tag{1.20}$$

From (1.18) we have substituted the scattering wave $\psi(2)$, which arrives at space–time point $(\boldsymbol{x}_2, t_2)$. In addition we have introduced the obvious abbreviations

$$\begin{aligned} (\boldsymbol{x}, t) &= x \quad , \\ \frac{1}{\hbar}V(\boldsymbol{x}_i, t_i) &= V(x_i) \quad . \end{aligned} \tag{1.21}$$

Note that we have absorbed the factor $1/\hbar$ in the potential, since the two always appear together according to (1.16). The first term in (1.20), which is proportional to $G_0(x'; x_2)V(x_2)G_0(x_2; x)\phi(x)$, represents a single scattering event at space–time point 2 and is illustrated in Fig. 1.3c. The second term in (1.20) is proportional to

$$G_0(x';x_2)V(x_2)G_0(x_2;x_1)V(x_1)G_0(x_1;x)\phi(x)$$

and represents a double scattering event at the potential at space–time points x_1 and x_2. This process is illustrated in Fig. 1.3d.

Now the total wave, which arrives at space–time point $(\boldsymbol{x}',t')$ after free propagation and single plus double scattering, is simply the sum of the partial waves (1.18) (this is the wave originating from free propagation and single scattering at $(\boldsymbol{x}_1,t_1)$) and (1.20) (this is the wave originating from single scattering at $(\boldsymbol{x}_2,t_2)$ as well as double scattering at $(\boldsymbol{x}_1,t_1)$ and $(\boldsymbol{x}_2,t_2)$). This yields

$$\begin{aligned}\psi(x') = \phi(x') &+ \int d^3x_1\, \Delta t_1\, G_0(x';x_1)V(x_1)\phi(x_1)\\ &+ \int d^3x_2\, \Delta t_2\, G_0(x';x_2)V(x_2)\phi(x_2)\\ &+ \int d^3x_1\, \Delta t_1\, d^3x_2\, \Delta t_2\, G_0(x';x_2)V(x_2)G_0(x_2;x_1)V(x_1)\phi(x_1) \quad . \end{aligned} \tag{1.22}$$

If we now switch on the potential V at n times $t_1 < t_2 < t_3 < \ldots < t_n$ for time intervals $\Delta t_1, \Delta t_2, \ldots$, then (1.22) must obviously be generalized to yield

$$\begin{aligned}\psi(x') = \phi(x') &+ \sum_i \int d^3x_i\, \Delta t_i\, G_0(x';x_i)V(x_i)\phi(x_i)\\ &+ \sum_{i,j;\,t_i>t_j} \int d^3x_i\, \Delta t_i\, d^3x_j\, \Delta t_j\, G_0(x';x_i)V(x_i)G_0(x_i;x_j)V(x_j)\phi(x_j)\\ &+ \sum_{i,j,k;\,t_i>t_j>t_k} \int d^3x_i\, \Delta t_i\, d^3x_j\, \Delta t_j\, d^3x_k\, \Delta t_k\, G_0(x';x_i)V(x_i)\\ &\times G_0(x_i;x_j)V(x_j)G_0(x_j;x_k)V(x_k)\phi(x_k)\\ &+ \ldots \quad . \end{aligned} \tag{1.23}$$

Note that one integrates over three-dimensional spatial coordinates in (1.22) and (1.23), e.g. $\int d^3x_i$. The summation runs over a grid of time values t_i, $\sum_i \int d^3x_i\, \Delta t_i = \sum_i \Delta t_i \int d^3x_i$. This is used in the following, when we take the limit $\Delta t_i \to 0$ and $n \to \infty$ so that $\sum_i \int d^3x_i\, \Delta t_i$ becomes a four-dimensional volume integral $\int d^4x_i$. If we express $\phi(x')$ and $\phi(x_i)$ in (1.23) by

$$\begin{aligned}\phi(x') &= \mathrm{i} \int d^3x\, G_0(x';x)\phi(x) \quad ,\\ \phi(x_i) &= \mathrm{i} \int d^3x\, G_0(x_i;x)\phi(x) \quad , \end{aligned} \tag{1.24}$$

we finally get

$$\psi(x') = \mathrm{i} \int d^3x \Bigg(G_0(x';x) + \sum_i \int d^3x_i \, \Delta t_i \, G_0(x';x_i) V(x_i) G_0(x_i;x)$$
$$+ \sum_{i,j;t_i>t_j} \int d^3x_1 \, \Delta t_i \, d^3x_j \, \Delta t_j \, G_0(x';x_i) V(x_i) G_0(x_i;x_j) V(x_j) G_0(x_j,x)$$
$$+ \sum_{i,j,k;t_i>t_j>t_k} \int d^3x_i \, \Delta t_i \, d^3x_j \, \Delta t_j \, d^3x_k \, \Delta t_k \, G_0(x';x_i) V(x_i)$$
$$\times G_0(x_i;x_j) V(x_j) G_0(x_j;x_k) V(x_k) G_0(x_k;x) + \dots \Bigg) \phi(x)$$
$$\equiv \mathrm{i} \int d^3x \, G(x';x) \phi(x) \quad . \tag{1.25}$$

The complete expression for the Green's function $G(x',x)$ including interactions results by comparing (1.25) with (1.2) or (1.4). Expanded in terms of the free Green function $G_0(x';x)$ the full Green's function reads

$$G(x';x) = G_0(x';x) + \sum_i \int d^3x_i \, \Delta t_i \, G_0(x';x_i) V(x_i) G_0(x_i;x)$$
$$+ \sum_{i,j;t_i>t_j} \int d^3x_1 \, \Delta t_i \, d^3x_j \, \Delta t_j \, G_0(x';x_i) V(x_i) G_0(x_i;x_j) V(x_j) G_0(x_j,x)$$
$$+ \sum_{i,j,k;t_i>t_j>t_k} \int d^3x_i \, \Delta t_i \, d^3x_j \, \Delta t_j \, d^3x_k \, \Delta t_k \, G_0(x';x_i) V(x_i)$$
$$\times G_0(x_i;x_j) V(x_j) \times G_0(x_j;x_k) V(x_k) G_0(x_k;x) + \dots \quad . \tag{1.26}$$

We have been careful to respect strict time ordering in the preceding expression. However it is possible to get rid of the constraints for the multiple sums if we introduce the retarded Green's function $G^+(x';x)$ (see (1.5)), which fulfills

$$G_0^+(\boldsymbol{x}',t';\boldsymbol{x},t) = \begin{cases} 0 & \text{for} \quad t'<t \\ G_0(\boldsymbol{x}',t';\boldsymbol{x},t) & \text{for} \quad t'>t \end{cases} \quad ,$$
$$G^+(\boldsymbol{x}',t';\boldsymbol{x},t) = \begin{cases} 0 & \text{for} \quad t'<t \\ G(\boldsymbol{x}',t';\boldsymbol{x},t) & \text{for} \quad t'>t \end{cases} \quad . \tag{1.27}$$

Furthermore, in the continuum limit $\Delta t_i \to 0$ etc., we can replace the sums over time intervals in (1.25) and (1.26) by time integrals $\int dt \dots$. This leads to the following series expansion for the retarded interacting Green's function

$$G^+(x';x) = G_0^+(x';x) + \int d^4x_1 \, G_0^+(x';x_1) V(x_1) G_0^+(x_1;x)$$
$$+ \int d^4x_1 \, d^4x_2 \, G_0^+(x';x_1) V(x_1) G_0^+(x_1;x_2) V(x_2) G_0^+(x_2;x)$$
$$+ \dots \quad , \tag{1.28}$$

where we used the abbreviation

$$d^4x = d^3x \, dt = d^3x \, dx_0 \quad . \tag{1.29}$$

In (1.28) the Green's function G with interaction is expanded as a *series of multiple scattering events*, where the propagation between single scattering events is determined by the free Green's function G_0. This multiple scattering series will be assumed to converge. We have also ignored complications arising from the possibility of bound states in the potential V.

It is possible to write down a closed expression for the interacting Green's function. This is achieved by formally summing the series (1.28) which leads to

$$G^+(x';x) = G_0^+(x';x) + \int d^4x_1\, G_0^+(x';x_1)V(x_1)G^+(x_1;x) \quad . \tag{1.30}$$

This is an integral equation for G^+. It is often called the *Lippmann–Schwinger equation*. As can be seen immediately, the multiple scattering series (1.28) can be generated by iterating the integral equation (1.30). Similarly the series (1.25) for the wave function $\psi(x')$ can be summed, resulting in

$$\begin{aligned}
\psi(x') &= \lim_{t\to-\infty} \mathrm{i} \int d^3x\, G^+(x';x)\phi(x) \\
&= \lim_{t\to-\infty} \mathrm{i} \int d^3x \left(G_0^+(x';x) + \int d^4x_1 G_0^+(x',x_1)V(x_1)G^+(x_1;x) \right) \phi(x) \\
&= \phi(x') + \lim_{t\to-\infty} \int d^4x_1\, G_0^+(x',x_1)V(x_1)\,\mathrm{i} \int d^3x\, G^+(x_1;x)\phi(x) \\
&= \phi(x') + \underbrace{\int d^4x_1 G_0^+(x',x_1)V(x_1)\psi(x_1)}_{\text{scattered wave}} \quad .
\end{aligned} \tag{1.31}$$

This is an *integral equation for* $\psi(x')$. One should realize that up to now nothing is solved, since one has to integrate over a still-unknown wave function ψ. However, in some sense the integral equations (1.30, 1.31) are more useful than the original differential equation (1.3). They allow a systematic approximation in the case of weak perturbations (that is a small perturbation potential V). Moreover, one can easily impose the correct boundary conditions (cf. the discussion in Sect. 1.5).

It should be noted that not only $G_0^+(x';x)$ vanishes for $t' < t$, but also $G^+(x';x)$. This property of the retarded Green's functions expresses the *principle of causality* in an elementary way through (1.31) and (1.25). For example, the expansion (1.26) means that an interaction with the potential V at time t_k can influence additional scattering interactions only if these occur later in time ($t_k < t_i, t_j$).

Let us return to the scattering expansion (1.28). If the infinite series is truncated after a finite number of terms (1.28) allows us to calculate G^+ as a functional of V and G_0. Given G^+ one can immediately solve the initial value problem. The wave function $\psi(\boldsymbol{x}', t')$ is obtained by a simple integration according to (1.31) if it is known at some former point in time.

1.5 Application to Scattering Problems

Let us consider a scattering problem. We know the incoming wave packet $\phi(\boldsymbol{x}, t)$; it describes a particle in the distant past moving towards the scattering centre. We want to construct the wave that originates from the interaction with the potential $V(\boldsymbol{x}, t)$, as it looks in the distant future. We idealize the scattering problem by assuming that no interaction is present at the initial time, i.e.

$$V(\boldsymbol{x}, t) \to 0 \quad \text{for} \quad t \to -\infty \quad ,$$

the initial wave ϕ is therefore a solution of the Schrödinger equation for free particles, which fulfills certain initial conditions.[3] The exact wave $\psi(\boldsymbol{x}, t)$ then approaches the incoming wave $\phi(\boldsymbol{x}, t)$ in the limit $t \to -\infty$:

$$\lim_{t \to -\infty} \psi(\boldsymbol{x}, t) = \phi(\boldsymbol{x}, t) \quad . \tag{1.32}$$

In the distant future the exact wave $\psi^{(+)}$ is, according to (1.25, 1.31) given by

$$\begin{aligned} \psi^{(+)}(\boldsymbol{x}', t') &= \lim_{t \to -\infty} \mathrm{i} \int d^3x \, G^{+}(\boldsymbol{x}', t'; \boldsymbol{x}, t) \phi(\boldsymbol{x}, t) \\ &= \phi(x') + \underbrace{\int d^4x_1 \, G_0^{+}(x'; x_1) V(x_1) \psi^{(+)}(x_1)}_{\text{scattered wave}} \quad . \end{aligned} \tag{1.33}$$

The $\psi^{(+)}(x_1)$ appearing in this equation is the exact wave that originates from the initial wave packets (1.32). The superscript (+) over ψ is meant to express the fact that we are dealing with a wave which propagates into the future. As has already been mentioned the second term in (1.33) represents the *scattered wave*. It includes all single and multiple scattering events. Now we assume that the potential $V(\boldsymbol{x}, t)$ vanishes after a certain time, i.e.

$$\lim_{t' \to \infty} V(\boldsymbol{x}', t') = 0 \quad . \tag{1.34}$$

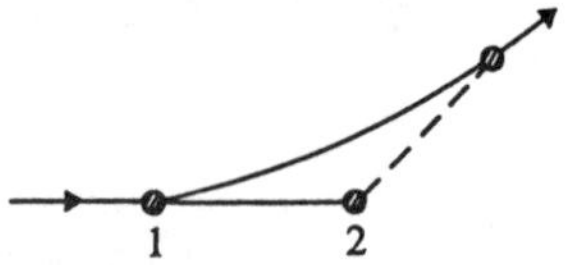

Fig. 1.4. The scattering of two particles: the interaction increases as they approach each other and decreases as they depart

If the interaction potential has a short range this condition will be met: Consider for example the mutual scattering of two particles, where the interaction increases from zero during their approach and decreases to zero when they depart (Fig. 1.4). If this condition is not fulfilled one may take recourse to the prescription of *adiabatic*

---

[3] The so-defined adiabatic approximation assumes that the solutions of Schrödinger's equation can be approximated by the stationary eigenfunctions of the instantaneous Hamiltonian, so that a certain eigenfunction at time t_1 transforms continuously into the corresponding eigenfunction at a later time. If we can solve the equation

$$H(t)\phi_n(t) = E_n(t)\phi_n(t)$$

at any time, then we expect that a system that assumes a discrete non-degenerate state $\phi_m(t)$ with energy $E_m(0)$ at time $t = 0$ will be in the state $\phi_m(t)$ with energy $E_m(t)$ at time t, provided that $H(t)$ varies slowly with time. However, this means that switching on or off $H(t)$ cannot cause excitations to other states $\phi_k(t)(k \neq m)$! The validity of the adiabatic approximation can be readily checked: if the typical excitation energies of a system are given by $\Delta E \approx E_m - E_k$, then the corresponding time scale is of the order $\Delta t \approx \hbar/\Delta E$. The switch-on time must be large compared to Δt!

switching. This means that the potential is forced to vanish asymptotically, e.g. by replacing

$$V(\boldsymbol{x},t) \quad \rightarrow \quad e^{-\lambda|t|}V(\boldsymbol{x},t) \quad .$$

The cutoff parameter λ has to be chosen small enough that the switching does not introduce spurious transient excitations.

We are now able to consider the exact wave $\psi^{(+)}(\boldsymbol{x}',t')$ in the distant future, i.e. in the limit $t' \rightarrow \infty$. All information about the scattered wave can be summarized in the probability amplitudes. Their squares express the probability that a particle from a given, free, initial state ϕ_i will be scattered into various final states ϕ_f in the limit $t' \rightarrow +\infty$. Since the potential is assumed to vanish for $t \rightarrow -\infty$ as well as for $t \rightarrow +\infty$, we can consider the ϕ's as plane waves.

$$\phi_f(\boldsymbol{x}',t') = \frac{1}{\sqrt{(2\pi\hbar)^3}} \exp\left[\mathrm{i}(\boldsymbol{k}_f\cdot\boldsymbol{x}' - \omega_f t')\right] \quad . \tag{1.35}$$

The plane wave (1.35) is subjected to the continuum normalization (or δ-function normalization). Alternatively one may use the box normalization, where the particle is confined to a box of volume V. Then the momentum variable becomes discretised and one has to replace

$$\frac{1}{\sqrt{(2\pi\hbar)^3}} \quad \rightarrow \quad \frac{1}{\sqrt{V}} \quad . \tag{1.36}$$

Dirac's δ function $\delta^3(\boldsymbol{k}_f - \boldsymbol{k}_i)$ is then replaced by Kronecker's delta

$$\delta_{\boldsymbol{k}_f,\boldsymbol{k}_i} = \begin{cases} 1 & \text{for} \quad \boldsymbol{k}_f = \boldsymbol{k}_i \\ 0 & \text{for} \quad \boldsymbol{k}_f \neq \boldsymbol{k}_i \end{cases} \quad .$$

The probability amplitudes are elements of Heisenberg's scattering matrix or S-matrix[4]

$$\begin{aligned}
S_{fi} &= \lim_{t'\rightarrow\infty} \left\langle \phi_f(\boldsymbol{x}',t') \middle| \psi_i^{(+)}(\boldsymbol{x}',t') \right\rangle \\
&= \lim_{t'\rightarrow\infty} \int d^3x' \phi_f^*(\boldsymbol{x}',t')\psi_i^{(+)}(\boldsymbol{x}',t') \\
&= \lim_{t'\rightarrow\infty} \lim_{t\rightarrow-\infty} \mathrm{i} \int d^3x' d^3x \int \phi_f^*(\boldsymbol{x}',t') G^+(\boldsymbol{x}',t';\boldsymbol{x},t)\phi_i(\boldsymbol{x},t) \\
&= \lim_{t'\rightarrow\infty} \int d^3x'\, \phi_f^*(\boldsymbol{x}',t') \left(\phi_i(\boldsymbol{x}',t') + \int d^4x G_0^+(\boldsymbol{x}',t';\boldsymbol{x},t)V(\boldsymbol{x},t)\psi_i^{(+)}(\boldsymbol{x},t) \right) \\
&= \delta^3(\boldsymbol{k}_f - \boldsymbol{k}_i) + \lim_{t'\rightarrow\infty} \int d^3x'\, d^4x \phi_f^*(\boldsymbol{x}',t') G_0^+(\boldsymbol{x}',t';\boldsymbol{x},t)V(\boldsymbol{x},t)\psi_i^{(+)}(\boldsymbol{x},t) \ .
\end{aligned} \tag{1.37}$$

$\psi_i^{(+)}$ is the solution (1.33) of the wave equation, which develops from the plane wave that emerges at $t \rightarrow -\infty$ and carries momentum $\boldsymbol{k}_i$ during the scattering process. The limit $t \rightarrow \pm\infty$ always means

[4] W. Heisenberg: Zeitschrift f. Naturforschung **1**, 608 (1946), see also C. Møller: Kgl. Danske Videnskab Selskab, Mat.-Fys. Medd. **23**, 1 (1948) and J.A. Wheeler: Phys. Rev. **52**, 1107 (1937).

$$t \to \text{large finite time } T \quad ,$$

at which the particles cease to interact. Typical times are, for example, the collision time or the production or detection time for particles.

Finally, if we insert ψ_i^+ from the iterated solution (1.33), we get an expansion of the S matrix in terms of multiple scattering events, namely

$$\begin{aligned} S_{fi} &= \delta^3(\boldsymbol{k}_f - \boldsymbol{k}_i) + \lim_{t' \to \infty} \int d^3x'\, d^4x\, \phi_f^*(\boldsymbol{x}', t') G_0^+(\boldsymbol{x}', t'; \boldsymbol{x}, t) V(\boldsymbol{x}, t) \phi_i(\boldsymbol{x}, t) \\ &\quad + \lim_{t' \to \infty} \int d^3x' d^4x_1\, d^4x\, \phi_f^*(\boldsymbol{x}', t') \\ &\quad \times G_0^+(\boldsymbol{x}', t'; \boldsymbol{x}_1, t_1) V(\boldsymbol{x}_1, t_1) G_0^+(\boldsymbol{x}_1, t_1; \boldsymbol{x}, t) V(\boldsymbol{x}, t) \phi_i(\boldsymbol{x}, t) \\ &\quad + \cdots \quad . \end{aligned} \tag{1.38}$$

The first term (the δ function) does not describe scattering but characterizes the particle flux without scattering. The second term represents single scattering, the third term double scattering, etc. Some of these contributions to the S matrix are illustrated in Fig. 1.3: they are coherently summed to give the total S matrix element.

One can obtain alternative expressions for the S matrix if one starts in a similar way to the above procedure, from the advanced Green's functions G_0^- and G^-, cf. (1.8) and (1.9). For example, the state $\phi_f(\boldsymbol{x}, t)$ corresponds to that wave function $\psi_f^{(-)}(\boldsymbol{x}, t)$, which becomes $\phi_f(\boldsymbol{x}, t)$ in the distant future ($t \to \infty$) after the interaction $V(\boldsymbol{x}, t)$ has vanished

$$\lim_{t \to \infty} \psi_f^{(-)}(\boldsymbol{x}, t) = \phi_f(\boldsymbol{x}, t) \quad . \tag{1.39}$$

Starting from (1.9) we can formulate this boundary condition precisely (now the primed $(\boldsymbol{x}', t')$ and unprimed $(\boldsymbol{x}, t)$ are exchanged):

$$\psi_f^{(-)}(\boldsymbol{x}, t) = \lim_{t' \to \infty} -\mathrm{i} \int d^3x G^-(\boldsymbol{x}, t; \boldsymbol{x}', t') \phi_f(\boldsymbol{x}', t') \quad . \tag{1.40}$$

We expect the S matrix element for the transition $i \to f$, i.e. S_{fi}, to be given by the scalar products of $\psi_f^{(-)}(\boldsymbol{x}, t)$ and $\phi_i(\boldsymbol{x}, t)$, calculated at a time t in the distant past ($\lim\limits_{t \to -\infty}$):

$$\begin{aligned} S_{fi} &= \lim_{t \to -\infty} \left\langle \psi_f^{(-)}(\boldsymbol{x}, t) \middle| \phi_i(\boldsymbol{x}, t) \right\rangle \\ &= \lim_{t \to -\infty} \lim_{t' \to \infty} \mathrm{i} \int\!\!\int d^3x d^3x' G^{-*}(\boldsymbol{x}, t; \boldsymbol{x}', t') \phi_f^*(\boldsymbol{x}', t') \phi_i(\boldsymbol{x}, t) \quad . \end{aligned} \tag{1.41}$$

The equivalence with the previous definition of the S-matrix is indeed immediately proved by means of the relation

$$G^+(\boldsymbol{x}', t'; \boldsymbol{x}, t) = G^{-*}(\boldsymbol{x}, t; \boldsymbol{x}', t') \tag{1.42}$$

(see Exercise 1.2), since then (1.41) becomes

$$S_{fi} = \lim_{t\to-\infty}\lim_{t'\to\infty} \mathrm{i}\int\int d^3x d^3x' G^+(\boldsymbol{x}',t';\boldsymbol{x},t)\phi_f^*(\boldsymbol{x}',t')\phi_i(\boldsymbol{x},t)$$
$$= \lim_{t'\to\infty}\left\langle \phi_f(\boldsymbol{x}',t') \middle| \psi_i^{(+)}(\boldsymbol{x}',t')\right\rangle \quad , \tag{1.43}$$

which agrees with the first line of (1.37). Here it is of crucial importance that the potential is real, which was used in the derivation of Exercise 1.2. The fact that (1.41) does not equal (1.37) for a complex potential V is easily understood physically, since an absorptive potential (having a negative imaginary part of $V(\boldsymbol{x},t)$) causes a reduction in the probability of finding the particle in a certain state. This means that $|\psi_i^{(+)}(\boldsymbol{x}',t')|^2$ for $t' \to \infty$ is in general smaller than $|\phi_i(\boldsymbol{x},t)|^2$ for $t \to -\infty$, where ϕ_i is the state from which $\psi_i^{(+)}$ originates. On the other hand, $|\psi_f^{(-)}(\boldsymbol{x},t)|^2$ for $t \to -\infty$ is in general larger than $|\phi_f(\boldsymbol{x}',t')|^2$ for $t' \to \infty$. Here ϕ_f is the state into which $\psi_i^{(-)}(\boldsymbol{x},t)$ changes for $t \to \infty$. Hence we expect that for absorptive potentials (1.41) is larger than (1.37).

Let us finally remark that the S matrix instead of using (1.37) or (1.43) can also be expressed in the following symmetrical way

$$S_{fi} = \left\langle \psi_f^{(-)}(\boldsymbol{x},t) \middle| \psi_i^{(+)}(\boldsymbol{x},t)\right\rangle \quad . \tag{1.44}$$

S_{fi} thus represents the overlap between the solutions $\psi_i^{(+)}$ satisfying the incoming and $\psi_f^{(-)}$ satisfying the outgoing boundary condition. The result is independent of the moment in time t at which the overlap is evaluated.

EXERCISE

1.2 Relation Between G^+ and G^-

Problem. a) Show the validity of the relation

$$G^+(\boldsymbol{x}',t';\boldsymbol{x},t) = G^{-*}(\boldsymbol{x},t;\boldsymbol{x}',t') \quad .$$

b) Prove the validity of (1.44).

Solution. a) We start from the integral equation (1.30) for G^+. In complete analogy to its derivation from Schrödinger's equation (cf. (1.10–19)) and considering the definition of G^- in (1.8) and (1.9), we find the integral equation for G^- which corresponds to (1.30), namely

$$G^-(\boldsymbol{x}',t';\boldsymbol{x},t) = G_0^-(\boldsymbol{x}',t';\boldsymbol{x},t) + \int d^4x_1 G_0^-(x';x_1)V(x_1)G^-(x_1,x) \quad . \tag{1}$$

Here the Green's functions G^+ and G_0^+ have obviously been replaced by G^- and G_0^- in contrast to (1.30). In Example 1.3 we show the validity of the relation

$$G_0^+(\boldsymbol{x}',t';\boldsymbol{x},t) = G_0^{-*}(\boldsymbol{x},t;\boldsymbol{x}',t') \tag{2}$$

for the free Green's functions. If we iterate the integral equation (1), i.e.

Exercise 1.2.

$$G^-(x';x) = G_0^-(x';x) + \int d^4x_1\, G_0^-(x';x_1)V(x_1)G_0^-(x_1;x)$$
$$+ \int d^4x_1\, d^4x_2\, G_0^-(x';x_1)V(x_1)G_0^-(x_1;x_2)V(x_2)G_0^-(x_2;x) + \cdots \quad , \tag{3}$$

it is obvious that by complex conjugation of (3) the following relation also holds:

$$\begin{aligned} G^{-*}(x';x) &= G_0^+(x;x') + \int d^4x_1\, G_0^+(x_1;x')V(x_1)G_0^+(x;x_1) \\ &\quad + \int d^4x_1\, d^4x_2\, G_0^+(x_1;x')V(x_1)G_0^+(x_2;x_1)V(x_2)G_0^+(x;x_2) + \cdots \\ &= G_0^+(x;x') + \int d^4x_1\, G_0^+(x;x_1)V(x_1)G_0^+(x_1;x') \\ &\quad + \int d^4x_1\, d^4x_2\, G_0^+(x;x_2)V(x_2)G_0^+(x_2;x_1)V(x_1)G_0^+(x_1;x') + \cdots \\ &= G^+(x;x') \quad . \end{aligned} \tag{4}$$

We have assumed here, that the potential V is real: $V(x) = V^*(x)$. This proves the validity of the assumption.

b) The claimed expression for the S matrix element can be expressed in terms of the advanced and retarded propagators as follows

$$\begin{aligned} S_{fi} = \int d^3x \Big[\lim_{t'\to+\infty} \mathrm{i} \int d^3x' G^{-*}(\boldsymbol{x},t;\boldsymbol{x}',t')\phi_f^*(\boldsymbol{x}',t') \Big] \\ \times \Big[\lim_{t\to-\infty} \mathrm{i} \int d^3x'' G^+(\boldsymbol{x},t;\boldsymbol{x}'',t'')\phi_i(\boldsymbol{x}'',t'') \Big] \quad . \end{aligned} \tag{5}$$

The spatial integral over d^3x can be solved using (4) and Exercise 1.1(a)

$$\begin{aligned} \int d^3x G^{-*}(\boldsymbol{x},t;\boldsymbol{x}',t')G^+(\boldsymbol{x},t;\boldsymbol{x}'',t'') &= \int d^3x G^+(\boldsymbol{x}',t';\boldsymbol{x},t)G^+(\boldsymbol{x},t;\boldsymbol{x}'',t'') \\ &= -\mathrm{i}G^+(\boldsymbol{x}',t';\boldsymbol{x}'',t'') \end{aligned} \tag{6}$$

Thus (5) becomes

$$S_{fi} = \lim_{t'\to+\infty} \lim_{t''\to-\infty} \mathrm{i} \iint d^3x' d^3x'' \phi_f^*(\boldsymbol{x}',t')G^+(\boldsymbol{x}',t';\boldsymbol{x}'',t'')\phi_i(\boldsymbol{x}'',t'') \; , \tag{7}$$

which agrees with the third line of (1.37). *Exercise 1.2.Exercise 1.2.*

EXAMPLE

1.3 The Free Green's Function and Its Properties

One can deduce an explicit expression for the free Green's function, i.e. in the case where $V = 0$.

As above, we denote the free Green's function by $G_0(\boldsymbol{x}',t';\boldsymbol{x},t)$ and remark that, according to the defining relation (1.2), arbitrary wave packets $\psi(\boldsymbol{x},t)$ develop in space and time corresponding to force free quantum-mechanical motion:

$$\psi(\boldsymbol{x}',t') = \mathrm{i}\int d^3x\; G_0(\boldsymbol{x}',t';\boldsymbol{x},t)\psi(\boldsymbol{x},t) \quad . \tag{1}$$

Example 1.3.

Initially we want to derive a general expression for the Green's function $G(\boldsymbol{x}',t';\boldsymbol{x},t)$ in the case of an arbitrary time-independent potential $V(\boldsymbol{x})$. Let us consider the complete and orthonormal set $(U_E(\boldsymbol{x}'))$ of eigensolutions of the stationary Schrödinger equation

$$\left(-\frac{\hbar^2}{2m}\nabla'^2 + V(\boldsymbol{x}')\right)U_E(\boldsymbol{x}') = EU_E(\boldsymbol{x}') \quad . \tag{2}$$

If the potential $V(\boldsymbol{x}')$ is time-independent, and provided that we know the solution of the following time-dependent Schrödinger equation

$$\mathrm{i}\hbar\frac{\partial\psi(\boldsymbol{x}',t')}{\partial t'} = \left(-\frac{\hbar^2}{2m}\nabla'^2 + V(\boldsymbol{x}')\right)\psi(\boldsymbol{x}',t') \tag{3}$$

at a certain time t, we can write down a formal expression for the solution $\psi(\boldsymbol{x}',t')$ at any time t'. To this end we expand $\psi(\boldsymbol{x}',t')$ at time t' in terms of the basis of energy eigenfunctions $U_E(\boldsymbol{x}')$, using the closure relation

$$\psi(\boldsymbol{x}',t') = \sum_E A_E(t')U_E(\boldsymbol{x}') \quad , \tag{4a}$$

$$A_E(t') = \int d^3x'\,U_E^*(\boldsymbol{x}')\psi(\boldsymbol{x}',t') \quad . \tag{4b}$$

Of course, the expansion coefficients $A_E(t')$ are time dependent. If we insert expansion (4) in (3) and consider (2), we find

$$\mathrm{i}\hbar\sum_E U_E(\boldsymbol{x}')\frac{d}{dt'}A_E(t') = \sum_E A_E(t')EU_E(\boldsymbol{x}') \quad ,$$

and, because of the orthonormality of the wave functions $U_E(\boldsymbol{x}')$,

$$\mathrm{i}\hbar\frac{d}{dt'}A_E(t') = EA_E(t') \quad .$$

This equation is solved by

$$A_E(t') = A_E(t)\exp\left[-\mathrm{i}E(t'-t)/\hbar\right] \quad . \tag{5}$$

It is well known that the probability $P(E)$ of finding the state $U_E(\boldsymbol{x}')$ as a part of $\psi(\boldsymbol{x}',t')$, $P(E) = |A_E(t)|^2$ is not dependent on time.

Hence, if $\psi(\boldsymbol{x}',t')$ is known at time $t' = t$, then the admixture coefficient $A_E(t)$ can be determined according to (4b) and its time dependence is given by (5). Then $\psi(\boldsymbol{x}',t')$ is known at any time t'. We obtain

$$\begin{aligned}\psi(\boldsymbol{x}',t') &= \sum_E A_E(t')U_E(\boldsymbol{x}')\\ &= \sum_E A_E(t)\exp\left[-\mathrm{i}E(t'-t)/\hbar\right]U_E(\boldsymbol{x}')\\ &= \int d^3x\left(\sum_E U_E^*(\boldsymbol{x})U_E(\boldsymbol{x}')\right)\exp\left[-\mathrm{i}E(t'-t)/\hbar\right]\psi(\boldsymbol{x},t) \quad .\end{aligned} \tag{6}$$

Example 1.3.

Comparing (6) with (1) yields as the result for the Green's function

$$G(\boldsymbol{x}',t';\boldsymbol{x},t) = -\mathrm{i}\sum_E U_E^*(\boldsymbol{x})U_E(\boldsymbol{x}')\exp\left[-\mathrm{i}E(t'-t)/\hbar\right] \quad . \tag{7}$$

For $t' = t$ the right-hand side becomes $-\mathrm{i}\delta^3(\boldsymbol{x}'-\boldsymbol{x})$ according to the completeness relation for the states U_E. Note that we derived the full Green's function G and not just G_0, since the above expression is valid if *any time-independent potential* $V(\boldsymbol{x})$ is present. As a special case for $V = 0$ we get the free Green's function G_0 if we insert the free stationary solutions (i.e. plane waves for $U_E(\boldsymbol{x})$. Using nonrelativistic plane waves for $U_E(\boldsymbol{x})$, i.e.

$$U_E(\boldsymbol{x}) = \frac{1}{(2\pi\hbar)^{3/2}}\exp(\mathrm{i}\boldsymbol{k}\cdot\boldsymbol{x}) \tag{8}$$

with wave vector $\boldsymbol{k} = \boldsymbol{p}/\hbar$ and $E = \boldsymbol{p}^2/2m$ leads to

$$\begin{aligned} G_0(\boldsymbol{x}',t';\boldsymbol{x},t) &= -\frac{\mathrm{i}}{(2\pi\hbar)^3}\int \exp\left[\mathrm{i}\boldsymbol{k}\cdot(\boldsymbol{x}'-\boldsymbol{x})\right]\exp\left[-\frac{\mathrm{i}}{\hbar}E(t'-t)\right]d^3p \\ &= -\frac{\mathrm{i}}{(2\pi\hbar)^3}\int \exp\left\{\frac{\mathrm{i}}{\hbar}\left[p_x(x'-x)+p_y(y'-y)+p_z(z'-z)\right]\right. \\ &\qquad \left. -\frac{\mathrm{i}}{\hbar}\frac{(p_x^2+p_y^2+p_z^2)}{2m}(t'-t)\right\}dp_x\,dp_y\,dp_z \quad , \end{aligned} \tag{9}$$

where we have used the Cartesian representation of the integral over d^3p. Further evaluation requires the elementary Gaussian integral formula

$$\int_{-\infty}^{\infty} dx\,\exp(-\mathrm{i}ax^2) = \sqrt{\frac{\pi}{\mathrm{i}a}} \quad . \tag{10}$$

First we consider the integration over p_x and complete the square of the exponent:

$$\begin{aligned} &-\frac{\mathrm{i}}{\hbar}\left[\frac{p_x^2(t'-t)}{2m} - p_x(x'-x)\right] \\ &= -\frac{\mathrm{i}}{\hbar}\left[\frac{p_x\sqrt{t'-t}}{\sqrt{2m}} - \frac{\sqrt{2m}(x'-x)}{2\sqrt{t'-t}}\right]^2 + \frac{\mathrm{i}}{\hbar}\frac{m(x'-x)^2}{2(t'-t)} \\ &= -\frac{\mathrm{i}}{\hbar}\frac{t'-t}{2m}\xi^2 + \frac{\mathrm{i}}{\hbar}\frac{m(x'-x)^2}{2(t'-t)} \quad , \end{aligned} \tag{11}$$

where

$$\xi = p_x - \frac{m(x'-x)}{t'-t} \quad . \tag{12}$$

With $a = (t'-t)/2m\hbar$ we consequently obtain for the first integral

$$\int_{-\infty}^{\infty} dp_x \exp\left\{\frac{\mathrm{i}}{\hbar}\left[p_x(x'-x)-\frac{p_x^2}{2m}(t'-t)\right]\right\}$$

$$= \exp\left[\frac{\mathrm{i}}{\hbar}\frac{m(x'-x)^2}{2(t'-t)}\right]\int_{-\infty}^{\infty} d\xi \exp\left[-\frac{\mathrm{i}}{\hbar}\frac{(t'-t)}{2m}\xi^2\right]$$

$$= \exp\left[\frac{\mathrm{i}}{\hbar}\frac{m(x'-x)^2}{2(t'-t)}\right]\sqrt{\frac{2\pi m\hbar}{\mathrm{i}(t'-t)}} \tag{13}$$

Example 1.3.

and finally, after triple application of the integral formula (10),

$$G_0(\boldsymbol{x}',t';\boldsymbol{x},t) = \frac{-\mathrm{i}}{(2\pi\hbar)^3}\left[\frac{2\pi m\hbar}{\mathrm{i}(t'-t)}\right]^{3/2}\exp\left[\frac{\mathrm{i}|\boldsymbol{x}'-\boldsymbol{x}|^2 2m\hbar}{4\hbar^2(t'-t)}\right]$$

$$= -\mathrm{i}\left(\frac{m}{2\pi\mathrm{i}\hbar(t'-t)}\right)^{3/2}\exp\left[\frac{\mathrm{i}m|\boldsymbol{x}'-\boldsymbol{x}|^2}{2\hbar(t'-t)}\right] \quad . \tag{14}$$

This is the unrestricted, free Green's function G_0 that describes the propagation into the future as well as back into the past. The retarded and advanced Green's functions are readily derived from (14) as follows:

$$G_0^+(\boldsymbol{x}',t';\boldsymbol{x},t) = +G_0(\boldsymbol{x}',t';\boldsymbol{x},t)\Theta(t'-t)$$

$$= -\mathrm{i}\left[\frac{m}{2\pi\mathrm{i}\hbar(t'-t)}\right]^{3/2}\exp\left[\frac{\mathrm{i}m|\boldsymbol{x}'-\boldsymbol{x}|^2}{2\hbar(t'-t)}\right]\Theta(t'-t) \tag{15}$$

and

$$G_0^-(\boldsymbol{x}',t';\boldsymbol{x},t) = -G_0(\boldsymbol{x}',t';\boldsymbol{x},t)\Theta(t-t')$$

$$= +\mathrm{i}\left[\frac{m}{2\pi\mathrm{i}\hbar(t'-t)}\right]^{3/2}\exp\left[\frac{\mathrm{i}m|\boldsymbol{x}'-\boldsymbol{x}|^2}{2\hbar(t'-t)}\right]\Theta(t-t') \quad . \tag{16}$$

The relation

$$G_0^+(\boldsymbol{x}',t';\boldsymbol{x},t) = G_0^{-*}(\boldsymbol{x},t;\boldsymbol{x}',t') \tag{17}$$

follows directly from (15) and (16). By the way, the result of Exercise 1.2 for the full Green's functions follows straight from the expansion (7).

In Example 1.5 we will get acquainted with the Green's function for diffusion and notice that this function is equivalent to the above Green's function (14) for the free motion of a quantum particle if we substitute $t' \to -\mathrm{i}t'$, $t \to -\mathrm{i}t$. This result is quite plausible because of the similarity between the diffusion equation and Schrödinger's equation.

1.6 The Unitarity of the S-Matrix

An important property of the S matrix is its unitarity, provided that the Hamiltonian operator is hermitian. To prove this we have to show that

$$\hat{S}\hat{S}^{\dagger} = \mathbb{1} = \hat{S}^{\dagger}\hat{S} \quad , \tag{1.45}$$

where $\hat{S} = \{S_{ik}\}$ denotes the full S matrix. We may use any of the above-discussed forms of $S_{fi} = \langle f|\hat{S}|i\rangle$ for the S matrix. In particular, we could base the proof on (1.41), which was established with the advanced Green's function G^-. However, we will use the form (1.37), where the retarded Green's function G^+ appears. An arbitrary matrix element of $\hat{S}\hat{S}^+$ then reads, inserting $\mathbb{1} = |\gamma\rangle\langle\gamma|$

$$\begin{aligned}
\langle\beta|\hat{S}\hat{S}^+|\alpha\rangle &= \sum_\gamma \langle\beta|\hat{S}|\gamma\rangle\langle\gamma|\hat{S}^+|\alpha\rangle \\
&= \sum_\gamma \langle\beta|\hat{S}|\gamma\rangle\langle\alpha|\hat{S}|\gamma\rangle^* \\
&= \lim_{t'\to\infty}\lim_{t\to-\infty}\sum_\gamma \int\!\!\int d^3x d^3x' \phi_\beta{}^*(\boldsymbol{x}',t')G^+(\boldsymbol{x}',t';\boldsymbol{x},t)\phi_\gamma(\boldsymbol{x},t) \\
&\quad \times \int\!\!\int d^3x'' d^3x''' \phi_\alpha(\boldsymbol{x}'',t')G^{+*}(\boldsymbol{x}'',t';\boldsymbol{x}''',t)\phi_\gamma^*(\boldsymbol{x}''',t) \quad .
\end{aligned}$$

Here it is advantageous to use the same time arguments t' and t in the matrix elements of both $\hat{S}$ and $\hat{S}^{\dagger}$, respectively. As previously mentioned, S matrix elements do not depend on these times in the limit $t' \to +\infty, t \to -\infty$, since the potential is assumed to decrease sufficiently fast. Since the ϕ_γ form a complete set of states,

$$\sum_\gamma \phi_\gamma(\boldsymbol{x},t)\phi_\gamma^*(\boldsymbol{x}''',t) = \delta^3(\boldsymbol{x}-\boldsymbol{x}''') \quad , \tag{1.46}$$

(1.45) transforms into

$$\begin{aligned}
\langle\beta|\hat{S}\hat{S}^+|\alpha\rangle &= \lim_{t'\to+\infty}\lim_{t\to-\infty}\int\!\!\int\!\!\int d^3x d^3x' d^3x'' \\
&\quad \times \phi_\beta{}^*(\boldsymbol{x}',t')G^+(\boldsymbol{x}',t';\boldsymbol{x},t)G^-(\boldsymbol{x},t;\boldsymbol{x}'',t')\phi_\alpha(\boldsymbol{x}'',t') \ .
\end{aligned} \tag{1.47}$$

Here we have exploited (1.42). Since

$$\int d^3x_1 G^+(\boldsymbol{x}',t';\boldsymbol{x}_1,t_1)G^-(\boldsymbol{x}_1,t_1;\boldsymbol{x}'',t') = \delta^3(\boldsymbol{x}'-\boldsymbol{x}'')$$

(see Exercise 1.1), it follows that

$$\begin{aligned}
\langle\beta|\hat{S}\hat{S}^{\dagger}|\alpha\rangle &= \lim_{t'\to+\infty}\int\!\!\int d^3x' d^3x'' \phi_\beta{}^*(\boldsymbol{x}',t')\delta^3(\boldsymbol{x}'-\boldsymbol{x}'')\phi_\alpha(\boldsymbol{x}'',t') \\
&= \lim_{t'\to+\infty}\int d^3x' \phi_\beta{}^*(\boldsymbol{x}',t')\phi_\alpha(\boldsymbol{x}',t') \\
&= \delta_{\beta\alpha} \quad .
\end{aligned} \tag{1.48}$$

Similarly one can show $\hat{S}^{\dagger}\hat{S} = \mathbb{1}$, which proves the unitarity of the S matrix.[5]

---

[5] This proof of the unitarity of the S matrix rests on the assumption that the potential $V(x)$ is switched on in the distant past and switched off in the distant future. The situation

1.7 Symmetry Properties of the S-Matrix

The S matrix possesses symmetries that reflect the symmetries of the corresponding Hamiltonian operator. In this context we refer to the detailed discussion in *Quantum Mechanics – Symmetries*, 2nd rev. ed., by W. Greiner, B. Müller (Springer, Berlin, Heidelberg 1994). We recognized there that symmetry operations can be represented by unitary operators $\hat{U}$ which act on the states of the Hilbert space. Let $\hat{U}$ be such an operator, which transforms the state $\phi_\beta(\boldsymbol{x},t)$ of a particle into another state

$$\phi_{\beta'}(\boldsymbol{x},t) = \hat{U}\phi_\beta(\boldsymbol{x},t) \quad . \tag{1.49}$$

Then $\phi_{\beta'}$ has to describe a possible free motion of the particles in the system, too, since $\hat{U}$ commutes with $\hat{H}_0$. Moreover, if $\hat{U}$ also commutes with $\hat{H}$, then the state ${\psi_\alpha}^{(+)}$ is transformed into ${\psi_{\alpha'}}^{(+)}$ by means of $\hat{U}$:

$$\psi_{\alpha'}{}^{(+)}(\boldsymbol{x},t) = \hat{U}\psi_\alpha{}^{(+)}(\boldsymbol{x},t) \quad , \tag{1.50}$$

where ${\psi_{\alpha'}}^{(+)}$ is also a possible state of the system including the interaction $V(\boldsymbol{x},t)$. This follows directly from the iterated equation (1.31) by using (1.49). The plus sign (+) characterises the development of the state into the future. We now obtain for the S matrix element between such transformed states, according to (1.37):

$$\begin{aligned}\langle\beta'|\hat{S}|\alpha'\rangle &= \left\langle\phi'_\beta\middle|\psi_{\alpha'}{}^{(+)}\right\rangle = \left\langle\hat{U}\phi_\beta\middle|\hat{U}\psi_\alpha{}^{(+)}\right\rangle \\ &= \left\langle\phi_\beta\middle|\hat{U}^\dagger\hat{U}\psi_\alpha{}^{(+)}\right\rangle = \left\langle\phi_\beta\middle|\psi_\alpha{}^{(+)}\right\rangle = \langle\beta|\hat{S}|\alpha\rangle \quad .\end{aligned} \tag{1.51}$$

The limit $t' \to \infty$ is implied for all these matrix elements. This equation means that the scattering amplitude between an arbitrary pair of states ϕ_α, ϕ_β and caused by, e.g., a spherically symmetric potential is numerically identical with the scattering amplitude between the "rotated" states $\phi_{\alpha'} = \hat{U}\phi_\alpha$ and $\phi_{\beta'} = \hat{U}\phi_\beta$. It is self-evident that both states have to be rotated by the same amount. If we explicitly insert the rotated states on the left-hand side of (1.51), we get

$$\langle\beta|\hat{S}|\alpha\rangle = \langle\hat{U}\phi_\beta|\hat{S}|\hat{U}\phi_\alpha\rangle = \langle\beta|\hat{U}^\dagger\hat{S}\hat{U}|\alpha\rangle = \langle\beta|\hat{S}|\alpha\rangle \quad , \tag{1.52}$$

i.e.

$$\hat{U}^\dagger\hat{S}\hat{U} = \hat{S} \quad \text{or} \quad [\hat{U},\hat{S}]_- = 0 \tag{1.53}$$

has to be valid. We therefore have the important statement: if the symmetry transformation operator $\hat{U}$ *commutes with* $\hat{H}$, then $\hat{U}$ also commutes with the $\hat{S}$ operator.

becomes more complicated if $V(x)$ is completely constant in time. In this case one has to work with wave packets. The switching on or off of the potential is caused by the fact that the wave packet is beyond the range of influence of the potential, when it is in the distant past and distant future. Further complications arise because the Hamiltonian might possess bound states. In this case one can show that the wave packets are orthogonal to these bound states and the set of wave packets is not closed. However, one can nevertheless demonstrate the unitarity of the S matrix for real potentials V, since bound states cannot be occupied because of energy conservation (therefore they are also called "closed channels"). These problems are discussed in texts on the formal theory of scattering, see e.g. M.L. Goldberger and K.M. Watson: *Scattering Theory of Waves and Particles* (McGraw-Hill, New York 1966).

For the anti-unitary operation of time-reversal[6] $\hat{T}$ the situation is more complicated. The time-reversal operator $\hat{T}$ can be written as $\hat{T} = \hat{U}\hat{K}$ where $\hat{U}$ is a unitary operator and $\hat{K}$ stands for complex conjugation. $\hat{T}$ transforms a free state $\phi_\beta(\boldsymbol{x}, t)$ into another state with reversed momentum and reversed angular momentum. The time-reversed state will be symbolically denoted by $\phi_{-\beta}$. However, one still has to pay attention to the fact that the operator $\hat{T}$ reverses the direction of time too, i.e. instead of (1.49) the transformation reads

$$\hat{T}\phi_\beta(\boldsymbol{x}, t) = \phi_{-\beta}(\boldsymbol{x}, -t) \quad . \tag{1.54}$$

If, for example,

$$\phi_\beta(\boldsymbol{x}, t) = N \exp\left[\mathrm{i}(\boldsymbol{k}_\beta \cdot \boldsymbol{x} - \omega_\beta t)\right]$$

is a plane wave for a spin-0 particle, then the state $\phi_{-\beta}(\boldsymbol{x}, t)$, according to (1.54), is given by

$$\phi_{-\beta}(\boldsymbol{x}, t) = N^* \exp\left[\mathrm{i}(-\boldsymbol{k}_\beta \cdot \boldsymbol{x} - \omega_\beta t)\right] \quad .$$

Note that $\phi_{-\beta}(\boldsymbol{x}, t) \neq \hat{T}\phi_\beta(\boldsymbol{x}, t)$, but rather $\phi_{-\beta}(\boldsymbol{x}, t) = \hat{T}\phi_\beta(\boldsymbol{x}, -t)$. This follows immediately from (1.54). Now let us consider the case of the time-reversal operator commuting with the Hamiltonian operator, i.e.

$$[\hat{T}, \hat{H}]_- = 0 \quad . \tag{1.55}$$

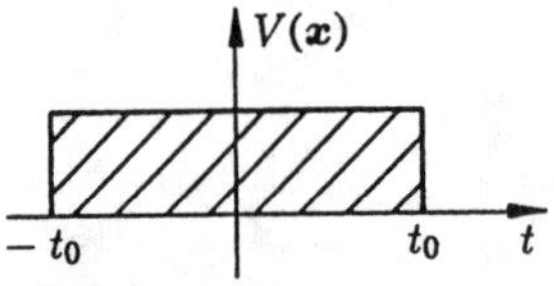

Fig. 1.5. A potential essentially constant in time, which is switched on and off at times $\pm t_0$

Then we speak of a time-reversal-invariant system. A simple example is a potential $V(\boldsymbol{x})$ that is constant in time but is switched on and off at times $-t_0$ and t_0, respectively, in a manner symmetrical in time (cf. Fig. 1.5). The time-reversal operator $\hat{T}$ then transforms the state $\psi_\alpha{}^{(+)}(\boldsymbol{x}, t)$ into another state that is also a possible state of motion of the system (because of its time-reversal-invariance). This state will have reversed momentum and reversed angular momentum compared to $\psi_\alpha{}^{(+)}$. In the distant future it converges towards $\phi_{-\alpha}(\boldsymbol{x}, -t)$, in contrast to $\psi_\alpha{}^{(+)}(\boldsymbol{x}, t)$, which originated from $\phi_\alpha(\boldsymbol{x}, t)$ in the distant past. We therefore call this time-reversed state $\psi_\alpha{}^{(-)}(\boldsymbol{x}, -t)$ and are able to write, because of (1.54),

$$\hat{T}\psi_\alpha{}^{(+)}(\boldsymbol{x}, t) = \psi_{-\alpha}{}^{(+)}(\boldsymbol{x}, -t) \quad . \tag{1.56}$$

The time-reversal operation $\hat{T}$ therefore reverses the direction of the time evolution ($t \to -t$), as we already know from (1.54). For a system invariant under time reversal and having a hermitian Hamiltonian, from (1.41):

$$\begin{aligned}
\left\langle -\alpha \left| \hat{S} \right| -\beta \right\rangle &= \lim_{t\to-\infty} \left\langle \psi_{-\alpha}{}^{(-)}(\boldsymbol{x}, t) \middle| \phi_{-\beta}(\boldsymbol{x}, t) \right\rangle \\
&= \lim_{t\to+\infty} \left\langle \psi_{-\alpha}{}^{(-)}(\boldsymbol{x}, -t) \middle| \phi_{-\beta}(\boldsymbol{x}, -t) \right\rangle \\
&= \lim_{t\to\infty} \left\langle \hat{T}\psi_\alpha{}^{(+)}(\boldsymbol{x}, t) \middle| \hat{T}\phi_\beta(\boldsymbol{x}, t) \right\rangle \quad ,
\end{aligned}$$

[6] See W. Greiner, B. Müller: *Quantum Mechanics – Symmetries*, 2nd rev. ed., (Springer, Berlin, Heidelberg 1994), Chap. 11 and Theoretical Physics, Vol. 3 by W. Greiner: *Relativistic Quantum Mechanics – Wave Equations* (Springer, Berlin, Heidelberg 1990), Chap. 12.

because of (1.54) and (1.56), and using $\hat{T} = \hat{U}\hat{K}$

$$\begin{aligned}\langle -\alpha | \hat{S} | -\beta \rangle &= \lim_{t\to\infty} \left\langle \hat{U}\hat{K}\psi_\alpha{}^{(+)}(\boldsymbol{x},t) \middle| \hat{U}\hat{K}\phi_\beta(\boldsymbol{x},t) \right\rangle \\ &= \lim_{t\to\infty} \left\langle \hat{K}\psi_\alpha{}^{(+)}(\boldsymbol{x},t) \middle| \hat{K}\phi_\beta(\boldsymbol{x},t) \right\rangle \\ &= \lim_{t\to\infty} \left\langle \phi_\beta(\boldsymbol{x},t) \middle| \psi_\alpha{}^{(+)}(\boldsymbol{x},t) \right\rangle = \langle \beta | \hat{S} | \alpha \rangle \quad . \end{aligned} \tag{1.57}$$

Equation (1.57) states that the scattering amplitude $S_{\beta\alpha}$ from an initial state $|\alpha\rangle$ into a final state $|\beta\rangle$ is numerically equal to the scattering amplitude from state $|-\beta\rangle$ (with momentum reversed compared to $|\beta\rangle$) into the state $|-\alpha\rangle$ (with momentum reversed as compared to $|\alpha\rangle$). This is, of course, only valid if the system is invariant under time reversal, since only then are the states $\hat{T}\psi_\alpha$ possible states of the system. The interesting relationship (1.57) is called the *reciprocity theorem*. One can even show its validity for complex potentials $V(\boldsymbol{x},t)$.[7]

1.8 The Green's Function in Momentum Representation and Its Properties

Until now we have emphasized the physical uses of the Green's functions. Now we want to develop the mathematical apparatus for practical calculations. We aim first at a differential equation for the Green's function and begin our discussion with the defining relation (1.7):

$$\Theta(t'-t)\psi(x') = \mathrm{i} \int d^3x \, G^+(x';x)\psi(x) \quad , \tag{1.58}$$

where $x = \{x_0, \boldsymbol{x}\}$ abbreviates the position 4-vector, and $\Theta(t'-t)$ is the step function introduced in (1.6). For further analysis it is useful to know the following integral representation of the step function:

$$\Theta(\tau) = -\frac{1}{2\pi\mathrm{i}} \lim_{\varepsilon\to 0} \int\limits_{-\infty}^{\infty} d\omega \frac{\mathrm{e}^{-\mathrm{i}\omega\tau}}{\omega + \mathrm{i}\varepsilon} \quad , \tag{1.59}$$

which we prove in Exercise 1.4.

[7] Cf., for example, L. Schiff: *Quantum Mechanics*, 3rd edition (McGraw-Hill, New York 1968), Chap. 20.

EXERCISE

1.4 An Integral Representation for the Step Function

Problem. Show that

$$\Theta(\tau) = -\frac{1}{2\pi\mathrm{i}} \lim_{\varepsilon\to 0} \int_{-\infty}^{\infty} d\omega \frac{\mathrm{e}^{-\mathrm{i}\omega\tau}}{\omega + \mathrm{i}\varepsilon} \tag{1}$$

is an integral representation of Heaviside's step function.

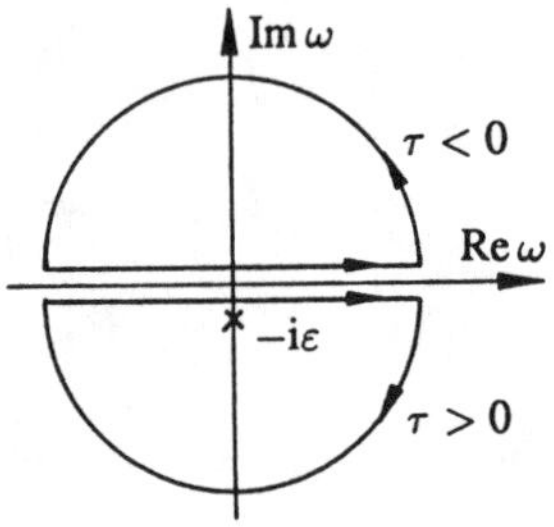

Fig. 1.6. Integration contours for $\tau < 0$ and $\tau > 0$

Solution. We perform the ω-integral as a contour integral in the complex ω plane (Fig. 1.6). There is a pole of first order at $\omega = -\mathrm{i}\varepsilon$. For $\tau < 0$ we close the contour in the upper half plane, since the contribution from the upper, infinitely distant half circle vanishes in this case. With $\omega = \varrho e^{\mathrm{i}\phi}$ the integrand reads

$$f(\varrho,\phi) = \frac{\mathrm{e}^{-\mathrm{i}\omega\tau}}{\omega} = \frac{\mathrm{e}^{-\mathrm{i}\varrho\tau(\cos\phi + \mathrm{i}\sin\phi)}}{\varrho e^{\mathrm{i}\phi}} = \mathrm{e}^{-\mathrm{i}\varrho\tau\cos\phi}\,\frac{\mathrm{e}^{+\varrho\tau\sin\phi}}{\varrho e^{\mathrm{i}\phi}} \quad .$$

For $\tau < 0$ the contribution from the upper half circle to the integral becomes smaller than

$$\pi\varrho|f(\varrho,\phi)| = \frac{\pi\varrho e^{-\varrho|\tau|\sin\phi}}{\varrho} \to 0 \quad (\varrho\to\infty) \quad .$$

According to Cauchy's integral theorem the complete integral over the contour closed in the upper half plane thus *vanishes*, since the pole lies outside the region enclosed by the integration boundaries. In the case $\tau > 0$ for similar reasons one can close the contour by means of an infinitely large half circle below the real axis. Then Cauchy's integral theorem states that the residue of the integrand at the pole determines the value of the integral. The clockwise direction of integration leads to a minus sign. Therefore, we obtain

$$\Theta(\tau > 0) = -\frac{1}{2\pi\mathrm{i}}(-1)2\pi\mathrm{i} \lim_{\varepsilon\to 0} \mathrm{Res}_{\omega=-\mathrm{i}\varepsilon}\left[\frac{\mathrm{e}^{-\mathrm{i}\omega\tau}}{\omega+\mathrm{i}\varepsilon}\right] = \mathrm{e}^{-\mathrm{i}\omega\tau}|_{\omega=0} = 1 \quad . \tag{2}$$

From this we get directly, by differentiating,

$$\frac{d\Theta(\tau)}{d\tau} = -\frac{1}{2\pi\mathrm{i}} \lim_{\varepsilon\to 0} \int_{-\infty}^{\infty} \frac{d}{d\tau}\frac{\mathrm{e}^{-\mathrm{i}\omega\tau}}{\omega+\mathrm{i}\varepsilon}\, d\omega = -\frac{1}{2\pi\mathrm{i}} \lim_{\varepsilon\to 0} \int_{-\infty}^{\infty} \frac{-\mathrm{i}\omega}{\omega+\mathrm{i}\varepsilon}\mathrm{e}^{-\mathrm{i}\omega\tau}\, d\omega$$

$$= \frac{1}{2\pi}\int_{-\infty}^{\infty} \mathrm{e}^{-\mathrm{i}\omega\tau}\, d\omega = \delta(\tau) \quad . \tag{1.60}$$

So the derivative of the step function yields Dirac's δ function. With the help of this relation one can specify a differential equation for the retarded Green's function $G^+(x';x)$ from (1.58) and deduce some of its other formal properties. We know that $\psi(x')$ fulfills Schrödinger's equation

$$\left(\mathrm{i}\hbar\frac{\partial}{\partial t'} - \hat{H}(x')\right)\psi(x') = 0 \quad . \tag{1.61}$$

Therefore, we apply the operator $\left(\mathrm{i}\hbar(\partial/\partial t') - \hat{H}(x')\right)$ to (1.58) from the left and get

$$\begin{aligned}&\left(\mathrm{i}\hbar\frac{\partial}{\partial t'} - \hat{H}(x')\right)\Theta(t'-t)\psi(x')\\&\quad = \mathrm{i}\int d^3x \left(\mathrm{i}\hbar\frac{\partial}{\partial t'} - \hat{H}(x')\right) G^+(x';x)\psi(x) \quad .\end{aligned} \tag{1.62}$$

The left-hand side is expanded to yield

$$\begin{aligned}&\left(\mathrm{i}\hbar\frac{\partial}{\partial t'}\Theta(t'-t)\right)\psi(x') + \Theta(t'-t)\left(\mathrm{i}\hbar\frac{\partial}{\partial t'} - \hat{H}(x')\right)\psi(x')\\&\quad = \left(\mathrm{i}\hbar\frac{\partial}{\partial t'}\Theta(t'-t)\right)\psi(x') = \mathrm{i}\hbar\delta(t'-t)\psi(x')\end{aligned} \tag{1.63}$$

so that (1.62) becomes

$$\mathrm{i}\int d^3x \left[\left(\mathrm{i}\hbar\frac{\partial}{\partial t'} - \hat{H}(x')\right) G^+(x';x) - \hbar\delta^3(\boldsymbol{x}'-\boldsymbol{x})\delta(t'-t)\right]\psi(x) = 0 \quad .$$

Since this equation has to be satisfied for arbitrary solutions $\psi(x)$, the term in square brackets must vanish, i.e.

$$\left(\mathrm{i}\hbar\frac{\partial}{\partial t'} - \hat{H}(x')\right) G^+(x';x) = \hbar\delta^4(x'-x) \quad , \tag{1.64}$$

where we have replaced $\delta^3(\boldsymbol{x}'-\boldsymbol{x})\delta(t'-t)$ by the four-dimensional δ function, $\delta^4(x'-x)$. This differential equation determines, together with the *boundary condition* for propagation forward in time, the retarded Green's function $G^+(x';x)$:

$$G^+(x';x) = 0 \quad \text{for} \quad t' < t \quad . \tag{1.65}$$

Obviously the Green's function is exactly the wave emitted from a point-like space–time source of strength

$$\hbar\delta^4(x'-x) = \hbar\delta^3(\boldsymbol{x}'-\boldsymbol{x})\delta(t'-t) \quad .$$

Equation (1.64) clearly illustrates how the Green's function technique can be used to solve an inhomogeneous linear differential equation. Given the Schrödinger equation

$$\left(\mathrm{i}\hbar\frac{\partial}{\partial t} - \hat{H}(x)\right)\psi(x) = \varrho(x) \quad , \tag{1.66}$$

with the source term $\varrho(x)$, we can immediately write down a solution

$$\psi(x) = \psi_0(x) + \frac{1}{\hbar}\int d^4x' G^+(x,x')\varrho(x') \quad \text{for} \quad t > t' \quad . \tag{1.67}$$

Here $\psi_0(x)$ is a solution of the homogeneous differential equation. Let us now calculate once more the propagator for *free particles*, but this time with the help of the differential equation (1.64) and the boundary condition (1.65). For free, nonrelativistic particles the Hamiltonian is

$$\hat{H}_0(x') = -\frac{\hbar^2}{2m}\nabla'^2 \quad . \tag{1.68}$$

In addition we note that $G_0{}^+(x';x)$ will depend only on the difference of the coordinates, $x' - x = \{\boldsymbol{x}', t'\} - \{\boldsymbol{x}, t\}$. This is because a wave emitted from the source at $\boldsymbol{x}$ at time t and arriving at $\boldsymbol{x}'$ at time t' depends only on the distance $\{\boldsymbol{x}' - \boldsymbol{x}, t' - t\}$. The Green's function, however, is precisely such a wave. Thus we are able to write

$$G_0{}^+(x';x) = G_0{}^+(x' - x) \quad . \tag{1.69}$$

Mathematically one readily appreciates this fact, since one can easily rewrite the differential equation analogous to (1.64) for the free propagator in a differential equation involving relative coordinates $z = x' - x = \{\boldsymbol{x}' - \boldsymbol{x}, t' - t\}$, because the Hamiltonian operator $\hat{H}_0$ is homogenous in spatial and time coordinates.

To proceed with the solution of (1.64) for free particles we consider the Fourier representation

$$G_0{}^+(x' - x) = \int \frac{d^3p\, dE}{(2\pi\hbar)^4} \exp\left[\frac{\mathrm{i}}{\hbar}\boldsymbol{p}\cdot(\boldsymbol{x}' - \boldsymbol{x})\right] \times \exp\left[-\frac{\mathrm{i}}{\hbar}E(t' - t)\right] G_0{}^+(\boldsymbol{p};E) \tag{1.70}$$

and determine with (1.68) and (1.64) the relation for the Fourier transform $G_0{}^+(\boldsymbol{p};E)$:

$$\begin{aligned}
&\left(\mathrm{i}\hbar\frac{\partial}{\partial t'} + \frac{\hbar^2}{2m}\nabla'^2\right) G_0{}^+(x' - x) \\
&= \int \frac{d^3p dE}{(2\pi\hbar)^4}\left(E - \frac{\boldsymbol{p}^2}{2m}\right) G_0{}^+(\boldsymbol{p};E) \exp\left[-\frac{\mathrm{i}}{\hbar}E(t' - t)\right] \exp\left[\frac{\mathrm{i}}{\hbar}\boldsymbol{p}(\boldsymbol{x}' - \boldsymbol{x})\right] \\
&\stackrel{!}{=} \hbar \int \frac{d^3p\, dE}{(2\pi\hbar)^4} \exp\left[-\frac{\mathrm{i}}{\hbar}E(t' - t)\right] \exp\left[\frac{\mathrm{i}}{\hbar}\boldsymbol{p}(\boldsymbol{x}' - \boldsymbol{x})\right] \quad .
\end{aligned} \tag{1.71}$$

The last term is the right-hand side of (1.64), i.e. $\hbar\delta^4(x' - x)$ in energy–momentum representation. Obviously one can immediately give the solution of the differential equation (1.64) in Fourier representation. For $E \neq \boldsymbol{p}^2/2m$ one obtains

$$G_0{}^+(\boldsymbol{p};E) = \frac{\hbar}{E - \frac{\boldsymbol{p}^2}{2m}} \quad . \tag{1.72}$$

This expression is still incomplete, since the treatment of the singularity at $E = \boldsymbol{p}^2/2m$ has not yet been determined. This is done using the retardation condition (1.65). We proceed as in Exercise 1.4 with the Fourier representation of the step function and add an infinitesimal, positive imaginary part $\mathrm{i}\varepsilon$ to the denominator of (1.72) and perform first of all the E integration in (1.70). As illustrated in Fig. 1.7 the singularity then lies below the real E axis. We obtain

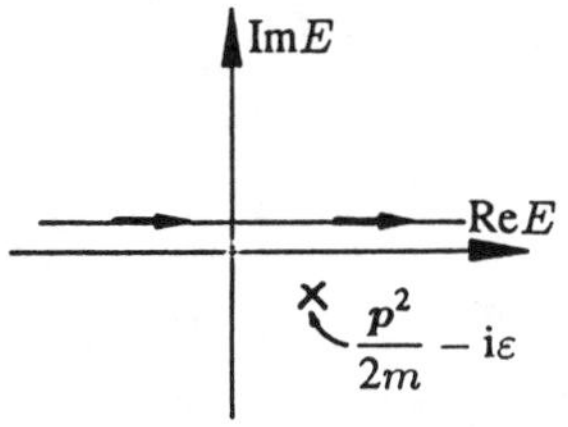

Fig. 1.7. The position of the singularity of $G^+(p;E)$ and the integration contour along the E axis

$$G_0{}^+(x'-x) = \hbar \int \frac{d^3p}{(2\pi\hbar)^3} \exp\left[\frac{\mathrm{i}}{\hbar}\boldsymbol{p}\cdot(\boldsymbol{x}'-\boldsymbol{x})\right] \times \int\limits_{-\infty}^{\infty} \frac{dE}{2\pi\hbar} \frac{\exp\left[-\frac{\mathrm{i}}{\hbar}E(t'-t)\right]}{E - \boldsymbol{p}^2/2m + \mathrm{i}\varepsilon} \quad . \tag{1.73}$$

With the substitution $E' = E - \boldsymbol{p}^2/2m$ the last integral changes into

$$\begin{aligned}
&\int\limits_{-\infty}^{\infty} \frac{dE'}{2\pi\hbar} \frac{\exp\left[-\frac{\mathrm{i}}{\hbar}(E'+\boldsymbol{p}^2/2m)(t'-t)\right]}{E'+\mathrm{i}\varepsilon} \\
&= \exp\left[-\frac{\mathrm{i}}{\hbar}\frac{\boldsymbol{p}^2}{2m}(t'-t)\right] \int_{-\infty}^{\infty} \frac{dE'}{2\pi\hbar} \frac{\exp\left[-\frac{\mathrm{i}}{\hbar}E'(t'-t)\right]}{E'+\mathrm{i}\varepsilon} \\
&= \exp\left[-\frac{\mathrm{i}}{\hbar}\frac{\boldsymbol{p}^2}{2m}(t'-t)\right]\left[-\frac{\mathrm{i}}{\hbar}\Theta\left(\frac{t'-t}{\hbar}\right)\right] \\
&= -\frac{\mathrm{i}}{\hbar}\exp\left[-\frac{\mathrm{i}}{\hbar}\frac{\boldsymbol{p}^2}{2m}(t'-t)\right]\Theta(t'-t) \quad .
\end{aligned}$$

In the last two steps we have exploited (1.59) and the property of the step function that $\Theta(\alpha x) = \Theta(x)$ for positive α . Now (1.73) transforms into

$$\begin{aligned}
G_0{}^+(x'-x) &= -\mathrm{i} \int \frac{d^3\boldsymbol{p}}{(2\pi\hbar)^3} \exp\left\{\frac{\mathrm{i}}{\hbar}\left[\boldsymbol{p}\cdot(\boldsymbol{x}'-\boldsymbol{x}) - \frac{\boldsymbol{p}^2}{2m}(t'-t)\right]\right\}\Theta(t'-t) \\
&= -\mathrm{i}\Theta(t'-t)\int d^3\boldsymbol{p}\,\phi_{\boldsymbol{p}}(\boldsymbol{x}',t')\phi_{\boldsymbol{p}}{}^*(\boldsymbol{x},t) \quad .
\end{aligned} \tag{1.74}$$

Here we have denoted the eigenfunctions of the free Hamiltonian $\hat{H}_0$, i.e. the plane waves, by

$$\begin{aligned}
\phi_{\boldsymbol{p}}(\boldsymbol{x},t) &= \frac{1}{\sqrt{(2\pi\hbar)^3}} \exp\left[\frac{\mathrm{i}}{\hbar}\left(\boldsymbol{p}\cdot\boldsymbol{x} - \frac{\boldsymbol{p}^2}{2m}t\right)\right] \\
&= \frac{1}{\sqrt{(2\pi\hbar)^3}} \exp\left[\mathrm{i}(\boldsymbol{k}\cdot\boldsymbol{x} - \omega t)\right] \,, \quad \hbar\omega = \boldsymbol{p}^2/2m \,, \quad \hbar\boldsymbol{k} = \boldsymbol{p} \quad .
\end{aligned} \tag{1.75}$$

Equation (1.74) is identical with the result of Example 1.3, (9) and (15). Further evaluation of the integral (1.74) proceeds then as above.

From this example we realize how useful it is to express the Green's function as a sum over a complete set of eigenfunctions of the corresponding Schrödinger equation. For such a set of functions $\psi_n(\boldsymbol{x},t)$ the closure relation reads

$$\sum_n \psi_n(\boldsymbol{x}',t)\psi_n^*(\boldsymbol{x},t) = \delta^3(\boldsymbol{x}'-\boldsymbol{x}) \quad . \tag{1.76}$$

Note that the *same time t* appears in $\psi_n(\boldsymbol{x}',t)$ and in $\psi_n^*(\boldsymbol{x},t)$.

Now one can easily verify that (1.64) together with boundary condition (1.65) is solved by

$$G^+(x';x) = -\mathrm{i}\Theta(t'-t)\sum_n \psi_n^*(x)\psi_n(x') \quad , \tag{1.77}$$

since

$$\begin{aligned}\left(\mathrm{i}\hbar\frac{\partial}{\partial t'} - \hat{H}(x')\right) G^+(x';x) &= \hbar\delta(t'-t)\sum_n \psi_n(\boldsymbol{x}',t)\psi_n^*(\boldsymbol{x},t)\\ &\quad - \mathrm{i}\Theta(t'-t)\sum_n \underbrace{\left[\left(\mathrm{i}\hbar\frac{\partial}{\partial t'} - \hat{H}(x')\right)\psi_n(x')\right]}_{=0}\psi_n^*(x)\\ &= \hbar\delta(t'-t)\delta^3(\boldsymbol{x}'-\boldsymbol{x})\\ &= \hbar\delta^4(x'-x) \quad .\end{aligned} \tag{1.78}$$

Next, we point out another important relationship: the same Green's function $G^+(x';x)$ that describes the evolution of a solution $\psi_n(\boldsymbol{x},t)$ of Schrödinger's equation forward in time also describes the propagation of the complex-conjugate solution $\psi_n^*(\boldsymbol{x},t)$ backward in time. From (1.77) we obtain on the one hand

$$\begin{aligned}\mathrm{i}\int d^3x\, G^+(x';x)\psi_n(x) &= \Theta(t'-t)\sum_m \psi_m(x')\underbrace{\int d^3x\,\psi_m^*(x)\psi_n(x)}_{\delta_{nm}}\\ &= \Theta(t'-t)\psi_n(x')\end{aligned} \tag{1.79a}$$

and on the other hand

$$\begin{aligned}\mathrm{i}\int d^3x'\,\psi_n^*(x')G^+(x';x) &= \Theta(t'-t)\sum_m \underbrace{\int d^3x'\psi_m(x')\psi_n^*(x')}_{\delta_{nm}}\,\psi_m^*(x)\\ &= \Theta(t'-t)\psi_n^*(x) \quad .\end{aligned} \tag{1.79b}$$

As stated above, (1.79a) expresses the propagation of $\psi_n(x)$ forward in time and (1.79b) the corresponding backward propagation of $\psi_n^*(x')$. The latter results may also be obtained by complex conjugating (1.9) and using Exercise 1.2. Conversely, starting from (1.9) and (1.79b) one easily proves the relations between G^+ and G^- outlined in Exercise 1.2.

1.9 Another Look at the Green's Function for Interacting Particles

Here we want to reconsider the iteration method for the Green's function $G^+(x';x)$ (cf. (1.28)) from a slightly different point of view. The starting point of our present discussion is the differential equation (1.64) for $G^+(x';x)$, which can be formulated with $\hat{H} = \hat{H}_0 + V$:

$$\left(\mathrm{i}\hbar\frac{\partial}{\partial t'} - \hat{H}_0(x')\right) G^+(x';x) = \hbar\delta^4(x'-x) + V(x')G^+(x';x) \quad . \tag{1.80}$$

The right-hand side can be interpreted as the source term in an inhomogeneous Schrödinger equation as in (1.66):

$$\left(\mathrm{i}\hbar\frac{\partial}{\partial t'} - \hat{H}_0(x')\right)\psi(x') = \varrho(x') \quad . \tag{1.81}$$

Using the free Green's function G_0 the solution of (1.81) is given by

$$\psi(x') = \frac{1}{\hbar}\int d^4x_1 G_0^+(x';x_1)\varrho(x_1) \quad . \tag{1.82}$$

Replacing $\psi(x')$ by $G^+(x',x)$ this leads immediately to the following integral equation for the interacting Green's function

$$\begin{aligned} G^+(x';x) &= \int d^4x_1\, G_0^+(x';x_1)\,\left(\,\delta^4(x_1-x) + V(x_1)G^+(x_1;x)\,\right) \\ &= G_0^+(x';x) + \int d^4x_1\, G_0^+(x';x_1)V(x_1)G^+(x_1;x) \quad . \end{aligned} \tag{1.83}$$

Here we again replaced $V(x)/\hbar \to V(x)$ as in (1.21). Equation (1.83) is identical with our earlier result (1.30). The iteration of (1.82) leads to the multiple scattering expansion (1.28) for the Green's function. This can be used to construct the S-matrix (1.37)

$$S_{fi} = \lim_{t\to-\infty}\lim_{t'\to+\infty} \mathrm{i}\int d^3x' d^3x\, \phi_f^*(x')G^+(x';x)\phi_i(x) \quad . \tag{1.84}$$

Using the equations (1.79a,b) for free particles

$$\int d^3x G_0^+(x_1,x)\phi_i(x) = -\mathrm{i}\phi_i(x_1) \quad \text{for } t_1 > t \quad , \tag{1.85a}$$

$$\int d^3x' \phi_f^*(x')G_0^+(x',x_1) = -\mathrm{i}\phi_f^*(x_1) \quad \text{for } t' > t_1 \quad , \tag{1.85b}$$

the x and x' integrations can be carried out and (1.28) leads to

$$\begin{aligned} S_{fi} = \delta_{fi} &- \mathrm{i}\int d^4x_1\; \phi_f^*(x_1)\; V(x_1)\; \phi_i(x_1) \\ &- \mathrm{i}\int d^4x_1\; d^4x_2\; \phi_f^*(x_1)\; V(x_1)G_0^+(x_1;x_2)V(x_2)\; \phi_i(x_2) \\ &- \mathrm{i}\int d^4x_1\; d^4x_2\; d^4x_3\; \phi_f^*(x_1)\; V(x_1)G_0^+(x_1;x_2)V(x_2)G_0^+(x_2;x_3)V(x_3)\; \phi_i(x_3) \\ &+ \cdots \quad . \end{aligned} \tag{1.86}$$

This is the expansion of the S matrix in terms of multiple scattering events. δ_{fi} describes the absence of scattering, the second term where V appears once describes single scattering, the next term double scattering, etc.

EXERCISE

1.5 Green's Function for Diffusion

Typical diffusion phenomena, e.g. heat conduction or two mutually permeating fluids, are determined solely by the gradients of the density. For example, a fluid of density $\varrho(\boldsymbol{x},t)$ tends to flow from a point, where the density is high, towards a region of low density. The current $\boldsymbol{J}$ is therefore assumed to be proportional to the gradient of the density:

$$\boldsymbol{J} = -D\nabla\varrho \quad , \tag{1}$$

where the constant D is called the diffusion constant. If we combine (1) with the continuity equation

$$\frac{\partial\varrho}{\partial t} = -\nabla\cdot\boldsymbol{J} \quad , \tag{2}$$

we obtain the diffusion equation

$$\frac{\partial\varrho}{\partial t} = D\,\nabla^2\,\varrho \quad . \tag{3}$$

In the case of heat conduction, ϱ is the "heat contents" per unit volume and is proportional to the temperature: $\varrho = CT$. The constant C is the specific heat capacity of the material. For later considerations we note that the transformation $t \to -\mathrm{i}t$ changes the diffusion equation (3) into a differential equation of Schrödinger type.

Now we want to construct explicitly the Green's function $G^+(\boldsymbol{x}',t';\boldsymbol{x},t)$ of the diffusion in an infinite three-dimensional region. Again we start with the defining differential equation

$$\nabla'^2 G \;-\; a^2\,\frac{\partial G}{\partial t'} = -4\pi\,\delta^3(\boldsymbol{x}'-\boldsymbol{x})\,\delta(t'-t) \quad , \tag{4}$$

with the constants $a^2 = 1/D$. The factor 4π on the right-hand side (instead of $\hbar$ as in (1.64)) is a matter of convention. This changes the Green's function only by a factor of $4\pi/\hbar$. Since the Green's function depends only on the time difference $t'-t$ and the spatial difference $\boldsymbol{x}'-\boldsymbol{x}$, we introduce the following abbreviations:

$$\begin{aligned} \tau &= t' - t \quad , \\ \boldsymbol{R} &= \boldsymbol{x}' - \boldsymbol{x} \quad . \end{aligned} \tag{5}$$

As an ansatz for the Green's function we write down the Fourier transform

$$G(\boldsymbol{x}',t';\boldsymbol{x},t) = \frac{1}{(2\pi)^3}\int d^3p\; \mathrm{e}^{\mathrm{i}\boldsymbol{p}\cdot(\boldsymbol{x}'-\boldsymbol{x})}\; g(\boldsymbol{p},\tau) \quad , \tag{6}$$

where the function $g(\boldsymbol{p},\tau)$ is not yet known. We insert (6) into the defining equation (4) and once again use the following three-dimensional Fourier representation of the δ function:

$$\delta^3(\boldsymbol{R}) = \frac{1}{(2\pi)^3}\int d^3p\; \mathrm{e}^{\mathrm{i}\boldsymbol{p}\cdot\boldsymbol{R}} \quad . \tag{7}$$

The left hand side of (4) becomes *Exercise 1.5.*

$$\nabla'^2 G - a^2 \frac{\partial G}{\partial \tau} = \frac{1}{(2\pi)^3} \int d^3p\, \mathrm{e}^{\mathrm{i}\boldsymbol{p}\cdot\boldsymbol{R}} \left(-p^2 g - a^2 \frac{\partial g}{\partial \tau} \right) \quad . \tag{8}$$

This results in a differential equation in time for g:

$$a^2 \frac{\partial g}{\partial \tau} + p^2 g = 4\pi\, \delta(\tau) \quad . \tag{9}$$

This differential equation has the following causal solution:

$$g = \frac{4\pi}{a^2}\, \mathrm{e}^{-(p^2\tau/a^2)}\, \Theta(\tau) \quad . \tag{10}$$

To prove this we insert solution (10) into (9) and use the relation

$$\frac{d\Theta(\tau)}{d\tau} = \delta(\tau) \tag{11}$$

between the step function Θ and the δ function:

$$\begin{aligned} &a^2 \frac{4\pi}{a^2} \left(\frac{-p^2}{a^2} \right) \mathrm{e}^{-(p^2\tau/a^2)}\, \Theta(\tau) + a^2 \frac{4\pi}{a^2}\, \mathrm{e}^{-(p^2\tau/a^2)} \frac{d\Theta}{d\tau} \\ &\quad + p^2 \frac{4\pi}{a^2}\, \mathrm{e}^{-(p^2\tau/a^2)}\, \Theta(\tau) = 4\pi\delta(\tau) \quad . \end{aligned} \tag{12}$$

Summarizing, we get

$$G(\boldsymbol{x}', t'; \boldsymbol{x}, t) = \left(\frac{4\pi}{(2\pi)^3 a^2} \right) \Theta(\tau) \int d^3p\, \exp(\mathrm{i}\boldsymbol{p}\cdot\boldsymbol{R}) \exp\left(-\frac{p^2\tau}{a^2} \right) \tag{13}$$

or explicitly in Cartesian coordinates

$$\begin{aligned} G(\boldsymbol{x}', t'; \boldsymbol{x}, t) &= \left(\frac{4\pi}{(2\pi)^3 a^2} \right) \Theta(\tau) \int\limits_{-\infty}^{\infty} dp_x \mathrm{e}^{\mathrm{i}p_x(x'-x)}\, \mathrm{e}^{-(p_x^2/a^2)\tau} \\ &\quad \times \int\limits_{-\infty}^{\infty} dp_y \mathrm{e}^{\mathrm{i}p_y(y'-y)}\, \mathrm{e}^{-(p_y^2/a^2)\tau} \int\limits_{-\infty}^{\infty} dp_z \mathrm{e}^{\mathrm{i}p_z(z'-z)}\, \mathrm{e}^{-(p_z^2/a^2)\tau} \, . \end{aligned} \tag{14}$$

Let us consider the exponent in the first integral:

$$\begin{aligned} \mathrm{i}p_x R_x - \frac{p_x^2}{a^2}\tau &= -\left(\frac{p_x}{a}\sqrt{\tau} - \frac{\mathrm{i}aR_x}{2\sqrt{\tau}} \right)^2 - \frac{a^2 R_x^2}{4\tau} \\ &= -\frac{\tau}{a^2}\,\xi^2 - \frac{a^2 R_x^2}{4\tau} \quad , \end{aligned} \tag{15}$$

with

$$\xi = p_x - \frac{\mathrm{i}a^2 R_x}{2\tau} \quad . \tag{16}$$

With this transformation the first integral in (14) can be rewritten as

Exercise 1.5.

$$\int\limits_{-\infty}^{\infty} dp_x \exp\left(-\frac{\tau\xi^2}{a^2} - \frac{a^2R_x^2}{4\tau}\right) dp_x = \exp\left(-\frac{a^2R_x^2}{4\tau}\right) \int\limits_{-\infty}^{\infty} d\xi \exp\left(-\frac{\tau\xi^2}{a^2}\right)$$
$$= \exp\left(-\frac{a^2R_x^2}{4\tau}\right) a\sqrt{\frac{\pi}{\tau}} \ . \tag{17}$$

In the derivation for the result (17) we have exploited the following elementary Gaussian integration formula:

$$\int\limits_{-\infty}^{\infty} dx \exp(-p^2x^2) = \frac{\sqrt{\pi}}{p} \ . \tag{18}$$

After performing all three integrals of (14) we obtain the Green's function for diffusion

$$G(\boldsymbol{x}',t';\boldsymbol{x},t) = \frac{1}{2\pi^2a^2}\Theta(\tau)\left(a\sqrt{\frac{\pi}{\tau}}\right)^3 \exp\left(-\frac{a^2R^2}{4\tau}\right)$$
$$= \frac{a}{2\tau^{3/2}\sqrt{\pi}} \exp\left(-\frac{a^2R^2}{4\tau}\right) \Theta(\tau) \ . \tag{19}$$

The transformation $t' \to -it'$ and $t \to -it$ changes the Green's function (19) into the free Green's function for Schrödinger's equation. This was already expected according to the relationship between both differential equations.

The Green's function for diffusion has a Gaussian shape with a maximum at $R = 0$. The width of the distribution grows with increasing τ. The quantity $\sqrt{4\tau/a^2}$ is a measure of this width. For $\tau = 0$ the width is still zero. In the case of heat conduction this means that all the heat is focussed at a point. As soon as τ becomes larger than zero, the temperature increases at $R > 0$ while it decreases continuously at $R = 0$. Eventually (for $\tau \to \infty$) the heat is uniformly distributed over the whole space.

EXAMPLE

1.6 Kirchhoff's Integral as an Example of Huygens' Principle in Electrodynamics

We start our examination with the solution of the wave equation in classical electrodynamics and the derivation of the corresponding Green's function. The defining equation for the potential follows from Maxwell's inhomogeneous equations:

$$\nabla^2\Phi + \frac{1}{c}\frac{\partial}{\partial t}\nabla\cdot\boldsymbol{A} = -4\pi\varrho \tag{1}$$

and

$$\nabla^2\boldsymbol{A} - \frac{1}{c^2}\frac{\partial^2\boldsymbol{A}}{\partial t^2} - \nabla\left(\nabla\cdot\boldsymbol{A} + \frac{1}{c}\frac{\partial\Phi}{\partial t}\right) = -\frac{4\pi}{c}\boldsymbol{J} \ . \tag{2}$$

Here Φ denotes the scalar potential, $\boldsymbol{A}$ the vector potential, ϱ a given charge distribution and $\boldsymbol{J}$ the current. In Lorentz gauge ($\nabla \cdot \boldsymbol{A} + (1/c)(\partial\Phi/\partial t) = 0$) these differential equations decouple:

Example 1.6.

$$\nabla^2\Phi - \frac{1}{c^2}\frac{\partial^2\Phi}{\partial t^2} = -4\pi\varrho \quad , \tag{3}$$

$$\nabla^2\boldsymbol{A} - \frac{1}{c^2}\frac{\partial^2\boldsymbol{A}}{\partial t^2} = -\frac{4\pi}{c}\boldsymbol{J} \quad . \tag{4}$$

On the other hand we get in Coulomb gauge ($\nabla \cdot \boldsymbol{A} = 0$)

$$\nabla^2\Phi = -4\pi\varrho \tag{5}$$

with the solution

$$\Phi(\boldsymbol{x}, t) = \int d^3x' \frac{\varrho(\boldsymbol{x}', t)}{|\boldsymbol{x} - \boldsymbol{x}'|} \tag{6}$$

and

$$\nabla^2\boldsymbol{A} - \frac{1}{c^2}\frac{\partial^2\boldsymbol{A}}{\partial t^2} = \frac{-4\pi}{c}\boldsymbol{J} + \frac{1}{c}\nabla\frac{\partial\Phi}{\partial t} \quad . \tag{7}$$

All of the wave equations (3), (4), and (7) have the form

$$\left(\nabla^2 - \frac{1}{c^2}\frac{\partial^2}{\partial t^2}\right)\psi = -4\pi f(\boldsymbol{x}, t) \quad , \tag{8}$$

where $f(\boldsymbol{x}, t)$ is a given source distribution. To solve (8) it is useful to introduce the Green's function G for the wave equation which is defined by

$$\left(\nabla_x^2 - \frac{1}{c^2}\frac{\partial^2}{\partial t^2}\right) G(\boldsymbol{x}, t; \boldsymbol{x}', t') = -4\pi\delta^3(\boldsymbol{x} - \boldsymbol{x}')\delta(t - t') \quad . \tag{9}$$

The Green's function depends only on the difference of the coordinates ($\boldsymbol{x} - \boldsymbol{x}'$) and times ($t - t'$). In further calculations we use the δ function in the Fourier representation:

$$\delta^3(\boldsymbol{x} - \boldsymbol{x}')\delta(t - t') = \frac{1}{(2\pi)^4}\int d^3k \int d\omega\, \mathrm{e}^{\mathrm{i}\boldsymbol{k}\cdot(\boldsymbol{x}-\boldsymbol{x}')}\,\mathrm{e}^{-\mathrm{i}\omega(t-t')} \quad . \tag{10}$$

First, we introduce the Fourier transform

$$G(\boldsymbol{x}, t; \boldsymbol{x}', t') = \int d^3k \int d\omega\, g(\boldsymbol{k}, \omega)\,\mathrm{e}^{\mathrm{i}\boldsymbol{k}\cdot(\boldsymbol{x}-\boldsymbol{x}')}\,\mathrm{e}^{-\mathrm{i}\omega(t-t')} \tag{11}$$

as an ansatz for G, where $g(\boldsymbol{k}, \omega)$ is not yet known. We insert the transform (11) into the defining equation (9) to obtain

$$\begin{aligned}(\nabla_x^2 - \frac{1}{c^2}\frac{\partial^2}{\partial t^2}) &\int d^3k \int d\omega\, g(\boldsymbol{k}, \omega)\,\mathrm{e}^{\mathrm{i}\boldsymbol{k}\cdot(\boldsymbol{x}-\boldsymbol{x}')}\,\mathrm{e}^{-\mathrm{i}\omega(t-t')} \\ &= -4\pi\frac{1}{(2\pi)^4}\int d^3k \int d\omega\, \mathrm{e}^{\mathrm{i}\boldsymbol{k}\cdot(\boldsymbol{x}-\boldsymbol{x}')}\,\mathrm{e}^{-\mathrm{i}\omega(t-t')} \quad .\end{aligned} \tag{12}$$

Application of the differential operators yields

Example 1.6.

$$-\int d^3k \int d\omega \left(\boldsymbol{k}^2 - \frac{\omega^2}{c^2}\right) g(\boldsymbol{k},\omega)\, \mathrm{e}^{\mathrm{i}\boldsymbol{k}\cdot(\boldsymbol{x}-\boldsymbol{x}')}\, \mathrm{e}^{-\mathrm{i}\omega(t-t')}$$
$$= -\frac{1}{4\pi^3}\int d^3k \int d\omega\, \mathrm{e}^{\mathrm{i}\boldsymbol{k}\cdot(\boldsymbol{x}-\boldsymbol{x}')}\, \mathrm{e}^{-\mathrm{i}\omega(t-t')} \quad . \tag{13}$$

Hence, $g(\boldsymbol{k},\omega)$ is determined as

$$g(\boldsymbol{k},\omega) = \frac{1}{4\pi^3}\,\frac{1}{\boldsymbol{k}^2 - \omega^2/c^2 - \mathrm{i}\varepsilon} \quad , \tag{14}$$

and we can express the Green's function as

$$G(\boldsymbol{x},t;\boldsymbol{x}',t') = \int d^3k \int d\omega\, \frac{1}{4\pi^3}\,\frac{1}{\boldsymbol{k}^2 - \omega^2/c^2 - \mathrm{i}\varepsilon}\, \mathrm{e}^{\mathrm{i}\boldsymbol{k}\cdot(\boldsymbol{x}-\boldsymbol{x}')}\, \mathrm{e}^{-\mathrm{i}\omega(t-t')} \ . \tag{15}$$

Using Cauchy's theorem the integrals can be solved in closed form with the result

$$G(\boldsymbol{x},t;\boldsymbol{x}',t') = \frac{\delta(t' - t\ +\ |\boldsymbol{x}-\boldsymbol{x}'|/c)}{|\boldsymbol{x}-\boldsymbol{x}'|} \quad . \tag{16}$$

This is the *retarded* Green's function which describes the propagation of a wave on the light cone $|\boldsymbol{x}'-\boldsymbol{x}| = c(t'-t)$ forward in time, $t > t'$. Causality was enforced by introducing the negative imaginary part $-\mathrm{i}\varepsilon$ in (14). The solution of the wave equation (8) can now be written as

$$\Psi(\boldsymbol{x},t) = \int d^3x'dt'G(\boldsymbol{x},t;\boldsymbol{x}',t')f(\boldsymbol{x}',t') \quad . \tag{17}$$

Indeed it follows that

$$\nabla\Psi - \frac{1}{c^2}\frac{\partial^2\Psi}{\partial t^2} = \int\left(\nabla_x^2 - \frac{1}{c^2}\frac{\partial^2}{\partial t^2}\right) G(\boldsymbol{x},t;\boldsymbol{x}',t')f(\boldsymbol{x}',t')\, d^3x'\, dt'$$
$$= -\int 4\pi\, \delta^3(\boldsymbol{x}-\boldsymbol{x}')\delta(t-t')f(\boldsymbol{x}',t')\, d^3x'\, dt'$$
$$= -4\pi f(\boldsymbol{x},t) \quad . \tag{18}$$

With the Green's function (16) we obtain for Ψ

$$\Psi(\boldsymbol{x},t) = \int \frac{\delta(t' + |\boldsymbol{x}-\boldsymbol{x}'|/c - t)}{|\boldsymbol{x}-\boldsymbol{x}'|} f(\boldsymbol{x}',t')\, d^3\boldsymbol{x}'\, dt' \quad . \tag{19}$$

The integration over t' can be performed and the result is the retarded solution

$$\Psi(\boldsymbol{x},t) = \int d^3x' \frac{[f(\boldsymbol{x}',t')]_{\mathrm{ret}}}{|\boldsymbol{x}-\boldsymbol{x}'|} \quad , \tag{20}$$

where $[\ \ ]_{\mathrm{ret}}$ means $t' = t - |\boldsymbol{x}-\boldsymbol{x}'|/c$. Hence we can specify the potentials $\boldsymbol{A}$ and Φ in closed form.

For further discussions we use Green's second theorem and integrate over time from $t' = t_0$ to $t' = t_1$:

$$\int_{t_0}^{t_1} dt' \int_V d^3x' \left(\Phi\nabla'^2\Psi - \Psi\nabla'^2\Phi\right) = \int_{t_0}^{t_1} dt' \oint_S da' \left(\Phi\frac{\partial\Psi}{\partial n'} - \Psi\frac{\partial\Phi}{\partial n'}\right) \quad , \tag{21}$$

Example 1.6.

where Φ and Ψ are – for the present – arbitrary scalar fields and $\partial\Psi/\partial n'$ is the normal derivative of Ψ on the surface S. For Ψ we now insert relation (20) and for Φ Green's function (16). Now we can rewrite the left hand side of (21) :

$$
\begin{aligned}
L &= \int_{t_0}^{t_1} dt' \int_V d^3x' \left(G\,\nabla'^2\Psi - \Psi\,\nabla'^2 G\right) \\
&= \int_{t_0}^{t_1} dt' \int_V d^3x' \left[G\left(-4\pi f(\boldsymbol{x}',t') + \frac{1}{c^2}\frac{\partial^2\Psi}{\partial t'^2}\right)\right. \\
&\quad \left. - \Psi\left(-4\pi\,\delta^3(\boldsymbol{x}'-\boldsymbol{x})\delta(t'-t) + \frac{1}{c^2}\frac{\partial^2 G}{\partial t'^2}\right)\right] \\
&= \int_{t_0}^{t_1} dt' \int_V d^3x' \left[4\pi\,\Psi(\boldsymbol{x}',t')\,\delta^3(\boldsymbol{x}'-\boldsymbol{x})\delta(t'-t) - 4\pi f(\boldsymbol{x}',t')G\right. \\
&\quad \left. + \frac{1}{c^2}\left(G\frac{\partial^2\Psi}{\partial t'^2} - \Psi\frac{\partial^2 G}{\partial t'^2}\right)\right] \quad ,
\end{aligned} \tag{22}
$$

where we have used the defining equations (8) and (9). The first term reduces to $4\pi\,\Psi(\boldsymbol{x},t)$, the second term will be kept, and the last two terms are integrated by parts with respect to time. This yields

$$
\begin{aligned}
L &= 4\pi\,\Psi(\boldsymbol{x},t) - 4\pi \int_{t_0}^{t_1} dt' \int_V d^3x'\, f(\boldsymbol{x}',t')\,G \\
&\quad + \frac{1}{c^2}\int_V d^3x' \left(G\frac{\partial\Psi}{\partial t'} - \Psi\frac{\partial G}{\partial t'}\right)\Bigg|_{t'=t_0}^{t'=t_1} \\
&\quad - \frac{1}{c^2}\int_V d^3x' \int_{t_0}^{t_1} dt' \left(\frac{\partial\Psi}{\partial t'}\frac{\partial G}{\partial t'} - \frac{\partial G}{\partial t'}\frac{\partial\Psi}{\partial t'}\right) \quad .
\end{aligned} \tag{23}
$$

The last term in (23) is zero. Since the retarded Green's function vanishes for $t' = t_1 > t$, the integral in the third term of (23) vanishes at the upper integration boundary. Now we combine (21) and (23) to get

$$
\begin{aligned}
\Psi(\boldsymbol{x},t) &= \frac{1}{4\pi c^2}\int d^3x' \left(G\frac{\partial\Psi}{\partial t'} - \Psi\frac{\partial G}{\partial t'}\right)\Bigg|_{t'=t_0} \\
&\quad + \frac{1}{4\pi}\int_{t_0}^{t_1} dt' \oint_S da' \left(G\frac{\partial\Psi}{\partial n'} - \Psi\frac{\partial G}{\partial n'}\right) \\
&\quad + \int_{t_0}^{t_1} dt' \int_V d^3x' f(\boldsymbol{x}',t')\,\frac{\delta(t' + |\boldsymbol{x}-\boldsymbol{x}'|/c - t)}{|\boldsymbol{x}-\boldsymbol{x}'|}
\end{aligned}
$$

Example 1.6.

$$= \int_V d^3x' \frac{\left[f(\boldsymbol{x}',t')\right]_{\text{ret}}}{|\boldsymbol{x}-\boldsymbol{x}'|} + \frac{1}{4\pi c^2}\int_V d^3x' \left(G\frac{\partial\Psi}{\partial t'} - \Psi\frac{\partial G}{\partial t'}\right)\bigg|_{t'=t_0}$$
$$+ \frac{1}{4\pi}\int_{t_0}^{t_1} dt' \oint_S da' \left(G\frac{\partial\Psi}{\partial n'} - \Psi\frac{\partial G}{\partial n'}\right) \quad . \tag{24}$$

Let us consider the so-called Kirchhoff representation of a field, which is expressed by the values of Ψ and $\partial\Psi/\partial n'$ in a surface S. To this end we assume that there are *no sources inside the volume* V and that the *initial values* of Ψ and $\partial\Psi/\partial t'$ vanish at $t' = t_0$. Then according to (24) the field is given by

$$\Psi(\boldsymbol{x},t) = \frac{1}{4\pi}\int_{t_0}^{t_1} dt' \oint_S da' \left(G\frac{\partial\Psi}{\partial n'} - \Psi\frac{\partial G}{\partial n'}\right) \quad . \tag{25}$$

We put

$$\boldsymbol{R} = \boldsymbol{x} - \boldsymbol{x}' \tag{26}$$

with

$$\nabla' R = -\frac{\boldsymbol{R}}{R} \quad . \tag{27}$$

We use the explicit form of the Green's function (16) to obtain, with

$$\frac{\partial}{\partial n'} = \boldsymbol{n}\cdot\nabla' \quad , \tag{28}$$

$$\nabla' G = \frac{\partial G}{\partial R}\nabla' R = -\frac{\boldsymbol{R}}{R}\frac{\partial}{\partial R}\left(\frac{\delta(t'+R/c-t)}{R}\right)$$
$$= -\frac{\boldsymbol{R}}{R}\left(-\frac{\delta(t'+R/c-t)}{R^2} + \frac{\delta'(t'+R/c-t)}{cR}\right) \quad . \tag{29}$$

$\boldsymbol{n}$ is the unit vector normal to the surface S. Here we have used the following relation:

$$\frac{\partial}{\partial R}\,\delta\left(t' + \frac{R}{c} - t\right) = \frac{1}{c}\,\delta'\left(t' + \frac{R}{c} - t\right) \quad , \tag{30}$$

where the prime denotes the derivative of the δ function with respect to its arguments. Furthermore, we make use of the relation for the derivative of the δ function

$$\int f(x)\,\delta'(x-a)\,dx = -f'(a) \quad , \tag{31}$$

which helps to give as a partial result

$$\oint_S da' \frac{\boldsymbol{n}\cdot\boldsymbol{R}}{cR^2}\int_{t_0}^{t_1} dt'\,\Psi\,\delta'\left(t' + \frac{R}{c} - t\right) = -\oint_S da' \frac{\boldsymbol{n}\cdot\boldsymbol{R}}{cR^2}\,\Psi'\left(\boldsymbol{x}', t' = t - \frac{R}{c}\right)$$
$$= \left[-\oint_S da' \frac{\boldsymbol{n}\cdot\boldsymbol{R}}{cR^2}\,\Psi'(\boldsymbol{x}',t')\right]_{\text{ret}} \quad . \tag{32}$$

It follows in conclusion that

$$\Psi(\boldsymbol{x},t) = \frac{1}{4\pi}\oint\limits_S da'\, \boldsymbol{n}\cdot\left[\frac{\nabla'\Psi(\boldsymbol{x}',t')}{R} - \frac{\boldsymbol{R}}{R^3}\Psi(\boldsymbol{x}',t') - \frac{\boldsymbol{R}}{cR^2}\frac{\partial\Psi(\boldsymbol{x}',t')}{\partial t'}\right]_{\text{ret}} . \tag{33}$$

We emphasize that this is *not* a solution for the field Ψ. It is only an integral representation for Ψ, expressing Ψ in terms of Ψ and its corresponding spatial and time derivatives on the surface S. However, these quantities cannot be chosen independently.

Kirchhoff's integral (33) is the mathematical expression of Huygens' principle which postulates that any point on a wave front behaves like a pointlike source that emits a spherical wave moving at the speed of light. Then the field at a given point and at a later time is a superposition of all fields emerging from these sources. The envelope of all these waves forms the next wave front. Kirchhoff's integral (33) serves e.g. as starting point for the discussion of problems related to optical diffraction.

1.10 Biographical Notes

FEYNMAN, Richard Phillip, American physicist. *11.5.1918 in New York, †15.2.1988 in Pasadena. F. studied at physics MIT and Caltech where he received his PhD in 1942. After working for the Manhattan project (Los Alamos) he became professor of physics at Cornell University (1946) and subsequently at the California Institute of Technology (since 1951). F. did pioneering work related to the formulation of quantum electrodynamics which earned him the 1965 Nobel prize in physics (with J. Schwinger and S. Tomonaga). He also explained the superfluidity in liquid helium and did fundamental work in the area of weak interactions. F. advanced the use of the path integral method in physics. In later years F. fostered the quark model by putting forward the parton picture to describe high-energy hadronic collisions.

SCHWINGER, Julian Seymour, American physicist. *12.2.1918 in New York, †16.7.1994. S. studied at Columbia University (NY) where he got his PhD in 1939. Subsequently he worked at the Universities of Berkeley, Cambridge (Mass.), Chicago and Boston and in 1947 became professor of physics at Harvard University. In 1972 he went to the University of California in Los Angeles. In 1948/49 S. developed a covariant formulation of quantum electrodynamics. Introducing charge and mass renormalization he was able to evaluate the anonmalous magnetic moment of the electron and the Lamb shift. For this work he received the Nobel prize in physics in 1965 (with R. Feynman and S. Tomonaga). S. made numerous important contributions to quantum field theory, many-body physics, gravitation theory etc.

STÜCKELBERG, Ernst Carl Gerlach S. von Breidenbach zu Breidenstein und Melsbach, Swiss physicist. *1.2.1905 in Basel, † 4.9.1984 in Geneva. S. studied physics in Basel and Munich (with A. Sommerfeld) and got his PhD at Basel University in 1927. He was research assistant and later professor at Princeton University (1927-32), and at the Universities of Zurich and Geneva (since 1935). In 1942-57 in addition he held a combined professorship at the University of Lausanne and the Federal Institute of Technology. The first theoretical work of S. was in the area of molecular physics. Later S. made pioneering contributions to quantum field theory. In 1942 he first conceived the idea that positrons can be interpreted as electrons running backward in time. The use of the causal propagator for calculating the scattering matrix was introduced in 1949 independently by R. Feynman and by S. (with his student D. Rivier). Later S. (with A. Petermann) developed the idea of the renormalization group. In his later years S. turned his main interest to the foundations of thermodynamics. In 1976 S. was honored with the Max Planck Medal of the German Physical Society.

TOMONAGA, Sin-itiro, Japanese physicist. *31.3.1906 in Kyoto, †8.7.1979. T. studied physics at Kyoto Imperial University where he graduated in 1929. Subsequently he worked at the Institute of Physico-Chemical Research in Tokyo in the laboratory of Y. Nishina. In 1937 he went to Leipzig University and joined the group of W. Heisenberg. In 1941 he became professor of physics at the Tokyo University of Education. In his later years he became involved in science administration, serving as president of his University and of the Science Council of Japan. The early work of T. was concerned with the creation and annihilation of positrons, with the stopping of cosmic-ray mesons in matter, and with the liquid-drop model of the nucleus. Later he turned to quantum field theory for which he developed a manifestly covariant formulation, incorporating renormalization. For this work he received the Nobel prize for physics in 1965 (with R. Feynman and J. Schwinger).

2. The Propagators for Electrons and Positrons

In the following we will generalize the nonrelativistic propagator theory developed in the previous chapter to the relativistic theory of electrons and positrons. We will be guided by the picture of the nonrelativistic theory where the propagator $G^+(x';x)$ is interpreted as the *probability amplitude for a particle wave originating at the space–time point* x *to propagate to the space–time point* x'. This amplitude can be decomposed as in (1.28) into a sum of partial amplitudes, the nth such partial amplitude being a product of factors illustrated in Fig. 2.1. According to (1.28), the probability amplitude consists of factors that describe the propagation of the particle between the particular scattering events (caused by the interaction $V(x)$) and when integrated over the space–time coordinates of the points of interaction represent the nth-order scattering process of the particle.

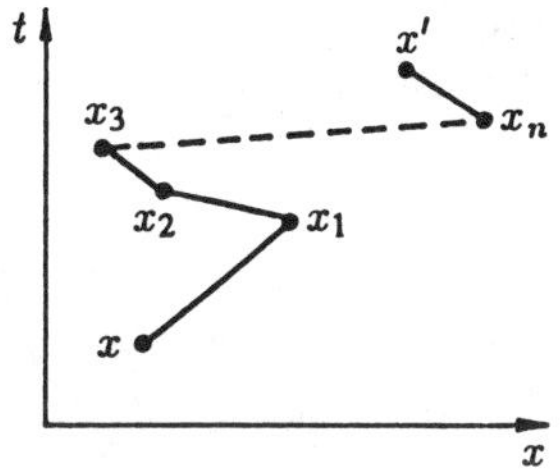

Fig. 2.1. Illustration of the nth-order contribution to the Green's function $G^+(x';x)$ which describes the probability amplitude for multiple scattering of a particle

Each line in Fig. 2.1 represents a Green's function; e.g. the line $\overline{x_{i-1}\quad x_i}$ signifies the Green's function $G_0^+(x_i,x_{i-1})$, i.e. the amplitude that a particle wave originating at the space–time point x_{i-1} propagates freely to the space–time point x_i. The space–time points where an interaction occurs are represented by small circles (•). At the point x_i the particle wave is scattered with the probability amplitude $V(x_i)$ per unit space–time volume. The resulting scattered wave then again propagates freely forward in time from the space–time point x_i towards the point x_{i+1} with the amplitude $G_0^+(x_{i+1},x_i)$ where the next interaction happens, and so on. The total amplitude is then given by the sum over contributions from all space–time points at which an interaction occurs. The particular space–time points at which the particle wave experiences an interaction are termed vertices. One may also describe the individual scattering processes by saying that *the interaction at the i-th vertex annihilates the particle that has propagated freely up to* x_i*, and creates a particle that propagates on to* x_{i+1}, with $t_{i+1} \geq t_i$.

This latter interpretation of scattering events is well suited for a generalization to relativistic hole theory since it contains the overall space–time structure of the scattering process and the interaction.

Our aim is now to develop, by analogy with the nonrelativistic propagator theory, methods to describe and calculate scattering processes mathematically within the framework of the Dirac hole theory. We need to focus on the new feature of *pair creation and annihilation processes* that are now contained in our relativistic picture of scattering processes. We shall adopt many of the calculation rules intuitively by requiring them to be consistent with the dynamics of the Dirac equation. A more rigorous mathematical justification of these rules can be given using the methods of quantum field theory. Some references on this subject are given in the appendix. In the following we shall use mainly heuristic arguments.

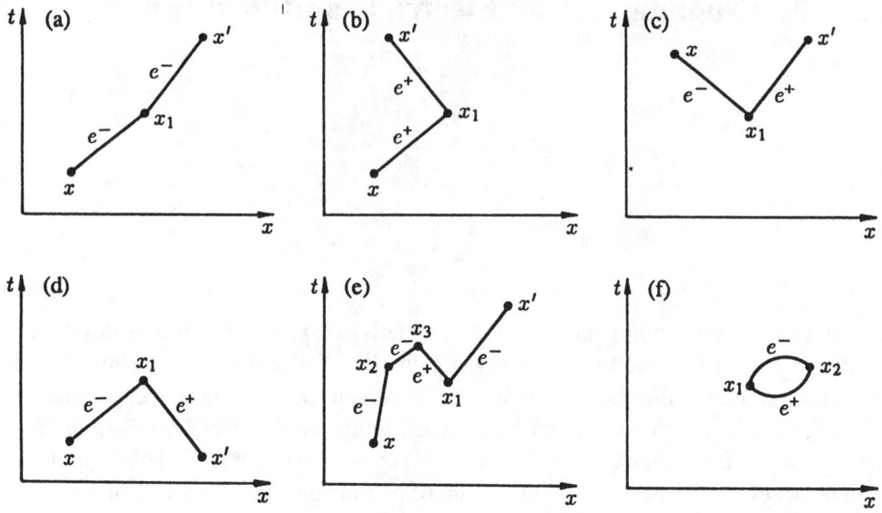

Fig. 2.2a–f. Some examples of processes encountered within the electron–positron theory. The diagrams represent: (**a**) electron scattering; (**b**) positron scattering; (**c**) electron–positron pair creation; (**d**) pair annihilation; (**e**) electron scattering that in addition includes an electron–positron pair creation process; and (**f**) a closed loop describing vacuum polarization

Let us now take a look at the typical processes that must be described within the relativistic theory. These are collected in Fig. 2.2, illustrated by diagrams that we shall learn to understand in the following.

In addition to the ordinary scattering processes of an electron (Fig. 2.2a) or a positron (Fig. 2.2b) there are also pair production and annihilation processes (Fig. 2.2c–f). Let us first take a look at the pair production illustrated in Fig. 2.2c: The electron–positron pair is created by a potential acting at space–time point x_1. The two particles then propagate freely forward in time, the positron to x' and the electron to x. Similarly, Fig. 2.2d shows the trajectories of an electron and a positron which start from the points x and x', respectively, and meet at the point x_1 where they annihilate.

Diagram 2.2e represents the scattering of an electron originating at x moving forward in time, experiencing several scatterings, and ending up at x'. Along its way from x to x' a pair is produced by a potential acting at x_1; the two created particles propagate forward in time. The positron of this pair and the initial electron converge at x_3 and are annihilated. The surviving electron of the pair then propagates to x'.

Diagram 2.2f shows a pair produced at x_1, propagating up to x_2, and being annihilated in the field there. It was only "virtually" present for a short intermediate time interval. Below we will recognize this process as the polarization of the vacuum.

These simple considerations already show that the relativistic electron–positron theory contains more ingredients than its nonrelativistic counterpart: we need to describe not only the amplitude for a particle (electron) to propagate from x_1 to x_2 but also the amplitude for the creation of a positron that propagates from one space–time point to another, where it is destroyed again. It is this *positron amplitude* we have to construct in the first place, enabling us then to find the total amplitude for the various processes illustrated in Fig. 2.2 by summing, or integrating, over all intermediate points (interaction events) that can contribute to the total process. In a scattering event (e.g. Fig. 2.2e) in general both electron and positron amplitudes will contribute.

The Dirac hole theory (see Theoretical Physics, Vol. 3 by W. Greiner: *Relativistic Quantum Mechanics – Wave Equations,* hereon referred to as *RQM*) interprets a positron as a hole in the Dirac sea, i.e. the absence of an electron with negative energy from the filled sea. Thus we may view the destruction of a positron at some space–time point as equivalent to the creation of an electron with negative energy at this point. This suggests the possibility, e.g. in Fig. 2.2e, that the amplitude for creating the positron at x_1 and destroying it at x_3 is related to the amplitude for creating a negative-energy electron at x_3 and destroying it at x_1 where $t_1 < t_3$. In this picture a pair creation process such as in Fig. 2.2e,c therefore leads to the following definition of positrons: *Positrons with positive energy moving forward in space–time are viewed within the propagator theory as electrons with negative energy travelling backward in space–time.*

This is the Stückelberg–Feynman definition of positrons, which we already encountered in the discussion of the time reversal and PCT-symmetries (see *RQM*, Chap. 12). *Electrons are represented by particle waves with positive energy propagating forward in space–time.* A process such as in Fig. 2.3 can therefore be interpreted using two different but equivalent languages as follows.

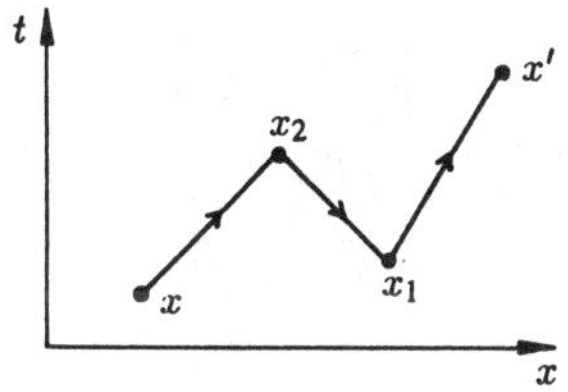

Fig. 2.3. Electron scattering with involving an intermediate pair creation process

An electron originating at x propagates forward in time, is scattered into a state of negative energy at x_2 by the interaction $V(x_2)$, propagates backward in time to x_1, where it is scattered again into a state of positive energy, and finally propagates forward in time to x'. Alternatively one may say that an electron originating at x moves forward in time up to x_2, where it is destroyed by the interaction $V(x_2)$ together with the positron of the $e^+ - e^-$ pair that has been created earlier at x_1 by $V(x_1)$. The electron of this pair propagates forward in time to x'.

Processes that are represented by closed loops as illustrated in Fig. 2.4 are interpreted in terms of an $e^+ - e^-$ pair being produced at x_1 by $V(x_1)$ that propagates forward in time to x_2, where it is destroyed again by $V(x_2)$. Equivalently, within the picture of the hole theory, we can say that the potential $V(x_1)$ at x_1 scatters an electron from the sea of negative-energy states into a state of positive energy leaving a hole behind; it is then scattered back into the sea, recombining with the hole at x_2 under the action of $V(x_2)$. Or, in propagator language, the electron created at x_1 is scattered back in time at x_2 to destroy itself at x_1.

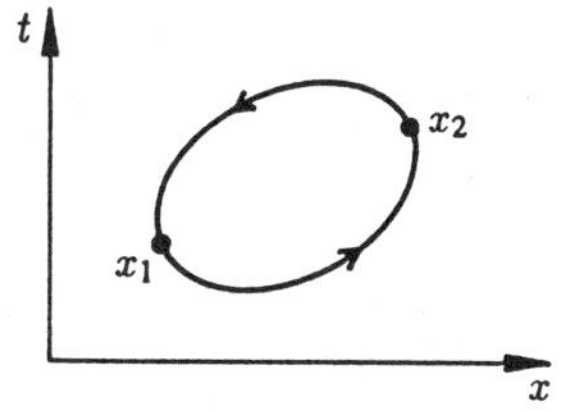

Fig. 2.4. A loop diagram

Our next aim is to find a unified mathematical description for the various processes making use of the relativistic propagator formalism. The first step is to construct the Green's function for electrons and positrons. It is known as the relativistic propagator[1]

$$S_{\mathrm{F}}(x',x;A) \tag{2.1}$$

and is required in analogy to the nonrelativistic propagator (1.64), to satisfy the following differential equation:

$$\sum_{\lambda=1}^{4}\left[\gamma_\mu\left(\mathrm{i}\hbar\frac{\partial}{\partial x'_\mu}-\frac{e}{c}A^\mu(x')\right)-m_0c\right]_{\alpha\lambda}(S_{\mathrm{F}})_{\lambda\beta}(x',x;A)=\hbar\delta_{\alpha\beta}\delta^4(x'-x)\ . \tag{2.2}$$

[1] The symbol S_{F} has been aptly chosen, bearing in mind that the originators of the relativistic propagator formalism were Stückelberg and Feynman: the propagator is commonly called the *Feynman propagator*. The original references are E.C.G. Stückelberg and D. Rivier: Helv. Phys. Acta **22**, 215 (1949) and R.P. Feynman: Phys. Rev. **76**, 749 (1949).

By means of this definition the propagator $S_F(x',x;A)$ is a 4×4 matrix corresponding to the dimension of the γ matrices. The third argument of S_F serves as a reminder that the propagator defined by (2.2) depends on the electromagnetic field A_μ.

In relativistic quantum theory usually one employs "natural units" and sets $\hbar = c = 1$, implying the substitutions

$$\frac{e}{\hbar c} \to e \quad , \quad \frac{m_0 c}{\hbar} \to m_0 \quad . \tag{2.3}$$

Thus in matrix notation with indices suppressed (2.2) becomes

$$(\mathrm{i}\nabla\!\!\!/' - eA\!\!\!/' - m_0)S_F(x',x;A) = \delta^4(x'-x)\mathbb{1} \quad . \tag{2.4}$$

Note that the definition of the relativistic propagator (2.2,2.4) differs from the nonrelativistic counterpart (1.64): the differential operator $\mathrm{i}\partial/\partial t' - \hat{H}(x')$ occurring in (1.64) has been multiplied by γ^0 in (2.2,2.4) in order to form the covariant operator $(\mathrm{i}\nabla\!\!\!/' - eA\!\!\!/' - m_0)$. The unit matrix in spinor space on the right-hand side of (2.4) is most commonly suppressed, i.e.

$$(\mathrm{i}\nabla\!\!\!/' - eA\!\!\!/' - m_0)S_F(x',x;A) = \delta^4(x'-x) \quad . \tag{2.5}$$

However, it must be kept in mind that (2.5) is a matrix equation so that the delta function in (2.5) is meant to be multiplied by $\mathbb{1}$.

The *free-particle propagator* must satisfy (2.5) with the interaction term $eA\!\!\!/'$ absent, i.e.

$$(\mathrm{i}\nabla\!\!\!/' - m_0)S_F(x',x) = \delta^4(x'-x) \quad . \tag{2.6}$$

As in the nonrelativistic case we compute $S_F(x',x)$ in momentum space, using the fact that $S_F(x',x)$ depends only on the distance vector $x'-x$. This property is a manifestation of the homogeneity of space and time and in general would not be valid for the interacting propagator $S_F(x',x;A)$. Fourier transformation to four-dimensional momentum space then yields for the free propagator

$$S_F(x',x) = S_F(x'-x) = \int \frac{d^4p}{(2\pi)^4} \exp\left[-\mathrm{i}p\cdot(x'-x)\right] S_F(p) \quad . \tag{2.7}$$

Inserting (2.7) into (2.6) we obtain an equation that determines the Fourier amplitude $S_F(p)$, namely

$$\int \frac{d^4p}{(2\pi)^4}(p\!\!\!/ - m_0)S_F(p)\exp\left[-\mathrm{i}p\cdot(x'-x)\right] = \int \frac{d^4p}{(2\pi)^4}\exp\left[-\mathrm{i}p\cdot(x'-x)\right] \ ,$$

which implies that

$$(p\!\!\!/ - m_0)S_F(p) = \mathbb{1} \tag{2.8}$$

or, in detail, restoring the indices,

$$\sum_{\lambda=1}^{4}(p\!\!\!/ - m_0)_{\alpha\lambda}(S_F(p))_{\lambda\beta} = \delta_{\alpha\beta} \quad . \tag{2.9}$$

Equation (2.8) can be solved for the Fourier amplitude $S_{\mathrm{F}}(p)$ by multiplying with $(\not{p} + m_0)$ from the left:

$$(\not{p} + m_0)(\not{p} - m_0)S_{\mathrm{F}}(p) = (\not{p} + m_0) \quad . \tag{2.10}$$

Since

$$\not{p}\not{p} = \gamma_\mu \gamma_\nu p^\mu p^\nu = \frac{1}{2}(\gamma_\mu\gamma_\nu + \gamma_\nu\gamma_\mu)p^\mu p^\nu = g_{\mu\nu}p^\mu p^\nu = p_\mu p^\mu = p^2 \ , \tag{2.11}$$

(2.10) becomes

$$(p^2 - m_0^2)S_{\mathrm{F}}(p) = (\not{p} + m_0) \tag{2.12}$$

or

$$S_{\mathrm{F}}(p) = \frac{\not{p} + m_0}{p^2 - m_0^2} \qquad \text{for} \quad p^2 \neq m_0^2 \quad . \tag{2.13}$$

In order to complete the definition of $S_{\mathrm{F}}(p)$ we must give a prescripton to handle the singularities at $p^2 = m_0^2$ which is just the mass-shell condition $p_0^2 - \boldsymbol{p}^2 = m_0^2$ or $p_0 = \pm\sqrt{m_0^2 + \boldsymbol{p}^2} = \pm E_p$. From the foregoing discussion of the nonrelativistic propagator formalism we know that this additional information comes from the boundary conditions that are imposed on $S_{\mathrm{F}}(x' - x)$. We will now put into practice the previous interpretation of positrons as negative-energy electrons moving backwards in time. In order to implement this concept we return to the Fourier representation (2.7) and the Fourier amplitude (2.13) and perform the energy integration (dp_0 integration) along the special contour C shown in Fig. 2.5. We obtain

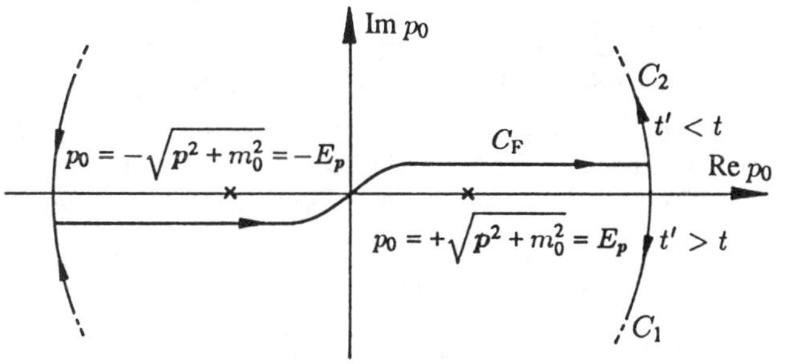

Fig. 2.5. Integration contour C_{F} that defines the Feynman propagator. The singularities are located on the real p_0 axis at $p_0 = -E_p$ and $p_0 = +E_p$

$$\begin{aligned} S_{\mathrm{F}}(x' - x) &= \int \frac{d^4p}{(2\pi)^4} S_{\mathrm{F}}(p) \exp\left[-\mathrm{i}p \cdot (x' - x)\right] \\ &= \int \frac{d^4p}{(2\pi)^4} S_{\mathrm{F}}(p) \exp\left\{-\mathrm{i}\left[p_0(t' - t) - \boldsymbol{p} \cdot (\boldsymbol{x}' - \boldsymbol{x})\right]\right\} \\ &= \int \frac{d^3p}{(2\pi)^3} \exp\left[\mathrm{i}\boldsymbol{p} \cdot (\boldsymbol{x}' - \boldsymbol{x})\right] \\ &\quad \times \int_{C_{\mathrm{F}}} \frac{dp_0}{2\pi} \frac{\exp\left[-\mathrm{i}p_0(t' - t)\right]}{p^2 - m_0^2} (\not{p} + m_0) \quad . \end{aligned} \tag{2.14}$$

For $t' > t$ we close the integration contour in the lower half plane, since in this case the integral along the lower semicircle, parametrized by $p_0 = \varrho e^{i\phi}$, does not contribute for $\varrho \to \infty$. By means of the residue theorem then only the positive energy pole at

$$p_0 = E_p = +\sqrt{\boldsymbol{p}^2 + m_0^2}$$

contributes to the p_0 integration. Hence, we obtain

$$\begin{aligned}
&\int_{C_F} \frac{dp_0}{2\pi} \frac{\exp\left[-\mathrm{i}p_0(t'-t)\right]}{p_0^2 - \boldsymbol{p}^2 - m_0^2}(\not{p} + m_0) \\
&= \int_{C_F + C_1} \frac{dp_0}{2\pi} \frac{\exp\left[-\mathrm{i}p_0(t'-t)\right]}{(p_0 - E_p)(p_0 + E_p)}(p_0\gamma^0 + p_i\gamma^i + m_0) \\
&= -2\pi\mathrm{i}\frac{\exp\left[-\mathrm{i}E_p(t'-t)\right]}{2\pi 2E_p}(E_p\gamma^0 - \boldsymbol{p}\cdot\boldsymbol{\gamma} + m_0) \quad ,
\end{aligned} \tag{2.15}$$

so that (2.14) yields

$$\begin{aligned}
S_F(x'-x) = -\mathrm{i}\int \frac{d^3p}{(2\pi)^3} &\exp\left[\mathrm{i}\boldsymbol{p}\cdot(\boldsymbol{x}'-\boldsymbol{x})\right]\exp\left[-\mathrm{i}E_p(t'-t)\right] \\
&\times \frac{(E_p\gamma^0 - \boldsymbol{p}\cdot\boldsymbol{\gamma} + m_0)}{2E_p} \qquad \text{for} \quad t' > t \quad .
\end{aligned} \tag{2.16}$$

The minus sign in (2.15) results from integrating along the contour in a mathematically negative (clockwise) sense. This propagator describes particle motion from x to x' *forward in time* ($t' > t$). At $x' = (\boldsymbol{x}', t')$ S_F contains positive-energy components only, since the energy factor occurring in the exponent of $\exp(-\mathrm{i}E_p(t'-t))$ is defined to be positive, $E_p = +\sqrt{\boldsymbol{p}^2 + m_0^2}$.

On the other hand, considering the particle propagation backward in time implies that $t' - t$ is negative so that the p_0 integration must be performed along the contour closed in the upper half plane in order to give a zero contribution along the semicircle for $\varrho \to \infty$. Then only the negative-energy pole at

$$p_0 = -E_p = -\sqrt{\boldsymbol{p}^2 + m_0^2}$$

contributes (Fig. 2.5). This yields

$$\begin{aligned}
&\int_{C_F + C_2} \frac{dp_0}{2\pi} \frac{\exp\left[-\mathrm{i}p_0(t'-t)\right]}{(p_0 - E_p)(p_0 + E_p)}(p_0\gamma^0 + p_i\gamma^i + m_0) \\
&\qquad = 2\pi\mathrm{i}\frac{\exp\left[-\mathrm{i}(-E_p)(t'-t)\right]}{2\pi(-2E_p)}(-E_p\gamma^0 - \boldsymbol{p}\cdot\boldsymbol{\gamma} + m_0) \quad .
\end{aligned} \tag{2.17}$$

Thus the propagator (2.14) for the case $t' < t$ reads

$$\begin{aligned}
S_F(x'-x) = -\mathrm{i}\int \frac{d^3p}{(2\pi)^3} &\exp\left[\mathrm{i}\boldsymbol{p}\cdot(\boldsymbol{x}'-\boldsymbol{x})\right]\exp\left[+\mathrm{i}E_p(t'-t)\right] \\
&\times \frac{(-E_p\gamma^0 - \boldsymbol{p}\cdot\boldsymbol{\gamma} + m_0)}{2E_p} \qquad \text{for} \quad t' < t \quad .
\end{aligned} \tag{2.18}$$

This propagator describes the propagation of negative-energy particle waves backward in time, as can be read off the factor $\exp\left(-\mathrm{i}(-E_p)(t'-t)\right)$. These negative-energy waves are absent in the nonrelativistic theory, since no solution of the energy–momentum relation at $p_0 = -E_p = -\sqrt{p^2 + m_0^2}$ exists. Here, in the relativistic case, they are unavoidable owing to the quadratic form of the energy–momentum dispersion relation.

We note that other choices of the integration contour C, e.g. as in Fig. 2.6, would lead to contributions from negative-energy waves propagating into the future (case a) or positive-energy waves into the past (case b). As we can see, the choice of the contour C_F according to Fig. 2.5 results in positive-energy waves moving forward in time and negative-energy waves backward in time, just as we required. *These negative-energy waves propagating backward in time we identify with positrons.*

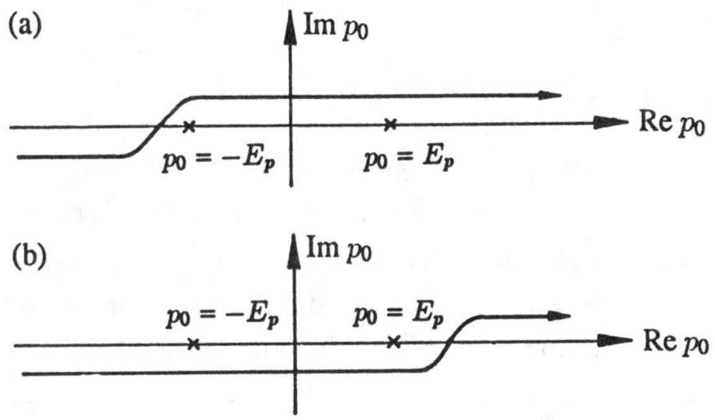

Fig. 2.6a,b. Possible alternative choices for the integration contour that lead to propagators with the wrong asymptotic behaviour

As we recall from hole theory it is the definition of the vacuum (specified by the position of the Fermi surface E_F) that prescribes which of the particle-wave states are to be interpreted as electrons and which as positrons. It is assumed that particle states with $E < E_\mathrm{F}$ are occupied and that the absence of such a particle state is interpreted in terms of a positron. The choice of the propagator is based on this definition of the vacuum, which determines the choice of the integration contour C, i.e. the transition of C from the lower to the upper complex p_0 half plane. For example, in supercritical fields (see Chap. 7) the vacuum carries charge. Consequently, some of the negative-energy states are to be interpreted as electrons propagating forward in time. For an atom the Fermi surface is usually located at a bound state. Hence, in this case the propagator is required to be evaluated along an integration contour that passes over from the lower half plane to the upper half plane at an energy slightly above the highest occupied bound state. Another example of a modified ground state (and thus a modified propagator) is the Fermi gas of electrons, which will be introduced in Exercise 2.2 below.

The integration contour C_F determining the propagator $S_\mathrm{F}(x'-x)$ may be alternatively characterized by adding a small positive imaginary part $+\mathrm{i}\varepsilon$ to the denominator in (2.14), where the limit $\varepsilon \to 0$ is to be taken at the end of the calculation:

$$S_F(x'-x) = \int \frac{d^4p}{(2\pi)^4} \frac{\exp\left[-ip\cdot(x'-x)\right]}{p^2 - m_0^2 + i\varepsilon}(\not{p} + m_0) \quad . \tag{2.19}$$

Then the singularities corresponding to positive-energy states,

$$p_0 = +\sqrt{\boldsymbol{p}^2 + m_0^2 - i\varepsilon} = +\sqrt{\boldsymbol{p}^2 + m_0^2} - i\eta(\varepsilon) \quad , \tag{2.20a}$$

lie below the real p_0 axis while the poles corresponding to negative-energy states,

$$p_0 = -\sqrt{\boldsymbol{p}^2 + m_0^2 - i\varepsilon} = -\sqrt{\boldsymbol{p}^2 + m_0^2} + i\eta(\varepsilon) \quad , \tag{2.20b}$$

are located above the p_0 axis, just as required for the contour C_F. The prescription of (2.19) is most easily remembered in the form of a rule: To ensure the correct boundary conditions, *the mass has to be given a small negative imaginary part.* The two propagators describing positive-energy particle waves (2.16) and negative-energy particle waves (2.18) moving forward and backward in time, respectively, may be combined by introducing the energy projection operators $\hat{\Lambda}_\pm(p)$(see *RQM*, Chap. 7)

$$\begin{aligned} \hat{\Lambda}_r(p) &= \frac{\varepsilon_r \not{p} + m_0}{2m_0} \quad , \\ \varepsilon_r &= \begin{cases} +1 & \text{for waves of positive energy} \\ -1 & \text{for waves of negative energy} \end{cases} \quad . \end{aligned} \tag{2.21}$$

Then, by changing the three-momentum $\boldsymbol{p}$ to $-\boldsymbol{p}$ in the propagator for negative-energy waves (2.18), which does not alter the result since the integral $\int d^3p$ includes all directions of the three-momentum, we can write

$$\begin{aligned} S_F(x'-x) &= -i \int \frac{d^3p}{(2\pi)^3} \Bigg\{ \exp\left[-i(+E_p)(t'-t)\right] \exp\left[+i\boldsymbol{p}\cdot(\boldsymbol{x}'-\boldsymbol{x})\right] \\ &\quad \times \frac{(+E_p\gamma^0 - \boldsymbol{p}\cdot\boldsymbol{\gamma} + m_0)}{2E_p} \Theta(t'-t) \\ &\quad + \exp\left[-i(-E_p)(t'-t)\right] \exp\left[-i\boldsymbol{p}\cdot(\boldsymbol{x}'-\boldsymbol{x})\right] \\ &\quad \times \frac{(-E_p\gamma^0 + \boldsymbol{p}\cdot\boldsymbol{\gamma} + m_0)}{2E_p} \Theta(t-t') \Bigg\} \\ &= -i \int \frac{d^3p}{(2\pi)^3} \frac{m_0}{E_p} \\ &\quad \times \Bigg(\frac{p_0\gamma^0 + p_i\gamma^i + m_0}{2m_0} \exp\left\{-i\left[p_0(t'-t) - \boldsymbol{p}\cdot(\boldsymbol{x}'-\boldsymbol{x})\right]\right\} \Theta(t'-t) \\ &\quad + \frac{-p_0\gamma^0 - p_i\gamma^i + m_0}{2m_0} \exp\left\{+i\left[p_0(t'-t) - \boldsymbol{p}\cdot(\boldsymbol{x}'-\boldsymbol{x})\right]\right\} \Theta(t-t') \Bigg) \\ &= -i \int \frac{d^3p}{(2\pi)^3} \frac{m_0}{E_p} \left\{ \frac{\not{p} + m_0}{2m_0} \exp\left[-ip\cdot(x'-x)\right] \Theta(t'-t) \right. \\ &\quad \left. + \frac{-\not{p} + m_0}{2m_0} \exp\left[+ip\cdot(x'-x)\right] \Theta(t-t') \right\} \\ &= -i \int \frac{d^3p}{(2\pi)^3} \frac{m_0}{E_p} \left\{ \hat{\Lambda}_+(p) \exp\left[-ip\cdot(x'-x)\right] \Theta(t'-t) \right. \\ &\quad \left. + \hat{\Lambda}_-(p) \exp\left[+ip\cdot(x'-x)\right] \Theta(t-t') \right\} \quad . \end{aligned} \tag{2.22}$$

Equivalently, by means of the normalized Dirac plane waves (see *RQM*, Chap. 6)

$$\psi_p^r(x) = \sqrt{\frac{m_0}{E_p}} \frac{1}{\sqrt{2\pi}^3} \, \omega^r(\boldsymbol{p}) \exp(-\mathrm{i}\varepsilon_r p \cdot x) \quad , \tag{2.23}$$

$S_F(x'-x)$ can be transcribed to the following form (cf. Exercise 2.1):

$$\begin{aligned} S_F(x'-x) = &-\mathrm{i}\Theta(t'-t) \int d^3p \sum_{r=1}^{2} \psi_p^r(x')\bar{\psi}_p^r(x) \\ &+ \mathrm{i}\Theta(t-t') \int d^3p \sum_{r=3}^{4} \psi_p^r(x')\bar{\psi}_p^r(x) \quad . \end{aligned} \tag{2.24}$$

This result is the relativistic generalization of the nonrelativistic Green's function (1.77). The propagator S_F now consists of two parts: the first describes the propagation of positive-energy states forward in time, the latter the propagation of negative-energy states backward in time. With the aid of (2.24) the following relations for positive-energy solutions ($\psi^{(+E)}$) and negative-energy solutions ($\psi^{(-E)}$) are easily verified:

$$\Theta(t'-t)\psi^{(+E)}(x') = \mathrm{i} \int d^3x S_F(x'-x)\gamma_0\psi^{(+E)}(x) \quad , \tag{2.25}$$

$$\Theta(t-t')\psi^{(-E)}(x') = -\mathrm{i} \int d^3x S_F(x'-x)\gamma_0\psi^{(-E)}(x) \quad . \tag{2.26}$$

In analogy to the nonrelativistic propagator theory (cf. (1.7) and (1.9)) the occurrence of an additional minus sign in (2.26) results from the difference of the direction of propagation in time between (2.25) and (2.26) corresponding to propagation of positive-energy solutions forward in time and negative-energy solutions backward in time, respectively. The validity of (2.25) can be seen by writing

$$\begin{aligned} &\mathrm{i} \int d^3x S_F(x'-x)\gamma_0\psi^{(+E)}(x) \\ &\quad = \Theta(t'-t) \int d^3p \sum_{r=1}^{2} \psi_p^r(x') \int d^3x \, \psi_p^{r\dagger}(x)\psi^{(+E)}(x) \\ &\quad - \Theta(t-t') \int d^3p \sum_{r=3}^{4} \psi_p^r(x') \int d^3x \, \psi_p^{r\dagger}(x)\psi^{(+E)}(x) \end{aligned} \tag{2.27}$$

and expanding the general positive-energy solution $\psi^{(+E)}(x)$ in terms of Dirac plane waves

$$\psi^{(+E)}(x) = \int d^3p \sum_{r=1}^{4} a_r(p)\psi_p^r(x) \quad . \tag{2.28}$$

The coefficients $a_r(p)$ vanish except for $r = 1, 2$, since by definition $\psi^{(+E_p)}$ describes a wave packet containing only positive energies or "frequencies". By means of the orthonormality relations of the $\psi_p^r(x)$

$$a_r(p) = \int d^3x \, \psi^{(+E)}(x) \psi_p^{r\dagger}(x) \neq 0 \qquad (r = 1, 2) \quad , \tag{2.29}$$

$$a_r(p) = \int d^3x \, \psi^{(+E)}(x) \psi_p^{r\dagger}(x) = 0 \qquad (r = 3, 4) \quad , \tag{2.30}$$

the second term in (2.27) vanishes, while the first term gives

$$\Theta(t' - t) \int d^3p \sum_{r=1}^{2} a_r(p) \psi_p^r(x') = \Theta(t' - t) \psi^{(+E)}(x') \quad . \tag{2.31}$$

Thus we have proved the relation (2.25).

Equation (2.26) can be verified in similar manner. Equations (2.25) and (2.26) explicitly express our interpretation of electrons and positrons in terms of positive-energy solutions propagating forward in time and negative-energy solutions moving backward in time, respectively.

EXERCISE

2.1 Plane-Wave Decomposition of the Feynman Propagator

Problem. Prove that the Stückelberg–Feynman propagator

$$S_{\mathrm{F}}(x' - x) = -\mathrm{i} \int \frac{d^3p}{(2\pi)^3} \frac{m_0}{E_p} \left\{ \hat{\Lambda}_+(p) \exp\left[-\mathrm{i}p \cdot (x' - x)\right] \Theta(t' - t) \right.$$
$$\left. + \hat{\Lambda}_-(p) \exp\left[\mathrm{i}p \cdot (x' - x)\right] \Theta(t - t') \right\}$$

may equivalently be represented as

$$S_{\mathrm{F}}(x' - x) = -\mathrm{i}\Theta(t' - t) \int d^3p \sum_{r=1}^{2} \psi_p^r(x') \bar{\psi}_p^r(x)$$
$$+ \mathrm{i}\Theta(t - t') \int d^3p \sum_{r=3}^{4} \psi_p^r(x') \bar{\psi}_p^r(x) \quad .$$

Solution. As we recall from *RQM*, Chap. 6, the Dirac plane waves satisfy the following relations:

$$(\hat{\not{p}} - m_0) \psi_p^r(x) = 0 \quad , \tag{1}$$

$$(\varepsilon_r \not{p} - m_0) \omega^r(\boldsymbol{p}) = 0 \quad \text{or} \quad (\not{p} - \varepsilon_r m_0) \omega^r(\boldsymbol{p}) = 0 \quad , \tag{2}$$

$$\bar{\omega}^r(\boldsymbol{p})(\not{p} - \varepsilon_r m_0) = 0 \quad \text{where} \quad \bar{\omega}^r(\boldsymbol{p}) = \omega^r(\boldsymbol{p})^\dagger \gamma^0 \quad , \tag{3}$$

$$\bar{\omega}^r(\boldsymbol{p}) \omega^{r'}(\boldsymbol{p}) = \delta_{rr'} \varepsilon_r \quad , \tag{4}$$

$$\sum_{r=1}^{4} \varepsilon_r \omega_\alpha^r(\boldsymbol{p}) \bar{\omega}_\beta^r(\boldsymbol{p}) = \delta_{\alpha\beta} \quad , \tag{5}$$

Exercise 2.1.

$$\omega^{r\dagger}(\varepsilon_r \boldsymbol{p})\omega^{r'}(\varepsilon_{r'}\boldsymbol{p}) = \frac{E_p}{m_0}\delta_{rr'} \quad . \tag{6}$$

Here r and r' can take on the values $1, 2, 3, 4$. With the aid of (2.23) we find that (bearing in mind that $\varepsilon_1 = \varepsilon_2 = +1, \varepsilon_3 = \varepsilon_4 = -1$)

$$\begin{aligned}
\sum_{r=1}^{2} \psi_p^r(x')\bar{\psi}_p^r(x) &= \frac{1}{(2\pi)^3}\frac{m_0}{E_p}\exp\left[-\mathrm{i}p\cdot(x'-x)\right]\sum_{r=1}^{2}\omega^r(\boldsymbol{p})\bar{\omega}^r(\boldsymbol{p}) \\
&= \frac{1}{(2\pi)^3}\frac{m_0}{E_p}\exp\left[-\mathrm{i}p\cdot(x'-x)\right]\sum_{r=1}^{4}\omega^r(\boldsymbol{p})\bar{\omega}^r(\boldsymbol{p})\frac{\not{p}+m_0}{2m_0} \\
&= \frac{1}{(2\pi)^3}\frac{m_0}{E_p}\exp\left[-\mathrm{i}p\cdot(x'-x)\right]\underbrace{\sum_{r=1}^{4}\varepsilon_r\omega^r(\boldsymbol{p})\bar{\omega}^r(\boldsymbol{p})}_{=\mathbf{1}\text{ (because of (5))}}\frac{\not{p}+m_0}{2m_0} \\
&= \frac{1}{(2\pi)^3}\frac{m_0}{E_p}\exp\left[-\mathrm{i}p\cdot(x'-x)\right]\frac{\not{p}+m_0}{2m_0} \\
&= \frac{1}{(2\pi)^3}\frac{m_0}{E_p}\exp\left[-\mathrm{i}p\cdot(x'-x)\right]\hat{\Lambda}_+(p) \quad .
\end{aligned}$$

This result is just the first term of the propagator $S_\mathrm{F}(x'-x)$ in its representation in terms of the projection operators $\hat{\Lambda}_\pm$. Similarly, for the second part one obtains

$$\begin{aligned}
\sum_{r=3}^{4} \psi_p^r(x')\bar{\psi}_p^r(x) &= \frac{1}{(2\pi)^3}\frac{m_0}{E_p}\exp\left[\mathrm{i}p\cdot(x'-x)\right]\sum_{r=3}^{4}\omega^r(\boldsymbol{p})\bar{\omega}(\boldsymbol{p}) \\
&= \frac{1}{(2\pi)^3}\frac{m_0}{E_p}\exp\left[+\mathrm{i}p\cdot(x'-x)\right]\sum_{r=1}^{4}\frac{-(\not{p}-m_0)}{2m_0}\omega^r(\boldsymbol{p})\bar{\omega}^r(\boldsymbol{p}) \\
&= \frac{1}{(2\pi)^3}\frac{m_0}{E_p}\exp\left[+\mathrm{i}p\cdot(x'-x)\right]\sum_{r=1}^{4}\frac{(-\not{p}+m_0)}{2m_0}(-\varepsilon_r)\omega^r(\boldsymbol{p})\bar{\omega}^r(\boldsymbol{p}) \\
&= -\frac{1}{(2\pi)^3}\frac{m_0}{E_p}\exp\left[\mathrm{i}p\cdot(x'-x)\right]\frac{(-\not{p}+m_0)}{2m_0} \\
&= -\frac{1}{(2\pi)^3}\frac{m_0}{E_p}\frac{\exp\left[\mathrm{i}p\cdot(x'-x)\right]}{(2\pi)^3}\hat{\Lambda}_-(p) \quad .
\end{aligned}$$

Thus we have verified the proposed equivalence between the two representations of $S_\mathrm{F}(x'-x)$.

Equations (2.22) or (2.24) determine the free-particle propagator of the electron-positron theory. In analogy to (1.83) and (1.86), respectively, we may now formally construct the complete Green's function and the S matrix for the electron-positron field interacting with an electromagnetic potential A. This will then enable us to calculate various scattering processes of electrons and positrons in the presence of external fields, as will be demonstrated in the following chapter. To accomplish the aim of constructing the exact propagator $S_F(x',x;A)$ we start from the differential equation (2.5) that determines $S_F(x',x;A)$ and transcribe it, paraphrasing the nonrelativistic treatment (cf. (1.80)), to the following form:

$$(\mathrm{i}\nabla\!\!\!/' - m_0)S_F(x',x;A) = \delta^4(x'-x) + eA\!\!\!/(x')S_F(x',x;A) \quad . \tag{2.32}$$

This can be viewed as an inhomogeneous Dirac equation of the form

$$(\mathrm{i}\nabla\!\!\!/ - m_0)\Psi(x) = \varrho(x) \quad , \tag{2.33}$$

which is solved by the Green's function technique as follows

$$\Psi(x) = \Psi_0(x) + \int d^4y S_F(x-y)\varrho(y) \quad . \tag{2.34}$$

In this way (2.32) leads to an integral equation for the Stückelberg–Feynman propagator

$$\begin{aligned} S_F(x',x;A) &= \int d^4y S_F(x'-y)\left[\delta^4(y-x) + eA\!\!\!/(y)S_F(y,x;A)\right] \\ &= S_F(x'-x) + e\int d^4y S_F(x',y)A\!\!\!/(y)S_F(y,x;A) \quad . \end{aligned} \tag{2.35}$$

This is the relativistic counterpart of the Lippmann–Schwinger equation (1.83). This integral equation determines the complete propagator $S_F(x',x;A)$ in terms of the free-particle propagator $S_F(x',x)$. Proceeding in analogy to the nonrelativistic treatment (cf. (1.28)) the iteration of the integral equation yields the following multiple scattering expansion:

$$\begin{aligned} S_F(x',x;A) = S_F(x'-x) &+ e\int d^4x_1 S_F(x'-x_1)A\!\!\!/(x_1)S_F(x_1-x) \\ &+ e^2\int d^4x_1 d^4x_2 S_F(x'-x_1)A\!\!\!/(x_1)S_F(x_1-x_2)A\!\!\!/(x_2)S_F(x_2-x) \\ &+ \dots \quad . \end{aligned} \tag{2.36}$$

In analogy to (1.31) the exact solution of the Dirac equation

$$(\mathrm{i}\nabla\!\!\!/_x - m_0)\Psi(x) = eA\!\!\!/(x)\Psi(x) \tag{2.37}$$

is completely determined in terms of S_F if one imposes the boundary condition of Feynman and Stückelberg, namely

$$\Psi(x) = \psi(x) + \int d^4y S_F(x-y)eA\!\!\!/(y)\Psi(y) \quad . \tag{2.38}$$

Here $\psi(x)$ is a solution of the free Dirac equation, i.e. of the homogeneous version of (2.37). The potential $V(x)$ occurring in (1.31) is now replaced by $eA\!\!\!/(x)$. The second term on the right-hand side represents the scattered wave. In accordance with the properties of the Stückelberg–Feynman propagator (2.24) this scattered wave contains only positive frequencies in the distant future and only negative frequencies in the distant past, since

$$\Psi(x) - \psi(x) \Rightarrow \int d^3p \sum_{r=1}^{2} \psi_p^r(x) \left(-\mathrm{ie} \int d^4y\, \bar{\psi}_p^r(y) A\!\!\!/(y) \Psi(y) \right) \quad \text{for } t \to +\infty \tag{2.39}$$

and

$$\Psi(x) - \psi(x) \Rightarrow \int d^3p \sum_{r=3}^{4} \psi_p^r(x) \left(+\mathrm{ie} \int d^4y\, \bar{\psi}_p^r(y) A\!\!\!/(y) \Psi(y) \right) \quad \text{for } t \to -\infty\,. \tag{2.40}$$

Notice that here x and y are to be identified with x' and x, respectively, in (2.24), and t in (2.37, 2.83) corresponds to t' in (2.24).

The result (2.39) expresses our formulation of the relativistic scattering problem, which is consistent with the requirements of hole theory. These requirements have been essentially built into the Stückelberg–Feynman propagator by the special choice of the integration contour and thus take into account the location of the Fermi border (cf. Fig. 2.5 and Exercise 2.2). Furthermore, according to (2.39), an electron cannot "fall into the sea" of (occupied) negative-energy states after scattering by an external field $A\!\!\!/(y)$, since only the unoccupied positive-energy states are available. In contrast, positrons interpreted in terms of negative-energy electrons travelling backward in time are scattered back to earlier times into other negative-energy states according to (2.40).

The *S-matrix elements* are defined in the same manner as in the nonrelativistic case (1.37). Terming $\psi_f(x)$ the final free wave with the quantum numbers f that is observed at the end of the scattering process, we infer from (2.38–40) with the aid of (2.24) that

$$\begin{aligned} S_{fi} &= \lim_{t\to\pm\infty} \langle \psi_f(x) | \Psi_i(x) \rangle \\ &= \lim_{t\to\pm\infty} \left\langle \psi_f(x) \middle| \psi_i(x) + \int d^4y S_\mathrm{F}(x-y) e A\!\!\!/(y) \Psi_i(y) \right\rangle \,. \end{aligned} \tag{2.41}$$

Here the limit $t \to +\infty$ is understood if $\psi_f(x)$ describes an electron and $t \to -\infty$ if $\psi_f(x)$ means a positron, since the latter is considered a negative-energy electron moving backward in time. For electron scattering we have

$$S_{fi} = \delta_{fi} - \mathrm{ie} \lim_{t\to+\infty} \left\langle \psi_f(x) \middle| \int d^3p \sum_{r=1}^{2} \psi_p^r(x) \int d^4y\, \bar{\psi}_p^r(y) A\!\!\!/(y) \Psi_i(y) \right\rangle, \tag{2.41a}$$

while positron scattering is described by

$$S_{fi} = \delta_{fi} + \mathrm{ie} \lim_{t\to-\infty} \left\langle \psi_f(x) \middle| \int d^3p \sum_{r=3}^{4} \psi_p^r(x) \int d^4y\, \bar{\psi}_p^r(y) A\!\!\!/(y) \Psi_i(y) \right\rangle. \tag{2.41b}$$

The $\int d^3x$ integral implied by the brackets projects out just that state $\psi_p^r(x)$ whose quantum numbers agree with $\psi_f(x)$. All other terms of the integral-sum $\int d^3p \sum_r$ do not contribute. This yields for (2.41a)

$$S_{fi} = \delta_{fi} - \mathrm{ie} \int d^4y\, \bar{\psi}_f(y) A\!\!\!/(y) \Psi_i(y)$$

and a similar expression for positron scattering. Both results can be combined by writing ($\varepsilon_f = +1$ for positive-energy waves in the future and $\varepsilon_f = -1$ for energy waves in the past)

$$S_{fi} = \delta_{fi} - \mathrm{i}e\varepsilon_f \int d^4y \, \bar{\psi}_f(y) A\!\!\!/(y) \Psi_i(y) \quad . \tag{2.42}$$

Depending on whether $\psi_f(x)$ represents an electron or a positron, the first or the second term, respectively, is nonzero. In (2.42) $\Psi_i(x)$ stands for the incoming wave, which either reduces at $y_0 \to -\infty$ to an incident positive-frequency wave $\psi_i(x)$ carrying the quantum numbers i or at $y_0 \to +\infty$ to an incident negative-frequency wave propagating into the past with quantum numbers i, according to the Stückelberg–Feynman boundary conditions.

To elucidate how the various scattering processes are contained in (2.42) we first consider the "ordinary" scattering of electrons. In this case

$$\Psi_i(y) \overset{y_0 \to \infty}{\Longrightarrow} \psi_i^{(+E)}(y) = \sqrt{\frac{m_0}{E_-}} \frac{1}{(2\pi)^{3/2}} u(p_-, s_-) \exp(\mathrm{i}p_- \cdot y) \tag{2.43}$$

reduces to an incoming electron wave with positive energy E_-, momentum p_- and spin s_-. The minus sign here designates the negative charge of the electron. The nth order contribution to the S-matrix element (2.42) is then

$$\begin{aligned} S_{fi}^{(n)} = -\mathrm{i}e^n \int & d^4y_1 \dots d^4y_n \, \bar{\psi}_f^{(+E)}(y_n) A\!\!\!/(y_n) S_\mathrm{F}(y_n - y_{n-1}) A\!\!\!/(y_{n-1}) \dots \\ & \times S_\mathrm{F}(y_2 - y_1) A\!\!\!/(y_1) \psi_i^{(+E)}(y_1) \quad . \end{aligned} \tag{2.44}$$

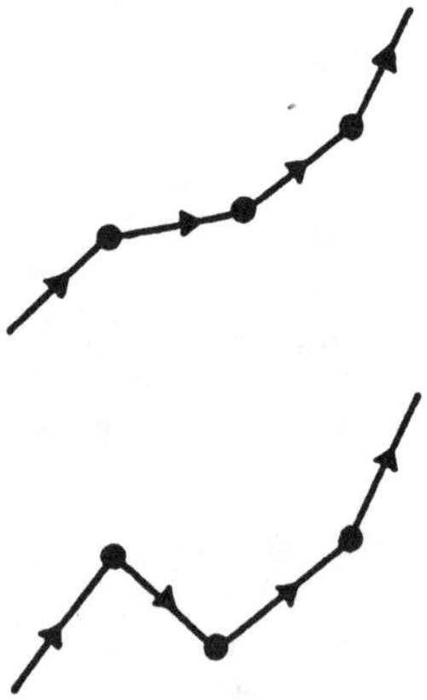

Fig. 2.7. Two graphs for third–order electron scattering. The lower graph involves an intermediate electron–positron pair

This expression contains both types of graphs shown in Fig. 2.7: That is, in addition to ordinary scattering intermediate pair creation and pair annihilation are included in the series, since the various d^4y integrations also allow for a reverse time ordering, $y_{n+1}^0 < y_n^0$. We therefore recognize that, inevitably, the second part of the propagator (2.24) also contributes.

Next we consider the *pair production* process. In accordance with the developed formalism, $\Psi_i(y)$ in this case at $y_0 \to +\infty$ reduces to a plane wave with negative energy. This particle state propagating backward in time then represents a positron. We use the notation $\boldsymbol{p}_-, s_-$ for three-momentum and spin corresponding to the physical electron and $\boldsymbol{p}_+, s_+$ for the physical positron where $p_\pm^0 > 0$. The physical positron state at $t \to \infty$ is described by a plane wave of negative energy with quantum numbers $-\boldsymbol{p}_+, -s_+, \varepsilon = -1$. This wave propagating backward in time enters into the vertex. That is,

$$\Psi_i(x) \overset{y_0 \to \infty}{\Longrightarrow} \psi_i^{(-E)}(y) = \sqrt{\frac{m_0}{E_+}} \frac{1}{(2\pi)^{3/2}} v(p_+, s_+) \exp\left(\mathrm{i}p_+ \cdot y\right) \quad . \tag{2.45}$$

This form of the wave function explicitly exhibits the negative energy and negative three–momentum of the particle wave. The positive sign in the exponent in (2.45) obviously expresses this property since a wave with positive energy and positive three-momentum carries a phase factor $\exp(-\mathrm{i}p_- \cdot y)$. The fact that the spin direction is reversed, i.e. $-s_+$, is taken into account by the definition of the spinor $v(p_+, s_+)$. As we recall from *RQM*, Chap. 6, the spinors have been defined according to

$$v(p_+, +1/2) = \omega^4(p_+) \quad \text{and} \quad v(p_+, -1/2) = \omega^3(p_+) \quad ,$$

where ω^4 is the spinor corresponding to a negative-energy electron with spin up and ω^3 a negative-energy electron with spin down.

The final wave function ψ_f in the case of the pair creation process is a positive-energy solution carrying the quantum numbers $\boldsymbol{p}_-, s_-, \varepsilon = +1$ and describes the electron.

To resume our previous considerations, from hole theory (see *RQM*, Chap. 12) we know that the absence of a negative-energy electron with four-momentum $-\boldsymbol{p}_+$ and spin $-s_+$ is interpreted in terms of a positron with four-momentum $+\boldsymbol{p}_+$ and polarization $+s_+$. Within the framework of the propagator formalism the probability amplitude for the creation of a positron at x propagating forward in space–time and emerging out of the interaction region into the final free state $(\boldsymbol{p}_+, s_+)$ at x' is calculated through the probability amplitude for the propagation of a negative-energy electron (four-momentum $-p_+$, spin $-s_+$) backward in time entering into the interaction region. Then, being scattered by the force field, it emerges out of the interaction volume as a positive-energy state propagating forward in time. The diagrams for the pair creation are illustrated in Fig. 2.8. We emphasize that the second-order amplitude consists of two diagrams corresponding to the second scattering of the positron. These two second-order diagrams are said to differ in the time ordering of the two scattering processes.

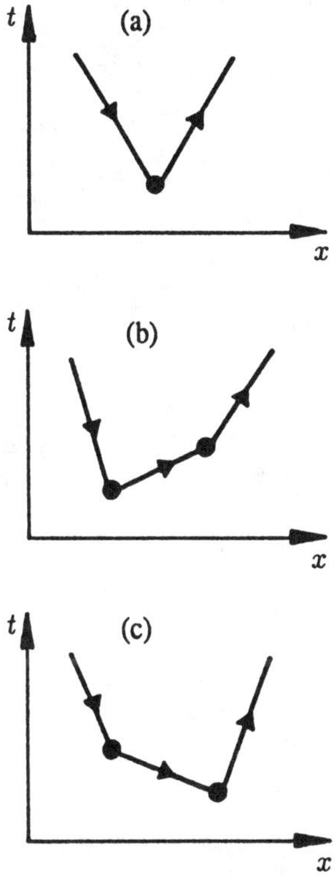

Fig. 2.8a–c. First- and second-order Feynman diagrams for electron–positron pair creation

Since the Feynman propagator according to (2.24) consists of two parts there is no need to deal explicitly with time orderings when calculating any process. The formula for the S-matrix automatically contains them all.

Now let us consider *pair annihilation*. This process in lowest order is represented by the graph of Fig. 2.9. In this case we insert for $\Psi_i(y)$ a solution of (2.38) that reduces to $\psi_i^{(+E)}(y)$ at $t \to -\infty$. This positive-energy solution represents an electron that propagates forward in time into the interaction volume, to be scattered backward in time and emerges into a negative energy state. According to (2.42) the nth-order amplitude that the electron scatters into a given final state $\psi_f^{(-E)}$, labelled by the physical quantum numbers $\boldsymbol{p}_+, \boldsymbol{s}_+, \varepsilon_f = -1$ (the corresponding formal quantum numbers entering the wave function, however, are $-\boldsymbol{p}_+, -s_+$; cf. the discussion following (2.45)), is given by

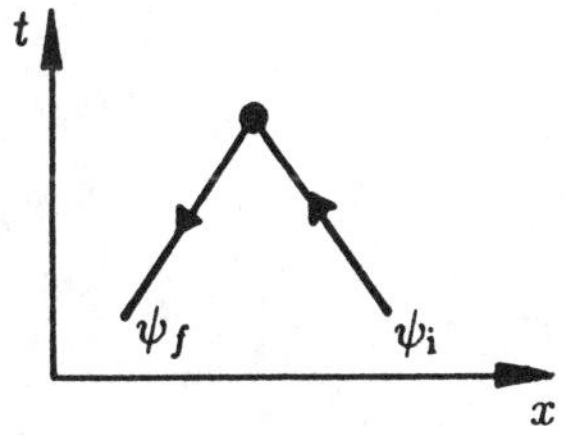

Fig. 2.9. The graph for pair annihilation

$$S_{fi}^{(n)} = \mathrm{i}e^n \int d^4y_1 \ldots \int d^4y_n \bar{\psi}_f^{(-E)}(y_n) A\!\!\!/(y_n) S_\mathrm{F}(y_n - y_{n-1}) \ldots A\!\!\!/(y_1) \psi_i^{(+E)}(y_1) \ . \tag{2.46}$$

In the language of hole theory this is the nth-order amplitude that a positive-energy electron is scattered into an electron state of negative energy, negative three-momentum $-\boldsymbol{p}_+$, and spin $-s_+$. This state must of course have been empty at $t \to -\infty$. That is, there must have been a hole or positron present with four–momentum p_+ and spin or polarization s_+.

Finally let us turn to *positron scattering*, which (in lowest order) is represented by either of the two equivalent graphs of Fig. 2.10. The incident wave is an electron of negative frequency (negative energy) labelled by the quantum numbers $-\boldsymbol{p}_+, -\boldsymbol{s}_+, \varepsilon_f = -1$. The final state (outgoing wave) is represented as a negative-energy electron too. Notice that the incoming electron of negative energy characterizes the outgoing positron of positive energy, and similarly the incoming

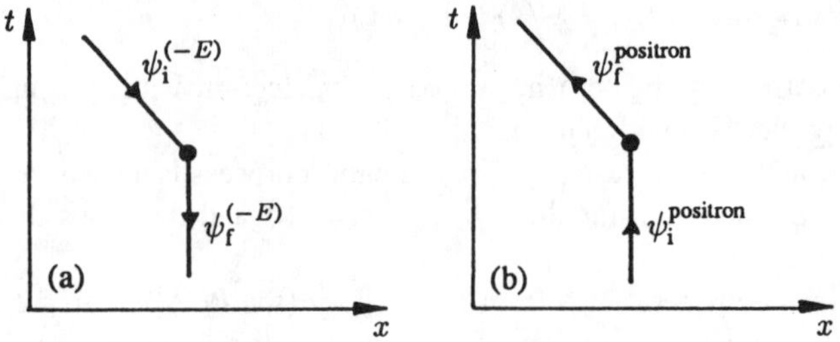

Fig. 2.10a,b. Positron scattering in lowest order. The emerging positron (ψ_f^{positron}, in (**b**)) corresponds to an incoming negative-energy electron ($\psi_i^{(-E)}$ in (**a**)). Similarly, the incident positron (ψ_i^{positron} in (**b**)) is represented in terms of an outcoming negative-energy electron ($\psi_f^{(-E)}$ in (**a**)). In other words, (**a**) describes the scattering process in accordance with our calculational techniques, whereas (**b**) illustrates the real physical picture of positron scattering

positron is represented as an outgoing negative-energy electron. In Chap. 3.4 we will elaborate this explicitly.

EXERCISE

2.2 Feynman Propagator for a Fermi Gas

Problem. Suppose in our formalism we replace the vacuum by a noninteracting Fermi gas of electrons with Fermi momentum k_F. How is the Stückelberg–Feynman propagator modified? Evaluate S_F in the low-density limit.

Solution. In a degenerate Fermi gas the levels in the positive-energy electron continuum are occupied up to the Fermi momentum k_F. These occupied states have to be treated like the negative-energy states of the Dirac sea. That is, the Feynman propagator is modified according to

$$\begin{aligned} \mathrm{i}S_F^G(x'-x) &= \Theta(t'-t) \sum_{\substack{\boldsymbol{k} \\ r=1,2}} \psi_{\boldsymbol{k}}^r(x')\bar{\psi}_{\boldsymbol{k}}^r(x)\Theta(k-k_F) \\ &\quad - \Theta(t-t')\sum_{\boldsymbol{k}} \left(\sum_{r=3,4} \psi_{\boldsymbol{k}}^r(x')\bar{\psi}_{\boldsymbol{k}}^r(x) \right. \\ &\quad \left. + \sum_{r=1,2} \psi_{\boldsymbol{k}}^r(x')\bar{\psi}_{\boldsymbol{k}}^r(x)\Theta(k_F-k) \right) \quad , \end{aligned} \tag{1}$$

where

$$\psi_{\boldsymbol{k}}^r(x) = (m_0/E_k)^{1/2}(2\pi)^{-3/2}\,\omega^r(\boldsymbol{k})\exp(-\mathrm{i}\varepsilon_r k\cdot x) \quad , \tag{2}$$

with $k_0 = E_k = \sqrt{\boldsymbol{k}^2 + m_0^2}$ are the normalized Dirac plane waves. For the special case $k_F = 0$ this expression reduces to the ordinary Feynman propagator. We recall the following representations of the Θ function:

Exercise 2.2.

$$\Theta(t'-t) = \mathrm{i}\int_{-\infty}^{+\infty} \frac{dp_0'}{2\pi} \exp\left[-\mathrm{i}p_0'(t'-t)\right] \frac{1}{p_0'+\mathrm{i}\varepsilon} \quad , \tag{3a}$$

$$\Theta(t-t') = -\mathrm{i}\int_{-\infty}^{+\infty} \frac{dp_0'}{2\pi} \exp\left[-\mathrm{i}p_0'(t'-t)\right] \frac{1}{p_0'-\mathrm{i}\varepsilon} \quad , \tag{3b}$$

where the second expression is obtained from the first by complex conjugation. Furthermore we need the relations

$$\begin{aligned} \sum_{r=1,2} \omega^r(\boldsymbol{k})\bar{\omega}^r(\boldsymbol{k}) &= \frac{\not{k}+m_0}{2m_0} = \hat{\Lambda}_+(k) \quad , \\ \sum_{r=3,4} \omega^r(\boldsymbol{k})\bar{\omega}^r(\boldsymbol{k}) &= \frac{\not{k}-m_0}{2m_0} = -\hat{\Lambda}_-(k) \quad . \end{aligned} \tag{4}$$

With the aid of (4), (1) yields

$$\begin{aligned} \mathrm{i}S_\mathrm{F}^G(x'-x) = {} & \Theta(t'-t)\int \frac{d^3k}{(2\pi)^3}\frac{m_0}{E_k}\hat{\Lambda}_+(k)\exp\left[-\mathrm{i}k\cdot(x'-x)\right]\Theta(k-k_\mathrm{F}) \\ & + \Theta(t-t')\int \frac{d^3k}{(2\pi)^3}\frac{m_0}{E_k}\hat{\Lambda}_-(k)\exp\left[\mathrm{i}k\cdot(x'-x)\right] \\ & - \Theta(t-t')\int \frac{d^3k}{(2\pi)^3}\frac{m_0}{E_k}\hat{\Lambda}_+(k)\exp\left[-\mathrm{i}k\cdot(x'-x)\right]\Theta(k_\mathrm{F}-k) \\ \equiv {} & I_1+I_2+I_3 \quad . \end{aligned} \tag{5}$$

Substituting the representation (3a) of the Θ function we find that

$$\begin{aligned} I_1 = {} & \mathrm{i}\int \frac{d^3k}{(2\pi)^3}\frac{m_0}{E_k}\hat{\Lambda}_+(k)\exp\left\{-\mathrm{i}\left[E_k(t'-t)-\boldsymbol{k}\cdot(\boldsymbol{x}'-\boldsymbol{x})\right]\right\} \\ & \times \int \frac{dk_0'}{2\pi}\exp\left\{-\mathrm{i}\left[k_0'(t'-t)\right]\right\}\frac{1}{k_0'+\mathrm{i}\varepsilon}\Theta(k-k_\mathrm{F}) \\ = {} & \mathrm{i}\int \frac{d^3k\,dk_0'}{(2\pi)^4}\frac{1}{2E_k}\exp\left\{-\mathrm{i}\left[(E_k+k_0')(t'-t)-\boldsymbol{k}\cdot(\boldsymbol{x}'-\boldsymbol{x})\right]\right\} \\ & \times \frac{E_k\gamma_0-\boldsymbol{k}\cdot\boldsymbol{\gamma}+m_0}{k_0'+\mathrm{i}\varepsilon}\Theta(k-k_\mathrm{F}) \quad . \end{aligned} \tag{6a}$$

Similarily, using (3b), we get

$$\begin{aligned} I_2 = {} & -\mathrm{i}\int \frac{d^3k\,dk_0'}{(2\pi)^4}\frac{1}{2E_k}\exp\left\{\mathrm{i}\left[(E_k-k_0')(t'-t)-\boldsymbol{k}\cdot(\boldsymbol{x}'-\boldsymbol{x})\right]\right\} \\ & \times \frac{-E_k\gamma_0+\boldsymbol{k}\cdot\boldsymbol{\gamma}+m_0}{k_0'-\mathrm{i}\varepsilon} \quad , \end{aligned} \tag{6b}$$

$$\begin{aligned} I_3 = {} & \mathrm{i}\int \frac{d^3k\,dk_0'}{(2\pi)^4}\frac{1}{2E_k}\exp\left\{-\mathrm{i}\left[(E_k+k_0')(t'-t)-\boldsymbol{k}\cdot(\boldsymbol{x}'-\boldsymbol{x})\right]\right\} \\ & \times \frac{E_k\gamma_0-\boldsymbol{k}\cdot\boldsymbol{\gamma}+m_0}{k_0'-\mathrm{i}\varepsilon}\Theta(k_\mathrm{F}-k) \quad . \end{aligned} \tag{6c}$$

In order to evaluate these integrals, we introduce the following substitutions:

Exercise 2.2.

$$k_0 = k_0' + E_k \quad \text{in} \quad I_1 \text{ and } I_3 \quad , \tag{7a}$$

$$k_0 = k_0' - E_k \quad \text{and} \quad \boldsymbol{k} \to -\boldsymbol{k} \quad \text{in} \quad I_2 \quad . \tag{7b}$$

In addition, in the integral I_2 we make use of the identity

$$1 = \Theta(k - k_{\text{F}}) + \Theta(k_{\text{F}} - k) \tag{8}$$

so that (6) becomes

$$\begin{aligned}
I_1 &= \mathrm{i} \int \frac{d^4k}{(2\pi)^4} \frac{1}{2E_k} \exp\left[-\mathrm{i}k \cdot (x' - x)\right] \frac{E_k\gamma_0 - \boldsymbol{k} \cdot \boldsymbol{\gamma} + m_0}{k_0 - E_k + \mathrm{i}\varepsilon} \Theta(k - k_{\text{F}}) \quad , \\
I_2 &= -\mathrm{i} \int \frac{d^4k}{(2\pi)^4} \frac{1}{2E_k} \exp\left[-\mathrm{i}k \cdot (x' - x)\right] \frac{-E_k\gamma_0 - \boldsymbol{k} \cdot \boldsymbol{\gamma} + m_0}{k_0 + E_k - \mathrm{i}\varepsilon} \\
&\quad \times \left[\Theta(k - k_{\text{F}}) + \Theta(k_{\text{F}} - k)\right] \quad , \\
I_3 &= \mathrm{i} \int \frac{d^4k}{(2\pi)^4} \frac{1}{2E_k} \exp\left[-\mathrm{i}k \cdot (x' - x)\right] \frac{E_k\gamma_0 - \boldsymbol{k} \cdot \boldsymbol{\gamma} + m_0}{k_0 - E_k - \mathrm{i}\varepsilon} \Theta(k_{\text{F}} - k) \quad .
\end{aligned} \tag{9}$$

In the next step we add I_1 to that part of I_2 which contains $\Theta(k - k_{\text{F}})$. The combined denominator of the two integrands is

$$\begin{aligned}
(k_0 - E_k + \mathrm{i}\varepsilon)(k_0 + E_k - \mathrm{i}\varepsilon) &= k_0^2 - E_k^2 + 2\mathrm{i}\varepsilon E_k + \varepsilon^2 \\
&= k_0^2 - E_k^2 + \mathrm{i}\varepsilon' = k^2 - m_0^2 + \mathrm{i}\varepsilon' \quad ,
\end{aligned} \tag{10}$$

since ε is an infinitesimal quantity and $E_k > 0$. This results in

$$\begin{aligned}
&\frac{E_k\gamma_0 - \boldsymbol{k} \cdot \boldsymbol{\gamma} + m_0}{k_0 - E_k + \mathrm{i}\varepsilon} - \frac{-E_k\gamma_0 - \boldsymbol{k} \cdot \boldsymbol{\gamma} + m_0}{k_0 + E_k - \mathrm{i}\varepsilon} \\
&\quad = \frac{(E_k\gamma_0 - \boldsymbol{k} \cdot \boldsymbol{\gamma} + m_0)(k_0 + E_k) - (-E_k\gamma_0 - \boldsymbol{k} \cdot \boldsymbol{\gamma} + m_0)(k_0 - E_k)}{k^2 - m_0^2 + \mathrm{i}\varepsilon'} \\
&\quad = \frac{2E_k(k_0\gamma_0 - \boldsymbol{k} \cdot \boldsymbol{\gamma} + m_0)}{k^2 - m_0^2 + \mathrm{i}\varepsilon'} \\
&\quad = \frac{2E_k(k \cdot \gamma + m_0)}{k^2 - m_0^2 + \mathrm{i}\varepsilon'} \quad .
\end{aligned} \tag{11}$$

Similarily the second part of I_2 is added to I_3. The combined denominator in this case is

$$\begin{aligned}
(k_0 + E_k - \mathrm{i}\varepsilon)(k_0 - E_k - \mathrm{i}\varepsilon) &= k_0^2 - E_k^2 - 2\mathrm{i}\varepsilon k_0 \\
&= k^2 - m_0^2 - \mathrm{i}\varepsilon' k_0 \quad .
\end{aligned} \tag{12}$$

Proceeding as in (11), we find that

$$\begin{aligned}
&\frac{E_k\gamma_0 - \boldsymbol{k} \cdot \boldsymbol{\gamma} + m_0}{k_0 - E_k - \mathrm{i}\varepsilon} - \frac{-E_k\gamma_0 - \boldsymbol{k} \cdot \boldsymbol{\gamma} + m_0}{k_0 + E_k - \mathrm{i}\varepsilon} \\
&\quad = \frac{(E_k\gamma_0 - \boldsymbol{k} \cdot \boldsymbol{\gamma} + m_0)(k_0 + E_k) - (-E_k\gamma_0 - \boldsymbol{k} \cdot \boldsymbol{\gamma} + m_0)(k_0 - E_k)}{k^2 - m_0^2 - \mathrm{i}\varepsilon' k_0} \\
&\quad = \frac{2E_k(k \cdot \gamma + m_0)}{k^2 - m_0^2 - \mathrm{i}\varepsilon' k_0} \quad .
\end{aligned} \tag{13}$$

We insert these expressions into (5) and obtain

Exercise 2.2.

$$S_{\mathrm{F}}^{G}(x'-x) = \int \frac{d^4k}{(2\pi)^4} \frac{\gamma \cdot k + m_0}{k^2 - m_0^2 + \mathrm{i}\varepsilon} \exp\left[-\mathrm{i}k \cdot (x'-x)\right] \Theta(k - k_{\mathrm{F}})$$
$$+ \int \frac{d^4k}{(2\pi)^4} \frac{\gamma \cdot k + m_0}{k^2 - m_0^2 - \mathrm{i}\varepsilon k_0} \exp\left[-\mathrm{i}k \cdot (x'-x)\right] \Theta(k_{\mathrm{F}} - k) \quad (14)$$

Instead of adding an infinitesimal $\mathrm{i}\varepsilon$ to the denominator of the propagators (14), one may alternatively perform the integrations along the contours in the complex k_0 plane as shown in Fig. 2.11:

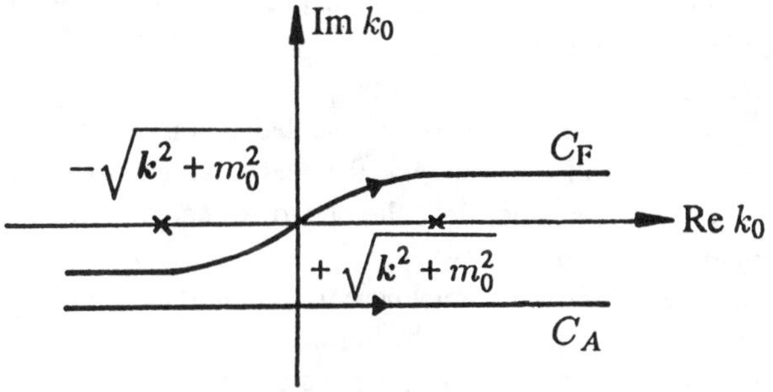

Fig. 2.11. The integration contours which define the Feynman propagator (C_{F}) and the advanced propagator (C_{A})

$$S_{\mathrm{F}}^{G}(x'-x) = \int_{C_{\mathrm{F}}} \frac{d^4k}{(2\pi)^4} \frac{\exp\left[-\mathrm{i}k \cdot (x'-x)\right]}{k \cdot \gamma - m_0} \Theta(k - k_{\mathrm{F}})$$
$$+ \int_{C_{\mathrm{A}}} \frac{d^4k}{(2\pi)^4} \frac{\exp\left[-\mathrm{i}k \cdot (x'-x)\right]}{k \cdot \gamma - m_0} \Theta(k_{\mathrm{F}} - k) \quad , \quad (15)$$

where we have introduced the symbolic notation

$$\frac{\gamma \cdot k + m_0}{k^2 - m_0^2} = \frac{\gamma \cdot k + m_0}{(\gamma \cdot k + m_0)(\gamma \cdot k - m_0)} = \frac{1}{\gamma \cdot k - m_0} \quad . \quad (16)$$

For $t' > t$ the second integral in (15),

$$\int_{C_A} \frac{d^4k}{(2\pi)^4} \frac{\exp\left\{-\mathrm{i}\left[k_0(t'-t) - \boldsymbol{k} \cdot (\boldsymbol{x}' - \boldsymbol{x})\right]\right\}}{k \cdot \gamma - m_0} \Theta(k_{\mathrm{F}} - k) \quad (17)$$

is evaluated along the contour C_{A} closed in the lower half plane so that it vanishes. This procedure yields the *advanced* propagator that transforms all solutions below the Fermi surface ($k < k_{\mathrm{F}}$) backward in time. The integration contour C_{F} in Fig. 2.11 is the ordinary contour in the vacuum, since the old vacuum remains unchanged above the Fermi momentum k_{F}. The corresponding "causal" propagator transforms particles (positive-energy solutions) to propagate forward in time. In Fig. 2.12 we have illustrated these properties:

Fig. 2.12. The integration contour C_{KF} crosses the real k_0 axis at the border between occupied and empty states

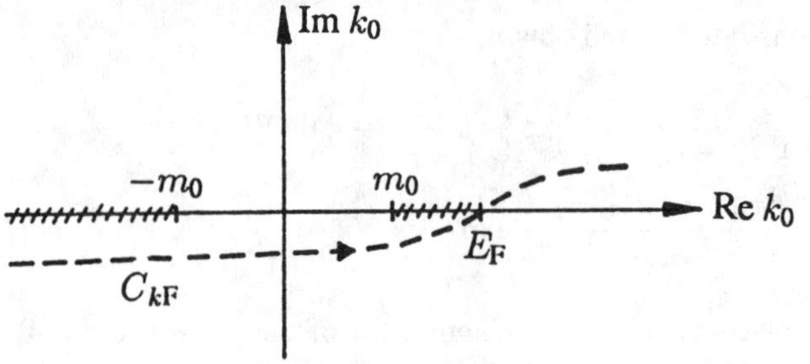

Solutions with a momentum $k < k_F$, i.e. with an energy below the corresponding Fermi energy $E_F = \sqrt{k_F^2 + m_0^2}$, propagate backward in time and are pictured as holes (hatched region). Particles, on the other hand, have energies larger than E_F and propagate forward in time.

We summarize the steps that led to this result. For particles of a Fermi gas the integration contour cuts the real k_0 axis just above the Fermi energy E_F. In the ordinary vacuum only the negative energy states are occupied. In this case one chooses $k_F = 0$, that is, $E_F \leq |E_k|$ for all k, and the point where the contour cuts the real k_0 axis lies somewhere in the interval $[-E_k, E_k]$, the precise position being irrelevant. In the case of the Fermi gas ($k_F > 0$) we have to distinguish between two alternatives. For $k > k_F$ the integration contour passes the same interval because $E_F < |E_k|$. In both cases the contour agrees with C_F. On the other hand, at low momenta $k < k_F$, implying $E_F > |E_k|$, i.e. $E_F > E_k > -E_k$, the integration has to be performed along the dashed contour, which is equivalent to C_A! This prescription is symbolically expressed as

$$S_F^G(x' - x) = \int\limits_{C_{KF}} \frac{d^4k}{(2\pi)^4} \frac{\exp\left[-ik \cdot (x' - x)\right]}{k \cdot \gamma - m_0} \quad , \tag{18}$$

where the contour C_{KF} depends on k and k_F, as explained above.

The extension of this prescription to the case of the Feynman propagator in the presence of an external field $A_\mu(\boldsymbol{x})$ is straightforward. For example, consider an atom with bound states (located within the interval $-m_0 < E < m_0$). In this case the integration contour in the complex k_0 plane has to be chosen such that it passes below the occupied and above the empty states.

In the low density limit the Fermi momentum k_F is directly related to the density of the electron gas. That is, with the normalization condition for a box of volume V the particle number is given by

$$N = \sum_{r=1}^{2} \sum_{k} \Theta(k_F - k) \quad \rightarrow 2V \int \frac{d^3k}{(2\pi)^3} \Theta(k_F - k) = \frac{V}{3\pi^2} k_F^3 \quad , \tag{19}$$

where the factor 2 accounts for the spin degeneracy. Thus in the low-density limit, $\varrho = N/V = k_F^3/3\pi^3 \rightarrow 0$, the Fermi momentum k_F approaches 0, so that the propagator S_F^G reduces to S_F.

Exercise 2.2.

Supplement. *Finite Temperatures.* The result (14) can be generalized to the case of a free-electron gas at finite temperature T. From statistical mechanics it is well known that a quantum-mechanical state with an energy E cannot definitely be said to be occupied or empty. Instead an *occupation probability* function $f(E)$ is introduced. The explicit form of this function depends on the type of particle considered; for particles with half-integer spin, Fermi–Dirac statistics requires $f(E)$ to be of the form

$$f(E) \equiv f(E, T, \mu) = \frac{1}{\exp\left[(E-\mu)/k_{\mathrm{B}}T + 1\right]} \quad , \tag{20}$$

where k_{B} is the Bolzmann constant $k_{\mathrm{B}} = 8.62 \times 10^{-11}$ MeV/K. The Fermi function contains two free parameters, the temperature T and the chemical potential μ. The latter is a generalization of the Fermi energy $E_{\mathrm{F}} = \sqrt{k_{\mathrm{F}}^2 + m_0^2}$, as becomes obvious in the limit $T \to 0$, when the Fermi function approaches the Θ function,

$$f(E, T, \mu) \to \Theta(\mu - E) \quad . \tag{21}$$

That is, below the chemical potential μ all states are occupied, whereas above μ all states are empty.

To generalize (1) to the case of finite temperature, we therefore replace the Θ functions $\Theta(k - k_{\mathrm{F}}) = \Theta(E_k - E_{\mathrm{F}})$ and $\Theta(k_{\mathrm{F}} - k) = \Theta(E_{\mathrm{F}} - E_k)$ by the occupation function (21). However, we must be careful to distinguish four different contributions: free-electron states (r=1,2) and occupied positron states (r=3,4) propagating forward in time, as well as occupied electron states and free-positron states, propagating backward in time. In contrast to (1), where all positron states were assumed empty, the electron gas also contains positrons owing to thermal excitation, as expressed by the Fermi function (20). This is depicted in Fig. 2.13. However, the temperature at which these contributions become important, i.e. where $k_{\mathrm{B}}T \approx 2m_0c^2$, is quite large, $T \approx 10^{10}$ K.

According to these considerations the temperature-dependent Feynman propagator must be of the following form:

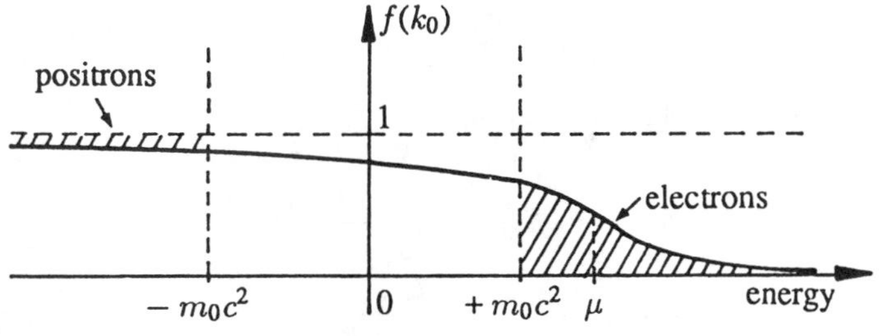

Fig. 2.13. The occupation probability for a hot electron gas. The hatched regions mark the occupied electron and positron states

Exercise 2.2.

$$
\begin{aligned}
\mathrm{i}S_{\mathrm{F}}^{G}(x'-x) = & \left[\sum_{\boldsymbol{k}}\sum_{r=1,2}(1-f(E_k))\,\psi_{\boldsymbol{k}}^{r}(x')\bar{\psi}_{\boldsymbol{k}}^{r}(x)\right.\\
& \left.+\sum_{\boldsymbol{k}}\sum_{r=3,4}(1-f(-E_k))\,\psi_{\boldsymbol{k}}^{r}(x')\bar{\psi}_{\boldsymbol{k}}^{r}(x)\right]\Theta(t'-t)\\
& -\left[\sum_{\boldsymbol{k}}\sum_{r=3,4}f(-E_k)\psi_{\boldsymbol{k}}^{r}(x')\bar{\psi}_{\boldsymbol{k}}^{r}(x)\right.\\
& \left.+\sum_{\boldsymbol{k}}\sum_{r=1,2}f(E_k)\psi_{\boldsymbol{k}}^{r}(x')\bar{\psi}_{\boldsymbol{k}}^{r}(x)\right]\Theta(t-t')\quad . \qquad (22)
\end{aligned}
$$

With the aid of the integral representation of the Θ function (3) and by employing (2) and (4), (22) yields

$$
\begin{aligned}
\mathrm{i}S_{\mathrm{F}}^{G}(x'-x) = & \,\mathrm{i}\int\frac{dk_0'}{(2\pi)}\frac{1}{k_0'+\mathrm{i}\varepsilon}\exp\left[-\mathrm{i}k_0'(t'-t)\right]\\
& \times\left\{\int\frac{d^3k}{(2\pi)^3}(1-f(E_k))\frac{m_0}{E_k}\frac{\not{k}+m_0}{2m_0}\exp\left[-\mathrm{i}k\cdot(x'-x)\right]\right.\\
& \left.-\int\frac{d^3k}{(2\pi)^3}(1-f(-E_k))\frac{m_0}{E_k}\frac{-\not{k}+m_0}{2m_0}\exp\left[+\mathrm{i}k\cdot(x'-x)\right]\right\}\\
& +\mathrm{i}\int\frac{dk_0'}{(2\pi)}\frac{1}{k_0'-\mathrm{i}\varepsilon}\exp\left[-\mathrm{i}k_0'(t'-t)\right]\\
& \times\left\{-\int\frac{d^3k}{(2\pi)^3}f(-E_k)\frac{m_0}{E_k}\frac{-\not{k}+m_0}{2m_0}\exp\left[-\mathrm{i}k\cdot(x'-x)\right]\right.\\
& \left.+\int\frac{d^3k}{(2\pi)^3}f(E_k)\frac{m_0}{E_k}\frac{\not{k}+m_0}{2m_0}\exp\left[-\mathrm{i}k\cdot(x'-x)\right]\right\}\quad . \qquad (23)
\end{aligned}
$$

To evaluate the four integrals we proceed as before by shifting the frequency variables and inverting the momentum variables, i.e. by carrying out the substitution (7a) in the first and last integrals and the substitution (7b) in the second and third integral. This gives

$$
S_{\mathrm{F}}^{G}(x'-x) = \int\frac{d^4k}{(2\pi)^4}S_{\mathrm{F}}^{G}(k)\exp\left[-\mathrm{i}k\cdot(x'-x)\right]\quad , \qquad (24)
$$

where

$$
\begin{aligned}
S_{\mathrm{F}}^{G}(k) = & \frac{1}{2E_k}\left[(1-f(E_k))\frac{E_k\gamma_0-\boldsymbol{k}\cdot\boldsymbol{\gamma}+m_0}{k_0-E_k+\mathrm{i}\varepsilon}\right.\\
& -(1-f(-E_k))\frac{-E_k\gamma_0-\boldsymbol{k}\cdot\boldsymbol{\gamma}+m_0}{k_0+E_k+\mathrm{i}\varepsilon}\\
& \left.-f(-E_K)\frac{-E_k\gamma_0-\boldsymbol{k}\cdot\boldsymbol{\gamma}+m_0}{k_0+E_k-\mathrm{i}\varepsilon}+f(E_k)\,\frac{E_k\gamma_0-\boldsymbol{k}\cdot\boldsymbol{\gamma}+m_0}{k_0-E_k+\mathrm{i}\varepsilon}\right]\quad . \qquad (25)
\end{aligned}
$$

We combine the four terms into two using the identity

Exercise 2.2.

$$\frac{1}{x-\mathrm{i}\varepsilon}=\frac{1}{x+\mathrm{i}\varepsilon}+2\pi\mathrm{i}\delta(x) \tag{26}$$

and obtain

$$S_{\mathrm{F}}^{G}(k)=\frac{\not k+m_0}{k^2-m_0^2+\mathrm{i}\varepsilon k_0}+\frac{1}{2E_k}(\not k+m_0)2\pi\mathrm{i}\left[f(E_k)\delta(k_0-E_k)\right.$$
$$\left.-f(-E_k)\delta(k_0+E_k)\right] \quad . \tag{27}$$

Finally, employing the relations

$$\begin{aligned}\Theta(k_0)\delta(k^2-m_0^2)&=\frac{1}{2E_k}\delta(k_0-E_k) \quad ,\\ \Theta(-k_0)\delta(k^2-m_0^2)&=\frac{1}{2E_k}\delta(k_0+E_k)\end{aligned} \tag{28}$$

we find that

$$\begin{aligned}S_{\mathrm{F}}^{G}(k)=&\frac{\not k+m_0}{k^2-m_0^2+\mathrm{i}\varepsilon k_0}+2\pi\mathrm{i}(\not k+m_0)\delta(k^2-m_0^2)\\ &\times\left[f(E_k)\Theta(k_0)-f(-E_k)\Theta(-k_0)\right] \quad .\end{aligned} \tag{29}$$

This expression may be transformed to a more symmetric form by separating off the free Feynman propagator according to the following identity:

$$\frac{\not k+m_0}{k^2-m_0^2+\mathrm{i}\varepsilon k_0}=\frac{\not k+m_0}{k^2-m_0^2+\mathrm{i}\varepsilon}+2\pi\mathrm{i}(\not k+m_0)\delta(k^2-m_0^2)\Theta(-k_0) \quad . \tag{30}$$

Hence, we have the final result

$$\begin{aligned}S_{\mathrm{F}}^{G}(k)=&S_{\mathrm{F}}(k)+2\pi\mathrm{i}(\not k+m_0)\delta(k^2-m_0^2)\\ &\times\left\{\frac{\Theta(k_0)}{\exp\left[(E_k-\mu)/k_{\mathrm{B}}T+1\right]}+\frac{\Theta(-k_0)}{\exp\left[(E_k+\mu)/k_{\mathrm{B}}T\right]+1}\right\} \quad .\end{aligned} \tag{31}$$

In the low-temperature limit $T\to 0$ and $\mu=E_{\mathrm{F}}>0$, (31) reduces to the previous expression (14) for the electron gas. This is easily proved by inserting (30) into (14).

EXERCISE

2.3 Nonrelativistic Limit of the Feynman Propagator

Problem. Show that $S_{\mathrm{F}}(x',x)$ reduces to the free retarded propagator for the Schrödinger equation in the nonrelativistic limit.

Exercise 2.3.

Solution. To solve the problem it is advantageous to change to momentum space. The representation of the propagators in configuration space is then obtained by Fourier transformation. However, to determine the propagators uniquely we need to give a prescription how the singularities have to be treated.
The Feynman propagator is

$$S_F(x'-x) = \int \frac{d^4p}{(2\pi)^4} \exp\left[-ip(x'-x)\right] S_F(p) \tag{1}$$

and the nonrelativistic retarded propagator is

$$G_0^+(x'-x) = \int \frac{d^3p}{(2\pi)^3} \exp\left[i\boldsymbol{p}\cdot(\boldsymbol{x}'-\boldsymbol{x})\right] \times \int_{-\infty}^{+\infty} \frac{d\omega}{2\pi} \exp\left[-i\omega(t'-t)\right] G_0^+(\boldsymbol{p},\omega) \quad . \tag{2}$$

From the previous discussion of the Feynman propagator we have learned that the appropriate boundary conditions correspond to shifting the poles by adding an infinitesimal imaginary constant, such that

$$S_F(p) = \frac{\not{p}+m_0}{p^2-m_0^2+i\varepsilon} \quad . \tag{3}$$

This form implies positive-energy solutions propagating forward in time and negative-energy solutions backward in time. In order to find the nonrelativistic limit of S_F we consider (3) in the approximation $|\boldsymbol{p}|/m_0 \ll 1$ and investigate the vicinity of the poles. We write

$$\frac{\not{p}+m_0}{p_0^2-\boldsymbol{p}^2-m_0^2+i\varepsilon} = \frac{p_0\gamma_0-\boldsymbol{p}\cdot\boldsymbol{\gamma}+m_0}{\left(p_0-\sqrt{\boldsymbol{p}^2+m_0^2}\right)\left(p_0+\sqrt{\boldsymbol{p}^2+m_0^2}\right)+i\varepsilon} \tag{4}$$

and obtain, using the approximation $\sqrt{\boldsymbol{p}^2+m_0} = m_0 + \boldsymbol{p}^2/2m_0 + O(\boldsymbol{p}^4/m_0^4)$,

$$S_F(p) \approx \frac{p_0\gamma_0-\boldsymbol{p}\cdot\boldsymbol{\gamma}+m_0}{\left(p_0-m_0-\frac{\boldsymbol{p}^2}{2m_0}\right)\left(p_0+m_0+\frac{\boldsymbol{p}^2}{2m_0}\right)+i\varepsilon} \quad . \tag{5}$$

Now we study the behaviour of the propagator in the vicinity of its positive-frequency pole. Introducing $\omega = p_0 - m_0$ we can reduce (5) to

$$S_F(p) \approx \frac{(\omega+m_0)\gamma_0-\boldsymbol{p}\cdot\boldsymbol{\gamma}+m_0}{\left(\omega-\frac{\boldsymbol{p}^2}{2m_0}\right)\left(\omega+2m_0+\frac{\boldsymbol{p}^2}{2m_0}\right)+i\varepsilon} \quad . \tag{6}$$

For the positive-frequency pole, ω lies in the vicinity of $\boldsymbol{p}^2/2m_0$. Therefore we have $\omega > 0$ and $(\omega + 2m_0 + \boldsymbol{p}^2/2m_0) \approx 2m_0 > 0$. Thus, within the approximation of small momenta, (5) can be transformed into

Exercise 2.3.

$$S_F(p) \approx \frac{1}{2m_0}\frac{m_0(\gamma_0+1)-\boldsymbol{p}\cdot\boldsymbol{\gamma}}{\left(\omega-\frac{p^2}{2m_0}\right)+\frac{\mathrm{i}\varepsilon}{2m_0}}$$
$$= \frac{\frac{1}{2}(\gamma_0+1)-\frac{\boldsymbol{p}\cdot\boldsymbol{\gamma}}{2m_0}}{\left(\omega-\frac{p^2}{2m_0}\right)+\mathrm{i}\varepsilon'} \quad , \tag{7}$$

where also ε' is a small imaginary constant. The first term

$$\frac{1}{2}(\gamma_0+1) = \begin{pmatrix} 1 & & & 0 \\ & 1 & & \\ & & 0 & \\ 0 & & & 0 \end{pmatrix}$$

selects the two upper components of a given bispinor. Since we have restricted our consideration to positive energy solutions by choosing the positive-energy pole, the two large spinor components are extracted. The second matrix

$$-\frac{\boldsymbol{p}\cdot\boldsymbol{\gamma}}{2m_0} = \begin{pmatrix} 0 & -\frac{\boldsymbol{p}\cdot\boldsymbol{\sigma}}{2m_0} \\ \frac{\boldsymbol{p}\cdot\boldsymbol{\sigma}}{2m_0} & 0 \end{pmatrix}$$

mixes the upper and lower components of the bispinor $\Psi = \begin{pmatrix} \varphi \\ \chi \end{pmatrix}$. Since $|\chi| \ll |\varphi|$ the contribution of this term is quadratic in $\boldsymbol{p}/m_0$, however, and can therefore be neglected within our small-momentum approximation. Thus the numerator of (7) reduces to unity (or, more precisely, to the unit matrix in spin space). We therefore have the result

$$S_F(p) \longrightarrow \frac{1}{\omega - \boldsymbol{p}^2/2m_0 + \mathrm{i}\varepsilon} = G_0^+(\boldsymbol{p},\omega) \quad . \tag{8}$$

Fourier transforming (8) back to coordinate space then yields the retarded propagator of the Schrödinger theory.

Remark. In the vicinity of the pole $p_0 = -\sqrt{\boldsymbol{p}^2+m_0^2}$ the procedure outlined above, but with the substitution $\omega = -p_0 - m_0$, would yield the same result (8). However, when Fourier transforming back to configuration space the energy-dependent exponent $\exp\left(\mathrm{i}p_0(t'-t)\right) = \exp\left(\mathrm{i}m(t'-t)\right)\exp\left(+\mathrm{i}\omega(t'-t)\right)$ produces a time dependence $\Theta(t-t')$. Thus, for antiparticles the Feynman propagator reduces to the advanced Green's function in the nonrelativistic limit.

EXERCISE

2.4 Time-Evolution of Dirac Wavefunctions

Problem. Prove the following identities:

$$\Theta(t'-t)\psi^{(+E)}(x') = \mathrm{i}\int d^3x S_F(x'-x)\gamma_0\psi^{(+E)}(x) \quad , \tag{1}$$

$$\Theta(t-t')\psi^{(-E)}(x') = -\mathrm{i}\int d^3x S_F(x'-x)\gamma_0\psi^{(-E)}(x) \quad , \tag{2}$$

and deduce similar relations for the adjoint solutions $\bar{\psi}^{(+E)}$ and $\bar{\psi}^{(-E)}$.

Exercise 2.4.

Solution. A wave packet of positive energy may be expressed in terms of a superposition of normalized plane waves:

$$\psi^{(+E)}(x) = \int \frac{d^3p}{(2\pi)^{3/2}} \sqrt{\frac{m_0}{E_p}} \sum_{r=1}^{2} b(p,r)\omega^r(\boldsymbol{p}) \exp(-\mathrm{i}\varepsilon_r p \cdot x) \quad , \tag{3}$$

where $E_p = \sqrt{p^2 + m_0^2}$ and $\varepsilon_1 = \varepsilon_2 = +1$. Similarly, for negative-energy wave packets we write $\varepsilon_3 = \varepsilon_4 = -1$.

$$\psi^{(-E)}(x) = \int \frac{d^3p}{(2\pi)^{3/2}} \sqrt{\frac{m_0}{E_p}} \sum_{r=3}^{4} d^*(p,r)\omega^r(\boldsymbol{p}) \exp(-\mathrm{i}\varepsilon_r p \cdot x) \quad . \tag{4}$$

In order to make use of the orthogonality condition for spinors

$$\omega^{r\dagger}(\varepsilon_r \boldsymbol{p})\omega^{r'}(\varepsilon_{r'}\boldsymbol{p}) = \frac{E_p}{m_0}\delta_{rr'} \tag{5}$$

we employ the plane-wave representation of the Feynman propagator

$$\begin{aligned} S_\mathrm{F}(x'-x) = &- \mathrm{i}\Theta(t'-t) \int d^3p \sum_{r=1}^{2} \psi_p^r(x')\bar{\psi}_p^r(x) \\ &+ \mathrm{i}\Theta(t-t') \int d^3p \sum_{r=3}^{4} \psi_p^r(x')\bar{\psi}_p^r(x) \quad , \end{aligned} \tag{6}$$

where the $\psi_p^r(x)$ are given by

$$\psi_p^r = \sqrt{\frac{m_0}{E_p}} \frac{1}{(2\pi)^{3/2}} \omega^r(\boldsymbol{p}) \exp(-\mathrm{i}\varepsilon_r p \cdot x) \quad . \tag{7}$$

Inserting (3), (6) and (7) into the right-hand side of (1) we then obtain

$$\begin{aligned} &\mathrm{i}\int d^3x\, S_\mathrm{F}(x'-x)\gamma_0\psi^{(+E)} \\ &= \Theta(t'-t) \int d^3x \int \frac{d^3p}{(2\pi)^3} \frac{m_0}{E_p} \\ &\quad \times \sum_{r=1}^{2} \omega^r(\boldsymbol{p})\bar{\omega}^r(\boldsymbol{p})\gamma_0 \exp\left[-\mathrm{i}\varepsilon_r p \cdot (x'-x)\right] \int \frac{d^3p'}{(2\pi)^{3/2}} \sqrt{\frac{m_0}{E_{p'}}} \\ &\quad \times \sum_{r'=1}^{2} b(p',r')\omega^{r'}(\boldsymbol{p}') \exp(-\mathrm{i}\varepsilon_{r'} p' \cdot x) \\ &\quad - \Theta(t-t') \int d^3x \int \frac{d^3p}{(2\pi)^3} \frac{m_0}{E_p} \sum_{r=3}^{4} \omega^r(\boldsymbol{p})\bar{\omega}^r(\boldsymbol{p})\gamma_0 \exp\left[-\mathrm{i}\varepsilon_r p \cdot (x'-x)\right] \\ &\quad \times \int \frac{d^3p'}{(2\pi)^{3/2}} \sqrt{\frac{m_0}{E_{p'}}} \sum_{r'=1}^{2} b(p',r')\omega^{r'}(\boldsymbol{p}') \exp(-\mathrm{i}\varepsilon_r p' \cdot x) \end{aligned}$$

Exercise 2.4.

$$
\begin{aligned}
&= \Theta(t'-t)\int \frac{d^3p\,d^3p'}{(2\pi)^{3/2}}\frac{m_0}{E_p}\sqrt{\frac{m_0}{E_{p'}}}\\
&\quad\times \sum_{\substack{r=1,2\\ r'=1,2}} \omega^r(\boldsymbol{p})\omega^{r\dagger}(\boldsymbol{p})\omega^{r'}(\boldsymbol{p}')b(p',r')\exp\left(-\mathrm{i}\varepsilon_r p\cdot x'\right)\\
&\quad\times \int\frac{d^3x}{(2\pi)^3}\exp\left[\mathrm{i}(\varepsilon_r p-\varepsilon_{r'}p')\cdot x\right] - \Theta(t-t')\int \frac{d^3p\,d^3p'}{(2\pi)^{3/2}}\frac{m_0}{E_p}\sqrt{\frac{m_0}{E_{p'}}}\\
&\quad\times \sum_{\substack{r=3,4\\ r'=1,2}} \omega^r(\boldsymbol{p})\omega^{r\dagger}(\boldsymbol{p})\omega^{r'}(\boldsymbol{p}')b(p',r')\exp\left(-\mathrm{i}\varepsilon_r p\cdot x'\right)\\
&\quad\times \int\frac{d^3x}{(2\pi)^3}\exp\left[\mathrm{i}(\varepsilon_r p-\varepsilon_{r'}p')\cdot x\right] \quad . \qquad (8)
\end{aligned}
$$

Performing the x integration in the $\Theta(t'-t)$ term yields

$$
\exp\left[\mathrm{i}(E_p-E_{p'})t\right]\delta^3(\boldsymbol{p}-\boldsymbol{p}') \quad , \qquad (9)
$$

where we have used $\varepsilon_r=\varepsilon_{r'}=1$, since $r,r'=1,2$. The $\Theta(t-t')$ term in the last line of (8) on the other hand produces a factor

$$
\exp\left[-\mathrm{i}(E_p+E_{p'})t\right]\delta^3(\boldsymbol{p}+\boldsymbol{p}') \quad , \qquad (10)
$$

since in this case $\varepsilon_r=-1$ ($r=3,4$) and $\varepsilon_{r'}=+1$ ($r'=1,2$). Integrating over d^3p and relabelling $\boldsymbol{p}'$ as $\boldsymbol{p}$ we find that

$$
\begin{aligned}
&\mathrm{i}\int d^3x\, S_\mathrm{F}(x'-x)\gamma_0\psi^{(+E)}(x)\\
&= \Theta(t'-t)\int\frac{d^3p}{(2\pi)^{3/2}}\left(\frac{m_0}{E_p}\right)^{3/2}\\
&\quad\times \sum_{\substack{r=1,2\\ r'=1,2}} \omega^r(\boldsymbol{p})\omega^{r\dagger}(\boldsymbol{p})\omega^{r'}(\boldsymbol{p})b(p,r')\exp\left(-\mathrm{i}\varepsilon_r p\cdot x'\right)\\
&\quad-\Theta(t-t')\int\frac{d^3p}{(2\pi)^{3/2}}\left(\frac{m_0}{E_p}\right)^{3/2}\sum_{\substack{r=3,4\\ r'=1,2}}\omega^r(-\boldsymbol{p})\omega^{r\dagger}(-\boldsymbol{p})\omega^{r'}(+\boldsymbol{p})b(p,r')\\
&\quad\times\exp\left(-\mathrm{i}\varepsilon_r\tilde{p}\cdot x'\right)\exp\left(-\mathrm{i}2E_p t\right) \qquad (11)
\end{aligned}
$$

where $\tilde{p}=(p_0,-\boldsymbol{p})$. Now we make use of the orthogonality relation (5), which reads, for $r,r'=1,2$,

$$
\omega^{r\dagger}(\boldsymbol{p})\omega^{r'}(\boldsymbol{p})=\omega^{r\dagger}(\varepsilon_r\boldsymbol{p})\omega^{r'}(\varepsilon_{r'}\boldsymbol{p})=\frac{E_p}{m_0}\delta_{rr'} \qquad (12)
$$

and, for $r=3,4$ and $r'=1,2$,

$$
\omega^{r\dagger}(-\boldsymbol{p})\omega^{r'}(\boldsymbol{p})=\omega^{r\dagger}(\varepsilon_r\boldsymbol{p})\omega^{r'}(\varepsilon_{r'}\boldsymbol{p})=0 \quad , \qquad (13)
$$

i.e. the second term in (11) vanishes. The remaining first term gives

Exercise 2.4.

$$\begin{aligned}
&\mathrm{i}\int d^3x\, S_F(x'-x)\gamma_0\psi^{(+E)}(x)\\
&\quad=\Theta(t'-t)\int\frac{d^3p}{(2\pi)^{3/2}}\sqrt{\frac{m_0}{E_p}}\sum_{r=1}^{2}b(p,r)\omega^r(\boldsymbol{p})\exp\left(-\mathrm{i}\varepsilon_r p\cdot x'\right)\\
&\quad=\Theta(t'-t)\psi^{(+E)}(x')\quad,
\end{aligned}\tag{14}$$

completing the proof of (1). The relation (2) is verified in an analogous manner. In this case, since $\psi^{(-E)}$ consists of spinors with $r=3,4$ only, the second part of $S_F(x'-x)$ contributes while the first term vanishes, thus yielding $-\Theta(t-t')\,\psi^{(-E)}(x')$.

Very similar relations can also be deduced for the propagation of the adjoint spinors $\bar{\psi}^{(+E)}(x)$, $\bar{\psi}^{(-E)}(x)$. Since the ordering of operators is inverted when performing Hermitian conjugation, the propagator S_F should now stand to the right of $\bar{\psi}$. Therefore we study the following integral

$$\begin{aligned}
&\mathrm{i}\int d^3x\,\bar{\psi}^{(+E)}(x)\gamma_0 S_F(x-x')\\
&\quad=\mathrm{i}\int d^3x\int\frac{d^3p'}{(2\pi)^{3/2}}\frac{m_0}{E_p}\sqrt{\frac{m_0}{E_{p'}}}\int\frac{d^3p}{(2\pi)^3}\\
&\qquad\times\sum_{r=1}^{2}b^*(p',r')\bar{\omega}^{r'}(\boldsymbol{p}')\exp\left(\mathrm{i}p'x\right)\gamma_0\\
&\qquad\times\left\{-\mathrm{i}\Theta(t-t')\sum_{r=1}^{2}\omega^r(\boldsymbol{p})\bar{\omega}^r(\boldsymbol{p})\exp\left[-\mathrm{i}p\cdot(x-x')\right]\right.\\
&\qquad\left.+\,\mathrm{i}\Theta(t'-t)\sum_{r=3}^{4}\omega^r(\boldsymbol{p})\bar{\omega}^r(\boldsymbol{p})\exp\left[+\mathrm{i}p\cdot(x-x')\right]\right\}\quad.
\end{aligned}\tag{15}$$

Now the calculation that led from (8) to (14) can be repeated, i.e. the x integration can be performed and the orthogonality relations for the unit spinors used. Then (15) reduces to the simple expression

$$\Theta(t-t')\int\frac{d^3p}{(2\pi)^{3/2}}\sqrt{\frac{m_0}{E_p}}\sum_{r=1}^{2}b^*(p,r)\bar{\omega}^r(\boldsymbol{p})\exp\left(\mathrm{i}p\cdot x\right)\quad.\tag{16}$$

This is the expansion of the adjoint spinor $\bar{\psi}^{(+E)}(x')$. Thus the ansatz (15) has indeed led to a propagation equation for the adjoint wave function, namely

$$\Theta(t-t')\bar{\psi}^{(+E)}(x')=\mathrm{i}\int d^3x\,\bar{\psi}^{(+E)}(x)\gamma_0 S_F(x-x')\quad.\tag{17}$$

In a similar manner one derives the relation

$$\Theta(t'-t)\bar{\psi}^{(-E)}(x')=-\mathrm{i}\int d^3x\,\bar{\psi}^{(-E)}(x)\gamma_0 S_F(x-x')\quad.\tag{18}$$

A comparison of (17,18) with (1,2) reveals that the order of the time arguments t and t' is interchanged. This is not surprising, since $\psi(x)$ describes an incoming wave and $\bar{\psi}(x)$ an outgoing wave.

EXERCISE

2.5 The Explicit Form of $S_F(x)$ in Coordinate Space

Problem. Derive a closed expression for the Feynman propagator in configuration space. How does it behave on the light cone, $x^2 \to 0$, and at large spacelike or timelike separations $x^2 \to \pm\infty$?

Solution. Our starting point is the integral representation of the Feynman propagator of the Dirac equation. The integral can be simplified by factorizing out the Dirac differential operator:

$$\begin{aligned} S_F(x) &= \int_{C_F} \frac{d^4p}{(2\pi)^4} \frac{e^{-ip\cdot x}}{p^2 - m^2} (\not{p} + m) \\ &= \int \frac{d^4p}{(2\pi)^4} \frac{i\partial_\mu \gamma^\mu + m}{p^2 - m^2 + i\varepsilon} e^{-ip\cdot x} \\ &= (i\gamma \cdot \partial + m) \int \frac{d^4p}{(2\pi)^4} \frac{e^{-ip\cdot x}}{p^2 - m^2 + i\varepsilon} \\ &= (i\gamma \cdot \partial + m) \Delta_F(x) \quad . \end{aligned} \tag{1}$$

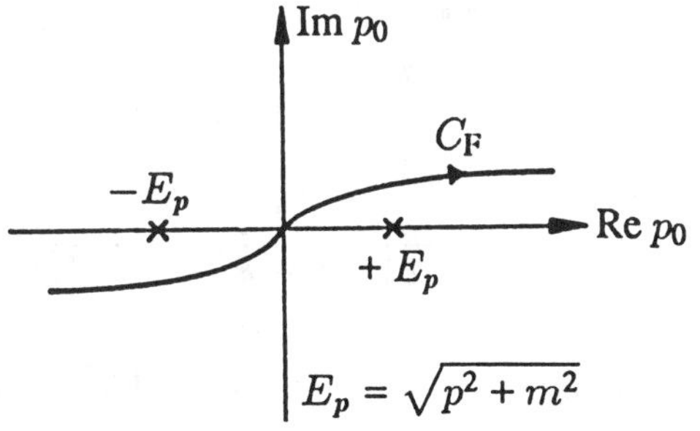

Fig. 2.14. The integration contour C_F

The Fig. 2.14 illustrates the integration contour C_F. Alternatively, the integration may be performed by shifting the poles by an infinitesimal constant $i\varepsilon$. The integral

$$\Delta_F(x) \equiv \int \frac{d^4p}{(2\pi)^4} \frac{e^{-ip\cdot x}}{p^2 - m^2 + i\varepsilon} = \int_{C_F} \frac{d^4p}{(2\pi)^4} \frac{e^{-ip\cdot x}}{p^2 - m^2} \quad , \tag{2}$$

which we introduced for the sake of mathematical simplification, also has a physical meaning. It is the *Feynman propagator of the Klein–Gordon field!*

The p_0-integration in $\Delta_F(x)$ can be evaluated by using the residue theorem, which determines the values of integrals along *closed* contours in the complex plane. Since the integrand carries a factor $\exp(-ip_0x_0 + i\boldsymbol{p} \cdot \boldsymbol{x})$, it is obvious that for $x_0 > 0$ the integrand vanishes asymptotically for large $|p_0|$ in the lower half plane. Thus, for $x_0 > 0$ an "infinite" semicircle in the lower half plane can be appended to the contour C_F without affecting the value of the integral. Since the integrand is regular everywhere except for the two poles, the path of integration

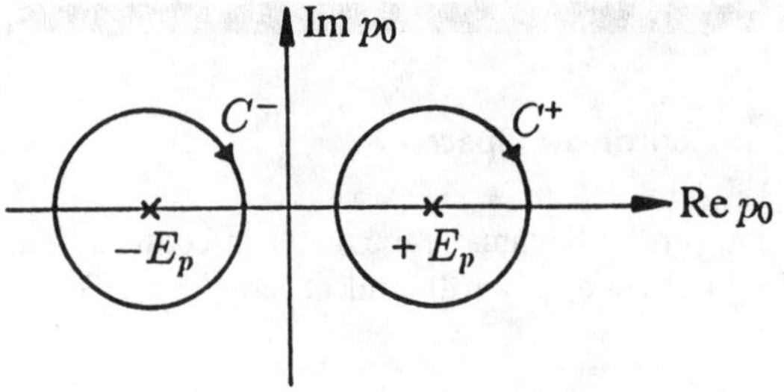

Fig. 2.15. Definition of the integration contours C^- and C^+

can be contracted to a contour C^+ which encircles the point $p_0 = +E_p$, as shown in the Fig. 2.15. Conversely, for $x_0 < 0$ the contour C_F needs to be closed in the upper half plane and we may integrate along the contour $-C^-$ (the direction of integration is essential). Thus, we obtain

$$\Delta_F(x) = \Theta(x_0)\Delta^+(x) - \Theta(-x_0)\Delta^-(x) \quad , \tag{3}$$

where

$$\Delta^\pm(x) = \oint_{C^\pm} \frac{d^4p}{(2\pi)^4} \frac{e^{-ip\cdot x}}{p^2 - m^2} \quad . \tag{4}$$

We proceed by rewriting the denominator as

$$\frac{1}{p^2 - m^2} = \frac{1}{2E_p}\left(\frac{1}{p_0 - E_p} - \frac{1}{p_0 + E_p}\right) \tag{5}$$

where $E_p = +\sqrt{\boldsymbol{p}^2 + m^2}$ to isolate the two poles and obtain

$$\begin{aligned} \Delta_F^\pm(x) &= \int \frac{d^3p}{(2\pi)^3} \exp(i\boldsymbol{p}\cdot\boldsymbol{x}) \frac{1}{2E_p} \oint_{C^\pm} \frac{dp_0}{2\pi} \exp(-ip_0x_0) \left(\frac{1}{p_0 - E_p} - \frac{1}{p_0 + E_p}\right) \\ &= \mp i \int \frac{d^3p}{(2\pi)^3} \frac{1}{2E_p} \exp\left[-i(\pm E_p x_0 - \boldsymbol{p}\cdot\boldsymbol{x})\right] \quad . \end{aligned} \tag{6}$$

Notice that the contours $C_\pm$ are directed in a negative mathematical sense.

Using this result both contributions to $\Delta_F(x)$ in (3) can be combined into a single expression

$$\Delta_F(x) = -i \int \frac{d^3p}{(2\pi)^3} \frac{1}{2E_p} \exp\left(-iE_p|x_0| + i\boldsymbol{p}\cdot\boldsymbol{x}\right) \quad . \tag{7}$$

In order to evaluate this three-dimensional integral we introduce spherical polar coordinates. The angular integrations can be carried out immediately

Exercise 2.5.

$$\begin{aligned}\Delta_{\mathrm{F}}(x) &= -\frac{\mathrm{i}}{(2\pi)^3}\int_0^\infty dp \int_{-1}^{+1} d\cos\theta \int_0^{2\pi} d\phi \frac{p^2}{2E_p}\exp\left(-\mathrm{i}E_p|x_0| + \mathrm{i}pr\cos\theta\right) \\ &= -\frac{\mathrm{i}}{(2\pi)^3}2\pi\int_0^\infty dp\frac{p^2}{2E_p}\frac{1}{\mathrm{i}pr}\exp\left(-\mathrm{i}E_p|x_0|\right)\left(\exp(\mathrm{i}pr) - \exp(-\mathrm{i}pr)\right) \\ &= -\frac{1}{8\pi^2 r}\int_0^\infty dp\frac{p}{E_p}\exp\left(-\mathrm{i}E_p|x_0|\right)\left(\exp(\mathrm{i}pr) - \exp(-\mathrm{i}pr)\right) \quad , \end{aligned} \tag{8}$$

where we have written $|\boldsymbol{p}| = p$ and $|\boldsymbol{x}| = r$. Substituting $p \to -p$ in the second term the two contributions in (8) can be combined into a single expression. Furthermore, the factor p under the integral can be replaced by a differentiation with respect to the parameter r

$$\begin{aligned}\Delta_{\mathrm{F}}(x) &= -\frac{1}{8\pi^2 r}\int_{-\infty}^{\infty} dp\frac{p}{E_p}\exp\left(-\mathrm{i}E_p|x_0|\right)\exp(\mathrm{i}pr) \\ &= \frac{\mathrm{i}}{8\pi^2 r}\frac{\partial}{\partial r}\int_{-\infty}^{\infty} dp\frac{\exp\left[-\mathrm{i}(E_p|x_0| - pr)\right]}{E_p} \quad . \end{aligned} \tag{9}$$

This integral can be brought into a more convenient form using the substitution

$$E_p = m\cosh\eta \quad , \quad p = m\sinh\eta \quad , \tag{10}$$

which obviously satisfies the relativistic energy momentum relation $E_p^2 - p^2 = m^2$. Now (9) takes the form

$$\begin{aligned}\Delta_{\mathrm{F}}(x) &= \frac{\mathrm{i}}{8\pi^2 r}\frac{\partial}{\partial r}\int_{-\infty}^{\infty} d\eta\frac{dp}{d\eta}\frac{\exp\left[-\mathrm{i}m(\cosh\eta|x_0| - \sinh\eta r)\right]}{m\cosh\eta} \\ &= \frac{\mathrm{i}}{8\pi^2 r}\frac{\partial}{\partial r}\int_{-\infty}^{\infty} d\eta\exp\left[-\mathrm{i}m(|x_0|\cosh\eta - r\sinh\eta)\right] \quad . \end{aligned} \tag{11}$$

The further evaluation of this integral depends on the relative size of the time and space arguments, $|x_0|$ and r. We will separately discuss the three possible cases.

Case 1: Timelike separation $x^2 > 0$, i.e. $|x_0| > r$. We substitute

$$\begin{aligned}|x_0| &= \sqrt{x_0^2 - r^2}\cosh\theta \quad , \\ r &= \sqrt{x_0^2 - r^2}\sinh\theta \quad , \end{aligned} \tag{12}$$

and use one of the addition theorems for the hyperbolic functions

$$\begin{aligned}|x_0|\cosh\eta - r\sinh\eta &= \sqrt{x_0^2 - r^2}(\cosh\theta\cosh\eta - \sinh\theta\sinh\eta) \\ &= \sqrt{x_0^2 - r^2}\cosh(\eta - \theta) \quad . \end{aligned} \tag{13}$$

Thus we have

Exercise 2.5.

$$\Delta_F(x) = \frac{i}{8\pi^2 r}\frac{\partial}{\partial r}\int_{-\infty}^{\infty} d\eta \exp\left(-im\sqrt{x_0^2 - r^2}\cosh(\eta-\theta)\right)$$
$$= \frac{i}{8\pi^2 r}\frac{\partial}{\partial r}\int_{-\infty}^{\infty} d\eta \exp\left(-im\sqrt{x_0^2 - r^2}\cosh\eta\right) \quad . \tag{14}$$

This integral can be solved in terms of Bessel functions[2] of zeroth order:

$$\int_{-\infty}^{\infty} d\eta \exp(-\mathrm{i}z\cosh\eta) = 2\int_0^{\infty} d\eta \cos(z\cosh\eta) - 2\mathrm{i}\int_0^{\infty} d\eta \sin(z\cosh\eta)$$
$$= -\mathrm{i}\pi J_0(z) - \pi N_0(z) = -\mathrm{i}\pi H_0^{(2)}(z) \quad . \tag{15}$$

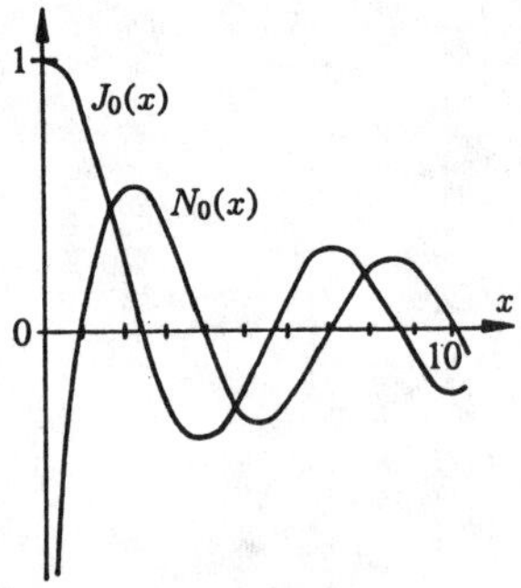

Fig. 2.16. Bessel function J_0 and Neumann function N_0 of zeroth order

J_0 and N_0 are the Bessel functions of first kind (often simply called the Bessel function) and of second kind (also known as the Neumann function). Both can be combined to yield the complex Hankel function $H_0^{(2)}(z)$ (Bessel function of third kind). The functions $J_0(z)$ and $N_0(z)$ are sketched in the Fig. 2.16. At $z \to 0$ $J_0(z)$ approaches 1 while $N_0(z)$ has a logarithmic singularity. Using the indentity

$$\frac{d}{dz}H_0^{(2)}(z) = -H_1^{(2)}(z) \tag{16}$$

we obtain the scalar Feynman propagator for $|x_0| > r$

$$\Delta_F(x) = \frac{1}{8\pi r}\frac{d(m\sqrt{x_0^2 - r^2})}{dr}\left[-H_1^{(2)}\left(m\sqrt{x_0^2 - r^2}\right)\right]$$
$$= \frac{m}{8\pi\sqrt{x_0^2 - r^2}}H_1^{(2)}\left(m\sqrt{x_0^2 - r^2}\right) \quad . \tag{17}$$

Case 2: Spacelike separation $x^2 < 0$, i.e. $|x_0| < r$. Here we substitute

$$|x_0| = \sqrt{r^2 - x_0^2}\sinh\theta \quad ,$$
$$r = \sqrt{r^2 - x_0^2}\cosh\theta \quad , \tag{18}$$

and use the addition theorem

$$\sinh\theta\cosh\eta - \cosh\theta\sinh\eta = -\sinh(\eta-\theta) \quad . \tag{19}$$

This leads to

$$\Delta_F(x) = \frac{i}{8\pi^2 r}\frac{\partial}{\partial r}\int_{-\infty}^{\infty} d\eta \exp\left(im\sqrt{r^2 - x_0^2}\sinh\eta\right)$$
$$= \frac{i}{8\pi^2 r}\frac{\partial}{\partial r}2\int_0^{\infty} d\eta \cos\left(m\sqrt{r^2 - x_0^2}\sinh\eta\right) \quad . \tag{20}$$

[2] See e.g. M. Abramowitz, I.A. Stegun: *Handbook of Mathematical Functions* (Dover, New York, 1965), Chap. 9.

Exercise 2.5.

The sin-term does not contribute, being an odd function in η. Here we encounter the integral representation of the modified Bessel function $K_0(z)$ (also known as the MacDonald function) which is related to the Hankel function of imaginary argument

$$2\int_0^\infty d\eta \cos(z\sinh\eta) = 2K_0(z)$$
$$= -\mathrm{i}\pi H_0^{(2)}(-\mathrm{i}z) \quad . \tag{21}$$

The MacDonald function has a logarithmic singularity at $z \to 0$ and falls off like $\sqrt{\pi/2z}\exp(-z)$ at $z \to \infty$, see Fig. 2.17. Using (21) we obtain for $r > |x_0|$

$$\Delta_\mathrm{F}(x) = \frac{1}{8\pi r}\frac{d\left(-m\mathrm{i}\sqrt{r^2-x_0^2}\right)}{dr}\left[-H_1^{(2)}\left(-\mathrm{i}m\sqrt{r^2-x_0^2}\right)\right]$$
$$= \frac{\mathrm{i}m}{8\pi\sqrt{r^2-x_0^2}}H_1^{(2)}\left(-\mathrm{i}m\sqrt{r^2-x_0^2}\right) \quad . \tag{22}$$

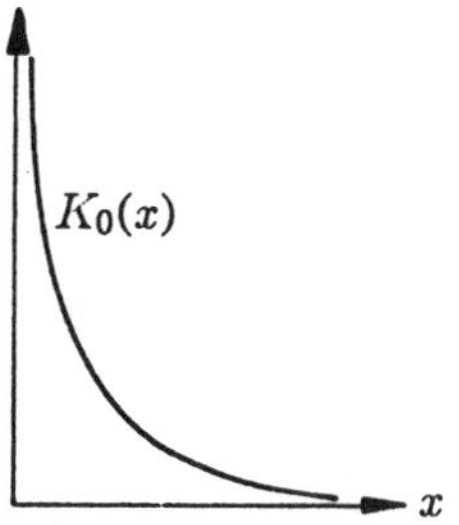

Fig. 2.17. Modified Bessel function of zeroth order $K_0(x)$

Obviously this is the analytical continuation of the result of case 1, (17).

Case 3: Lightlike separation $x^2 = 0$, i.e. $|x_0| = r$. This case has to be treated with special care since here the integral (8) is divergent. For large values of p the integrand approaches

$$\lim_{p\to\infty}\frac{p}{E_p}\exp\left(-\mathrm{i}E_p r\right)\left(\exp(\mathrm{i}pr) - \exp(\mathrm{i}pr)\right)$$
$$= \lim_{p\to\infty}\left(1 - \exp(-2\mathrm{i}pr)\right) \quad . \tag{23}$$

Since the first term approaches a constant (instead of oscillating, which would be the case for $|x_0| \neq r$) the integral will diverge. A certain singular behaviour of $S_\mathrm{F}(x)$ is already apparent when the results (17) or (22) are continued to the argument $|x_0| \to r$. In addition, however, also a singular distribution might contribute which has its support solely on the light cone $|x_0| = r$ and thus does not emerge when one studies the limit just mentioned.

It is easy to see that this indeed is the case. Let us study the divergent part of the integral (8) explicitly. For this it is justified to replace $E_p \to p$. Then we find

$$\Delta_\mathrm{F}(x)|_{x_0\to r} \simeq -\frac{1}{8\pi^2 r}\int_0^\infty dp\left\{\exp\left[-\mathrm{i}p(|x_0|-r)\right] - \exp\left[-\mathrm{i}p(|x_0|+r)\right]\right\}$$
$$\simeq -\frac{1}{8\pi^2 r}\left(\frac{1}{2}\,2\pi\delta(|x_0|-r) - \frac{1}{2}2\pi\delta(|x_0|+r)\right) \quad . \tag{24}$$

This calculation has taken into account only the delta-function contribution. Both terms in (24) can be combined to yield

$$\Delta_\mathrm{F}(x)|_{x_0\to r} \simeq -\frac{1}{4\pi}\delta(x_0^2 - r^2) \quad . \tag{25}$$

We have to add this singular contribution to our earlier result. The final result for the Feynman propagator Δ_F for the Klein–Gordon field then reads (using $x^2 = x_0^2 - r^2$)

Exercise 2.5.

$$\Delta_F(x) = -\frac{1}{4\pi}\delta(x^2) + \frac{m\Theta(x^2)}{8\pi\sqrt{x^2}} H_1^{(2)}\left(m\sqrt{x^2}\right) + \frac{im\Theta(-x^2)}{8\pi\sqrt{-x^2}} H_1^{(2)}\left(-im\sqrt{-x^2}\right) \quad . \tag{26}$$

As an important special case of $\Delta_F(x)$, let us consider the limit $m \to 0$. Since $H_1^{(2)}(z) \sim 2i/\pi z$ for $z \to 0$ (Abramowitz, Stegun, p. 360, No. 9.1.9) it follows that

$$\begin{aligned} D_F(x) &\equiv \lim_{m\to 0} \Delta_F(x) \\ &= -\frac{1}{4\pi}\delta(x^2) + \lim_{m\to 0}\left(\frac{\Theta(x^2)2im}{8\pi^2 mx^2} + \frac{\Theta(-x^2)2imi}{8\pi^2 imx^2}\right) \\ &= -\frac{1}{4\pi}\delta(x^2) + \frac{i}{4\pi^2}\frac{1}{x^2} \\ &= \frac{i}{4\pi^2}\frac{1}{x^2 - i\varepsilon} \quad . \end{aligned} \tag{27}$$

Up to a constant factor this agrees with the *photon propagator*, which will be discussed in Sect. 3.2.

Let us return to the Feynman propagator of the Dirac equation $S_F(x)$ which is related to $\Delta_F(x)$ by (1)

$$S_F(x) = (i\gamma\cdot\partial + m)\Delta_F(x) = m\,\Delta_F(x) + i\gamma\cdot\partial\Delta_F(x) \quad , \tag{28}$$

where the first term tacitly contains the unit matrix in spinor space. Often it is sufficient to work with this representation of the propagator. For completeness, however, we will derive the explicit form of $S_F(x)$ which calls for an evaluation of the derivative of $\Delta_F(x)$ given in (26). We proceed by employing the following identities:

$$\partial_\mu\Theta(x^2) = 2x_\mu\delta(x^2) = -\partial_\mu\Theta(-x^2) \quad , \tag{29}$$

$$\begin{aligned} \partial_\mu(x^2)^{1/2} &= x_\mu(x^2)^{-1/2} \quad , \\ \partial_\mu(x^2)^{-1/2} &= -x_\mu(x^2)^{-3/2} \quad , \end{aligned} \tag{30}$$

and also (Abramowitz, Stegun, p.361, No. 9.1.27)

$$\frac{d}{dz}H_1^{(2)}(z) = \frac{1}{2}\left(H_0^{(2)}(z) - H_2^{(2)}(z)\right) \quad . \tag{31}$$

The last term in (28) has the form

$$i\gamma\cdot\partial\Delta_F(x) = -i\frac{1}{4\pi}\gamma\cdot\partial\delta(x^2) + \frac{m}{8\pi}i\gamma\cdot\partial\left[\frac{\Theta(x^2)}{\sqrt{x^2}}H_1^{(2)}\left(m\sqrt{x^2}\right) + \frac{i\Theta(-x^2)}{\sqrt{-x^2}}H_1^{(2)}\left(-im\sqrt{-x^2}\right)\right] \quad . \tag{32}$$

We evaluate the derivative of the term in square brackets by using (29–31) and obtain

Exercise 2.5.

$$
\begin{aligned}
&\frac{m}{8\pi}\mathrm{i}\gamma^\mu x_\mu\left\{2\frac{\delta(x^2)}{(x^2)^{1/2}}H_1^{(2)}\left(m\sqrt{x^2}\right)-\frac{\Theta(x^2)}{(x^2)^{3/2}}H_1^{(2)}\left(m\sqrt{x^2}\right)\right.\\
&+\frac{m\Theta(x^2)}{2x^2}\left[H_0^{(2)}\left(m\sqrt{x^2}\right)-H_2^{(2)}\left(m\sqrt{x^2}\right)\right]\\
&-2\mathrm{i}\frac{\delta(x^2)}{(-x^2)^{1/2}}H_1^{(2)}\left(-\mathrm{i}m\sqrt{-x^2}\right)+\frac{\mathrm{i}\Theta(-x^2)}{(-x^2)^{3/2}}H_1^{(2)}\left(-\mathrm{i}m\sqrt{-x^2}\right)\\
&\left.+\frac{m\Theta(-x^2)}{2x^2}\left[H_0^{(2)}\left(-\mathrm{i}m\sqrt{-x^2}\right)-H_2^{(2)}\left(-\mathrm{i}m\sqrt{-x^2}\right)\right]\right\}\quad . \qquad (33)
\end{aligned}
$$

The two factors that are multiplied by $\delta(x^2)$ can be combined. Then we have

$$
\begin{aligned}
&\lim_{|x^2|\to 0}\left[\frac{H_1^{(2)}\left(m\sqrt{|x^2|}\right)}{\sqrt{|x^2|}}-\mathrm{i}\frac{H_1^{(2)}\left(-\mathrm{i}m\sqrt{|x^2|}\right)}{\sqrt{|x^2|}}\right]\\
&=\lim_{|x^2|\to 0}\left[\frac{\mathrm{i}}{\pi}\frac{2}{m\left(\sqrt{|x^2|}\right)^2}-\frac{\mathrm{i}}{\pi}\frac{2\mathrm{i}}{(-\mathrm{i}m)\left(\sqrt{|x^2|}\right)^2}\right]\\
&=\lim_{|x^2|\to 0}\left[\frac{4\mathrm{i}}{\pi m}\frac{1}{|x^2|}\right]\quad , \qquad (34)
\end{aligned}
$$

where we have used the asymptotic expansion of the Hankel functions for small arguments

$$
H_\nu^{(2)}(z)\sim\frac{\mathrm{i}}{\pi}\Gamma(\nu)\left(\frac{2}{z}\right)^\nu\quad ,\quad \nu>0\quad . \qquad (35)
$$

Thus the explicit expression for the Feynman propagator in coordinate space reads

$$
\begin{aligned}
S_{\mathrm{F}_{\alpha\beta}}(x)&=m\delta_{\alpha\beta}\Delta_{\mathrm{F}}(x)-\frac{\mathrm{i}}{4\pi}\gamma^\mu_{\alpha\beta}\partial_\mu\delta(x^2)-\frac{1}{\pi^2}\gamma^\mu_{\alpha\beta}x_\mu\frac{\delta(x^2)}{|x^2|}\\
&+\frac{\mathrm{i}m}{8\pi}\gamma^\mu_{\alpha\beta}x_\mu\left\{\Theta(x^2)\left[-\frac{1}{(x^2)^{3/2}}H_1^{(2)}\left(m\sqrt{x^2}\right)\right.\right.\\
&\left.+\frac{m}{2x^2}\left(H_0^{(2)}\left(m\sqrt{x^2}\right)-H_2^{(2)}\left(m\sqrt{x^2}\right)\right)\right]\\
&+\mathrm{i}\Theta(-x^2)\left[\frac{1}{(-x^2)^{(3/2)}}H_1^{(2)}\left(-\mathrm{i}m\sqrt{-x^2}\right)\right.\\
&\left.\left.-\mathrm{i}\frac{m}{2x^2}\left(H_0^{(2)}\left(-\mathrm{i}m\sqrt{-x^2}\right)-H_2^{(2)}\left(-\mathrm{i}m\sqrt{-x^2}\right)\right)\right]\right\}\quad , \qquad (36)
\end{aligned}
$$

where $\Delta_{\mathrm{F}}(x)$ is given by (26). We emphasize that propagators like $\Delta_{\mathrm{F}}(x)$ and $S_{\mathrm{F}}(x)$, looked upon mathematically are *distributions*, that is, they only make sense in integrals when multiplied with suitable "well-behaved" test functions.

Asymptotic Behaviour. 1) x^2 *small.* $\Delta_{\mathrm{F}}(x)$ and - even more so - $S_{\mathrm{F}}(x)$ exhibit several kinds of *singularities on the light cone* $x^2\to 0$. A study of the asymptotic behaviour of the scalar Feynman propagator, (26), leads to

Exercise 2.5.

$$\Delta_F(x) \approx -\frac{1}{4\pi}\delta(x^2) + \frac{i}{4\pi^2}\frac{1}{x^2} - \frac{im^2}{8\pi^2}\ln\left(\frac{m\sqrt{x^2}}{2}\right) - \frac{m^2}{16\pi^2}\Theta(x^2) \quad . \tag{37}$$

The leading singularity is contained in the first two terms, namely

$$\Delta_F(x) \approx \frac{i}{4\pi^2}\frac{1}{x^2 - i\varepsilon} + O(m^2) \quad . \tag{38}$$

Note that this result agrees with the massless propagator $\Delta_F(x)$ given in (27). This coincidence is quite reasonable since the singularity at $x^2 \to 0$ in momentum space is related to the divergence of integrals at $p \to \infty$. In this region the mass can be neglected. The singular nature of the propagators is the cause of great concern when integrals involving the product of several propagators have to be evaluated. In general the "collision" of singularities will render the integral divergent. The elaborate formalism of renormalization theory is required to extract meaningful results from these infinite quantities, see Chap. 5. These calculations, however, are more easily performed using momentum space propagators $\Delta_F(p)$.

2) x^2 *large*. The Hankel function behaves for *large* arguments $|z|$ as

$$H_\nu^{(2)}(z) \sim \sqrt{\frac{2}{\pi z}}\exp\left[-i\left(z - \frac{\pi\nu}{2} - \frac{1}{4}\pi\right)\right] \quad \text{for} \quad |z| \to \infty \quad . \tag{39}$$

Applying this relation to (26) we deduce the following asymptotic behaviour of the scalar Feynman propagator

$$\Delta_F(x) \to \text{const}\left(x^2\right)^{-\frac{3}{4}}\exp\left(-im\sqrt{x^2}\right) \quad \text{for} \quad x^2 \to \infty \quad , \tag{40a}$$

$$\Delta_F(x) \to \text{const}\,|x^2|^{-\frac{3}{4}}\exp\left(-m\sqrt{|x^2|}\right) \quad \text{for} \quad x^2 \to -\infty \quad . \tag{40b}$$

Thus for large *timelike* distances ($x^2 \to +\infty$) the propagator is an oscillationg function slowly decreasing in amplitude owing to the power-law factor. On the other hand, for large spacelike distances ($x^2 \to -\infty$) the propagator rapidly falls to zero according to the exponential function in (40b). The scale is set by the inverse mass of the particle, i.e. by it's Compton wavelength. These conclusions remain valid also for the Spin-1/2 Feynman propagator $S_F(x)$ given in (36).

Figure 2.18 illustrates the qualitative behaviour of the propagators. This result can be understood quite easily if one thinks of the propagation of a wave $\Psi(x) \to \Psi(x')$ in terms of Huygens' principle. Classically, from each point x elementary waves emanate which can propagate with velocities up to the velocity of light, i.e. inside the forward light cone $(x'-x)^2 \geq 0$. The fact that the propagator is nonzero (albeit rapidly decreasing) also in the region of spacelike distances is a quantum mechanical tunnelling phenomenon caused by the difficulty to localize a particle on a scale smaller than its Compton wavelength. This apparent violation of causality vanishes in the classical limit $m \to \infty$ (or, formally, $\hbar \to 0$).

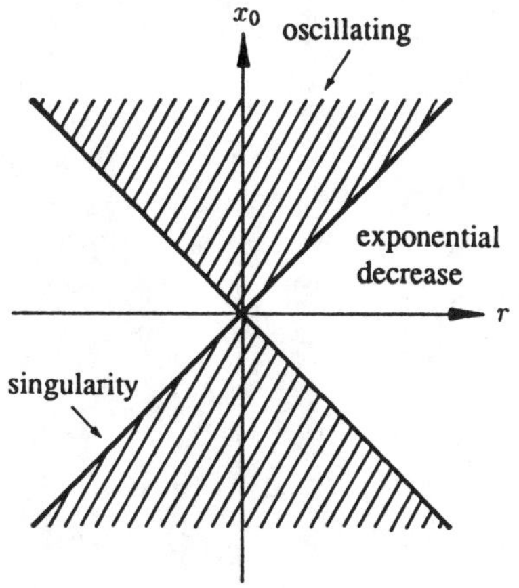

Fig. 2.18. The propagators $\Delta_F(x)$ and $S_F(x)$ are oscillating functions inside the light cone and fall off outside the light cone. On the light cone they are singular distributions

3. Quantum-Electrodynamical Processes

In this chapter we will gain some practical abilities in calculating various interesting quantum-electrodynamical processes that are of great importance. Thus the following chapter consists mainly of examples and problems. First, we start by applying the propagator formalism to problems related to electron-positron scattering. We shall proceed by considering more complicated processes including photons and other particles. As in the original publications of Feynman[1] we shall derive general rules for the practical calculation of transition probabilities and cross sections of any process involving electrons, positrons, and photons. These rules, although derived in a nonrigorous fashion, provide a correct and complete description of QED processes. The same set of "Feynman rules" results from a systematic treatment within the framework of quantum field theory.

3.1 Coulomb Scattering of Electrons

We calculate the Rutherford scattering of an electron at a fixed Coulomb potential. The appropriate S-matrix element is given by (2.41a) and (2.42) and can be used directly. For $f \neq i$ and renaming the integration variable $y \to x$ one gets

$$S_{fi} = -\mathrm{i}e \int d^4x \; \bar{\psi}_f(x) \not{A}(x)\, \Psi_i(x) \qquad (f \neq i) \quad . \tag{3.1}$$

Here $e < 0$ is the charge of the electron. In order to discuss (3.1) in an approximation that is solvable in practice we calculate the process in *lowest order of perturbation theory*. Then $\Psi_i(x)$ is approximated by the incoming plane wave $\psi_i(x)$ of an electron with momentum p_i and spin s_i :

$$\psi_i(x) = \sqrt{\frac{m_0}{E_i V}}\, u(p_i, s_i)\, \mathrm{e}^{-\mathrm{i}p_i \cdot x} \quad . \tag{3.2}$$

V denotes the normalization volume, i.e. ψ_i is normalized to probability 1 in a box with volume V. Similarly $\bar{\psi}_f(x)$ is given by

$$\bar{\psi}_f(x) = \sqrt{\frac{m_0}{E_f V}}\, \bar{u}(p_f, s_f)\, \mathrm{e}^{\mathrm{i}p_f \cdot x} \quad . \tag{3.3}$$

The Coulomb potential $A_0(x)$ is generated by a static point charge $-Ze$; thus we have

[1] R.P. Feynman: Phys. Rev. **76** 749, 769 (1949).

$$A_0(x) = A_0(\boldsymbol{x}) = -\frac{Ze}{|\boldsymbol{x}|} \quad , \qquad \boldsymbol{A}(x) = 0 \quad . \tag{3.4}$$

With these assumptions the S-matrix element (3.1) reads

$$S_{fi} = \mathrm{i}Ze^2 \frac{1}{V} \sqrt{\frac{m_0^2}{E_f E_i}} \; \bar{u}(p_f, s_f)\, \gamma^0\, u(p_i, s_i) \int d^4x \, \mathrm{e}^{\mathrm{i}(p_f - p_i)\cdot x} \, \frac{1}{|\boldsymbol{x}|} \quad . \tag{3.5}$$

The integral over the time coordinate can be separated yielding

$$\int_{-\infty}^{\infty} dx_0 \, \mathrm{e}^{\mathrm{i}(E_f - E_i)x_0} = 2\pi\, \delta(E_f - E_i) \quad . \tag{3.6}$$

This result expresses the fact that energy is conserved for the scattering in a time-independent potential. The remaining three-dimensional Fourier transform of the Coulomb potential (3.4)

$$A_0(\boldsymbol{q}) = -Ze \int d^3x \, \frac{1}{|\boldsymbol{x}|} \, \mathrm{e}^{-\mathrm{i}\boldsymbol{q}\cdot\boldsymbol{x}} \quad ,$$

with the momentum transfer $\boldsymbol{q} = \boldsymbol{p}_f - \boldsymbol{p}_i$, can easily be solved with the help of the following trick based on partial integration:

$$\begin{aligned} \int d^3x \, \frac{1}{|\boldsymbol{x}|} \, \mathrm{e}^{-\mathrm{i}\boldsymbol{q}\cdot\boldsymbol{x}} &= -\frac{1}{\boldsymbol{q}^2} \int d^3x \, \frac{1}{|\boldsymbol{x}|} \, \Delta\, \mathrm{e}^{-\mathrm{i}\boldsymbol{q}\cdot\boldsymbol{x}} \\ &= -\frac{1}{\boldsymbol{q}^2} \int d^3x \left(\Delta \frac{1}{|\boldsymbol{x}|} \right) \mathrm{e}^{-\mathrm{i}\boldsymbol{q}\cdot\boldsymbol{x}} \\ &= -\frac{1}{\boldsymbol{q}^2} \int d^3x \, (-4\pi\delta^3(\boldsymbol{x}))\, \mathrm{e}^{-\mathrm{i}\boldsymbol{q}\cdot\boldsymbol{x}} = \frac{4\pi}{\boldsymbol{q}^2} \quad . \end{aligned} \tag{3.7}$$

Thus the S-matrix element (3.5) follows:

$$S_{fi} = \mathrm{i}Ze^2 \frac{1}{V} \sqrt{\frac{m_0^2}{E_f E_i}} \; \bar{u}(p_f, s_f)\, \gamma^0\, u(p_i, s_i) \, \frac{4\pi}{\boldsymbol{q}^2} \, 2\pi\delta(E_f - E_i) \quad . \tag{3.8}$$

Now we need the number of final states dN_f within the range of momentum d^3p_f. It is given by

$$dN_f = V \frac{d^3p_f}{(2\pi)^3} \quad . \tag{3.9}$$

This can be understood by considering the following inset.

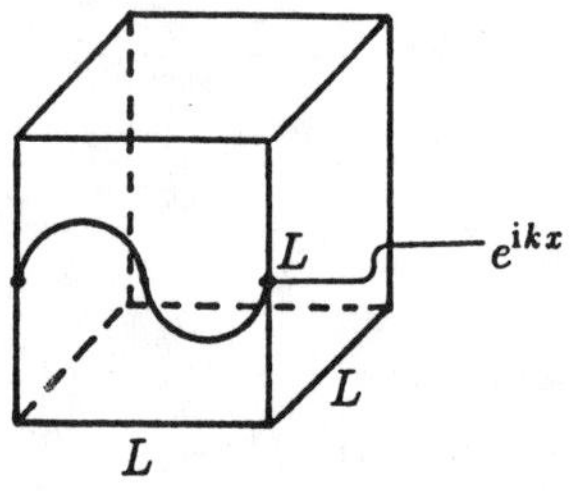

Fig. 3.1. Wave functions are normalized within a cubic box of side length L

Standing waves in a cubical box of volume $V = L^3$ (Fig. 3.1) require

$$\begin{aligned} k_x L &= n_x\, 2\pi \quad , \\ k_y L &= n_y\, 2\pi \quad , \\ k_z L &= n_z\, 2\pi \quad , \end{aligned}$$

with integer numbers n_x, n_y, n_z. For large L the discrete set of $\boldsymbol{k}$-values approaches a continuum. The number of states is

$$\begin{aligned} dN &= dn_x dn_y dn_z \\ &= \frac{1}{(2\pi)^3} L^3 \, dk_x dk_y dk_z \\ &= \frac{V}{(2\pi)^3} d^3k = \frac{V}{(2\pi)^3} d^3p \end{aligned}$$

($\hbar$ set to 1).

At this point we can already state the *transition probability per particle* into these final states which is obtained by multiplying the squared S matrix element by the density of final states

$$\begin{aligned} dW &= |S_{fi}|^2 \frac{V d^3p_f}{(2\pi)^3} \\ &= \frac{Z^2 (4\pi\alpha)^2 m_0^2}{E_i V} \frac{|\bar{u}(p_f, s_f)\gamma^0 u(p_i, s_i)|^2}{|q|^4} \frac{d^3p_f}{(2\pi)^3 E_f} (2\pi\delta(E_f - E_i))^2 \quad . \quad (3.10) \end{aligned}$$

Here we have set $e^2 = e^2/\hbar c = \alpha$, $\alpha \simeq 1/137$ being the fine-structure constant. In Sect. 4.2 the system of units we employ will be discussed in more detail.

In (3.10) the square of the δ-function enters. This is a mathematically not well defined divergent quantity and has to be specified by a limiting procedure. Instead of (3.6) which refers to an infinite time interval $-\infty \le t \le \infty$ we now assume that the transition takes place only within a finite time interval $-\frac{T}{2} \le t \le \frac{T}{2}$. Then, instead of a δ-function we get a function that is 'smeared out' in energy:

$$\begin{aligned} 2\pi\delta(E_f - E_i) &\Longrightarrow \int_{-T/2}^{T/2} dt \; e^{i(E_f - E_i)t} \\ &= \frac{1}{i(E_f - E_i)} e^{i(E_f - E_i)t} \Big|_{-T/2}^{T/2} = \frac{2\sin(E_f - E_i)T/2}{E_f - E_i} \quad . \quad (3.11) \end{aligned}$$

Thus the square of the δ-function is replaced by

$$(2\pi\delta(E_f - E_i))^2 \Longrightarrow 4 \frac{\sin^2(E_f - E_i)T/2}{(E_f - E_i)^2} \quad . \quad (3.12)$$

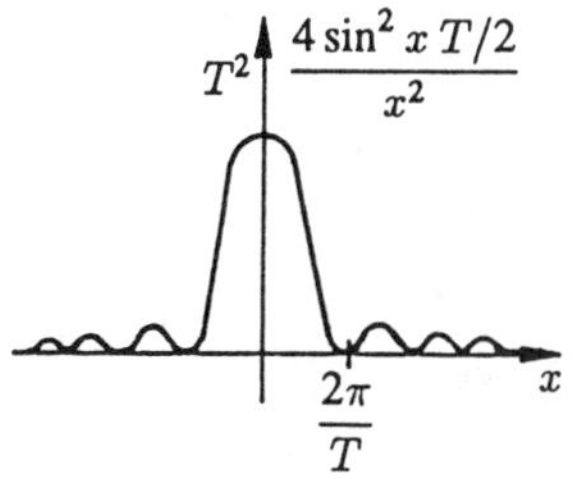

Fig. 3.2. The function under the integral of (3.13)

In Exercise 3.1 we show that the area under this function is

$$\int_{-\infty}^{\infty} dE_f \, 4 \frac{\sin^2(E_f - E_i)\,T/2}{(E_f - E_i)^2} = 2\pi T \quad . \quad (3.13)$$

This result can be understood by inspecting the graph of the function $4\sin^2(xT/2)/x^2$ (see Fig. 3.2). The area can be approximated by a triangle with height T^2 and length of the basis $4\pi/T$:

$$\int_{-\infty}^{\infty} dx\, 4\frac{\sin^2(x\,T/2)}{x^2} \simeq \frac{1}{2}T^2\frac{4\pi}{T} = 2\pi T \quad , \tag{3.14}$$

which incidentally gives the exact result. For increasing T the shape of the function $4\sin^2(xT/2)/x^2$ approaches a δ-function, the area under the function having the value $2\pi T$. Therefore we may give meaning to the square of the energy-preserving δ-function:

$$\begin{aligned}(2\pi\delta(E_f - E_i))^2 &= 2\pi\delta(0)\,2\pi\delta(E_f - E_i)\\ &\overset{!}{=} 2\pi\,T\,\delta(E_f - E_i) \quad .\end{aligned} \tag{3.15a}$$

This identification ensures that the integration over dE_f yields $2\pi T$ according to (3.13), and we are led to the following rule of replacement

$$2\pi\delta(0) \Rightarrow T \quad . \tag{3.15b}$$

This result can be made plausible by another approach. It is

$$2\pi\delta(E_f - E_i) = \lim_{T\to\infty} \int_{-T/2}^{T/2} dt\ \mathrm{e}^{\mathrm{i}(E_f - E_i)t} \quad .$$

For $E_f = E_i$ it follows that

$$2\pi\delta(0) = \lim_{T\to\infty} \int_{-T/2}^{T/2} dt = \lim_{T\to\infty} T \quad . \tag{3.16}$$

Inserting (3.15) into the expression for the transition probabilities per particle (3.10), we can now state the *transition probabilities per particle and per unit of time with final states within the momentum range* d^3p_f.

We denote this rate dR,

$$dR = \frac{dW}{T} = \frac{4Z^2\alpha^2 m_0^2}{E_i V}\frac{|\bar{u}(p_f, s_f)\gamma^0 u(p_i, s_i)|^2}{|\boldsymbol{q}|^4}\frac{d^3p_f}{E_f}\,\delta(E_f - E_i) \quad . \tag{3.17}$$

The *scattering cross section* can be defined as the transition probability per particle and per unit of time divided by the incoming current of particles

$$J^a_{\text{inc.}} = c\,\bar{\psi}_i(x)\gamma^a\psi_i(x) \quad . \tag{3.18}$$

The upper index 'a' determines the component of the current vector in the direction of the velocity of the incoming particles

$$\boldsymbol{v}_i = \frac{\boldsymbol{p}_i}{E_i} \quad . \tag{3.19}$$

Taking the spinors (3.2) with spin polarization in the direction of the z-axis and using (6.30) from *RQM* we can determine the current:

$$
\begin{aligned}
J^a_{\text{inc.}} &= c\,\bar{\psi}_i(x)\gamma^a\psi_i(x) = c\,\frac{m_0c^2}{E_iV}\,\bar{u}(p_i,s_i)\,\gamma^3\,u(p_i,s_i)\\
&= c\,\frac{m_0c^2}{E_iV}\,\frac{(E_i+m_0c^2)}{2m_0c^2}\left(1\,0\,\frac{p_ic}{E_i+m_0c^2}\,0\right)\gamma^0\gamma^3\begin{pmatrix}1\\0\\\frac{p_ic}{E_i+m_0c^2}\\0\end{pmatrix}\\
&= c\,\frac{m_0c^2}{E_iV}\,\frac{(E_i+m_0c^2)}{2m_0c^2}\left(1\,0\,\frac{p_ic}{E_i+m_0c^2}\,0\right)\begin{pmatrix}0&\sigma_3\\\sigma_3&0\end{pmatrix}\begin{pmatrix}1\\0\\\frac{p_ic}{E_i+m_0c^2}\\0\end{pmatrix}\\
&= c\,\frac{m_0c^2}{E_iV}\,\frac{(E_i+m_0c^2)}{2m_0c^2}\left(1\,0\,\frac{p_ic}{E_i+m_0c^2}\,0\right)\begin{pmatrix}\frac{p_ic}{E_i+m_0c^2}\\0\\1\\0\end{pmatrix}\\
&= \frac{p_ic^2}{E_i}\,\frac{1}{V}\quad. \qquad (3.20)
\end{aligned}
$$

(3.20) is just the ratio of velocity and volume

$$
|\boldsymbol{J}_{\text{inc.}}| = \frac{|\boldsymbol{v}_i|}{V}\quad. \qquad (3.21)
$$

Performing the calculation in (3.20) we assumed without restriction of generality the direction of velocity to be parallel to the z-axis. Further we used the relation $\gamma^0\gamma^3 = \alpha^3 = \begin{pmatrix}0&\sigma^3\\\sigma^3&0\end{pmatrix}$. The result (3.21) is plausible and could have been written down without any calculation. It can also be derived in a simple way for velocities whose direction is not parallel to the z-axis. In this case one has to take the more general spinors ((6.32) in *RQM*). The differential cross section $d\sigma$ can now be determined with the help of (3.21) and (3.17)

$$
d\sigma = \frac{dR}{J_{\text{inc.}}} = \frac{4Z^2\alpha^2m_0^2}{E_iV\frac{|\boldsymbol{v}_i|}{V}}\,\frac{|\bar{u}(p_f,s_f)\gamma^0u(p_i,s_i)|^2}{|\boldsymbol{q}|^4}\,\frac{\boldsymbol{p}_f^2d|\boldsymbol{p}_f|}{E_f}\,d\Omega_f\,\delta(E_f-E_i)\,. \qquad (3.22)
$$

The *differential cross section* per unit solid angle $d\Omega_f$ of the scattered particle follows:

$$
\frac{d\sigma}{d\Omega_f} = \int_{\Delta p_f}\frac{4Z^2\alpha^2m_0^2}{E_i|\boldsymbol{v}_i|}\,\frac{|\bar{u}(p_f,s_f)\gamma^0u(p_i,s_i)|^2}{|\boldsymbol{q}|^4}\,\frac{\boldsymbol{p}_f^2d|\boldsymbol{p}_f|}{E_f}\,\delta(E_f-E_i)\quad. \qquad (3.23)
$$

Here the momentum space volume element $d^3p_f = \boldsymbol{p}_f^2d|\boldsymbol{p}_f|d\Omega_f$ was used. We introduced the integral since in every practical case one has to integrate over a small interval Δp_f (uncertainty of measurement). The integration has the effect that the apparently singular behaviour arising from the $\delta(E_f-E_i)$-function (being infinitely large for $E_i = E_f$) vanishes. Because

$$
E_f^2 = \boldsymbol{p}_f^2 + m_0^2
$$

we get

$$E_f dE_f = |\boldsymbol{p}_f| d|\boldsymbol{p}_f|$$

and thus

$$\frac{d\sigma}{d\Omega_f} = \frac{4Z^2\alpha^2 m_0^2}{|\boldsymbol{q}|^4} |\bar{u}(p_f, s_f)\gamma^0 u(p_i, s_i)|^2 \int_{\Delta p_f} \frac{|\boldsymbol{p}_f| E_f \delta(E_f - E_i)}{E_i\, |\boldsymbol{v}_i|\, E_f} dE_f$$

$$= \frac{4Z^2\alpha^2 m_0^2}{|\boldsymbol{q}|^4} |\bar{u}(p_f, s_f)\gamma^0 u(p_i, s_i)|^2 \quad . \tag{3.24}$$

Here we used the relation $|\boldsymbol{p}_f|/E_i|\boldsymbol{v}_i| = |\boldsymbol{p}_i|/E_i|\boldsymbol{v}_i| = |\boldsymbol{v}_i|/|\boldsymbol{v}_i| = 1$ resulting from (3.19) and the δ-function. In the nonrelativistic limit it holds that

$$|\bar{u}(p_f, s_f)\gamma^0 u(p_i, s_i)|^2 \rightarrow \left| (1 \quad 0) \begin{pmatrix} 1 & 0 \\ 0 & 1 \end{pmatrix} \begin{pmatrix} 1 \\ 0 \end{pmatrix} \right|^2 = 1 \quad ,$$

and (3.24) reduces to the well-known *Rutherford scattering cross section*.

The differential cross section (3.24) can in principle be applied to calculate the scattering of a particle from the initial polarization (s_i) to the final (s_f). This will be treated in Sect. 3.5. However, in most experiments neither the polarization s_f of the scattered particle nor the initial polarization s_i is measured. Therefore the various possible *initial polarization states have the same probability*. That is, the actually measured cross section is given by *summing* the cross section (3.24) over the final polarizations s_f and then *averaging* over the initial polarizations s_i. Thus the unpolarized scattering cross section reads

$$\frac{d\bar{\sigma}}{d\Omega} = \frac{4Z^2\alpha^2 m_0^2}{|\boldsymbol{q}|^4} \frac{1}{2} \sum_{s_f, s_i} |\bar{u}(p_f, s_f)\gamma^0 u(p_i, s_i)|^2 \quad . \tag{3.25}$$

The problem of calculating polarization sums of this kind is very frequently encountered when dealing with processes involving Dirac particles. Fortunately an elegant mathematical technique has been developed which avoids the explicit handling of the unit spinors $u(p, s)$. The double sum over the polarizations (spins) can be rewritten as

$$\sum_{s_f, s_i} \bar{u}_\alpha(p_f, s_f)\gamma^0_{\alpha\beta} u_\beta(p_i, s_i) u^\dagger_\lambda(p_i, s_i)\gamma^{0\dagger}_{\lambda\delta}\gamma^{0\dagger}_{\delta\sigma} u_\sigma(p_f, s_f)$$

$$= \sum_{s_f, s_i} \bar{u}_\alpha(p_f, s_f)\gamma^0_{\alpha\beta} u_\beta(p_i, s_i)\bar{u}_\delta(p_i, s_i)\gamma^0_{\delta\sigma} u_\sigma(p_f, s_f) \quad , \tag{3.26}$$

using the Hermiticity $\gamma^{0\dagger} = \gamma^0$. Here the summation over doubly occurring Dirac indices is implied. For an arbitrary operator $\hat{\Gamma}$ we have the general rule

$$|\bar{u}(f)\hat{\Gamma} u(i)|^2 = \left(\bar{u}(f)\hat{\Gamma} u(i)\right) \left(\bar{u}(i)\hat{\bar{\Gamma}} u(f)\right) \quad , \tag{3.27}$$

with the "barred" operator

$$\hat{\bar{\Gamma}} = \gamma^0 \hat{\Gamma}^\dagger \gamma^0 \quad . \tag{3.28}$$

This can be easily shown:

$$\begin{aligned}&\left(\bar{u}_\alpha(f)\,\hat{\Gamma}_{\alpha\beta}\,u_\beta(i)\right)\left(u^\dagger_\gamma(i)\,\hat{\Gamma}^\dagger_{\gamma\delta}\,\gamma^{0\dagger}_{\delta\tau}\,u_\tau(f)\right)\\&\quad=\left(\bar{u}_\alpha(f)\,\hat{\Gamma}_{\alpha\beta}\,u_\beta(i)\right)\left(u^\dagger_\gamma(i)\,\gamma^0_{\gamma\varepsilon}\,\gamma^0_{\varepsilon\mu}\hat{\Gamma}^\dagger_{\mu\delta}\,\gamma^{0\dagger}_{\delta\tau}\,u_\tau(f)\right)\\&\quad=\left(\bar{u}(f)\,\hat{\Gamma}\,u(i)\right)\left(\bar{u}(i)\,\bar{\hat{\Gamma}}\,u(f)\right)\quad.\end{aligned}$$

Here we used $\gamma^{0\dagger}=\gamma^0$, $(\gamma^0)^2=1$, and, further, the identity of the conjugate complex and the Hermitian conjugate of the number (matrix element) $\left(\bar{u}(f)\,\hat{\Gamma}\,u(i)\right)^*=\left(\bar{u}(f)\,\hat{\Gamma}u(i)\right)^\dagger$. The barred matrices $\bar{\hat{\Gamma}}$ can be directly calculated for a number of operators $\hat{\Gamma}$, for instance

$$\bar{\gamma}^\mu=\gamma^0\gamma^{\mu\dagger}\gamma^0=\gamma^\mu \tag{3.29a}$$

because $\gamma^0\gamma^{0\dagger}\gamma^0=\gamma^0\gamma^0\gamma^0=\gamma^0$ and $\gamma^0\gamma^{i\dagger}\gamma^0=-\gamma^0\gamma^i\gamma^0=\gamma^i\gamma^0\gamma^0=\gamma^i$, since γ^0 is Hermitian and γ^i is anti-Hermitian. For the γ^5 matrix we find

$$\overline{\mathrm{i}\gamma^5}=\mathrm{i}\gamma^5\quad, \tag{3.29b}$$

because

$$\mathrm{i}\gamma^5=-\gamma^0\gamma^1\gamma^2\gamma^3$$

and

$$\overline{\mathrm{i}\gamma^5}=-\gamma^0\gamma^{3\dagger}\gamma^{2\dagger}\gamma^{1\dagger}\gamma^{0\dagger}\gamma^0=+\gamma^0\gamma^3\gamma^2\gamma^1=-\gamma^0\gamma^1\gamma^2\gamma^3=\mathrm{i}\gamma^5\quad.$$

In a similar way we get

$$\overline{\gamma^\mu\gamma^5}=\gamma^\mu\gamma^5 \tag{3.29c}$$

and from (3.29a)

$$\overline{\not{a}\not{b}\not{c}\cdots\not{p}}=\not{p}\cdots\not{c}\not{b}\not{a}\quad. \tag{3.29d}$$

In order to calculate the sum over spins in expressions like (3.26) or (3.27) in a direct and simple way we have to learn some new calculational techniques. These sums can be *reduced to calculating traces*.[2] Here we use the energy projection operators (see *RQM*, Chap. 7)

$$\hat{\Lambda}_\pm=\frac{\pm\not{p}+m_0}{2m_0}\quad\text{or}\quad\hat{\Lambda}_r=\frac{\varepsilon_r\not{p}+m_0}{2m_0}\quad. \tag{3.30}$$

As an example we calculate

$$\begin{aligned}\sum_{s_i}u_\beta(p_i,s_i)\,\bar{u}_\delta(p_i,s_i)&=\sum_{\gamma,r=1}^{4}\varepsilon_r\,\omega^r_\beta(\boldsymbol{p}_i)\,\bar{\omega}^r_\gamma(\boldsymbol{p}_i)\left(\frac{\not{p}_i+m_0}{2m_0}\right)_{\gamma\delta}\\&=\left(\frac{\not{p}_i+m_0}{2m_0}\right)_{\beta\delta}=\left(\hat{\Lambda}_+(p_i)\right)_{\beta\delta}\quad.\end{aligned} \tag{3.31}$$

[2] This elegant technique was introduced in H.B.G. Casimir, Helv. Phys. Acta **6**, 287 (1933).

We have used here the relation (cf. *RQM*, (6.33))

$$\bar{\omega}^r(\not{p} - \varepsilon_r m_0) = 0 \tag{3.32a}$$

to extend the range of summation to $r = 1, \dots, 4$. Subsequently we used

$$\sum_{r=1}^{4} \varepsilon_r\, \omega^r_\beta(\boldsymbol{p}_i)\, \bar{\omega}^r_\gamma(\boldsymbol{p}_i) = \delta_{\beta\gamma} \quad . \tag{3.32b}$$

The last expression is the completeness relation (*RQM*, (6.41)). Similarly we calculate the spin sum in (3.26). We write explicitly

$$\begin{aligned}
&\sum_{\alpha,\sigma,\beta,\delta} \sum_{s_f} \bar{u}_\alpha(p_f,s_f)\, \gamma^0_{\alpha\beta} \underbrace{\left(\sum_{s_i} u_\beta(p_i,s_i)\, \bar{u}_\delta(p_i,s_i) \right)}_{\left(\frac{\not{p}_i + m_0}{2m_0}\right)_{\beta\delta} \text{ because of (3.31)}} \gamma^0_{\delta\sigma}\, u_\sigma(p_f,s_f) \\
&= \sum_{\alpha,\sigma} \sum_{s_f} \bar{u}_\alpha(p_f,s_f) \left(\gamma^0 \frac{\not{p}_i + m_0}{2m_0} \gamma^0 \right)_{\alpha\sigma} u_\sigma(p_f,s_f) \\
&= \sum_{\alpha,\sigma} \sum_{r=1}^{2} \bar{\omega}^r_\alpha(\boldsymbol{p}_f) \left(\gamma^0 \frac{\not{p}_i + m_0}{2m_0} \gamma^0 \right)_{\alpha\sigma} \omega^r_\sigma(\boldsymbol{p}_f) \\
&= \sum_{\alpha,\sigma,\tau} \sum_{r=1}^{4} \varepsilon_r\, \bar{\omega}^r_\alpha(\boldsymbol{p}_f) \left(\gamma^0 \frac{\not{p}_i + m_0}{2m_0} \gamma^0 \right)_{\alpha\sigma} \left(\frac{\not{p}_f + m_0}{2m_0} \right)_{\sigma\tau} \omega^r_\tau(\boldsymbol{p}_f) \\
&= \sum_{\alpha,\sigma} \left(\gamma^0 \frac{\not{p}_i + m_0}{2m_0} \gamma^0 \right)_{\alpha\sigma} \left(\frac{\not{p}_f + m_0}{2m_0} \right)_{\sigma\alpha} \\
&= \mathrm{Tr} \left[\gamma^0 \frac{\not{p}_i + m_0}{2m_0} \gamma^0 \frac{\not{p}_f + m_0}{2m_0} \right] \quad .
\end{aligned} \tag{3.33a}$$

In the last line but two we used the relation (cf. *RQM*, (6.33))

$$(\not{p} - \varepsilon_r m_0)\, \omega^r(\boldsymbol{p}) = 0 \quad . \tag{3.32c}$$

The reasoning leading to (3.33a) equally applies to the spin summation of the general squared matrix element (3.27). The result is

$$\sum_{s_f\, s_i} |\bar{u}(p_f,s_f)\, \hat{\Gamma}\, u(p_i,s_i)|^2 = \mathrm{Tr} \left[\hat{\Gamma} \frac{\not{p}_i + m_0}{2m_0} \hat{\bar{\Gamma}} \frac{\not{p}_f + m_0}{2m_0} \right] \quad , \tag{3.33b}$$

where the barred operator $\hat{\bar{\Gamma}}$ has been introduced in (3.28). Using this result the unpolarized differential cross section (3.25) can be elegantly written as

$$\frac{d\bar{\sigma}}{d\Omega_f} = \frac{4Z^2\alpha^2 m_0^2}{2|\boldsymbol{q}|^4}\, \mathrm{Tr} \left[\gamma^0 \frac{\not{p}_i + m_0}{2m_0} \gamma^0 \frac{\not{p}_f + m_0}{2m_0} \right] \quad . \tag{3.34}$$

To proceed in the calculation we now use several relations that will be discussed in Mathematical Supplement 3.3. Since the trace of an odd number of γ-matrices vanishes (Theorem 1, Mathematical Supplement 3.3), (3.34) can be reduced to

$$\frac{d\bar{\sigma}}{d\Omega_f} = \frac{Z^2\alpha^2}{2|\boldsymbol{q}|^4}\left[\mathrm{Tr}(\gamma^0 \not{p}_i \gamma^0 \not{p}_f) + m_0^2\,\mathrm{Tr}(\gamma^0)^2\right] \quad . \tag{3.35}$$

With $\mathrm{Tr}\,(\gamma^0)^2 = \mathrm{Tr}\,\mathbb{1} = 4$ and $\mathrm{Tr}(\gamma^0\not{p}_i\gamma^0\not{p}_f) = \mathrm{Tr}(\not{a}\,\not{p}_i\,\not{a}\,\not{p}_f)$ for $a = (1,0,0,0)$ and furthermore using Theorem 3, Mathematical Supplement 3.3,

$$\mathrm{Tr}(\gamma^0\not{p}_i\gamma^0\not{p}_f) = a\cdot p_i\,\mathrm{Tr}\not{a}\not{p}_f - a\cdot a\,\mathrm{Tr}\not{p}_i\not{p}_f + a\cdot p_f\,\mathrm{Tr}\not{p}_i\not{a} \quad .$$

Using $\mathrm{Tr}\,\not{a}\,\not{b} = 4\,a\cdot b$ (Theorem 2, Mathematical Supplement 3.3) we get

$$\begin{aligned}\mathrm{Tr}(\gamma^0\not{p}_i\gamma^0\not{p}_f) &= 4\,(a\cdot p_i)(a\cdot p_f) - (a\cdot a)\,4\,(p_i\cdot p_f) + 4\,(a\cdot p_f)(a\cdot p_i)\\ &= 4\,E_iE_f - 4\,(E_iE_f - \boldsymbol{p}_i\cdot\boldsymbol{p}_f) + 4\,E_iE_f\\ &= 8\,E_iE_f - 4\,p_i\cdot p_f = 4\,E_iE_f + 4\,\boldsymbol{p}_i\cdot\boldsymbol{p}_f \quad .\end{aligned} \tag{3.36}$$

The $\delta(E_i - E_f)$-function in (3.23) ensures energy conservation $E_i = E_f$ and thus $E_i^2 = E_f^2$, yielding

$$m_0^2 + \boldsymbol{p}_i^2 = m_0^2 + \boldsymbol{p}_f^2 \quad \text{or} \quad |\boldsymbol{p}_i| = |\boldsymbol{p}_f| = |\boldsymbol{p}| \quad .$$

As a function of the scattering angle θ we can write for the scalar product of initial and final momentum

$$\begin{aligned}\boldsymbol{p}_i\cdot\boldsymbol{p}_f = |\boldsymbol{p}|^2\cos\theta &= |\boldsymbol{p}|^2\left(\cos^2\frac{\theta}{2} - \sin^2\frac{\theta}{2}\right) = |\boldsymbol{p}|^2\left(1 - 2\sin^2\frac{\theta}{2}\right)\\ &= \beta^2E^2\left(1 - 2\sin^2\frac{\theta}{2}\right) \quad ,\end{aligned} \tag{3.37}$$

with $|\boldsymbol{p}| = |\boldsymbol{v}|E = \beta E$.

Taking this result and the momentum transfer (see Fig. 3.3)

$$|\boldsymbol{q}| = |\boldsymbol{p}_f - \boldsymbol{p}_i| = 2\,|\boldsymbol{p}|\sin\frac{\theta}{2} \quad , \tag{3.38}$$

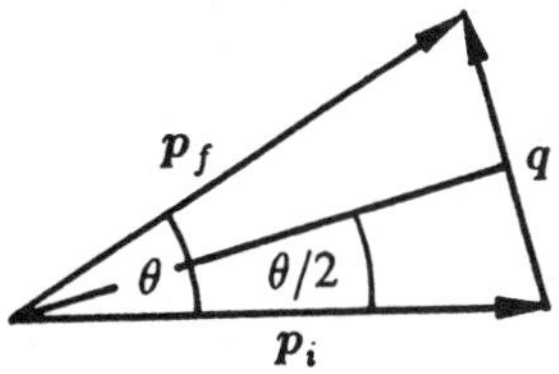

Fig. 3.3. Sketch of the momentum transfer $\boldsymbol{q}$

the differential cross section (3.35) can be written in the form

$$\begin{aligned}\frac{d\bar{\sigma}}{d\Omega_f} &= \frac{Z^2\alpha^2}{2(\boldsymbol{q}^2)^2}\left[8\,E_iE_f - 4\,p_i\cdot p_f + 4\,m_0^2\right]\\ &= \frac{Z^2\alpha^2}{2\times 16|\boldsymbol{p}|^4\sin^4\frac{\theta}{2}}\left[8\,E_iE_f - 4\,p_i\cdot p_f + 4\,m_0^2\right]\\ &= \frac{Z^2\alpha^2}{32|\boldsymbol{p}|^4\sin^4\frac{\theta}{2}}\left[4\,E_iE_f + 4\,\boldsymbol{p}_i\cdot\boldsymbol{p}_f + 4\,m_0^2\right]\\ &= \frac{Z^2\alpha^2}{8|\boldsymbol{p}|^4\sin^4\frac{\theta}{2}}\left[E^2 + |\boldsymbol{p}|^2(1 - 2\sin^2\frac{\theta}{2}) + m_0^2\right]\\ &= \frac{Z^2\alpha^2\left[E^2 - \beta^2E^2\sin^2\frac{\theta}{2}\right]}{4\beta^4E^4\sin^4\frac{\theta}{2}}\\ &= \frac{Z^2\alpha^2\left(1 - \beta^2\sin^2\frac{\theta}{2}\right)}{4\beta^4E^2\sin^4\frac{\theta}{2}} = \frac{Z^2\alpha^2\left(1 - \beta^2\sin^2\frac{\theta}{2}\right)}{4\beta^2|\boldsymbol{p}|^2\sin^4\frac{\theta}{2}} \quad .\end{aligned} \tag{3.39}$$

Here we used $\beta^2E^2 = m^2c^2v^2 = c^2|\boldsymbol{p}|^2$ with $\beta = v/c$, yielding $\beta^2E^2 = |\boldsymbol{p}|^2$ for $c = 1$. Equation (3.39) is just the well-known **Mott** *scattering formula*, which reduces to *Rutherford's scattering formula* in the limit $\beta \to 0$ (small velocities)

$$\frac{d\bar{\sigma}}{d\Omega_f} = \frac{Z^2\alpha^2}{4\beta^2|\mathbf{p}|^2 \sin^4\frac{\theta}{2}} \quad . \tag{3.40}$$

In addition to the correct treatment of the relativistic kinematics, (3.39) differs from the Rutherford formula for another reason: the Dirac electron has a *magnetic moment* interacting with the magnetic field of the scattering centre (viewed in the rest frame of the electron!). For small velocities this effect is negligible.

EXERCISE

3.1 Calculation of a Useful Integral

Problem. Show that

$$I = \int_{-\infty}^{\infty} dE_f \, \frac{4\,\sin^2(E_f - E_i)T/2}{(E_f - E_i)^2} = 2\pi T \quad .$$

Solution. In a first step we introduce the variable $x := (E_f - E_i)\frac{T}{2}$; then

$$I = \int_{-\infty}^{\infty} dx \, 4\, \frac{\sin^2 x}{\frac{4}{T^2}x^2}\,\frac{2}{T} = 2\,T \int_{-\infty}^{\infty} dx \, \frac{\sin^2 x}{x^2} \quad . \tag{1}$$

Since $\sin^2 x/x^2|_{x=0} = 1$, the integrand is continuous and bounded everywhere. By partial integration this expression can be simplified to

$$\begin{aligned} I &= 2\,T\left[-\sin^2 x\,\frac{1}{x}\right]_{-\infty}^{\infty} + 2\,T\int_{-\infty}^{\infty} dx\,\frac{2\,\sin x\,\cos x}{x} \\ &= 2\,T\int_{-\infty}^{\infty} dx\,\frac{\sin 2x}{x} = 2\,T\int_{-\infty}^{\infty} dy\,\frac{\sin y}{y} \quad . \end{aligned} \tag{2}$$

This is the so-called 'Dirichlet integral'. It can be easily calculated using the residue theorem. The function $\sin y/y$ with the extension $\sin y/y|_{y=0} = 1$ is holomorphic in the finite plane. The integral I does not change its value if the path of integration is deformed near the origin to the contour C (dashed line in Fig. 3.4). It follows that

$$I = 2\,T\left(\int_C dz\,\frac{e^{iz}}{2iz} - \int_C dz\,\frac{e^{-iz}}{2iz}\right) \quad . \tag{3}$$

The first integral can be performed by closing the path of integration in the upper half plane, the second in the lower half plane:

$$I = 2\,T\left(\int_{C_1}\frac{e^{iz}}{2iz}\,dz - \int_{C_2}\frac{e^{-iz}}{2iz}\,dz\right) \quad . \tag{4}$$

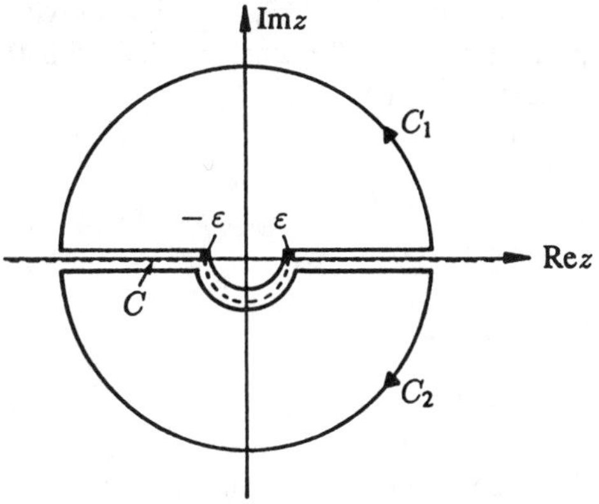

Fig. 3.4. Integration contours in the complex plane

The residue theorem states that

$$I = 2T\,2\pi\mathrm{i}\left(\frac{\mathrm{e}^{0\mathrm{i}}}{2\mathrm{i}} - 0\right) = 2\pi T \quad . \tag{5}$$

EXERCISE

3.2 Lorentz Transformation of Plane Waves

Problem. The plane waves in box normalization are given by

$$\begin{aligned} \psi(x) &= \sqrt{\frac{m_0}{EV}}\,u(p,s)\,\mathrm{e}^{-\mathrm{i}p\cdot x} \quad , \\ \bar{\psi}(x) &= \sqrt{\frac{m_0}{EV}}\,\bar{u}(p,s)\,\mathrm{e}^{\mathrm{i}p\cdot x} \quad . \end{aligned} \tag{1}$$

V is the normalization volume and

$$\begin{aligned} u(p,s) &= \sqrt{\frac{E+m_0}{2m_0}}\begin{pmatrix} \chi_s \\ \frac{\sigma\cdot p}{E+m_0}\chi_s \end{pmatrix} \quad , \\ \chi_{1/2} &= \begin{pmatrix} 1 \\ 0 \end{pmatrix} \quad , \quad \chi_{-1/2} = \begin{pmatrix} 0 \\ 1 \end{pmatrix} \quad . \end{aligned} \tag{2}$$

Show that $\psi(x)$ has the right properties under Lorentz transformations, i.e. that the bilinear quantity $\bar{\psi}(x)\psi(x)$ is a Lorentz scalar and $\psi^\dagger(x)\psi(x)$ is the time component of a four-vector.

Solution. The action of a Lorentz transformation on the box volume V has to be taken into account. The length contraction is a simple kinematical consequence of the Lorentz transformation yielding a modification of the observed volume.

First a measuring unit in the rest frame S' of the electron is given with end points on the z-axis z_1' and z_2'; the length of the unit is $l_0 = z_2' - z_1'$. An observer in

Exercise 3.2.

frame S measures its length at time t. Without restriction of generality we choose the z-axis in the S-frame, moving with the velocity $-v$ with respect to S', to coincide with the z'-axis.

With $\gamma = (1-v^2)^{-1/2}$ it follows that

$$z_2' = \gamma(z_2 - vt) \quad , \tag{3a}$$
$$z_1' = \gamma(z_1 - vt) \quad , \tag{3b}$$

and thus

$$l_0 = z_2' - z_1' = \gamma(z_2 - z_1) = \gamma l \quad , \tag{4}$$

which gives the length l measured in the observer's frame S,

$$l = \frac{1}{\gamma} l_0 \tag{5}$$

(length contraction). The invariant volume V_0 as 'seen' by an electron in its rest frame changes to

$$V = \frac{1}{\gamma} V_0 \tag{6}$$

in the observer's frame depending on the relative velocity of observer and electron.

The coordinate $\boldsymbol{x}$ of the wave function (1) refers to a specific choice of the origin – here given by the observer's position. The normalization volume as seen by the observer

$$V = \int d^3x \quad , \tag{7}$$

thus depends on the velocity of the electron. However,

$$V_0 = V\gamma \tag{8}$$

is independent of the velocity, and thus it is invariant.

Using (1) and (2) we get for the scalar density

$$\bar{\psi}(x)\,\psi(x) = \frac{E+m_0}{2EV}\left[\chi_s^\dagger \chi_s - \chi_s^\dagger \frac{(\boldsymbol{\sigma}\cdot\boldsymbol{p})^\dagger(\boldsymbol{\sigma}\cdot\boldsymbol{p})}{(E+m_0)^2}\chi_s\right] \quad . \tag{9}$$

We write explicitly

$$\begin{aligned}(\boldsymbol{\sigma}\cdot\boldsymbol{p})^\dagger(\boldsymbol{\sigma}\cdot\boldsymbol{p}) &= \begin{pmatrix} p_3 & p_1 - ip_2 \\ p_1 + ip_2 & -p_3 \end{pmatrix}^\dagger \begin{pmatrix} p_3 & p_1 - ip_2 \\ p_1 + ip_2 & -p_3 \end{pmatrix} \\ &= \begin{pmatrix} p_3^2 + p_1^2 + p_2^2 & 0 \\ 0 & p_1^2 + p_2^2 + p_3^2 \end{pmatrix} = \boldsymbol{p}^2\,\mathbb{1} \quad , \end{aligned} \tag{10}$$

which can be used to simplify (9) to the result

Exercise 3.2.

$$
\begin{aligned}
\bar{\psi}(x)\,\psi(x) &= \frac{E+m_0}{2EV}\left[1-\frac{p^2}{(E+m_0)^2}\right]\\
&= \frac{1}{2EV}\,\frac{E^2+2\,Em_0+m_0^2-E^2+m_0^2}{E+m_0}\\
&= \frac{2m_0\,(E+m_0)}{2EV\,(E+m_0)}=\frac{m_0}{E}\,\frac{1}{V}\quad .
\end{aligned}
\tag{11}
$$

Since E/m_0 is just the Lorentz factor γ we find

$$
\bar{\psi}(x)\,\psi(x) = \frac{1}{\gamma V} = \frac{1}{V_0}\quad ,
\tag{12}
$$

which is a Lorentz-invariant expression. On the other hand the 0-component of the 4-current density is

$$
\begin{aligned}
\psi^\dagger(x)\;\psi(x) &= \frac{E+m_0}{2EV}\left[\chi_s^\dagger\chi_s+\chi_s^\dagger\,\frac{(\boldsymbol{\sigma}\cdot\boldsymbol{p})^\dagger\,(\boldsymbol{\sigma}\cdot\boldsymbol{p})}{(E+m_0)^2}\,\chi_s\right]\\
&= \frac{E+m_0}{2EV}\left[1+\frac{p^2}{(E+m_0)^2}\right]\\
&= \frac{1}{2EV}\,\frac{E^2+2Em_0+m_0^2+E^2-m_0^2}{E+m_0}\\
&= \frac{2E\,(E+m_0)}{2EV\,(E+m_0)}=\frac{1}{V}=\frac{\gamma}{V_0}\quad .
\end{aligned}
\tag{13}
$$

Since γ is the time component of the four-velocity, the transformation properties of the result are obvious.

MATHEMATICAL SUPPLEMENT

3.3 Traces and Identities Involving γ-Matrices

When calculating Feynman diagrams and the resulting physically measurable cross sections, one is confronted with the task of evaluating traces of special combinations of γ-matrices. In Sect. 3.1 the importance of calculating traces has already been discussed, and this will be a recurring theme also in the following examples. Very useful techniques have been developed to simplify these calculations. It is not necessary to use the explicit form of the γ-matrices; usually it is sufficient to exploit the commutator algebra of the γ-matrices. We collect these properties in several theorems.

Theorem 1. The trace of an odd number of γ-matrices vanishes.

Proof. We make use of the matrix γ_5 which anticommutes with all other γ-matrices and satisfies $\gamma_5\gamma_5 = 1\!\!1$. For arbitrary four-vectors $a_1, \ldots, a_n$ we have

$$
\mathrm{Tr}\not{a}_1\cdots\not{a}_n = \mathrm{Tr}\not{a}_1\cdots\not{a}_n\,\gamma_5\gamma_5 = \mathrm{Tr}\,\gamma_5\not{a}_1\cdots\not{a}_n\,\gamma_5
$$

because of the cyclic permutation within a trace, i.e. $\mathrm{Tr}\,AB = \mathrm{Tr}\,BA$. We use the relation $\gamma_\mu\gamma_5 + \gamma_5\gamma_\mu = 0$ and commute the first γ_5 to the right. This yields n minus signs, and thus we get

Mathematical Supplement 3.3.

$$\begin{aligned}\mathrm{Tr}\,\not{a}_1 \cdots \not{a}_n &= (-1)^n\, \mathrm{Tr}\,\not{a}_1 \cdots \not{a}_n \gamma_5\gamma_5 \\ &= (-1)^n \mathrm{Tr}\,\not{a}_1 \cdots \not{a}_n \quad .\end{aligned}$$

Obviously the trace vanishes for odd n.

Theorem 2. $\mathrm{Tr}\,\not{a}\not{b} = 4a\cdot b \quad .$

Proof.

$$\begin{aligned}\mathrm{Tr}\,\not{a}\not{b} = \mathrm{Tr}\,\not{b}\not{a} &= \tfrac{1}{2}\,\mathrm{Tr}\,(\not{a}\not{b} + \not{b}\not{a}) \\ &= \frac{1}{2}\,\mathrm{Tr}\,(\gamma^\mu\gamma^\nu + \gamma^\nu\gamma^\mu)\, a_\mu b_\nu \\ &= a_\mu b_\nu \,\mathrm{Tr}\, g^{\mu\nu}\,\mathbb{1} \\ &= a\cdot b\ \mathrm{Tr}\,\mathbb{1} = 4\, a\cdot b \quad .\end{aligned}$$

Theorem 3.

$$\begin{aligned}\mathrm{Tr}\,\not{a}_1 \cdots \not{a}_n = a_1\cdot a_2\, \mathrm{Tr}\,\not{a}_3 \cdots \not{a}_n - a_1\cdot a_3\, \mathrm{Tr}\,\not{a}_2\not{a}_4 \cdots \not{a}_n + \dots \\ + a_1\cdot a_n\, \mathrm{Tr}\,\not{a}_2 \cdots \not{a}_{n-1} \quad .\end{aligned}$$

This theorem is very useful in calculating traces involving many γ-matrices. A special case is

$$\mathrm{Tr}\,\not{a}_1\not{a}_2\not{a}_3\not{a}_4 = 4\left(a_1\cdot a_2\, a_3\cdot a_4 - a_1\cdot a_3\, a_2\cdot a_4 + a_1\cdot a_4\, a_2\cdot a_3\right) \quad .$$

Proof. By using $\not{a}_1\not{a}_2 = -\not{a}_2\not{a}_1 + 2a_1\cdot a_2\,\mathbb{1}$ we shift $\not{a}_1$ to the right-hand side of $\not{a}_2$, that is

$$\mathrm{Tr}\,\not{a}_1 \cdots \not{a}_n = 2a_1\cdot a_2\, \mathrm{Tr}\,\not{a}_3 \cdots \not{a}_n - \mathrm{Tr}\,\not{a}_2\,\not{a}_1\,\not{a}_3 \cdots \not{a}_n \quad .$$

Repeating this procedure we get (remember that n must be even according to Theorem 1)

$$\begin{aligned}\mathrm{Tr}\,\not{a}_1 \cdots \not{a}_n = 2\, a_1\cdot a_2\, \mathrm{Tr}\,\not{a}_3 \cdots \not{a}_n - \dots + 2\, a_1\cdot a_n\, \mathrm{Tr}\,\not{a}_2 \cdots \not{a}_{n-1} \\ - \mathrm{Tr}\,\not{a}_2 \cdots \not{a}_n\not{a}_1 \quad .\end{aligned}$$

Finally we use the invariance of the trace under cyclic permutations to move $\not{a}_1$ back to the left hand side of the expression. This yields our theorem. In particular, we get

$$\begin{aligned}\mathrm{Tr}\,\not{a}_1\not{a}_2\,\not{a}_3\not{a}_4 &= a_1\cdot a_2\, \mathrm{Tr}\,\not{a}_3\not{a}_4 - a_3\cdot a_1\, \mathrm{Tr}\,\not{a}_2\not{a}_4 + a_1\cdot a_4\, \mathrm{Tr}\,\not{a}_2\not{a}_3 \\ &= 4\, a_1\cdot a_2\, a_3\cdot a_4 - 4\, a_1\cdot a_3\, a_2\cdot a_4 + 4\, a_1\cdot a_4\, a_2\cdot a_3 \quad .\end{aligned}$$

Here we have applied Theorem 2.

Theorem 4. $\mathrm{Tr}\,\gamma^5 = 0 \quad .$

Mathematical Supplement 3.3

Proof. In our representation the matrix

$$\gamma_5 = \gamma^5 = \mathrm{i}\gamma^0\gamma^1\gamma^2\gamma^3$$

has the explicit form

$$\gamma_5 = \begin{pmatrix} 0 & 1\!\!1 \\ 1\!\!1 & 0 \end{pmatrix} ,$$

which is obviously traceless.

The same result can be derived from the algebra of the γ-matrices without using any special representation. With $\gamma^\mu\gamma^5 + \gamma^5\gamma^\mu = 0$ and thus in particular $\gamma^0\gamma^5 = -\gamma^5\gamma^0$ we get

$$\begin{aligned} \operatorname{Tr}\gamma^5 &= \operatorname{Tr}\gamma^5(\gamma^0)^2 = -\operatorname{Tr}\gamma^0\gamma^5\gamma^0 \\ &= -\operatorname{Tr}\gamma^5(\gamma^0)^2 = -\operatorname{Tr}\gamma^5 = 0 \quad . \end{aligned}$$

We again used the cyclic permutation of matrices in a trace.

Theorem 5. $\operatorname{Tr}\gamma^5 \not{a}\not{b} = 0.$

Proof. We have to show that $\operatorname{Tr}(\gamma^5\gamma^\mu\gamma^\nu) = 0$. If the indices are equal, $\mu = \nu$, the assertion follows from Theorem 4 since $(\gamma^\mu)^2 = g^{\mu\mu}1\!\!1$. In the case $\mu \neq \nu$ we choose an index λ that differs from μ and from ν and proceed as follows

$$\begin{aligned} \operatorname{Tr}\gamma^5\gamma_\mu\gamma_\nu &= \operatorname{Tr}\gamma^5\gamma_\mu\gamma_\nu\gamma_\lambda^{-1}\gamma_\lambda = \operatorname{Tr}\gamma_\lambda\gamma^5\gamma_\mu\gamma_\nu\gamma_\lambda^{-1} \\ &= (-1)^3\operatorname{Tr}\gamma^5\gamma_\mu\gamma_\nu\gamma_\lambda\gamma_\lambda^{-1} \\ &= -\operatorname{Tr}\gamma^5\gamma_\mu\gamma_\nu = 0 \quad . \end{aligned}$$

Theorem 6. $\operatorname{Tr}\gamma^5\not{a}\not{b}\not{c}\not{d} = -4\mathrm{i}\,\varepsilon^{\alpha\beta\gamma\delta}\,a_\alpha b_\beta c_\gamma d_\delta$.

Here ε is the completely antisymmetric unit tensor: $\varepsilon^{\alpha\beta\gamma\delta} = +1$ if $(\alpha, \beta, \gamma, \delta)$ is an even permutation of $(0, 1, 2, 3)$, $\varepsilon^{\alpha\beta\gamma\delta} = -1$ for an odd permutation, and $\varepsilon^{\alpha\beta\gamma\delta} = 0$ if any two indices are identical.

Proof. We have to evaluate

$$\operatorname{Tr}\gamma^5\not{a}\not{b}\not{c}\not{d} = a_\alpha b_\beta c_\gamma d_\delta \operatorname{Tr}\gamma^5\gamma^\alpha\gamma^\beta\gamma^\gamma\gamma^\delta \quad ,$$

where summation over all repeated indices is implied. Most of the $4^4 = 256$ terms in this sum do not contribute. Indeed, if any two of the indices $\alpha, \beta, \gamma, \delta$ take on equal values, the trace will vanish. Let, say, the first and third indices be equal. Using the commutation relations the number of γ-matrices under the trace can be reduced by two:

$$\begin{aligned} \operatorname{Tr}\gamma^5\gamma^\alpha\gamma^\beta\gamma^\alpha\gamma^\delta &= \operatorname{Tr}\gamma^5\gamma^\alpha(-\gamma^\alpha\gamma^\beta + 2g^{\alpha\beta}1\!\!1)\gamma^\delta \\ &= \operatorname{Tr}\gamma^5(-\gamma^\alpha\gamma^\alpha\gamma^\beta + 2\gamma^\alpha g^{\alpha\beta}1\!\!1)\gamma^\delta \\ &= -g^{\alpha\alpha}\operatorname{Tr}\gamma^5\gamma^\beta\gamma^\delta + 2g^{\alpha\beta}\operatorname{Tr}\gamma^5\gamma^\alpha\gamma^\delta \\ &= 0 \end{aligned}$$

using Theorem 5.

Mathematical Supplement 3.3.

Thus only indices $(\alpha, \beta, \gamma, \delta)$ that are a permutation of the numbers $(0, 1, 2, 3)$ can contribute. We only have to evaluate the trace

$$\begin{aligned}\mathrm{Tr}\gamma^5\gamma^0\gamma^1\gamma^2\gamma^3 &= \mathrm{Tr}\gamma^5(-\mathrm{i}\gamma^5) = -\mathrm{i}\,\mathrm{Tr}\mathbb{1}\\ &= -4\mathrm{i} = -4\mathrm{i}\varepsilon^{0123} \quad .\end{aligned}$$

Since the four γ^μ matrices are mutually anticommuting, an odd permutation of the indices $(0, 1, 2, 3)$ introduces an additional minus sign, which completes the proof of the theorem.

Theorem 7. $\mathrm{Tr}\,\not{a}_1\not{a}_2 \cdots \not{a}_{2n} = \mathrm{Tr}\,\not{a}_{2n} \cdots \not{a}_1$.

Proof. We take advantage of the matrix $\hat{C}$, which was introduced in the discussion of charge conjugation. $\hat{C}$ has the property $\hat{C}\,\gamma_\mu\,\hat{C}^{-1} = -\gamma_\mu^T$. It follows that

$$\begin{aligned}\mathrm{Tr}\,\not{a}_1\not{a}_2 \cdots \not{a}_{2n} &= \mathrm{Tr}\,\hat{C}\,\not{a}_1\hat{C}^{-1}\,\hat{C}\,\not{a}_2\,\hat{C}^{-1} \cdots \hat{C}\,\not{a}_{2n}\,\hat{C}^{-1}\\ &= (-1)^{2n}\,\mathrm{Tr}\,\not{a}_1^T\not{a}_2^T \cdots \not{a}_{2n}^T\\ &= \mathrm{Tr}\,[\not{a}_{2n} \cdots \not{a}_1]^T\\ &= \mathrm{Tr}\,\not{a}_{2n} \cdots \not{a}_1 \quad .\end{aligned}$$

Theorem 8. The following useful identities hold for contracted products of γ-matrices:

$$\begin{aligned}
&\text{a)} & \gamma_\mu\gamma^\mu &= 4\mathbb{1} \;,\\
&\text{b)} & \gamma_\mu\not{a}\gamma^\mu &= -2\not{a} \;,\\
&\text{c)} & \gamma_\mu\not{a}\not{b}\gamma^\mu &= 4a\cdot b\,\mathbb{1} \;,\\
&\text{d)} & \gamma_\mu\not{a}\not{b}\not{c}\gamma^\mu &= -2\not{c}\not{b}\not{a} \;,\\
&\text{e)} & \gamma_\mu\not{a}\not{b}\not{c}\not{d}\gamma^\mu &= 2\not{d}\not{a}\not{b}\not{c} + 2\not{c}\not{b}\not{a}\not{d} \quad .
\end{aligned}$$

Proof. These identities all follow from the anticommutation relations of the γ-matrices.

$$\text{a)}\quad \gamma_\mu\gamma^\mu = \frac{1}{2}(\gamma_\mu\gamma^\mu + \gamma_\mu\gamma^\mu) = \frac{1}{2}2g^\mu{}_\mu\mathbb{1} = 4\mathbb{1} \;,$$

$$\begin{aligned}\text{b)}\quad \gamma_\mu\not{a}\gamma^\mu &= \gamma_\mu\gamma^\nu a_\nu\gamma^\mu = a_\nu\gamma_\mu(2g^{\mu\nu}\mathbb{1} - \gamma^\mu\gamma^\nu)\\ &= 2\not{a} - 4\not{a} = -2\not{a} \;,\end{aligned}$$

$$\begin{aligned}\text{c)}\quad \gamma_\mu\,\not{a}\,\not{b}\gamma^\mu &= \gamma_\mu\gamma^\nu\,a_\nu\,\gamma^\lambda\,b_\lambda\gamma^\mu\\ &= \gamma_\mu\,\not{a}\,2g^{\lambda\mu}\,b_\lambda - \gamma_\mu\,\not{a}\,\gamma^\mu\not{b} \;=\; 2\not{b}\not{a} + 2\not{a}\not{b}\\ &= 4\,a\cdot b\mathbb{1} - 2\not{a}\not{b} + 2\not{a}\not{b} = 4\,a\cdot b\mathbb{1} \;,\end{aligned}$$

$$\begin{aligned}\text{d)}\quad \gamma_\mu\,\not{a}\,\not{b}\,\not{c}\,\gamma^\mu &= \gamma_\mu\,\not{a}\,\not{b}\,2g^{\nu\mu}\,c_\nu - \gamma_\mu\,\not{a}\,\not{b}\,\gamma^\mu\,\not{c}\\ &= 2\,\not{c}\,\not{a}\,\not{b} - 4\,a\cdot b\,\not{c}\\ &= 4\,\not{c}\,a\cdot b - 2\not{c}\,\not{b}\,\not{a} - 4\,a\cdot b\,\not{c}\\ &= -2\,\not{c}\,\not{b}\,\not{a} \;,\end{aligned}$$

$$\begin{aligned}\text{e)}\quad \gamma_\mu\,\not{a}\,\not{b}\,\not{c}\,\not{d}\,\gamma^\mu &= \gamma_\mu\,\not{a}\,\not{b}\,\not{c}\,2g^{\nu\mu}\,d_\nu - \gamma_\mu\,\not{a}\,\not{b}\,\not{c}\,\gamma^\mu\,\not{d}\\ &= 2\,\not{d}\,\not{a}\,\not{b}\,\not{c} + 2\,\not{c}\,\not{b}\,\not{a}\,\not{d} \quad .\end{aligned}$$

EXAMPLE

3.4 Coulomb Scattering of Positrons

In the discussion of electron scattering at a Coulomb potential we found that the scattering matrix element depends quadratically on e. Therefore we expect the Coulomb scattering of positrons to yield the same result in that order of e. This can be seen by denoting the matrix element explicitly

$$S_{fi} = \mathrm{i}e \int d^4x \, \bar{\psi}_f(x) \not{A}(x) \Psi_i^{(-E)}(x) \quad . \tag{1}$$

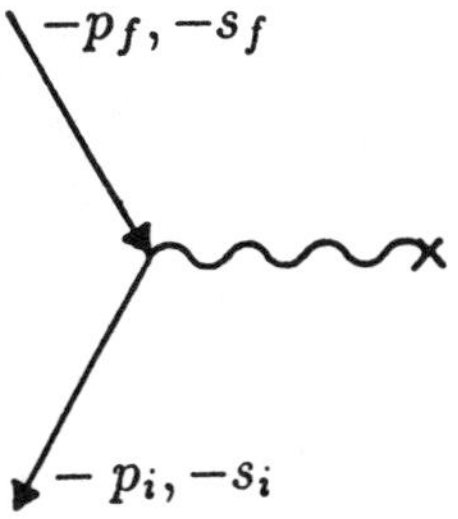

Fig. 3.5. Scattering of a positron at an external potential (×) to lowest order. The incoming positron with momentum p_i and spin s_i is described by an outgoing electron with negative energy, with momentum $-p_i$ and spin $-s_i$. For the outgoing positron the treatment is analogous

Here the overall sign is positive, because we scatter waves with negative frequency (cf. (2.39) and (2.40)). The incoming state corresponds to the future and is treated as an electron with negative energy and four-momentum $-p_f$. This electron moves backward in time. The scattering process is illustrated in Fig. 3.5. If as a lowest-order approximation we insert a plane-wave solution, the corresponding wave function (incoming electron with negative energy) is

$$\psi_i^{(\text{electron})}(-p_f, -s_f) = \sqrt{\frac{m_0}{E_f V}} \, v(p_f, s_f) \, \mathrm{e}^{+\mathrm{i} p_f \cdot x} \quad . \tag{2a}$$

For the outgoing state we write equivalently (outgoing electron with negative energy)

$$\psi_f^{(\text{electron})}(-p_i, -s_i) = \sqrt{\frac{m_0}{E_i V}} \, v(p_i, s_i) \, \mathrm{e}^{+\mathrm{i} p_i \cdot x} \quad . \tag{2b}$$

Here we have adopted the language of electrons on the left-hand side and the language of positrons on the right-hand side. An incoming electron with negative energy moving backward in time corresponds to an outgoing positron with positive energy moving forward in time ($\psi_i^{(\text{electron})}(-p_f, -s_f) \leftrightarrow v(p_f, s_f)$). Similarly an outgoing electron with negative energy moving backward in time corresponds to an incoming positron with positive energy moving forward in time.

By using the spinors $v(p,s)$ we have taken care of the fact that the spin of electrons with negative energy is $-s$. Here s denotes the spin of the positron. This redefinition, which we performed in *RQM*, Chap. 6, is now obvious. A positron is described by an electron with negative energy, negative momentum, and negative spin as demanded by the hole theory. The negative momentum is automatically included in the spinors $\omega^r(\boldsymbol{p})$ with $r = 3, 4$, since these solutions belong to the plane waves $\psi^r(x) = \omega^r(\boldsymbol{p}) \exp(-\mathrm{i}\varepsilon_r p \cdot x)$. (For further information we refer to *RQM*, Chap. 6). We always construct the appropriate graph in terms of electrons. By doing this we retain the clarity of our calculations and attain a well-defined procedure to treat electrons with positive and negative energies avoiding possible errors.

As a consequence of our considerations we arrive at a result that is completely analogous to electron scattering:

$$S_{fi} = -\mathrm{i} Z e^2 \frac{1}{V} \sqrt{\frac{m_0^2}{E_f E_i}} \, \bar{v}(p_i, s_i) \gamma^0 v(p_f, s_f) \int d^4x \, \mathrm{e}^{\mathrm{i}(p_f - p_i)\cdot x} \frac{1}{|\boldsymbol{x}|} \quad . \tag{3}$$

Example 3.4.

Differences only occur regarding the total sign and the spinors v. We do not have to repeat the steps leading to the cross section since they have already been presented for the case of electron scattering. The unpolarized differential cross section follows as

$$\frac{d\bar{\sigma}_{e^+}}{d\Omega} = \frac{2Z^2\alpha^2 m_0^2}{|\boldsymbol{q}|^4} \sum_{s_f, s_i} |\bar{v}(p_i, s_i)\gamma^0 v(p_f, s_f)|^2 \quad . \tag{4}$$

Again the sums over the spins can be reduced to a trace. We rewrite

$$\sum_{s_i} v_\alpha(p_i, s_i)\bar{v}_\beta(p_i, s_i) = \sum_{r=1}^{4} \varepsilon_r \, \omega^r_\alpha(\boldsymbol{p}_i)\bar{\omega}^r_\gamma(\boldsymbol{p}_i)(-)\left(\frac{-\not{p}_i + m_0}{2m_0}\right)_{\gamma\beta} \quad . \tag{5}$$

This follows from the Dirac equation for the adjoint spinor $\bar{\omega}^r(\boldsymbol{p})(\not{p} - \varepsilon_r m_0) = 0$, which for $r = 1, 2$ gives

$$\bar{\omega}^r(\boldsymbol{p}_i)(\not{p}_i - m_0) = 0 \quad , \tag{6}$$

whereas for $r = 3, 4$ we have

$$\varepsilon_r \, \bar{\omega}^r(\boldsymbol{p}_i)\left(\frac{\not{p}_i - m_0}{2m_0}\right) = \bar{\omega}^r(\boldsymbol{p}_i) \quad . \tag{7}$$

We make use of the following closure relation:

$$\sum_{r=1}^{4} \varepsilon_r \, \omega^r_\alpha(\boldsymbol{p})\bar{\omega}^r_\beta(\boldsymbol{p}) = \delta_{\alpha\beta} \tag{8}$$

and derive the result

$$\begin{aligned} \sum_{s_i} v_\alpha(p_i, s_i)\bar{v}_\beta(p_i, s_i) &= -\delta_{\alpha\gamma}\left(\frac{-\not{p}_i + m_0}{2m_0}\right)_{\gamma\beta} \\ &= \left(\frac{\not{p}_i - m_0}{2m_0}\right)_{\alpha\beta} \\ &= \left(\Lambda_-(p_i)\right)_{\alpha\beta} \quad . \end{aligned} \tag{9}$$

Thus we finally get the scattering cross section

$$\begin{aligned} \frac{d\bar{\sigma}_{e^+}}{d\Omega} &= \frac{Z^2\alpha^2}{2|\boldsymbol{q}|^4} \operatorname{Tr}\left[\gamma^0(\not{p}_i - m_0)\gamma^0(\not{p}_f - m_0)\right] \\ &= \frac{Z^2\alpha^2}{2|\boldsymbol{q}|^4}\left(\operatorname{Tr}\left[\gamma^0\not{p}_i\gamma^0\not{p}_f\right] + m_0^2 \operatorname{Tr}(\gamma^0)^2\right) \quad . \end{aligned} \tag{10}$$

This result is *identical with the formula for electron scattering*, (3.31), and thus yields the same angular distribution. As is the case in the classical theory the cross section for Coulomb scattering is independent of the sign of the charge.

Yet, if we take care of *higher terms* in the expansion of the S-matrix, this statement is no longer valid. The first-order (proportional to $e(-Ze)$) and second-order (proportional to $e^2(-Ze)^2$) contributions to the transition amplitude have different signs for electron and positron scattering, resulting in differing cross sections. The

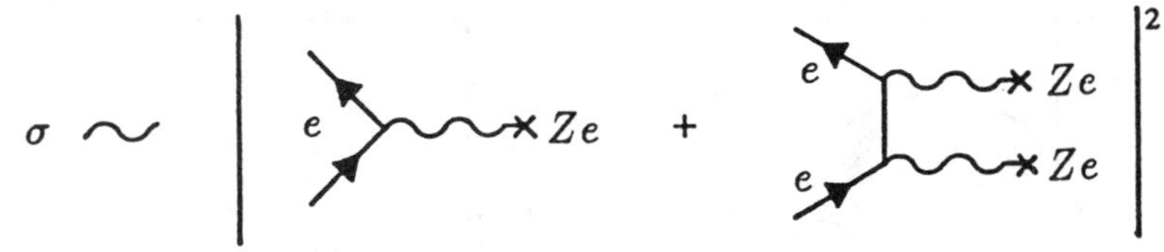

Fig. 3.6. The first-order and second-order scattering amplitudes can interfere

source of this interference term, which gives a contribution to the cross section proportional to $e^3(-Ze)^3$ is schematically illustrated in Fig. 3.6.

Owing to the infinite range of the Coulomb interaction, a more complicated calculation has to be done, since higher-order contributions to the S-matrix diverge. Yet a thorough analysis[3] shows how to collect the divergent parts in a physically irrevelant phase factor, which drops out when squaring the S-matrix element. Instead of using plane waves for describing the charged particle one can employ *Coulomb waves* which include the distortion caused by the $\frac{1}{r}$-potential. The asymptotic form of these distorted waves is

$$\mathrm{e}^{-\mathrm{i}\boldsymbol{p}\cdot\boldsymbol{r}+\mathrm{i}\eta\ln(pr-\boldsymbol{p}\cdot\boldsymbol{r})} = \mathrm{e}^{-\mathrm{i}pr\cos\theta+\mathrm{i}\eta\ln(2pr\sin^2\theta/2)} \quad , \tag{11}$$

where $\eta = Z\alpha/\beta$. The calculation using Coulomb waves in principle is exact, i.e. it is equivalent to summing up all orders of the perturbation series. However, no closed analytical expression can be given for the scattering cross section.

In the nonrelativistic limit ($\beta \to 0$) the exact cross section (in all orders) reduces to Rutherford's result again! It is interesting to study the next order of the expansion in $Z\alpha$. Without presenting the lengthy calculation we just quote the result[4]

$$\frac{d\bar{\sigma}_{e^\mp}}{d\Omega_f} = \frac{Z^2\alpha^2}{4|\boldsymbol{p}|^2\beta^2}\frac{1}{\sin^4\theta/2}\left[1-\beta^2\sin^2\frac{\theta}{2} \pm \pi Z\alpha\beta\sin\frac{\theta}{2}\left(1-\sin\frac{\theta}{2}\right)\right] \quad . \tag{12}$$

The effect of the interference term can be seen immediately. It leads to an increase of electron scattering at the positively charged nucleus (upper sign) and a decrease of positron scattering. As expected the term contains the factor $e(Ze)e^2(Ze)^2 = (Z\alpha)^3$. As an illustration of (12) Fig. 3.7 shows the differential cross section for scattering of electrons (left) and positrons (right) divided by the Rutherford cross section. The charge number is $Z = 40$, for still higher charges the truncation of the series expansion in $Z\alpha$ becomes noticeable, i.e. (12) deviates considerably from the exact results.

[3] R. H. Dalitz: Proc. Roy. Soc. **A206**, 509 (1951).
[4] W.A. McKinley and H. Feshbach: Phys. Rev. **74**, 1759 (1948).

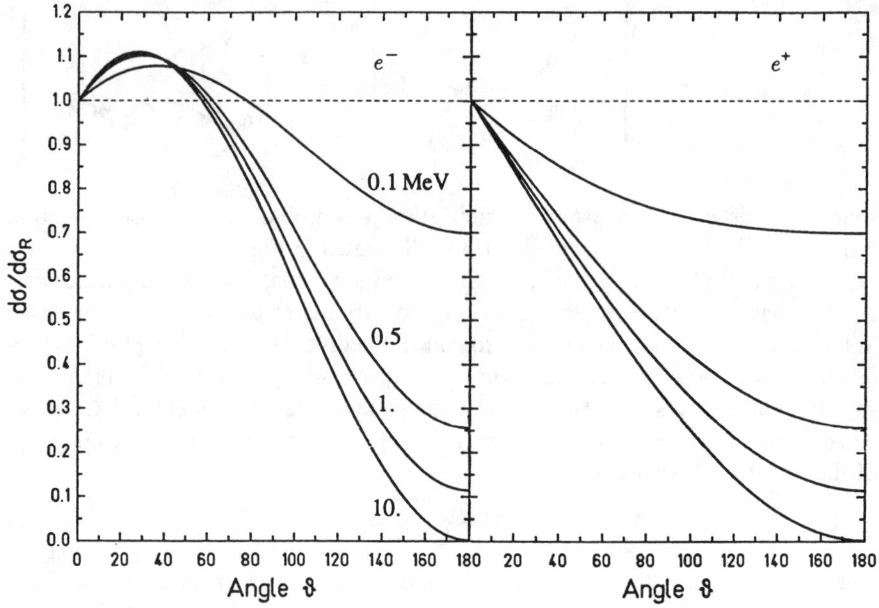

Fig. 3.7. Differential cross section (12) for the scattering of electrons (left) and positrons (right) off a nucleus with charge $Z = 40$, normalized to the Rutherford cross section. The values of the incident kinetic energies are $E_{\text{kin}} = 0.1, 0.5, 1., 10.$ MeV

3.2 Scattering of an Electron off a Free Proton: The Effect of Recoil

In contrast to Sect. 3.1, where the scattering centre was assumed to be fixed, we now consider electron scattering off a freely moveable nucleus. To be specific we choose a proton as the target, i.e. a spin-$\frac{1}{2}$ particle. In a first approximation this will be treated as a structureless Dirac particle. One should expect a result different to the one derived in Sect. 3.1, since now also recoil effects are present.

In order to solve this problem we proceed in two steps. Let us assume we know the proton current $J^\mu(x)$. Then, with the help of the Maxwell equations we can determine the *electromagnetic field* $A^\mu(x)$ produced by the current. This field $A^\mu(x)$ can be inserted into the S-matrix (3.1)

$$S_{fi} = -\mathrm{i}e \int d^4x\, \bar{\psi}_f(x) A\!\!\!/(x) \Psi_i(x) \quad . \tag{3.41}$$

If we replace Ψ_i and ψ_f by the plane Dirac waves, (3.2), we have constructed the scattering amplitude of the electron in the field produced by the proton to lowest order of α. Analogously to Sect. 3.1 this result leads to the lowest-order transition amplitude and cross section.

In the first step we calculate the four-potential $A^\mu(x)$ produced by the Dirac proton. This is achieved by solving the inhomogeneous wave equation with the proton current $J^\mu(x)$ as a source term,

$$\Box A^\mu(x) = 4\pi J^\mu(x) \tag{3.42a}$$

with the wave operator $\Box = \partial_\mu \partial^\mu = \partial^2/\partial t^2 - \nabla^2$. One should keep in mind that for this equation to be valid the *Lorentz gauge* $\partial_\mu A^\mu(x) = 0$ has to be chosen. As we know from classical electrodynamics otherwise the differential operator would be more complicated, namely

$$\Box A^{\mu}(x) - \partial^{\mu}\partial_{\nu}A^{\nu}(x) = 4\pi J^{\mu}(x) \quad . \tag{3.42b}$$

Since we are free to choose the most convenient gauge, the following calculations will be based on (3.42a).[5] The solution of this equation again can be most clearly and systematically formulated by using the appropriate Green's function (propagator) which will be called $D_F(x-y)$. As in the electron case the photon propagator is defined by the equation

$$\Box D_F(x-y) = 4\pi\delta^4(x-y) \quad . \tag{3.43}$$

The Fourier-transformed propagator is defined by

$$D_F(x-y) = \int \frac{d^4q}{(2\pi)^4} \exp\left[-iq\cdot(x-y)\right] D_F(q^2) \quad . \tag{3.44}$$

Using

$$\delta^4(x-y) = \int \frac{d^4q}{(2\pi)^4} \exp\left[-iq\cdot(x-y)\right] \tag{3.45}$$

we obviously get

$$D_F(q^2) = -\frac{4\pi}{q^2} \quad \text{for} \quad q^2 \neq 0 \quad . \tag{3.46}$$

As in the fermionic case (cf. (2.9 – 2.19)) the pole of $D_F(q^2)$ at $q^2 = 0$ has to be treated carefully. As before we add an infinitesimally small positive imaginary number $\mathrm{i}\varepsilon$, i.e. we write[6]

$$D_F(q^2) = -\frac{4\pi}{q^2 + \mathrm{i}\varepsilon} \quad . \tag{3.47}$$

We may also say that we provide the photon with a small negative imaginary mass. This prescription of treating the pole guarantees the causality principle. Photons with positive frequency (i.e. positive energy) only can propagate forward in time. Contributions to the A^{μ}-field which have negative frequency move backward in time. Since the photon (in contrast to the electron) carries no charge and "is its own antiparticle" these two processes are physically identical. There is no need to speak of photons with negative energy.

The causal behaviour of $D_F(q^2)$ (3.47) can be seen mathematically by substituting (3.47) into (3.44). This yields

$$D_F(x-y) = -4\pi \int \frac{d^3q}{(2\pi)^3} \exp\left[+i\boldsymbol{q}\cdot(\boldsymbol{x}-\boldsymbol{y})\right] \int \frac{dq_0}{(2\pi)} \frac{\exp[-iq_0(x_0-y_0)]}{q_0^2 - \boldsymbol{q}^2 + \mathrm{i}\varepsilon} \quad . \tag{3.48}$$

---

[5] Other gauges will be briefly discussed in Chap. 4.

[6] The factor of 4π arises from our use of the Gaussian system of units. When 'rationalized' units are used the numerator in (3.47) is replaced by 1, see Sect. 4.2.

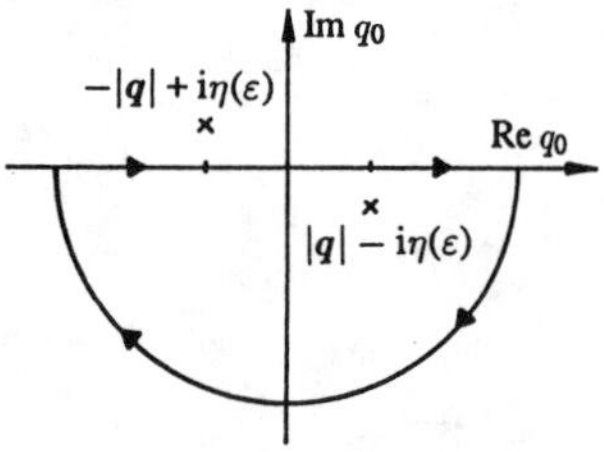

Fig. 3.8. Integration contour in the complex q_0-plane used to evaluate the Feynman propagator in the case $x_0 > y_0$

The path of integration in the complex q_0-plane (see Fig. 3.8) has to be closed in the lower half plane for $x_0 > y_0$, which yields a contribution at

$$q_0 = +|\boldsymbol{q}| \quad . \tag{3.49a}$$

This result simply states that only waves with positive energy ($q_0 > 0$) move into the future (from y_0 to x_0). Similarly for $y_0 > x_0$ the pole at

$$q_0 = -|\boldsymbol{q}| \tag{3.49b}$$

contributes to the photon propagator. This can be interpreted as a positive energy photon ($-q_0 > 0$) moving from x_0 to y_0.

With the help of the Feynman propagator for photons,

$$D_{\mathrm{F}}(x-y) = \int \frac{d^4q}{(2\pi)^4} \exp\left[-\mathrm{i}q\cdot(x-y)\right]\left(\frac{-4\pi}{q^2+\mathrm{i}\varepsilon}\right) \quad , \tag{3.50}$$

the four-potential $A^\mu(x)$ solving (3.42a) is

$$A^\mu(x) = \int d^4y\, D_{\mathrm{F}}(x-y)J^\mu(y) \quad . \tag{3.51a}$$

Note that in more general gauges (3.51a) will be replaced by

$$A^\mu(x) = \int d^4y\, D_{\mathrm{F}}^{\mu\nu}(x-y)J_\nu(y) \quad . \tag{3.51b}$$

In our case the tensor $D_{\mathrm{F}}^{\mu\nu}$ is just proportional to $g^{\mu\nu}$ so that the tensor indices can be discarded for convenience, $D_{\mathrm{F}}^{\mu\nu} = g^{\mu\nu} D_{\mathrm{F}}$.

Using (3.51a) in lowest order the S-matrix element (3.41) is given by

$$S_{fi} = -\mathrm{i}\int d^4x d^4y \left[e\bar{\psi}_f(x)\gamma_\mu\psi_i(x)\right] D_{\mathrm{F}}(x-y)J^\mu(y) \quad . \tag{3.52}$$

The term inside the brackets represents the current of the electron. It is a matrix element of the current operator between an initial and a final state and thus is called a *transition current*.[7] Up to now nothing is known about the proton's current. However, since the electron and the proton play equivalent roles in the scattering process, the proton's current has to be of the same form as the electronic current. Therefore we make the replacement

$$J^\mu(y) \to J_{fi}^\mu(y) = e_p\bar{\psi}_f^{\mathrm{P}}(y)\gamma^\mu\psi_i^{\mathrm{P}}(y) \quad . \tag{3.53}$$

$e_p = -e > 0$ is the proton's charge and $\psi_f^{\mathrm{P}}(y)$ and $\psi_i^{\mathrm{P}}(y)$ are the free final and initial states (i.e. plane Dirac waves) of the proton. They have the same form as the electron waves (3.2) and (3.3):

[7] As a historical remark we mention that already Heisenberg, developing his formulation of nonrelativistic quantum mechanics, used the transition matrix element of the current as a source for $A^\mu(x)$. In particular, he adopted this procedure for describing electronic transitions entering the calculations of atomic spectra in the framework of his matrix mechanics.

$$\psi_i^{\mathrm{p}}(y) = \sqrt{\frac{M_0}{E_i^{\mathrm{p}}V}}\, u(P_i,S_i)\exp(-\mathrm{i}P_i\cdot y) \quad ,$$
$$\psi_f^{\mathrm{p}}(y) = \sqrt{\frac{M_0}{E_f^{\mathrm{p}}V}}\, u(P_f,S_f)\exp\left(-\mathrm{i}P_f\cdot y\right) \quad . \tag{3.54}$$

Here P_i and P_f denote the four-momentum of the proton, S_i, S_f and $E_i^{\mathrm{p}}, E_f^{\mathrm{p}}$ denote its spin and energy, respectively. M_0 is the rest mass of the proton. Thus the proton's transition current can be written in the form

$$J_{fi}^{\mu}(y) = -\sqrt{\frac{M_0^2}{E_f^{\mathrm{p}}E_i^{\mathrm{p}}}}\frac{e}{V}\exp\left[\mathrm{i}(P_f-P_i)\cdot y\right]\bar{u}(P_f,S_f)\gamma^{\mu}u(P_i,S_i) \quad . \tag{3.55}$$

Insertion of the transition current (3.53) or (3.55) into equation (3.41) defines the so-called **Møller** *potential*[8] of the Dirac proton.

Now we insert (3.55) and the analogous expression for the electronic current into (3.52) and get

$$\begin{aligned} S_{fi} = &+\mathrm{i}\frac{e^2}{V^2}\sqrt{\frac{m_0^2}{E_fE_i}}\sqrt{\frac{M_0^2}{E_f^{\mathrm{p}}E_i^{\mathrm{p}}}}\left[\bar{u}(p_f,s_f)\gamma_{\mu}u(p_i,s_i)\right] \\ &\times\int d^4x d^4y\frac{d^4q}{(2\pi)^4}\exp\left[-\mathrm{i}q\cdot(x-y)\right]\exp\left[\mathrm{i}(p_f-p_i)\cdot x\right]\exp\left[\mathrm{i}(P_f-P_i)\cdot y\right] \\ &\times\left(-\frac{4\pi}{q^2+\mathrm{i}\varepsilon}\right)\left[\bar{u}(P_f,S_f)\gamma^{\mu}u(P_i,S_i)\right] \quad . \end{aligned} \tag{3.56}$$

The x- and y-integrations can be performed immediately yielding

$$\int d^4x\exp\left(\mathrm{i}(p_f-p_i-q)\cdot x\right) = (2\pi)^4\delta^4(p_f-p_i-q) \quad ,$$
$$\int d^4y\exp\left(\mathrm{i}(P_f-P_i+q)\cdot y\right) = (2\pi)^4\delta^4(P_f-P_i+q) \quad . \tag{3.57}$$

Now the q-integration is easily done:

$$\begin{aligned} &\int\frac{d^4q}{(2\pi)^4}(2\pi)^4\delta^4(p_f-p_i-q)(2\pi)^4\,\delta^4(P_f-P_i+q)\left[-\frac{4\pi}{q^2+\mathrm{i}\varepsilon}\right] \\ &= (2\pi)^4\delta^4(P_f-P_i+p_f-p_i)\left(-\frac{4\pi}{(p_f-p_i)^2+\mathrm{i}\varepsilon}\right) \quad , \end{aligned} \tag{3.58}$$

and the total S-matrix element (3.56) reads

$$\begin{aligned} S_{fi} = &\frac{-\mathrm{i}e^2}{V^2}(2\pi)^4\delta^4(P_f-P_i+p_f-p_i)\sqrt{\frac{m_0^2}{E_fE_i}}\sqrt{\frac{M_0^2}{E_f^{\mathrm{p}}E_i^{\mathrm{p}}}} \\ &\times\left[\bar{u}(p_f,s_f)\gamma_{\mu}u(p_i,s_i)\right]\frac{4\pi}{(p_f-p_i)^2+\mathrm{i}\varepsilon}\left[\bar{u}(P_f,S_f)\gamma^{\mu}u(P_i,S_i)\right] \quad . \end{aligned} \tag{3.59}$$

We notice that the electron and proton enter this equation in a completely symmetric way. This symmetry is necessary, since there is no physical difference between

[8] C. Møller: Ann. Phys. **14**, 531 (1932).

electron scattering at a proton field and proton scattering at a field generated by the electronic current.

If we compare this expression with the result (3.8) of Sect. 3.1, we recognize the difference between electron scattering at an external Coulomb field and at a proton. It is given by the substitutions

$$\frac{Z\gamma^0}{|\boldsymbol{q}|^2} \Rightarrow \gamma_\mu \left(\frac{-1}{q^2+\mathrm{i}\varepsilon}\right) \sqrt{\frac{M_0^2}{E_f^{\mathrm{p}} E_i^{\mathrm{p}}}}\, \bar{u}(P_f,S_f)\gamma^\mu u(P_i,S_i) \tag{3.60a}$$

and

$$V \Rightarrow (2\pi)^3 \delta^3(\boldsymbol{P}_f - \boldsymbol{P}_i + \boldsymbol{p}_f - \boldsymbol{p}_i) \quad . \tag{3.60b}$$

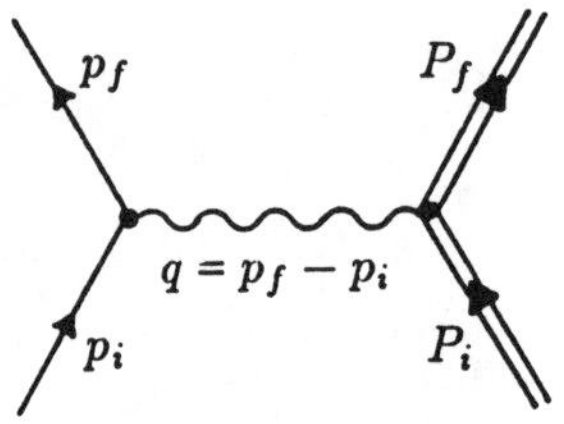

Fig. 3.9. Graph of lowest-order electron–proton scattering. The thin line represents the electron, the double line represents the proton. The wavy line symbolizes the exchanged photon

The last replacement guarantees momentum conservation which was not accounted for in the earlier calculation.

The S-matrix element (3.59) describes electron-proton scattering in lowest order; for higher orders the currents in (3.59) would change. Figure 3.9 shows the process (3.50) graphically. The electromagnetic interaction is expressed by a wavy line. It enters the matrix element (3.59) as the inverse square of the transferred momentum, $1/q^2 = 1/(p_f - p_i)^2$. This can be viewed as the reciprocal □-operator (3.42) in momentum space. We note that the wavy line represents a *virtual photon* being exchanged between electron and proton. The four-momentum of the photon is

$$q = p_f - p_i = P_i - P_f \tag{3.61}$$

(compare the δ-functions in (3.58)). The factor

$$-\frac{4\pi}{q^2+\mathrm{i}\varepsilon} \tag{3.62}$$

in a way represents the amplitude for the propagation of a photon with momentum q. The points where the photon starts and ends are called *vertices* . According to (3.59) the factors $e\gamma_\mu$ act at these points, enclosed between spinors of the form $\sqrt{m_0/E}\, u(p,s)$. The spinors describe the free, ingoing and outgoing Dirac particles, which can be observed as real particles. As we see, each line and vertex of the graph corresponds to a specific factor in the S-matrix element. In addition, the S-matrix element (3.59) contains *the four-dimensional δ-function, ensuring the conservation of total energy and momentum in the scattering process.*

Now we calculate the scattering cross section, beginning with the *transition rate per unit volume.* To that end we divide $|S_{fi}|^2$ by the time interval T of observation and by the space volume of the reaction (normalization volume of Dirac waves), which gives

$$W_{fi} = \frac{|S_{fi}|^2}{VT} = \frac{[(2\pi)^4\delta^4(P_f + p_f - P_i - p_i)]^2}{VT} \frac{1}{V^4} \frac{m_0^2}{E_f E_i} \frac{M_0^2}{E_f^{\mathrm{p}} E_i^{\mathrm{p}}} |M_{fi}|^2 \quad . \tag{3.63}$$

Here,

$$M_{fi} = \left[\bar{u}(p_f,s_f)\gamma_\mu u(p_i,s_i)\right] \frac{4\pi e e_{\mathrm{p}}}{q^2+\mathrm{i}\varepsilon} \left[\bar{u}(P_f,S_f)\gamma^\mu u(P_i,S_i)\right] \tag{3.64}$$

is the so-called *invariant amplitude*. The choice of this name is quite natural since the matrix element (3.64) consists of a scalar product of 4-vectors which is *Lorentz invariant.*

As in Sect. 3.1 we have to consider the square of the δ^4-function. Again we will make use of the relation (3.15a)

$$(2\pi\delta(E_f - E_i))^2 = 2\pi\delta(0)\, 2\pi\delta(E_f - E_i) \Rightarrow T\, 2\pi\, \delta(E_f - E_i) \quad . \tag{3.65}$$

This is valid for the one-dimensional δ-function. The four-dimensional δ-function

$$\delta^4(x-y) = \delta(x_0 - y_0)\, \delta(x_1 - y_1)\, \delta(x_2 - y_2)\, \delta(x_3 - y_3) \tag{3.66}$$

is just the product of four one-dimensional δ-functions. By denoting the time and spatial intervals by T and L – or $L^3 = V$, respectively – the following four-dimensional generalization of (3.65) suggests itself:

$$\begin{aligned}[(2\pi)^4\delta^4(p_f - p_i)]^2 &= (2\pi)^4\delta^4(0)(2\pi)^4\delta^4(p_f - p_i) = (2\pi)^4 T\, L^3\, \delta^4(p_f - p_i) \\ &\Rightarrow T\, V\, (2\pi)^4\delta^4(p_f - p_i) \quad . \end{aligned} \tag{3.67}$$

In the case of the δ^4-function occurring in (3.63) we get

$$\begin{aligned}[(2\pi)^4\delta^4(P_f + p_f - P_i - p_i)]^2 &= (2\pi)^4\delta^4(0)\, (2\pi)^4\, \delta^4(P_f + p_f - P_i - p_i) \\ &\Rightarrow T\, V\, (2\pi)^4\delta^4(P_f + p_f - P_i - p_i) \quad . \end{aligned} \tag{3.68}$$

With that result the transition rate per unit volume (3.63) reads

$$W_{fi} = (2\pi)^4\delta^4(P_f + p_f - P_i - p_i)\frac{1}{V^4}\frac{m_0^2}{E_f E_i}\frac{M_0^2}{E_f^{\mathrm{p}} E_i^{\mathrm{p}}}\, |M_{fi}|^2 \quad . \tag{3.69}$$

In order to determine the cross section we have to divide W_{fi} by the *flux of the incoming particles* $|\boldsymbol{J}_{\mathrm{inc.}}|$ and by *the number of target particles per unit volume.*

The latter is given by $1/V$, since the normalization of the wave functions was performed in such a way that there is just one particle in the normalization volume V. Furthermore we have to sum (integrate) over the possible final states of the electron and the proton to obtain the cross section. For a given spin the number of final states in the momentum interval $d^3p_f d^3P_f$ is given by

$$V^2\frac{d^3p_f}{(2\pi)^3}\frac{d^3P_f}{(2\pi)^3} \tag{3.70}$$

(cf. the discussion of Sect. 3.1, (3.9)). Now we can write down the sixfold differential scattering cross section

$$\begin{aligned} d\sigma &= V^2\frac{d^3p_f}{(2\pi)^3}\frac{d^3P_f}{(2\pi)^3}\frac{1}{|\boldsymbol{J}_{\mathrm{inc.}}|}\frac{1}{1/V}W_{fi} \\ &= \left(\frac{m_0}{E_i}\frac{M_0}{E_i^{\mathrm{p}}}\frac{1}{|\boldsymbol{J}_{\mathrm{inc.}}|V}\right)|M_{fi}|^2(2\pi)^4\delta^4(P_f + p_f - P_i - p_i) \\ &\quad\times\frac{m_0}{E_f}\frac{d^3p_f}{(2\pi)^3}\frac{M_0}{E_f^{\mathrm{p}}}\frac{d^3P_f}{(2\pi)^3} \quad . \end{aligned} \tag{3.71}$$

To compare with measurements (3.71) has to be integrated over an appropriate range of the momentum variables determined by the experimental setup. E.g., if only the scattered electron is observed, one has to integrate over all values of the proton momentum d^3P_f. In the cross section (3.71) the initial and final polarizations (spin directions) of the scattering particles are fixed. This can be seen directly in the invariant amplitude (3.64). If polarizations are not measured, i.e. if one determines the cross section using an unpolarized beam and detectors not sensitive to polarization, one has to sum over the final spin states and to average over the spin states of the initial particles. The expression (3.71) for the cross section exhibits some general features, which are worth discussing. These general features are common to all scattering processes.

The square of the invariant amplitudes $|M_{fi}|$ incorporates the essential physics of the process. The conservation of total energy and momentum is guaranteed by the factor $(2\pi)^4\delta^4(P_f + p_f - P_i - p_i)$. Furthermore there are exactly four factors of the type m_0/E. In general there occurs a factor m_0/E for every external fermion line of the corresponding graph of the process. Since these factors result from the Dirac particles involved in the process (compare e.g. (3.54)), every Dirac particle entering and leaving the interaction yields such a factor in the cross section. In addition each particle yields a *phase-space factor* $d^3p_f/(2\pi)^3$. We can say *each particle leaving the scattering contributes a factor*

$$\frac{m_0}{E_f}\frac{d^3p_f}{(2\pi)^3} \tag{3.72}$$

to the cross section. This factor is Lorentz-invariant. It is just the three-dimensional Lorentz-invariant volume in momentum space which can be written in four-dimensional form as

$$\frac{d^3p}{2E} = \int_0^\infty dp_0\; \delta(p^2 - m_0^2)\; d^3p \quad . \tag{3.73}$$

This is derived as follows

$$\begin{aligned}\int_0^\infty dp_0\, \delta(p_0^2 - \boldsymbol{p}^2 - m_0^2)\; d^3p &= \int_0^\infty dp_0\, \delta(p_0^2 - E^2)\; d^3p \\ &= \int_0^\infty dp_0\, \delta\left[(p_0 - E)(p_0 + E)\right]\; d^3p = \frac{d^3p}{2E} \quad .\end{aligned}$$

Here we used the well-known formula $\int dx\; \delta(f(x)) = \sum_k 1/|\frac{df}{dx}|_{x_k}$, x_k being the roots of $f(x)$ within the interval of integration. The right-hand side of (3.73) can further be transformed to

$$\frac{d^3p}{2E} = \int_{-\infty}^{\infty} d^4p\; \delta(p^2 - m_0^2)\Theta(p_0) \quad , \tag{3.74}$$

with

$$\Theta(p_0) = \begin{cases} 1 & \text{for} \quad p_0 > 0 \\ 0 & \text{for} \quad p_0 < 0 \end{cases}$$

being the Lorentz-invariant step function with respect to energy. $\Theta(p_0)$ is Lorentz-invariant since Lorentz transformations always transform time-like four-vectors (like p^μ) into time-like vectors, and correspondingly for space-like vectors. In our case p is a time-like four-vector in the forward light cone (because $p^2 = m_0^2$ and therefore $p_0^2 > \boldsymbol{p}^2$) independent of the specific Lorentz frame. Thus it is obvious that because of (3.74) $d^3p/2E$ is a Lorentz-invariant factor.

Now we have to consider the factor in brackets in (3.71). The flux is given by the number of particles per unit area that come together in a unit of time. Denoting the velocities of electrons $\boldsymbol{v}_i$ and protons $\boldsymbol{V}_i$ we see that (cf. Fig. 3.10)

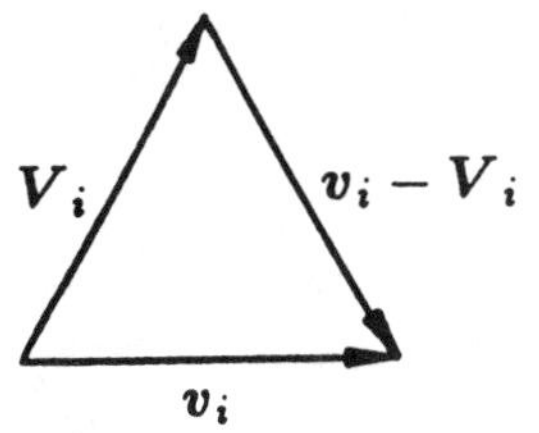

Fig. 3.10. The relative velocity $\boldsymbol{v}_i - \boldsymbol{V}_i$ is the relevant quantity determining the incoming particle current

$$|\boldsymbol{J}_{\text{inc.}}| = \frac{1}{V}|\boldsymbol{v}_i - \boldsymbol{V}_i| = \text{ particle density} \times \text{relative velocity} \quad . \tag{3.75}$$

We will now show that the factor $1/V|\boldsymbol{J}_{\text{inc.}}|$ when combined with the remaining factors m_0/E_i, M_0/E_i^{p} is nearly – but not exactly – Lorentz-invariant. However, the correct Lorentz-invariant flux factor can then be guessed easily. With

$$\boldsymbol{v}_i = \frac{\boldsymbol{p}_i}{E_i} \qquad \text{and} \qquad \boldsymbol{V}_i = \frac{\boldsymbol{P}_i}{E_i^{\text{p}}}$$

the intuitive expression (3.75) leads to

$$\begin{aligned}\frac{m_0}{E_i}\frac{M_0}{E_i^{\text{p}}}\frac{1}{V|\boldsymbol{J}_{\text{inc.}}|} &= \frac{m_0 M_0}{E_i E_i^{\text{p}}|\boldsymbol{v}_i - \boldsymbol{V}_i|} = \frac{m_0 M_0}{E_i E_i^{\text{p}}\sqrt{\boldsymbol{v}_i^2 + \boldsymbol{V}_i^2 - 2\boldsymbol{v}_i \cdot \boldsymbol{V}_i}} \\ &= \frac{m_0 M_0}{\sqrt{\boldsymbol{p}_i^2 E_i^{\text{p}2} + \boldsymbol{P}_i^2 E_i^2 - 2\boldsymbol{p}_i \cdot \boldsymbol{P}_i E_i E_i^{\text{p}}}} \quad . \end{aligned} \tag{3.76}$$

As we will see in the following this result is nearly identical to the Lorentz scalar

$$\frac{m_0 M_0}{\sqrt{(p_i \cdot P_i)^2 - m_0^2 M_0^2}} \quad ,$$

because

$$\begin{aligned}\frac{m_0 M_0}{\sqrt{(p_i \cdot P_i)^2 - m_0^2 M_0^2}} &= \frac{m_0 M_0}{\sqrt{(E_i E_i^{\text{p}} - \boldsymbol{p}_i \cdot \boldsymbol{P}_i)^2 - m_0^2 M_0^2}} \\ &= \frac{m_0 M_0}{\sqrt{E_i^2 E_i^{\text{p}2} - 2E_i E_i^{\text{p}}\, \boldsymbol{p}_i \cdot \boldsymbol{P}_i + (\boldsymbol{p}_i \cdot \boldsymbol{P}_i)^2 - m_0^2 M_0^2}} \\ &= \frac{m_0 M_0}{\sqrt{(m_0^2 + \boldsymbol{p}_i^2)(M_0^2 + \boldsymbol{P}_i^2) - 2E_i E_i^{\text{p}}\, \boldsymbol{p}_i \cdot \boldsymbol{P}_i + (\boldsymbol{p}_i \cdot \boldsymbol{P}_i)^2 - m_0^2 M_0^2}} \\ &= \frac{m_0 M_0}{\sqrt{\boldsymbol{p}_i^2 E_i^{\text{p}2} + m_0^2 \boldsymbol{P}_i^2 - 2E_i E_i^{\text{p}}\, \boldsymbol{p}_i \cdot \boldsymbol{P}_i + (\boldsymbol{p}_i \cdot \boldsymbol{P}_i)^2}} \\ &\approx \frac{m_0 M_0}{\sqrt{\boldsymbol{p}_i^2 E_i^{\text{p}2} + m_0^2 \boldsymbol{P}_i^2 - 2E_i E_i^{\text{p}}\, \boldsymbol{p}_i \cdot \boldsymbol{P}_i + \boldsymbol{p}_i^2 \cdot \boldsymbol{P}_i^2}} \\ &= \frac{m_0 M_0}{\sqrt{\boldsymbol{p}_i^2 E_i^{\text{p}2} + \boldsymbol{P}_i^2 E_i^2 - 2E_i E_i^{\text{p}}\, \boldsymbol{p}_i \cdot \boldsymbol{P}_i}} \quad . \end{aligned} \tag{3.77}$$

In the last step but one we had to assume that $(\boldsymbol{p}_i \cdot \boldsymbol{P}_i)^2 = \boldsymbol{p}_i^2 \boldsymbol{P}_i^2$ which requires that the velocity vectors are collinear. Thus we have deduced the relation

$$\frac{m_0 M_0}{E_i E_i^{\mathrm{p}} |\boldsymbol{v}_i - \boldsymbol{V}_i|} = \frac{m_0 M_0}{\sqrt{(p_i \cdot P_i)^2 - m_0^2 M_0^2}} \tag{3.78}$$

which is only valid for *collinear collisions*. However, Lorentz invariance has the higher priority. Consequently the naive flux factor (3.76) in the cross section (3.71) has in general to be replaced by the Lorentz-invariant flux factor (3.77). In the case of collinear collisions both results are identical.

By using the just-derived Lorentz-invariant flux factor the total cross section (3.71) becomes Lorentz-invariant. We write it in an *invariant form*:

$$d\sigma = \frac{m_0 M_0}{\sqrt{(p_i \cdot P_i)^2 - m_0^2 M_0^2}} |M_{fi}|^2 (2\pi)^4 \, \delta^4(P_f - P_i + p_f - p_i) \frac{m_0 d^3 p_f}{(2\pi)^3 E_f} \frac{M_0 d^3 P_f}{(2\pi)^3 E_f^{\mathrm{p}}} \,. \tag{3.79}$$

Every factor that occurs has a general meaning and has always to appear in this form: the first factor represents the reciprocal incoming particle flux per unit area and time, the second is the squared invariant amplitude (which describes the physics involved beyond pure kinematics), the third incorporates energy and momentum conservation, and the last factor describes the densities of the final states. Note that the normalization volume V in the final result (3.79) has disappeared, as it should.

In a short excursion we shall discuss how to treat noncollinear collisions, which, for instance, occur in scattering processes in a plasma. In this case it is most convenient to calculate the *number of events per unit time dN/dt* directly. Using (3.69) we get

$$\begin{aligned} \frac{dN}{dt} &= \int d^3x \; \varrho_e(\boldsymbol{x}, t) \varrho_p(\boldsymbol{x}, t) \\ &\times \int \frac{m_0}{E_i} \frac{M_0}{E_i^{\mathrm{p}}} |M_{fi}|^2 \, (2\pi)^4 \, \delta^4(P_f - P_i + p_f - p_i) \frac{m_0 d^3 p_f}{(2\pi)^3 E_f} \frac{M_0 d^3 P_f}{(2\pi)^3 E_f^{\mathrm{p}}} \,. \end{aligned} \tag{3.80}$$

$\varrho_{\mathrm{e}}(\boldsymbol{x}, t)$ and $\varrho_{\mathrm{p}}(\boldsymbol{x}, t)$ denote the densities of the electrons and protons, respectively. They both contain a factor $1/V$, while two additional powers of $1/V$ have cancelled after multiplication with the final state densities (3.70).

As mentioned before we have to sum over the final states and to average over the initial ones if we do not consider polarization effects. Since in (3.79) the spin quantum numbers occur solely in the squared invariant amplitude, we define the average squared invariant amplitude

$$\overline{|M_{fi}|^2} = \frac{1}{4} \sum_{S_f, S_i, s_f, s_i} \left| \bar{u}(p_f, s_f) \gamma^\mu u(p_i, s_i) \frac{e e_{\mathrm{p}} (4\pi)}{q^2 + \mathrm{i}\varepsilon} \bar{u}(P_f, S_f) \gamma_\mu u(P_i, S_i) \right|^2 \,. \tag{3.81}$$

This expression can be calculated in the same way as (3.25) in Sect. 3.1, yet one has to take into account that according to Einstein's sum convention (3.81) contains a sum over μ which has to be squared. Therefore we cannot directly take over (3.33). We discuss the right-hand side of (3.81) in more detail by explicitly distinguishing the terms that occur. Terms with the form adjoint spinor $\times$ matrix $\times$ spinor are complex numbers for which the operations of complex conjugation and taking the adjoint are identical:

$$\begin{aligned}
\left[\bar{u}(p_f,s_f)\gamma^\mu u(p_i,s_i)\right]^* &= \left[\bar{u}(p_f,s_f)\gamma^\mu u(p_i,s_i)\right]^\dagger = \left[u^\dagger(p_f,s_f)\gamma^0\gamma^\mu u(p_i,s_i)\right]^\dagger \\
&= u^\dagger(p_i,s_i)\gamma^{\mu\dagger}\gamma^{0\dagger}u^{\dagger\dagger}(p_f,s_f) = \bar{u}(p_i,s_i)\gamma^0\gamma^{\mu\dagger}\gamma^0 u(p_f,s_f) \\
&= \bar{u}(p_i,s_i)\gamma^\mu u(p_f,s_f) = \sum_{\delta,\varepsilon=1}^{4} \bar{u}_\delta(p_i,s_i)\,\gamma^\mu_{\delta\varepsilon}\,u_\varepsilon(p_f,s_f) \quad .
\end{aligned} \tag{3.82}$$

This yields for the spin sum in (3.81)

$$\begin{aligned}
&\sum_{S_f,S_i,s_f,s_i} \left|\sum_{\mu=0}^{3} \left[\bar{u}(p_f,s_f)\gamma^\mu u(p_i,s_i)\right]\left[\bar{u}(P_f,S_f)\gamma_\mu u(P_i,S_i)\right]\right|^2 \\
&= \sum_{S_f,S_i,s_f,s_i} \left\{\sum_{\mu=0}^{3} \left[\bar{u}(p_f,s_f)\gamma^\mu u(p_i,s_i)\right]\left[\bar{u}(P_f,S_f)\gamma_\mu u(P_i,S_i)\right]\right\} \\
&\quad \times \left\{\sum_{\nu=0}^{3} \left[\bar{u}(p_f,s_f)\gamma^\nu u(p_i,s_i)\right]^*\left[\bar{u}(P_f,S_f)\gamma_\nu u(P_i,S_i)\right]^*\right\} \\
&= \sum_{\mu,\nu=0}^{3}\sum_{S_f,S_i,s_f,s_i} \left[\bar{u}(p_f,s_f)\gamma^\mu u(p_i,s_i)\right]\left[\bar{u}(p_f,s_f)\gamma^\nu u(p_i,s_i)\right]^\dagger \\
&\quad \times \left[\bar{u}(P_f,S_f)\gamma_\mu u(P_i,S_i)\right]\left[\bar{u}(P_f,S_f)\gamma_\nu u(P_i,S_i)\right]^\dagger \\
&= \sum_{\mu,\nu=0}^{3}\sum_{S_f,S_i,s_f,s_i}\sum_{\alpha,\beta,\delta,\varepsilon=1}^{4} \bar{u}_\alpha(p_f,s_f)\gamma^\mu_{\alpha\beta}u_\beta(p_i,s_i)\,\bar{u}_\delta(p_i,s_i)\,\gamma^\nu_{\delta\varepsilon}\,u_\varepsilon(p_f,s_f) \\
&\quad \times \sum_{\varrho,\sigma,\tau,\lambda=1}^{4} \bar{u}_\varrho(P_f,S_f)\gamma_{\mu\,\varrho\sigma}u_\sigma(P_i,S_i)\,\bar{u}_\tau(P_i,S_i)\gamma_{\nu\,\tau\lambda}\,u_\lambda(P_f,S_f) \\
&= \sum_{\mu,\nu=0}^{3}\sum_{\alpha,\beta,\delta,\varepsilon=1}^{4}\sum_{s_f}\bar{u}_\alpha(p_f,s_f)\gamma^\mu_{\alpha\beta}\underbrace{\sum_{s_i}u_\beta(p_i,s_i)\bar{u}_\delta(p_i,s_i)}_{\left(\frac{\not p_i+m_0}{2m_0}\right)_{\beta\delta}}\ \gamma^\nu_{\delta\varepsilon}\,u_\varepsilon(p_f,s_f) \\
&\quad \times \sum_{\varrho,\sigma,\tau,\lambda=1}^{4}\sum_{S_f}\bar{u}_\varrho(P_f,S_f)\gamma_{\mu\,\varrho\sigma}\underbrace{\sum_{S_i}u_\sigma(P_i,S_i)\bar{u}_\tau(P_i,S_i)}_{\left(\frac{\not P_i+M_0}{2M_0}\right)_{\sigma\tau}}\ \gamma_{\nu\,\tau\lambda}\,u_\lambda(P_f,S_f) \\
&= \sum_{\mu,\nu=0}^{3}\sum_{\alpha,\varepsilon=1}^{4}\sum_{s_f}\bar{u}_\alpha(p_f,s_f)u_\varepsilon(p_f,s_f)\left(\gamma^\mu\frac{\not p_i+m_0}{2m_0}\gamma^\nu\right)_{\alpha\varepsilon} \\
&\quad \times \sum_{\varrho,\lambda=1}^{4}\sum_{S_f}\bar{u}_\varrho(P_f,S_f)u_\lambda(P_f,S_f)\left(\gamma_\mu\frac{\not P_i+M_0}{2M_0}\gamma_\nu\right)_{\varrho\lambda}
\end{aligned}$$

$$= \sum_{\mu,\nu=0}^{3} \sum_{\alpha,\varepsilon=1}^{4} \left(\frac{\not p_f + m_0}{2m_0}\right)_{\varepsilon\alpha}$$
$$\times \left(\gamma^\mu \frac{\not p_i + m_0}{2m_0} \gamma^\nu\right)_{\alpha\varepsilon} \sum_{\varrho,\lambda=1}^{4} \left(\frac{\not P_f + M_0}{2M_0}\right)_{\lambda\varrho} \left(\gamma_\mu \frac{\not P_i + M_0}{2M_0} \gamma_\nu\right)_{\varrho\lambda}$$
$$= \sum_{\mu,\nu=0}^{3} \mathrm{Tr}\left[\frac{\not p_f + m_0}{2m_0} \gamma^\mu \frac{\not p_i + m_0}{2m_0} \gamma^\nu\right] \mathrm{Tr}\left[\frac{\not P_f + M_0}{2M_0} \gamma_\mu \frac{\not P_i + M_0}{2M_0} \gamma_\nu\right] \ . \tag{3.83}$$

This calculation, which we have spelled out in great detail, thus leads to the following result for the averaged squared matrix element

$$\overline{|M_{fi}|^2} = \frac{1}{4} \frac{e^2 e_\mathrm{p}^2 (4\pi)^2}{(q^2)^2} \mathrm{Tr}\left[\frac{\not p_f + m_0}{2m_0} \gamma^\mu \frac{\not p_i + m_0}{2m_0} \gamma^\nu\right]$$
$$\times \mathrm{Tr}\left[\frac{\not P_f + M_0}{2M_0} \gamma_\mu \frac{\not P_i + M_0}{2M_0} \gamma_\nu\right] \ . \tag{3.84}$$

Note that the squaring of the amplitude which contained the scalar product of two Lorentz scalars has lead to the contraction of two tensors, i.e. a double sum. One often abbreviates this as

$$\overline{|M_{fi}|^2} = \frac{e^2 e_\mathrm{p}^2 (4\pi)^2}{(q^2)^2} L^{\mu\nu} H_{\mu\nu} \quad , \tag{3.85}$$

where $L^{\mu\nu}$ is the *lepton* (i.e. electron) *tensor* and $H_{\mu\nu}$ the *hadron* (i.e. proton) *tensor*.

$$L^{\mu\nu} = \frac{1}{2} \sum_{s_i s_f} \bar u(p_f, s_f) \gamma^\mu u(p_i, s_i)\ \bar u(p_i, s_i) \gamma^\nu u(p_f, s_f)$$
$$= \frac{1}{2} \mathrm{Tr}\left[\frac{\not p_f + m_0}{2m_0} \gamma^\mu \frac{\not p_i + m_0}{2m_0} \gamma^\nu\right] \quad , \tag{3.86a}$$

and similarly

$$H_{\mu\nu} = \frac{1}{2} \mathrm{Tr}\left[\frac{\not P_f + M_0}{2M_0} \gamma_\mu \frac{\not P_i + M_0}{2M_0} \gamma_\nu\right] \ . \tag{3.86b}$$

The factorisation of (3.85) remains meaningful as long as a single virtual photon is exchanged in the scattering process, even if the transition currents becomes more complicated than those used in (3.86). By applying the theorems derived in Mathematical Supplement 3.3 we can quite easily evaluate the lepton tensor (3.86a) since traces of odd numbers of γ-matrices vanish it immediately is simplified to

$$L^{\mu\nu} = \frac{1}{2} \frac{1}{4m_0^2} \mathrm{Tr}\left[\not p_f \gamma^\mu \not p_i \gamma^\nu + m_0^2\, \gamma^\mu \gamma^\nu\right] \ . \tag{3.87}$$

This result can be written in the 'slash' notation used by the above-mentioned theorems. We introduce two unit four-vectors with a 1 in the component μ and ν respectively. The other components are zero:

$$A = (0, \underbrace{1}_{\text{comp. } \mu}, 0, 0) \quad \text{and} \quad B = (0, 0, \underbrace{1}_{\text{comp. } \nu}, 0) \quad .$$

We can now write

$$\begin{aligned}
\mathrm{Tr}\, \not{p}_f \gamma^\mu \not{p}_i \gamma^\nu &= \mathrm{Tr}\, \not{p}_f \not{A} \not{p}_i \not{B} \\
&= p_f \cdot A\, \mathrm{Tr}\, \not{p}_i \not{B} - p_f \cdot p_i\, \mathrm{Tr}\, \not{A}\not{B} + p_f \cdot B\, \mathrm{Tr}\, \not{A}\not{p}_i \\
&= 4 p_f \cdot A\, p_i \cdot B - 4 p_f \cdot p_i\, A \cdot B + 4 p_f \cdot B\, A \cdot p_i \\
&= 4 p_f^\mu p_i^\nu - 4 p_f \cdot p_i\, g^{\mu\nu} + 4 p_f^\nu p_i^\mu \quad ,
\end{aligned} \tag{3.88}$$

and

$$\mathrm{Tr}\, \gamma^\mu \gamma^\nu = \mathrm{Tr}\, \not{A}\not{B} = 4 A \cdot B = 4 g^{\mu\nu} \quad . \tag{3.89}$$

Then the lepton tensor (3.87) reads

$$L^{\mu\nu} = \frac{1}{2} \frac{1}{m_0^2} \left[p_f^\mu p_i^\nu + p_i^\mu p_f^\nu - g^{\mu\nu}(p_f \cdot p_i - m_0^2) \right] \tag{3.90}$$

and analogously for the hadron tensor $H_{\mu\nu}$ (3.86b). In this case we just have to replace the small letters by capitals and exchange covariant by contravariant indices. Multiplying both and summing over μ and ν yields the squared invariant amplitude (see (3.85))

$$\begin{aligned}
\overline{|M_{fi}|^2} &= \frac{e^2 e_\mathrm{p}^2 (4\pi)^2}{4 m_0^2 M_0^2 (q^2)^2} \left[p_f^\mu p_i^\nu + p_i^\mu p_f^\nu - g^{\mu\nu}(p_f \cdot p_i - m_0^2) \right] \\
&\quad \times \left[P_{f\mu} P_{i\nu} + P_{i\mu} P_{f\nu} - g_{\mu\nu}(P_f \cdot P_i - M_0^2) \right] \\
&= \frac{e^2 e_\mathrm{p}^2 (4\pi)^2}{4 m_0^2 M_0^2 (q^2)^2} \Big[(p_f \cdot P_f)(p_i \cdot P_i) + (p_f \cdot P_i)(p_i \cdot P_f) \\
&\quad + (p_f \cdot P_i)(p_i \cdot P_f) + (p_i \cdot P_i)(p_f \cdot P_f) - 2(p_i \cdot p_f)(P_f \cdot P_i - M_0^2) \\
&\quad - 2(P_i \cdot P_f)(p_f \cdot p_i - m_0^2) + 4(p_f \cdot p_i - m_0^2)(P_f \cdot P_i - M_0^2) \Big] \\
&= \frac{e^2 e_\mathrm{p}^2 (4\pi)^2}{4 m_0^2 M_0^2 (q^2)^2} \Big[2(p_i \cdot P_i)(p_f \cdot P_f) + 2(p_i \cdot P_f)(p_f \cdot P_i) \\
&\quad - 4(p_i \cdot p_f)(P_i \cdot P_f) + 2(p_i \cdot p_f) M_0^2 + 2(P_i \cdot P_f) m_0^2 + 4(p_i \cdot p_f)(P_i \cdot P_f) \\
&\quad + 4 m_0^2 M_0^2 - 4(p_i \cdot p_f) M_0^2 - 4(P_i \cdot P_f) m_0^2 \Big] \\
&= \frac{e^2 e_\mathrm{p}^2 (4\pi)^2}{2 m_0^2 M_0^2 (q^2)^2} \Big[(p_i \cdot P_i)(p_f \cdot P_f) + (p_i \cdot P_f)(p_f \cdot P_i) \\
&\quad - (p_i \cdot p_f) M_0^2 - (P_i \cdot P_f) m_0^2 + 2 m_0^2 M_0^2 \Big] \quad .
\end{aligned} \tag{3.91}$$

This average squared invariant amplitude has to be inserted into (3.79). To evaluate the scattering cross section any further the frame of reference has to be specified. Usually calculations take their simplest form in the centre-of-mass system. However, electron-proton scattering experiments mostly are performed using a fixed target in the laboratory frame. Therefore we will evaluate (3.79) *in the rest frame of the incoming proton*. We define

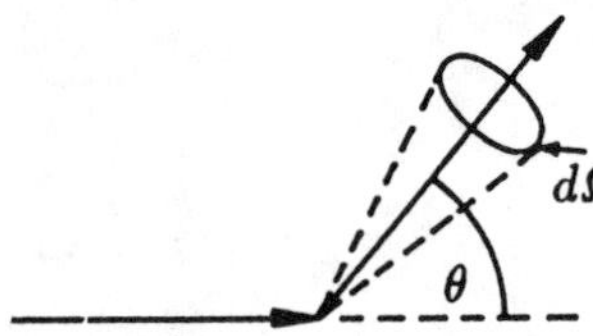

Fig. 3.11. The electron is scattered into a solid-angle element $d\Omega'$

$$\begin{aligned} p_f &= (E',\boldsymbol{p}') \equiv p' \quad , \\ p_i &= (E,\boldsymbol{p}) \equiv p \quad , \\ P_i &= (M_0,\boldsymbol{0}) \quad . \end{aligned} \tag{3.92}$$

We want to calculate the differential cross section for electron scattering into a given solid-angle element $d\Omega'$ centered around the scattering angle θ (cf. Fig. 3.11). Therefore the differential quantity (3.79) has to be integrated over all momentum variables except for the direction of $\boldsymbol{p}_f$. The volume element can be written as

$$d^3p_f = d^3p' = |\boldsymbol{p}'|^2\, d|\boldsymbol{p}'|\, d\Omega' = |\boldsymbol{p}'|\, E'\, dE'\, d\Omega' \tag{3.93}$$

because $E'^2 = \boldsymbol{p}'^2 + m_0^2$ and thus $|\boldsymbol{p}'|d|\boldsymbol{p}'| = E'dE'$. Using (3.92) the invariant flux factor reduces to

$$\frac{m_0 M_0}{\sqrt{(p_i\cdot P_i)^2 - m_0^2 M_0^2}} = \frac{m_0 M_0}{\sqrt{E^2 M_0^2 - m_0^2 M_0^2}} = \frac{m_0}{\sqrt{E^2 - m_0^2}} = \frac{m_0}{|\boldsymbol{p}|} \quad , \tag{3.94}$$

and with the help of (3.74) we get

$$\begin{aligned} d\bar{\sigma} &= \frac{m_0}{|\boldsymbol{p}|}\, \overline{|M_{fi}|^2}\, (2\pi)^4\, \delta^4(P_f + p' - P_i - p) \\ &\quad \times \frac{m_0}{(2\pi)^3} |\boldsymbol{p}'|\, dE'\, d\Omega'\, \frac{2M_0}{(2\pi)^3}\, d^4P_f\, \delta(P_f^2 - M_0^2)\, \Theta(P_f^0) \quad . \end{aligned} \tag{3.95}$$

Thus by integrating over dE' and d^4P_f the differential cross section becomes

$$\begin{aligned} \frac{d\bar{\sigma}}{d\Omega'} &= \frac{2}{|\boldsymbol{p}|} \int \frac{m_0^2 M_0 |\boldsymbol{p}'| dE'}{(2\pi)^2}\, \overline{|M_{fi}|^2}\, d^4P_f\, \delta(P_f^2 - M_0^2)\, \Theta(P_f^0)\, \delta^4(P_f + p' - P_i - p) \\ &= \frac{2m_0^2 M_0}{|\boldsymbol{p}| 4\pi^2} \int |\boldsymbol{p}'|\, dE'\, \overline{|M_{fi}|^2}\, \delta\big((p' - P_i - p)^2 - M_0^2\big)\, \Theta\,(P_i^0 + E - E') \\ &= \frac{m_0^2 M_0}{|\boldsymbol{p}| 2\pi^2} \int_{m_0}^{M_0+E} |\boldsymbol{p}'|\, dE'\, \overline{|M_{fi}|^2}\, \delta\big(2m_0^2 - 2(E' - E)M_0 - 2E'E + 2|\boldsymbol{p}||\boldsymbol{p}'|\cos\theta\big) \,. \end{aligned} \tag{3.96}$$

In the last step we used the fact that the energy E' of the scattered electron is bounded by $E' \leq M_0 + E$ because of the step function. The argument of the step function has to be larger than zero, otherwise the integrand vanishes, i.e. $E' \leq P_i^0 + E = M_0 + E$. Of course E' also has to be larger or equal to m_0. Furthermore the argument of the δ-function in (3.96) was expressed in terms in the kinematical variables in the laboratory frame:

$$\begin{aligned} (p' - P_i - p)^2 - M_0^2 &= p'^2 + P_i^2 + p^2 - 2p'\cdot P_i - 2p'\cdot p + 2P_i\cdot p - M_0^2 \\ &= m_0^2 + M_0^2 + m_0^2 - 2E'M_0 - 2(E'E - \boldsymbol{p}'\cdot\boldsymbol{p}) + 2M_0E - M_0^2 \\ &= 2m_0^2 - 2M_0(E' - E) - 2E'E + 2|\boldsymbol{p}||\boldsymbol{p}'|\cos\theta \quad . \end{aligned} \tag{3.97}$$

The remaining integral over E' in (3.96) can be solved by using the familiar formula

$$\delta\big(f(x)\big) = \sum_k \frac{\delta(x - x_k)}{|\frac{df}{dx}|_{x_k}} \quad , \tag{3.98}$$

x_i being the roots of $f(x)$ contained in the interval of integration. Thus we get

$$\frac{d\bar{\sigma}}{d\Omega'} = \frac{m_0^2 M_0}{4\pi^2} \frac{|\boldsymbol{p}'|}{|\boldsymbol{p}|} \frac{\overline{|M_{fi}|^2}}{M_0 + E - |\boldsymbol{p}|(E'/|\boldsymbol{p}'|)\cos\theta} \quad , \tag{3.99}$$

where we have used $|\boldsymbol{p}'|d|\boldsymbol{p}'| = E'dE'$ The argument of the δ-function in (3.96) leads to the following condition for energy conservation:

$$E'(M_0 + E) \; - \; |\boldsymbol{p}||\boldsymbol{p}'|\cos\theta \; = \; EM_0 + m_0^2 \quad . \tag{3.100}$$

For a given scattering angle θ the final energy E' of the electrons can be determined as a function of E and θ using (3.100) which is a quadratic equation in E'. The resulting E' and the corresponding $|\boldsymbol{p}'|^2 = E'^2 - m_0^2$ have to be inserted into (3.99). In order to understand the scattering formula (3.99) we check the limit $E/M_0 \ll 1$, where the *electron energy is small compared to the rest mass of the proton*. In this limit we should approach the limiting case of scattering at a fixed Coulomb potential (Sect. 3.1). Applying this approximation we can reduce (3.100) to

$$E'M_0 = EM_0 \quad , \quad \text{and thus} \quad E' = E \quad , \tag{3.101}$$

implying completely elastic scattering of the electron. Because $E^2 = |\boldsymbol{p}|^2 + m_0^2 = E'^2 = |\boldsymbol{p}'|^2 + m_0^2$ it follows that $|\boldsymbol{p}| = |\boldsymbol{p}'|$ and (3.99) yields

$$\frac{d\bar{\sigma}}{d\Omega'} = \frac{m_0^2}{4\pi^2} \overline{|M_{fi}|^2} \quad \text{for} \quad \frac{E}{M_0} \ll 1 \quad . \tag{3.102}$$

This result corresponds to Mott's scattering cross section known from Sect. 3.1, (3.25). To see the complete agreement we consider the square of the invariant amplitude (3.91) in the same approximation. The exact expression evaluated in the laboratory frame is

$$\begin{aligned}
\overline{|M_{fi}|^2} = {} & \frac{e^2 e_{\mathrm{p}}^2 (4\pi)^2}{2m_0^2 M_0^2 (q^2)^2} \Big\{ (M_0 E)\big[p_f\cdot(P_i + p_i - p_f)\big] + \big[p_i\cdot(P_i + p_i - p_f)(E'M_0)\big] \\
& - (p_i\cdot p_f)M_0^2 - M_0(M_0 + E - E')m_0^2 + 2m_0^2 M_0^2 \Big\} \\
= {} & \frac{e^2 e_{\mathrm{p}}^2 (4\pi)^2}{2m_0^2 M_0^2 (q^2)^2} \Big\{ M_0 E\big[M_0 E' + p_f\cdot p_i - m_0^2\big] + M_0 E'\big[M_0 E + m_0^2 - p_f\cdot p_i\big] \\
& - M_0^2 p_f\cdot p_i - M_0^2 m_0^2 - M_0 E m_0^2 + M_0 E' m_0^2 + 2m_0^2 M_0^2 \Big\} \\
= {} & \frac{e^2 e_{\mathrm{p}}^2 (4\pi)^2}{2m_0^2 M_0^2 (q^2)^2} \left\{ 2M_0^2 EE' - p_f\cdot p_i\big[M_0^2 + M_0(E' - E)\big] + m_0^2 M_0^2 \right\} \quad .
\end{aligned} \tag{3.103}$$

In the limit $E/M_0 \ll 1$ only the terms proportional to M_0^2 have to be kept. With $E = E'$ this leads to

$$\overline{|M_{fi}|^2} \approx \frac{e^2 e_{\mathrm{p}}^2 (4\pi)^2}{2m_0^2 (q^2)^2} \left(2E^2 - p_f\cdot p_i + m_0^2\right) \quad , \quad \text{for} \quad \frac{E}{M_0} \ll 1 \quad . \tag{3.104}$$

Inserting the result into (3.101) and taking into account that in the limit considered the momentum transfer q has no 0-component, i.e. $q^2 = -\boldsymbol{q}^2$, we finally get

$$\frac{d\bar{\sigma}}{d\Omega'} = \frac{2\alpha^2}{(\boldsymbol{q}^2)^2}\left(2E^2 - p_f\cdot p_i + m_0^2\right) \quad , \quad \frac{E}{M_0} \ll 1 \quad . \tag{3.105}$$

This is just the expected result we derived in Sect. 3.1, (3.39). In that limit the proton does not recoil. It can be considered as the source of a static external field. Remember that we have chosen units such that $\hbar = c = 1$, $e^2 = e^2/\hbar c = \alpha$.

Another interesting limit in which the quite complicated general expression for the scattering cross section simplifies is the case of *ultrarelativistic electrons*. In this case $m_0/E \ll 1$, and the recoil of the proton should modify the scattering formula considerably. Inspecting (3.99) we note that

$$\frac{|\boldsymbol{p}'|}{|\boldsymbol{p}|} = \frac{\sqrt{E'^2 - m_0^2}}{\sqrt{E^2 - m_0^2}} \quad \to \quad \frac{E'}{E} \quad \text{for} \quad \frac{m_0}{E}, \frac{m_0}{E'} \ll 1 \quad ,$$

and thus (3.99) becomes

$$\begin{aligned}\frac{d\bar{\sigma}}{d\Omega'} &\approx \frac{m_0^2}{4\pi^2}\frac{\frac{E'}{E}}{1 + \frac{E}{M_0} - \frac{E}{M_0}\cos\theta}\overline{|M_{fi}|^2} \\ &= \frac{m_0^2}{4\pi^2}\frac{\frac{E'}{E}}{1 + \frac{2E}{M_0}\sin^2\frac{\theta}{2}}\overline{|M_{fi}|^2} \quad , \quad \frac{m_0}{E} \ll 1 \quad .\end{aligned} \tag{3.106}$$

In order to calculate $\overline{|M_{fi}|^2}$ we consider (3.103) and express the scalar product $p_f\cdot p_i$ in terms of the squared momentum transfer through

$$q^2 = (p_f - p_i)^2 = p_f^2 + p_i^2 - 2p_f\cdot p_i = 2\left[m_0^2 - (p_f\cdot p_i)\right] \quad . \tag{3.107}$$

This yields

$$\begin{aligned}\overline{|M_{fi}|^2} &= \frac{(4\pi)^2\alpha^2}{2m_0^2M_0^2(q^2)^2}\left\{2M_0^2EE' + \frac{q^2}{2}\left[M_0^2 + M_0(E' - E)\right] + m_0^2M_0(E' - E)\right\} \\ &= \frac{(4\pi)^2\alpha^2EE'}{m_0^2(q^2)^2}\left\{1 + \frac{q^2}{4EE'}\left(1 + \frac{E' - E}{M_0}\right) + \frac{m_0^2}{2EE'}\frac{(E' - E)}{M_0}\right\} \quad ,\end{aligned} \tag{3.108}$$

which is still exact. In the ultrarelativistic limit E/m_0, $E'/m_0 \gg 1$, energy and momentum become equal and the squared momentum transfer (3.107) is related to the scattering angle in a simple way

$$\begin{aligned}q^2 &= 2\left(m_0^2 - EE' + \boldsymbol{p}'\cdot\boldsymbol{p}\right) = 2\left(m_0^2 - EE' + |\boldsymbol{p}'||\boldsymbol{p}|\cos\theta\right) \\ &\approx -2EE'(1 - \cos\theta) = -4EE'\sin^2\frac{\theta}{2} \quad .\end{aligned} \tag{3.109}$$

Furthermore, the condition of energy conservation (3.100) simplifies to

$$\begin{aligned}M_0(E - E') &= E'E - |\boldsymbol{p}||\boldsymbol{p}'|\cos\theta - m_0^2 \\ &\approx EE'(1 - \cos\theta) = 2EE'\sin^2\frac{\theta}{2}\end{aligned} \tag{3.110}$$

or, using (3.109)

$$\frac{E'-E}{M_0} \approx -\frac{2EE'}{M_0^2}\sin^2\frac{\theta}{2} \approx \frac{q^2}{2M_0^2} \quad . \tag{3.111}$$

Thus (3.108) can be reduced to

$$\begin{aligned}\overline{|M_{fi}|^2} &\approx \frac{\pi^2\alpha^2}{m_0^2 EE'\sin^4\frac{\theta}{2}}\Bigg(\underbrace{1+\frac{q^2}{4EE'}}_{1-\sin^2(\theta/2)}-\frac{q^2}{4EE'}\frac{2EE'}{M_0^2}\sin^2\frac{\theta}{2}\Bigg)\\ &= \frac{\pi^2\alpha^2}{m_0^2 EE'\sin^4\frac{\theta}{2}}\left(\cos^2\frac{\theta}{2}-\frac{q^2}{2M_0^2}\sin^2\frac{\theta}{2}\right) \quad , \quad E,E'\gg m_0 \quad .\end{aligned} \tag{3.112}$$

We insert this result into (3.106) and get

$$\frac{d\bar{\sigma}}{d\Omega'} = \frac{\alpha^2}{4E^2}\frac{1}{\sin^4\frac{\theta}{2}}\,\frac{\cos^2\frac{\theta}{2}-\frac{q^2}{2M_0^2}\sin^2\frac{\theta}{2}}{1+\frac{2E}{M_0}\sin^2\frac{\theta}{2}} \quad \text{for} \quad E,E'\gg m_0 \quad . \tag{3.113}$$

This formula determines the scattering cross section in the ultrarelativistic case under the assumption that the proton behaves like a heavy electron with mass M_0. Equation (3.113) can be compared with the Mott scattering formula (3.39) in the limit $\beta \to 1$. Two deviations are found. The denominator in (3.113) originates from the *recoil of the target* as we see from (3.110) which can be written as

$$E' = E\frac{1}{1+\frac{2E}{M_0}\sin^2\frac{\theta}{2}} \quad . \tag{3.114}$$

Furthermore the angular dependence of the numerator in (3.113) is more involved compared to Mott scattering. The q^2-dependent second term is found to originate from the fact that *the target is a spin-$\frac{1}{2}$ particle*. This term is absent when the collision of electrons with spin-0 particles is considered.

We finally remark that (3.113) does not provide a realistic description of electron-proton collisions at high energies since the de Broglie wavelength of the electron is so small that the substructure of the proton becomes detectable. This fact has not been considered in (3.113), where we assumed the proton to be a point-like Dirac particle without internal structure. In addition the proton's anomalous magnetic moment has to be considered in that case. We remark, then, that in a complete treatment for very high energies (several 100 MeV) formula (3.113) has to be modified by introducing electric and magnetic *form factors* representing the internal structure of the proton. This yields the so-called **Rosenbluth** formula (see Exercise 3.5). Equation (3.113) would apply with great accuracy, however, to the *scattering of electrons and muons* which both are structureless Dirac particles.

EXERCISE

3.5 Rosenbluth's Formula

The realistic description of the scattering of an electron at a spin-$\frac{1}{2}$ hadron has to take into account the internal structure and anomalous magnetic moment of the hadron. To that end one replaces the transition current in momentum space which originates from the Dirac equation with the more general bilinear expression

$$\bar{u}(P')\gamma_\mu u(P) \to \bar{u}(P')\Gamma_\mu(P',P)u(P) \quad . \tag{1}$$

Problem.

a) Show that the most general expression for a transition current that fulfils the conditions of Lorentz covariance, Hermiticity, and gauge invariance can be written as

$$\bar{u}(P')\Gamma_\mu(P',P)u(P) = \bar{u}(P')\Big(\gamma_\mu F_1(q^2) + \mathrm{i}\frac{1}{2M_0}F_2(q^2)q^\nu\sigma_{\mu\nu}\Big)u(P) \quad . \tag{2}$$

Here $q = P' - P$ is the momentum transfer and $F_1(q^2)$, $F_2(q^2)$ are unspecified real functions ("form factors"), cf. Fig. 3.12.

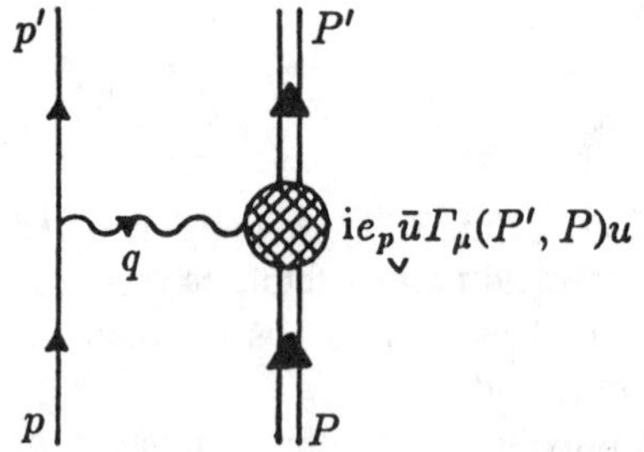

Fig. 3.12. Feynman diagram for the scattering of a point-like Dirac particle at an extended target, symbolized by the hatched blob

b) What is the physical meaning of $F_1(0)$ and $F_2(0)$? This can be deduced by studying the interaction energy with static electromagnetic fields in the nonrelativistic limit.

c) Calculate the unpolarized cross sections of electron scattering at a hadron with the vertex function (2) in the ultrarelativistic limit.

Solution. a) In order to construct the vertex function $\Gamma_\mu(P',P)$ we have at our disposal the two kinematic quantities P^μ and P'^μ. Since the proton moves freely before and after the collision (it is "on the mass shell"), there is only a single independent scalar variable, because $P^2 = P'^2 = M_0^2$, which we choose as the square of momentum transfer $q^2 = (P' - P)^2$. Since $\Gamma_\mu(P',P)$ has to be a Lorentz vector the most general ansatz can be directly noted with the help of the known bilinear convariants of the Dirac theory (cf. *RQM*, Chap. 5) :

$$\begin{aligned}\bar{u}(P')\Gamma_\mu(P',P)u(P) = \bar{u}(P')\big(&A(q^2)\gamma_\mu + B(q^2)P'_\mu + C(q^2)P_\mu \\ &+ \mathrm{i}D(q^2)P'^\nu\sigma_{\mu\nu} + \mathrm{i}E(q^2)P^\nu\sigma_{\mu\nu}\big)u(P) \quad .\end{aligned} \tag{3}$$

Exercise 3.5.

$A(q^2), \dots, E(q^2)$ are undetermined scalar functions of the variable q^2 and $\sigma_{\nu\mu} = (\mathrm{i}/2)(\gamma_\mu\gamma_\nu - \gamma_\nu\gamma_\mu)$. Since we demanded Hermiticity, they are real functions. Further restrictions follow from our requiring gauge invariance, which takes the form

$$q^\mu\, \bar{u}(P')\Gamma_\mu(P',P)u(P) = 0 \quad . \tag{4}$$

This follows from the condition of current conservation applied to the electromagnetic transition current of the hadron, i.e. $\partial_\mu J^\mu_{P',P}(x) = 0$. In momentum space the operator ∂_μ is replaced by a factor $-\mathrm{i}P'_\mu$ or $\mathrm{i}P_\mu$, resp., when acting on the final (inital) wavefunction. This leads to the condition (4).

A short reminder on gauge invariance and current conservation: Gauge invariance implies

$$\int d^4x\, j_\mu(x)A^\mu(x) = \int d^4x\, j_\mu(x)(A^\mu(x) - \partial^\mu\chi(x))$$

which means $\int d^4x\, j_\mu(x)\partial^\mu\chi(x) = 0$. Assuming that surface terms do not contribute (the function $\chi(x)$ can be chosen appropriately) the latter condition can be rewritten as $\int d^4x\, \partial^\mu j_\mu(x)\chi(x) = 0$. Since $\chi(x)$ is an arbitrary function this implies $\partial^\mu j_\mu(x) = 0$.

The first term in (3) is just the usual Dirac current, which obviously fulfils condition (4). This can be easily shown with the help of the Dirac equation. For the subsequent terms we get

$$(P'^\mu - P^\mu)(BP'_\mu + CP_\mu) = (B - C)(M_0^2 - P'\cdot P) = 0 \tag{5}$$

and

$$\begin{aligned} \mathrm{i}(P'^\mu - P^\mu)(DP'^\nu + EP^\nu)\sigma_{\mu\nu} &= -\mathrm{i}DP^\mu P'^\nu\sigma_{\mu\nu} + \mathrm{i}EP'^\mu P^\nu\sigma_{\mu\nu} \\ &= \mathrm{i}(D + E)P'^\mu P^\nu\sigma_{\mu\nu} = 0 \quad , \end{aligned} \tag{6}$$

taking into account the antisymmetry of the tensor $\sigma_{\mu\nu}$. Thus we deduce that $C = B$ and $E = -D$, i.e.

$$\begin{aligned} \bar{u}(P')\Gamma_\mu(P',P)u(P) = \bar{u}(P')\big[&A(q^2)\gamma_\mu + B(q^2)(P' + P)_\mu \\ &+ \mathrm{i}D(q^2)(P' - P)^\nu\sigma_{\mu\nu}\big]u(P) \quad . \end{aligned} \tag{7}$$

Because of the Gordon decomposition (cf. *RQM*, Chap. 8)

$$\bar{u}\gamma_\mu u = \frac{1}{2M_0}(P' + P)_\mu\, \bar{u}u + \frac{\mathrm{i}}{2M_0}q^\nu\, \bar{u}\sigma_{\mu\nu}u \tag{8}$$

these three terms are not linearly independent. Therefore one of the terms, for instance $B(q^2)(P' + P)_\mu$, can be eliminated. This yields (2).

b) In order to understand the significance of the form factors $F_1(0)$ and $F_2(0)$ we consider the energy of a nonrelativistic hadron with charge e_p in a static external electromagnetic field $A^\mu(\boldsymbol{x})$.

Exercise 3.5.

$$\begin{aligned}
W &= e_p \int d^3x\, A^\mu(\boldsymbol{x}) J_\mu(\boldsymbol{x}) \\
&= e_p \int d^3x\, A^\mu(\boldsymbol{x})\, \bar{\psi}_{P'}(\boldsymbol{x}) \Gamma_\mu(P',P) \psi_P(\boldsymbol{x}) \\
&= \frac{e_p}{V}\sqrt{\frac{M_0^2}{E'E}} \int d^3x\, \mathrm{e}^{-\mathrm{i}(\boldsymbol{P}'-\boldsymbol{P})\cdot\boldsymbol{x}}\, A^\mu(\boldsymbol{x}) \bar{u}(P')\Gamma_\mu(P',P)u(P) \quad . \qquad (9)
\end{aligned}$$

In the case of a pure *electrostatic field* $A^0(\boldsymbol{x})$ it is advantageous to rewrite the vertex function $\Gamma_\mu(P',P)$ using the Gordon decomposition (8):

$$\Gamma_\mu(P',P) = \gamma_\mu\big(F_1(q^2) + F_2(q^2)\big) - \frac{1}{2M_0}(P'+P)_\mu F_2(q^2) \quad . \qquad (10)$$

This leads to the following expression for the integrand of (9), given in the rest frame of the hadron

$$\begin{aligned}
&A^0(\boldsymbol{x})\bar{u}(P')\Gamma_0(P',P)u(P) \\
&\quad = A^0(\boldsymbol{x})\,\bar{u}(P')\Big[\gamma_0(F_1(q^2)+F_2(q^2)) - \frac{1}{2M_0}(P'+P)_0 F_2(q^2)\Big]u(P) \\
&\quad = A^0(\boldsymbol{x})\,\bar{u}(P')\Big[\gamma_0 F_1(q^2) + \frac{2M_0\gamma_0 - E' - M_0}{2M_0}F_2(q^2)\Big]u(P) \quad . \qquad (11)
\end{aligned}$$

In the nonrelativistic limit the lower components of the Dirac spinors can be neglected and $\bar{u}u \approx \bar{u}\gamma^0 u \approx 1$ which leads to

$$A^0(\boldsymbol{x})\,\bar{u}(P')\Gamma_0(P',P)u(P) \approx A^0(\boldsymbol{x})\Big(F_1(q^2) + \frac{q^2}{4M_0^2}F_2(q^2)\Big) \quad . \qquad (12)$$

Here the identity

$$q^2 = (P'-P)^2 = 2M_0^2 - 2P'\cdot P = 2M_0(M_0 - E') \qquad (13)$$

was used. In the static limit $q^2 \to 0$ (considering an external potential which is constant or slowly varying) the interaction energy simply gets

$$\begin{aligned}
W &\simeq e_p F_1(0)\frac{1}{V}\int d^3x\, \mathrm{e}^{-\mathrm{i}\boldsymbol{q}\cdot\boldsymbol{x}}\, A^0(\boldsymbol{x}) \\
&\simeq e_p F_1(0) A^0 \quad . \qquad (14)
\end{aligned}$$

This result is obviously the electrostatic energy of a particle with charge $e_p F_1(0)$ in a potential A^0, from which we conclude that for the proton

$$F_1(0) = 1 \quad . \qquad (15)$$

In the case of a *magnetic field* given by the vector potential $A^k(\boldsymbol{x})$ we use (2) and get

$$\begin{aligned}
&A^k(\boldsymbol{x})\bar{u}(P')\Gamma_k(P',P)u(P) \\
&\quad = A^k(\boldsymbol{x})\,\bar{u}(P')\Big[\frac{1}{2M_0}(P'_k + P_k)F_1(q^2) + \frac{\mathrm{i}}{2M_0}(F_1(q^2)+F_2(q^2))q^l\sigma_{kl}\Big]u(P) \\
&\quad \simeq -\bar{u}(P')\Big[\boldsymbol{A}\cdot\frac{\boldsymbol{P}}{M_0}F_1(q^2) + \frac{\mathrm{i}}{2M_0}(F_1(q^2)+F_2(q^2))\boldsymbol{q}\times\boldsymbol{A}\cdot\boldsymbol{\Sigma}\Big]u(P) \quad , \qquad (16)
\end{aligned}$$

Exercise 3.5.

since

$$\sigma_{kl} = \varepsilon_{klm}\,\Sigma^m \qquad \text{with} \qquad \boldsymbol{\Sigma} = \begin{pmatrix} \boldsymbol{\sigma} & 0 \\ 0 & \boldsymbol{\sigma} \end{pmatrix} \quad .$$

The first term in (16) just describes the interaction of a moving charge with the magnetic field yielding $W \simeq -e_\mathrm{p} F_1(0) \boldsymbol{v}\cdot\boldsymbol{A}\,\bar{u}u$. The second term yields (again considering the limit $q \to 0$)

$$\begin{aligned} W &\simeq \frac{e_\mathrm{p}}{V}\int d^3x\; \mathrm{e}^{-\mathrm{i}\boldsymbol{q}\cdot\boldsymbol{x}}\,\frac{1}{2M_0}\big(F_1(q^2)+F_2(q^2)\big)\big(-\mathrm{i}\boldsymbol{q}\times\boldsymbol{A}(\boldsymbol{x})\big)\cdot\bar{u}(P')\boldsymbol{\Sigma}u(P) \\ &\simeq -e_\mathrm{p}\frac{1}{2M_0}\big(F_1(0)+F_2(0)\big)\big(\bar{u}\,\boldsymbol{\Sigma}u\big)\frac{1}{V}\int d^3x\; \mathrm{e}^{-\mathrm{i}\boldsymbol{q}\cdot\boldsymbol{x}}\,\nabla\times\boldsymbol{A}(\boldsymbol{x}) \\ &\simeq -\frac{e_\mathrm{p}}{2M_0}\big(1+F_2(0)\big)\;2\langle\boldsymbol{s}\rangle\cdot\boldsymbol{B} \quad . \end{aligned} \tag{17}$$

This is the energy of a particle with spin-$\frac{1}{2}$ having the gyromagnetic ratio $g = 2(1+F_2(0))$ in a homogeneous magnetic field. The number $F_2(0)$ therefore describes the *anomalous magnetic moment* of the particle. In the case of proton and neutron experiment gives the following form factors at zero momentum transfer:

$$\begin{aligned} &F_1^\mathrm{p}(0) = 1 \quad , \quad F_1^n(0) = 0 \quad , \\ &F_2^\mathrm{p}(0) = 1.79284 \quad , \quad F_2^n(0) = -1.91304 \quad . \end{aligned} \tag{18}$$

c) The calculation of the electron–hadron cross section can be done as Sect. 3.2; we just have to insert the more complex vertex operator (2). In fact, the calculation simplifies if we use the equivalent expression (10). The squared spin-averaged transition matrix element is

$$\begin{aligned} \overline{|M_{fi}|^2} &= \frac{1}{4}\sum_{\text{spin}}\Big|\bar{u}(p',s')\gamma^\mu u(p,s)\,\frac{4\pi e e_\mathrm{p}}{q^2} \\ &\quad\times\bar{u}(P',S')\Big[\gamma_\mu(F_1+F_2)-\frac{1}{2M_0}(P'+P)_\mu F_2\Big]u(P,S)\Big|^2 \\ &= \frac{(4\pi)^2e^2e_\mathrm{p}^2}{(q^2)^2}\,\frac{1}{4}\sum_{\text{spin}}u^\dagger(P,S)\Big[\gamma_\mu^\dagger(F_1+F_2) \\ &\quad-\frac{1}{2M_0}(P'+P)_\mu F_2\Big]\gamma_0 u(P',S') \\ &\quad\times u^\dagger(p,s)\gamma^{\mu\dagger}\gamma^0 u(p',s')\,\bar{u}(p',s')\gamma^\nu u(p,s) \\ &\quad\times\bar{u}(P',S')\Big[\gamma_\nu(F_1+F_2)-\frac{1}{2M_0}(P'+P)_\nu F_2\Big]u(P,S) \quad . \end{aligned} \tag{19}$$

With the help of (3.29) and (3.31) of Sect. 3.1 this expression can be rewritten as a product of two traces. As in Sect. 3.2, (3.85), it is given by

$$\overline{|M_{fi}|^2} = \frac{(4\pi)^2e^2e_\mathrm{p}^2}{(q^2)^2}\,L^{\mu\nu}H_{\mu\nu} \quad , \tag{20}$$

with the lepton tensor

Exercise 3.5.

$$L^{\mu\nu} = \frac{1}{2}\mathrm{Tr}\left[\gamma^{\mu}\frac{\not{p}+m_0}{2m_0}\gamma^{\nu}\frac{\not{p}'+m_0}{2m_0}\right] \tag{21}$$

and the hadron tensor

$$\begin{aligned} H_{\mu\nu} = \frac{1}{4M_0^2}\frac{1}{2}\mathrm{Tr}\Big[&(\not{P}+M_0)\big(\gamma_\mu(F_1+F_2) - \frac{F_2}{2M_0}(P'_\mu+P_\mu)\big) \\ &\times(\not{P}'+M_0)\big(\gamma_\nu(F_1+F_2) - \frac{F_2}{2M_0}(P'_\nu+P_\nu)\big)\Big] \quad . \end{aligned} \tag{22}$$

Expanding the product we get 16 traces from which 8 vanish, since they contain an odd number of γ-matrices. The remaining terms are

$$\begin{aligned} 8M_0^2 H_{\mu\nu} = &(F_1+F_2)^2\,\mathrm{Tr}\big[\not{P}\gamma_\mu\not{P}'\gamma_\nu\big] - (F_1+F_2)M_0\frac{F_2}{2M_0}(P'_\nu+P_\nu)\,\mathrm{Tr}\big[\not{P}\gamma_\mu\big] \\ &+ \frac{F_2}{2M_0}(P'_\mu+P_\mu)\,\frac{F_2}{2M_0}(P'_\nu+P_\nu)\,\mathrm{Tr}\big[\not{P}\not{P}'\big] \\ &- \frac{F_2}{2M_0}(P'_\mu+P_\mu)\,M_0\,(F_1+F_2)\,\mathrm{Tr}\big[\not{P}\gamma_\nu\big] \\ &- M_0(F_1+F_2)\frac{F_2}{2M_0}(P'_\nu+P_\nu)\,\mathrm{Tr}\big[\gamma_\mu\not{P}'\big] \\ &+ M_0(F_1+F_2)\,M_0\,(F_1+F_2)\,\mathrm{Tr}\big[\gamma_\mu\gamma_\nu\big] \\ &- M_0\frac{F_2}{2M_0}(P'_\mu+P_\mu)(F_1+F_2)\,\mathrm{Tr}\big[\not{P}'\gamma_\nu\big] \\ &+ M_0\frac{F_2}{2M_0}(P'_\mu+P_\mu)\,M_0\frac{F_2}{2M_0}(P'_\nu+P_\nu)\,\mathrm{Tr}[\mathbb{1}] \quad . \end{aligned} \tag{23}$$

After inserting the values of all these traces we can sum the result to the form

$$\begin{aligned} H_{\mu\nu} &= \frac{1}{8M_0^2}\Big\{4(F_1+F_2)^2\big[P_\mu P'_\nu + P'_\mu P_\nu - (P\cdot P' - M_0^2)\,g_{\mu\nu}\big] \\ &\qquad + \big[-4(F_1+F_2)F_2 + F_2^2(P\cdot P'/M_0^2+1)\big](P_\mu+P'_\mu)(P_\nu+P'_\nu)\Big\} \\ &\equiv H^{(1)}_{\mu\nu} + H^{(2)}_{\mu\nu} \quad . \end{aligned} \tag{24}$$

$H^{(1)}_{\mu\nu}$ is already known from previous discussions and apart from the factor $(F_1+F_2)^2$ it again yields the cross section of (3.113) when inserted into (20). In addition we get

$$\begin{aligned} \overline{|M_{fi}(2)|^2} = &\frac{(4\pi)^2 e^2 e_\mathrm{p}^2}{(q^2)^2}\frac{1}{16M_0^2 m_0^2}\big[p^\mu p'^\nu + p'^\mu p^\nu - (p\cdot p' - m_0^2)\,g^{\mu\nu}\big] \\ &\times\big[(P_\mu+P'_\mu)(P_\nu+P'_\nu)\big]\big[-4(F_1+F_2)F_2 + F_2^2(P\cdot P'/M_0^2+1)\big] \quad , \end{aligned} \tag{25}$$

where we have taken the lepton tensor from (3.90). Using energy–momentum conservation

$$p + P = p' + P' \quad , \tag{26}$$

we will now eliminate the final momentum of the hadron and collect the momentum dependent terms in (25). In the *ultrarelativistic approximation* (neglecting the electron rest mass) we simply get

Exercise 3.5.

$$
\begin{aligned}
[\ldots][\ldots] &= 2p'\cdot(P+P')\,p\cdot(P+P') - (P+P')\cdot(P+P')(p\cdot p' - m_0^2) \\
&\approx 4(2p'\cdot P\; p\cdot P - M_0^2\, p\cdot p') \\
&\approx 4\left[2E'M_0\, EM_0 - M_0^2|\boldsymbol{p}||\boldsymbol{p}'|(1-\cos\theta)\right] \\
&\approx 8M_0^2 E'E\ \cos^2\frac{\theta}{2}
\end{aligned}
\qquad (27)
$$

in the laboratory system. Further, using (3.111) the product $P\cdot P'$ in (23) can be expressed in terms of the momentum transfer q

$$
P\cdot P' = P\cdot(P+p-p') = M_0^2 + M_0(E-E') \simeq M_0^2\left(1-\frac{q^2}{2M_0^2}\right) \quad . \qquad (28)
$$

q^2 is related to the scattering angle θ through (3.109)

$$
q^2 \approx -4EE'\sin^2\frac{\theta}{2} \quad . \qquad (29)
$$

By adding $\overline{|M_{fi}(1)|^2}$ from (3.112) we finally get

$$
\begin{aligned}
\overline{|M_{fi}|^2} &= \frac{(4\pi)^2 e^2 e_\mathrm{p}^2}{(q^2)^2}\frac{E'E}{m_0^2}\left\{(F_1+F_2)^2\left[1-\sin^2\frac{\theta}{2}\left(1+\frac{q^2}{2M_0^2}\right)\right]\right. \\
&\quad \left. +\frac{1}{2}\cos^2\frac{\theta}{2}\left[-4(F_1+F_2)F_2 + 2F_2^2\left(1-\frac{q^2}{4M_0^2}\right)\right]\right\} \\
&= \frac{(4\pi)^2 e^2 e_\mathrm{p}^2}{(q^2)^2}\frac{E'E}{m_0^2}\left[\left(F_1^2 - \frac{q^2}{4M_0^2}F_2^2\right)\cos^2\frac{\theta}{2}\right. \\
&\quad \left. -(F_1+F_2)^2\frac{q^2}{2M_0}\sin^2\frac{\theta}{2}\right] \quad .
\end{aligned}
\qquad (30)
$$

With (3.106) the spin-averaged scattering cross section in the laboratory system results as

$$
\begin{aligned}
\frac{d\bar{\sigma}}{d\Omega} &= \frac{e^2 e_\mathrm{p}^2}{4E^2\sin^4\frac{\theta}{2}\left(1+\frac{2E}{M_0}\sin^2\frac{\theta}{2}\right)} \\
&\quad \times\left[\left(F_1^2-\frac{q^2}{4M_0^2}F_2^2\right)\cos^2\frac{\theta}{2} - (F_1+F_2)^2\frac{q^2}{2M_0^2}\sin^2\frac{\theta}{2}\right] \quad .
\end{aligned}
\qquad (31)
$$

This result is known as *Rosenbluth's formula*.[9]

Additional Remarks. Instead of the functions $F_1(q^2)$ and $F_2(q^2)$ one often introduces the so-called electric and magnetic "Sachs form factors"

$$
\begin{aligned}
G_\mathrm{E}(q^2) &= F_1(q^2) + \frac{q^2}{4M_0^2}F_2(q^2) \quad , \\
G_\mathrm{M}(q^2) &= F_1(q^2) + F_2(q^2) \quad .
\end{aligned}
\qquad (32)
$$

---

[9] M. N. Rosenbluth: Phys. Rev. **79**, 615 (1950).

Exercise 3.5.

These combinations emerge in a natural way from the interaction energy with electric and magnetic fields, (12) and (17). Expressed in terms of the Sachs form factors the Rosenbluth formula (31) becomes

$$\frac{d\bar{\sigma}}{d\Omega} = \left(\frac{d\bar{\sigma}}{d\Omega}\right)_{\text{Mott}} \left[\frac{G_E^2(q^2) + \tau G_M^2(q^2)}{1+\tau} + 2\tau G_M^2(q^2)\tan^2\frac{\theta}{2}\right] \quad . \tag{33}$$

with the abbreviation $\tau = -q^2/(4M_0^2) > 0$. Experimentally the two form factors can be determined by varying E and θ at a given momentum transfer

$$q^2 = -4EE'\sin^2\frac{\theta}{2} = -4E^2 \frac{\sin^2\frac{\theta}{2}}{1+\frac{2E}{M_0}\sin^2\frac{\theta}{2}} \quad . \tag{34}$$

Figure 3.13 shows some experimental data[10] on the form factors $G_E(q^2)$ and $G_M(q^2)$ for protons. It was found that at not too large values of q^2 a good description of both G_E and $\frac{1}{\kappa}G_M$ (κ is the magnetic moment of the proton) is given by the so-called "dipole fit" formula

$$G_D(q^2) = \frac{1}{\left(1+\frac{|q^2|}{Q^2}\right)^2} \quad , \tag{35}$$

with the empirical parameter (remember the choice of units $\hbar c$ = 0.197 GeV fm =1)

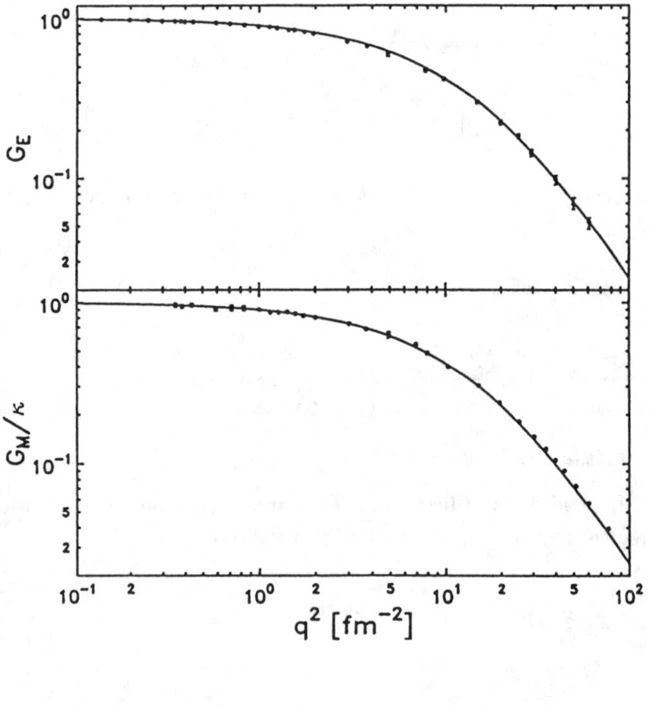

Fig. 3.13. The electric and the normalized magnetic form factor of the proton as a function of the squared momentum transfer q^2. The experimental data are well described by the "dipole fit" (35)

[10] G. Simon, Ch. Schmitt, F. Borkowski, V.H. Walther: Nucl. Phys. **A333**, 381 (1980).

Exercise 3.5.

$$Q^2 = 0.71\,\mathrm{GeV}^2 = 18.2\,\mathrm{fm}^{-2} \quad . \tag{36}$$

In nonrelativistic quantum mechanical scattering theory a form factor $F(\boldsymbol{q})$ is introduced which relates the scattering at extended and point-like targets:

$$\left(\frac{d\sigma}{d\Omega}\right)_{\text{extended}} = \left(\frac{d\sigma}{d\Omega}\right)_{\text{point}} |F(\boldsymbol{q})|^2 \quad . \tag{37}$$

$F(\boldsymbol{q})$ has a simple interpretation:[11]

It is the spatial *Fourier transform of the density distribution* $\varrho(\boldsymbol{x})$ of the extended scattering centre

$$F(\boldsymbol{q}) = \int d^3x\, \varrho(\boldsymbol{x})\, \mathrm{e}^{\mathrm{i}\boldsymbol{q}\cdot\boldsymbol{x}} \quad . \tag{38}$$

By analogy one may identify $G_{\mathrm{E}}(q^2)$ and $G_{\mathrm{M}}(q^2)$ with the Fourier transforms of the hadron's charge and magnetic moment density distribution. At relativistic energies this interpretation becomes problematic since it depends on the frame of reference. It is valid in the so called "**Breit**-frame" where no energy is transferred to the nucleon, $q_0 = 0$, which means $P = (E, \boldsymbol{P})$, $P' = (E, -\boldsymbol{P})$.

With this caveat, the density distribution of the proton charge derived from the dipole formula (35) by inverse Fourier transformation is simply

$$\varrho_{\mathrm{p}}(\boldsymbol{x}) = \int \frac{d^3q}{(2\pi)^3} G_{\mathrm{D}}(-\boldsymbol{q}^2)\, \mathrm{e}^{-\mathrm{i}\boldsymbol{q}\cdot\boldsymbol{x}} = \frac{Q^3}{8\pi}\mathrm{e}^{-Q|\boldsymbol{x}|} \quad . \tag{39}$$

The size of the nucleon may be defined by the second moment of this distribution

$$\langle R^2\rangle = \int d^3x\, |\boldsymbol{x}|^2 \varrho_{\mathrm{p}}(\boldsymbol{x}) = \frac{12}{Q^2} \quad . \tag{40}$$

This gives the mean *proton radius*

$$R_{\mathrm{p}} = \sqrt{\langle R^2\rangle} = \sqrt{\frac{12}{18.2\ \mathrm{fm}^{-2}}} = 0.81\ \mathrm{fm} \quad . \tag{41}$$

More detailed information on the internal structure of hadrons can be obtained from the study of *inelastic* electron scattering. In this way the presence of pointlike constituents (partons) inside the hadrons was revealed.

[11] See e.g. R. Hofstadter, Ann. Rev. Nucl. Sci. **7**, 231 (1957).

EXAMPLE

3.6 Higher-Order Electron-Proton Scattering

In Sect. 3.2 we calculated electron–proton scattering to lowest order in $\alpha = e^2$. Now we shall discuss corrections arising at the next higher order of the perturbation expansion. To that end we refer to the general expression of the nth order contribution to the electron scattering matrix (2.44). The amplitude of second-order electron–proton interaction is given by

$$S_{fi}^{(2)} = -\mathrm{i}e^2 \int d^4x \, d^4y \, \bar{\psi}_f(x) \not{A}(x) S_{\mathrm{F}}(x-y) \not{A}(y) \psi_i(y) \quad . \tag{1}$$

As in Sect. 3.2, the electromagnetic potential $A(x)$ is produced by the proton current. However, for us to be consistent the proton current has also to be treated in second order. To do this we again consider (1), describing the interaction of the electron current with the $A_\mu(x)\, A_\nu(y)$-fields in second order. Again, we require the total expression for $S_{fi}^{(2)}$ to be symmetric with respect to the electron and proton currents. From (1), it is obvious that the second-order electron current is given by

$$J_{\mu\nu}^{(2)}(x,y) = \mathrm{i}e^2 \, \bar{\psi}_f(x)\, \gamma_\mu S_{\mathrm{F}}(x-y) \gamma_\nu \, \psi_i(y) \, . \tag{2}$$

The factor i here has been introduced for convenience. Namely, inserting the electron propagator $S_{\mathrm{F}}(x-y)$ explicitly this becomes

$$\begin{aligned} J_{\mu\nu}^{(2)}(x,y) = e^2 \bar{\psi}_f(x) \gamma_\mu \Bigg(& \sum_{n;p_0>0} \Theta(x_0 - y_0) \psi_n(x) \bar{\psi}_n(y) \\ & - \sum_{n;p_0<0} \Theta(y_0 - x_0) \psi_n(x) \bar{\psi}_n(y) \Bigg) \gamma_\nu \psi_i(y) \quad . \end{aligned} \tag{3}$$

The two indices (and the two arguments x, y) indicate the second-order structure of the transition current. Here we used the Stückelberg–Feynman propagator (2.24). The factor i introduced in (2) cancels the factor minus i included in the propagator $S_{\mathrm{F}}(x-y)$, enabling us to represent the second-order current $J^{\mu\nu}(x,y)$ as a sum $(\sum_n)$ of products consisting of first-order transition currents (cf. Sect. 3.2, (3.53)) of the form

$$\big(J_\mu(x)\big)_{fn} = e \, \bar{\psi}_f(x)\, \gamma_\mu \, \psi_n(x) \quad , \tag{4}$$

which yields

$$\begin{aligned} J_{\mu\nu}^{(2)}(x,y) = & \sum_{n,p_0>0} \big(J_\mu(x)\big)_{fn} \big(J_\nu(y)\big)_{ni} \, \Theta(x_0 - y_0) \\ & - \sum_{n,p_0<0} \big(J_\mu(x)\big)_{fn} \big(J_\nu(y)\big)_{ni} \, \Theta(y_0 - x_0) \quad . \end{aligned} \tag{5}$$

The single index and the single argument express the first-order structure of these transition currents $J_\mu(x)$. According to Sect. 3.2, (3.51), each first-order transition current produces a vector potential

Example 3.6.

$$A^{\mu}(x) = \int d^4y\, D_{\mathrm{F}}(x-y)\, J^{\mu}(y) \quad .$$

Generalizing this relation we can conjecture that the following expression describes the electromagnetic fields produced by the proton:

$$\begin{aligned}
&A_{\mu}(x)A_{\nu}(y)\\
&= \int d^4X d^4Y\, D_{\mathrm{F}}(x-X) D_{\mathrm{F}}(y-Y) J^{\mathrm{p}(2)}_{\mu\nu}(X,Y)\\
&= \sum_{n;p_0>0} \int d^4X d^4Y\, D_{\mathrm{F}}(x-X) D_{\mathrm{F}}(y-Y) \left(J^{\mathrm{p}}_{\mu}(X)\right)_{fn} \left(J^{\mathrm{p}}_{\nu}(Y)\right)_{ni} \Theta(X_0-Y_0)\\
&\quad - \sum_{n;p_0<0} \int d^4X d^4Y\, D_{\mathrm{F}}(x-X) D_{\mathrm{F}}(y-Y) \left(J^{\mathrm{p}}_{\mu}(X)\right)_{fn} \left(J^{\mathrm{p}}_{\nu}(Y)\right)_{ni} \Theta(Y_0-X_0)\\
&= e_p^2 \int d^4X d^4Y\, D_{\mathrm{F}}(x-X) D_{\mathrm{F}}(y-Y)\\
&\quad \times \bar{\psi}^{\mathrm{p}}_f(X)\gamma_{\mu} \left(\sum_{n;p_0>0} \Theta(X_0-Y_0)\, \psi^{\mathrm{p}}_n(X)\, \bar{\psi}^{\mathrm{p}}_n(Y) \right.\\
&\quad \left. - \sum_{n;p_0<0} \Theta(Y_0-X_0)\, \psi^{\mathrm{p}}_n(X)\, \bar{\psi}^{\mathrm{p}}_n(Y) \right) \gamma_{\nu}\psi^{\mathrm{p}}_i(Y)\\
&= \mathrm{i}e_p^2 \int d^4X d^4Y\, D_{\mathrm{F}}(x-X) D_{\mathrm{F}}(y-Y) \bar{\psi}^{\mathrm{p}}_f(X)\gamma_{\mu} S^{\mathrm{p}}_{\mathrm{F}}(X-Y)\gamma_{\nu}\psi^{\mathrm{p}}_i(Y) \quad . \qquad (6)
\end{aligned}$$

Each single first-order transition current occurring in the products (5) creates a photon field like the one produced by the transition current in (3.51). Note the factor +i occurring in the last step that cancels the factor −i contained in the proton propagator $S^{\mathrm{p}}_{\mathrm{F}}(X-Y)$. Equation (6) can be substantiated by considering the differential equation for the two-photon field $A_{\mu\nu}(x,y)$:

$$\Box_x \Box_y\, A_{\mu\nu}(x,y) = (4\pi)^2 J^{\mathrm{p}(2)}_{\mu\nu}(x,y) \quad , \qquad (7)$$

which is analogous to (3.42). The expression (6) obviously fulfils this differential equation. This result can be represented graphically as shown in Fig. 3.14.

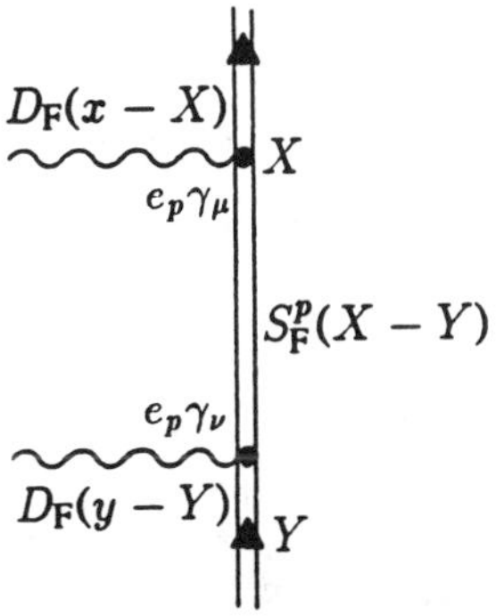

Fig. 3.14. The second-order proton transition current involves two photons. One propagates between the space-time points x and X, the second one between y and Y. Depending on the time ordering of the arguments of $D_{\mathrm{F}}(x-X)$ and $D_{\mathrm{F}}(y-Y)$ the photons are either absorbed or emitted. The propagation of the proton between the vertices at X and Y is described by the Feynman propagator $S^{\mathrm{p}}_{\mathrm{F}}(X-Y)$

The two photons are emitted at the space–time points Y, X. At each vertex a factor $e_p\gamma_\mu$ or $e_p\gamma_\nu$, respectively, enters into the calculation of the fields. The proton line between Y and X is called an *internal proton line*. It represents the propagation of the proton between the points of interaction with the photons according to the proton propagator $S^{\mathrm{p}}_{\mathrm{F}}(X-Y)$. The photons emitted by the proton at Y and X travel to the electron which absorbs them at the space–time points y and x. This process is shown in Fig. 3.15. It represents the second-order S-matrix element obtained by inserting (6) into (1), i.e.

$$\begin{aligned}
S^{(2)}_{fi}(\mathrm{dir.}) = e^2 e_p^2 &\int d^4x\, d^4y\, d^4X\, d^4Y\, \left[\bar{\psi}_f(x)\gamma^{\mu}\, S_{\mathrm{F}}(x-y)\, \gamma^{\nu}\, \psi_i(y)\right]\\
&\times D_{\mathrm{F}}(x-X)\, D_{\mathrm{F}}(y-Y) \left[\bar{\psi}^{\mathrm{p}}_f(X)\, \gamma_{\mu}\, S^{\mathrm{p}}_{\mathrm{F}}(X-Y)\, \gamma_{\nu}\, \psi^{\mathrm{p}}_i(Y)\right] \quad , \qquad (8)
\end{aligned}$$

which for reasons to become clear soon is called the direct amplitude. Here, in contrast to (1), no factor (−i) occurs, because it is compensated by the factor +i of the last expression in (6).

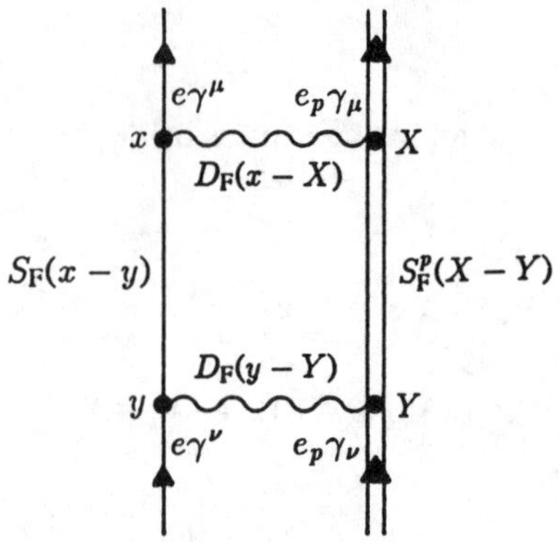

Fig. 3.15. The graph for the direct second-order electron–proton scattering amplitude

We see very clearly how to translate the various lines in the graph directly into mathematical expressions. However, (8) does not yet represent the full second-order scattering amplitude, since the two photons emitted by the proton current cannot be distinguished. The electron at x does not 'know' whether the photon being absorbed at that point has been emitted at space–time point Y or X. The *indistinguishability of the photons* then implies the contribution of a second graph (see Fig. 3.16). According to the principles of quantum mechanics the complete electron-proton scattering amplitude is then given by coherently adding the contribution of an *exchange diagram* shown in Fig. 3.16. This yields

$$\begin{aligned} S_{fi}^{(2)} &= S_{fi}^{(2)}(\text{dir.}) + S_{fi}^{(2)}(\text{exch.}) \\ &= e^2 e_p^2 \int d^4x\, d^4y\, d^4X\, d^4Y \left[\bar{\psi}_f(x)\,\gamma^\mu\, S_F(x-y)\,\gamma^\nu\, \psi_i(y)\right] \\ &\quad \times \left\{ D_F(x-X)\, D_F(y-Y) \left[\bar{\psi}_f^{\mathrm{p}}(X)\,\gamma_\mu\, S_F^{\mathrm{p}}(X-Y)\,\gamma_\nu\, \psi_i^{\mathrm{p}}(Y)\right] \right. \\ &\quad \left. + D_F(x-Y)\, D_F(y-X) \left[\bar{\psi}_f^{\mathrm{p}}(X)\,\gamma_\nu\, S_F^{\mathrm{p}}(X-Y)\,\gamma_\mu\, \psi_i^{\mathrm{p}}(Y)\right] \right\} \quad . \end{aligned} \tag{9}$$

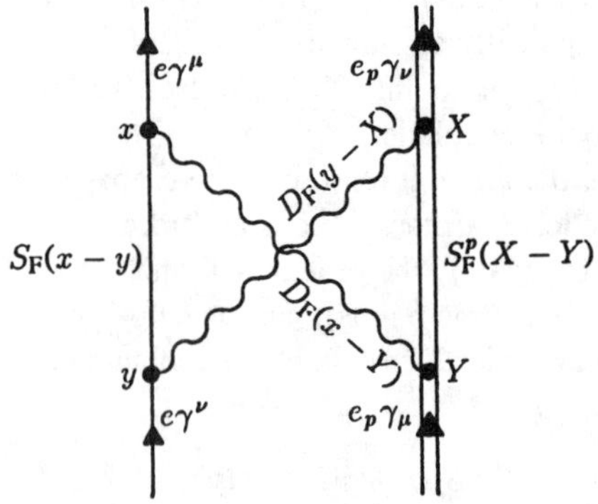

Fig. 3.16. The exchange graph corresponding to Fig. 3.15. The photon lines are crossed

The second term in the curly brackets is the exchange term. We notice also that the indices μ and ν are exchanged with respect to the direct term. This can be easily understood, because the photon at Y propagates to x and thus the γ^μ at x has to have a corresponding γ_μ at Y. The current–current interaction enforces this

Example 3.6.

structure. This point can also be understood by adding the exchange term to the direct term of the two-photon field in (6):

$$A_\mu(x)A_\nu(y) = A_\mu(x)A_\nu(y)|_{\text{dir}} + A_\mu(x)A_\nu(y)|_{\text{exch}} \quad , \tag{10}$$

where the first contribution agrees with (6). The second contribution arises from the possibility that the photon described by the vector potential $A_\mu(x)$ originates from the proton at a point Y prior to the interaction point X of the second photon which produces the field $A_\nu(y)$. Of course the names given to the coordinates X and Y are unimportant and could be interchanged since these are integration variables. What does matter, however, is the sequential ordering of the two interaction vertices. Therefore the exchange contribution

$$\begin{aligned} A_\mu(x)A_\nu(x)|_{\text{exch}} = \mathrm{i}e_p^2 \int d^4X\, d^4Y\, D_{\mathrm{F}}(x-X)\, D_{\mathrm{F}}(y-Y) \\ \times \bar{\psi}_f^p(X)\,\gamma_\nu\, S_{\mathrm{F}}^p(X-Y)\,\gamma_\mu\,\psi_i^p(Y) \end{aligned} \tag{11}$$

differs from the "direct" term (6) in that the matrices γ_μ and γ_ν are interchanged. Both contributions have been added in (9).

Note that (9) contains an integration over four time variables x_0, y_0, X_0, Y_0, each integration extending over the whole real axis. This means that all $4! = 24$ possible time orderings are allowed and will contribute to the amplitude. This is automatically taken care of by the use of the covariant Feynman propagators S_{F} and D_{F} which describe the propagation in both directions of the relative time.

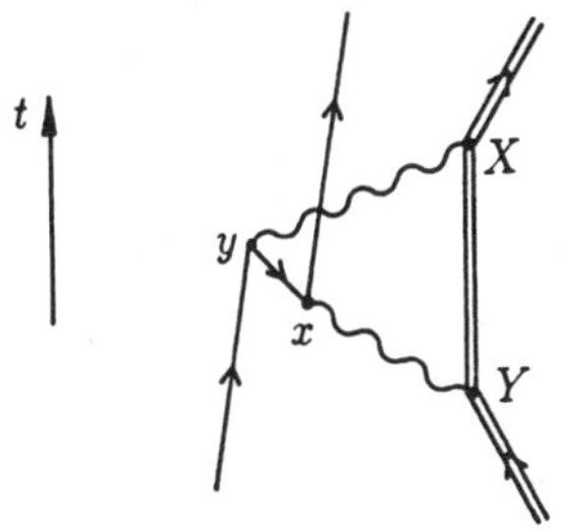

Fig. 3.17. One of several time orderings of the general exchange graph depicted in Fig. 3.16. In contrast to the usual convention for Feynman graphs here the relative position of the vertices has been assumed to have a physical meaning

As an example let us look at the special time ordering $X_0 > y_0 > x_0 > Y_0$. The corresponding exchange graph is depicted in Fig. 3.17. The first photon is emitted by the proton at Y and travels to point x where it creates an e^+e^--pair.[12] The second photon originates from pair annihilation at point y and later is absorbed by the proton at X. In all one could draw $2 \cdot 24 = 48$ different graphs which, however, can be condensed in one direct and one exchange graph. Thus it is clear that the absolute positions of the vertices when drawing a Feynman diagram in coordinate space is not intended to imply a specific time ordering. All that matters is the topological structure of the graph.

In order to develop a general rule we will now consider the factors i that occur in the S-matrix elements. The nondiagonal S-matrix elements generally contain a factor $-\mathrm{i}$ (cf. (1.86) and (2.42)). Also, we introduced a factor i in the expression of the electron propagator in (3) in order to factorize the second-order transition currents. This should also be valid in higher orders, and therefore we generally incorporate a factor i in the electron propagator, writing $\mathrm{i}S_{\mathrm{F}}(x-y)$. For the proton propagator we do the same, i.e. $S_{\mathrm{F}}^p(x-y) \to \mathrm{i}S_{\mathrm{F}}^p(x-y)$. In order to preserve symmetry we require this procedure for every fermion propagator. This can in fact be done if we substitute *simultaneously* $\mathrm{i}S_{\mathrm{F}}(x-y)$ for each electron line *and* $(-\mathrm{i}\not{A})$ for each photon field $\not{A}$. Then the total n^{th} order scattering amplitude does not change, because (cf. e.g. (1) or (2.44)) when following an electron line in a Feynman graph we may write

[12] Also graphs involving the virtual production of proton-antiproton pairs will occur. However, this implies energies where it is not justified to treat the proton as an elementary Dirac particle.

Example 3.6.

$$-\mathrm{i}e\,A\!\!\!/\,S_\mathrm{F}\,e\,A\!\!\!/\,S_\mathrm{F}\ldots e\,A\!\!\!/=(-\mathrm{i}e\,A\!\!\!/)\,\mathrm{i}\,S_\mathrm{F}\,(-\mathrm{i}e\,A\!\!\!/)\ldots(-\mathrm{i}e\,A\!\!\!/)\quad. \tag{12}$$

Note, that the overall factor $(-\mathrm{i})$ is included in the extra factor $A\!\!\!/$ in the first position of the amplitude. Alternatively we may say, that at every electron vertex (point of creation or absorption of a photon A) we write a factor $-\mathrm{i}$ in addition to γ_μ. For symmetry reasons we will have to attach factors $-\mathrm{i}$ also to each proton vertex and factors i to each proton propagator S_F^p. This will lead to a consistent description if a factor i is attached also to each photon propagator D_F. We can easily check that this procedure indeed gives the right overall factor for the two-photon field of (6):

$$\begin{aligned}A_\mu(x)A_\nu(y)|_\mathrm{dir}&=\int d^4X\,d^4Y\,\mathrm{i}D_\mathrm{F}(x-X)\,\mathrm{i}D_\mathrm{F}(y-Y)\\&\quad\times\bar{\psi}_f(X)(-\mathrm{i}e_p)\gamma_\mu\,\mathrm{i}S_\mathrm{F}^p(X-Y)(-\mathrm{i}e_p)\gamma_\nu\,\psi_i^p(Y)\quad.\end{aligned}\tag{13}$$

Generalizing from this result we arrive at a *general rule for the* i*-factors*: At every vertex in a graph (electron or proton vertex) there is a factor $-\mathrm{i}$, for every line (no matter whether electron, proton, or photon line) we have a factor $+\mathrm{i}$. It can be easily seen that this rule does not change the amplitude $S_{fi}^{(2)}$ in (9).

We shall apply this rule in all forthcoming calculations, so we do not have to bother any more about the i-factors and thus about the phase of the graph.

In practice it is convenient to calculate in momentum space. Therefore we determine the Fourier transforms of the expressions occurring in (8). The external particles (i.e. the incoming and outgoing electron and proton) are described by plane waves, as we did before (cf. (3.2), (3.3) and (3.54)). The first term of (9) becomes

$$\begin{aligned}S_{fi}^{(2)}(\mathrm{dir.})&=\frac{(4\pi)^2e^4}{V^2}\int d^4x\,d^4y\,d^4X\,d^4Y\\&\quad\times\sqrt{\frac{m_0^2}{E_fE_i}}\sqrt{\frac{M_0^2}{E_f^\mathrm{p}E_i^\mathrm{p}}}\int\frac{d^4q_1}{(2\pi)^4}\frac{d^4q_2}{(2\pi)^4}\frac{d^4p}{(2\pi)^4}\frac{d^4P}{(2\pi)^4}\frac{\mathrm{e}^{-\mathrm{i}q_1\cdot(x-X)}}{q_1^2+\mathrm{i}\varepsilon}\frac{\mathrm{e}^{-\mathrm{i}q_2\cdot(y-Y)}}{q_2^2+\mathrm{i}\varepsilon}\\&\quad\times\left[\mathrm{e}^{\mathrm{i}p_f\cdot x}\,\bar{u}(p_f,s_f)\,\gamma^\mu\,\frac{\mathrm{e}^{-\mathrm{i}p\cdot(x-y)}}{p\!\!\!/-m_0+\mathrm{i}\varepsilon}\,\gamma^\nu\,u(p_i,s_i)\,\mathrm{e}^{-\mathrm{i}p_i\cdot y}\right]\\&\quad\times\left[\mathrm{e}^{\mathrm{i}P_f\cdot X}\bar{u}(P_f,S_f)\gamma_\mu\frac{\mathrm{e}^{-\mathrm{i}P\cdot(X-Y)}}{P\!\!\!/-M_0+\mathrm{i}\varepsilon}\gamma_\nu u(P_i,S_i)\mathrm{e}^{-\mathrm{i}P_i\cdot Y}\right]\quad.\end{aligned}\tag{14}$$

Here we have used the expressions for electron and proton propagators from (2.7), (2.13) and (2.19) and further the Fourier representation of the photon propagators known from (3.50). We can easily perform the integration over the space coordinates:

$$\begin{aligned}&\int d^4x\,d^4y\,d^4X\,d^4Y\;\mathrm{e}^{-\mathrm{i}q_1\cdot(x-X)}\,\mathrm{e}^{-\mathrm{i}q_2\cdot(y-Y)}\,\mathrm{e}^{\mathrm{i}p_f\cdot x}\\&\qquad\times\mathrm{e}^{-\mathrm{i}p\cdot(x-y)}\mathrm{e}^{-\mathrm{i}p_i\cdot y}\,\mathrm{e}^{\mathrm{i}P_f\cdot X}\,\mathrm{e}^{-\mathrm{i}P\cdot(X-Y)}\,\mathrm{e}^{-\mathrm{i}P_i\cdot Y}\\&\quad=(2\pi)^4\,\delta^4(q_1+p-p_f)\,(2\pi)^4\,\delta^4(q_2-p+p_i)\\&\qquad\times(2\pi)^4\delta^4(-q_2-P+P_i)(2\pi)^4\delta^4(-q_1+P-P_f)\quad.\end{aligned}\tag{15}$$

Each δ^4-function expresses the energy–momentum conservation at one of the four vertices. Now we can integrate over the momentum variables q_2, p, and P occurring in (12):

Example 3.6.

$$\int \frac{d^4q_1}{(2\pi)^4}\frac{d^4q_2}{(2\pi)^4}\frac{d^4p}{(2\pi)^4}\frac{d^4P}{(2\pi)^4}(2\pi)^4\,\delta^4(q_1+p-p_f)$$
$$\times\,(2\pi)^4\,\delta^4(q_2-p+p_i)(2\pi)^4\,\delta^4(-q_2-P+P_i)$$
$$\times\,(2\pi)^4\delta^4(-q_1+P-P_f)\frac{1}{q_1^2+\mathrm{i}\varepsilon}\frac{1}{q_2^2+\mathrm{i}\varepsilon}$$
$$\times\left[\bar{u}(p_f,s_f)\,\gamma^\mu\frac{1}{\not{p}-m_0+\mathrm{i}\varepsilon}\gamma^\nu\,u(p_i,s_i)\right]$$
$$\times\left[\bar{u}(P_f,S_f)\,\gamma_\mu\frac{1}{\not{P}-M_0+\mathrm{i}\varepsilon}\gamma_\nu\,u(P_i,S_i)\right]$$
$$=(2\pi)^4\,\delta^4(P_f+p_f-P_i-p_i)\int\frac{d^4q_1}{(2\pi)^4}\frac{1}{q_1^2+\mathrm{i}\varepsilon}\frac{1}{(q-q_1)^2+\mathrm{i}\varepsilon}$$
$$\times\left[\bar{u}(p_f,s_f)\,\gamma^\mu\frac{1}{\not{p}_f-\not{p}_1-m_0+\mathrm{i}\varepsilon}\gamma^\nu\,u(p_i,s_i)\right]$$
$$\times\left[\bar{u}(P_f,S_f)\,\gamma_\mu\frac{1}{\not{P}_f+\not{p}_1-M_0+\mathrm{i}\varepsilon}\gamma_\nu\,u(P_i,S_i)\right]\quad, \tag{16}$$

with

$$q\;\equiv\;p_f-p_i\;=\;-(P_f-P_i) \tag{17}$$

being the fixed momentum transfer to the electron. According to (14) the contribution to the second-order S-matrix element in the direct scattering graph is given by

$$S_{fi}^{(2)}(\mathrm{dir.})=\frac{(4\pi)^2e^4}{V^2}\sqrt{\frac{m_0^2}{E_fE_i}}\sqrt{\frac{M_0^2}{E_f^{\mathrm{p}}E_i^{\mathrm{p}}}}(2\pi)^4\delta^4(P_f+p_f-P_i-p_i)$$
$$\times\int\frac{d^4q_1}{(2\pi)^4}\frac{1}{q_1^2+\mathrm{i}\varepsilon}\frac{1}{(q-q_1)^2+\mathrm{i}\varepsilon}$$
$$\times\left[\bar{u}(p_f,s_f)\,\gamma^\mu\frac{1}{\not{p}_f-\not{p}_1-m_0+\mathrm{i}\varepsilon}\gamma^\nu\,u(p_i,s_i)\right]$$
$$\times\left[\bar{u}(P_f,S_f)\,\gamma_\mu\frac{1}{\not{P}_f+\not{p}_1-M_0+\mathrm{i}\varepsilon}\gamma_\nu\,u(P_i,S_i)\right]\quad. \tag{18}$$

Again $\delta^4(P_f+p_f-P_i-p_i)$ guarantees energy–momentum conservation. It is very satisfying to recognize that these conservation laws follow automatically. We have further to consider the integral $\int d^4q_1/(2\pi)^4\ldots$ over the four-momentum q_1. q_1 represents the four-momentum "circling around" in a closed loop, which appears in the graph (see Fig. 3.18) corresponding to the process (18) in Fourier space. As we have seen in (15), energy–momentum conservation is fulfilled at each vertex, for instance the upper right vertex yields

$$P_f\;+\;q_1\;-\;q_1\;=\;P_f\quad,$$

the upper left

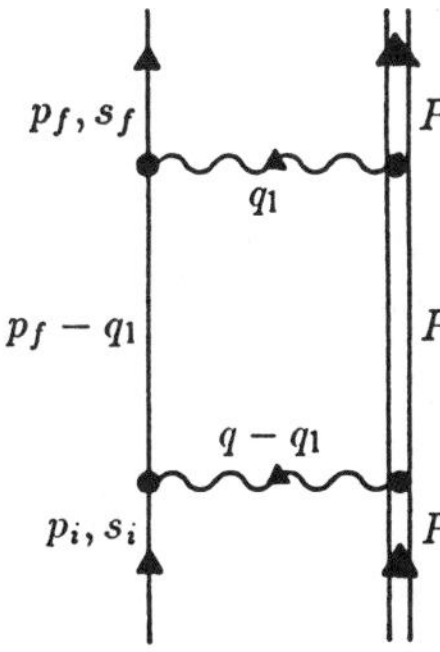

Fig. 3.18. The second-order direct graph for electron–proton scattering in momentum space

Example 3.6.

$$p_f - q_1 + q_1 = p_f \quad ,$$

the lower left

$$p_i + (q - q_1) = p_f - q_1 \quad ,$$

and finally the lower right

$$P_i - (q - q_1) = P_f + q_1 \quad .$$

According to (18) we have to integrate over all intermediate momenta q_1. The intermediate momenta in the loop of Fig. 3.18 have just the appropriate value to conserve the total momentum relation

$$P_f + p_f = P_i + p_i$$

for every value of q_1.

In addition each vertex contributes a factor of the form $-ie\gamma_\mu$ to (18), whereas each external particle yields a factor $\sqrt{m_0/E}$. Note the matrix factors in (18) of the form

$$\frac{i}{\not{p} - m + i\varepsilon}$$

being inserted between the γ-matrices that represent the vertices. These factors represent the propagator of the internal virtual fermionic lines.

We shall soon have had enough practice to be able to write down directly the correct mathematical expressions corresponding to a given Feynman diagram. For instance, we can easily draw the exchange graph corresponding to the direct graph (Fig. 3.18). Formulated in momentum space it is depicted in Fig. 3.19. It contributes to the same amplitude as the direct graph. Again we have individual energy–momentum conservation at each vertex, which leads to conservation of total energy–momentum. We can write down at once the contribution of this graph to the amplitude $S_{fi}^{(2)}$ according to the rules we just derived:

Fig. 3.19. The exchange graph in momentum space

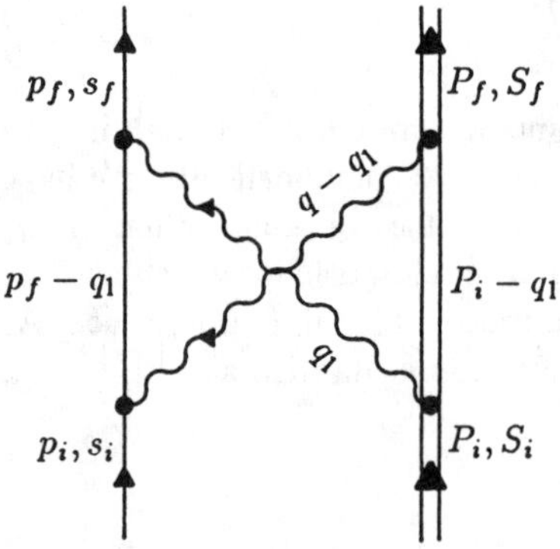

Example 3.6.

$$S_{fi}^{(2)}(\text{exch.}) = \frac{(4\pi)^2 e^4}{V^2}\sqrt{\frac{m_0^2}{E_f E_i}}\sqrt{\frac{M_0^2}{E_f^{\mathrm{p}} E_i^{\mathrm{p}}}}(2\pi)^4\delta^4(P_f + p_f - P_i - p_i)$$
$$\times \int \frac{d^4 q_1}{(2\pi)^4}\frac{1}{q_1^2 + \mathrm{i}\varepsilon}\frac{1}{(q-q_1)^2 + \mathrm{i}\varepsilon}$$
$$\times \left[\bar{u}(p_f, s_f)\gamma^\mu \frac{1}{\not{p}_f - \not{q}_1 - m_0 + \mathrm{i}\varepsilon}\gamma^\nu u(p_i, s_i)\right]$$
$$\times \left[\bar{u}(P_f, S_f)\gamma_\nu \frac{1}{\not{P}_i - \not{q}_1 - M_0 + \mathrm{i}\varepsilon}\gamma_\mu u(P_i, S_i)\right] \quad . \qquad (19)$$

Performing the four-dimensional integrals occurring in (18) and (19) is a difficult task. In the limit of a static point-like charged proton at rest this was done by **Dalitz**[13] for the first time. We shall discuss this problem further in Exercise 3.7, but we remark here that even in this special case problems occur owing to the infinite range of the Coulomb potential.

EXERCISE

3.7 Static Limit of the Two-Photon Exchange

Problem. Write down the scattering amplitude of a two-photon exchange between electron and proton corresponding to the two graphs of Fig. 3.20. Show that in the static limit (the proton has infinite mass) the result coincides with the amplitude of electron scattering at a Coulomb potential in second order Born approximation.

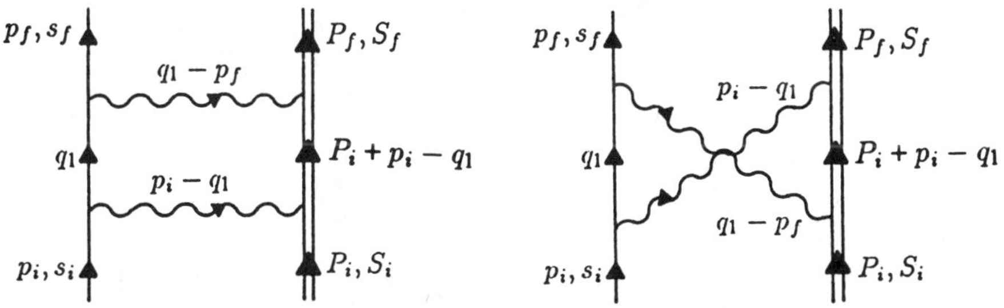

Fig. 3.20. The direct and exchange two-photon scattering graphs

Solution. By use of the Feynman rules the sum of both graphs can be written as (the phase factor resulting from four internal lines and four vertices is $(\mathrm{i})^4(-\mathrm{i})^4 = +1$)

$$S_{fi}^{(2)} = \frac{e^4}{V^2}\sqrt{\frac{m_0^2}{E_f E_i}}\sqrt{\frac{M_0^2}{E_f^{\mathrm{p}} E_i^{\mathrm{p}}}}(2\pi)^4\,\delta^4(P_f + p_f - P_i - p_i)$$
$$\times \int \frac{d^4 q_1}{(2\pi)^4}\frac{4\pi}{(q_1 - p_f)^2 + \mathrm{i}\varepsilon}\frac{4\pi}{(p_i - q_1)^2 + \mathrm{i}\varepsilon}$$

[13] R.H. Dalitz: Proc. Roy. Soc. (London) **A206**, 509 (1951).

Exercise 3.7.

$$\times \left[\bar{u}(p_f,s_f)\gamma_\mu \frac{1}{\not p_1 - m_0 + \mathrm{i}\varepsilon}\gamma_\nu u(p_i,s_i)\right]$$
$$\times \bar{u}(P_f,S_f)\left[\gamma^\mu \frac{1}{\not P_i + \not p_i - \not p_1 - M_0 + \mathrm{i}\varepsilon}\gamma^\nu \right.$$
$$\left. + \gamma^\nu \frac{1}{\not P_f - \not p_i + \not p_1 - M_0 + \mathrm{i}\varepsilon}\gamma^\mu\right] u(P_i,S_i) \quad , \tag{1}$$

where the capital letters refer to the proton and the lower-case ones to the electron. (1) differs from the corresponding expressions (16) and (17) in Example 3.6 in the way the momentum variables are labeled. Both results agree if q_1 is replaced by $p_f - q_1$, which is admissible since q_1 is only an integration variable. In the limit of very large proton mass ($M_0 \gg E_i, E_f$) the recoil energy becomes negligible, i.e.,

$$E_f^{\mathrm{p}} \approx E_i^{\mathrm{p}} \to M_0 \quad ,$$
$$\frac{M_0^2}{E_f^{\mathrm{p}}E_i^{\mathrm{p}}} \to 1 \quad ,$$
$$\delta(E_f^{\mathrm{p}} + E_f - E_i^{\mathrm{p}} - E_i) \to \delta(E_f - E_i) \quad ,$$
$$u(P_f,S_f) \to u(0,S_f) \quad ,$$
$$u(P_i,S_i) \to u(0,S_i) \quad .$$

Taking only the leading powers of M_0 the matrix element involving the proton spinors (1) reduces to

$$P^{\mu\nu}$$
$$\to \bar{u}(0,S_f)\left[\gamma^\mu \frac{M_0\gamma_0 + \not p_i - \not p_1 + M_0}{(P_i + p_i - q_1)^2 - M_0^2 + \mathrm{i}\varepsilon}\gamma^\nu + \right.$$
$$\left. + \gamma^\nu \frac{M_0\gamma_0 - \not p_i + \not p_1 + M_0}{(P_f - p_i + q_1)^2 - M_0^2 + \mathrm{i}\varepsilon}\gamma^\mu\right] u(0,S_i)$$
$$\to \bar{u}(0,S_f)\left[\gamma^\mu \frac{M_0(\gamma_0 + 1)}{M_0^2 + 2M_0(E_i - q_1^0) - M_0^2 + \mathrm{i}\varepsilon}\gamma^\nu \right.$$
$$\left. + \gamma^\nu \frac{M_0(\gamma_0 + 1)}{M_0^2 - 2M_0(E_i - q_1^0) - M_0^2 + \mathrm{i}\varepsilon}\gamma^\mu\right] u(0,S_i)$$
$$= \bar{u}(0,S_f)\left[\gamma^\mu \frac{\gamma_0 + 1}{2(E_i - q_1^0) + \mathrm{i}\varepsilon}\gamma^\nu + \gamma^\nu \frac{\gamma_0 + 1}{-2(E_i - q_1^0) + \mathrm{i}\varepsilon}\gamma^\mu\right] u(0,S_i) \quad . \tag{2}$$

Now we have

$$\gamma_0 + 1 = 2\begin{pmatrix} 1 & 0 \\ 0 & 0 \end{pmatrix} \quad ,$$

and $u(0,S)$ has just upper components. Therefore,

$$u(0,S) = \begin{pmatrix} \chi_s \\ 0 \end{pmatrix} \quad , \quad \gamma_0\, u(0,S) = u(0,S) \quad .$$

On the other hand,

$$\boldsymbol{\gamma}\, u(0,S) = \begin{pmatrix} 0 \\ -\boldsymbol{\sigma}\chi_s \end{pmatrix}$$

Exercise 3.7.

has only nonvanishing lower components, and thus

$$(\gamma_0 + 1)\,\boldsymbol{\gamma}\,u(0,S) = 0 \quad .$$

Therefore, the only non vanishing term in $P^{\mu\nu}$ is

$$P^{\mu\nu} = g^{\mu 0} g^{\nu 0} u^\dagger(0,S_f)\,u(0,S_i)\left(\frac{1}{E_i - q_1^0 + \mathrm{i}\varepsilon} - \frac{1}{E_i - q_1^0 - \mathrm{i}\varepsilon}\right) \quad . \tag{3}$$

The two pole terms can be combined to yield a delta function which is seen from the identity

$$\frac{1}{x \pm \mathrm{i}\varepsilon} = P\left(\frac{1}{x}\right) \mp \mathrm{i}\pi\delta(x) \quad . \tag{4}$$

Thus we finally get

$$P^{\mu\nu} = -2\pi\mathrm{i}\; g^{\mu 0} g^{\nu 0} \delta_{S_f S_i}\,\delta(E_i - q_1^0) \quad . \tag{5}$$

By inserting P into S_{fi} and considering that the proton's degrees of freedom (spin, momentum) are not to be observed and thus in the cross section can be summed over (final values $\boldsymbol{P}_f, S_f$) or averaged (initial spin S_i), we can substitute

$$\begin{aligned} &(2\pi)^3\delta^3(\boldsymbol{P}_f + \boldsymbol{p}_f - \boldsymbol{P}_i - \boldsymbol{p}_i) \to V \quad , \\ &\delta_{S_f S_i} \to 1 \quad . \end{aligned} \tag{6}$$

This can be understood by applying the heuristically deduced relation $(2\pi)^3\delta^3(\boldsymbol{0}) \to V$ (Sect. 3.2) to determine the cross section. Namely, the spin averaging and momentum integration yields

$$\begin{aligned} &\frac{1}{2}\sum_{S_i S_f}\int V\frac{d^3P_f}{(2\pi)^3}\left[(2\pi)^3\delta^3(\boldsymbol{P}_f + \boldsymbol{p}_f - \boldsymbol{P}_i - \boldsymbol{p}_i)\right]^2\delta_{S_i S_f} \\ &= \frac{1}{2}2\int V\frac{d^3P_f}{(2\pi)^3}(2\pi)^3\delta^3(\boldsymbol{P}_f + \boldsymbol{p}_f - \boldsymbol{P}_i - \boldsymbol{p}_i)\underbrace{[(2\pi)^3\delta^3(\boldsymbol{0})]}_{V} \to V^2 \quad . \end{aligned}$$

The same effect is achieved if one substitutes (6) in the scattering matrix element and forgets about the proton altogether. After integrating over q_1^0 we get

$$\begin{aligned} S_{fi}^{(2)} \to\; &\frac{-\mathrm{i}e^4}{V}\sqrt{\frac{m_0^2}{E_f E_i}}\,2\pi\delta(E_f - E_i)\int\frac{d^3q}{(2\pi)^3}\frac{4\pi}{|\boldsymbol{q} - \boldsymbol{p}_f|^2}\frac{4\pi}{|\boldsymbol{q} - \boldsymbol{p}_i|^2} \\ &\times\left[\bar{u}(p_f, s_f)\frac{E_i\gamma_0 + \boldsymbol{q}\cdot\boldsymbol{\gamma} + m_0}{|\boldsymbol{p}|^2 - |\boldsymbol{q}|^2 + \mathrm{i}\varepsilon}u(p_i, s_i)\right] \quad , \end{aligned} \tag{7}$$

which is valid in the limit $M_0 \to \infty$. (7) is just the amplitude of electron scattering at a Coulomb potential with charge e in second-order Born approximation. As mentioned in Example 3.4 the integral is divergent owing to the long range of the Coulomb interaction.

3.3 Scattering of Identical Fermions

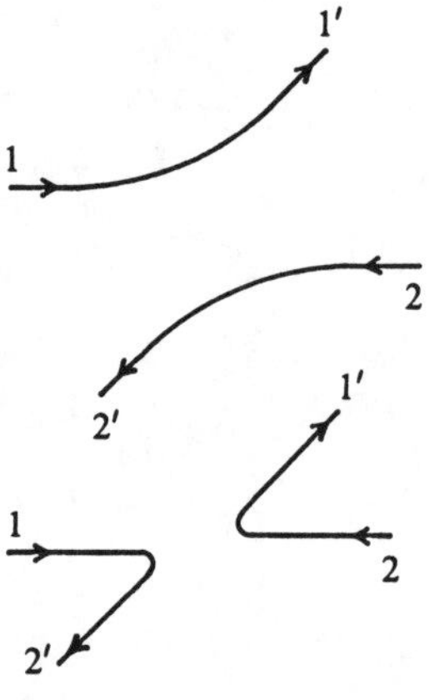

Fig. 3.21. Classical scattering trajectories of two identical particles leading to the same final state. In quantum theory both processes can not be distinguished

Using the example of electron–electron and electron–positron scattering we shall discuss the general aspects of scattering of identical fermions and particle–antiparticle scattering. To this end we can take over many of the results of electron–proton scattering (Sect. 3.2). However, slight complications arise from the fact that the scattering particles are of the same type. There is no way to tell which of the two emerging electrons is the "incident" and which is the "target" particle. This ambiguity is present already in classical mechanics as depicted in Fig. 3.21. Classically, however, one can trace the trajectories and distinguish between the two alternatives. In quantum physics this is no longer possible and therefore the two processes can interfere. Therefore the amplitudes of two indistinguishable processes, which differ in the association of momentum and spin variables (p_1', s_1') and (p_2', s_2') with the outgoing electron lines, have to be added coherently.

Figure 3.22 shows the two lowest order Feynman graphs for *electron–electron scattering*. The figure also illustrates the kinematic conditions for this scattering process. We note the scattering amplitude of electron–electron scattering in momentum space directly (compare (3.59) in Sect. 3.2)

$$
\begin{aligned}
S_{fi} &= S_{fi}(\text{dir.}) + S_{fi}(\text{exch.}) \\
&= -e^2 \frac{1}{V^2} \sqrt{\frac{m_0^2}{E_1 E_2}} \sqrt{\frac{m_0^2}{E_1' E_2'}}\, (2\pi)^4 \delta^4(p_1 + p_2 - p_1' - p_2') \\
&\quad \times \Bigg\{ + \left[\bar{u}(p_1', s_1')(-\mathrm{i}\gamma_\mu) u(p_1, s_1)\right] \frac{\mathrm{i}4\pi}{(p_1 - p_1')^2} \left[\bar{u}(p_2', s_2')(-\mathrm{i}\gamma^\mu) u(p_2, s_2)\right] \\
&\quad - \left[\bar{u}(p_2', s_2')(-\mathrm{i}\gamma_\mu) u(p_1, s_1)\right] \frac{\mathrm{i}4\pi}{(p_1 - p_2')^2} \left[\bar{u}(p_1', s_1')(-\mathrm{i}\gamma^\mu) u(p_2, s_2)\right] \Bigg\} \quad .
\end{aligned}
\tag{3.115}
$$

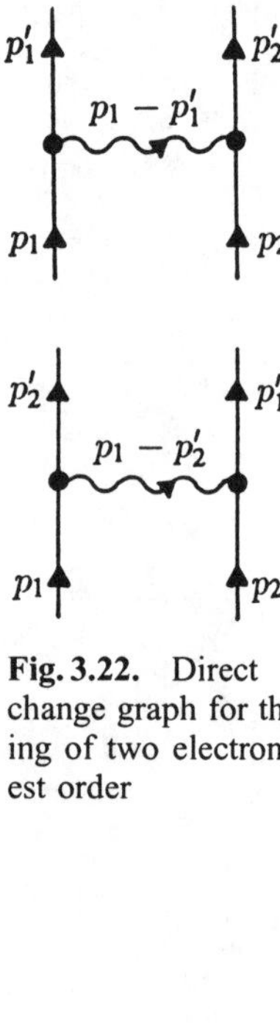

Fig. 3.22. Direct and exchange graph for the scattering of two electrons in lowest order

Since electrons obey Fermi statistics, *the exchange graph has a minus sign.* This yields an antisymmetric scattering amplitude with respect to electron exchange in the final state $(p_1' \leftrightarrow p_2')$ or the initial state $(p_1 \leftrightarrow p_2)$. If we were to calculate the scattering of identical Bose particles, the total amplitude would be symmetric with respect to exchange of these identical bosons. Section 3.7 will illustrate this, since there the amplitude for Compton scattering is symmetric with respect to photon exchange, photons of course being bosons.

Important remarks: (i) The matrix elements $\bar{u}\gamma_\mu u$ are numbers (one for every μ) and thus their order can be exchanged

$$
\bar{u}(1)\gamma_\mu u(1)\, \bar{u}(2)\gamma_\mu u(2) = \bar{u}(2)\gamma_\mu u(2)\, \bar{u}(1)\gamma^\mu u(1) \quad .
$$

(ii) There are no additional normalization factors in (1). In principle one could think of factors like $\frac{1}{\sqrt{2}}$ or $\frac{1}{2}$ – they do not enter here. Also the rules of calculating the differential cross section are not changed by the occurrence of identical particles. We have merely to take care of introducing a factor $\frac{1}{2}$ in the calculation of the total cross section in order to prevent double-counting of the identical particles in the final state. On the other hand, there is no further factor due to the presence of identical particles in the initial state, because the incident flux stays the same whether there are identical particles or not. We can control the validity of our considerations in the case of electron–electron scattering by considering the scattering amplitude (1) for forward scattering. Forward scattering implies small momentum transfer $(p_1 - p_1')^2$. Therefore in (1) the direct term is large compared to the exchange term so that the latter can be neglected. In this case (1) is reduced to the amplitude of electron–proton scattering (to lowest order) already derived in Sect. 3.2, (3.59).

Now we determine the differential cross section of scattering of unpolarized electrons. We start with

$$d\sigma = \int \frac{|S_{fi}|^2}{T \cdot V \cdot \frac{1}{V}} \cdot \frac{1}{\frac{1}{V} v_{\mathrm{rel}}} \frac{V d^3 p_1'}{(2\pi)^3} \frac{V d^3 p_2'}{(2\pi)^3} \quad . \tag{3.116}$$

The first factor denotes the transition rate per unit volume and time per electron $(1/V)^{-1}$, the second one describes the division by the incident electron flux, and the last two factors give the number of final states in phase space. We will calculate $d\sigma$ in the *centre-of-mass frame* of the incoming electrons. In this frame both particles have the same energy $E_1 = E_2 = E$. The centre-of-mass energy does not change in the collision, i.e. it is $E_1' = E_2' = E$. In the centre-of-mass frame momentum conservation yields

$$\boldsymbol{p}_1 + \boldsymbol{p}_2 = \boldsymbol{p}_1' + \boldsymbol{p}_2' = 0 \quad , \quad \text{and thus} \quad \boldsymbol{p}_1' = -\boldsymbol{p}_2' \quad .$$

From this the energies follow:

$$E_1' = \sqrt{m_0^2 + \boldsymbol{p}_1'^2} = E_2' = \sqrt{m_0^2 + \boldsymbol{p}_2'^2} \quad .$$

Energy conservation demands that

$$E_1 + E_2 = E_1' + E_2' \quad ,$$

which gives

$$E_1 = E_2 = E_1' = E_2' = E \quad .$$

With β being the velocity of one electron with respect to the centre-of-mass frame, the relative velocity of the colliding electrons in the centre-of-mass frame is given by

$$v_{\mathrm{rel}} = 2\beta \quad . \tag{3.117}$$

For relativistic energies this relative velocity approaches twice the velocity of light. On first sight one might suspect a contradiction with special relativity. This, in fact, is not the case, because the velocity v of one electron as seen from the other one is given by the well-known equation for the addition of velocities:

$$v = \frac{v_1 + v_2}{1 + v_1 v_2 / c^2} = \frac{\beta + \beta}{1 + \beta \cdot \beta} c \quad ,$$

which can at best reach the velocity of light ($v \to c$) for $\beta \to 1$. By observing the electrons from the centre-of-mass frame, however, their relative velocity is just the difference of their velocities in this system, which is expressed by (3.117). This also follows from the covariant flux factor which was introduced in (3.77)

$$\frac{m_0}{E_1}\frac{m_0}{E_2}\frac{1}{|\boldsymbol{v}_{\rm rel}|} = \frac{m_0^2}{\sqrt{(p_1\cdot p_2)^2 - m_0^4}} = \frac{m_0^2}{\sqrt{(E^2+\boldsymbol{p}^2)^2 - m_0^4}}$$
$$= \frac{m_0^2}{E^2}\frac{1}{2|\boldsymbol{p}|/E} \quad . \tag{3.118}$$

From (3.115) and (3.116) we get the spin-averaged differential scattering cross section

$$d\bar{\sigma} = \frac{m_0^4}{(2\pi)^2 E^4 2\beta}\int \overline{|M_{fi}|^2}\,\delta^4(p_1'+p_2'-p_1-p_2)d^3p_1'\,d^3p_2' \quad , \tag{3.119}$$

with the squared invariant matrix element

$$\overline{|M_{fi}|^2} = e^4(4\pi)^2\frac{1}{4}\sum_{s_1's_1}\sum_{s_2's_2}\left|\bar{u}(p_1',s_1')\gamma_\mu u(p_1,s_1)\frac{1}{(p_1-p_1')^2}\bar{u}(p_2',s_2')\gamma^\mu u(p_2,s_2)\right.$$
$$\left. -\bar{u}(p_2',s_2')\gamma_\mu u(p_1,s_1)\frac{1}{(p_1-p_2')^2}\bar{u}(p_1',s_1')\gamma^\mu u(p_2,s_2)\right|^2 \quad . \tag{3.120}$$

The averaging over the initial spins s_1 and s_2 is responsible for the factor $\frac{1}{4}$. We can easily determine the sums over the spins and convert them into traces with the help of the identity (3.33) in Sect. 3.1. Since $\bar{\gamma}_\mu = \gamma_\mu$ (3.29a), we obtain

$$\overline{|M_{fi}|^2} = e^4(4\pi)^2\frac{1}{4}\left\{\frac{1}{(p_1-p_1')^4}\,\mathrm{Tr}\left[\frac{\not{p}_1'+m_0}{2m_0}\gamma_\mu\frac{\not{p}_1+m_0}{2m_0}\gamma_\nu\right]\right.$$
$$\times\,\mathrm{Tr}\left[\frac{\not{p}_2'+m_0}{2m_0}\gamma^\mu\frac{\not{p}_2+m_0}{2m_0}\gamma^\nu\right]$$
$$-\frac{1}{(p_1-p_1')^2(p_1-p_2')^2}\mathrm{Tr}\left[\frac{\not{p}_1'+m_0}{2m_0}\gamma_\mu\frac{\not{p}_1+m_0}{2m_0}\gamma_\nu\frac{\not{p}_2'+m_0}{2m_0}\gamma^\mu\frac{\not{p}_2+m_0}{2m_0}\gamma^\nu\right]$$
$$\left.+(p_1'\leftrightarrow p_2')\right\} \quad . \tag{3.121}$$

By writing $(p_1'\leftrightarrow p_2')$ we have abbreviated the terms that can be generated from those written down by exchanging p_1' and p_2'. (3.121) contains a sum of four trace terms which are produced by squaring an amplitude consisting of two terms. Let us consider these traces, their origin, and their evaluation in more detail. We start with the first trace term in (3.121) which already was encountered in Sect. 3.2, (3.33). We will repeat the calculation here to gain familiarity with the trace technique. The spin sum to be calculated to obtain the first contribution to (3.121) obviously is

$$\sum_{s_1's_1}\sum_{s_2's_2}\left(\bar{u}(p_1',s_1')\gamma_\mu u(p_1,s_1)\,\bar{u}(p_2',s_2')\gamma^\mu u(p_2,s_2)\right)$$
$$\times\left(\bar{u}(p_1',s_1')\gamma_\nu u(p_1,s_1)\,\bar{u}(p_2',s_2')\gamma^\nu u(p_2,s_2)\right)^*$$

$$
\begin{aligned}
&= \left[\sum_{s_1' s_1} \left(\bar{u}(p_1',s_1')\gamma_\mu u(p_1,s_1)\right) \left(\bar{u}(p_1',s_1')\gamma_\nu u(p_1,s_1)\right)^* \right] \\
&\times \left[\sum_{s_2' s_2} \left(\bar{u}(p_2',s_2')\gamma^\mu u(p_2,s_2)\right) \left(\bar{u}(p_2',s_2')\gamma^\nu u(p_2,s_2)\right)^* \right] \quad . \qquad (3.122)
\end{aligned}
$$

It is sufficient to consider just the first sum over s_1', s_1, since the second sum s_2', s_2 has the same structure. The complex conjugate product of spinors can be rewritten as

$$
\begin{aligned}
\left(\bar{u}(p_2',s_2')\gamma_\nu u(p_2,s_2)\right)^\dagger &= u^\dagger(p_2,s_2)\gamma_\nu^\dagger \gamma_0^\dagger u(p_2',s_2') \\
&= \bar{u}(p_2,s_2)\gamma_0\gamma_\nu^\dagger\gamma_0 u(p_2',s_2') \\
&= \bar{u}(p_2,s_2)\bar{\gamma}_\nu u(p_2',s_2') \quad , \qquad (3.123)
\end{aligned}
$$

with the barred gamma matrix $\bar{\gamma}_\nu = \gamma_\nu$ (see (3.29a)). The evaluation of the spin summation as usual makes use of the projection operators (3.31):

$$
\sum_{s_1} u_\beta(p_1,s_1)\bar{u}_\lambda(p_1,s_1) = \left(\frac{\not{p}_1 + m_0}{2m_0}\right)_{\beta\lambda} \quad . \qquad (3.124)
$$

The sum over s_1 and s_1' in this way leads to

$$
\begin{aligned}
&\sum_{s_1 s_1'} \left(\bar{u}(p_1',s_1')\gamma_\mu u(p_1,s_1)\right) \left(\bar{u}(p_1',s_1')\gamma_\nu u(p_1,s_1)\right)^* \\
&\quad = \sum_{s_1' s_1} \left(\bar{u}(p_1',s_1')\gamma_\mu u(p_1,s_1)\right) \left(\bar{u}(p_1,s_1)\bar{\gamma}_\nu u(p_1',s_1')\right) \\
&\quad = \sum_{s_1'} \bar{u}(p_1',s_1')\gamma_\mu \left(\frac{\not{p}_1 + m_0}{2m_0}\right) \bar{\gamma}_\nu u(p_1',s_1') \\
&\quad = \sum_{\alpha\beta} \left(\gamma_\mu \frac{\not{p}_1 + m_0}{2m_0} \bar{\gamma}_\nu\right)_{\alpha\beta} \sum_{s_1'} u_\beta(p_1',s_1')\bar{u}_\alpha(p_1',s_1') \\
&\quad = \mathrm{Tr}\left[\gamma_\mu \frac{\not{p}_1 + m_0}{2m_0} \bar{\gamma}_\nu \frac{\not{p}_1' + m_0}{2m_0}\right] \quad . \qquad (3.125)
\end{aligned}
$$

Analogously the sum $\sum_{s_2' s_2}$ in (3.122) can be calculated yielding the total result

$$
\begin{aligned}
&\sum_{s_1' s_1} \sum_{s_2' s_2} \left| \left(\bar{u}(p_1',s_1')\gamma_\mu u(p_1,s_1)\right) \left(\bar{u}(p_2',s_2')\gamma^\mu u(p_2,s_2)\right) \right|^2 \\
&\quad = \mathrm{Tr}\left[\gamma_\mu \frac{\not{p}_1 + m_0}{2m_0} \bar{\gamma}_\nu \frac{\not{p}_1' + m_0}{2m_0}\right] \mathrm{Tr}\left[\gamma^\mu \frac{\not{p}_2 + m_0}{2m_0} \bar{\gamma}^\nu \frac{\not{p}_2' + m_0}{2m_0}\right] \quad . \qquad (3.126)
\end{aligned}
$$

Since $\bar{\gamma}_\nu = \gamma_\nu$, this is just the result stated in (3.121). Now we consider the more complicated mixed terms in the square $|\cdots|^2$ of (3.120). The first of these is

$$
\begin{aligned}
&\sum_{s_1' s_1}\sum_{s_2' s_2}\left[\left(\bar{u}(p_1',s_1')\gamma_\mu u(p_1,s_1)\right)\left(\bar{u}(p_2',s_2')\gamma^\mu u(p_2,s_2)\right)\right]\\
&\quad\times\left[\left(\bar{u}(p_2',s_2')\gamma_\nu u(p_1,s_1)\right)\left(\bar{u}(p_1',s_1')\gamma^\nu u(p_2,s_2)\right)\right]^*\\
&=\sum_{s_1' s_1}\sum_{s_2' s_2}\left(\bar{u}(p_1',s_1')\gamma_\mu u(p_1,s_1)\right)\left(\bar{u}(p_1,s_1)\bar{\gamma}_\nu u(p_2',s_2')\right)\\
&\quad\times\left(\bar{u}(p_2',s_2')\gamma^\mu u(p_2,s_2)\right)\left(\bar{u}(p_2,s_2)\bar{\gamma}^\nu u(p_1',s_1')\right)\\
&=\sum_{s_1'}\sum_{s_2'}\left(\bar{u}(p_1',s_1')\gamma_\mu\frac{\not p_1+m_0}{2m_0}\bar{\gamma}_\nu u(p_2',s_2')\right)\\
&\quad\times\left(\bar{u}(p_2',s_2')\gamma^\mu\frac{\not p_2+m_0}{2m_0}\bar{\gamma}^\nu u(p_1',s_1')\right)\\
&=\sum_{s_1'}\bar{u}(p_1',s_1')\gamma_\mu\frac{\not p_1+m_0}{2m_0}\bar{\gamma}_\nu\frac{\not p_2'+m_0}{2m_0}\gamma^\mu\frac{\not p_2+m_0}{2m_0}\bar{\gamma}^\nu u(p_1',s_1')\\
&=\mathrm{Tr}\left[\gamma_\mu\frac{\not p_1+m_0}{2m_0}\bar{\gamma}_\nu\frac{\not p_2'+m_0}{2m_0}\gamma^\mu\frac{\not p_2+m_0}{2m_0}\bar{\gamma}^\nu\frac{\not p_1'+m_0}{2m_0}\right]\quad. \qquad (3.127)
\end{aligned}
$$

In the course of this derivation the identity (3.124) has been used four times. With $\bar{\gamma}_\nu=\gamma_\nu$ we see that this trace is identical with the second trace in (3.121). It represents the *interference term of direct and exchange scattering.*

In order to evaluate the traces we refer to the theorems we proved in the Mathematical Supplement 3.3. Let us consider the trace containing eight γ-matrices originating from the interference term in (3.121). To make life easier here we consider only the *ultrarelativistic limit* where $E\gg m_0$ (the complete result is derived in Exercise 3.8). Then we may neglect terms proportional to m_0^2 and only one term in the expansion of (3.127) remains. Using $\gamma^\nu\not p_1'\gamma_\mu\not p_1\gamma_\nu=-2\not p_1\gamma_\mu\not p_1'$ (cf. Mathematical Supplement 3.3, Theorem (8d)) this trace becomes

$$
\begin{aligned}
\mathrm{Tr}\left[\not p_1'\gamma_\mu\not p_1\gamma_\nu\not p_2'\gamma^\mu\not p_2\gamma^\nu\right]&=-2\,\mathrm{Tr}\left[\not p_1\gamma_\mu\not p_1'\not p_2'\gamma^\mu\not p_2\right]\\
&=-8p_1'\cdot p_2'\mathrm{Tr}\left[\not p_1\not p_2\right]\\
&=-32(p_1'\cdot p_2')(p_1\cdot p_2)\quad, \qquad (3.128)
\end{aligned}
$$

where we have applied Theorems (8c) and (2) in the last steps. In the same ultrarelativistic approximation the traces occurring in the direct term in (3.121) can be simplified to

$$
\begin{aligned}
&\mathrm{Tr}\left[\not p_1'\gamma_\mu\not p_1\gamma_\nu\right]\ \mathrm{Tr}\left[\not p_2'\gamma^\mu\not p_2\gamma^\nu\right]\\
&\quad=4\left[(p_1')_\mu(p_1)_\nu+(p_1')_\nu(p_1)_\mu+p_1'\cdot p_1 g_{\mu\nu}\right]\\
&\quad\quad\times 4\left[(p_2')^\mu(p_2)^\nu+(p_2')^\nu(p_2)^\mu+p_2'\cdot p_2 g^{\mu\nu}\right]\\
&\quad=32\left[(p_1'\cdot p_2')(p_1\cdot p_2)+(p_1'\cdot p_2)(p_1\cdot p_2')\right]\quad. \qquad (3.129)
\end{aligned}
$$

This is a repetition of the calculation in Sect. 3.2, (3.91). The traces were converted to the "slash"-notation by introducing two unit vectors $A^\sigma=g^\sigma{}_\mu$ and $B^\sigma=g^\sigma{}_\nu$ so that $\gamma_\mu=\not A$ and $\gamma_\nu=\not B$. This gives $p_1'\cdot A=(p_1')_\sigma A^\sigma=(p_1')_\mu$ etc. and $A\cdot B=A^\sigma B_\sigma=g_{\mu\nu}$. Furthermore, in the calculation of (3.129) we used Theorems 2 and 3 of Mathematical Supplement 3.3.

Putting together (3.128) and (3.129) the squared invariant matrix element (3.121) reads in the ultrarelativistic limit

$$\begin{aligned}
&\overline{|M_{fi}|^2_{\mathrm{UR}}} \\
&= e^4(4\pi)^2\frac{1}{4}\left\{\frac{32\left[(p_1'\cdot p_2')(p_1\cdot p_2)+(p_1'\cdot p_2)(p_1\cdot p_2')\right]}{(p_1-p_1')^4(2m_0)^4}\right. \\
&\quad -\frac{\left[-32(p_1'\cdot p_2')(p_1\cdot p_2)\right]}{(p_1-p_2')^2(p_1-p_1')^2(2m_0)^4}+\frac{32\left[(p_1'\cdot p_2')(p_1\cdot p_2)+(p_2'\cdot p_2)(p_1'\cdot p_1)\right]}{(p_1-p_2')^4(2m_0)^4} \\
&\quad \left.-\frac{\left[-32(p_1'\cdot p_2')(p_1\cdot p_2)\right]}{(p_1-p_2')^2(p_1-p_1')^2(2m_0)^4}\right\} \quad . \qquad (3.130)
\end{aligned}$$

The two interference terms, being complex conjugates of each other and at the same time real-valued, are found to agree. The scattering cross section in the centre-of-mass frame is given by (3.119):

$$d\bar{\sigma}_{\mathrm{UR}}=\frac{m_0^4}{(2\pi)^2E^42\beta}\int\overline{|M_{fi}|^2_{\mathrm{UR}}}\,\delta^4(p_1'+p_2'-p_1-p_2)\,4E_1'E_2'\frac{d^3p_1'}{2E_1'}\frac{d^3p_2'}{2E_2'}\,. \qquad (3.131)$$

In the integral we introduced the factor $4E_1'E_2'$ in order to get the usual invariant phase-space factor $d^3p'/2E'$.

In the centre-of-mass frame we have

$$\begin{aligned}
p_1&=(E_1,\boldsymbol{p}_1)=(E,\boldsymbol{p}) \quad , \quad p_2=(E_2,\boldsymbol{p}_2)=(E,-\boldsymbol{p}) \quad , \\
p_1'&=(E_1',\boldsymbol{p}_1')=(E',\boldsymbol{p}') \quad , \quad p_2'=(E_2',\boldsymbol{p}_2')=(E',-\boldsymbol{p}') \quad .
\end{aligned} \qquad (3.132)$$

In the second line we already made use of momentum conservation which gives $\boldsymbol{p}_1'=-\boldsymbol{p}_2'$. In the ultrarelativistic limit $E\gg m_0$, i.e. $\boldsymbol{p}^2=E^2-m_0^2\approx E^2$, the scalar products needed in (3.130) are

$$\begin{aligned}
p_1\cdot p_2&=E^2-\boldsymbol{p}_1\cdot\boldsymbol{p}_2=E^2+\boldsymbol{p}^2\approx E^2+E^2=2E^2 \quad , \\
p_1'\cdot p_2'&=E'^2-\boldsymbol{p}_1'\cdot\boldsymbol{p}_2'=E'^2+\boldsymbol{p}'^2\approx E'^2+E'^2=2E'^2 \quad , \\
p_1\cdot p_2'&=EE'-\boldsymbol{p}_1\cdot\boldsymbol{p}_2'=EE'+\boldsymbol{p}\cdot\boldsymbol{p}'=EE'+|\boldsymbol{p}|\cdot|\boldsymbol{p}'|\cos\theta \\
&\approx EE'(1+\cos\theta)=2EE'\cos^2\frac{\theta}{2} \quad , \qquad (3.133) \\
p_1'\cdot p_2&=EE'-\boldsymbol{p}_1{}'\cdot\boldsymbol{p}_2=EE'+\boldsymbol{p}\cdot\boldsymbol{p}'\approx 2EE'\cos^2\frac{\theta}{2} \quad , \\
p_1\cdot p_1'&=EE'-\boldsymbol{p}_1\cdot\boldsymbol{p}_1'=EE'-|\boldsymbol{p}|\cdot|\boldsymbol{p}'|\cos\theta\approx 2EE'\sin^2\frac{\theta}{2} \quad ,
\end{aligned}$$

where θ denotes the scattering angle (cf. Fig. 3.23). The integral in (17) has the structure

$$I=\int\frac{d^3p_1'}{2E'}\frac{d^3p_2'}{2E'}\,\delta^4(p_1'+p_2'-p_1-p_2)\,f(p_1',p_2') \quad , \qquad (3.134)$$

with $f(p_1',p_2')$ containing the squared invariant matrix element (3.130) and the factor $4E'^2$. According to Sect. 3.2, (3.74), the integral over p_2' can be extended from three to four dimensions:

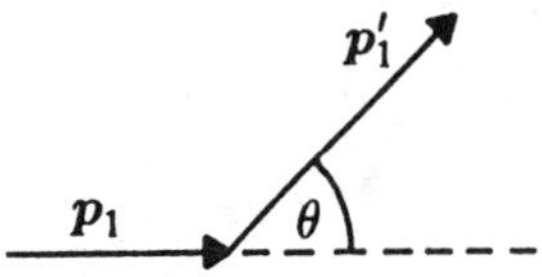

Fig. 3.23. Definition of the scattering angle

$$\int \frac{d^3p_2'}{2E'} = \int_{-\infty}^{\infty} d^4p_2' \, \delta(p_2' \cdot p_2' - m_0^2) \, \Theta((p_2')_0) \quad . \tag{3.135}$$

This integration “eats” the four-dimensional delta function in (3.134) leading to

$$\begin{aligned} I &= \int \frac{|\boldsymbol{p}_1'|^2 d|\boldsymbol{p}_1'| d\Omega_1'}{2E'} \, \delta\left[(p_1 + p_2 - p_1')^2 - m_0^2\right] \\ &\quad \times \Theta(2E - E') f(p_1', p_2' = p_1 + p_2 - p_1') \\ &= \int_0^{2E} \frac{1}{2} |\boldsymbol{p}_1'| dE' d\Omega_1' \, \delta\left[(p_1 + p_2)^2 - 2(p_1 + p_2)\cdot p_1'\right] f(p_1', p_2' = p_1 + p_2 - p_1') \\ &= \frac{d\Omega_1'}{2} \int_0^{2E} |\boldsymbol{p}_1'| dE' \, \delta\left[(2E)^2 - 2(2E)E'\right] f(p_1', p_2' = p_1 + p_2 - p_1') \\ &= \frac{d\Omega_1'}{2} \frac{|\boldsymbol{p}_1'|}{|-4E|} f(p_1', p_2' = p_1 + p_2 - p_1')\Big|_{E'=E} \quad . \end{aligned} \tag{3.136}$$

If we insert this intermediate result into (3.131) we get the differential cross section in the ultrarelativistic limit:

$$\begin{aligned} \left(\frac{d\bar{\sigma}}{d\Omega_1'}\right)_{\mathrm{UR}} &= \frac{m_0^4}{4\pi^2 E^4 2\beta} \frac{|\boldsymbol{p}_1'|}{E} \frac{1}{8} 4E^2 \, \overline{|M_{\mathrm{fi}}|^2_{\mathrm{UR}}} \\ &= \frac{e^4 (4\pi)^2}{4\pi^2 E^4 2} \frac{1}{16} \frac{1}{8} E^2 32 \left[\frac{(p_1' \cdot p_2')(p_1 \cdot p_2) + (p_1' \cdot p_2)(p_1 \cdot p_2')}{(p_1 - p_1')^4} \right. \\ &\quad + \frac{(p_1' \cdot p_2')(p_1 \cdot p_2) + (p_2' \cdot p_2)(p_1' \cdot p_1)}{(p_1 - p_2')^4} \\ &\quad \left. + 2 \, \frac{(p_1' \cdot p_2')(p_1 \cdot p_2)}{(p_1 - p_2')^2 (p_1 - p_1')^2} \right] \Bigg|_{p_2' = p_1 + p_2 - p_1', E' = E} \quad . \end{aligned} \tag{3.137}$$

For an explicit evaluation of this result in terms of the scattering angle θ we insert the scalar products of (3.133) and at various places neglect terms proportional to the electron mass $m_0 \ll E$. In particular, the momentum-transfer denominators are given by

$$\begin{aligned} (p_1 - p_1')^2 &= p_1^2 + p_1'^2 - 2p_1 \cdot p_1' \\ &= 2m_0^2 - 2(E^2 - \boldsymbol{p} \cdot \boldsymbol{p}') \approx -4E^2 \sin^2 \frac{\theta}{2} \quad , \\ (p_1 - p_2')^2 &= p_1^2 + p_2'^2 - 2p_1 \cdot p_2' \\ &= 2m_0^2 - 2(E^2 + \boldsymbol{p} \cdot \boldsymbol{p}') \approx -4E^2 \cos^2 \frac{\theta}{2} \quad . \end{aligned} \tag{3.138}$$

This leads to the final expression

$$
\begin{aligned}
\left(\frac{d\bar{\sigma}}{d\Omega_1'}\right)_{\mathrm{UR}} &= \frac{(4\pi)^2 e^4}{(4\pi)^2 E^2 2}\left[\frac{2E^2 2E^2 + (2E^2\cos^2\frac{\theta}{2})^2}{(4E^2\sin^2\frac{\theta}{2})^2}\right.\\
&\qquad \left. + \frac{2E^2 2E^2 + (2E^2\sin^2\frac{\theta}{2})^2}{(4E^2\cos^2\frac{\theta}{2})^2} + 2\frac{2E^2 2E^2}{4E^2\cos^2\frac{\theta}{2}\,4E^2\sin^2\frac{\theta}{2}}\right]\\
&= \frac{\alpha^2}{8E^2}\left(\frac{1+\cos^4\frac{\theta}{2}}{\sin^4\frac{\theta}{2}} + \frac{1+\sin^4\frac{\theta}{2}}{\cos^4\frac{\theta}{2}} + \frac{2}{\sin^2\frac{\theta}{2}\cos^2\frac{\theta}{2}}\right) \quad . \qquad (3.139)
\end{aligned}
$$

The first and second terms originate from the squares of the direct matrix element and the exchange matrix element, respectively, whereas the last term is the *interference contribution*. The elastic scattering of electrons is known as *Møller scattering*, named after the author who first treated this process correctly using the Dirac equation.[14] Thus (3.139) represents the ultrarelativistic limit of the Møller formula in the centre-of-mass frame.

3.4 Electron–Positron Scattering: Bhabha Scattering and Muon Pair Creation

The scattering of particles and antiparticles is closly related to the process of particle-particle scattering. In fact, we will find that the knowledge of the scattering cross section for one of the processes is sufficient to deduce the corresponding other cross section.

In Example 3.4 we have obtained the amplitude for positron scattering at a Coulomb potential. It was related to the corresponding process of electron scattering by the replacement of an incoming electron spinor $u(p_i, s_i)$ with an outgoing positron spinor $\bar{v}(p_f, s_f)$ and vice versa. In the same way the amplitude for electron-positron scattering can be constructed. Written in momentum space the *direct amplitude* (compare (3.115) in the previous section) becomes

$$
\begin{aligned}
S_{fi}(\mathrm{dir.}) &= +e^2 \frac{1}{V^2}\sqrt{\frac{m_0^2}{E_1\bar{E}_2'}}\sqrt{\frac{m_0^2}{E_1'\bar{E}_2}}\,(2\pi)^4\,\delta^4(p_1 - \bar{p}_2' - p_1' + \bar{p}_2)\\
&\quad \times \left[\bar{u}(p_1', s_1')(-\mathrm{i}\gamma_\mu)u(p_1, s_1)\right]\frac{\mathrm{i}4\pi}{(p_1 - p_1')^2}\left[\bar{v}(\bar{p}_2, \bar{s}_2)(-\mathrm{i}\gamma^\mu)v(\bar{p}_2', \bar{s}_2')\right] \quad , \qquad (3.140)
\end{aligned}
$$

where the bar over a variable is meant to indicate that this quantity refers to an antiparticle. I.e. $(\bar{p}_2, \bar{s}_2)$, $(\bar{p}_2', \bar{s}_2')$ are the momentum and spin of the incoming (outgoing) positron. The corresponding Feynman graph (Fig. 3.24a) contains the labels $-\bar{p}_2'$ and $-\bar{s}_2$ at the positron line which is drawn in the reversed direction when compared to Fig. 3.22. As in the case of electron-electron scattering there also exists an *exchange amplitude*. The corresponding Feynman graph is shown in Fig. 3.24b).

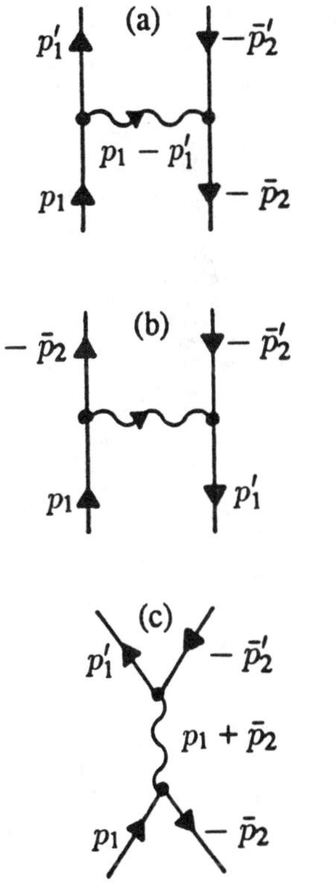

Fig. 3.24. Direct **(a)** and exchange **(b)** diagrams describing electron–positron scattering ("Bhabha scattering"). The indices $-\bar{p}_2'$ and $-\bar{p}_2$ of the positron lines show that we treat the positrons in the language of electrons. Then the positron is an electron with negative four-momentum, in particular with negative energy. The exchange graph usually is drawn as in part **(c)**

[14] C. Møller: Ann. Phys. **14**, 531 (1932).

Here the outgoing electron (p_1') with positive energy is exchanged with the outgoing electron ($-\bar{p}_2$) with negative energy. Graph (b) can also be drawn as shown in (c). Thus it can be interpreted as representing an incoming electron (p_1) with positive energy which by emitting a photon is scattered into an electron ($-\bar{p}_2$) with negative energy moving backwards in time. The photon is absorbed by a negative energy electron ($-\bar{p}_2'$) moving backwards in time. This electron is scattered into a state with positive energy moving forwards in time. This is the interpretation purely in terms of electrons. Adopting the picture involving both electrons and positrons we may say that an incoming electron (p_1) and an incoming positron ($\bar{p}_2$) are annihilated into a photon; the photon is again annihilated ("leptonized") by creating an electron (p_1')– positron ($\bar{p}_2'$) pair. Both nomenclatures, the one using positive and negative energy electrons and the other using electrons and positrons, are equivalent.

The exchange diagram translates into the following 'annihilation amplitude' in momentum space

$$S_{fi}(\text{exch.}) = -e^2 \frac{1}{V^2} \sqrt{\frac{m_0^2}{E_1 \bar{E}_2'}} \sqrt{\frac{m_0^2}{E_1' \bar{E}_2}} (2\pi)^4 \, \delta^4(p_1 - \bar{p}_2' - p_1' + \bar{p}_2)$$
$$\times \left[\bar{v}(\bar{p}_2, \bar{s}_2)(-\mathrm{i}\gamma_\mu) u(p_1, s_1)\right] \frac{\mathrm{i}4\pi}{(p_1 + \bar{p}_2)^2} \left[\bar{u}(p_1', s_1')(-\mathrm{i}\gamma^\mu) v(\bar{p}_2', \bar{s}_2')\right] , \tag{3.141}$$

which has a *relative minus sign* compared with the direct amplitude (3.140). This can be understood in the following way. The initial state (before interaction) consists of an electron (p_1) with positive energy and a sea of electrons with negative energy containing a hole with four-momentum $-\bar{p}_2$. The occupied Dirac sea contains an electron with four-momentum $-\bar{p}_2'$. According to Fermi statistics, the initial state, containing the electrons p_1 and $-\bar{p}_2'$, has to be antisymmetric with respect to the exchange $p_1 \leftrightarrow -\bar{p}_2'$. In the final state we have the analogous antisymmetry concerning $p_1' \leftrightarrow -\bar{p}_2$. In other words: Within our adopted "electron language" the outgoing electrons in Fig. 3.24a with momenta p_1' and $-\bar{p}_2$ have to be antisymmetric. We can immediately check that this antisymmetry is fulfilled by the amplitude (3.141). Of course there is also antisymmetry with respect to all other particles in the Dirac sea, but these do not show up in the scattering process. Thus we can neglect them in our considerations. Again this discussion demonstrates the advantage and clarity of the electron formalism.

Apart from the exchange of spinors, the annihilation amplitude corresponding to Fig. 3.24c has one qualitative difference compared to the scattering processes we have studied up to now: *The virtual photon is timelike*, i.e. its momentum has the property $q^2 > 0$. This is most easily seen in the centre-of-mass frame where $p_1 = (E, \boldsymbol{p})$, $\bar{p}_2 = (E, -\boldsymbol{p})$, $q = (2E, \boldsymbol{0})$. In ordinary scattering processes the exchanged photons are spacelike, $q^2 < 0$. Although being closely related to each other, processes involving timelike and spacelike photons may have quite different properties.

To evaluate the cross section for elastic electron-positron scattering we can proceed exactly as in Sect. 3.3. Equation (3.118) remains valid and (3.120) for the squared spin-averaged invariant matrix element becomes

$$
\begin{aligned}
\overline{|M_{fi}|^2} = e^4(4\pi)^2\frac{1}{4}\Bigg\{ & \frac{1}{(p_1-p_1')^4} \\
& \times \mathrm{Tr}\left[\frac{\not p_1'+m_0}{2m_0}\gamma_\mu\frac{\not p_1+m_0}{2m_0}\gamma_\nu\right]\mathrm{Tr}\left[\frac{-\bar{\not p}_2+m_0}{2m_0}\gamma^\mu\frac{-\bar{\not p}_2'+m_0}{2m_0}\gamma^\nu\right] \\
& -\frac{1}{(p_1-p_1')^2(p_1+\bar p_2)^2} \\
& \times \mathrm{Tr}\left[\frac{\not p_1'+m_0}{2m_0}\gamma_\mu\frac{\not p_1+m_0}{2m_0}\gamma_\nu\frac{-\bar{\not p}_2+m_0}{2m_0}\gamma^\mu\frac{-\bar{\not p}_2'+m_0}{2m_0}\gamma^\nu\right] \\
& +(p_1'\leftrightarrow -\bar p_2)\Bigg\} \quad . \qquad (3.142)
\end{aligned}
$$

Table 3.1. Electron-electron scattering and electron-positron scattering are related by crossing symmetry

e^-+e^-	e^-+e^+
p_1	p_1
p_1'	p_1'
p_2	$-\bar p_2'$
p_2'	$-\bar p_2$

Here we can make a very important observation. The result (3.142) could have been obtained from the corresponding expression (3.120) for electron–electron scattering simply by using the translation Table 3.1 for the momentum variables. This *substitution rule* has very general validity. It is a special case of the law of *crossing symmetry*. This tells us that the S–matrix elements of the processes which are related to each other by an exchange of incoming particles and outgoing antiparticles etc. are essentially the same. One only has to exchange the corresponding momentum variables in the expression for the S–matrix element. For example from the amplitude of a two–body reaction $A+B \to C+D$ we can obtain the corresponding expression involving antiparticles, e.g. $A+\bar D \to C+\bar B$ (see Fig. 3.25) by simply exchanging the momentum variables $p_B \to -p_D$, $p_D \to -p_B$. This rule *holds for the exact amplitudes* as well as in any order of perturbation theory. Also processes which differ in the grouping of incoming and outgoing particles are related to each other. E.g. the matrix element of the three–body decay $A \to \bar B + C + D$ can be derived from that of the two–body scattering process.[15]

If we have obtained an expression for one of the matrix elements we get the corresponding results for the other processes related by crossing symmetry for free, i.e. by an analytic continuation in the momentum variables.

Using the substitution rule ($p_2 \to -\bar p_2'$, $p_2' \to -\bar p_2$) the results from electron–electron scattering can be immediately translated to the case of electron–positron scattering. Thus the cross section for unpolarized elastic e^+e^- scattering in the ultrarelativistic limit (see (3.137)) becomes

$$
\begin{aligned}
\left(\frac{d\bar\sigma}{d\Omega_1'}\right)_{\mathrm{UR}} &= \frac{e^4}{2E^2}\Bigg[\frac{(p_1'\cdot\bar p_2)(p_1\cdot\bar p_2')+(p_1'\cdot\bar p_2')(p_1\cdot\bar p_2)}{(p_1-p_1')^4} \\
&\quad + \frac{(p_1'\cdot\bar p_2)(p_1\cdot\bar p_2')+(\bar p_2\cdot\bar p_2')(p_1'\cdot p_1)}{(p_1+\bar p_2)^4} \\
&\quad + 2\,\frac{(p_1'\cdot\bar p_2)(p_1\cdot\bar p_2')}{(p_1+\bar p_2)^2(p_1-p_1')^2}\Bigg]\Bigg|_{\bar p_2'=p_1+\bar p_2-p_1',\,E'=E} \\
&= \frac{\alpha^2}{8E^2}\left(\frac{1+\cos^4\frac{\theta}{2}}{\sin^4\frac{\theta}{2}}+\frac{1+\cos^2\theta}{2}-\frac{2\cos^4\frac{\theta}{2}}{\sin^2\frac{\theta}{2}}\right) \quad . \qquad (3.143)
\end{aligned}
$$

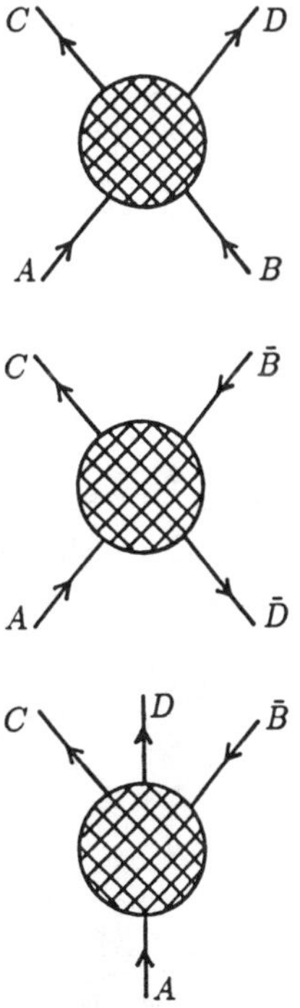

Fig. 3.25. Three processes which are related by crossing symmetry

[15] Note, however, that a process may be forbidden by the energy–momentum conservation law. I.e. the delta function by which the squared amplitude is multiplied to get the cross section (or decay rate) may vanish.

We have used $p_1 \cdot \bar{p}_2 = p_1{}' \cdot \bar{p}_2' = 2E^2$, $p_1 \cdot \bar{p}_2' = p_1{}' \cdot \bar{p}_2 = 2E^2 \cos^2 \frac{\theta}{2}$, and $p_1 \cdot p_1{}' = \bar{p}_2 \cdot \bar{p}_2' = 2E^2 \sin^2 \frac{\theta}{2}$. The full result[16] valid for arbitrary energies will be derived in Exercise 3.8.

The process of elastic electron–positron scattering is also known under the name of **Bhabha** *scattering*. The validity of the Bhabha formula has been verified in many experiments, mainly at high energies. It describes the "elastic background" in all experiments performed at electron–positron colliders. An example[17] is shown in Fig. 3.26. It is essential to include all three contributions (the direct, annihilation and interference terms) to get agreement with the measurements.

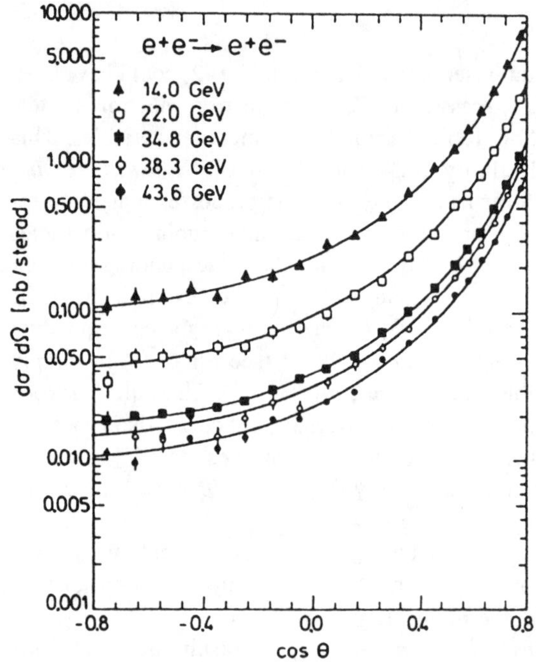

Fig. 3.26. The differential cross section for Bhabha scattering as a function fo the cosine of the scattering angle. The measurements were performed at five different energies $\sqrt{s} = \sqrt{(p_1 + p_2)^2} = 2E$ between 14 GeV and 43.6 GeV. The full curves are the prediction of QED, (3.143)

There are, however, processes where only the annihilation diagram can contribute. This is the case if the initial and final states consist of particle–antiparticle pairs of different type. The simplest example is the process of *muon pair creation* $e^+ + e^- \to \mu^+ + \mu^-$. Since muons differ from electrons just by their much higher mass ($m_\mu = 105.6584\,\text{MeV}$ compared to $m_e = 0.510999\,\text{MeV}$) we can simply take that part of the Bhabha formula which originates from the annihilation diagram. In the ultrarelativistic limit ($E \gg m_\mu$) the differential cross section for muon pair creation thus becomes

$$\left(\frac{d\bar{\sigma}}{d\Omega_1'}\right)_{e^+ + e^- \to \mu^+ + \mu^-} = \frac{\alpha^2}{16E^2}\left(1 + \cos^2\theta\right) \quad . \tag{3.144}$$

Integration over the solid angle gives the total cross section

[16] H.J. Bhabha: Proc. Roy. Soc. (London) **A154**, 195 (1936).

[17] TASSO Collaboration, W. Braunschweig et al.: Z. Physik **C37**, 171 (1988).

$$\bar{\sigma}_{e^+ + e^- \to \mu^+ + \mu^-} = \frac{\pi}{3}\frac{\alpha^2}{E^2} \quad . \tag{3.145}$$

Finally we note that Quantum Electrodynamics no longer gives a correct description of annihilation processes of the type $e^+ + e^- \to l^+ + l^-$ (where l refers to any leptons) if the available energy comes close to the mass of the neutral intermediate vector boson Z, which is $m_Z = 91.16$ GeV. This particle can be produced "on the mass shell" as a *resonance* (the width is $\Gamma_Z = 2.53$ GeV) and then completely dominates over the contribution from the virtual photon.[18] The theory of electroweak interactions gives a unified description of both contributions. (See W. Greiner and B. Müller: Theoretical Physics, Vol. 5, *Gauge Theory of Weak Interactions* (Springer, Berlin, Heidelberg, New York 1993))

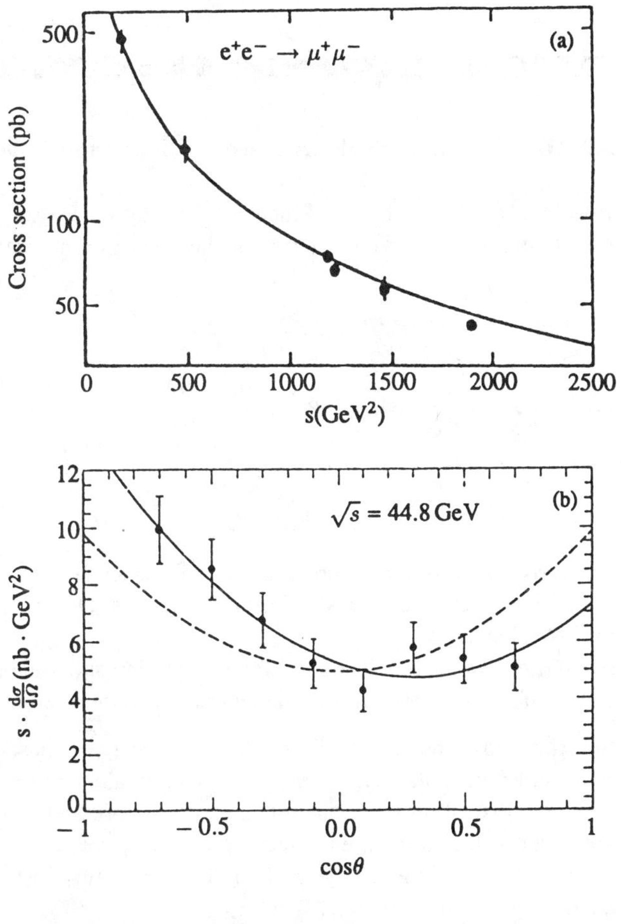

Fig. 3.27. (a) Total cross section for the process $e^+ + e^- \to \mu^+ + \mu^-$ as a function of the squared centre-of-mass energy $s = (2E)^2$, compared with the prediction of (3.145). **(b)** Differential cross section for muon pair creation at centre-of- mass energy $s = 44.8$ GeV. The angular distribution deviates from the QED prediction of (3.144) (dashed line). The fit to the data (full line) shows an asymmetry which results from the influence of the weak interaction

Figure 3.27 shows the measured total muon pair cross section[19] which agrees well with the prediction of (3.145). Already well below the Z-resonance energy,

[18] Recently there has been interest in the search for new particles via their signature as resonances in Bhabha scattering in the MeV energy region. See the review A. Scherdin, J. Reinhardt, W. Greiner, B. Müller: Rep. Prog. Phys. **54** 1 (1991).

[19] TASSO Collaboration, W. Braunschweig et al.: Z. Physik **C40**, 163 (1988).

however, the measured angular distribution[20] shows a notable deviation from the prediction of pure QED (dashed line).

If the theoretically well understood contribution from the weak interaction is included the predictions of QED compare well with experimental data. (In order to get quantitative agreement radiative corrections of the order α^3 have to be taken into account). From this we conclude that the electron is a point-like elementary particle. An extended composite object would be described by a momentum-dependent form-factor $F(q^2)$ as discussed in Exercise 3.5 for the case of electron–nucleon-scattering. Experiments tell us that for the electron $F(q^2) = F(0)$ up to momentum transfers of several hundred GeV. This implies that *the electron is a point-like object down to a distance of at least* $r_e < 10^{-16}$ cm! The same conclusion also holds for the heavy leptons μ and τ.

EXERCISE

3.8 Mandelstam Variables Applied to Møller and Bhabha Scattering

Problem. (a) Show that the kinematics of any binary scattering process $A + B \to C + D$ can be expressed in terms of the three Lorentz-invariant Mandelstam variables

$$\begin{aligned} s &= (p_a + p_b)^2 = (p_c + p_d)^2 \quad , \\ t &= (p_c - p_a)^2 = (p_d - p_b)^2 \quad , \\ u &= (p_c - p_b)^2 = (p_d - p_a)^2 \quad . \end{aligned} \tag{1}$$

Prove the identity

$$s + t + u = m_a^2 + m_b^2 + m_c^2 + m_d^2 \quad . \tag{2}$$

(b) Derive the differential cross sections for electron-electron and electron-positron scattering in terms of the Mandelstam variables. Do not neglect the electron mass in this calculation.

(c) Write down explicit results for the Møller and Bhabha cross sections in the centre-of-mass system and in the laboratory system.

Solution. (a) The two-body scattering process is described by the 16 four-momentum variables p_a, p_b, p_c, p_d. The equivalence principle of special relativity demands that observable quantities can be expressed in terms of Lorentz invariants. Out of the four Lorentz vectors $p_i (i = a, b, c, d)$ one can construct 10 scalar products $p_i \cdot p_j$ where $j \geq i$. Four of these are constrained by the relativistic energy-momentum relations

$$p_i^2 = m_i^2 \quad , \qquad (i = a, b, c, d) \quad . \tag{3}$$

The remaining six degrees of freedom still are interdependent since energy-momentum conservation must be satisfied by any scattering process

[20] JADE Collaboration, W. Bartel et al.: Z. Physik **C30**, 371 (1986).

$$p_a + p_b = p_c + p_d \quad . \tag{4}$$

This gives four additional constraints, leading to the conclusion that two independent kinematic variables are sufficient to describe a two-body scattering process (if no polarizations are involved). From (1) it is obvious that any of the 6 scalar products $p_i \cdot p_j$ can be expressed in terms of one of the Mandelstam variables, e.g. $p_a \cdot p_b = (1/2)(s - m_a^2 - m_b^2)$ etc. (Fig. 3.28). The equivalence of the two expressions given for each of the Mandelstam variables (1) is an immediate consequence of (4). The relation (2) between s, t, and u follows from

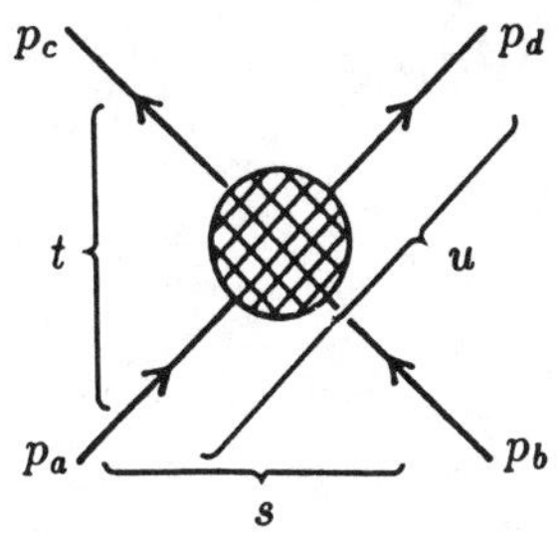

Fig. 3.28. Definition of the Mandelstam variables s, t, u in a two-body scattering process

$$\begin{aligned} s + t + u &= (p_a + p_b)^2 + (p_c - p_a)^2 + (p_d - p_a)^2 \\ &= 3p_a^2 + p_b^2 + p_c^2 + p_d^2 + 2p_a \cdot (p_b - p_c - p_d) \\ &= 3p_a^2 + p_b^2 + p_c^2 + p_d^2 - 2p_a^2 \\ &= m_a^2 + m_b^2 + m_c^2 + m_d^2 \quad . \end{aligned} \tag{5}$$

Although the third variable can be eliminated by means of (2), results often take a more symmetric form if s, t, and u are used simultaneously.

(b) In the centre-of-mass frame of two equal-mass particles (e.g. $e^- + e^- \to e^- + e^-$) the momenta are $p_1 = (E, \boldsymbol{p})$, $p_2 = (E, -\boldsymbol{p})$, $p_1' = (E, \boldsymbol{p}')$, $p_2' = (E, -\boldsymbol{p}')$, and thus the Mandelstam invariants have the values

$$s = (p_1 + p_2)^2 = 4E^2 \quad , \tag{6a}$$

$$\begin{aligned} t &= (p_1' - p_1)^2 = -(\boldsymbol{p}' - \boldsymbol{p})^2 \\ &= -2|\boldsymbol{p}|^2(1 - \cos\theta) = -4|\boldsymbol{p}|^2 \sin^2\frac{\theta}{2} \quad , \end{aligned} \tag{6b}$$

$$\begin{aligned} u &= (p_2' - p_1)^2 = -(-\boldsymbol{p}' - \boldsymbol{p})^2 \\ &= -2|\boldsymbol{p}|^2(1 + \cos\theta) = -4|\boldsymbol{p}|^2 \cos^2\frac{\theta}{2} \quad , \end{aligned} \tag{6c}$$

which obviously satisfies (2).

According to (3.137) in Sect. 3.3 the Møller cross section in the centre-of-mass frame reads (this is valid for arbitrary velocities)

$$\left(\frac{d\bar{\sigma}}{d\Omega}\right)_{\text{cm}} = \frac{m_0^4}{16\pi^2 E^2}\overline{|M_{fi}|^2} = \frac{m_0^4}{4\pi^2 s}\overline{|M_{fi}|^2} \quad . \tag{7}$$

A Lorentz-invariant differential scattering cross section $d\bar{\sigma}/dt$ can be defined if we express the angle between $\boldsymbol{p}_1$ and $\boldsymbol{p}_1'$ (the scattering angle θ_1') in terms of the squared momentum transfer $t = q^2$. In the centre-of-mass frame the transformation is given by (writing $\theta_1' = \pi - \theta_2' = \theta$)

$$\frac{d\cos\theta}{dt} = \left(\frac{dt}{d\cos\theta}\right)^{-1} = (2|\boldsymbol{p}|^2)^{-1} = \frac{2}{s - 4m_0^2} \tag{8}$$

so that (7) becomes

$$\begin{aligned} \frac{d\bar{\sigma}}{dt} &= \frac{d\cos\theta}{dt}\left(\frac{d\bar{\sigma}}{d\cos\theta}\right)_{\text{cm}} = \frac{2}{s - 4m_0^2}\,2\pi\left(\frac{d\bar{\sigma}}{d\Omega}\right)_{\text{cm}} \\ &= \frac{m_0^4}{\pi(s - 4m_0^2)s}\overline{|M_{fi}|^2} \quad . \end{aligned} \tag{9}$$

Exercise 3.8.

We have made use of the fact that the spin-averaged cross section does not depend on the azimuthal angle φ. This would not in general be true in the case of polarized scattering. Although the derivation of (9) was made in the centre-of-mass frame the result is expressed in terms of Lorentz-invariant quantities only and thus is valid in any frame of reference.

Let us evaluate the invariant matrix element $\overline{|M_{fi}|^2}$ for *electron-electron scattering* explicitly, expressed in terms of the Mandelstam variables. According to (3.121) in Sect. 3.3 we have

$$\overline{|M_{fi}|^2}_{\text{Moller}} = e^4(4\pi)^2 \frac{1}{4}\frac{1}{2m_0^2}\left(\frac{1}{t^2}A_{\text{dir}} + \frac{1}{u^2}A_{\text{ex}} - \frac{2}{tu}A_{\text{int}}\right) \quad , \tag{10}$$

where the direct, exchange and interference terms A_{dir}, A_{ex}, A_{int} are functions of the variables s, t, and u. They are given by the traces (see (3.121))

$$\begin{aligned}
A_{\text{dir}} &= \frac{1}{8}\text{Tr}\left[(\not{p}_1' + m_0)\gamma_\mu(\not{p}_1 + m_0)\gamma_\nu\right]\text{Tr}\left[(\not{p}_2' + m_0)\gamma^\mu(\not{p}_2 + m_0)\gamma^\nu\right] \quad , \\
A_{\text{ex}} &= \frac{1}{8}\text{Tr}\left[(\not{p}_2' + m_0)\gamma_\mu(\not{p}_1 + m_0)\gamma_\nu\right]\text{Tr}\left[(\not{p}_1' + m_0)\gamma^\mu(\not{p}_2 + m_0)\gamma^\nu\right] \quad , \\
A_{\text{int}} &= \frac{1}{8}\text{Tr}\left[(\not{p}_1' + m_0)\gamma_\mu(\not{p}_1 + m_0)\gamma_\nu(\not{p}_2' + m_0)\gamma^\mu(\not{p}_2 + m_0)\gamma^\nu\right] \quad .
\end{aligned} \tag{11}$$

According to (3.90) we find for the direct term

$$\begin{aligned}
A_{\text{dir}} &= \frac{1}{8}4\left[p_{1\mu}' p_{1\nu} + p_{1\mu} p_{1\nu}' - g_{\mu\nu}(p_1\cdot p_1' - m_0^2)\right] \\
&\quad \times 4\left[p_2'^{\mu} p_2^{\nu} + p_2^{\mu} p_2'^{\nu} - g^{\mu\nu}(p_2\cdot p_2' - m_0^2)\right] \\
&= 4\left[(p_1\cdot p_2)^2 + (p_1\cdot p_2')^2 - 2m_0^2\, p_1\cdot p_1' + 2m_0^4\right] \quad .
\end{aligned} \tag{12}$$

The scalar products can be expressed in terms of the Mandelstam variables according to

$$\begin{aligned}
p_1\cdot p_2 &= p_1'\cdot p_2' = \tfrac{1}{2}(s - 2m_0^2) \quad , \\
p_1\cdot p_1' &= p_2\cdot p_2' = -\tfrac{1}{2}(t - 2m_0^2) \quad , \\
p_1\cdot p_2' &= p_1'\cdot p_2 = -\tfrac{1}{2}(u - 2m_0^2) \quad .
\end{aligned} \tag{13}$$

Therefore,

$$A_{\text{dir}} = (s - 2m_0^2)^2 + (u - 2m_0^2)^2 + 4m_0^2 t \quad . \tag{14a}$$

The exchange term results from the replacement $p_1' \leftrightarrow p_2'$ which corresponds to the exchange of the variables $u \leftrightarrow t$, i.e.

$$A_{\text{ex}} = (s - 2m_0^2)^2 + (t - 2m_0^2)^2 + 4m_0^2 u \quad . \tag{14b}$$

The interference term consists of a long trace (involving 8 gamma matrices) which can be evaluated using the theorems of the Mathematical Supplement 3.3. Without writing out all intermediate steps the result is

Exercise 3.8.

$$
\begin{aligned}
A_{\mathrm{int}} &= \frac{1}{8}\Big[-32(p_1\cdot p_2)(p_1'\cdot p_2') - 32m_0^4 \\
&\quad + 16m_0^2(p_1\cdot p_2 + p_1\cdot p_1' + p_1\cdot p_2' + p_1'\cdot p_2 + p_1'\cdot p_2' + p_2\cdot p_2')\Big] \\
&= -\left(s^2 - 8m_0^2 s + 12m_0^4\right) \\
&= -(s - 2m_0^2)(s - 6m_0^2) \quad . \qquad (14c)
\end{aligned}
$$

Using the crossing symmetry it is very simple to obtain the corresponding results for *electron–positron scattering*. All we have to do is to replace $p_2 \to -\bar{p}_2'$, $p_2' \to -\bar{p}_2$. This means for the Mandelstam variables:

$$
\begin{aligned}
s &= (p_1 + p_2)^2 \to (p_1 - \bar{p}_2')^2 = u \quad , \\
t &= (p_1' - p_1)^2 \qquad \text{unchanged} \quad , \qquad (15) \\
u &= (p_2' - p_1)^2 \to (\bar{p}_2 + p_1)^2 = s \quad .
\end{aligned}
$$

The Mandelstam variables on the r.h.s. refer to "antiparticle language" where $\bar{p}_2$ is the physical four-momentum of the incoming positron and thus the centre-of-mass energy variable s in Bhabha scattering is identified with $s = (p_1 + \bar{p}_2)^2$. Thus according to (15) the transition from particle-particle to particle-antiparticle scattering simply amounts to the *exchange of the variables s and u*. One often refers to these processes as scattering "in the s-channel" (Møller) and "in the u-channel" (Bhabha).

Thus the invariant matrix element becomes (the bar denotes Bhabha scattering)

$$
\overline{|M_{fi}|^2}_{\mathrm{Bhabha}} = e^4(4\pi)^2 \frac{1}{4}\frac{1}{2m_0^2}\left(\frac{1}{t^2}\bar{A}_{\mathrm{dir}} + \frac{1}{s^2}\bar{A}_{\mathrm{ex}} - \frac{2}{st}\bar{A}_{\mathrm{int}}\right) \quad , \qquad (16)
$$

with

$$
\begin{aligned}
\bar{A}_{\mathrm{dir}} &= (u - 2m_0^2)^2 + (s - 2m_0^2)^2 + 4m_0^2 t = A_{\mathrm{dir}} \quad , \qquad (17a) \\
\bar{A}_{\mathrm{ex}} &= (u - 2m_0^2)^2 + (t - 2m_0^2)^2 + 4m_0^2 s \quad , \qquad (17b) \\
\bar{A}_{\mathrm{int}} &= -(u - 2m_0^2)(u - 6m_0^2) \quad . \qquad (17c)
\end{aligned}
$$

Using the redefinition (15) the relation between s, t, u and the variables E and θ is restored, i.e. (6a-c) remains true for both particle-particle and particle-antiparticle scattering. Note that the formula relating the cross section to the invariant matrix element does not undergo the exchange $s \leftrightarrow u$, i.e. (8) remains unchanged.

(c) Using (6), (7), and (14) the complete *Møller cross section* in the centre-of-mass frame is

$$
\begin{aligned}
\left(\frac{d\bar{\sigma}}{d\Omega}\right)_{\mathrm{cm}} &= \frac{\alpha^2}{8E^2}\left(\frac{1}{t^2}A_{\mathrm{dir}} + \frac{1}{u^2}A_{\mathrm{ex}} - \frac{2}{tu}A_{\mathrm{int}}\right) \\
&= \frac{\alpha^2}{8E^2\boldsymbol{p}^4} \\
&\quad \times \left\{\frac{1}{(1-\cos\theta)^2}\left[(2E^2 - m_0^2)^2 + (\boldsymbol{p}^2(1+\cos\theta) + m_0^2)^2 - 2m_0^2\boldsymbol{p}^2(1-\cos\theta)\right]\right. \\
&\quad + \frac{1}{(1+\cos\theta)^2}\left[(2E^2 - m_0^2)^2 + (\boldsymbol{p}^2(1-\cos\theta) + m_0^2)^2 - 2m_0^2\boldsymbol{p}^2(1+\cos\theta)\right] \\
&\quad \left. + \frac{2}{(1-\cos\theta)(1+\cos\theta)}(2E^2 - m_0^2)(2E^2 - 3m_0^2)\right\} \quad . \qquad (18)
\end{aligned}
$$

Exercise 3.8.

A more compact expression can be deduced with the help of a few elementary trigonometric transformations:

$$\left(\frac{d\bar{\sigma}}{d\Omega}\right)_{\mathrm{cm}} = \frac{\alpha^2}{4E^2}\frac{1}{(E^2-m_0^2)^2}\left[\frac{4(2E^2-m_0^2)^2}{\sin^4\theta} - \frac{8E^4-4E^2m_0^2-m_0^4}{\sin^2\theta} + (E^2-m_0^2)^2\right] \quad . \tag{19}$$

In the ultrarelativistic limit $E \gg m_0$ this can be shown to approach the result (3.139) of Sect. 3.3 In the nonrelativistic limit $E \simeq m_0, \boldsymbol{p}^2 \ll m_0^2$ (18) approaches

$$\left(\frac{d\bar{\sigma}}{d\Omega}\right)_{\mathrm{nr}} = \frac{\alpha^2}{m_0^2}\frac{1}{16\beta^4}\left(\frac{1}{\sin^4\theta/2} + \frac{1}{\cos^4\theta/2} - \frac{1}{\sin^2\theta/2\cos^2\theta/2}\right) \quad . \tag{20}$$

This is a symmetrized version of the Rutherford cross section, including an interference term.

The cross section for *Bhabha scattering* in the center-of-mass frame follows from (16), (17). Contrary to (19) it can not be expressed in terms of powers of $1/\sin^2\theta$. This is plausible since now it is possible to distinguish between forward ($\theta < \pi/2$) and backward ($\theta > \pi/2$) scattering so that the cross section will not be symmetric unter the exchange $\theta \to \pi - \theta$. The Bhabha cross section can be written as

$$\begin{aligned}\left(\frac{d\bar{\sigma}}{d\Omega}\right)_{\mathrm{cm}} = \frac{\alpha^2}{16E^2}\Bigg\{ & \frac{1}{\boldsymbol{p}^4\sin^4\theta/2}\left[m_0^4 + 4\boldsymbol{p}^2m_0^2\cos^2\frac{\theta}{2} + 2\boldsymbol{p}^4(1+\cos^4\frac{\theta}{2})\right] \\ & + \frac{1}{E^4}\left[3m_0^4 + 4\boldsymbol{p}^2m_0^2 + \boldsymbol{p}^4(1+\cos^2\theta)\right] \\ & - \frac{1}{E^2\boldsymbol{p}^2\sin^2\theta/2}\left(3m_0^4 + 8\boldsymbol{p}^2m_0^2\cos^2\frac{\theta}{2} + 4\boldsymbol{p}^4\cos^4\frac{\theta}{2}\right)\Bigg\} \quad .\end{aligned} \tag{21}$$

The ultrarelativistic limit ((3.143) in Sect. 3.4) can be read off immediately. The nonrelativistic limit is

$$\left(\frac{d\bar{\sigma}}{d\Omega}\right)_{\mathrm{nr}} = \frac{\alpha^2}{m_0^2}\frac{1}{16\beta^4}\frac{1}{\sin^4\theta/2} \quad , \tag{22}$$

which just agrees with the Rutherford cross section. The contribution of the annihilation graph is suppressed by a factor $O(\beta^2)$ at low energies.

In the *laboratory system* the target particle is initially at rest, i.e.

$$\begin{aligned} p_1 &= (E, \boldsymbol{p}) \quad , \quad p_2 = (m_0, \boldsymbol{0}) \quad , \\ p_1' &= (E_1', \boldsymbol{p}_1') \quad , \quad p_2' = (E_2', \boldsymbol{p}_2') \quad . \end{aligned} \tag{23}$$

The Mandelstam invariants take the form

$$s = (p_1+p_2)^2 = 2m_0^2 + 2p_1\cdot p_2 = 2m_0(E+m_0) \quad , \tag{24a}$$

$$t = (p_1'-p_1)^2 = 2m_0^2 - 2p_1'\cdot p_1 = 2m_0^2 - 2E_1'E + 2|\boldsymbol{p}_1'||\boldsymbol{p}|\cos\theta_1' \quad , \tag{24b}$$

$$u = (p_1'-p_2)^2 = 2m_0^2 - 2p_1'\cdot p_2 = -2m_0(E_1'-m_0) \quad . \tag{24c}$$

Since the scattering process is uniquely specified by fixing the incident energy E and the scattering angle θ_1' the variable E_1' is redundant and can be expressed as

Exercise 3.8.

a function of E and θ_1'. We can make use of the relation (2) which can be solved for E_1'

$$\begin{aligned} 0 &= s + t + u - 4m_0^2 \\ &= -2(E + m_0)(E_1' - m_0) + \sqrt{E_1'^2 - m_0^2}\sqrt{E^2 - m_0^2}\cos\theta_1' \quad , \end{aligned} \tag{25}$$

with the result

$$E_1' = m_0 \frac{E + m_0 + (E - m_0)\cos^2\theta_1'}{E + m_0 - (E - m_0)\cos^2\theta_1'} \tag{26}$$

and consequently

$$|\boldsymbol{p}_1'| = \frac{2|\boldsymbol{p}|m_0\cos\theta_1'}{E + m_0 - (E - m_0)\cos^2\theta_1'} \quad . \tag{27}$$

This leads to the following explicit expressions for the Mandelstam variables

$$t = -\frac{2m_0(E^2 - m_0^2)\sin^2\theta_1'}{E + m_0 - (E - m_0)\cos^2\theta_1'} \quad , \tag{28a}$$

$$u = -\frac{4m_0^2(E - m_0)\cos^2\theta_1'}{E + m_0 - (E - m_0)\cos^2\theta_1'} \quad . \tag{28b}$$

We also can derive an explicit expression for the scattering angle in the centre-of-mass frame which now will be denoted by θ^*. Using (6) we get

$$\cos\theta^* = \frac{t - u}{4\sqrt{E^{*2} - m_0^2}} \quad , \tag{29}$$

which becomes, after inserting (28),

$$\cos\theta^* = \frac{-(E + m_0) + (E + 3m_0)\cos^2\theta_1'}{E + m_0 - (E - m_0)\cos^2\theta_1'} \quad . \tag{30}$$

The centre-of-mass energy E^* is related to the laboratory energy E through (equating the expressions for s)

$$E^* = \sqrt{\tfrac{1}{2}m_0(E + m_0)} \quad . \tag{31}$$

The scattering cross section in the laboratory frame follows from the invariant cross section by

$$\left(\frac{d\bar{\sigma}}{d\Omega}\right)_{\mathrm{lab}} = \frac{1}{2\pi}\frac{dt}{d\cos\theta_1'}\frac{d\bar{\sigma}}{dt} \quad . \tag{32}$$

Differentiating (28a) with respect to $\cos\theta_1'$ we get

$$\begin{aligned} \frac{dt}{d\cos\theta_1'} &= 2\cos\theta_1'\frac{dt}{d\cos^2\theta_1'} \\ &= \frac{8m_0^2(E^2 - m_0^2)\cos\theta_1'}{[E + m_0 - (E - m_0)\cos^2\theta_1']^2} \quad . \end{aligned} \tag{33}$$

Exercise 3.8.

Using (9) this leads to

$$\left(\frac{d\bar{\sigma}}{d\Omega}\right)_{\text{lab}} = \frac{m_0^4}{\pi^2} \frac{\cos\theta_1'}{[E+m_0-(E-m_0)\cos^2\theta_1']^2} \overline{|M_{fi}|^2} \quad . \tag{34}$$

The invariant matrix elements have been constructed earlier in terms of the Mandelstam variables, which in turn have been expressed through the laboratory variables E and θ_1'. The results (10) with (14) for Møller scattering and (16) with (17) for Bhabha scattering are quite complicated expressions which can not be simplified significantly when laboratory variables are used.

3.5 Scattering of Polarized Dirac Particles

Up to now all calculations have been performed under the assumption that the spin of the electron (or positron) is not observed. In this example we will learn a technique to calculate the scattering of polarized fermions. This will be applied to the simplest possible case, i.e. Coulomb scattering of electrons as discussed in Sect. 3.1.

First let us briefly review the description of spin polarization (see Chaps. 6 and 7 in *RQM* for details). Free electrons with momentum p and spin s are described by spinors $u(p,s)$. s^μ is a Lorentz vector which is properly defined in the rest system of the particle where it reduces to a spatial unit vector

$$(s^\mu)_{\text{R.S.}} = \left(0, \boldsymbol{s}'\right) \quad . \tag{3.146}$$

The components of s^μ in a frame in which the particle moves with momentum $\boldsymbol{p}$ are obtained by a Lorentz boost with the result

$$s^\mu = \left(\frac{\boldsymbol{p}\cdot\boldsymbol{s}'}{m_0}, \boldsymbol{s}' + \frac{\boldsymbol{s}'\cdot\boldsymbol{p}}{m_0(E+m_0)}\boldsymbol{p}\right) \quad . \tag{3.147}$$

It is easily checked that (3.147) satisfies the normalization and orthogonality relations

$$s^2 = -1 \quad , \quad p\cdot s = 0 \quad , \tag{3.148}$$

which in the rest frame are trivially fulfilled. In the rest frame the unit spinors are eigenstates of the operator $\boldsymbol{\Sigma}\cdot\boldsymbol{s}'$:

$$\boldsymbol{\Sigma}\cdot\boldsymbol{s}' u(\boldsymbol{0},\pm\boldsymbol{s}') = \pm u(\boldsymbol{0},\pm\,\boldsymbol{s}') \quad . \tag{3.149}$$

Here $\boldsymbol{\Sigma} = \gamma_5\gamma^0\boldsymbol{\gamma}$ which in the standard representation is the “double” Pauli matrix

$$\boldsymbol{\Sigma} = \begin{pmatrix} \boldsymbol{\sigma} & 0 \\ 0 & \boldsymbol{\sigma} \end{pmatrix} \quad . \tag{3.150}$$

The condition (3.149) has the covariant generalisation

$$\gamma_5 \not{s}\, u(p, \pm s) = \pm u(p, \pm s) \quad . \tag{3.151}$$

Here an extra factor γ^0 was included in order to make (3.151) valid also for positron spinors $v(p, \pm s)$. Using (3.151) one introduces the *spin projection operator*

$$\hat{\Sigma}(s) = \frac{1}{2}(1 + \gamma_5 \not{s}) \tag{3.152}$$

with the property

$$\hat{\Sigma}(s)\, u(p, +s) = u(p, +s) \quad , \quad \hat{\Sigma}(s)\, u(p, -s) = 0 \quad . \tag{3.153}$$

The formalism also applies to *helicity states* where the spin points in the direction of the momentum. Here one has to choose

$$s'_\lambda = \lambda \frac{\boldsymbol{p}}{|\boldsymbol{p}|} \quad , \quad \text{where} \quad \lambda = \pm 1 \quad , \tag{3.154}$$

which according to (3.147) leads to the spin 4-vector

$$s^\mu_\lambda = \lambda \left(\frac{|\boldsymbol{p}|}{m_0}, \frac{\boldsymbol{p}}{|\boldsymbol{p}|} + \frac{\boldsymbol{p}^2}{m_0(E + m_0)} \frac{\boldsymbol{p}}{|\boldsymbol{p}|} \right) = \lambda \left(\frac{|\boldsymbol{p}|}{m_0}, \frac{E}{m_0} \frac{\boldsymbol{p}}{|\boldsymbol{p}|} \right) \quad . \tag{3.155}$$

We call an electron with spin s parallel to p *right-handed* and conversely one with spin direction opposite to the momentum *left-handed*. Alternatively one speaks of *positive and negative helicity* ($\lambda = +1$ or $\lambda = -1$).

After these preliminary remarks we look at the cross section for Coulomb scattering of an electron of momentum p_i and initial spin s_i. If also the final spin s_f is measured the result is (see Sect. 3.1, (3.25)):

$$\frac{d\sigma}{d\Omega}(s_i, s_f) = \frac{4Z^2\alpha^2 m_0^2}{|\boldsymbol{q}|^4} \left| \bar{u}(p_f, s_f)\gamma^0 u(p_i, s_i) \right|^2 \quad . \tag{3.156}$$

This expression in principle can be evaluated by choosing a particular representation of the Dirac matrices and inserting the corresponding explicit expressions for the spinors $u(p, s)$. However , there is a more elegant way to proceed. We can make use of the trace techniques developed in the previous sections by introducing auxiliary summations over the spin orientations s_i and s_f. This can be done with the help of the spin projection operator $\Sigma(s)$ which according to (3.153) suppresses the "wrong" spin state $u(p, -s)$. By use of these operators the Coulomb scattering cross section (3.156) can be expressed as follows:

$$\begin{aligned} \frac{d\sigma}{d\Omega}(s_i, s_f) &= \frac{4Z^2\alpha^2 m_0^2}{|\boldsymbol{q}|^4} \left(\bar{u}(p_f, s_f)\gamma_0 u(p_i, s_i)\right) \left(u^\dagger(p_i, s_i)\gamma_0^\dagger \gamma_0^\dagger u(p_f, s_f)\right) \\ &= \frac{4Z^2\alpha^2 m_0^2}{|\boldsymbol{q}|^4} \sum_{s'_f, s'_i} \left(\bar{u}(p_f, s'_f)\gamma_0 \hat{\Sigma}(s_i) u(p_i, s'_i)\right) \left(\bar{u}(p_i, s'_i)\gamma_0 \hat{\Sigma}(s_f) u(p_f, s_f)\right) \quad . \end{aligned} \tag{3.157}$$

The projection $\hat{\Sigma}(s_i)u(p_i, s'_i) = \delta_{s_i s'_i} u(p_i, s_i)$ ensures that the sum over s'_i yields just one term with $s'_i = s_i$. The same holds true for the sum over s'_f. It is obviously sufficient to introduce the projection operators into one of the two matrix elements.

The double sum (3.157) can be transformed into a trace as in Sect. 3.1, (3.33). Thus we get

$$\begin{aligned}\frac{d\sigma}{d\Omega}(s_i,s_f) &= \frac{4Z^2\alpha^2m_0^2}{|q|^4}\operatorname{Tr}\left[\gamma_0\hat{\Sigma}(s_i)\frac{\not{p}_i+m_0}{2m_0}\gamma_0\hat{\Sigma}(s_f)\frac{\not{p}_f+m_0}{2m_0}\right]\\ &= \frac{4Z^2\alpha^2m_0^2}{|q|^4}\operatorname{Tr}\left[\gamma_0\frac{1+\gamma_5\not{s}_i}{2}\frac{\not{p}_i+m_0}{2m_0}\gamma_0\frac{1+\gamma_5\not{s}_f}{2}\frac{\not{p}_f+m_0}{2m_0}\right],\end{aligned} \tag{3.158}$$

Before discussing this result let us look at a scattering process where either the incoming beam is *unpolarized* or the polarization of the outgoing particles is *not observed.* Then only one of the projection operators will be present in the trace. E.g. if the incoming beam is unpolarized (3.158) has to be replaced by

$$\frac{d\sigma}{d\Omega}(s_f) = \frac{1}{2}\frac{4Z^2\alpha^2m_0^2}{|q|^4}\operatorname{Tr}\left[\frac{1+\gamma_5\not{s}_f}{2}\gamma_0\frac{\not{p}_i+m_0}{2m_0}\gamma_0\frac{\not{p}_f+m_0}{2m_0}\right] \quad . \tag{3.159}$$

The factor $1/2$ originates from averaging over the initial spins.

In the last step the traces containing $\gamma_5\not{s}_f$ are found to vanish. From Theorem 1 of Mathematical Supplement 3.3 we know that the trace of a product of γ_5 and an odd number of $\not{a}$-matrices vanishes:

$$\begin{aligned}\operatorname{Tr}\left[\gamma_5\not{a}_1\cdots\not{a}_n\right] &= (-1)^n\operatorname{Tr}\left[\not{a}_1\cdots\not{a}_n\gamma_5\right]\\ &= (-1)^n\operatorname{Tr}\left[\gamma_5\not{a}_1\cdots\not{a}_n\right] = 0 \qquad \text{for odd } n \quad .\end{aligned} \tag{3.160}$$

After expanding (3.159) two traces containing γ_5 remain, which can be easily transformed by anticommutation of the matrix γ_0 into terms of the form (Theorem 5)

$$\operatorname{Tr}\left[\gamma_5\not{a}\not{b}\right] = 0 \quad . \tag{3.161}$$

Thus the cross section (3.159) is *independent of the final spin* and agrees with half the unpolarized Mott scattering cross section of (3.39), Sect. 3.1,

$$\frac{d\sigma}{d\Omega}(s_f) = \frac{1}{2}\frac{d\bar{\sigma}}{d\Omega} \quad . \tag{3.162}$$

Thus it appears that Coulomb scattering of electrons does not lead to a polarization of the beam. It has to be stressed, however, that this is true only in the lowest order of perturbation theory! If Coulomb–Dirac wave functions are used instead of plane waves one finds that the outgoing particles are polarized in the direction orthogonal to the scattering plane. The degree of polarization is of the order $Z\alpha$ and can become quite large for heavy nuclei.

Even in a first-order calculation we may expect that an incident polarized electron escapes the scattering process with a final polarization depending on its scattering angle. We shall now calculate these final polarizations. We assume the spin of the incoming electron to be parallel to its direction of motion i.e. we start from a state with a well–defined helicity $\lambda_i = +1$. In the final state with some specific probability the spin of the particle will be parallel or antiparallel to its direction of motion (helicity $\lambda_f = \pm 1$) (see Fig. 3.29). Thus (3.158) has to be evaluated inserting the polarization 4-vectors

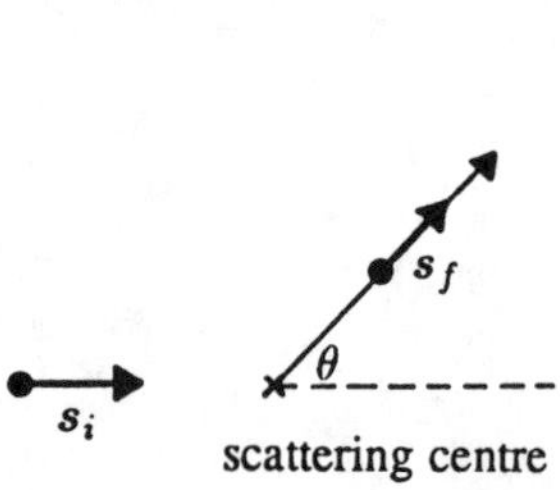

Fig. 3.29. A scattering process with positive helicities in the initial and final channel

$$
\begin{aligned}
s_{i_{\lambda_i}} &= \lambda_i \left(\frac{|\boldsymbol{p}|}{m_0} \quad , \frac{E}{m_0} \frac{\boldsymbol{p}_i}{|\boldsymbol{p}|} \right) \equiv \lambda_i s_i \quad , \\
s_{f_{\lambda_f}} &= \lambda_f \left(\frac{|\boldsymbol{p}|}{m_0} \quad , \frac{E}{m_0} \frac{\boldsymbol{p}_f}{|\boldsymbol{p}|} \right) \equiv \lambda_f s_f \quad .
\end{aligned} \tag{3.163}
$$

The polarized scattering cross section then takes the form

$$
\begin{aligned}
\frac{d\sigma}{d\Omega}(\lambda_f, \lambda_i) &= \frac{4Z^2\alpha^2 m_0^2}{|q|^4} \operatorname{Tr} \left[\gamma_0 \frac{1+\lambda_i \gamma_5 \not{s}_i}{2} \frac{\not{p}_i + m_0}{2m_0} \gamma_0 \frac{1+\lambda_f \gamma_5 \not{s}_f}{2} \frac{\not{p}_f + m_0}{2m_0} \right] \\
&= \frac{4Z^2\alpha^2 m_0^2}{|q|^4} \frac{1}{4} \frac{1}{(2m_0)^2} \Big\{ \operatorname{Tr}\big[\gamma_0(\not{p}_i + m_0)\gamma_0(\not{p}_f + m_0)\big] \\
&\quad + \lambda_i \lambda_f \operatorname{Tr}\big[\gamma_0 \gamma_5 \not{s}_i (\not{p}_i + m_0) \gamma_0 \gamma_5 \not{s}_f (\not{p}_f + m_0)\big] \Big\} \quad .
\end{aligned} \tag{3.164}
$$

Here we dropped those terms which contain only a single $\gamma_5 \not{s}$–factor since the traces vanish, see (3.159)–(3.162). As a consequence the scattering cross section depends on the product of helicities $\lambda_i \lambda_f$. In many experiments the *degree of polarization* of the scattered particles is measured. It is defined as the difference between counting rates for positive and negative helicities, normalized to the total countig rate:

$$
P = \frac{d\sigma(\lambda_f = +1) - d\sigma(\lambda_f = -1)}{d\sigma(\lambda_f = +1) + d\sigma(\lambda_f = -1)} \quad . \tag{3.165}
$$

If the initial state is fully polarized, e.g. $\lambda_i = +1$, the final degree of polarization becomes, using (3.164)

$$
P = \frac{\operatorname{Tr}\big[\gamma_0 \gamma_5 \not{s}_i (\not{p}_i + m_0) \gamma_0 \gamma_5 \not{s}_f (\not{p}_f + m_0)\big]}{\operatorname{Tr}\big[\gamma_0(\not{p}_i + m_0)\gamma_0(\not{p}_f + m_0)\big]} \quad . \tag{3.166}
$$

The evaluation of the traces (see Exercise 3.9) leads to the result

$$
P = 1 - \frac{2\sin^2\frac{\theta}{2}}{\left(\frac{E}{m_0}\right)^2 \cos^2\frac{\theta}{2} + \sin^2\frac{\theta}{2}} \quad . \tag{3.167}
$$

In the nonrelativistic limit $E \longrightarrow m_0$ this reduces to

$$
P \simeq 1 - 2\sin^2\frac{\theta}{2} = \cos\theta \quad . \tag{3.168}
$$

This is just the geometric overlap between the initial and final quantization axes, $\cos\theta = \boldsymbol{p}_i \cdot \boldsymbol{p}_f / |\boldsymbol{p}^2|$, and indicates, that the spin is not influenced at all by the collision, when viewed from a fixed system. When the collision becomes relativistic the degree of polarization becomes less strongly angle dependent and approaches a constant value at $P = 1$ as $E \rightarrow \infty$. This is shown in Fig. 3.30.

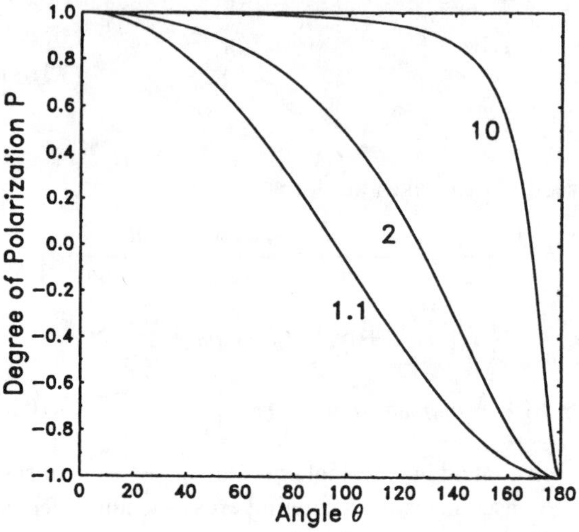

Fig. 3.30. The degree of polarisation according to (3.167) for various incident kinetic energies E/m_0 of the electron

EXERCISE

3.9 Degree of Polarization

Problem. Calculate the differential scattering cross section $(d\sigma/d\Omega)(\lambda_f,\lambda_i)$ for Coulomb scattering of electrons with longitudinal polarization. Derive the degree of polarization P as a function of the scattering angle.

Solution. According to Sect. 3.5, (3.164) the polarized Coulomb scattering cross section is given by

$$\frac{d\sigma}{d\Omega}(\lambda_f,\lambda_i) = \frac{4Z^2\alpha^2 m_0^2}{|q|^4}\frac{1}{16m_0^2}(T_1+\lambda_i\lambda_f T_2) \quad , \tag{1}$$

with the traces

$$\begin{aligned} T_1 &= \mathrm{Tr}\left[\gamma_0(\not{p}_i+m_0)\gamma_0(\not{p}_f+m_0)\right] \quad , \\ T_2 &= \mathrm{Tr}\left[\gamma_0\gamma_5\not{s}_i(\not{p}_i+m_0)\gamma_0\gamma_5\not{s}_f(\not{p}_f+m_0)\right] \quad . \end{aligned} \tag{2}$$

The first trace can be easily determined (cf. Sect. 3.1, (3.36):

$$\begin{aligned} T_1 &= \mathrm{Tr}\left[\gamma_0\not{p}_i\gamma_0\not{p}_f\right] + m_0^2\,\mathrm{Tr}\left[\gamma_0\gamma_0\right] \\ &= 4\,E_iE_f + 4E_fE_i - 4(p_i\cdot p_f) + 4m_0^2 \\ &= 4\left(E^2 + p^2\cos\theta + m^2\right) \\ &= 4\left(E^2(1+\cos\theta) + m_0^2(1-\cos\theta)\right) \\ &= 8\left(E^2\cos^2\frac{\theta}{2} + m_0^2\sin^2\frac{\theta}{2}\right) \quad , \end{aligned} \tag{3}$$

Exercise 3.9.

where we have taken care of energy conservation in Mott scattering, i.e. $E_i = E_f = E$ and $|\boldsymbol{p}_i| = |\boldsymbol{p}_f| = |\boldsymbol{p}|$. Since γ_5 anticommutes with all gamma matrices γ^μ, and since $\gamma_5^2 = 1$, the second trace can be written in the form

$$\begin{aligned} T_2 &= \mathrm{Tr}\left[\gamma_0 \not{s}_i(-\not{p}_i + m_0)\gamma_0 \not{s}_f(\not{p}_f + m_0)\right] \\ &= -\mathrm{Tr}\left[\gamma_0 \not{s}_i \not{p}_i \gamma_0 \not{s}_f \not{p}_f\right] + \mathrm{Tr}\left[\gamma_0 \not{s}_i \gamma_0 \not{s}_f\right] m_0^2 \quad . \end{aligned} \tag{4}$$

Again, all other traces vanish since they contain an odd number of γ-matrices. Theorem 3 of Mathematical Supplement 3.3 implies that

$$\begin{aligned} &\mathrm{Tr}\left[\not{a}_1 \not{a}_2 \not{a}_3 \not{a}_4 \not{a}_5 \not{a}_6\right] \\ &\quad = 4\big[\, (12)(34)(56) + (12)(36)(45) - (12)(35)(46) \\ &\quad - (13)(24)(56) - (13)(26)(45) + (13)(25)(46) \\ &\quad + (14)(23)(56) + (14)(26)(35) - (14)(25)(36) \\ &\quad - (15)(23)(46) - (15)(26)(34) + (15)(24)(36) \\ &\quad + (16)(23)(45) + (16)(25)(34) - (16)(24)(35)\big] \quad , \end{aligned} \tag{5}$$

with (ij) denoting $a_i \cdot a_j$. The 4-spin vectors for positive helicity are given by (see (3.163) in Sect. 3.5)

$$\begin{aligned} s_i &= \left(\frac{|\boldsymbol{p}|}{m_0}, \frac{E}{m_0}\frac{\boldsymbol{p}_i}{|\boldsymbol{p}|}\right) \quad , \\ s_f &= \left(\frac{|\boldsymbol{p}|}{m_0}, \frac{E}{m_0}\frac{\boldsymbol{p}_f}{|\boldsymbol{p}|}\right) \quad . \end{aligned} \tag{6}$$

The following scalar products are needed in (5)

$$p_i \cdot s_i = p_f \cdot s_f = 0 \quad , \tag{7a}$$

$$p_i \cdot s_f = p_f \cdot s_i = \frac{|\boldsymbol{p}|}{m_0}(1 - \cos\theta) \quad , \tag{7b}$$

$$s_i \cdot s_f = \frac{1}{m_0^2}\left(\boldsymbol{p}^2 - E^2 \cos\theta\right) \quad , \tag{7c}$$

$$p_i \cdot p_f = E^2 - \boldsymbol{p}^2 \cos\theta \quad . \tag{7d}$$

Inserting the vectors $a_1 = a_4 = (1, 0, 0, 0)$, $a_2 = s_i$, $a_3 = p_i$, $a_5 = s_f$, $a_6 = p_f$ the result of the first term of the sum (4) is

$$\begin{aligned} &\mathrm{Tr}\left[\gamma_0 \not{s}_i \not{p}_i \gamma_0 \not{s}_f \not{p}_f\right] \\ &\quad = 4\left[s_i^0 E_i (s_f \cdot p_f) + s_i^0 (p_i \cdot p_f) s_f^0 - s_i^0 (p_i \cdot s_f) E_f\right. \\ &\quad - E_i s_i^0 (s_f \cdot p_f) - E_i (s_i \cdot p_f) s_f^0 + E_i (s_i \cdot s_f) E_f \\ &\quad + (s_i \cdot p_i)(s_f \cdot p_f) + (s_i \cdot p_f)(p_i \cdot s_f) - (s_i \cdot s_f)(p_i \cdot p_f) \\ &\quad - s_f^0 (s_i \cdot p_i) E_f - s_f^0 (s_i \cdot p_f) E_i + s_f^0 s_i^0 (p_i \cdot p_f) \\ &\quad \left. + E_f (s_i \cdot p_i) s_f^0 + E_f (s_i \cdot s_f) E_i - E_f s_i^0 (p_i \cdot s_f)\right] \quad . \end{aligned} \tag{8}$$

Because $s_f \cdot p_f = s_i \cdot p_i = 0$, in every line the first term of the sum vanishes. Inserting the scalar products from (7) the remaining part yields

Exercise 3.9.

$$\begin{aligned}
\mathrm{Tr}\left[\gamma_0 \not{s}_i \not{p}_i \gamma_0 \not{s}_f \not{p}_f\right] &= 4\left[\frac{\boldsymbol{p}^2}{m_0^2}(E^2 - \boldsymbol{p}^2\cos\theta) - \frac{\boldsymbol{p}^2E^2}{m_0^2}(1-\cos\theta)\right.\\
&\quad - \frac{\boldsymbol{p}^2E^2}{m_0^2}(1-\cos\theta) + \frac{E^2}{m_0^2}(\boldsymbol{p}^2 - E^2\cos\theta) + \frac{\boldsymbol{p}^2E^2}{m_0^2}(1-\cos\theta)^2\\
&\quad - \frac{1}{m_0^2}(\boldsymbol{p}^2 - E^2\cos\theta)(E^2 - \boldsymbol{p}^2\cos\theta) - \frac{\boldsymbol{p}^2E^2}{m_0^2}(1-\cos\theta)\\
&\quad \left. + \frac{\boldsymbol{p}^2}{m_0^2}(E^2 - \boldsymbol{p}^2\cos\theta) + \frac{E^2}{m_0^2}(\boldsymbol{p}^2 - E^2\cos\theta) - \frac{\boldsymbol{p}^2E^2}{m_0^2}(1-\cos\theta)\right]\\
&= 4\left[\frac{2\boldsymbol{p}^2}{m_0^2}(E^2 - \boldsymbol{p}^2\cos\theta) - \frac{4\boldsymbol{p}^2E^2}{m_0^2}(1-\cos\theta) + \frac{2E^2}{m_0^2}(\boldsymbol{p}^2 - E^2\cos\theta)\right.\\
&\quad \left. + \frac{\boldsymbol{p}^2E^2}{m_0^2}(1-\cos\theta)^2 - \frac{1}{m_0^2}(\boldsymbol{p}^2 - E^2\cos\theta)(E^2 - \boldsymbol{p}^2\cos\theta)\right]\\
&= \frac{4}{m_0^2}\left(-\boldsymbol{p}^4\cos\theta - E^4\cos\theta + 2\boldsymbol{p}^2E^2\cos\theta\right)\\
&= -\frac{4}{m_0^2}\left(\boldsymbol{p}^2 - E^2\right)^2\cos\theta = -4m_0^2\cos\theta \quad .
\end{aligned} \tag{9}$$

With the abbreviation $\tilde{s}_i = (s_i^0, -\boldsymbol{s}_i)$ the second trace in (4) yields

$$\mathrm{Tr}\left[\gamma_0 \not{s}_i \gamma_0 \not{s}_f\right] m_0^2 = m_0^2\,\mathrm{Tr}\left[\not{\tilde{s}}_i \not{s}_f\right] = 4m_0^2(\tilde{s}_i \cdot s_f) = 4\left(\boldsymbol{p}^2 + E^2\cos\theta\right) \quad . \tag{10}$$

The sum of the traces in (4) then is

$$\begin{aligned}
T_2 &= 4m_0^2\cos\theta + 4(\boldsymbol{p}^2 - E^2\cos\theta) = 4E^2(1+\cos\theta) - 4m_0^2(1-\cos\theta)\\
&= 8\left(E^2\cos^2\frac{\theta}{2} - m_0^2\sin^2\frac{\theta}{2}\right) \quad .
\end{aligned} \tag{11}$$

The polarized scattering cross section (1) thus becomes

$$\begin{aligned}
\frac{d\sigma}{d\Omega}(\lambda_f, \lambda_i) = \frac{4Z^2\alpha^2m_0^2}{|\boldsymbol{q}|^4}\frac{1}{2}&\left[E^2\cos^2\frac{\theta}{2} + m_0^2\sin^2\frac{\theta}{2}\right.\\
&\left. + \lambda_i\lambda_f\left(E^2\cos^2\frac{\theta}{2} - m_0^2\sin^2\frac{\theta}{2}\right)\right] \quad .
\end{aligned} \tag{12}$$

Special cases are the *helicity–flip* cross section $\lambda_f = -\lambda_i$

$$\left(\frac{d\sigma}{d\Omega}\right)_{\text{flip}} = \left(\frac{d\sigma}{d\Omega}\right)_{\text{Ruth}}\frac{m_0^2}{E^2}\sin^2\frac{\theta}{2} \quad , \tag{13}$$

and the *non–flip* cross section $\lambda_f = +\lambda_i$

$$\left(\frac{d\sigma}{d\Omega}\right)_{\text{non-flip}} = \left(\frac{d\sigma}{d\Omega}\right)_{\text{Ruth}}\cos^2\frac{\theta}{2} \quad . \tag{14}$$

We have factorized out the Rutherford cross section

$$\left(\frac{d\sigma}{d\Omega}\right)_{\text{Ruth}} = \frac{4Z^2\alpha^2E^2}{|\boldsymbol{q}|^4} = \frac{Z^2\alpha^2}{4p^2\beta^2\sin^4\frac{\theta}{2}} \quad . \tag{15}$$

Exercise 3.9.

The unpolarized Mott cross section derived in Sect. 3.1, (3.39), is given by the sum of (13) and (14)

$$\begin{aligned}\left(\frac{d\sigma}{d\Omega}\right)_{\text{Mott}} &= \left(\frac{d\sigma}{d\Omega}\right)_{\text{flip}} + \left(\frac{d\sigma}{d\Omega}\right)_{\text{non-flip}} \\ &= \left(\frac{d\sigma}{d\Omega}\right)_{\text{Ruth}} \left(1 - \beta^2 \sin^2\frac{\theta}{2}\right) \quad . \end{aligned} \tag{16}$$

The correction factor obviously results from the suppression of helicity-flip transitions at high energies in (13).

Finally the degree of polarization can be obtain by taking the difference between the non-flip and flip cross sections and dividing by the sum of both quantities. This leads to the result

$$\begin{aligned} P &= \frac{E^2\cos^2\theta/2 - m_0^2\sin^2\theta/2}{E^2\cos^2\theta/2 + m_0^2\sin^2\theta/2} \\ &= 1 - \frac{2m_0^2\sin^2\theta/2}{E^2\cos^2\theta/2 + m_0^2\sin^2\theta/2} \quad . \end{aligned} \tag{17}$$

3.6 Bremsstrahlung

When electrons scatter at protons or in the field of a nucleus, they can emit real photons. This process is called *bremsstrahlung* because it involves an acceleration or deceleration (in German: "bremsen") of the projectile. The emitted *real photons* fulfill the Einstein relation

$$q^2 = 0 \quad . \tag{3.169}$$

Bremsstrahlung can be described by Feynman graphs, similar to those we have already encountered, with the difference that now the photon line does not end at a vertex. In this case the corresponding particle either travels into the future as a free photon (emission) or it emerges from the past (absorption). Generally the difference between real and virtual particles is given by the fact that the graphical lines of the former have an open end which signals an emission or absorption process whereas those of the latter both start and end at a vertex (see Fig. 3.31).

In order to study the interaction of particles with the electromagnetic field in the case of bremsstrahlung we start with the *four-potential* of "one" photon with momentum $k^\mu = (\frac{\omega}{c}, \boldsymbol{k})$ and polarization ε_μ. The vector potential A_μ is given by a plane wave:

$$A^\mu(x,k) = \varepsilon^\mu N_k\left(\mathrm{e}^{-\mathrm{i}k\cdot x} + \mathrm{e}^{\mathrm{i}k\cdot x}\right) \quad , \tag{3.170}$$

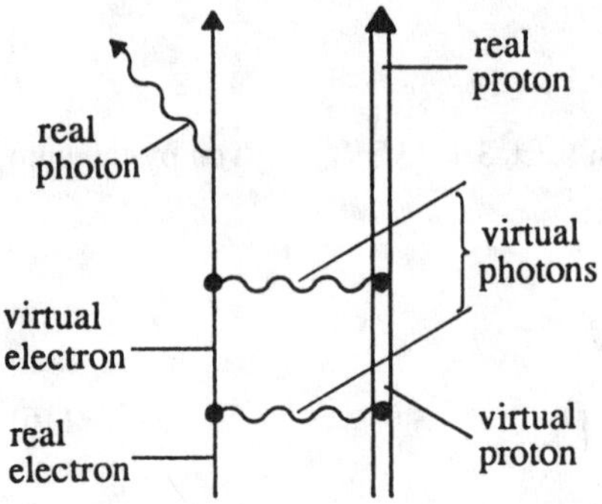

Fig. 3.31. A Feynman graph involving several real and virtual particles

where the normalization constant N_k will be determined later. With $\hbar = c = 1$ we have $\omega = |\boldsymbol{k}|$ and the dispersion relation reads

$$k_\mu k^\mu = 0 \quad . \tag{3.171}$$

In order to understand the polarization vector $\varepsilon_\mu(\boldsymbol{k}, \lambda)$ we have to study the gauge dependence of the A_μ-field more closely. In an arbitrary gauge the vector potential of the free electromagnetic field satisfies the wave equation

$$\Box A^\mu - \partial^\mu(\partial_\nu A^\nu) = 0 \quad , \tag{3.172}$$

which is a consequence of Maxwell's equations. A_μ is a four-dimensional vector field and thus appears to have four degrees of freedom. However, all observables are invariant under gauge transformations $A^\mu \to A^\mu + \partial^\mu \Lambda(x)$ with an arbitrary function $\Lambda(x)$. It is always possible to find a function such that the transformed potential satisfies the *Lorentz gauge*

$$\partial_\mu A^\mu = 0 \tag{3.173}$$

so that the second term in (3.172) vanishes. Even within this restricted class it is still possible to make further gauge transformations, provided that the function $\Lambda(x)$ satisfies the d'Alembert equation $\Box \Lambda(x) = 0$. One possible choice is to set

$$A^0 = 0 \quad , \quad \nabla \cdot \boldsymbol{A} = 0 \quad , \tag{3.174}$$

which is called the "radiation gauge". In this way the number of degrees of freedom has been reduced twice by imposing constraints on the the A_μ-field. Thus we have derived the well-known fact that photons can have only two polarization states, $\lambda = 1, 2$, which both are transversal. The condition (3.174) of course is not covariant and will be valid only in one particular Lorentz frame. In this frame the polarization vectors are purely space-like.

$$\varepsilon^\mu = \big(0, \boldsymbol{\varepsilon}(\boldsymbol{k}, \lambda)\big) \quad , \quad \lambda = 1, 2 \quad , \tag{3.175}$$

with the two transverse three-vectors

$$\boldsymbol{k} \cdot \boldsymbol{\varepsilon}(\boldsymbol{k}, \lambda) = 0 \quad , \tag{3.176}$$

which are normalized to unit length

$$\varepsilon\cdot\varepsilon = 1 \quad . \tag{3.177}$$

By performing a Lorentz boost transformation the three-vectors ε can be generalized into 4-vectors ε^μ which satisfy the covariant conditions

$$\varepsilon^\mu\varepsilon_\mu = -1 \quad , \tag{3.178}$$
$$k^\mu\varepsilon_\mu = 0 \quad . \tag{3.179}$$

Let us discuss the normalization of the photon field in (3.170). The constant N_k will be chosen in such a way that the energy of the wave A^μ is just equal to ω ($\hbar = 1!$), i.e. it is equal to the energy of a single photon. The energy of the electromagnetic field in Gaussian units is given by

$$E_{\text{photon}} = \frac{1}{8\pi}\int d^3x \; < \boldsymbol{E}^2 + \boldsymbol{B}^2 > \; = \frac{1}{4\pi}\int d^3x \; < \boldsymbol{B}^2 > \quad , \tag{3.180}$$

since $\boldsymbol{E}^2 = \boldsymbol{B}^2$ on the average. As

$$\boldsymbol{B} = \nabla\times\boldsymbol{A} = \mathrm{i}N_k\,\boldsymbol{k}\times\boldsymbol{\varepsilon}\left(\mathrm{e}^{-\mathrm{i}k\cdot x} - \mathrm{e}^{\mathrm{i}k\cdot x}\right) = 2N_k\,\boldsymbol{k}\times\boldsymbol{\varepsilon}\,\sin(k\cdot x) \quad . \tag{3.181}$$

The square of the cross product becomes

$$\begin{aligned}(\boldsymbol{k}\times\boldsymbol{\varepsilon})\cdot(\boldsymbol{k}\times\boldsymbol{\varepsilon}) &= \boldsymbol{\varepsilon}\cdot\boldsymbol{\varepsilon}\,\boldsymbol{k}\cdot\boldsymbol{k} - (\boldsymbol{k}\cdot\boldsymbol{\varepsilon})^2 = (\varepsilon_0^2 - \varepsilon\cdot\varepsilon)\boldsymbol{k}^2 - (k_0\varepsilon_0 - k\cdot\varepsilon)^2 \\ &= \varepsilon_0^2\boldsymbol{k}^2 + \boldsymbol{k}^2 - \varepsilon_0^2k_0^2 = \boldsymbol{k}^2 = \omega^2 \quad ,\end{aligned} \tag{3.182}$$

where the conditions (3.178) and (3.179) have been used. We find the energy (3.180) to be

$$E_{\text{photon}} = \frac{4\omega^2}{4\pi}N_k^2\int d^3x \; < \sin^2(\omega\cdot t - \boldsymbol{k}\cdot\boldsymbol{x}) > \; = \frac{2\omega^2}{4\pi}N_k^2V \quad . \tag{3.183}$$

The condition $E_{\text{photon}} = \omega$ leads to the normalisation constant[21]

$$N_k = \sqrt{\frac{4\pi}{2\omega V}} \quad . \tag{3.184}$$

Now, our task is to determine the scattering amplitude for emitting such a photon during electron scattering. In order to simplify this problem we first consider the electron scattering at an external (static) Coulomb field as in Sect. 3.1. The corresponding Feynman diagrams for emitting a photon in lowest-order scattering are shown in Fig. 3.32.

As we can see, *bremsstrahlung is a second-order process.* The emission of a photon by a free electron in the presence of an external field to first order does not happen, since in that case energy and momentum conservation could not be fulfilled simultaneously. The graph of this forbidden process is shown in Fig. 3.33. The conservation law in this case would require

$$k = p_f - p_i \quad , \quad \text{i.e.} \quad k^2 = (p_f - p_i)^2 \quad . \tag{3.185}$$

For a real photon, $k^2 = 0$. On the other hand, it follows that

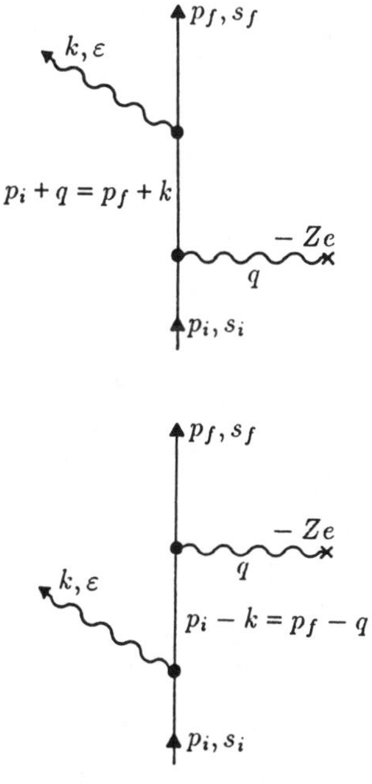

Fig. 3.32. The two Feynman diagrams describing the emission of a photon in lowest-order electron scattering at a static Coulomb field (bremsstrahlung)

[21] When rationalized instead of Gaussian units are used the factor 4π in (3.184) is absent.

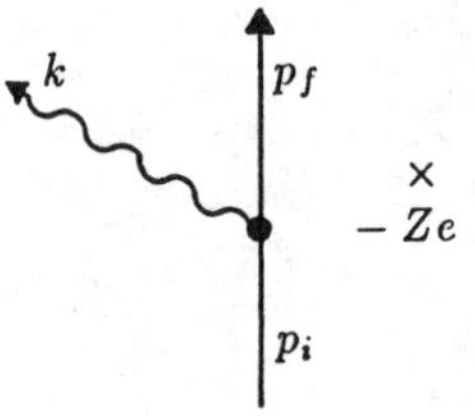

Fig. 3.33. Ficticious graph of first order bremsstrahlung. This process is forbidden (the corresponding amplitude vanishes, because energy and momentum conservation cannot be fulfilled simultaneously)

$$
\begin{aligned}
(p_f - p_i)^2 &= p_f{}^2 + p_i{}^2 - 2p_f \cdot p_i = 2m_0^2 - 2E_f E_i + 2\boldsymbol{p}_f \cdot \boldsymbol{p}_i \\
&= 2\left(m_0^2 + \boldsymbol{p}_f \cdot \boldsymbol{p}_i - \sqrt{m_0^2 + \boldsymbol{p}_f^2}\sqrt{m_0^2 + \boldsymbol{p}_i^2} \right) < 0 \quad . \qquad (3.186)
\end{aligned}
$$

Thus (3.185) yields a contradiction.

The S-matrix element of the 2nd order processes shown in Fig. 3.32 can be directly noted by applying our rules deduced before:

$$
\begin{aligned}
S_{fi} = e^2 \int d^4x d^4y\, \bar{\psi}_f(x) \Big[&(-\mathrm{i}\not{A}(x,k))\, \mathrm{i}S_{\mathrm{F}}(x-y)(-\mathrm{i}\gamma^0) A_0^{\mathrm{coul}}(y) \\
&+ (-\mathrm{i}\gamma^0) A_0^{\mathrm{coul}}(x) \mathrm{i}S_{\mathrm{F}}(x-y)(-\mathrm{i}\not{A}(y,k)) \Big] \psi_i(y) \quad , \qquad (3.187)
\end{aligned}
$$

with

$$
A_0^{\mathrm{coul}}(x) = -\frac{Ze}{|\boldsymbol{x}|} \qquad (3.188)
$$

being the Coulomb potential known from Sect. 3.1. Since it is impossible to distinguish whether the photon is emitted from the incoming or outgoing electron both amplitudes have been added coherently in (3.187).

As before, it is convenient to transform the S-matrix element (3.187) into momentum space. We use the Fourier representation of the Coulomb potential

$$
-\frac{Ze}{|\boldsymbol{x}|} = -Ze\, 4\pi \int \frac{d^3q}{(2\pi)^3} \frac{1}{|\boldsymbol{q}|^2} \mathrm{e}^{+\mathrm{i}\boldsymbol{q}\cdot\boldsymbol{x}} \qquad (3.189)
$$

taken from Sect. 3.1, (3.7). With this we get

$$
\begin{aligned}
S_{fi} = &-\frac{Ze^3 4\pi}{V^{3/2}} \sqrt{\frac{4\pi}{2\omega}} \sqrt{\frac{m_0^2}{E_f E_i}} \int d^4x d^4y \frac{d^3q}{(2\pi)^3} \frac{d^4p}{(2\pi)^4} \\
&\times \bar{u}(p_f, s_f)\, \mathrm{e}^{\mathrm{i}p_f\cdot x} \Bigg[-\mathrm{i}\not{\epsilon}\left(\mathrm{e}^{-\mathrm{i}k\cdot x} + \mathrm{e}^{+\mathrm{i}k\cdot x}\right) \frac{\mathrm{i}\mathrm{e}^{-\mathrm{i}p\cdot(x-y)}}{\not{p} - m_0 + \mathrm{i}\varepsilon} (-\mathrm{i}\gamma^0) \frac{\mathrm{e}^{+\mathrm{i}\boldsymbol{q}\cdot\boldsymbol{y}}}{|\boldsymbol{q}|^2} \\
&- \mathrm{i}\gamma^0 \frac{\mathrm{e}^{+\mathrm{i}\boldsymbol{q}\cdot\boldsymbol{x}}}{|\boldsymbol{q}|^2} \frac{\mathrm{i}\mathrm{e}^{-\mathrm{i}p\cdot(x-y)}}{\not{p} - m_0 + \mathrm{i}\varepsilon} (-\mathrm{i}\not{\epsilon})\left(\mathrm{e}^{-\mathrm{i}k\cdot y} + \mathrm{e}^{\mathrm{i}k\cdot y}\right) \Bigg] u(p_i, s_i)\, \mathrm{e}^{-\mathrm{i}p_i\cdot y} \\
= &-\frac{Ze^3 4\pi}{V^{3/2}} \sqrt{\frac{4\pi}{2\omega}} \sqrt{\frac{m_0^2}{E_f E_i}} \int \frac{d^3q}{(2\pi)^3} \frac{d^4p}{(2\pi)^4} \\
&\times \Bigg\{ \left[(2\pi)^4 \delta^4(p_f - k - p) + (2\pi)^4 \delta^4(p_f + k - p)\right] (2\pi)^4 \delta^4(p - q - p_i) \\
&\times \bar{u}(p_f, s_f)(-\mathrm{i}\not{\epsilon}) \frac{\mathrm{i}}{\not{p} - m_0 + \mathrm{i}\varepsilon} (-\mathrm{i}\gamma^0) \frac{1}{|\boldsymbol{q}|^2} u(p_i, s_i) \\
&+ \left[(2\pi)^4 \delta^4(p - k - p_i) + (2\pi)^4 \delta^4(p + k - p_i)\right] (2\pi)^4 \delta^4(p_f - q - p) \\
&\times \bar{u}(p_f, s_f)(-\mathrm{i}\gamma^0) \frac{1}{|\boldsymbol{q}|^2} \frac{\mathrm{i}}{\not{p} - m_0 + \mathrm{i}\varepsilon} (-\mathrm{i}\not{\epsilon}) u(p_i, s_i) \Bigg\} \quad . \qquad (3.190)
\end{aligned}
$$

Here we have introduced the four-vector $q = (0, \boldsymbol{q})$ into the arguments of some of the δ-functions. We recognize that in the direct as well as in the exchange term two contributions appear originating from the factors $\mathrm{e}^{-\mathrm{i}k\cdot x}$ and $\mathrm{e}^{+\mathrm{i}k\cdot x}$ in the photon field (3.170). Let us first consider the direct term. The d^4p and d^3q integrations break down leading to

$$\begin{aligned}&\int \frac{d^3q}{(2\pi)^3}\frac{d^4p}{(2\pi)^4}(2\pi)^4\delta^4(p_f \pm k - p)(2\pi)^4\delta^4(p - q - p_i)f(p,|\boldsymbol{q}|)\\ &= \int \frac{d^3q}{(2\pi)^3}(2\pi)^4\delta^4(p_f \pm k - q - p_i)f(p,|\boldsymbol{q}|)\\ &= 2\pi\delta(E_f - E_i \pm \omega)f(p,|\boldsymbol{q}|) \quad , \end{aligned} \tag{3.191}$$

where $q = p_f \pm k - p_i$ and $p = p_f \pm k$.

Equation (3.190) containes a sum over two contributions where the photon either has momentum $+k$ or $-k$. However, for a given process the energies E_f and E_i are fixed and therefore only one of the two delta functions in (3.191) will contribute. Since we want to describe *photon emission* the electron looses energy, $E_f < E_i$, and the positive sign has to be taken in (3.191) which corresponds to $E_f = E_i - \omega$. The alternate possibility $E_f = E_i + \omega$ corresponds to a process where the electron gains energy from the radiation field during the scattering. This process of *photon absorption* can be represented by the Feynman graph in Fig. 3.34.

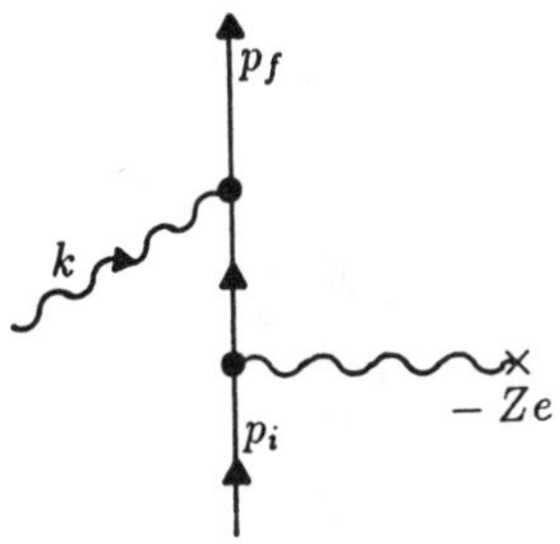

Fig. 3.34. Absorption of a real photon during scattering

In a bremsstrahlung-type process there are no incoming photons and the emitted photons are observed. Therefore we take only these parts of the scattering amplitude into account. Equation (3.190) gives

$$\begin{aligned}S_{fi} = &-Ze^3 2\pi\delta(E_f + \omega - E_i)\sqrt{\frac{4\pi}{2\omega V}}\sqrt{\frac{m_0^2}{E_f E_i V^2}}\frac{4\pi}{|\boldsymbol{q}|^2}\bar{u}(p_f, s_f)\\ &\times \left[(-\mathrm{i}\not{\varepsilon})\frac{\mathrm{i}}{\not{p}_f + \not{k} - m_0}(-\mathrm{i}\gamma_0) + (-\mathrm{i}\gamma_0)\frac{\mathrm{i}}{\not{p}_i - \not{k} - m_0}(-\mathrm{i}\not{\varepsilon})\right]u(p_i, s_i) \quad .\end{aligned} \tag{3.192}$$

Here,

$$\boldsymbol{q} = \boldsymbol{p}_f + \boldsymbol{k} - \boldsymbol{p}_i$$

is the momentum transfer to the nucleus. Since the vectors $\boldsymbol{p}_i$, $\boldsymbol{p}_f$, and $\boldsymbol{k}$ are fixed experimentally also $\boldsymbol{q}$ is a fixed vector. There is no energy transfer to the nucleus since the latter was assumed to be infinitely heavy.

Guided by the construction of the bremsstrahlung amplitude (3.192) we adopt the following general rule: at each vertex, where a free photon with polarization vector ε_μ is emitted, a factor $(-\mathrm{i}\not{\varepsilon})$ occurs and the normalization factor (3.184) of the photon, i. e. $\sqrt{4\pi/2\omega V}$ enters.

To simplify the notation we split off the normalization and kinematical factors in (3.192) according to

$$S_{fi} = \mathrm{i}Ze^3\, 2\pi\delta(E_f + \omega - E_i)\sqrt{\frac{4\pi}{2\omega V}}\sqrt{\frac{m_0^2}{E_f E_i V^2}}\frac{4\pi}{|\boldsymbol{q}|^2}\varepsilon^\mu M_\mu(k) \quad , \tag{3.193}$$

with the matrix element

$$M_\mu(k) = \bar{u}(p_f, s_f)\left[\gamma_\mu \frac{1}{\not{p}_f + \not{k} - m_0 + \mathrm{i}\varepsilon}\gamma_0 + \gamma_0 \frac{1}{\not{p}_i - \not{k} - m_0 + \mathrm{i}\varepsilon}\gamma_\mu\right]u(p_i, s_i) \, . \tag{3.194}$$

Using the energy–momentum relations $p_i^2 = p_f^2 = m_0^2$, $k^2 = 0$, this can be rewritten as

$$M_\mu(k) = \bar{u}(p_f, s_f)\left[\gamma_\mu \frac{\not p_f + \not k + m_0}{2p_f\cdot k + \mathrm{i}\varepsilon}\gamma_0 + \gamma_0 \frac{\not p_i - \not k + m_0}{-2p_i\cdot k + \mathrm{i}\varepsilon}\gamma_\mu\right] u(p_i, s_i)\ . \quad (3.195)$$

The cross section of bremsstrahlung is given by the square of the scattering amplitude $|S_{fi}|^2$ per incoming electron flux (v_i/V) and time $(T = 2\pi\delta(0))$, cf. (3.16). Furthermore we have to sum over the final states of the photons $(V\ d^3k/(2\pi)^3)$ and the electrons $(V\ d^3p_f/(2\pi)^3)$. This yields the total *phase-space factor*

$$V^2 \frac{d^3k}{(2\pi)^3}\frac{d^3p_f}{(2\pi)^3}\ . \quad (3.196)$$

Thus the bremsstrahlung cross section is given by

$$\begin{aligned} d\sigma &= \frac{1}{\frac{|v_i|}{V}T}|S_{fi}|^2 \frac{Vd^3k}{(2\pi)^3}\frac{Vd^3p_f}{(2\pi)^3} \\ &= \frac{Z^2e^6}{|v_i|}\frac{4\pi}{2\omega}\frac{m_0^2}{E_fE_i}\frac{(4\pi)^2}{|q|^4}\left|\varepsilon^\mu M_\mu(k)\right|^2 2\pi\delta(E_f + \omega - E_i)\frac{d^3k}{(2\pi)^3}\frac{d^3p_f}{(2\pi)^3}\ . \end{aligned} \quad (3.197)$$

In the following we want to evaluate this expression. We restrict ourselves to the case $k \to 0$, i.e. we consider *the emission of very soft photons*. The general case is known as the **Bethe**–*Heitler formula*[22] which we discuss in Example 3.11.

In the limit $k \to 0$ the matrix element $\varepsilon^\mu M_\mu$ in (3.193) can be approximated in the following manner:

$$\begin{aligned} \varepsilon^\mu M_\mu(k) \simeq \bar{u}(p_f, s_f)&\left[\frac{2\varepsilon\cdot p_f - (\not p_f - m_0)\not\varepsilon}{2k\cdot p_f + \mathrm{i}\varepsilon}\gamma_0\right. \\ &\left. + \gamma_0\frac{2\varepsilon\cdot p_i - \not\varepsilon(\not p_i - m_0)}{-2k\cdot p_i + \mathrm{i}\varepsilon}\right] u(p_i, s_i)\ . \end{aligned} \quad (3.198)$$

Here we have suppressed terms linear in $\not k$ in the numerator. Furthermore the order of the $\not p$ and $\not\varepsilon$ factors has been changed using the anticommutation relation

$$\not\varepsilon\not p = \varepsilon_\mu\gamma^\mu p_\nu\gamma^\nu = \varepsilon_\mu p_\nu(2g^{\mu\nu}\mathbb{1} - \gamma^\nu\gamma^\mu) = 2\varepsilon\cdot p - \not p\not\varepsilon\ . \quad (3.199)$$

This makes it possible to simplify (3.198) since the unit spinors satisfy the free Dirac equation

$$\bar{u}(p_f, s_f)(\not p_f - m_0) = 0 \quad \text{and} \quad (\not p_i - m_0)u(p_i, s_i) = 0\ . \quad (3.200)$$

Thus the term in brackets occurring in (3.198) can be reduced to

$$\varepsilon^\mu M_\mu \simeq \bar{u}(p_f, s_f)\gamma_0 u(p_i, s_i)\left(\frac{\varepsilon\cdot p_f}{k\cdot p_f} - \frac{\varepsilon\cdot p_i}{k\cdot p_i}\right)\ . \quad (3.201)$$

The first factor is just the elastic scattering amplitude (see (3.8)). Thus in the limit $k \to 0$ we get the plausible result that the matrix element describing soft

[22] See e.g. Walter Heitler: *The Quantum Theory of Radiation* (Oxford University Press, Oxford 1957).

bremsstrahlung is proportional to the elastic scattering amplitude. Using (3.197) and (3.201) the cross section for bremsstrahlung becomes

$$d\sigma = \frac{Z^2 e^6 (4\pi)^3 m_0^2}{2\omega |\boldsymbol{v}_i| E_f E_i} \left(\frac{\varepsilon \cdot p_f}{k \cdot p_f} - \frac{\varepsilon \cdot p_i}{k \cdot p_i}\right)^2 \frac{|\bar{u}(p_f, s_f)\gamma_0 u(p_i, s_i)|^2}{|\boldsymbol{q}|^4}$$
$$\times\, 2\pi\delta(E_f + \omega - E_i)\, \frac{d^3k d^3p_f}{(2\pi)^6} \quad . \qquad (3.202)$$

This result can be compared to the cross section for elastic electron scattering (Sect. 3.1, (3.23)):

$$d\sigma_{\text{elastic}} = \frac{4Z^2 e^4 m_0^2}{|\boldsymbol{v}_i| E_f E_i} \frac{|\bar{u}(p_f, s_f)\gamma_0 u(p_i, s_i)|^2}{|\boldsymbol{q}|^4} \delta(E_f - E_i)\, d^3p_f \quad , \qquad (3.203)$$

which is obviously contained as a factor in (3.202). Thus neglecting the effect of the soft photon's energy and momentum in (3.202), $\boldsymbol{q} = \boldsymbol{p}_f - \boldsymbol{p}_i + \boldsymbol{k} \approx \boldsymbol{p}_f - \boldsymbol{p}_i$, $E_f = E_i - \omega \approx E_i$, the differential cross section for bremsstrahlung can be written as

$$\frac{d\sigma}{d\Omega_f d\Omega_k d\omega} = \left(\frac{d\sigma}{d\Omega_f}\right)_{\text{elastic}} \frac{e^2\omega}{(2\pi)^2} \left(\frac{\varepsilon \cdot p_f}{k \cdot p_f} - \frac{\varepsilon \cdot p_i}{k \cdot p_i}\right)^2 \Theta(E_i - m_0 - \omega)\,. \qquad (3.204)$$

This is the cross section of bremsstrahlung for an electron scattered into the solid angle $d\Omega_f$. The soft photons ($k \to 0$) with polarization $\boldsymbol{\varepsilon}$ and momentum $\boldsymbol{k}$ are observed in the interval $d\omega d\Omega_k$. It is very natural that the bremsstrahlung of soft photons is proportional to the scattering cross section of the decelerating electrons at the same energy and scattering angle. Indeed, the amount of energy and momentum carried off by the photon is so small that the "trajectory" of the electron remains nearly undisturbed.

If the cross section of unpolarized electrons is to be calculated, one has to sum over the final spin states of the electrons and to average over the initial spin states. Owing to the factorisation property (3.204) this is easily achieved. One merely has to replace the elastic cross section by the unpolarized expression $(d\bar{\sigma}/d\Omega_f)_{\text{elastic}}$ which was derived in Sect. 3.1, (3.39).

The result of (3.204) is more general than one might expect. It has been shown that in the limit $k \to 0$ *the amplitude for any process leading to photon emission can factorized* according to

$$\lim_{k\to 0} M(k) = \sqrt{4\pi} e \left(\frac{\varepsilon \cdot p_f}{k \cdot p_f} - \frac{\varepsilon \cdot p_i}{k \cdot p_i}\right) M_0 \quad , \qquad (3.205)$$

where M_0 is the amplitude for the same process without photon emission. This result is true for any kind of process, irrespective of the spin or internal structure of the charged particle! The expression (3.205) is divergent in the limit $k \to 0$. It can be viewed as the leading term of an expansion in powers of k. According to the *soft-photon theorem*[23] the first two terms of this expansion are *universal expressions* which depend only on the charge, mass, and magnetic moment of the particle. Loosely speaking one may say that photons with a long wavelength can not

[23] F. E. Low: Phys. Rev. **96**, 1428 (1954) and Phys. Rev. **110**, 974 (1958).

resolve the detailed structure of the radiating source. Similar *low-energy theorems* also hold for the emission of other kinds of bosonic field quanta, in particular for *pion emission* in nuclear collisions.

Let us consider in more detail the energy spectrum of the photons emitted according to (3.204). The probability that soft quanta are emitted is obviously proportional to

$$\frac{dW_{\text{photon}}}{d\Omega_k} \sim \frac{d\,|\boldsymbol{k}|}{|\boldsymbol{k}|} \quad , \tag{3.206}$$

which tends to infinity for $k \to 0$. This behaviour is known as the *infrared catastrophe*. In the following we will discuss how to cure this unphysical divergence by a detailed examination of the measuring process of bremsstrahlung. It is important to realize that the electron and photon detectors have only a finite energy resolution. Therefore, if photons with momenta $|\boldsymbol{k}| \approx 0$ are measured, not only inelastically but also elastically scattered electrons ($|\boldsymbol{k}| = 0$) are detected. In a comparison of theory and experiment we have consequently to consider elastic *and* inelastic cross sections, both up to order e^2. In other words, since the bremsstrahlung cross section (3.204) is of the order e^2 with respect to the elastic scattering cross section of electrons, one has also to include the so-called radiative corrections to $(d\sigma/d\Omega_f)_{\text{elast.}}$ up to the same order. There exist two types of corrections shown in Fig. 3.35.

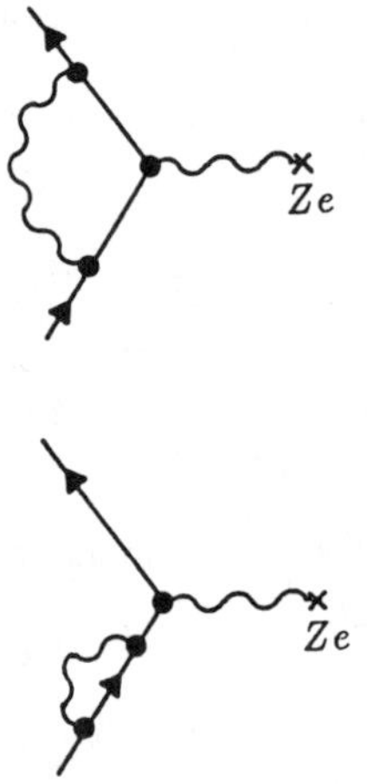

Fig. 3.35. The two types of lowest-order radiative corrections that occur in the Coulomb scattering of electrons

Both diagrams in the figure contain a *virtual photon* being emitted and reabsorbed by the same electron. This differs from the two-photon exchange we considered in Example 3.6. There both photons are emitted by the electron and both are absorbed by the proton. In contrast in the case of Fig. 3.35 the electron interacts with itself via the radiative field. Later on in Example 5.7 within the (quite complicated) calculation of these processes we shall see that these graphs produce a divergent contribution which just cancels the infrared divergence (3.206).

For the time being we shall continue the calculation of the cross section for emitting soft bremsstrahlung and ignore the infrared divergence. First we sum over the different polarizations of the photon. This can be done very elegantly[24] if one makes use of the *gauge invariance* property of the electromagnetic field. The interaction of any electromagnetic current $J_\mu(x)$ with the vector potential $A_\mu(x)$ is given by $\int d^4x\, J_\mu(x) A^\mu(x)$. This integral must be invariant under the gauge transformation

$$A^\mu(x) \to A^\mu(x) + \frac{\partial \Lambda(x)}{\partial x_\mu} \quad . \tag{3.207}$$

Integrating by part this implies the condition

$$\int d^4x\, J_\mu(x) \frac{\partial \Lambda(x)}{\partial x_\mu} = 0 = \int d^4x\, \frac{\partial J_\mu(x)}{\partial x_\mu}\, \Lambda(x) \quad . \tag{3.208}$$

Since $\Lambda(x)$ is an arbitrary function this yields the condition of current conservation

$$\frac{\partial J_\mu(x)}{\partial x_\mu} = 0 \quad , \tag{3.209}$$

[24] R.P. Feynman: Phys. Rev. **76**, 769 (1949).

which can be written in momentum space as

$$k_\mu J^\mu(k) = 0 \quad , \tag{3.210}$$

since $J_\mu(x) = \int d^4x\, J_\mu(k) \mathrm{e}^{-\mathrm{i}k\cdot x}$. This property is shared also by quantum mechanical transition currents. Thus we can expect that the matrix element $M_\mu(k)$ introduced in (3.194) satisfies

$$k^\mu M_\mu(k) = 0 \quad , \tag{3.211}$$

since $M_\mu(x)$ up to a numerical factor is the transition current for bremsstrahlung in lowest order perturbation theory. Using $\not{k}\not{p} = -\not{p}\not{k} + 2p\cdot k$ and the Dirac equation (3.200) this is easily verified explicitly:

$$\begin{aligned}
k^\mu M_\mu(k) &= \bar{u}(p_f, s_f)\left[\not{k}\frac{\not{p}_f + \not{k} + m_0}{2p_f\cdot k + \mathrm{i}\varepsilon}\gamma_0 + \gamma_0\frac{\not{p}_i - \not{k} + m_0}{-2p_i\cdot k + \mathrm{i}\varepsilon}\not{k}\right] u(p_i, s_i) \\
&= \bar{u}(p_f, s_f)\left[\frac{-(\not{p}_f - m_0)\not{k} + 2p_f\cdot k + k^2}{2p_f\cdot k + \mathrm{i}\varepsilon}\gamma_0\right. \\
&\quad \left. + \gamma_0\frac{-\not{k}(\not{p}_i - m_0) + 2p_i\cdot k - k^2}{-2p_i\cdot k + \mathrm{i}\varepsilon}\right] u(p_i, s_i) \\
&= \bar{u}(p_f, s_f)\left[\frac{2p_f\cdot k}{2p_f\cdot k + \mathrm{i}\varepsilon}\gamma_0 + \gamma_0\frac{2p_i\cdot k}{-2p_i\cdot k + \mathrm{i}\varepsilon}\right] u(p_i, s_i) \\
&= 0 \quad .
\end{aligned} \tag{3.212}$$

Now we are ready to perform the summation over the photon polarizations characterized by the polarization vectors $\varepsilon_\mu(\boldsymbol{k}, \lambda)$ with $\lambda = 1, 2$. The quantity of interest is

$$\begin{aligned}
\overline{|\varepsilon\cdot M|^2} &= \sum_{\lambda=1,2} |\varepsilon_\mu(\boldsymbol{k}, \lambda) M^\mu(k)|^2 \\
&= \sum_{\lambda=1,2} \varepsilon_\mu(\boldsymbol{k}, \lambda)\varepsilon^*_\nu(\boldsymbol{k}, \lambda)\, M^\mu(k) M^{*\nu}(k) \quad .
\end{aligned} \tag{3.213}$$

To simplify the following calculation the coordinate system will be chosen such that the momentum vector $\boldsymbol{k}$ points into the z-direction

$$k^\mu = \omega(1\,, 0\,, 0\,, 1) \quad . \tag{3.214}$$

The two transverse polarization vectors have to satisfy (3.178) and (3.179). We choose them to be the purely spatial real unit vectors

$$\begin{aligned}
\varepsilon(\boldsymbol{k}, 1) &= (0\,, 1\,, 0\,, 0) \quad , \\
\varepsilon(\boldsymbol{k}, 2) &= (0\,, 0\,, 1\,, 0) \quad .
\end{aligned} \tag{3.215}$$

Of course this choice is valid only in a particular Lorentz frame and it also implies a particular gauge for the vector potential, namely the "radiation gauge"

$$A^0(x) = 0 \quad . \tag{3.216}$$

However, this is no serious drawback since the final result will be Lorentz- and gauge-invariant. Using (3.215) we obtain

$$\overline{|\varepsilon \cdot M|^2} = M^1 M^{*1} + M^2 M^{*2} \quad . \tag{3.217}$$

Now we make use of the condition of current conservation (3.210) which reduces to

$$k_\mu M^\mu = \omega (M^0 - M^3) = 0 \quad , \tag{3.218}$$

and thus implies $M^0 = M^3$. Then we can transform (3.217) into a four dimensional scalar product by adding a vanishing contribution

$$\overline{|\varepsilon \cdot M|^2} = M^1 M^{*1} + M^2 M^{*2} + M^3 M^{*3} - M^0 M^{*0} = -M_\mu M^{*\mu} \quad . \tag{3.219}$$

Obviously this result is covariant. Comparing this with (3.123) we see that mathematically what we have proven is the *completeness relation of the polarization vectors* which can be written as

$$\sum_{\lambda=1,2} \varepsilon_\mu(\boldsymbol{k}, \lambda) \varepsilon_\nu(\boldsymbol{k}, \lambda) = -g_{\mu\nu} + \text{gauge terms} \quad . \tag{3.220}$$

The additional gauge terms need not to be specified in detail. They are proportional to k_μ or k_ν and thus do not contribute to any observable quantity since (3.220) will be multiplied with conserved currents which satisfy $k \cdot J = 0$. Nevertheless these terms have to be present in (3.220) since a complete basis in the 4-dimensional space of Lorentz vectors has to contain 4 elements. The contributions of *longitudinal*, $\varepsilon_\mu(\boldsymbol{k}, 3)$, and *scalar*, $\varepsilon_\mu(\boldsymbol{k}, 0)$, photons to the completeness relation make their appearance on on the r.h.s. of (3.220). They do not correspond to physical photons, however.

We apply the completeness relation (3.220) to the cross section of bremsstrahlung (3.204), which we integrate over the photon angle $d\Omega_k$ and photon energies in the interval

$$0 < \omega_{\min} \leq \omega \leq \omega_{\max} \ll E_i \tag{3.221}$$

in order to circumvent the infrared catastrophe. The summation over the polarization states of the photon leads to

$$\sum_{\lambda=1,2} \left(\frac{\varepsilon \cdot p_f}{k \cdot p_f} - \frac{\varepsilon \cdot p_i}{k \cdot p_i} \right)^2 = -\left(\frac{p_f}{k \cdot p_f} - \frac{p_i}{k \cdot p_i} \right)^2 \quad . \tag{3.222}$$

This yields

$$\begin{aligned} \frac{d\bar{\sigma}}{d\Omega_f} &= \left(\frac{d\bar{\sigma}}{d\Omega_f} \right)_{\text{elastic}} \frac{4\pi\alpha}{2 \cdot (2\pi)^3} \int_{\omega_{\min}}^{\omega_{\max}} \omega d\omega \int d\Omega_k \\ &\quad \times \left[\frac{2 p_f \cdot p_i}{(k \cdot p_f)(k \cdot p_i)} - \frac{m_0^2}{(k \cdot p_f)^2} - \frac{m_0^2}{(k \cdot p_i)^2} \right] \\ &= \left(\frac{d\bar{\sigma}}{d\Omega_f} \right)_{\text{elastic}} \frac{4\pi\alpha}{(2\pi)^2} \ln \left(\frac{\omega_{\max}}{\omega_{\min}} \right) \int \frac{d\Omega_k}{4\pi} \\ &\quad \times \left[\frac{2(1 - \boldsymbol{\beta}_f \cdot \boldsymbol{\beta}_i)}{(1 - \hat{\boldsymbol{k}} \cdot \boldsymbol{\beta}_f)(1 - \hat{\boldsymbol{k}} \cdot \boldsymbol{\beta}_i)} - \frac{m_0^2}{E_f^2 (1 - \hat{\boldsymbol{k}} \cdot \boldsymbol{\beta}_f)^2} - \frac{m_0^2}{E_i^2 (1 - \hat{\boldsymbol{k}} \cdot \boldsymbol{\beta}_i)^2} \right] \quad . \end{aligned} \tag{3.223}$$

We have (remember $c = 1$)

$$\boldsymbol{\beta}_i = \boldsymbol{v}_i = \frac{\boldsymbol{p}_i}{E_i} \quad \text{and} \quad \boldsymbol{\beta}_f = \boldsymbol{v}_f = \frac{\boldsymbol{p}_f}{E_f} \quad , \tag{3.224}$$

which are the initial and final velocities of the electron, $\hat{\boldsymbol{k}} = \boldsymbol{k}/|\boldsymbol{k}|$ being the unit vector in the direction of the photon momentum. If the emitted bremsstrahlung photons are very soft, the initial and final energies of the electron are almost the same and we get

$$|\boldsymbol{\beta}_i| = |\boldsymbol{\beta}_f| \equiv \beta \tag{3.225}$$

and thus

$$\boldsymbol{\beta}_i \cdot \boldsymbol{\beta}_f = \beta^2 \cos\Theta \quad , \tag{3.226}$$

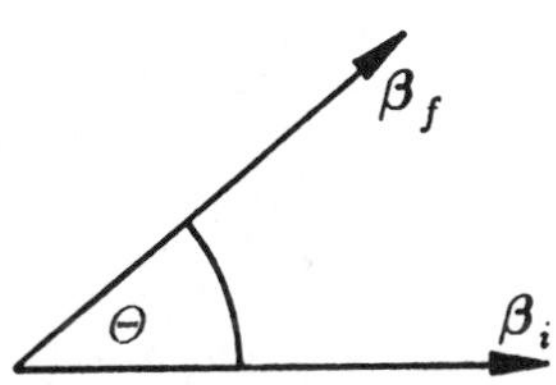

Fig. 3.36. Θ is the angle between the initial and final direction of the electron

with the scattering angle Θ (see Fig. 3.36). The angular integrations of the last two terms in (3.223) can be performed by elementary means:

$$\begin{aligned}
\int \frac{d\Omega_k}{4\pi} \frac{m_0^2}{E^2(1-\boldsymbol{\beta}\cdot\hat{\boldsymbol{k}})^2} &= \frac{m_0^2}{E^2} \int\limits_{-1}^{1} \frac{d\cos\theta}{2(1-\beta\cos\theta)^2} = \frac{m_0^2}{E^2} \int\limits_{-1}^{1} \frac{dz}{2(1-\beta z)^2} \\
&= \frac{m_0^2}{E^2}\left(-\frac{1}{\beta}\right) \int\limits_{1+\beta}^{1-\beta} \frac{dx}{2x^2} = -\frac{m_0^2}{\beta E^2}\frac{1}{2}\left(-\frac{1}{x}\right)\Bigg|_{1+\beta}^{1-\beta} \\
&= \frac{m_0^2}{1-\beta^2}\frac{1}{E^2} = 1 \quad .
\end{aligned} \tag{3.227}$$

θ is the angle between electron and photon. The first integral in (3.223) is more difficult to evaluate. It can be calculated with the help of a trick also developed by Feynman. The two factors in the denominator of the integrand can be combined by introducing the following auxiliary integral

$$\begin{aligned}
\int\limits_0^1 \frac{dx}{[ax+b(1-x)]^2} &= \int\limits_0^1 \frac{dx}{[(a-b)x+b]^2} = \frac{1}{a-b}\int\limits_b^a \frac{dz}{z^2} \\
&= \frac{1}{a-b}\left(-\frac{1}{a}+\frac{1}{b}\right) = \frac{1}{a\,b} \quad .
\end{aligned} \tag{3.228}$$

Expressing the denominator through this identity we see that the first integral in (3.223) follows as

$$\begin{aligned}
I &= \int \frac{d\Omega_k}{4\pi} \frac{1}{(1-\hat{\boldsymbol{k}}\cdot\boldsymbol{\beta}_f)(1-\hat{\boldsymbol{k}}\cdot\boldsymbol{\beta}_i)} \\
&= \int_0^1 dx \int \frac{d\Omega_k}{4\pi} \frac{1}{[(1-\hat{\boldsymbol{k}}\cdot\boldsymbol{\beta}_f)\,x + (1-\hat{\boldsymbol{k}}\cdot\boldsymbol{\beta}_i)(1-x)]^2} \\
&= \int_0^1 dx \int \frac{d\Omega_k}{4\pi} \frac{1}{[1-\hat{\boldsymbol{k}}\cdot(\boldsymbol{\beta}_f x + \boldsymbol{\beta}_i(1-x))]^2} \\
&= \int_0^1 dx \int_{-1}^1 \frac{d\varphi d\cos\vartheta}{4\pi} \frac{1}{[1-|\boldsymbol{\beta}_f x + \boldsymbol{\beta}_i(1-x)|\cos\vartheta]^2} \\
&= \int_0^1 dx \frac{1}{2}\left(\frac{-1}{u}\right) \int_{1+u}^{1-u} \frac{dz}{z^2} \qquad \text{(where } u = |\boldsymbol{\beta}_f x + \boldsymbol{\beta}_i(1-x)|) \\
&= \int_0^1 dx \frac{1}{1-u^2} = \int_0^1 \frac{dx}{1-|\boldsymbol{\beta}_f x + \boldsymbol{\beta}_i(1-x)|^2} \\
&= \int_0^1 \frac{dx}{1-\beta^2+4\beta^2 x(1-x)\sin^2(\Theta/2)} \quad . \qquad (3.229)
\end{aligned}$$

This integral with a quadratic polynomial in the denominator can be solved in closed form with the result

$$I = \frac{1}{2\beta\sin(\Theta/2)\sqrt{1-\beta^2\cos^2(\Theta/2)}} \ln\left(\frac{\sqrt{1-\beta^2\cos^2(\Theta/2)}+\beta\sin(\Theta/2)}{\sqrt{1-\beta^2\cos^2(\Theta/2)}-\beta\sin(\Theta/2)}\right) . \qquad (3.230)$$

This expression simplifies in the nonrelativistic limit ($\beta \ll 1$) where the Taylor series has the leading terms

$$I_{\mathrm{NR}} \approx 1 + \beta^2 - \frac{2}{3}\beta^2 \sin^2\frac{\Theta}{2} + O(\beta^4) \quad . \qquad (3.231)$$

In the opposite ultrarelativistic limit ($\beta = 1-\delta$, $\delta \ll 1$) the Taylor expansion with respect to the small parameter δ leads to

$$I_{\mathrm{UR}} \approx \frac{1}{2\sin^2\frac{\Theta}{2}} \ln\left(\frac{2\sin^2\frac{\Theta}{2}}{\delta}\right)\Big(1+O(\delta)\Big) \quad . \qquad (3.232)$$

Using $E = m_0\gamma = m_0(1-\beta^2)^{-1/2} \approx m_0/\sqrt{2\delta}$ the argument of the logarithm can be expressed in terms of the momentum transfer (cf. (3.109) in Sect. 3.2)

$$q^2 = (p_f - p_i)^2 \simeq -4E^2\sin^2\frac{\Theta}{2} = -m_0^2\left(\frac{2\sin^2\frac{\Theta}{2}}{\delta}\right) \quad . \qquad (3.233)$$

Using these results (3.223) leads to the following cross section of soft bremsstrahlung in the nonrelativistic and ultrarelativistic limit

$$\frac{d\bar{\sigma}}{d\Omega_f} = \left(\frac{d\sigma}{d\Omega_f}\right)_{\text{elastic}} \frac{\alpha}{\pi} \ln\frac{\omega_{\text{max}}}{\omega_{\text{min}}} \left[2(1-\beta^2\cos\theta)I - 2\right]$$

$$= \left(\frac{d\bar{\sigma}}{d\Omega_f}\right)_{\text{elastic}} \frac{2\alpha}{\pi} \ln\frac{\omega_{\text{max}}}{\omega_{\text{min}}} \begin{cases} \frac{4}{3}\beta^2\sin^2\frac{\Theta}{2} & \text{nonrelativistic limit} \\ \ln(\frac{-q^2}{m_0^2}) - 1 & \text{ultrarelativistic limit} \end{cases} . \quad (3.234)$$

As explained above, the infrared divergence has been cut off by using ω_{min} as the lower limit in the momentum integration. In the limit $\omega_{\text{min}} \to 0$ we have to include the radiative corrections entering the calculation of electron scattering, i.e. $\left(\frac{d\sigma}{d\Omega_f}\right)_{\text{elastic}}$, in order to get a finite result.

EXERCISE

3.10 Static Limit of Bremsstrahlung

Problem. Derive the S-matrix element for bremsstrahlung in electron-proton collisions treating the target as a finite-mass Dirac particle. Show that in the static limit it is reduced to the amplitude given in (3.192)

$$S_{fi} = \frac{\mathrm{i}e^3}{V^{3/2}} 2\pi\delta(E_f + \omega - E_i)\sqrt{\frac{4\pi}{2\omega}}\sqrt{\frac{m_0^2}{E_i E_f}} \frac{4\pi}{|\boldsymbol{p}_f + \boldsymbol{k} - \boldsymbol{p}_i|^2}$$
$$\times \bar{u}(p_f, s_f)\left[\not{\epsilon}\frac{1}{\not{p}_f + \not{k} - m_0}\gamma_0 + \gamma_0\frac{1}{\not{p}_i - \not{k} - m_0}\not{\epsilon}\right] u(p_i, s_i) \quad , \quad (1)$$

which describes electron bremsstrahlung in an external Coulomb field. Further show that the same relation between the two problems holds as in the case of elastic scattering, i.e. that the replacements

$$\gamma^\mu \left(\frac{-1}{q^2 + \mathrm{i}\varepsilon}\right)\sqrt{\frac{M_0^2}{E_f E_i}}\,\bar{u}(p_f, s_f)\gamma_\mu u(p_i, s_i) \to \gamma_0 \frac{1}{|\boldsymbol{q}|^2} \quad , \quad (2a)$$

where $q = p_f + k - p_i$ and

$$(2\pi)^3\delta^3(\boldsymbol{P}_f + \boldsymbol{p}_f + \boldsymbol{k} - \boldsymbol{P}_i - \boldsymbol{p}_i) \to V \quad (2b)$$

have to be made when going to the static limit.

Solution. The amplitude to be determined is represented by the two graphs of Fig. 3.37. Graph (a) yields

$$S_{fi}^{(a)} = \frac{1}{V^2}\sqrt{\frac{m_0^2}{E_i E_f}}\sqrt{\frac{M_0^2}{E_i^{\mathrm{p}} E_f^{\mathrm{p}}}}(2\pi)^4\delta^4(P_f + p_f + k - P_i - p_i)$$
$$\times \left[\bar{u}(p_f, s_f)(-\mathrm{i}e\gamma^\rho)\frac{\mathrm{i}}{\not{p}_f + \not{k} - m_0 + \mathrm{i}\varepsilon}(-\mathrm{i}e\gamma^\nu)u(p_i, s_i)\right]$$
$$\times \left[\bar{u}(P_f, S_f)(+\mathrm{i}e\gamma^\mu)u(P_i, S_i)\right]\frac{(-\mathrm{i}g_{\mu\nu})4\pi}{(P_f - P_i)^2 + \mathrm{i}\varepsilon}\sqrt{\frac{4\pi}{2\omega V}}\,\varepsilon_\rho(\boldsymbol{k}, \lambda) \quad (3a)$$

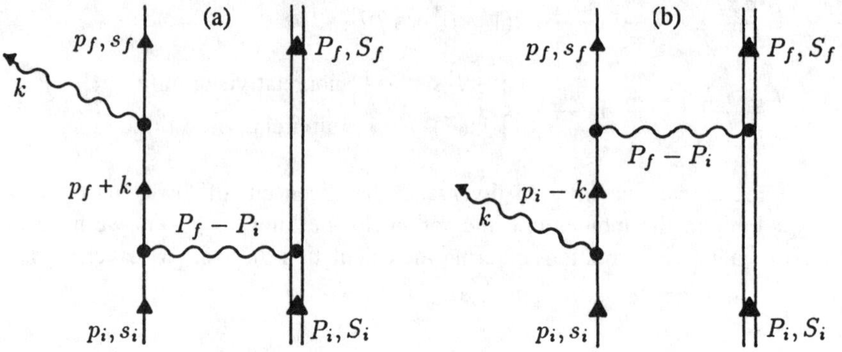

Fig. 3.37a,b. The two Feynman diagrams describing bremsstrahlung in electron–proton collisions

and graph (b) yields

$$
\begin{aligned}
S_{fi}^{(b)} = {} & \frac{1}{V^2}\sqrt{\frac{m_0^2}{E_iE_f}}\sqrt{\frac{M_0^2}{E_i^{\mathrm{p}}E_f^{\mathrm{p}}}}(2\pi)^4\delta^4(P_f+p_f+k-P_i-p_i) \\
& \times\left[\bar u(p_f,s_f)(-\mathrm{i}e\gamma^\rho)\frac{\mathrm{i}}{\not p_i-\not k-m_0+\mathrm{i}\varepsilon}(-\mathrm{i}e\gamma^\nu)u(p_i,s_i)\right] \\
& \times\left[\bar u(P_f,S_f)(+\mathrm{i}e\gamma^\mu)u(P_i,S_i)\right]\frac{(-\mathrm{i}g_{\mu\rho})4\pi}{(P_f-P_i)^2+\mathrm{i}\varepsilon}\sqrt{\frac{4\pi}{2\omega V}}\,\varepsilon_\nu(\boldsymbol{k},\lambda) \quad . \qquad (3b)
\end{aligned}
$$

The sum of both terms is

$$
\begin{aligned}
S_{fi} = {} & \frac{-\mathrm{i}e^3}{V^{5/2}}\sqrt{\frac{m_0^2}{E_iE_f}}\sqrt{\frac{M_0^2}{E_i^{\mathrm{p}}E_f^{\mathrm{p}}}}\sqrt{\frac{4\pi}{2\omega}}(2\pi)^4 \\
& \times\delta^4(P_f+p_f+k-P_i-p_i)\left[\bar u(P_f,S_f)\gamma^\mu u(P_i,S_i)\right] \\
& \times\left[\bar u(p_f,s_f)\left(\not\varepsilon\frac{1}{\not p_f+\not k-m_0+\mathrm{i}\varepsilon}\gamma_\mu+\gamma_\mu\frac{1}{\not p_i-\not k-m_0+\mathrm{i}\varepsilon}\not\varepsilon\right)u(p_i,s_i)\right] \\
& \times\frac{4\pi}{(p_f+k-p_i)^2+\mathrm{i}\varepsilon} \quad . \qquad (4)
\end{aligned}
$$

If we compare (4) and (1) the relation of the factors is obvious.

Note that in principle there are two additional bremsstrahlung graphs in which the photon is emitted by the proton instead of the electron. The corresponding amplitudes, however, involve the propagator of the proton and are suppressed by the large proton mass in the denominator. This corresponds to the fact that classically the acceleration of the recoiling proton is smaller by a factor m_0/M_0 compared to that of the electron.

Now we assume that the proton mass M_0 is large compared to the kinetic energies. In the rest frame of the proton we have $P_i = (M_0, \mathbf{0})$ and $P_f^0 \approx M_0$, i.e. $M_0 = E_i^{\mathrm{p}} \approx E_f^{\mathrm{p}}$ and thus

$$
\sqrt{\frac{M_0^2}{E_i^{\mathrm{p}}E_f^{\mathrm{p}}}} \approx 1 \quad . \qquad (5)
$$

Exercise 3.10.

Furthermore,

$$(2\pi)^4\delta^4(P_f + p_f + k - p_i - P_i)$$
$$= (2\pi)\delta(E_f + \omega - E_i)(2\pi)^3\delta^3(\boldsymbol{P}_f + \boldsymbol{p}_f + \boldsymbol{k} - \boldsymbol{p}_i) \quad , \tag{6}$$

with $\boldsymbol{P}_f$ being the final momentum of the proton. The δ-function contributes only for $\boldsymbol{P}_f + \boldsymbol{p}_f + \boldsymbol{k} - \boldsymbol{p}_i = 0$. If the proton is infinitely heavy it can gain an arbitrary high momentum without violating energy conservation. Then the momentum balance $\boldsymbol{P}_f = \boldsymbol{p}_i - \boldsymbol{p}_f - \boldsymbol{k}$ puts no constraint on the vectors $\boldsymbol{p}_i$, $\boldsymbol{p}_f$ and $\boldsymbol{k}$. In this case we can make the replacement

$$(2\pi)^3\delta^3(\boldsymbol{P}_f + \boldsymbol{p}_f + \boldsymbol{k} - \boldsymbol{p}_i) = (2\pi)^3\,\delta^3(\boldsymbol{0}) \; = \; V \quad . \tag{7}$$

As in Exercise 3.7 in the infinite-mass limit, inner degrees of freedom of the proton, e.g. spin-flip transitions, can be neglected, which leads to

$$\lim_{M_0\to\infty} \bar{u}(P_f, S_f)\gamma^\mu u(P_i, S_i) = \bar{u}(0, S)\gamma^\mu u(0, S) = g^{\mu 0} \quad . \tag{8}$$

Inserting (5), (6), (8), and (8) into (4) yields

$$\begin{aligned} S_{fi} = {} & \frac{-\mathrm{i}e^3}{V^{3/2}}\sqrt{\frac{m_0^2}{E_i E_f}}\sqrt{\frac{4\pi}{2\omega}}(2\pi)\delta(E_f + \omega - E_i) \\ & \times \frac{4\pi}{(E_f + \omega - E_i)^2 - |\boldsymbol{p}_f + \boldsymbol{k} - \boldsymbol{p}_i|^2 + \mathrm{i}\varepsilon} \\ & \times \bar{u}(p_f, s_f)\left[\not{\epsilon}\frac{1}{\not{p}_f + \not{k} - m_0 + \mathrm{i}\varepsilon}\gamma_0 + \gamma_0\frac{1}{\not{p}_i - \not{k} - m_0 + \mathrm{i}\varepsilon}\not{\epsilon}\right] u(p_i, s_i) \quad . \end{aligned} \tag{9}$$

The δ-function makes the energy part of the photon propagator vanish. The result is identical with (1).

EXAMPLE

3.11 The Bethe–Heitler Formula for Bremsstrahlung

We shall determine the unpolarized differential cross section of bremsstrahlung emission in electron scattering at a fixed Coulomb potential up to order $\alpha(Z\alpha)^2$. According to Sect. 3.6, (3.197) the unpolarized bremsstrahlung cross section is given by

$$d\bar{\sigma} = \frac{Z^2e^6(4\pi)^3m_0^2}{2\omega|\boldsymbol{v}_i|E_f E_i}\int\frac{d^3k}{(2\pi)^3}\frac{d^3p_f}{(2\pi)^3}2\pi\delta(E_f + \omega - E_i)\frac{1}{|\boldsymbol{q}|^4}F(p_i, p_f; k) \quad , \tag{1}$$

with the abbreviation

$$\begin{aligned} F = {} & \frac{1}{2}\sum_\lambda\sum_{s_i,s_f}\left|\bar{u}(p_f, s_f)\left[\not{\epsilon}\frac{\not{p}_f + \not{k} + m_0}{(p_f + k)^2 - m_0^2}\gamma_0\right.\right. \\ & \left.\left. + \gamma_0\frac{\not{p}_i - \not{k} + m_0}{(p_i - k)^2 - m_0^2}\not{\epsilon}\right]u(p_i, s_i)\right|^2 \quad . \end{aligned} \tag{2}$$

Example 3.11.

The differential cross section with respect to the solid angle of the scattered electron and the photon energy and solid angle can be determined by integrating over dE_f:

$$d\bar{\sigma} = \frac{Z^2 e^6 m_0^2}{\pi^2} \frac{|\boldsymbol{p}_f|}{|\boldsymbol{p}_i|} \omega d\omega d\Omega_k d\Omega_e \, \Theta(E_i - m_0 - \omega) \frac{1}{|\boldsymbol{q}|^4} F(p_i, p_f; k) \quad . \tag{3}$$

The averaging over the initial spin $\boldsymbol{s}_i$ and the summation over the final electron spin $\boldsymbol{s}_f$ can be reduced to the calculation of a trace in the usual manner. In order to sum over the two polarizations of the photon we use the completeness relation (3.220) for the photon polarization vectors $\varepsilon_\mu(\boldsymbol{k}, \lambda)$. From this it follows that

$$\begin{aligned} F = -\frac{1}{2} \mathrm{Tr} \Bigg[& \left(\gamma^\mu \frac{\not{p}_f + \not{k} + m_0}{2p_f \cdot k} \gamma^0 + \gamma^0 \frac{\not{p}_i - \not{k} + m_0}{-2p_i \cdot k} \gamma^\mu \right) \left(\frac{\not{p}_i + m_0}{2m_0} \right) \\ & \times \left(\gamma^0 \frac{\not{p}_f + \not{k} + m_0}{2p_f \cdot k} \gamma_\mu + \gamma_\mu \frac{\not{p}_i - \not{k} + m_0}{-2p_i \cdot k} \gamma^0 \right) \left(\frac{\not{p}_f + m_0}{2m_0} \right) \Bigg] \quad . \end{aligned} \tag{4}$$

We introduce the more convenient notation

$$F = -\frac{1}{32 m_0^2} \left(\frac{1}{(p_f \cdot k)^2} F_1 + \frac{1}{(p_i \cdot k)^2} F_2 + \frac{1}{-(p_i \cdot k)(p_f \cdot k)} (F_3 + F_4) \right) \quad . \tag{5}$$

with

$$\begin{aligned} F_1 &= \mathrm{Tr}\big[\gamma^\mu(\not{p}_f + \not{k} + m_0)\gamma^0(\not{p}_i + m_0)\gamma^0(\not{p}_f + \not{k} + m_0)\gamma_\mu(\not{p}_f + m_0)\big] \quad , \\ F_2 &= \mathrm{Tr}\big[\gamma^0(\not{p}_i - \not{k} + m_0)\gamma^\mu(\not{p}_i + m_0)\gamma_\mu(\not{p}_i - \not{k} + m_0)\gamma^0(\not{p}_f + m_0)\big] \quad , \\ F_3 &= \mathrm{Tr}\big[\gamma^0(\not{p}_i - \not{k} + m_0)\gamma^\mu(\not{p}_i + m_0)\gamma^0(\not{p}_f + \not{k} + m_0)\gamma_\mu(\not{p}_f + m_0)\big] \quad , \\ F_4 &= \mathrm{Tr}\big[\gamma^\mu(\not{p}_f + \not{k} + m_0)\gamma^0(\not{p}_i + m_0)\gamma_\mu(\not{p}_i - \not{k} + m_0)\gamma^0(\not{p}_f + m_0)\big] \quad . \end{aligned} \tag{6}$$

It suffices to calculate two of these complicated traces. That is, by substituting $p_i \leftrightarrow -p_f$ and cyclic permutation in the trace we get

$$\begin{aligned} F_1(p_i \leftrightarrow -p_f) &= \mathrm{Tr}\big[\gamma^0(-\not{p}_i + \not{k} + m_0)\gamma_\mu(-\not{p}_i + m_0) \\ &\qquad \times \gamma^\mu(-\not{p}_i + \not{k} + m_0)\gamma^0(-\not{p}_f + m_0)\big] \\ &= \mathrm{Tr}\big[\gamma^0(\not{p}_i - \not{k} - m_0)\gamma^\mu(\not{p}_i - m_0) \\ &\qquad \times \gamma_\mu(\not{p}_i - \not{k} - m_0)\gamma^0(\not{p}_f - m_0)\big] \\ &= F_2 \quad . \end{aligned} \tag{7}$$

Here we made use of the fact that the F_i have to be functions even in m_0. It can be easily seen that contributions with m_0 and m_0^3 contain an odd number of γ-matrices in the trace and therefore vanish. Analogously we find

$$F_3(p_i \leftrightarrow -p_f) = F_4 \tag{8}$$

by applying Theorem 7 from Mathematical Supplement 3.3 permitting the reversal of the γ-matrices in a trace

$$\mathrm{Tr}\,[\not{a}_1 \cdots \not{a}_n] = \mathrm{Tr}\,[\not{a}_n \cdots \not{a}_1] \quad .$$

In order to determine F_1 we introduce *Example 3.11.*

$$\gamma^0 \not{a}\, \gamma^0 = \tilde{\not{a}} \quad \text{with} \quad \tilde{\not{a}} = a_0\gamma_0 + \boldsymbol{a}\cdot\boldsymbol{\gamma} = 2a_0\gamma^0 - \not{a} \tag{9}$$

and use

$$\gamma^\mu \gamma_\mu = 4 \quad , \tag{10}$$

$$\gamma^\mu \not{a} \gamma_\mu = -2\not{a} \quad , \tag{11}$$

yielding

$$F_1 = \mathrm{Tr}\big[(\not{p}_f + \not{k} + m_0)(\tilde{\not{p}}_i + m_0)(\not{p}_f + \not{k} + m_0)(-2\not{p}_f + 4m_0)\big] \quad . \tag{12}$$

This expression can be split up into a sum of traces containing 0, 2, and 4 γ-matrices, respectively. Using

$$\mathrm{Tr}\{\not{a}\not{b}\} = 4\, a\cdot b \quad , \tag{13}$$

$$\mathrm{Tr}\{\not{a}\not{b}\not{c}\not{d}\} = 4[a\cdot b\; c\cdot d - a\cdot c\; b\cdot d + a\cdot d\; b\cdot c] \tag{14}$$

we finally get

$$F_1 = 16\big(m_0^2\, p_i\cdot\tilde{p}_f - k\cdot\tilde{p}_i\, k\cdot p_f + m_0^2\, k\cdot\tilde{p}_i + m_0^2\, k\cdot p_f + m_0^4\big) \quad . \tag{15}$$

The calculation of the interference term F_3 is more complicated. With the help of (11) and

$$\gamma^\mu \not{a}\not{b}\gamma_\mu = 4a\cdot b \tag{16}$$

as well as

$$\gamma^\mu \not{a}\not{b}\not{c}\gamma_\mu = -2\not{c}\not{b}\not{a} \tag{17}$$

we find

$$\begin{aligned}&\gamma^\mu(\not{p}_i + m_0)\gamma^0(\not{p}_f + \not{k} + m_0)\gamma_\mu \\ &\quad = -2(\not{p}_f + \not{k})\gamma^0\not{p}_i + 4m_0(p_i^0 + p_f^0 + k^0) - 2m_0^2\gamma^0 \quad .\end{aligned}$$

Applying (9) we can eliminate the γ^0-factors giving

$$\begin{aligned}F_3 = \mathrm{Tr}\big\{&(\not{p}_i - \not{k} + m_0)\big[-2(\not{p}_f + \not{k})\tilde{\not{p}}_i \\ &+ 4m_0(p_i^0 + p_f^0 + k^0)\gamma^0 - 2m_0^2\big](\tilde{\not{p}}_f + m_0)\big\} \quad .\end{aligned} \tag{18}$$

This expression expands into a sum of traces which contain at most four γ-matrices and can be calculated using (13) and (14). The result is

$$\begin{aligned}F_3 = 8\Big[&-(p_i\cdot p_f)^2 + k\cdot p_f\, p_i\cdot p_f - k\cdot p_i\, p_i\cdot p_f \\ &+ p_i\cdot\tilde{p}_i\, p_f\cdot\tilde{p}_f - 2E_f^2 k\cdot\tilde{p}_i + 2E_i^2 k\cdot\tilde{p}_f \\ &- (p_i\cdot\tilde{p}_f)^2 + k\cdot\tilde{p}_f\, p_i\cdot\tilde{p}_f - p_i\cdot\tilde{p}_f\, k\cdot\tilde{p}_i \\ &- 2p_i\cdot\tilde{p}_f m_0^2 - m_0^4 + 2m_0^2\big((E_i + E_f)^2 - \omega^2\big)\Big] \quad .\end{aligned} \tag{19}$$

Example 3.11.

This expression is invariant with respect to the exchange $p_i \leftrightarrow -p_f$ (note that $a \cdot \tilde{b} = \tilde{a} \cdot b$), so it follows that $F_4 = F_3$ from (8).

Now we can add the various contributions according to (5) in order to get the complete trace $\mathrm{Tr}\, F$. Further we notice that owing to energy conservation (the Coulomb centre does not absorb energy) the relation

$$\omega \equiv k^0 = E_i - E_f \tag{20}$$

is valid, and we introduce the momentum transfer

$$q = p_f + k - p_i \quad , \tag{21}$$

with the square

$$q^2 = 2(m_0^2 + p_f \cdot k - p_i \cdot k - p_i \cdot p_f) = -|\boldsymbol{q}|^2 \quad . \tag{22}$$

After some lengthy rewriting we get

$$\begin{aligned} F = &-\frac{1}{4m_0^2} \frac{1}{(p_i \cdot k)^2 (p_f \cdot k)^2} \Big[4m_0^2 (p_f \cdot k\, E_f - p_i \cdot k\, E_i)^2 \\ &+ \left((p_f \cdot k)^2 + (p_i \cdot k)^2\right)\left(2 p_i \cdot k\, p_f \cdot k + q^2 m_0^2\right) \\ &+ 2 p_i \cdot k\, p_f \cdot k\, q^2 \left(E_i^2 + E_f^2 - p_i \cdot p_f\right)\Big] \quad . \end{aligned} \tag{23}$$

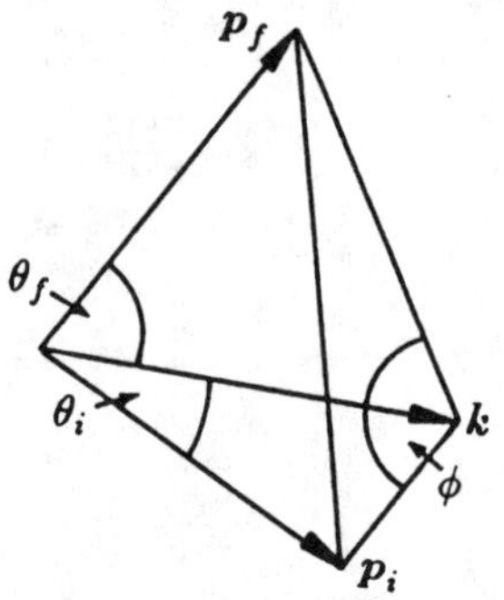

Fig. 3.38. The three momentum vectors $\boldsymbol{p}_i$, $\boldsymbol{p}_f$ and $\boldsymbol{k}$ form a spherical triangle

Inserting this result into (3) yields the bremsstrahlung cross section. For a comparison with experimental data it is convenient to put in the scalar products of the four-vectors explicitly. As variables we use the absolute values of the electron and photon momenta (or energies) before and after the scattering, the polar angle of $\boldsymbol{p}_i$ and $\boldsymbol{p}_f$ with respect to the direction $\boldsymbol{k}$ of the photon, and the angle ϕ between the planes $(\boldsymbol{p}_i, \boldsymbol{k})$ and $(\boldsymbol{p}_f, \boldsymbol{k})$ (see Fig. 3.38). From a formula from spherical geometry the angle between $\boldsymbol{p}_i$ and $\boldsymbol{p}_f$ is given by

$$\cos(\boldsymbol{p}_i, \boldsymbol{p}_f) = \cos\theta_i \cos\theta_f + \sin\theta_i \sin\theta_f \cos\phi \quad .$$

Then the square of the momentum transfer follows as

$$\begin{aligned} q^2 &= -\boldsymbol{q}^2 \\ &= -\boldsymbol{p}_i^2 - \boldsymbol{p}_f^2 - \omega^2 - 2\omega|\boldsymbol{p}_f|\cos\theta_f + 2\omega|\boldsymbol{p}_i|\cos\theta_i \\ &\quad + 2|\boldsymbol{p}_i||\boldsymbol{p}_f|(\cos\theta_i\cos\theta_f + \sin\theta_i\sin\theta_f\cos\phi) \quad . \end{aligned} \tag{24}$$

Furthermore,

$$\begin{aligned} p_i \cdot k &= \omega(E_i - |\boldsymbol{p}_i|\cos\theta_i) \quad , \\ p_f \cdot k &= \omega(E_f - |\boldsymbol{p}_f|\cos\theta_f) \quad . \end{aligned} \tag{25}$$

Together with (3) and (23) this finally yields the *Bethe–Heitler formula*[25] for the bremsstrahlung cross section:

[25] H. Bethe and W. Heitler: Proc. Roy. Soc. **A146**, 83 (1934).

Example 3.11.

$$
\begin{aligned}
d\bar{\sigma} = {} & \frac{Z^2\alpha^3}{(2\pi)^2}\frac{|\boldsymbol{p}_f|}{|\boldsymbol{p}_i|}\frac{d\omega}{\omega}\frac{d\Omega_e d\Omega_k}{|\boldsymbol{q}|^4}\,\Theta(E_i - m_0 - \omega) \\
& \times \left[\frac{\boldsymbol{p}_f^2\sin^2\theta_f}{(E_f - |\boldsymbol{p}_f|\cos\theta_f)^2}(4E_i^2 - \boldsymbol{q}^2) + \frac{\boldsymbol{p}_i^2\sin^2\theta_i}{(E_i - |\boldsymbol{p}_i|\cos\theta_i)^2}(4E_f^2 - \boldsymbol{q}^2)\right. \\
& + 2\omega^2\frac{\boldsymbol{p}_i^2\sin^2\theta_i + \boldsymbol{p}_f^2\sin^2\theta_f}{(E_i - |\boldsymbol{p}_i|\cos\theta_i)(E_f - |\boldsymbol{p}_f|\cos\theta_f)} \\
& \left. - 2\frac{|\boldsymbol{p}_i||\boldsymbol{p}_f|\sin\theta_i\sin\theta_f\cos\phi}{(E_i - |\boldsymbol{p}_i|\cos\theta_i)(E_f - |\boldsymbol{p}_f|\cos\theta_f)}(2E_i^2 + 2E_f^2 - \boldsymbol{q}^2)\right] \quad . \qquad (26)
\end{aligned}
$$

The validity of this formula can most easily be shown by tracing it back to (23). An extended discussion of (26) can be found in Heitler's book.[26] There also the analytical result of the integration over electron and photon angles, Ω_e and Ω_k , is stated.

In Exercise 3.16 we shall show that the Bethe–Heitler formula with some minor modifications also applies to the creation of electron–positron pairs.

Although the steps leading from the basic S-matrix element to the Bethe–Heitler-formula have been "exact" one should keep in mind that the result has its limitations before comparing with experiments.[27]

i) The derivation is based on perturbation theory using plane waves for the electron. If the criterion for the validity of the Born approximation

$$
\frac{Ze^2}{\hbar|\boldsymbol{v}|} \ll 1 \qquad (27)
$$

is violated for the initial and/or final velocity, Coulomb waves should be used instead of plane waves.

ii) The nuclear Coulomb potential in a neutral atom is *screened by the electron cloud.* This will lead to a reduction of the bremsstrahlung cross section in such cases where a significant contribution would arise from distances larger than the atomic radius. This happens at high electron energies as can be seen from the following qualitative argument. In momentum space the largest contribution to the radiation cross section originates from the region where the momentum transfer $\boldsymbol{q} = \boldsymbol{p}_f + \boldsymbol{k} - \boldsymbol{p}_i$ is smallest. This happens at

$$
|q_{\min}| = |\boldsymbol{p}_i| - |\boldsymbol{p}_f| - |\boldsymbol{k}| \quad . \qquad (28)
$$

Insertion of the relativistic energy-momentum relation yields in the limit $E_i/m_0 \gg 1$, $E_f/m_0 \gg 1$

$$
\begin{aligned}
|\boldsymbol{q}_{\min}| &= \sqrt{E_i^2 - m_0^2} - \sqrt{E_f^2 - m_0^2} - \omega \\
&\approx E_i\left(1 - \frac{m_0^2}{2E_i^2}\right) - E_f\left(1 - \frac{m_0^2}{2E_f^2}\right) - \omega \\
&= \frac{m_0^2}{2}\left(\frac{1}{E_f} - \frac{1}{E_i}\right) = \frac{m_0^2\omega}{2E_fE_i} \quad . \qquad (29)
\end{aligned}
$$

[26] W. Heitler: *The Quantum Theory of Radiation*, (Oxford University Press, Oxford 1957); see also H.W. Koch, J.W. Motz: Rev. Mod Phys. **31**, 920 (1959).

[27] For a recent review see, e.g., R.H. Pratt and I.J. Feng in *Atomic inner-shell Physics*, Ed. B. Craseman (Plenum Publishing Corporation, New York 1985).

Example 3.11.

In coordinate space this corresponds to a distance from the nucleus of the order

$$R = \frac{1}{|q_{\min}|} = \frac{2E_f E_i}{m_0^2 \omega} \quad . \tag{30}$$

This value has to be compared with the extension of the atomic shell $a(Z)$. According to the Thomas–Fermi model $a(Z)$ is of the order $Z^{-\frac{1}{3}}$ times the hydrogenic Bohr radius:

$$a(Z) = \frac{1}{m_0 \alpha} Z^{-\frac{1}{3}} \quad . \tag{31}$$

From (30) and (31) one deduces that atomic screening will significantly reduce the radiation intensity at energies exceeding

$$E > \frac{m_0}{Z^{\frac{1}{3}} \alpha} \tag{32}$$

taking $E_i \sim E_f \sim \omega$.

3.7 Compton Scattering – The Klein-Nishina Formula

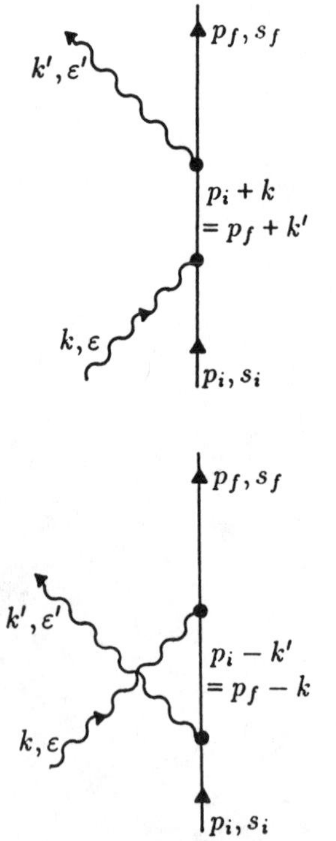

Fig. 3.39. The direct and exchange diagram describing Compton scattering

The name Compton scattering refers to the scattering of photons by free electrons. In the language of quantum electrodynamics an incoming photon with four-momentum k and polarization vector ε is absorbed by an electron (or another charged particle) and a second photon with four-momentum k' and polarization vector ε' is emitted. The corresponding Feynman diagrams are shown in Fig. 3.39.

We describe the incoming photon as a plane wave (see Sect. 3.6, (3.170)):

$$A_\mu(x, k) = \sqrt{\frac{4\pi}{2\omega V}}\, \varepsilon_\mu(\boldsymbol{k}, \lambda)\left(\mathrm{e}^{-\mathrm{i}k\cdot x} + \mathrm{e}^{\mathrm{i}k\cdot x}\right) \quad , \tag{3.235}$$

and the outgoing (scattered) photon by

$$A'_\mu(x', k') = \sqrt{\frac{4\pi}{2\omega' V}}\, \varepsilon_\mu(\boldsymbol{k}', \lambda')\left(\mathrm{e}^{-\mathrm{i}k'\cdot x'} + \mathrm{e}^{\mathrm{i}k'\cdot x'}\right) \quad . \tag{3.236}$$

Figure 3.39 shows that the Compton process is of second order and differs from bremsstrahlung by the fact that here we have an *incoming real photon* instead of the virtual photon exchanged with a recoiling charged particle. As a consequence the amplitude for Compton scattering can be obtained from that of bremsstrahlung (Sect. 3.6, (3.187)) just by the replacement $\gamma_0 A_0(y) \to \gamma^\mu A_\mu(y, k')$. In coordinate space the S-matrix element of Compton scattering is therefore given by

$$\begin{aligned} S_{fi} = e^2 \int d^4x d^4y\, \bar{\psi}_f(x)\Big[&\big(-\mathrm{i}\not{A}(x, k')\big)\mathrm{i}S_\mathrm{F}(x-y)\big(-\mathrm{i}\not{A}(y, k)\big) \\ &+ \big(-\mathrm{i}\not{A}(x, k)\big)\mathrm{i}S_\mathrm{F}(x-y)\big(-\mathrm{i}\not{A}(y, k')\big)\Big]\psi_i(y) \quad . \end{aligned} \tag{3.237}$$

Also in momentum space the amplitude can be directly written down with the help of the Feynman rules:

$$\begin{aligned} S_{fi} &= \frac{e^2}{V^2}\sqrt{\frac{m_0^2}{E_i E_f}}\sqrt{\frac{(4\pi)^2}{2\omega 2\omega'}}\,(2\pi)^4\delta^4(p_f+k'-p_i-k) \\ &\times \bar{u}(p_f,s_f)\left[(-\mathrm{i}\not{\varepsilon}')\frac{\mathrm{i}}{\not{p}_i+\not{k}-m_0}(-\mathrm{i}\not{\varepsilon})\right. \\ &\left.+\,(-\mathrm{i}\not{\varepsilon})\frac{\mathrm{i}}{\not{p}_i-\not{k}'-m_0}(-\mathrm{i}\not{\varepsilon}')\right]u(p_i,s_i) \quad . \end{aligned} \tag{3.238}$$

Here we have chosen the appropriate boundary conditions for the photon field. Owing to the plane wave factors $\exp(\pm \mathrm{i}k\cdot x)$ and $\exp(\pm\mathrm{i}k'\cdot x')$ in (3.235), (3.236) there are four possible sign combinations when passing from (3.237) to (3.238). Each of them gives rise to a delta function which constrains the four-momenta of the particles:

$$+k+p_i = +k'+p_f \quad , \tag{3.239a}$$

$$-k+p_i = +k'+p_f \quad , \tag{3.239b}$$

$$+k+p_i = -k'+p_f \quad , \tag{3.239c}$$

$$-k+p_i = -k'+p_f \quad . \tag{3.239d}$$

(3.239a) describes the correct conservation relation for energy and momentum in Compton scattering. The energy–momentum conditions of processes b) and c) can not be fulfilled; therefore their contributions vanish (see Problem 3.12, where this is shown in case c); the argument for b) is similar). Physically (3.239b) and (3.239c) describe the emission or absorption of two photons by a free electron which is kinematically impossible. In process d) the photons k and k' are exchanged with respect to the process a) discussed here. This corresponds to an incoming photon with momentum k' and a scattered (outgoing) photon with momentum k. The kinematical conditions fixed by the experiment are those noted in (3.239a). Thus the relation (3.239d) is not compatible with the prescribed experimental conditions, i.e. process (3.239d) does not have to be considered.

Here the situation is similar to the bremsstrahlung case (cf. the discussion in Sect. 3.6, (3.191)). Not every term that occurs in the scattering amplitude is physically relevant for the process considered. The remaining term in the Compton scattering amplitude stems from the part $\exp(-\mathrm{i}k\cdot x)$ of the photon field in (2.235) describing the absorption of a photon with four-momentum k_μ by the electron at x and from the part $\exp(+\mathrm{i}k'\cdot x')$ of the photon field in (2.236). The latter describes a photon with four–momentum k'_μ emitted by the electron at x'.

Inspecting (3.238) we note that the scattering amplitude has a symmetry property: obviously it is invariant under the exchange

$$k,\varepsilon \leftrightarrow -k',\varepsilon' \quad . \tag{3.240}$$

This is a new example for the *crossing symmetry* which we first encountered in Sect. 3.4. In our case the crossing symmetry implies that the amplitude for absorbing the photon k,ε and emitting the photon k',ε' is the same as that for absorbing a photon k', ε' and emitting a photon with k, ε. In general crossing relates ingoing particles to outgoing antiparticles and vice versa. In the case of photons this distinction does not arise since the photon "is its own antiparticle".

It is useful to split the S-matrix element (3.238) into two parts:

$$S_{fi} = -\mathrm{i}\frac{e^2}{V^2}\sqrt{\frac{m_0^2}{E_i E_f}}\sqrt{\frac{(4\pi)^2}{2\omega 2\omega'}}(2\pi)^4\delta^4(p_f + k' - p_i - k) \\ \times \varepsilon^\mu(\boldsymbol{k}', \lambda')\varepsilon^\nu(\boldsymbol{k}, \lambda)\, M_{\mu\nu} \quad . \tag{3.241}$$

Here we have introduced the *Compton tensor*

$$M_{\mu\nu} = \bar{u}(p_f, s_f)\left[\gamma_\mu \frac{1}{\not{p}_i + \not{k} - m_0}\gamma_\nu + \gamma_\nu \frac{1}{\not{p}_i - \not{k}' - m_0}\gamma_\mu\right] u(p_i, s_i) \ , \tag{3.242}$$

which is a function of the four-momenta p_i, p_f, k, k' subject to the condition $k + p_i = k' + p_f$. Using the on-shell conditions for the momentum vectors (3.242) can be written as

$$M_{\mu\nu} = \bar{u}(p_f, s_f)\left[\gamma_\mu \frac{\not{p}_i + \not{k} + m_0}{2p_i \cdot k}\gamma_\nu + \gamma_\nu \frac{\not{p}_i - \not{k}' + m_0}{-2p_i \cdot k'}\gamma_\mu\right] u(p_i, s_i) . \tag{3.243}$$

The tensor $M_{\mu\nu}$ is an obvious generalization of the matrix element M_μ that we introduced when discussing bremsstrahlung in Sect. 3.6, (3.194). The relation between both quantities is given by

$$M_{\mu 0}(p_i, p_f; k, k') = M_\mu(p_i, p_f; k') \quad . \tag{3.244}$$

Since both photons interact with conserved currents the Compton tensor is a *gauge-invariant* object, characterized by the property

$$k'^\mu M_{\mu\nu} = k^\nu M_{\mu\nu} = 0 \quad . \tag{3.245}$$

The proof of these two relations can be copied from the analogous case of bremsstrahlung, (3.212), and will not be repeated here.

Now we shall calculate the photon scattering cross section using the rules we have derived; only the spinor algebra will be somewhat more complicated than in the previous examples. The cross section results as

$$d\sigma = \int \frac{|S_{fi}|^2}{T|\boldsymbol{v}_{\mathrm{rel}}|/V} V\frac{d^3p_f}{(2\pi)^3} V\frac{d^3k'}{(2\pi)^3} \quad , \tag{3.246}$$

with

$$\frac{|S_{fi}|^2}{T} = \frac{|S_{fi}|^2}{VT/V} = \frac{|S_{fi}|^2}{(2\pi)^4\delta^4(0)1/V}$$

being the transition rate per volume and normalized to one electron per volume (see (3.63–3.71)). $|\boldsymbol{v}_{\mathrm{rel}}|/V$ is the incoming photon flux. $\boldsymbol{v}_{\mathrm{rel}} = \boldsymbol{c} - \boldsymbol{v}_{\mathrm{e}}$ is the relative velocity of the photons with respect to the electron and $1/V$ is again the number of electrons per unit volume. An additional factor $1/V$ originates from the number of photons per unit volume. The phase space volume elements of the final electron and photon in (3.246) are been given by $Vd^3p_f/(2\pi)^3$ and $Vd^3k'/(2\pi)^3$. This yields:

$$
\begin{aligned}
d\sigma &= \frac{e^4}{V^4}\frac{m_0^2}{E_iE_f}\frac{1}{(|\boldsymbol{v}_{\text{rel}}|/V)(1/V)} \\
&\times \int (2\pi)^4\delta^4(p_f+k'-p_i-k)\frac{(4\pi)^2}{2\omega 2\omega'}\left|\varepsilon'^{\mu}M_{\mu\nu}\varepsilon^{\nu}\right|^2 V^2\frac{d^3p_f}{(2\pi)^3}\frac{d^3k'}{(2\pi)^3} \\
&= \frac{e^4}{(2\pi)^2}\frac{m_0(4\pi)^2}{E_i|\boldsymbol{v}_{\text{rel}}|2\omega} \\
&\times \int \delta^4(p_f+k'-p_i-k)\left|\varepsilon'^{\mu}M_{\mu\nu}\varepsilon^{\nu}\right|^2\frac{m_0d^3p_f}{E_f}\frac{d^3k'}{2\omega'} \quad .
\end{aligned}
\tag{3.247}
$$

In the following we want to evaluate this expression in the *laboratory frame* where the electron is at rest initially, $p_i = (m_0, \mathbf{0})$. Hence we have $E_i = m_0$ in that frame and also $|\boldsymbol{v}_{\text{rel}}| = |\boldsymbol{c} - \boldsymbol{v}_{\text{e}}| = |\boldsymbol{c}| = 1$. Now we use the covariant expression for the density of final states known from Sect. 3.2, (3.74),

$$
\frac{d^3p}{2E} = \int_{-\infty}^{+\infty} d^4p\;\delta(p^2-m_0^2)\,\Theta(p_0) \quad , \tag{3.248}
$$

and perform the integral over the recoil electrons (d^3p_f) and the absolute value of the photon momentum $|\boldsymbol{k}'| = \omega'$.

$$
\begin{aligned}
&\int \frac{|\boldsymbol{k}'|^2d|\boldsymbol{k}'|}{2|\boldsymbol{k}'|}\int\frac{m_0d^3p_f}{E_f}\,\delta^4(p_f+k'-p_i-k) \\
&= m_0\int_0^{\infty}|\boldsymbol{k}'|d|\boldsymbol{k}'|\int d^4p_f\,\delta(p_f^2-m_0^2)\,\Theta(p_{f0})\,\delta^4(p_f+k'-p_i-k) \\
&= m_0\int_0^{\infty}\omega'd\omega'\,\delta\left((p_i+k-k')^2-m_0^2\right)\Theta(m_0+\omega-\omega') \\
&= m_0\int_0^{\omega+m_0}\omega'd\omega'\,\delta\left(2m_0(\omega-\omega')-2\omega\omega'(1-\cos\theta)\right) \\
&= m_0\frac{\omega'}{|-2m_0-2\omega(1-\cos\theta)|} = \frac{\omega'}{2|1+\frac{\omega}{m_0}(1-\cos\theta)|} \\
&= \frac{\omega'^2}{2\omega} \quad .
\end{aligned}
\tag{3.249}
$$

Here we have used

$$
\begin{aligned}
(k+p_i-k')^2 &= k^2+k'^2+p_i^2+2k\cdot p_i-2k\cdot k'-2k'\cdot p_i \\
&= m_0^2+2m_0(\omega-\omega')-2\omega\omega'(1-\cos\theta)
\end{aligned}
\tag{3.250}
$$

and we applied the familiar formula

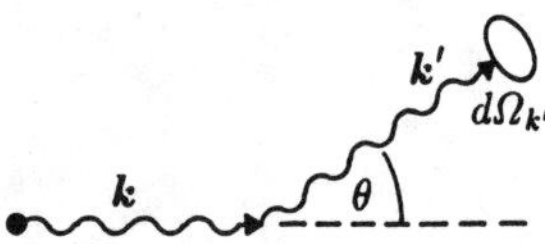

Fig. 3.40. The outgoing photon (k') is scattered with respect to the incoming one (k') by an angle θ into the spherical angle element $d\Omega_{k'}$

$$\int dx\, \delta(f(x))g(x) = \sum \left. \frac{g(x)}{\left|\frac{df}{dx}\right|} \right|_{\text{zero of } f(x)} \quad . \tag{3.251}$$

θ is the scattering angle of the photon, see Fig. 3.40. The energies of the incident and scattered photons are connected by

$$\omega' = \frac{\omega}{1 + \frac{\omega}{m_0}(1 - \cos\theta)} \quad . \tag{3.252}$$

This relation follows from the root of the delta function that occurred in the derivation of (3.249) and thus is a simple consequence of the laws of energy and momentum conservation. Equation (3.252) takes a particularly simple form if one looks at the wavelength $\lambda = 2\pi/\omega$ which leads to *Compton's formula*

$$\lambda' = \lambda + 2\pi \frac{1}{m_0}(1 - \cos\theta) \quad . \tag{3.253}$$

The wavelength of the scattered photon is increased by an amount of the order $1/m_0$ which, of course, is the Compton wavelength $\hbar/m_0c$.

From (3.243) and (3.247) the differential photon scattering cross section results as

$$\begin{aligned} \frac{d\sigma}{d\Omega_{k'}} &= \alpha^2 \frac{\omega'^2}{\omega^2} \left| \varepsilon'^{\mu} M_{\mu\nu} \varepsilon^{\nu} \right|^2 \\ &= \alpha^2 \frac{\omega'^2}{\omega^2} \left| \bar{u}(p_f, s_f) \left(\frac{\not{\varepsilon}'(\not{p}_i + \not{k} + m_0)\not{\varepsilon}}{2p_i \cdot k} \right.\right. \\ &\quad \left.\left. + \frac{\not{\varepsilon}(\not{p}_i - \not{k}' + m_0)\not{\varepsilon}'}{-2p_i \cdot k'} \right) u(p_i, s_i) \right|^2 \quad . \end{aligned} \tag{3.254}$$

In the following we will be interested in the case of *unpolarized electrons*[28] but keep, for the time being, the photon polarizations λ and λ'. Thus (3.254) has to be averaged over the initial spin and summed over the final spin of the electron:

$$\frac{d\bar{\sigma}}{d\Omega_{k'}}(\lambda', \lambda) = \frac{1}{2} \sum_{s_i, s_f} \frac{d\sigma}{d\Omega_{k'}}(s_f, s_i; \lambda', \lambda) \quad . \tag{3.255}$$

Using the familiar trace technique to eliminate the electron spinors this leads to

$$\frac{d\bar{\sigma}}{d\Omega_{k'}}(\lambda', \lambda) = \alpha^2 \frac{\omega'^2}{\omega^2} \frac{1}{2} \mathrm{Tr}\left[\frac{\not{p}_f + m_0}{2m_0} \Gamma \frac{\not{p}_i + m_0}{2m_0} \bar{\Gamma} \right] \quad , \tag{3.256}$$

where

$$\Gamma = \frac{\not{\varepsilon}'(\not{p}_i + \not{k} + m_0)\not{\varepsilon}}{2p_i \cdot k} + \frac{\not{\varepsilon}(\not{p}_i - \not{k}' + m_0)\not{\varepsilon}'}{-2p_i \cdot k'} \tag{3.257}$$

and

$$\bar{\Gamma} = \gamma^0 \Gamma^{\dagger} \gamma^0 = \Gamma(\varepsilon \leftrightarrow \varepsilon') \tag{3.258}$$

---

[28] Polarization effects in Compton scattering have been discussed by F.W. Lipps and H.A. Tolhoek: Physica **20**, 85 (1954).

using $\bar{\gamma}^\mu = \gamma^\mu$. The expression for Γ can be simplified by anticommuting the factor $\not{p}_i$ to the right:

$$\Gamma = \frac{2p_i\cdot\varepsilon\,\not{\varepsilon}' + \not{\varepsilon}'\not{k}\not{\varepsilon} - \not{\varepsilon}'\not{\varepsilon}(\not{p}_i - m_0)}{2p_i\cdot k} + \frac{2p_i\cdot\varepsilon'\,\not{\varepsilon} - \not{\varepsilon}\not{k}'\not{\varepsilon}' - \not{\varepsilon}\not{\varepsilon}'(\not{p}_i - m_0)}{-2p_i\cdot k'} \quad . \tag{3.259}$$

The last term in the numerators can be discarded since it is orthogonal to the energy projection operator in (3.256):

$$(\not{p}_i - m_0)(\not{p}_i + m_0) = p_i^2 - m_0^2 = 0 \quad . \tag{3.260}$$

Thus instead of (3.257) we will use

$$\Gamma \to \frac{2p_i\cdot\varepsilon\,\not{\varepsilon}' + \not{\varepsilon}'\not{k}\not{\varepsilon}}{2p_i\cdot k} + \frac{2p_i\cdot\varepsilon'\,\not{\varepsilon} - \not{\varepsilon}\not{k}'\not{\varepsilon}'}{-2p_i\cdot k'} \quad . \tag{3.261a}$$

and similarly (commuting $\not{p}_i$ to the left)

$$\bar{\Gamma} \to \frac{2p_i\cdot\varepsilon\,\not{\varepsilon}' + \not{\varepsilon}\not{k}\not{\varepsilon}'}{2p_i\cdot k} + \frac{2p_i\cdot\varepsilon'\,\not{\varepsilon} - \not{\varepsilon}'\not{k}'\not{\varepsilon}}{-2p_i\cdot k'} \quad . \tag{3.261b}$$

The Dirac matrices (3.261) are still quite complicated expressions. The following calculations can be simplified considerably, however, if we choose a convenient gauge in which the polarization vectors are orthogonal to the initial electron momentum p_i:

$$\varepsilon\cdot p_i = 0 \quad , \qquad \varepsilon'\cdot p_i = 0 \quad . \tag{3.262}$$

In the laboratory frame where $p_i^\mu = (m_0, \mathbf{0})$ this amounts to the "radiation gauge" in which the electromagnetic potential has no 0-component, i.e. $\varepsilon^\mu = (0, \boldsymbol{\varepsilon})$. However, the condition (3.262) can be imposed in any given frame of reference. Starting from an arbitrary set of polarization vectors ε, ε' we can perform a gauge transformation

$$\begin{aligned} \tilde{\varepsilon}^\mu &= \varepsilon^\mu - \frac{p_i\cdot\varepsilon}{p_i\cdot k}\,k^\mu \quad , \\ \tilde{\varepsilon}'^\mu &= \varepsilon'^\mu - \frac{p_i\cdot\varepsilon'}{p_i\cdot k'}\,k'^\mu \quad , \end{aligned} \tag{3.263}$$

so that the new polarization vectors $\tilde{\varepsilon}$ are orthogonal to p_i. The normalization and transversality conditions (3.178), (3.179) are not affected by the transformation (3.263):

$$\begin{aligned} \tilde{\varepsilon}\cdot\tilde{\varepsilon} &= \tilde{\varepsilon}'\cdot\tilde{\varepsilon}' = -1 \quad , \\ \tilde{\varepsilon}\cdot k &= \tilde{\varepsilon}'\cdot k' = 0 \quad , \end{aligned} \tag{3.264}$$

which immediately follows from $k^2 = k'^2 = 0$. Thus without restricting the generality of our calculation we will impose the condition (3.262). In the remainder of this section for simplicity we will continue to write ε instead of $\tilde{\varepsilon}$.

Using (3.262) we finally have to evaluate the trace in (3.256) with

$$\Gamma = \frac{\not{\varepsilon}'\not{k}\not{\varepsilon}}{2k\cdot p_i} + \frac{\not{\varepsilon}\not{k}'\not{\varepsilon}'}{2k'\cdot p_i} \tag{3.265a}$$

and

$$\bar{\Gamma} = \frac{\not{\varepsilon}\not{k}\not{\varepsilon}'}{2k\cdot p_i} + \frac{\not{\varepsilon}'\not{k}'\not{\varepsilon}}{2k'\cdot p_i} \quad . \tag{3.265b}$$

The calculation of the trace in (3.256) is not easily done since products of up to 8 γ-matrices are involved. Two terms are identical, that is the two mixed terms with a denominator proportional to $(k\cdot p_i)(k'\cdot p_i)$:

$$\begin{aligned}
&\frac{1}{16m_0^2(k\cdot p_i)(k'\cdot p_i)} \mathrm{Tr}\Big[(\not{p}_f + m_0)\big(\not{\varepsilon}'\not{k}\not{\varepsilon}(\not{p}_i + m_0)\not{\varepsilon}'\not{k}'\not{\varepsilon} + \not{\varepsilon}\not{k}'\not{\varepsilon}'(\not{p}_i + m_0)\not{\varepsilon}\not{k}\not{\varepsilon}'\big)\Big] \\
&= \frac{2}{16m_0^2(k\cdot p_i)(k'\cdot p_i)} \mathrm{Tr}\big[(\not{p}_f + m_0)\not{\varepsilon}'\not{k}\not{\varepsilon}(\not{p}_i + m_0)\not{\varepsilon}'\not{k}'\not{\varepsilon}\big] \quad .
\end{aligned} \tag{3.266}$$

This follows from Theorem 7 in the Mathematical Supplement 3.3 according to which the trace remains unchanged when the order of the factors is reversed.

We have to evaluate

$$\begin{aligned}
\frac{d\bar{\sigma}}{d\Omega_{k'}}(\lambda', \lambda) &= \alpha^2 \frac{\omega'^2}{\omega^2} \frac{1}{2} \mathrm{Tr}\left[\frac{\not{p}_f + m_0}{2m_0}\left(\frac{\not{\varepsilon}'\not{\varepsilon}\not{k}}{2k\cdot p_i} + \frac{\not{\varepsilon}\not{\varepsilon}'\not{k}'}{2k'\cdot p_i}\right)\right. \\
&\quad \times \left.\frac{\not{p}_i + m_0}{2m_0}\left(\frac{\not{k}\not{\varepsilon}\not{\varepsilon}'}{2k\cdot p_i} + \frac{\not{k}'\not{\varepsilon}'\not{\varepsilon}}{2k'\cdot p_i}\right)\right] \\
&= \alpha^2 \frac{\omega'^2}{\omega^2} \frac{1}{2} \frac{1}{4m_0^2}\left(\frac{S_1}{(2k\cdot p_i)^2} + \frac{S_2}{(2k'\cdot p_i)^2} + \frac{2S_3}{(2k\cdot p_i)(2k'\cdot p_i)}\right) \quad .
\end{aligned} \tag{3.267}$$

Note that in (3.267) the factors have been slightly reordered using $\not{k}\not{\varepsilon} = -\not{\varepsilon}\not{k}$, $\not{k}'\not{\varepsilon}' = -\not{\varepsilon}'\not{k}'$. Each of the terms S_1, S_2, S_3 contains a trace involving 8 gamma matrices.[29] Without further simplifications each trace would evaluated to a sum over $3\cdot 5\cdot 7 = 105$ terms, each consisting of a product of four scalar products. This follows from the expansion rule for traces, Theorem 3 in the Mathematical Supplement 3.3. Fortunately the final result of the calculation will be much simpler since in our case many of the scalar products vanish

$$k\cdot k = k'\cdot k' = \varepsilon\cdot k = \varepsilon'\cdot k' = \varepsilon\cdot p_i = \varepsilon'\cdot p_i = 0 \tag{3.268a}$$

or are trivial

$$\varepsilon\cdot\varepsilon = \varepsilon'\cdot\varepsilon' = -1 \quad . \tag{3.268b}$$

Thus we might write down the fully expanded trace and then simplify the general expression with the help of (3.268). However, we can avoid this tedious calculation

[29] Complicated trace calculations in QED and other field theories nowadays are routinely performed with the help of the computer, using symbolic-algebra programs like REDUCE, MACSYMA, FORM, Mathematica, and others.

by making use of the fact that the arguments of the traces contain repeated factors (like $\not{\varepsilon}$ in (3.266). We shall anticommute these factors until they stand next to each other. In this way two gamma matrices are eliminated since $\not{a}\not{a} = a \cdot a$ is proportional to the unit matrix. Let us apply this strategy to the evaluation of the three traces in (3.267).

(a) The first trace can be expressed by

$$\begin{aligned} S_1 &= \mathrm{Tr}\left[(\not{p}_f + m_0)\not{\varepsilon}'\not{\varepsilon}\not{k}(\not{p}_i + m_0)\not{k}\not{\varepsilon}\not{\varepsilon}'\right] \\ &= \mathrm{Tr}\left[\not{p}_f\not{\varepsilon}'\not{\varepsilon}\not{k}\not{p}_i\not{k}\not{\varepsilon}\not{\varepsilon}'\right] + m_0^2\,\mathrm{Tr}\left[\not{\varepsilon}'\not{\varepsilon}\not{k}\not{k}\not{\varepsilon}\not{\varepsilon}'\right] \\ &= 2k\cdot p_i\,\mathrm{Tr}\left[\not{p}_f\not{\varepsilon}'\not{\varepsilon}\not{k}\not{\varepsilon}\not{\varepsilon}'\right] - \mathrm{Tr}\left[\not{p}_f\not{\varepsilon}'\not{\varepsilon}\not{p}_i\not{k}\not{k}\not{\varepsilon}\not{\varepsilon}'\right] \\ &= -2k\cdot p_i\,\mathrm{Tr}\left[\not{p}_f\not{\varepsilon}'\not{k}\not{\varepsilon}\not{\varepsilon}\not{\varepsilon}'\right] = 2k\cdot p_i\,\mathrm{Tr}\left[\not{p}_f\not{\varepsilon}'\not{k}\not{\varepsilon}'\right] \\ &= 2k\cdot p_i\left\{2(k\cdot\varepsilon')\mathrm{Tr}\left[\not{p}_f\not{\varepsilon}'\right] - \mathrm{Tr}\left[\not{p}_f\not{\varepsilon}'\not{\varepsilon}'\not{k}\right]\right\} \\ &= 2k\cdot p_i\left[2(k\cdot\varepsilon')4(p_f\cdot\varepsilon') + 4p_f\cdot k\right] \\ &= 8(k\cdot p_i)\left[k\cdot p_f + 2(k\cdot\varepsilon')(p_f\cdot\varepsilon')\right] \quad . \end{aligned} \tag{3.269}$$

In the course of these transformations we have used $k^2 = 0$, $\varepsilon^2 = \varepsilon'^2 = -1$, $\varepsilon\cdot k = 0$. To eliminate the dependence on the final momentum p_f we use

$$k\cdot p_f = k'\cdot p_i \quad , \quad \varepsilon'\cdot p_f = \varepsilon'\cdot k \quad . \tag{3.270}$$

The first identity follows from squaring the four-momentum conservation relation in the form $p_f - k = p_i - k'$ and using $p_i^2 = p_f^2 = m_0^2$ and $k^2 = k'^2 = 0$. Similarly the second identity (3.270) is obtained by multiplying the energy-momentum relation by ε' and using $\varepsilon'\cdot p_i = \varepsilon'\cdot k' = 0$. Thus the final result for the first trace is

$$S_1 = 8(k\cdot p_i)\left[k'\cdot p_i + 2(k\cdot\varepsilon')^2\right] \quad . \tag{3.271}$$

(b) Now we calculate the second trace in (3.267):

$$S_2 = \mathrm{Tr}\left[(\not{p}_f + m_0)\not{\varepsilon}\not{\varepsilon}'\not{k}'(\not{p}_i + m_0)\not{k}'\not{\varepsilon}'\not{\varepsilon}\right] \quad . \tag{3.272}$$

The comparison with (3.269) shows that S_2 results from S_1 if we replace $\varepsilon \leftrightarrow \varepsilon'$ and $k \leftrightarrow k'$ Thus from (3.271) the result is

$$S_2 = 8(k'\cdot p_i)\left[k\cdot p_i - 2(k'\cdot\varepsilon)^2\right] \quad . \tag{3.273}$$

(c) Finally we have to calculate the trace S_3:

$$\begin{aligned} S_3 &= \mathrm{Tr}\left[(\not{p}_f + m_0)\not{\varepsilon}'\not{\varepsilon}\not{k}(\not{p}_i + m_0)\not{k}'\not{\varepsilon}'\not{\varepsilon}\right] \\ &= \mathrm{Tr}\left[(\not{p}_i + m_0)\not{\varepsilon}'\not{\varepsilon}\not{k}(\not{p}_i + m_0)\,\not{k}'\not{\varepsilon}'\not{\varepsilon}\right] + \mathrm{Tr}\left[(\not{k} - \not{k}')\not{\varepsilon}'\not{\varepsilon}\not{k}\not{p}_i\not{k}'\not{\varepsilon}'\not{\varepsilon}\right] \\ &\equiv S_3^a + S_3^b \end{aligned} \tag{3.274}$$

(since $p_f = p_i + k - k'$).

In the first trace, S_3^a, we can anticommute the factor $(\not{p}_i + m_0)$ to the right. This leads to

$$(\not{p}_i + m_0)\not{\varepsilon}'\not{\varepsilon}\not{k} = \not{\varepsilon}'\not{\varepsilon}(\not{p}_i + m_0)\not{k} = 2p_i\cdot k\,\not{\varepsilon}'\not{\varepsilon} + \not{\varepsilon}'\not{\varepsilon}\not{k}(-\not{p}_i + m_0) \quad .$$

The second term drops out since $(-\not{p}_i + m_0)\,(\not{p}_i + m_0) = p_i^2 - m_0^2 = 0$ so that we are left with

$$\begin{aligned}
S_3^a &= 2p_i\cdot k\,\mathrm{Tr}\big[\not\varepsilon'\not\varepsilon\not p_i\not k'\not\varepsilon'\not\varepsilon\big] \\
&= 2p_i\cdot k\left\{2\varepsilon\cdot\varepsilon'\,\mathrm{Tr}\big[\not p_i\not k'\not\varepsilon'\not\varepsilon\big]-\mathrm{Tr}\big[\not\varepsilon\not\varepsilon'\not p_i\not k'\not\varepsilon'\not\varepsilon\big]\right\} \\
&= 2p_i\cdot k\left\{2\varepsilon\cdot\varepsilon'\,\mathrm{Tr}\big[\not p_i\not k'\not\varepsilon'\not\varepsilon\big]-\mathrm{Tr}\big[\not p_i\not k'\big]\right\} \quad . \qquad (3.275)
\end{aligned}$$

The second term in (3.274) can be simplified in the following way:

$$\begin{aligned}
S_3^b &= \mathrm{Tr}\big[(\not k-\not k')\not\varepsilon'\not\varepsilon\not k\not p_i\not k'\not\varepsilon'\not\varepsilon\big] \\
&= 2\varepsilon'\cdot k\,\mathrm{Tr}\big[\not\varepsilon\not k\not p_i\not k'\not\varepsilon'\not\varepsilon\,\big]-\mathrm{Tr}\big[\not\varepsilon'(\not k-\not k')\not\varepsilon\not k\not p_i\not k'\not\varepsilon'\not\varepsilon\big] \\
&= 2\varepsilon'\cdot k\mathrm{Tr}\big[\not\varepsilon\not\varepsilon\not k\not p_i\not k'\not\varepsilon'\big]-\mathrm{Tr}\big[\not\varepsilon'(-\not\varepsilon\not k-\not k'\not\varepsilon)\not k\not p_i\not k'\not\varepsilon'\not\varepsilon\big] \\
&= -2\varepsilon'\cdot k\ \mathrm{Tr}\big[\not k\not p_i\not k'\not\varepsilon'\big]+\mathrm{Tr}\big[\not\varepsilon'\not k'\not\varepsilon\not k\not p_i\not k'\not\varepsilon'\not\varepsilon\big] \\
&= -2\varepsilon'\cdot k\ \mathrm{Tr}\big[\not k\not p_i\not k'\not\varepsilon'\big]+\mathrm{Tr}\big[\not k'\not\varepsilon'\not\varepsilon\not k\not p_i\not\varepsilon'\not k'\not\varepsilon\big] \\
&= -2\varepsilon'\cdot k\ \mathrm{Tr}\big[\not k\not p_i\not k'\not\varepsilon'\big]+2k'\cdot\varepsilon\ \mathrm{Tr}\big[\not k'\not\varepsilon'\not\varepsilon\not k\not p_i\not\varepsilon'\big]-\mathrm{Tr}\big[\not k'\not\varepsilon'\not\varepsilon\not k\not p_i\not\varepsilon'\not\varepsilon\not k'\big] \\
&= -2\varepsilon'\cdot k\ \mathrm{Tr}\big[\not k\not p_i\not k'\not\varepsilon'\big]-2k'\cdot\varepsilon\ \mathrm{Tr}\big[\not\varepsilon'\not k'\not\varepsilon\not k\not p_i\not\varepsilon'\big] \\
&= -2\varepsilon'\cdot k\ \mathrm{Tr}\big[\not k\not p_i\not k'\not\varepsilon'\big]+2k'\cdot\varepsilon\ \mathrm{Tr}\big[\not k'\not\varepsilon\not k\not p_i\big] \quad . \qquad (3.276)
\end{aligned}$$

The remaining traces in (3.275) and (3.276) can not be simplified further and have to be expanded explicitly. This leads to

$$\begin{aligned}
S_3 &= 8p_i\cdot k\Big[2\varepsilon\cdot\varepsilon'\big((p_i\cdot k')(\varepsilon'\cdot\varepsilon)-(p_i\cdot\varepsilon')(k'\cdot\varepsilon)+(p_i\cdot\varepsilon)(k'\cdot\varepsilon)\big)-p_i\cdot k'\Big] \\
&\quad -8k\cdot\varepsilon'\Big[(k\cdot p_i)(k'\cdot\varepsilon')-(k\cdot k')(p_i\cdot\varepsilon')+(k\cdot\varepsilon')(p_i\cdot k)\Big] \\
&\quad +8k'\cdot\varepsilon\Big[(k'\cdot\varepsilon)(k\cdot p_i)-(k'\cdot k)(\varepsilon\cdot p_i)+(k'\cdot p_i)(\varepsilon\cdot k)\Big] \\
&= 8(k\cdot p_i)(k'\cdot p_i)\,\big[2(\varepsilon'\cdot\varepsilon)^2-1\big]-8(k\cdot\varepsilon')^2(p_i\cdot k')+8(k'\cdot\varepsilon)^2(p_i\cdot k) \quad .
\end{aligned} \qquad (3.277)$$

where again $k\cdot\varepsilon=k'\cdot\varepsilon'=\varepsilon\cdot p_i=\varepsilon'\cdot p_i=0$ has been used.

Now we can finally construct the differential photon scattering cross section (3.267) using the three traces just calculated, i.e. S_1 (3.271), S_2 (3.273), and S_3 (3.277). We arive at

$$\begin{aligned}
\frac{d\bar\sigma}{d\Omega_{k'}}(\lambda',\lambda) &= \frac{1}{2}\alpha^2\frac{\omega'^2}{\omega^2}\frac{1}{4m_0^2}\left(\frac{8(k\cdot p_i)\,\big[k'\cdot p_i+2(k\cdot\varepsilon')^2\big]}{4(k\cdot p_i)^2}\right. \\
&\quad +\frac{8(k'\cdot p_i)\,\big[k\cdot p_i-2(k'\cdot\varepsilon)^2\big]}{4(k'\cdot p_i)^2} \\
&\quad \left.+\frac{2\left\{8(k\cdot p_i)(k'\cdot p_i)\,\big[2(\varepsilon'\cdot\varepsilon)^2-1\big]-8(k\cdot\varepsilon')^2(p_i\cdot k')+8(k'\cdot\varepsilon)^2(p_i\cdot k)\right\}}{4(k\cdot p_i)(k'\cdot p_i)}\right) \\
&= \frac{1}{2}\alpha^2\frac{\omega'^2}{\omega^2}\frac{1}{4m_0^2}\,2\left[\frac{k'\cdot p_i}{k\cdot p_i}+\frac{k\cdot p_i}{k'\cdot p_i}+4(\varepsilon'\cdot\varepsilon)^2-2\right] \quad , \qquad (3.278)
\end{aligned}$$

where some of the terms have cancelled each other pairwise. While the trace calculation has been fully covariant we now insert the kinematics of the laboratory frame, i.e. $k{\cdot}p_i = \omega m_0$, $k'{\cdot}p_i = \omega' m_0$. This leads to the well-known *Klein–***Nishina** *formula*[30] which describes Compton scattering of photons:

$$\frac{d\bar{\sigma}}{d\Omega_{k'}}(\lambda,\lambda') = \alpha^2 \frac{1}{4m_0^2}\frac{\omega'^2}{\omega^2}\left[\frac{\omega'}{\omega} + \frac{\omega}{\omega'} + 4(\varepsilon{\cdot}\varepsilon')^2 - 2\right] \quad . . \tag{3.279}$$

ω' depends on ω and on the photon scattering angle according to (3.252). For small energies ($\omega \to 0$) we have $\omega' = \omega$ and the photon scattering cross section (3.279) is reduced to the cross section of *Thomson's scattering formula*:

$$\left(\frac{d\bar{\sigma}}{d\Omega_{k'}}(\lambda',\lambda)\right)_{\omega\to 0} = \alpha^2 \frac{1}{m_0^2}(\varepsilon{\cdot}\varepsilon')^2 \quad . \tag{3.280}$$

This result contains no quantum effects, it can be derived from classical electrodynamics. The classical nature of the cross section (3.280) is rather obvious since it does not depend on Planck's constant $\hbar$. Writing out the constants $\hbar$ and c we find that the cross section is proportional to the square of the length

$$r_0 = \alpha \frac{1}{m_0} = \frac{e^2}{\hbar c}\frac{\hbar}{m_0 c} = \frac{e^2}{m_0 c^2} \simeq 2.8 \times 10^{-13} \text{ cm} \quad . \tag{3.281}$$

This quantity sometimes is called the "classical radius of the electron". This name originates from the (incorrect) notion that the rest mass of the electron can be explained in terms of the electrostatic energy of an extended charged sphere. Equation (3.280) can be written in the form

$$\left(\frac{d\bar{\sigma}}{d\Omega_{k'}}(\lambda',\lambda)\right)_{\omega\to 0} = r_0^2\,(\varepsilon{\cdot}\varepsilon')^2 \quad . \tag{3.282}$$

Equation (3.252) shows that the classical limit $\omega' = \omega$ also applies if the photon scattering angle becomes small. In forward scattering ($\theta \to 0$), therefore, the exact Compton cross section reduces to the Thomson cross section (3.282).

Finally, we shall sum over the polarizations λ' of the scattered photon and average over the initial polarizations λ of the incoming photon thus obtaining the *unpolarized cross section*:

$$\frac{d\bar{\sigma}}{d\Omega_{k'}} = \frac{1}{2}\sum_{\lambda,\lambda'=1}^{2} \frac{d\bar{\sigma}}{d\Omega_{k'}}(\lambda',\lambda) \quad . \tag{3.283}$$

We could have performed these summations from the outset, i.e. by applying the completeness relation (3.220) of the photon polarization vectors to the squared matrix element in (3.254). This would have eliminated the dependence on the ε vectors, thus slightly simplifying the trace calculations.

Instead we will start from the polarization dependent result (3.279) and explicitly sum over λ and λ'. We will use the radiation gauge where the polarization vectors are purely space-like, $\varepsilon = (0, \boldsymbol{\varepsilon})$, so that

$$\varepsilon{\cdot}\varepsilon' = \varepsilon^\mu(\boldsymbol{k},\lambda)\varepsilon_\mu(\boldsymbol{k}',\lambda') = -\boldsymbol{\varepsilon}(\boldsymbol{k},\lambda){\cdot}\boldsymbol{\varepsilon}(\boldsymbol{k}',\lambda') \quad . \tag{3.284}$$

[30] O. Klein and Y. Nishina: Z. Phys. **52**, 853 (1929).

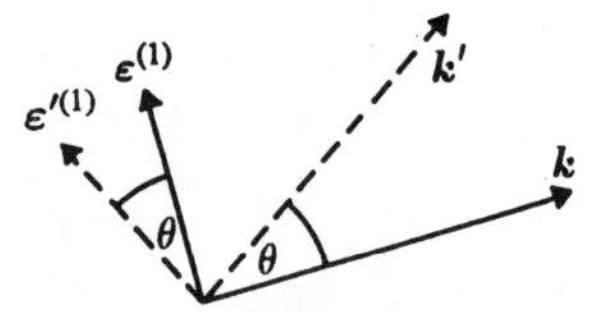

Fig. 3.41. The angle between the vectors of polarization $\varepsilon^{(1)}$ and $\varepsilon'^{(1)}$ is the same as the one between $\boldsymbol{k}$ and $\boldsymbol{k}'$. All four vectors $\varepsilon^{(1)}$, $\varepsilon'^{(1)}$, $\boldsymbol{k}$ and $\boldsymbol{k}'$ are chosen to lie in the same plane. $\varepsilon^{(2)} = \varepsilon'^{(2)}$ is a unit vector orthogonal to this plane

The spatial vectors $\varepsilon(\boldsymbol{k},1)$, $\varepsilon(\boldsymbol{k},2)$, $\boldsymbol{k}$ form an orthogonal system, the same holds for the primed quantities. Now without restricting generality we can choose $\varepsilon(\boldsymbol{k},1)$ and $\varepsilon(\boldsymbol{k},2)$ to lie in the plane spanned by $\boldsymbol{k}$ and $\boldsymbol{k}'$, see Fig. 3.41. Then $\varepsilon(\boldsymbol{k},2)$ and $\varepsilon'(\boldsymbol{k}',2)$ are perpendicular to this plane and thus identical. We have

$$\begin{aligned} \varepsilon^{(1)}\cdot\varepsilon'^{(1)} &= \cos\theta \quad , \\ \varepsilon^{(2)}\cdot\varepsilon'^{(2)} &= 1 \quad , \\ \varepsilon^{(1)}\cdot\varepsilon'^{(2)} &= \varepsilon^{(2)}\cdot\varepsilon'^{(1)} = 0 \quad . \end{aligned} \tag{3.285}$$

The averaged polarization dependent term in (3.279) becomes

$$\frac{1}{2}\sum_{\lambda,\lambda'=1}^{2}\left|\varepsilon^{(\lambda)}\cdot\varepsilon'^{(\lambda')}\right|^2 = \frac{1}{2}\left(\cos^2\theta + 1\right) \quad . \tag{3.286}$$

Using this result and (3.279) the *unpolarized cross section for Compton scattering* becomes

$$\frac{d\bar{\sigma}}{d\Omega_{k'}} = \alpha^2\frac{1}{2m_0^2}\frac{\omega'^2}{\omega^2}\left(\frac{\omega'}{\omega}+\frac{\omega}{\omega'}-\sin^2\theta\right) \quad . \tag{3.287}$$

The classical limit of this result ($\omega \to 0$ or $\theta \to 0$) is the unpolarized Thomson cross section

$$\left(\frac{d\bar{\sigma}}{d\Omega_{k'}}\right)_{\text{class}} = r_0^2\,\frac{1}{2}\left(1+\cos^2\theta\right) \quad . \tag{3.288}$$

In the *ultrarelativistic limit* $\omega, \omega' \gg m_0$ Compton's formula (3.252) reduces to

$$\omega' = \frac{\omega}{1+\frac{\omega}{m_0}(1-\cos\theta)} \simeq \frac{m_0}{1-\cos\theta} = \frac{m_0}{2\sin^2\frac{\theta}{2}} \quad , \tag{3.289}$$

which is valid for not too small scattering angles $\theta^2 \gg \frac{2m_0}{\omega}$. In this limit the Klein–Nishina cross section approaches

$$\left(\frac{d\bar{\sigma}}{d\Omega_{k'}}\right)_{\text{UR}} = r_0^2\,\frac{m_0}{\omega}\frac{1}{4}\frac{1}{\sin^2\frac{\theta}{2}} \qquad \text{for} \quad \theta^2 \gg \frac{2m_0}{\omega} \quad . \tag{3.290}$$

Figure 3.42 shows how the exact result (3.287) with (3.252) interpolates between the limiting cases (3.288) and (3.290). At high energies the angular distribution gets concentrated in a narrow cone in the forward direction.

Lastly we integrate over the photon solid angle $d\Omega_{k'}$ in order to derive the *total cross section*. Here we have to take into account relation (2.252), because a dependence on the scattering angle of the photon is also hidden in ω'. This yields

$$\begin{aligned} \bar{\sigma} &= \alpha^2\frac{1}{2m_0^2}\int \sin\theta\, d\theta\, d\phi \frac{1}{\left(1+\frac{\omega}{m_0}(1-\cos\theta)\right)^2}\left\{\frac{1}{1+\frac{\omega}{m_0}(1-\cos\theta)} + \right. \\ &\quad \left. + \left[1+\frac{\omega}{m_0}(1-\cos\theta)\right] - \sin^2\theta\right\} \\ &= \alpha^2\frac{\pi}{m_0^2}\int_{-1}^{1} dz \left\{\frac{1}{\left[1+\frac{\omega}{m_0}(1-z)\right]^3} + \frac{1}{\left[1+\frac{\omega}{m_0}(1-z)\right]} - \frac{(1-z^2)}{\left[1+\frac{\omega}{m_0}(1-z)\right]^2}\right\}, \end{aligned} \tag{3.291}$$

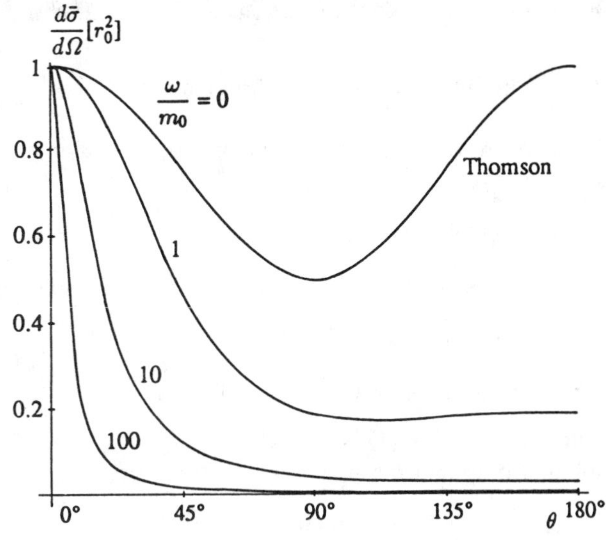

Fig. 3.42. The differential cross section of unpolarized Compton scattering as a function of the scattering angle θ for various photon energies ω

where we have set $z = \cos\theta$. This integral can be calculated in a closed form. Simple expressions can be derived for high and low photon energies.

(a) Small photon energies ($\omega \ll m_0$):

$$\bar{\sigma} \simeq \alpha^2 \frac{\pi}{m_0^2} \left[z + z - \left(z - \frac{z^3}{3} \right) \right]_{-1}^{1} = \alpha^2 \frac{8\pi}{3} \frac{1}{m_0^2} = \frac{8\pi}{3} r_0^2 \quad . \tag{3.292}$$

This is just the classical Thomson cross section.

(b) High photon energies ($\omega \gg m_0$):

$$\begin{aligned} \bar{\sigma} &\simeq \alpha^2 \frac{\pi}{m_0^2} \int\limits_{-1}^{1-m_0/\omega} dz \left[\frac{1}{\left(\frac{\omega}{m_0}\right)^3 (1-z)^3} + \frac{1}{\frac{\omega}{m_0}(1-z)} - \frac{(1-z^2)}{\left(\frac{\omega}{m_0}\right)^2 (1-z)^2} \right] \\ &\simeq \alpha^2 \frac{\pi}{\omega m_0} \left[\ln \frac{2\omega}{m_0} + \frac{1}{2} + O\left(\frac{m_0}{\omega} \ln \frac{\omega}{m_0} \right) \right] \quad . \end{aligned} \tag{3.293}$$

In order to estimate the value of the integral the 1 in the denominators of (3.291) was neglected compared to $(\omega/m_0)(1-z)$, which is valid if $z < 1 - m_0/\omega$. This was accounted for by lowering the upper boundary of integration. The dominating logarithm in (3.293) results from the second term in the integrand.

For completeness we also quote the result of the exact angular integration (3.291). This leads to the total Klein–Nishina cross section, valid for all values of ω

$$\bar{\sigma} = 2\pi \frac{\alpha^2}{m_0^2} \left[\frac{1+\gamma}{\gamma^3} \left(\frac{2\gamma(1+\gamma)}{1+2\gamma} - \ln(1+2\gamma) \right) + \frac{1}{2\gamma} \ln(1+2\gamma) - \frac{1+3\gamma}{(1+2\gamma)^2} \right], \tag{3.294}$$

with the abbreviation $\gamma = \omega/m_0$.

EXERCISE

3.12 Relations of Energy and Momentum

Problem. Assume that the four-momenta p_i, p_f of an electron and k, k' of photons satisfy the relation

$$p_f = p_i + k + k' \quad . \tag{1}$$

Show that this can only be satisfied by the trivial solution

$$p_f = p_i \quad \text{and} \quad k = k' = 0 \tag{2}$$

if the momenta are on the mass shell.

Solution. Equation (1) is equivalent to $p_f - p_i = k + k'$. Squaring this relation and splitting it up into space and time parts we get

$$\omega\omega' + E_i E_f - m_0^2 = \boldsymbol{p}_i \cdot \boldsymbol{p}_f + \boldsymbol{k} \cdot \boldsymbol{k}' \quad , \tag{3}$$

where we used the on-shell conditions

$$p_i^2 = p_f^2 = m_0^2 \quad , \quad k^2 = k'^2 = 0 \quad . \tag{4}$$

The absolute value of the r.h.s. of (3) is limited by the following inequality

$$|\boldsymbol{p}_i \cdot \boldsymbol{p}_f + \boldsymbol{k} \cdot \boldsymbol{k}'| \le |\boldsymbol{p}_i||\boldsymbol{p}_f| + |\boldsymbol{k}||\boldsymbol{k}'| = \sqrt{E_i^2 - m_0^2}\sqrt{E_f^2 - m_0^2} + \omega\omega' \quad . \tag{5}$$

We insert this result into (3):

$$E_i E_f - m_0^2 \le \sqrt{E_i^2 - m_0^2}\sqrt{E_f^2 - m_0^2} \quad . \tag{6}$$

Squaring yields (since $E_i E_f - m_0^2 \ge 0$)

$$E_i^2 E_f^2 - 2m_0^2 E_i E_f + m_0^4 \le E_i^2 E_f^2 - m_0^2(E_i^2 + E_f^2) + m_0^4 \quad . \tag{7}$$

which rewritten is

$$(E_i - E_f)^2 \le 0 \quad . \tag{8}$$

This implies that

$$E_i = E_f \tag{9}$$

and, owing to the time component of (1),

$$\omega = -\omega' \quad , \tag{10}$$

which means that $\omega = \omega' = 0$. Since $\omega = |\boldsymbol{k}|$ and $\omega' = |\boldsymbol{k}'|$, also the photon momentum four-vectors vanish and (2) is proven.

3.8 Annihilation of Particle and Antiparticle

The annihilation of matter and antimatter is a conceptually very interesting process. We shall study the process by considering the example of the annihilation of an electron–positron pair into two photons. The corresponding Feynman graphs are shown in Fig. 3.43.

We immediately notice that these graphs are very similar to those describing Compton scattering. In fact, Fig. 3.43 becomes identical to Fig. 3.39 in Sect. 3.7, if the direction of the time axis is rotated by 90°! In the experiment, of course, both processes appear to be quite different. In pair annihilation two photons are emitted, whereas in Compton scattering one photon is absorbed and one is emitted, together with an electron.

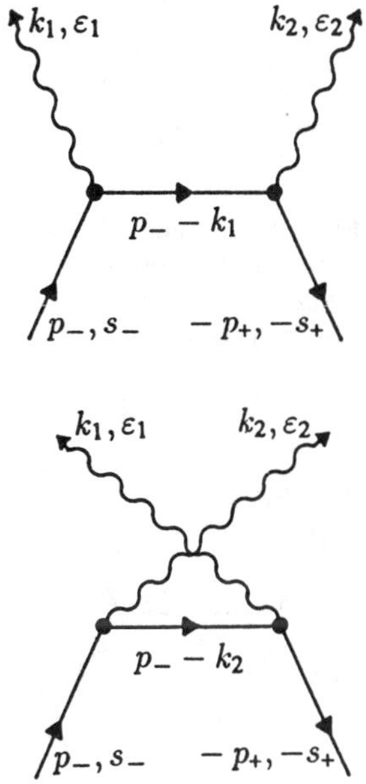

Fig. 3.43. Graph and exchange graph of pair annihilation into two photons

The S-matrix element, which corresponds to the processes in Fig. 3.43, can be easily written down by applying the usual rules. We have to consider the correct kinematics: two particles (electron and positron) enter and two photons leave. In coordinate space the S-matrix element reads

$$S_{fi} = e^2 \int d^3x d^3y \, \bar{\psi}_+(x) \Big[\big(-\mathrm{i}\not{A}(x, k_2)\big) \mathrm{i}S_\mathrm{F}(x-y) \big(-\mathrm{i}\not{A}(y, k_1)\big) + \big(-\mathrm{i}\not{A}(x, k_1)\big) \mathrm{i}S_\mathrm{F}(x-y) \big(-\mathrm{i}\not{A}(y, k_2)\big) \Big] \psi_-(y) \quad , \tag{3.295}$$

which leads to the following expression in momentum space:

$$\begin{aligned} S_{fi} = {} & \frac{e^2}{V^2} \sqrt{\frac{m_0^2}{E_+ E_-}} \sqrt{\frac{(4\pi)^2}{2\omega_1\omega_2}} (2\pi)^4 \delta^4(k_1 + k_2 - p_+ - p_-) \\ & \times \bar{v}(p_+, s_+) \left[(-\mathrm{i}\not{\varepsilon}_2) \frac{\mathrm{i}}{\not{p}_- - \not{k}_1 - m_0} (-\mathrm{i}\not{\varepsilon}_1) \right. \\ & \left. + (-\mathrm{i}\not{\varepsilon}_1) \frac{\mathrm{i}}{\not{p}_- - \not{k}_2 - m_0} (-\mathrm{i}\not{\varepsilon}_2) \right] u(p_-, s_-) \quad . \end{aligned} \tag{3.296}$$

Here for the description of both photons outgoing plane waves have been used. Obviously the S-matrix element is symmetric with respect to photon exchange, as it should be according to Bose-statistics. The coherent summation of both graphs is necessary to preserve this symmetry. We will first interpret this process.

An electron with positive energy and momentum and spin (p_-, s_-) was produced in the past and moves forward in time. It is scattered into a state with negative energy and four-momentum $-p_+$ moving backwards in time. The wave function of an electron with negative energy and momentum $-\boldsymbol{p}_+$ is given by $v(p_+, s_+) \exp(\mathrm{i}p_+ \cdot x)$ (see *RQM*, Chap. 6). During the scattering it converts its energy into radiation by emitting two photons.

As usual the δ^4-function in (3.296) expresses energy-momentum conservation between ingoing and outgoing particles. Only the electron enters the reaction: the two photons and the positron exit the zone of reaction. However, the positron has a four-momentum $-p_+$ because it is represented by an electron with negative energy. Therefore

$$p_- = k_1 + k_2 + (-p_+) = k_1 + k_2 - p_+ \quad . \tag{3.297}$$

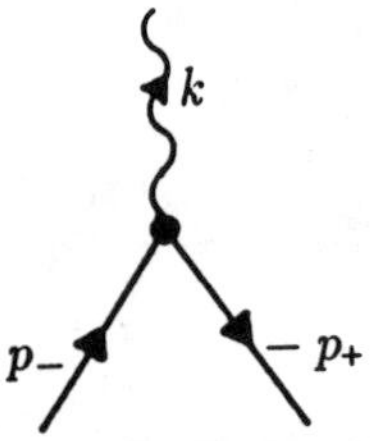

Fig. 3.44. The annihilation of a free electron–positron pair into a single photon is kinematically forbidden

It can be easily understood that the annihilation of an electron–positron pair into a single photon is kinematically forbidden. This hypothetical process is shown in Fig. 3.44, which gives $p_- = k + (-p_+)$ or $p_- + p_+ = k$ and, conclusively, $(p_- + p_+)^2 = k^2 = 0$. This, however, is not possible, as can be seen most easily by going to the center-of-mass frame where $\boldsymbol{p}_+ + \boldsymbol{p}_- = 0$. Note that this argument is only valid if the electron is a free particle. The one-photon annihilation of positrons and bound atomic electrons is indeed possible. In this case the second photon is absorbed by the nucleus.

Before beginning with the evaluation of the annihilation cross section we come back to the close relation between pair annihilation and Compton scattering (Table 3.2). Both processes are related by the *crossing symmetry* which we first encountered when we compared electron-electron and electron–positron scattering in Sect. 3.4. There we noted that processes of the type $A + B \longrightarrow C + D$ and $A + \bar{D} \rightarrow C + \bar{B}$ are related to each other by a *substitution rule*, i.e. by replacing the momentum variables $p_B \longrightarrow -p_D$ and $p_D \longrightarrow -p_B$. In the case of pair annihilation, $e^+ + e^- \rightarrow \gamma + \gamma$, we can identify A with the electron, B with the positron, and C, D with the two photons. Then crossing leads to the process of Compton scattering, $e^- + \gamma \rightarrow \gamma + e^-$. We can identify $D = \bar{D}$ since the photon is its own antiparticle.

To complete the picture we note that the crossing symmetry can be applied a second time. The Compton process thus is related to $\bar{C} + \bar{D} \longrightarrow \bar{A} + \bar{B}$ which means $\gamma + \gamma \rightarrow e^+ + e^-$. This is the process of *electron–positron pair creation* by two photons (see Exercise 3.15). The corresponding Feynman graphs are shown in Fig. 3.45. The following table summarizes how these three processes are related to each other according to the substitution rule. Thus all of the three processes essentially are governed by the same physics. Only the kinematical conditions are different. We again stress that the crossing symmetry is exact, not being restricted to a particular order of perturbation theory.

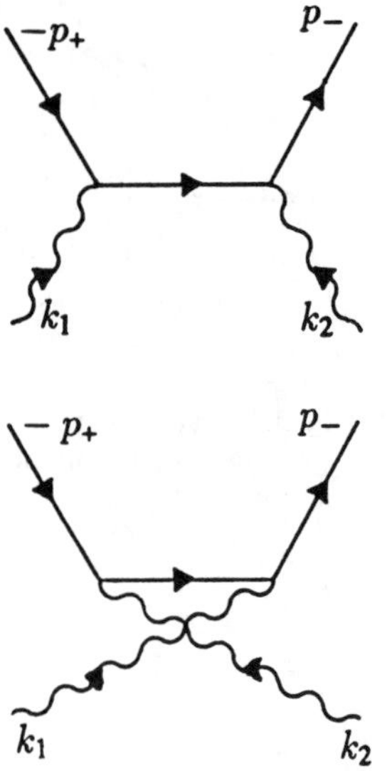

Fig. 3.45. The process of pair production by photons is closely related to that of pair annihilation into photons, Fig. 3.43

Table 3.2. Three processes which are related by crossing symmetry

Pair annihilation $e^- + e^+ \rightarrow \gamma + \gamma$	Compton scattering $e^- + \gamma \rightarrow \gamma + e^-$	Pair creation $\gamma + \gamma \rightarrow e^+ + e^-$
p_-	p_i	$-p_+$
p_+	$-p_f$	$-p_-$
k_1	$-k$	$-k_1$
k_2	k'	$-k_2$

We now will determine the cross section for pair annihilation. As usual it is related to the S-matrix element by

$$d\sigma = \int \frac{|S_{fi}|^2}{T \cdot V \cdot \frac{|v|}{V} \cdot \frac{1}{V}} V \frac{d^3k_1}{(2\pi)^3} V \frac{d^3k_2}{(2\pi)^3} \quad . \tag{3.298}$$

Insertion of (3.296) leads to

$$d\sigma = \frac{e^4}{(2\pi)^2} \frac{m_0^2}{E_+ E_-} \frac{(4\pi)^2}{|v_{\rm rel}|} \int \delta^4(k_1 + k_2 - p_+ - p_-) \, |\varepsilon_2^\mu M_{\mu\nu} \varepsilon_1^\nu|^2 \frac{d^3k_1}{2\omega_1} \frac{d^3k_2}{2\omega_2} . \tag{3.299}$$

Here we have introduced the tensor

$$M_{\mu\nu} = \bar{v}(p_+, s_+)\left[\gamma_\mu \frac{1}{\not{p}_- - \not{k}_1 - m_0}\gamma_\nu + \gamma_\nu \frac{1}{\not{p}_- - \not{k}_2 - m_0}\gamma_\mu\right] u(p_-, s_-)$$
$$= \bar{v}(p_+, s_+)\left[\gamma_\mu \frac{\not{p}_- - \not{k}_1 + m_0}{-2p_- \cdot k_1}\gamma_\nu + \gamma_\nu \frac{\not{p}_- - \not{k}_2 + m_0}{-2p_- \cdot k_2}\gamma_\mu\right] u(p_-, s_-) \quad . \tag{3.300}$$

In accordance with crossing symmetry we note that this "annihilation tensor" is identical with the "Compton tensor" defined in Sect. 3.7, (3.242) when the replacements indicated in the Table are made. Therefore most of the calculations in Sect. 3.7 can be taken over. The kinematics of the collision, however, is different since we now have two photons in the final state and two massive particles in the incoming channel.

The six-dimensional two-body phase space is reduced to two dimensions because of the delta function. Let us consider the integral

$$I = \int \frac{d^3k_1}{2\omega_1}\frac{d^3k_2}{2\omega_2}\,\delta^4(k_1 + k_2 - p_+ - p_-)f(k_1, k_2) \quad , \tag{3.301}$$

where $f(k_1, k_2)$ stands for the momentum-dependent integrand of (3.299). We now integrate out the variable k_2, once again making use of the formula (3.74)

$$\frac{d^3k_2}{2\omega_2} = \int_{-\infty}^{\infty} d^4k_2\,\delta(k_2 \cdot k_2 - 0)\,\Theta(k_{2_0}) \quad . \tag{3.302}$$

Equation (3.301) becomes

$$I = \int_0^\infty \frac{1}{2}\omega_1 d\omega_1 d\Omega_{k_1}\delta\left[(p_+ + p_- - k_1)^2\right]$$
$$\times\,\Theta(E_+ + E_- - \omega_1)f(k_1, k_2 = p_+ + p_- - k_1)$$
$$= \frac{d\Omega_{k_1}}{2}\int_0^{E_+ + E_-} \omega_1 d\omega_1\delta\left[(p_+ + p_-)^2 - 2k_1\cdot(p_+ + p_-)\right]$$
$$\times f(k_1, k_2 = p_+ + p_- - k_1) \quad . \tag{3.303}$$

To evaluate this integral we have to specify the frame of reference. The following calculation will be done *in the rest frame of the electron* where $p_- = (m_0, \mathbf{0})$. In this frame (3.303) becomes

$$I = \frac{d\Omega_{k_1}}{2}\int_0^{E_+ + m_0} \omega_1 d\omega_1\,\delta\left(2m_0^2 + 2m_0E_+ - 2\omega_1(m_0 + E_+ - |\boldsymbol{p}_+|\cos\theta)\right)$$
$$\times f(k_1, k_2 = p_+ + p_- - k_1)$$
$$= \frac{d\Omega_{k_1}}{2}\left.\frac{\omega_1 f(k_1, k_2 = p_+ + p_- - k_1)}{2|m_0 + E_+ - |\boldsymbol{p}_+|\cos\theta|}\right|_{\omega_1}$$
$$= \frac{d\Omega_{k_1}}{4}\frac{m_0(m_0 + E_+)f(k_1, k_2 = p_+ + p_- - k_1)}{[m_0 + E_+ - |\boldsymbol{p}_+|\cos\theta]^2} \quad , \tag{3.304}$$

where the photon energy $\omega_1 = |\boldsymbol{k}_1|$ is determined by

$$\omega_1 = \frac{m_0(m_0 + E_+)}{m_0 + E_+ - |\boldsymbol{p}_+| \cos\theta} \quad . \tag{3.305}$$

Here θ denotes the angle between the momenta of the first photon $\boldsymbol{k}_1$ and the incoming positron $\boldsymbol{p}_+$ (compare Fig. 3.46). In (3.304) every photon energy ω_1 occuring in the function $f(k_1, \dots)$ is replaced by (3.305).

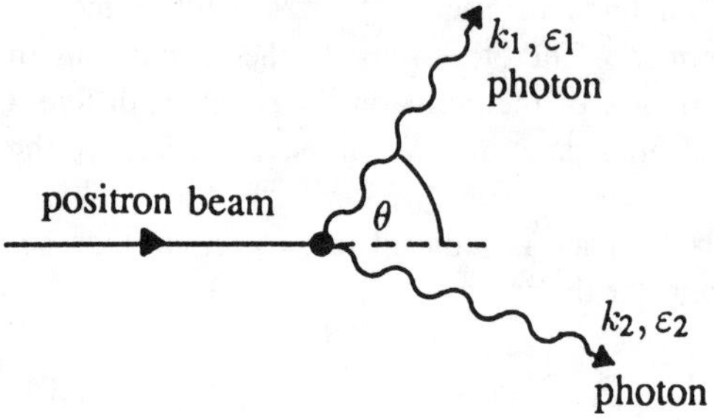

Fig. 3.46. Definition of the angle θ

In the laboratory frame where $|\boldsymbol{v}_{\text{rel}}| = |\boldsymbol{v}_+| = |\boldsymbol{p}_+|/E_+$ the differential cross section (3.299) for pair annihilation becomes

$$\frac{d\sigma}{d\Omega_{k_1}} = \alpha^2 \frac{\omega_1^2}{m_0(m_0 + E_+)} \frac{m_0}{|\boldsymbol{p}_+|} \left| \varepsilon_2^\mu \, M_{\mu\nu} \, \varepsilon_1^\nu \right|^2 \quad , \tag{3.306}$$

with ω_1 given by (3.305). Note the different kinematical dependence when comparing (3.306) with the corresponding formula for Compton scattering (3.254).

The further evaluation of (3.306) is straightforward. Averaging over the electron and positron spins

$$\frac{d\bar{\sigma}}{d\Omega_{k_1}}(\lambda_2, \lambda_1) = \frac{1}{4} \sum_{s_-, s_+} \frac{d\sigma}{d\Omega_{k_1}}(s_+, s_-; \lambda_2, \lambda_1) \tag{3.307}$$

leads to

$$\begin{aligned} \frac{d\bar{\sigma}}{d\Omega_{k_1}}(\lambda_2, \lambda_1) &= \frac{1}{4}\alpha^2 \frac{\omega_1^2}{m_0(m_0 + E_+)} \frac{m_0}{|\boldsymbol{p}_+|} (-) \\ &\times \mathrm{Tr}\left[\frac{-\not{p}_+ + m_0}{2m_0} \left(\frac{\not{\varepsilon}_2 \not{k}_1 \not{\varepsilon}_1}{2p_- \cdot k_1} + \frac{\not{\varepsilon}_1 \not{k}_2 \not{\varepsilon}_2}{2p_- \cdot k_2} \right) \frac{\not{p}_- + m_0}{2m_0} \left(\frac{\not{\varepsilon}_1 \not{k}_1 \not{\varepsilon}_2}{2p_- \cdot k_1} + \frac{\not{\varepsilon}_2 \not{k}_2 \not{\varepsilon}_1}{2p_- \cdot k_2} \right) \right] . \end{aligned} \tag{3.308}$$

The trace in this expression coincides with the result for Compton scattering if the translation of the momentum variables according to the substitution rule is made (see (3.256) and (3.265)). Note the extra minus sign in (3.308) which arises from the summation over positron spinors

$$\sum_s v_\alpha(p, s) \bar{v}_\beta(p, s) = -\left(\frac{-\not{p} + m_0}{2m_0} \right)_{\alpha\beta} \quad . \tag{3.309}$$

As in Sect. 3.7 the special tranverse gauge condition

$$\varepsilon_1\cdot p_- = \varepsilon_2\cdot p_- = 0 \tag{3.310}$$

has been imposed to simplify (3.308). According to (3.278) the calculation of the trace leads to

$$\mathrm{Tr}\big[\cdots\big] = \frac{1}{4m_0^2}\,2\Big[\frac{k_2\cdot p_-}{-k_1\cdot p_-} + \frac{-k_1\cdot p_-}{k_2\cdot p_-} + 4(\varepsilon_2\cdot\varepsilon_1)^2 - 2\Big] \quad , \tag{3.311}$$

where the substitution $p_i \to p_-$, $k \to -k_1$, $k' \to k_2$ has been performed. Then the cross section for pair annihilation in the electron rest frame becomes

$$\begin{aligned}\frac{d\bar\sigma}{d\Omega_{k_1}}(\lambda_2,\lambda_1) &= \frac{1}{8}\frac{\alpha^2}{m_0^2}\frac{\omega_1^2}{m_0(m_0+E_+)}\frac{m_0}{|\boldsymbol{p}_+|}\Big[\frac{k_2\cdot p_-}{k_1\cdot p_-} + \frac{k_1\cdot p_-}{k_2\cdot p_-} - 4(\varepsilon_2\cdot\varepsilon_1)^2 + 2\Big] \\ &= \frac{1}{8}\frac{\alpha^2}{m_0^2}\frac{m_0(m_0+E_+)}{(m_0+E_+-|\boldsymbol{p}_+|\cos\theta)^2}\frac{m_0}{|\boldsymbol{p}_+|}\Big[\frac{\omega_2}{\omega_1} + \frac{\omega_1}{\omega_2} + 2 - 4(\varepsilon_2\cdot\varepsilon_1)^2\Big] \ ,\end{aligned} \tag{3.312}$$

where $p_- = (m_0, \boldsymbol{0})$. The photon energy ω_1 depends on the photon angle θ according to (3.305) while ω_2 follows from energy conservation:

$$\begin{aligned}\omega_2 &= m_0 + E_+ - \omega_1 = m_0 + E_+ - \frac{m_0(m_0+E_+)}{m_0+E_+-|\boldsymbol{p}_+|\cos\theta} \\ &= (m_0+E_+)\left(1 - \frac{m_0}{m_0+E_+-|\boldsymbol{p}_+|\cos\theta}\right) \\ &= \frac{E_+ - |\boldsymbol{p}_+|\cos\theta}{m_0}\omega_1 \quad .\end{aligned} \tag{3.313}$$

If the photon polarizations are not observed (3.312) has to be summed over λ_1, λ_2. Using (3.286) this leads to the *unpolarized cross section for pair annihilation*

$$\begin{aligned}\frac{d\bar\sigma}{d\Omega_{k_1}} &= \sum_{\lambda_1,\lambda_2=1}^{2}\frac{d\bar\sigma}{d\Omega_{k_1}}(\lambda_2,\lambda_1) \\ &= \frac{1}{2}\frac{\alpha^2}{m_0^2}\frac{m_0(m_0+E_+)}{(m_0+E_+-|\boldsymbol{p}_+|\cos\theta)^2}\frac{m_0}{|\boldsymbol{p}_+|}\left(\frac{\omega_2}{\omega_1} + \frac{\omega_1}{\omega_2} + \sin^2\tilde\theta\right) \quad .\end{aligned} \tag{3.314}$$

Here $\tilde\theta$ is the angle between the momentum vectors $\boldsymbol{k}_1$ and $\boldsymbol{k}_2$ of the two photons. In the case of Compton scattering $\tilde\theta$ happened to coincide with the scattering angle θ.

In the *nonrelativistic limit* $E_+ - m_0 \ll m_0$ the created photons have equal energies $\omega_1 \to m_0$, $\omega_2 \to m_0$ and are emitted back-to-back, $\tilde\theta \to \pi$. Since the incoming positron momentum is negligible in this case the angular distribution of photons becomes isotropic and (3.314) reduces to

$$\left(\frac{d\bar\sigma}{d\Omega_{k_1}}\right)_{\mathrm{nr}} = \frac{1}{2}\frac{\alpha^2}{m_0^2}\frac{1}{\beta_+} \quad , \tag{3.315}$$

where $\beta_+ = |\boldsymbol{p}_+|/m_0$ is the positron velocity. If the incoming positron has a large momentum the angular distribution (3.314) becomes peaked in the forward direction. In this case nearly all the energy E_+ is carried by the photon that is emitted in the beam direction. Figure 3.47 shows the differential cross section (3.314) for some typical values of the positron energy.

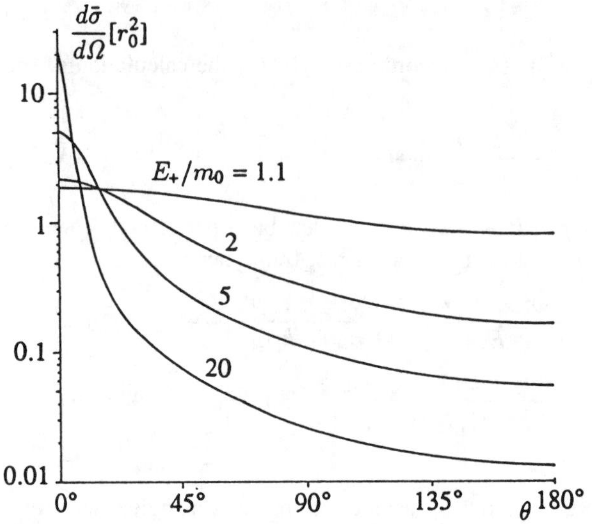

Fig. 3.47. Logarithmic plot of the differential pair annihilation cross section in units of $r_0^2 = \alpha^2/m_0^2$ as a function of the photon angle θ, drawn for various values of the positron energy E_+

The *total annihilation cross section* is obtained by integrating (3.314) over the solid angle $d\Omega_{k_1}$. When performing this integration we have to keep in mind that there are *two identical particles* in the final state. This point is already implemented in the cross section (3.314) originating from the symmetry of the scattering amplitude (3.296) with respect to the exchange of two photons $(k_1, \varepsilon_1 \leftrightarrow k_2, \varepsilon_2)$. Thus (3.314) gives the cross section that one of the photons is scattered into the angle $d\Omega_{k_1}$. Owing to the indistinguishability of the photons either of them could be so scattered. We therefore double-count the photons when we integrate $d\bar{\sigma}/d\Omega_{k_1}$ over the full solid angle 4π; i.e., we would count four – not two – photons per scattering event. To correct for this double counting the cross section has to be multiplied by $1/2$:

$$\bar{\sigma} = \frac{1}{2}\int d\Omega_{k_1} \frac{d\bar{\sigma}}{d\Omega_{k_1}} \quad . \tag{3.316}$$

This integration is carried out in Exercise 3.13.

EXERCISE

3.13 The Total Cross Section of Pair Annihilation

Problem. An unpolarized positron with four-momentum $(E, \boldsymbol{p})$ hits an equally unpolarized electron at rest, annihilating into two photons with momenta $\boldsymbol{k}_1$ and $\boldsymbol{k}_2$ and polarizations λ_1 and λ_2. The angle between positron momentum and $\boldsymbol{k}$ is denoted by θ, the angle between $\boldsymbol{k}_1$ and $\boldsymbol{k}_2$ by $\tilde{\theta}$ (see Fig. 3.48). Determine the total cross section of pair annihilation.

Solution. The differential cross section for pair annihilation has been given in (3.314)

$$\frac{d\bar{\sigma}}{d\Omega} = \frac{\alpha^2}{2|\boldsymbol{p}|}\frac{m_0+E}{(x+m_0)^2}\left(\frac{m_0^2+x^2}{m_0 x}+\sin^2\tilde{\theta}\right) \quad . \tag{1}$$

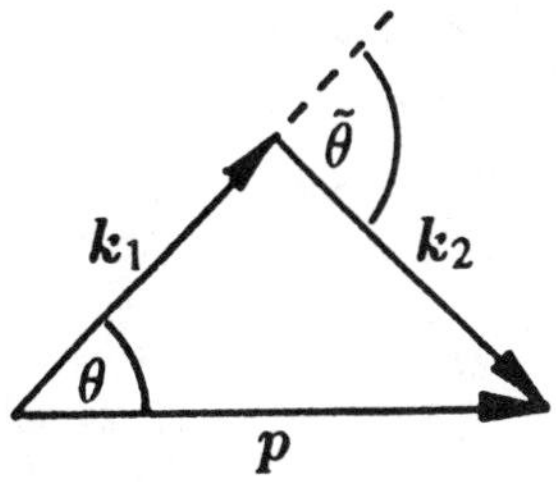

Fig. 3.48. Definition of the angles θ and $\tilde{\theta}$

Here we have introduced the variable x which depends on the photon angle and is defined by

$$x = \frac{\boldsymbol{p}\cdot\boldsymbol{k}_1}{|\boldsymbol{k}_1|} = E - |\boldsymbol{p}|\cos\theta \qquad \text{or} \qquad \cos\theta = \frac{E-x}{|\boldsymbol{p}|} \quad . \tag{2}$$

From kinematics we get the photon energies

$$\omega_1 = m_0\frac{E+m_0}{x+m_0} \qquad \text{and} \qquad \omega_2 = \frac{x}{m_0}\omega_1 \quad . \tag{3}$$

To evaluate (1) the dependence $\tilde{\theta} = \tilde{\theta}(\theta)$ is needed. The sine rule applied to the triangle spanned by $\boldsymbol{k}_1$, $\boldsymbol{k}_2$, and $\boldsymbol{p}$ leads to

$$\frac{\sin\theta}{|\boldsymbol{k}_2|} = \frac{\sin(\pi-\tilde{\theta})}{|\boldsymbol{p}|} = \frac{\sin\tilde{\theta}}{|\boldsymbol{p}|} \tag{4}$$

and thus, using (3)

$$\sin\tilde{\theta} = \frac{|\boldsymbol{p}|}{E+m_0}\frac{x+m_0}{x}\sin\theta \quad . \tag{5}$$

From

$$\sin^2\theta = 1-\cos^2\theta = 1-\frac{(E-x)^2}{|\boldsymbol{p}|^2} = -\frac{x^2-2Ex+m_0^2}{|\boldsymbol{p}|^2} \tag{6}$$

it follows that

$$\sin^2\tilde{\theta} = -\frac{1}{(E+m_0)^2}(x^2-2Ex+m_0^2)\left(\frac{m_0+x}{x}\right)^2 \quad . \tag{7}$$

The total cross section, corrected for the presence of two identical particles in the final state, is given by

$$\begin{aligned}\bar{\sigma} &= \frac{1}{2}\int d\Omega\,\frac{d\bar{\sigma}}{d\Omega}\\ &= \pi\int_{-1}^{+1} d\cos\theta\,\frac{d\bar{\sigma}}{d\Omega} = \frac{\pi}{|\boldsymbol{p}|}\int_{E-|\boldsymbol{p}|}^{E+|\boldsymbol{p}|} dx\,\frac{d\bar{\sigma}}{d\Omega} \quad .\end{aligned} \tag{8}$$

We insert (1) and (7) into (8) and obtain

Exercise 3.13.

$$\bar{\sigma} = \frac{\alpha^2\pi}{2m_0|\boldsymbol{p}|^2} \int\limits_{E-|\boldsymbol{p}|}^{E+|\boldsymbol{p}|} dx \left[(E+m_0)\frac{m_0^2+x^2}{x(m_0+x)^2} - \frac{m_0}{m_0+E}\frac{1}{x^2}(x^2-2Ex+m_0^2)\right] \quad . \tag{9}$$

We can immediately derive an asymptotic expression for the *nonrelativistic limit* from this integral. Replacing the variable x by the constant value $x \sim E \sim m_0$ everywhere in the integral yields the estimate

$$\bar{\sigma}_{\mathrm{nr}} \simeq \frac{\alpha^2\pi}{2m_0|\boldsymbol{p}|^2} 2|\boldsymbol{p}| \left[2\frac{2}{2^2} - \frac{1}{2}(1-2+1)\right] = \frac{\alpha^2\pi}{m_0|\boldsymbol{p}|} \quad . \tag{10}$$

This can also be written as

$$\bar{\sigma}_{\mathrm{nr}} \simeq \pi r_0^2 \frac{1}{\beta} \quad , \tag{11}$$

where $r_0 = e^2/m_0c^2 \simeq 2.8$ fm is the "classical radius of the electron" and $\beta = |\boldsymbol{v}|/c$ denotes the incoming positron velocity.

It is not difficult to go beyond this approximation and solve the integral (9) exactly. We write

$$\bar{\sigma} = \frac{\alpha^2\pi}{2m_0|\boldsymbol{p}|^2}(I_1+I_2) \quad , \tag{12}$$

with

$$I_1 = (E+m_0) \int\limits_{E-|\boldsymbol{p}|}^{E+|\boldsymbol{p}|} dx \, \frac{m_0^2+x^2}{x(m_0+x)^2} \tag{13}$$

and

$$I_2 = -\frac{m_0}{m_0+E} \int\limits_{E-|\boldsymbol{p}|}^{E+|\boldsymbol{p}|} dx \left(1 - \frac{2E}{x} + \frac{m_0^2}{x^2}\right) \quad . \tag{14}$$

The integrals can be calculated analytically:

$$\begin{aligned} I_1 &= (E+m_0)\left(\frac{2m_0}{m_0+x} + \ln\frac{x}{m_0}\right)\bigg|_{E-|\boldsymbol{p}|}^{E+|\boldsymbol{p}|} \\ &= (E+m_0)\left(\frac{2m_0}{m_0+E+|\boldsymbol{p}|} - \frac{2m_0}{m_0+E-|\boldsymbol{p}|} + \ln\frac{E+|\boldsymbol{p}|}{E-|\boldsymbol{p}|}\right) \\ &= (E+m_0)\left[\frac{2m_0(-2|\boldsymbol{p}|)}{(m_0+E)^2-|\boldsymbol{p}|^2} + \ln\frac{(E+|\boldsymbol{p}|)^2}{E^2-|\boldsymbol{p}|^2}\right] \\ &= -2|\boldsymbol{p}| + 2(E+m_0)\ln\frac{E+|\boldsymbol{p}|}{m_0} \end{aligned} \tag{15}$$

and

Exercise 3.13.

$$I_2 = \frac{m_0}{m_0+E}\left(-x+2E\ln x+\frac{1}{x}\right)\Bigg|_{E-|\boldsymbol{p}|}^{E+|\boldsymbol{p}|}$$
$$= \frac{4m_0}{m_0+E}\left(E\ln\frac{E+|\boldsymbol{p}|}{m_0}-|\boldsymbol{p}|\right) \quad . \tag{16}$$

This leads to the following exact expression for the total pair annihilation cross section[31]

$$\bar{\sigma} = \frac{\alpha^2\pi}{m_0|\boldsymbol{p}|^2(E+m_0)}\left[(E^2+4m_0E+m_0^2)\ln\frac{E+|\boldsymbol{p}|}{m_0}-(E+3m_0)|\boldsymbol{p}|\right] \quad . \tag{17}$$

Now we derive two limiting cases.

(a) *The nonrelativistic limit* ($|\boldsymbol{p}|\to 0$, $E\to m_0$). In the case $|\boldsymbol{p}|/m_0 \ll 1$ we have the expansions

$$E(|\boldsymbol{p}|) = m_0 + O(\boldsymbol{p}^2) \quad ,$$
$$\frac{m_0}{E+m_0} = \frac{1}{2}\frac{1}{1+\frac{E-m_0}{2m_0}} = \frac{1}{2}+O(\boldsymbol{p}^2) \quad ,$$
$$\ln\frac{E+|\boldsymbol{p}|}{m_0} = \frac{1}{2}\ln\frac{1+|\boldsymbol{p}|/E}{1-|\boldsymbol{p}|/E} = \frac{|\boldsymbol{p}|}{E}+O(|\boldsymbol{p}|^3) = \frac{|\boldsymbol{p}|}{m_0}+O(|\boldsymbol{p}|^3) \quad ,$$

yielding

$$\bar{\sigma}_{\text{nr}} = \frac{\alpha^2\pi}{m_0|\boldsymbol{p}|}\left(1+O(\boldsymbol{p}^2)\right) \tag{18}$$

in accordance with (11).

(b) *The ultrarelativistic limit* ($|\boldsymbol{p}|\to\infty$, $E\to\infty$). For $E/m_0 \gg 1$ we approximate

$$|\boldsymbol{p}| = E\sqrt{1-\frac{m_0^2}{E^2}} = E+O(E^{-1}) \quad ,$$
$$\frac{m_0^2}{\boldsymbol{p}^2} = \frac{m_0^2}{E^2}\frac{1}{1-m_0^2/E^2} = \frac{m_0^2}{E^2}+O(E^{-4}) \quad ,$$
$$\frac{m_0}{m_0+E} = \frac{m_0}{E}\frac{1}{1+m_0/E} = \frac{m_0}{E}+O(E^{-2}) \quad .$$

Furthermore, the Taylor expansion of the logarithm gives

$$\ln\frac{E+|\boldsymbol{p}|}{m_0} = \ln\frac{2E+(|\boldsymbol{p}|-E)}{m_0} = \ln\frac{2E}{m_0}+O\left(\frac{1}{E}(|\boldsymbol{p}|-E)\right)$$
$$= \ln\frac{2E}{m_0}+O(E^{-2}) \quad .$$

[31] P.A.M. Dirac: Proc. Camb. Phil. Soc. **26**, 361 (1930).

Thus we get the asymptotic formula for the cross section

$$\bar{\sigma}_{\mathrm{ur}} = \frac{\alpha^2\pi}{m_0 E}\left[\ln\frac{2E}{m_0} - 1 + O\left(\frac{\ln 2E/m_0}{E/m_0}\right)\right] \quad . \tag{19}$$

EXERCISE

3.14 Electron–Positron Annihilation in the Centre-of-Mass Frame

Problem. Derive the differential and total unpolarized cross section for pair annihilation $e^+ + e^- \to \gamma + \gamma$ in the centre-of-mass frame.

Solution. The general expression for the differential cross section of pair annihilation has been given in (3.299), (3.300):

$$\begin{aligned} d\bar{\sigma} &= \frac{\alpha^2}{|\boldsymbol{v}_{\mathrm{rel}}|}\frac{m_0}{E_+}\frac{m_0}{E_-}\overline{|\varepsilon_2\cdot M_{fi}\cdot\varepsilon_1|^2} \\ &\quad\times (2\pi)^4\delta^4(k_1+k_2-p_+-p_-)\frac{4\pi d^3k_1}{(2\pi)^3 2\omega_1}\frac{4\pi d^3k_2}{(2\pi)^3 2\omega_2} \quad . \end{aligned} \tag{1}$$

where

$$\begin{aligned} \overline{|\varepsilon_2\cdot M_{fi}\cdot\varepsilon_1|^2} &= \frac{1}{4}\sum_{s_+,s_-}\sum_{\lambda_1,\lambda_2}|\varepsilon_2\cdot M_{fi}\cdot\varepsilon_1|^2 \\ &= \frac{1}{2}\frac{1}{4m_0^2}\sum_{\lambda_1,\lambda_2}\left(\frac{k_2\cdot p_-}{k_1\cdot p_-} + \frac{k_1\cdot p_-}{k_2\cdot p_-} - 4(\tilde{\varepsilon}_2\cdot\tilde{\varepsilon}_1)^2 + 2\right) \quad . \end{aligned} \tag{2}$$

In the derivation of (2) the special gauge condition $\tilde{\varepsilon}_1\cdot p_- = \tilde{\varepsilon}_2\cdot p_- = 0$ was chosen, which we indicate by the tilde. This was convenient for calculations in the electron rest frame where $p_- = (m_0, \mathbf{0})$. Here the polarization vectors could be chosen to be purely spacelike vectors $\varepsilon = (0, \boldsymbol{\varepsilon})$ transverse to the photon momentum $\boldsymbol{k}$ and the λ-summation was easily performed. In an arbitrary frame of reference the special gauge condition loses its simplicity and complicated $\tilde{\varepsilon}$-vectors have to be constructed. It would be more appealing to use the general completeness relation (3.220) for photon polarization vectors

$$\sum_{\lambda=1}^{2}\varepsilon_\mu(k,\lambda)\varepsilon_\nu(k,\lambda) \to -g_{\mu\nu} \quad . \tag{3}$$

This replacement, however, may be applied only to *gauge-invariant* expressions, a condition which is not fulfilled by (2). Nevertheless we can use (3) if we reinstall the gauge invariance of (2). This is achieved by going back from the special polarization vectors $\tilde{\varepsilon}$ to the general case ε with the help of

$$\begin{aligned} \tilde{\varepsilon}_1 &= \varepsilon_1 - \frac{\varepsilon_1\cdot p_-}{k_1\cdot p_-}k_1 \quad , \\ \tilde{\varepsilon}_2 &= \varepsilon_2 - \frac{\varepsilon_2\cdot p_-}{k_2\cdot p_-}k_2 \quad , \end{aligned} \tag{4}$$

Exercise 3.14.

see (3.263) in Sect. 3.7. It is obvious that the expression

$$\tilde{\varepsilon}_1 \cdot \tilde{\varepsilon}_2 = \left(\varepsilon_1 - \frac{\varepsilon_1 \cdot p_-}{k_1 \cdot p_-} k_1\right) \cdot \left(\varepsilon_2 - \frac{\varepsilon_2 \cdot p_-}{k_2 \cdot p_-} k_2\right) \tag{5}$$

is invariant with respect to gauge transformations of the form $\varepsilon_1 \to \varepsilon_1' + f_1(k)k_1$, $\varepsilon_2 \to \varepsilon_2' + f_2(k)k_2$. Thus (2) has been made gauge-invariant so that (3) can be applied.

We introduce the abbreviations

$$\sum_{\lambda_1,\lambda_2} (\tilde{\varepsilon}_1 \cdot \tilde{\varepsilon}_2)^2 = A^{\mu\nu} B_{\mu\nu} \quad , \tag{6}$$

where

$$A^{\mu\nu} = \sum_{\lambda_1} \left(\varepsilon_1 - \frac{\varepsilon_1 \cdot p_-}{k_1 \cdot p_-} k_1\right)^\mu \left(\varepsilon_1 - \frac{\varepsilon_1 \cdot p_-}{k_1 \cdot p_-} k_1\right)^\nu \tag{7}$$

and a similar expression defines $B_{\mu\nu}$. Using (3) we get

$$\begin{aligned} A^{\mu\nu} &= \sum_{\lambda_1} \left(\varepsilon_1^\mu \varepsilon_1^\nu - \varepsilon_1^\alpha \varepsilon_1^\nu \frac{p_{-\alpha} k_1^\mu}{k_1 \cdot p_-} - \varepsilon_1^\mu \varepsilon_1^\alpha \frac{p_{-\alpha} k_1^\nu}{k_1 \cdot p_-} + \varepsilon_1^\alpha \varepsilon_1^\beta \frac{p_{-\alpha} p_{-\beta} k_1^\mu k_1^\nu}{(k_1 \cdot p_-)^2}\right) \\ &= -g^{\mu\nu} + \frac{k_1^\mu p_-^\nu + p_-^\mu k_1^\nu}{k_1 \cdot p_-} - \frac{m_0^2 k_1^\mu k_1^\nu}{(k_1 \cdot p_-)^2} \end{aligned} \tag{8a}$$

and

$$B_{\mu\nu} = -g_{\mu\nu} + \frac{k_{2\mu} p_{-\nu} + p_{-\mu} k_{2\nu}}{k_2 \cdot p_-} - \frac{m_0^2 k_{2\mu} k_{2\nu}}{(k_2 \cdot p_-)^2} \quad . \tag{8b}$$

The contraction of these tensors leads to (using the on-shell condition $k_1^2 = k_2^2 = 0$)

$$\sum_{\lambda_1,\lambda_2} (\tilde{\varepsilon}_1 \cdot \tilde{\varepsilon}_2)^2 = 2 - 2m_0^2 \frac{k_1 \cdot k_2}{k_1 \cdot p_- \, k_2 \cdot p_-} + m_0^4 \frac{(k_1 \cdot k_2)^2}{(k_1 \cdot p_-)^2 (k_2 \cdot p_-)^2} \quad . \tag{9}$$

Inserting this into (2) leads to the unpolarized squared invariant matrix element

$$\begin{aligned} \overline{|\varepsilon_2 \cdot M_{fi} \cdot \varepsilon_1|^2} = \frac{1}{2} \frac{1}{4m_0^2} 4 &\left(\frac{k_2 \cdot p_-}{k_1 \cdot p_-} + \frac{k_1 \cdot p_-}{k_2 \cdot p_-} + 2m_0^2 \frac{k_1 \cdot k_2}{k_1 \cdot p_- \, k_2 \cdot p_-} \right. \\ &\left. - m_0^4 \frac{(k_1 \cdot k_2)^2}{(k_1 \cdot p_-)^2 (k_2 \cdot p_-)^2} \right) \quad . \end{aligned} \tag{10}$$

The differential cross section $d\bar{\sigma}/d\Omega_1$ is obtained by integrating (1) over d^3k_2 and $d\omega_1$. With the help of

$$\int \frac{d^3 k_2}{2\omega_2} = \int d^4 k_2 \, \delta(k_2^2) \Theta(k_{20}) \tag{11}$$

Exercise 3.14.

a brief calculation similar to that in Sect. 3.7 leads to

$$\frac{d\bar{\sigma}}{d\Omega_1} = \frac{\alpha^2}{|\boldsymbol{v}_{\rm rel}|}\frac{m_0^2}{E_+E_-}2\int\limits_0^{E_++E_-} d\omega_1\,\omega_1\,\delta\big[(p_++p_--k_1)^2\big]\,\overline{|\varepsilon_2\cdot M_{fi}\cdot\varepsilon_1|^2} \tag{12}$$

to be evaluated at $k_2 = p_++p_--k_1$.

While (12) is still completely general we now select the centre-of-mass system where $p_+ = (E,\boldsymbol{p})$, $p_- = (E,-\boldsymbol{p})$. Then the delta function becomes

$$\begin{aligned}\delta\big[(p_++p_--k_1)^2\big] &= \delta\big[(2E-\omega_1)^2-\boldsymbol{k}_1^2\big]\\ &= \delta\big[4E(E-\omega_1)\big]\end{aligned} \tag{13}$$

so that (12) reads

$$\frac{d\bar{\sigma}}{d\Omega_1} = \frac{\alpha^2}{|\boldsymbol{v}_{\rm rel}|}\frac{m_0^2}{E^2}2\frac{E}{4E}\overline{|\varepsilon_2\cdot M_{fi}\cdot\varepsilon_1|^2} \quad . \tag{14}$$

The scalar products of (10) in the centre-of-mass frame reduce to

$$\begin{aligned}k_1\cdot p_- &= E^2(1-v\cos\theta) \quad ,\\ k_2\cdot p_- &= E^2(1+v\cos\theta) \quad ,\\ k_1\cdot k_2 &= 2E^2 \quad ,\end{aligned} \tag{15}$$

with the velocity $v = |\boldsymbol{p}|/E = \sqrt{E^2-m_0^2}/E$. θ is the angle between the incoming and outgoing momenta $\boldsymbol{p}$ and $\boldsymbol{k}$. With $|\boldsymbol{v}_{\rm rel}| = 2v$ the final result for the *differential cross section of pair annihilation in the centre-of-mass frame* is

$$\begin{aligned}\frac{d\bar{\sigma}}{d\Omega_1} = \frac{1}{8}\frac{\alpha^2}{m_0^2}\frac{m_0^2}{E^2}\frac{1}{v}\bigg(&\frac{1+v\cos\theta}{1-v\cos\theta}+\frac{1-v\cos\theta}{1+v\cos\theta}\\ &+4\frac{m_0^2}{E^2}\frac{1}{1-v^2\cos^2\theta}-4\frac{m_0^4}{E^4}\frac{1}{(1-v^2\cos^2\theta)^2}\bigg) \quad .\end{aligned} \tag{16}$$

Using $m_0^2/E^2 = 1-v^2$ this result can also be written as

$$\frac{d\bar{\sigma}}{d\Omega_1} = \frac{1}{4}\frac{\alpha^2}{m_0^2}\frac{m_0^2}{E^2}\frac{1}{v}\frac{1+2v^2-2v^4-2v^2(1-v^2)\cos^2\theta-v^4\cos^4\theta}{(1-v^2\cos^2\theta)^2} \quad . \tag{17}$$

It is an elementary task to integrate this expression over $d\cos\theta$ to obtain the *total cross section for pair annihilation*. The result is

$$\begin{aligned}\bar{\sigma} &= \frac{1}{2}\int d\Omega_1\frac{d\bar{\sigma}}{d\Omega_1}\\ &= \frac{\pi}{4}\frac{\alpha^2}{m_0^2}\frac{1-v^2}{v^2}\left[(3-v^4)\ln\frac{1+v}{1-v}-2v(2-v^2)\right] \quad .\end{aligned} \tag{18}$$

Since the total cross section is invariant under Lorentz transformations (in the beam direction) (18) should agree with the result of Exercise 3.13. which was derived in the rest frame of the electron.

Exercise 3.14.

$$\bar{\sigma} = \frac{\pi\alpha^2}{m_0|\boldsymbol{p}_\mathrm{L}|^2(E_\mathrm{L}+m_0)}\left[\left(E_\mathrm{L}^2+4m_0E_\mathrm{L}+m_0^2\right)\ln\frac{E_\mathrm{L}+|\boldsymbol{p}_\mathrm{L}|}{m_0} - (E_\mathrm{L}+3m_0)|\boldsymbol{p}_\mathrm{L}|\right] , \tag{19}$$

where the subscript L refers to the laboratory frame. To express E_L in terms of the centre-of-mass velocity v we relate both quantities to the Mandelstam invariant s, i.e.

$$\begin{aligned} s &= (p_{+\mathrm{L}}+p_{-\mathrm{L}})^2 = (E_\mathrm{L}+m_0, \boldsymbol{p}_\mathrm{L})^2 \\ &= E_\mathrm{L}^2 + 2m_0E_\mathrm{L} + m_0^2 - \boldsymbol{p}_\mathrm{L}^2 = 2m_0(E_\mathrm{L}+m_0) \end{aligned} \tag{20}$$

and

$$s = (p_+ + p_-)^2 = (E+E, \boldsymbol{0})^2 = 4E^2 \quad . \tag{21}$$

The velocity is given by

$$v = \frac{|\boldsymbol{p}|}{E} = \sqrt{\frac{s-4m_0^2}{s}} \quad , \quad \text{thus} \quad s = \frac{4m_0^2}{1-v^2} \quad . \tag{22}$$

This leads to

$$E_\mathrm{L} = \frac{s}{2m_0} - m_0 = m_0\frac{1+v^2}{1-v^2} \, , \quad |\boldsymbol{p}_\mathrm{L}| = m_0\frac{2v}{1-v^2} \quad . \tag{23}$$

Inserting (23) into (19) confirms the result (18).

Finally we come back to (17) and note the *ultrarelativistic limit* of the cross section, $v \longrightarrow 1$:

$$\left(\frac{d\bar{\sigma}}{d\Omega_1}\right)_\mathrm{ur} \simeq \frac{1}{4}\frac{\alpha^2}{E^2}\frac{1-\cos^4\theta}{\sin^4\theta} = \frac{\alpha^2}{s}\frac{1+\cos\theta}{\sin^2\theta} \quad . \tag{24}$$

This result has been tested experimentally at various electron–positron storage-ring accelerators. As an example Fig. 3.49 shows data taken with the JADE detector at the PETRA collider.[32] Within the experimental accuracy the prediction of QED is fully confirmed.

[32] W. Bartel et al. (JADE collaboration): Z. Physik **C19**, 197 (1983).

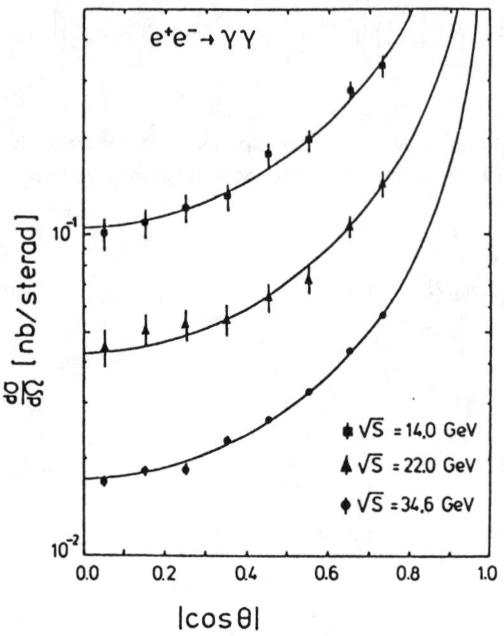

Fig. 3.49. The differential cross section $d\bar{\sigma}/d\Omega_1$ for the process $e^+ + e^- \to \gamma + \gamma$ for three different centre-of-mass energies $\sqrt{s}$. The experimental data are in good agreement with the QED prediction of (24)

EXERCISE

3.15 Pair Creation by Two Photons

Problem. Derive the total unpolarized cross section for the creation of an electron–positron pair by two colliding photons, $\gamma + \gamma \to e^+ + e^-$. Express the result in terms of the velocity of the produced particles in the centre-of-mass frame. Hint: Use the result of Exercise 3.13 for the pair annihilation cross section.

Solution. The differential cross section for pair creation according to the graphs of Fig. 3.45 is given by

$$d\sigma_{\text{pair}} = \frac{\alpha^2}{2}\frac{4\pi}{2\omega_1}\frac{4\pi}{2\omega_2}\left|M_{fi}^{\text{pair}}\right|^2(2\pi)^4\delta^4(p_+ + p_- - k_1 - k_2)\frac{m_0 d^3p_+}{(2\pi)^3E_+}\frac{m_0 d^3p_-}{(2\pi)^3E_-}\,, \quad (1)$$

where the factor 1/2 results from the photon flux factor $1/|\boldsymbol{v}_1 - \boldsymbol{v}_2|$. The relative velocity of two collinearly colliding photons is 2 in any frame of reference. The invariant amplitude is given by

$$M_{fi}^{\text{pair}} = \bar{u}(p_-, s_-)\left[\not{\varepsilon}_2\frac{1}{-\not{p}_+ + \not{k}_1 - m_0}\not{\varepsilon}_1 + \not{\varepsilon}_1\frac{1}{-\not{p}_+ + \not{k}_2 - m_0}\not{\varepsilon}_2\right]v(p_+, s_+)\quad . \quad (2)$$

The corresponding expressions for pair annihilation $e^+ + e^- \to \gamma + \gamma$ was derived in Sect. 3.8:

$$d\sigma_{\text{anni}} = \frac{\alpha^2}{|\boldsymbol{v}_{\text{rel}}|}\frac{m_0}{E_+}\frac{m_0}{E_-}\left|M_{fi}^{\text{anni}}\right|^2 (2\pi)^4\delta^4(k_1+k_2-p_+ + p_-)\frac{4\pi d^3k_1}{(2\pi)^3 2\omega_1}\frac{4\pi d^3k_2}{(2\pi)^3 2\omega_2}\ , \tag{3}$$

Exercise 3.15.

with

$$M_{fi}^{\text{anni}} = \bar{v}(p_+,s_+)\left[\not{\varepsilon}_2\frac{1}{\not{p}_- - \not{k}_1 - m_0}\not{\varepsilon}_1 + \not{\varepsilon}_1\frac{1}{\not{p}_- - \not{k}_2 - m_0}\not{\varepsilon}_2\right]u(p_-,s_-)\ . \tag{4}$$

According to the principle of crossing symmetry the invariant amplitudes can be transformed into each other if the momenta are substituted according to the Table in Sect 3.8, namely $k_1 \to -k_1$, $k_2 \to -k_2$, $p_- \to -p_+$, $p_+ \to -p_-$. In fact the values of (2) and (4) are equal in magnitude

$$\left|M_{fi}^{\text{pair}}\right| = \left|M_{fi}^{\text{anni}}\right|\ . \tag{5}$$

This is easily verified by calculating $\left(M_{fi}^{\text{pair}}\right)^* = \left(M_{fi}^{\text{pair}}\right)^\dagger$. Using $\gamma^0\not{a}^\dagger\gamma^0 = \not{a}$, $-p_+ + k_2 = p_- - k_1$, $-p_+ + k_1 = p_- - k_2$ this quantity is found to agree with M_{fi}^{anni}. Thus the differential cross sections (1) and (3) are equal, except for the phase space volumes and the flux factor.

To get the total cross section for pair creation (1) will be integrated over d^3p_- and subsequently over dE_+. Using the familiar identity

$$\frac{d^3p_-}{2E_-} = \int d^4p_-\,\delta(p_-^2 - m_0^2)\,\Theta(p_{-0}) \tag{6}$$

we obtain

$$\begin{aligned}\sigma_{\text{pair}} &= \frac{1}{2}\int d^3p_+ d^3p_-\,\delta^4(p_+ + p_- - k_1 - k_2)\,F\\ &= \frac{1}{2}\int d^3p_+\,2E_-\delta\left[(k_1+k_2-p_+)^2 - m_0^2\right]F\end{aligned} \tag{7}$$

to be evaluated at the electron momentum $p_- = k_1 + k_2 - p_+$. $F \equiv F(p_+,p_-;k_1,k_2)$ is an obvious abbreviation for the factors which are common to (1) and (3). The analogous expression for the total annihilation cross section reads

$$\begin{aligned}\sigma_{\text{anni}} &= \frac{1}{2}\frac{1}{|\boldsymbol{v}_{\text{rel}}|}\int d^3k_1 d^3k_2\,\delta[k_1 + k_2 - p_+ - p_-]\,F\\ &= \frac{1}{2}\frac{1}{|\boldsymbol{v}_{\text{rel}}|}\int d^3k_1\,2\omega_2\,\delta\left((p_+ + p_- - k_1)^2\right)F\end{aligned} \tag{8}$$

to be taken at $k_2 = p_+ + p_- - k_1$. The factor $1/2$ was introduced in (7) to account for the presence of two identical particles in the final states (see the discussion at the end of Sect. 3.8).

The results (7) and (8) are most easily compared in the centre-of-mass frame. Since the total cross section is Lorentz-invariant this choice does not restrict generality. In this frame we have $p_+ = (E,\boldsymbol{p})$, $p_- = (E,-\boldsymbol{p})$, and $k_1 = (\omega,\boldsymbol{k})$, $k_2 = (\omega,-\boldsymbol{k})$ where energy conservation demands $\omega = E$.

Exercise 3.15.

The delta function in (7) becomes

$$\begin{aligned}\delta\left((k_1+k_2-p_+)^2-m_0^2\right) &= \delta\left((2\omega-E_+)^2-p_+^2-m_0^2\right)\\ &= \delta(4\omega(\omega-E_+))\\ &= \frac{1}{4\omega}\delta(\omega-E_+)\end{aligned} \tag{9}$$

leading to

$$\begin{aligned}\sigma_{\text{pair}} &= \frac{1}{2}\int d\Omega_+\, 2E_-\frac{1}{4\omega}|\boldsymbol{p}_+|E_+\,F\\ &= \frac{1}{2}\frac{1}{2}E|\boldsymbol{p}|\int d\Omega_+\,F \quad .\end{aligned} \tag{10}$$

The same reasoning leads to the total cross section for two-photon annihilation

$$\begin{aligned}\sigma_{\text{anni}} &= \frac{1}{2}\frac{1}{|\boldsymbol{v}_{\text{rel}}|}\int d\Omega_1\, 2\omega_2\frac{1}{4E}\,\omega_1^2\,F\\ &= \frac{1}{2}\frac{1}{|\boldsymbol{v}_{\text{rel}}|}\frac{1}{2}E^2\int d\Omega_1\,F \quad .\end{aligned} \tag{11}$$

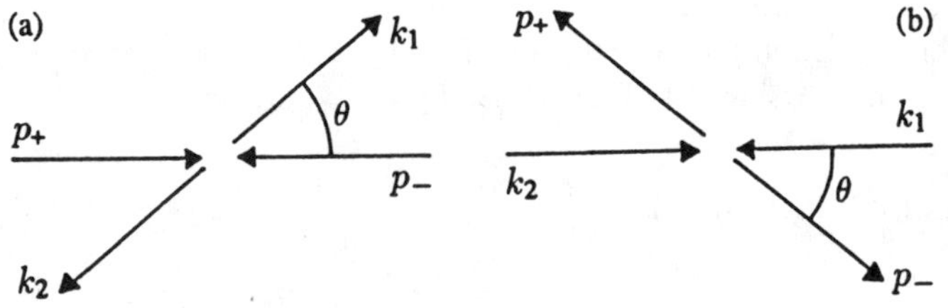

Fig. 3.50a,b. The momentum balance of two-photon pair annihilation **(a)** and pair production by two photons **(b)** in the centre-of-mass frame

The angular integrals in (10) and (11) are identical since both extend over the relative angle θ between $\boldsymbol{p}$ and $\boldsymbol{k}$, see Fig. 3.50. Inserting $|\boldsymbol{v}_{\text{rel}}| = |\boldsymbol{v}_+ - \boldsymbol{v}_-| = 2v$ where $v = |\boldsymbol{p}|/E$, the comparison of (10) and (11) leads to the simple relation

$$\sigma_{\text{anni}} = \frac{1}{2v^2}\,\sigma_{\text{pair}} \quad . \tag{12}$$

Therefore, we can use the result for the total unpolarized cross section for pair annihilation from Exercise 3.14, (18) and obtain the cross section for pair creation

$$\bar{\sigma}_{\text{pair}} = \frac{\pi}{2}\frac{\alpha^2}{m_0^2}(1-v^2)\left[(3-v^4)\ln\frac{1+v}{1-v} - 2v(2-v^2)\right] \quad . \tag{13}$$

This result is known as the **Breit-Wheeler** *formula*.[33] The nonrelativistic limit ($v \ll 1$) is

$$\bar{\sigma}^{\text{nr}}_{\text{pair}} \simeq \pi\frac{\alpha^2}{m_0^2}\,v \tag{14}$$

and the ultrarelativistic limit $v = \sqrt{(E^2-m_0^2)/E^2} \to 1$ becomes

[33] G. Breit and J.A. Wheeler: Phys. Rev. **46**, 1087 (1934).

$$\bar{\sigma}^{\mathrm{ur}}_{\mathrm{pair}} \simeq \pi \frac{\alpha^2}{m_0^2}\left(\frac{m_0}{E}\right)^2\left(\ln\frac{2E}{m_0} - 1\right) \quad . \tag{15}$$

Exercise 3.15.

The pair production cross section thus is suppressed at the threshold (owing to the vanishing phase space volume), rises to a maximum (at $v \simeq 0.701$), and at high energies falls of again according to (15).

Remark. The cross section (13) is reasonably large, being of the order of the squared "classical electron radius" r_0 as in the case of, e.g., Compton scattering. However, pair production by two real photons has not been observed experimentally since it is difficult to prepare two colliding beams of high-energy photons. Nevertheless the graph of Fig. 3.45 can be tested in the collision of *charged particles*. The graph of Fig. 3.51a can be interpreted as describing the collision of two *virtual photons* which produce an electron–positron pair.[34]

Note that this process competes with the graph in Fig. 3.51b where a single virtual bremsstrahlung photon can be split into an e^+e^- pair since its momentum is off the mass shell.

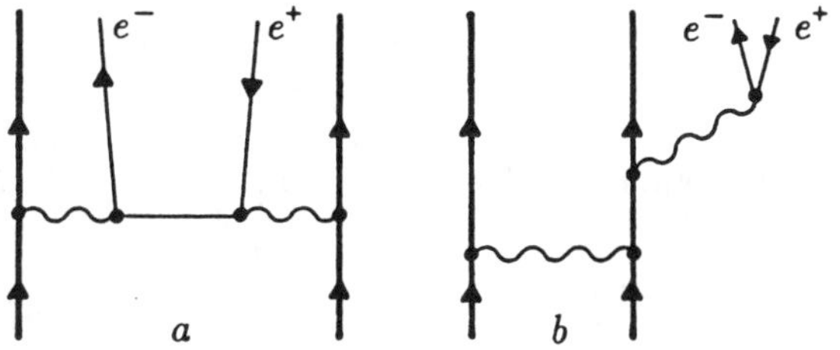

Fig. 3.51a,b. Lowest-order Feynman graphs for pair production in the collision of charged particles (thick lines). **(a)** Collision of two virtual photons. **(b)** Pair conversion of a virtual bremsstrahlung photon

EXERCISE

3.16 Pair Creation in the Field of an Atomic Nucleus

Problem. Calculate the cross section of electron–positron pair creation by an incoming photon in the field of a heavy nucleus with charge $-Ze$. Hint: The calculation can be considerably simplified by exploiting crossing symmetry which relates pair creation and bremsstrahlung. After a simple substitution the results from Sect. 3.6 and Example 3.11 can be used.

Solution. To lowest order the pair creation can be represented by the graphs of Fig. 3.52. They differ from the graphs of bremsstrahlung (Fig. 3.32, Sect. 3.6) just by the interpretation of the external lines (incoming photon $\leftrightarrow$ outgoing photon, incoming electron $\leftrightarrow$ outgoing positron).

[34] See e.g. V.M. Budnev, I.F. Ginzburg, G.V. Meledin, V.G. Serbo, Phys. Rep. **15**, 181 (1975); C. Bottcher, M.R. Strayer: Phys. Rev. **D39**, 1330 (1989).

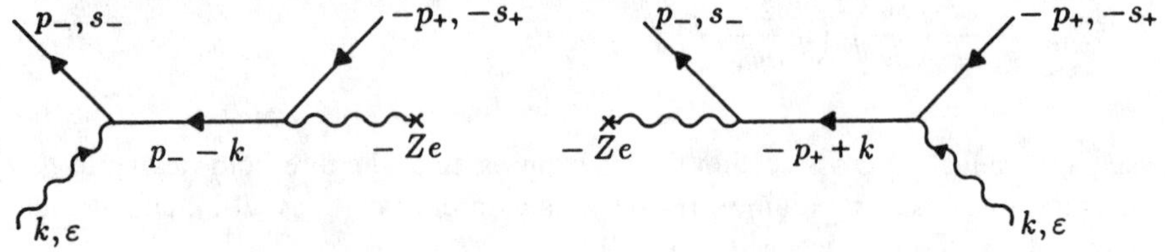

Fig. 3.52. The graphs for pair production by a photon in an external Coulomb field

The second-order S-matrix element in coordinate space reads

$$\begin{aligned} S_{fi} = & -e^2 \int d^4x\, d^4y \sqrt{\frac{m_0^2}{E_+E_-V^2}} \sqrt{\frac{4\pi}{2\omega V}}\, \bar{u}(p_-, s_-) e^{ip_-\cdot x} \\ & \times \Big[(-i\not{\varepsilon})\left(e^{-ik\cdot x} + e^{ik\cdot x}\right) iS_F(x-y)(-i\gamma^0)A^{coul}(y) \\ & + (-i\gamma^0)A_0^{coul}(x) iS_F(x-y)(-i\not{\varepsilon})\left(e^{-ik\cdot y} + e^{+ik\cdot y}\right)\Big] v(p_+, s_+) e^{ip_+\cdot y} \quad . \end{aligned} \tag{1}$$

Using the static Coulomb potential of the nucleus

$$A_0^{coul}(x) = \frac{-Ze}{|\boldsymbol{x}|} = -4\pi Ze \int \frac{d^3q}{(2\pi)^3} \frac{e^{-i\boldsymbol{q}\cdot\boldsymbol{x}}}{|\boldsymbol{q}|^2} \tag{2}$$

and performing the Fourier integrations we get the S-matrix element

$$\begin{aligned} S_{fi} = & Ze^3 2\pi\delta(E_- + E_+ - \omega) \sqrt{\frac{4\pi}{2\omega V}} \sqrt{\frac{m_0^2}{E_+E_-V^2}} \frac{4\pi}{|\boldsymbol{q}|^2} \\ & \times \bar{u}(p_-, s_-) \left[(-i\not{\varepsilon}) \frac{i}{\not{p}_- - \not{k} - m_0}(-i\gamma^0) \right. \\ & \left. + (-i\gamma^0) \frac{i}{-\not{p}_+ + \not{k} - m_0}(-i\not{\varepsilon})\right] v(p_+, s_+) \quad . \end{aligned} \tag{3}$$

Here $p_+ = (E_+, \boldsymbol{p}_+)$, $p_- = (E_-, \boldsymbol{p}_-)$ and $k = (\omega, \boldsymbol{k})$ denote the four-momenta of positron, electron and photon, respectively.

$$\boldsymbol{q} = \boldsymbol{p}_+ + \boldsymbol{p}_- - \boldsymbol{k} \tag{4}$$

is the momentum transferred to the nucleus.

Apart from the sign, (3) is identical with (3.192) if we substitute

$$\begin{aligned} p_i, s_i \quad & \rightarrow \quad -p_+, s_+ \quad , \\ p_f, s_f \quad & \rightarrow \quad p_-, s_- \quad , \\ k, \lambda \quad & \rightarrow \quad -k, \lambda \quad , \end{aligned} \tag{5}$$

and if we replace the electron spinor $u(p_i, s_i)$ by the positron spinor $v(p_+, s_+)$. All further calculations can be traced back to the case of bremsstrahlung. There is a slight difference when noting the cross section. One has to divide by the flux of the incoming photons $\frac{c}{V} \equiv \frac{1}{V}$ instead of $\frac{|\boldsymbol{v}_i|}{V}$. In addition the phase space is changed, since a positron is emitted instead of a photon. Thus we get the cross section

Exercise 3.16.

$$d\sigma = \int \frac{|S_{fi}|^2}{T\frac{1}{V}} V\frac{d^3p_+}{(2\pi)^3} V\frac{d^3p_-}{(2\pi)^3} \quad . \tag{6}$$

Integration over the electron energy E_- gives the cross section, which is five-fold differential with respect to positron energy and solid angles $d\Omega_+$ and $d\Omega_-$. Averaging over the photon polarization λ and summing over the spin directions s_+ and s_- we obtain the unpolarized cross section

$$\begin{aligned} d\bar{\sigma} = &\frac{Z^2\alpha^3}{(2\pi)^2}\frac{m_0^2}{\omega E_+E_-}4\frac{1}{|q|^4}|\boldsymbol{p}_+|^2\, d|\boldsymbol{p}_+|\, d\Omega_+\, |\boldsymbol{p}_-|\, E_-\, d\Omega_- \\ &\times \Theta(\omega - E_+ - m_0)F'(p_+,p_-;k) \quad . \end{aligned} \tag{7}$$

Here the function

$$\begin{aligned} F' = \frac{1}{2}\sum_{\lambda}\sum_{s_+,s_-} \Bigg| &\bar{u}(p_-,s_-)\left[\not{\epsilon}\frac{1}{\not{p}_- - \not{k} - m_0}\gamma_0\right. \\ &\left. + \gamma_0\frac{1}{-\not{p}_- + \not{k} - m_0}\not{\epsilon}\right] v(p_+,s_+)\Bigg|^2 \end{aligned} \tag{8}$$

was introduced which can be expressed as a trace over up to 8 gamma matrices. We do not have to evaluate this complicated expression because $F'(p_+,p_-;k)$ is connected with the function $F(p_i,p_f;k)$ known from (2), Example 3.11 (bremsstrahlung), namely

$$F'(p_+,p_-;k) = -F(-p_+,p_-;-k) \quad , \tag{9}$$

the sign originating from the sum over the spin of the positron, since

$$\sum_{s_+} v_\alpha(p_+,s_+)\bar{v}_\beta(p_+,s_+) = -\left(\frac{-\not{p}_+ + m_0}{2m_0}\right)_{\alpha\beta} \quad . \tag{10}$$

The final result can again be expressed as a function of the angle between the momentum vectors. Let θ_+ (θ_-) be the angle between $\boldsymbol{p}_+$ ($\boldsymbol{p}_-$) and $\boldsymbol{k}$, and ϕ the angle between the planes ($\boldsymbol{p}_+, \boldsymbol{k}$) and ($\boldsymbol{p}_-, \boldsymbol{k}$). Then our rule of substitution (5) yields the relation

$$\theta_i \rightarrow \theta_+ \quad , \quad \theta_f \rightarrow \pi - \theta_- \quad , \quad \phi \rightarrow \phi - \pi \quad , \tag{11}$$

with the angles as defined in Example 3.11. Then the explicit cross section of pair creation follows from (26) in Example 3.11 (the Bethe–Heitler formula):

$$\begin{aligned} d\bar{\sigma} = &\frac{Z^2\alpha^3}{(2\pi)^2}\frac{|\boldsymbol{p}_+||\boldsymbol{p}_-|}{|q|^4}\frac{dE_+\, d\Omega_+\, d\Omega_-}{\omega^3}\Theta(\omega - E_+ - m_0) \\ &\times \left[-\frac{\boldsymbol{p}_-^2\sin^2\theta_-}{(E_- - |\boldsymbol{p}_-|\cos\theta_-)^2}(4E_+^2 - q^2) - \frac{\boldsymbol{p}_+^2\sin^2\theta_+}{(E_+ - |\boldsymbol{p}_+|\cos\theta_+)^2}(4E_-^2 - q^2)\right. \\ &+ 2\omega^2\frac{\boldsymbol{p}_+^2\sin^2\theta_+ + \boldsymbol{p}_-^2\sin^2\theta_-}{(E_+ - |\boldsymbol{p}_+|\cos\theta_+)(E_- - |\boldsymbol{p}_-|\cos\theta_-)} \\ &\left. + 2\frac{|\boldsymbol{p}_+||\boldsymbol{p}_-|\sin\theta_+\sin\theta_-\cos\phi}{(E_+ - |\boldsymbol{p}_+|\cos\theta_+)(E_- - |\boldsymbol{p}_-|\cos\theta_-)}(2E_+^2 + 2E_-^2 - q^2)\right] \quad . \end{aligned} \tag{12}$$

Exercise 3.16.

To get the signs right, note that the substitution $k \to -k$ implies $\omega \to -\omega$ but of course $|\boldsymbol{k}| \to |-\boldsymbol{k}| = +|\boldsymbol{k}|$. Therefore the denominators in (12) should be treated as follows: $p_f \cdot k = \omega E_f - |\boldsymbol{k}|\,|\boldsymbol{p}_f|\cos\theta_f = \omega(E_f - |\boldsymbol{p}_f|\cos\theta_f) \to -\omega E_- - |\boldsymbol{k}|\,|\boldsymbol{p}_-|\cos(\pi - \theta_-) = -\omega(E_- - |\boldsymbol{p}_-|\cos\theta_-) = -p_- \cdot k$.

One should mention that the result (12), being based on the lowest-order graphs (the plane wave Born approximation) has only a limited range of validity. For high nuclear charges Z or low velocities v_+, v_- the interaction of the produced charged particles with the nuclear Coulomb field becomes important. This can be taken into account by replacing the plane waves by Coulomb distorted waves. This calculation, however, can no longer be performed fully analytically. The criterion of validity of the Born approximation is

$$\frac{Ze^2}{\hbar|\boldsymbol{v}_\pm|} \ll 1 \quad .$$

For heavy nuclei this condition is no longer satisfied and the cross section changes. In particular the complete symmetry of (12) with respect to the interchange $e^+ \leftrightarrow e^-$ will be lost since the electrons (positrons) feel the attraction (repulsion) by the Coulomb potential of the nucleus.

Additional Remarks. Despite its complicated appearance the differential cross section (12) can be integrated analytically with respect to the electron and positron solid angles $d\Omega_-$, $d\Omega_+$. Since this calculation is lengthy and not very illuminating we merely quote the result which already was derived by Bethe and Heitler in their original publication:

$$\begin{aligned}\frac{d\bar{\sigma}}{dE_+} = \frac{Z^2\alpha^3}{m_0^2}\frac{|\boldsymbol{p}_+|\,|\boldsymbol{p}_-|}{\omega^3}\Bigg\{&-\frac{4}{3} - 2E_+E_-\frac{\boldsymbol{p}_+^2 + \boldsymbol{p}_-^2}{\boldsymbol{p}_+^2\boldsymbol{p}_-^2} + \frac{m_0E_+\eta_-}{\boldsymbol{p}_-^2}\\ &+ \frac{m_0E_-\eta_+}{\boldsymbol{p}_+^2} - \eta_+\eta_- + L\Bigg[\frac{\omega^2}{|\boldsymbol{p}_+|^3|\boldsymbol{p}_-|^3}(E_+^2E_-^2 + \boldsymbol{p}_+^2\boldsymbol{p}_-^2) - \frac{8}{3}\frac{E_+E_-}{|\boldsymbol{p}_+||\boldsymbol{p}_-|}\\ &- \frac{m_0^2\omega}{2|\boldsymbol{p}_+||\boldsymbol{p}_-|}\left(\frac{E_+E_- - \boldsymbol{p}_-^2}{\boldsymbol{p}_-^2 m_0}\eta_- + \frac{E_+E_- - \boldsymbol{p}_+^2}{\boldsymbol{p}_+^2 m_0}\eta_+ + \frac{2\omega E_+E_-}{\boldsymbol{p}_+^2\boldsymbol{p}_-^2}\right)\Bigg]\Bigg\} \quad ,\end{aligned} \tag{13}$$

where the abbreviations

$$\begin{aligned}\eta_\pm &= 2\frac{m_0}{|\boldsymbol{p}_\pm|}\ln\frac{E_\pm + |\boldsymbol{p}_\pm|}{m_0} \quad ,\\ L &= 2\ln\frac{E_+E_- + |\boldsymbol{p}_+||\boldsymbol{p}_-| + m_0^2}{m_0\omega}\end{aligned} \tag{14}$$

have been introduced. Figure 3.53 shows the energy distribution of created positrons as a function of the kinetic energy $E_+ - m_0$, normalized to the total available energy $\omega - 2m_0$. The cross section $d\bar{\sigma}/dE_+$ was multiplied by $\omega - 2m_0$ so that the area under the curve represents the total cross section $\bar{\sigma}$. The latter is found to rise slowly with energy.

In the *ultrarelativistic limit* $\omega \gg m_0$ (13) simplifies and can be integrated over dE_+ analytically, leading to a logarithmically rising cross section

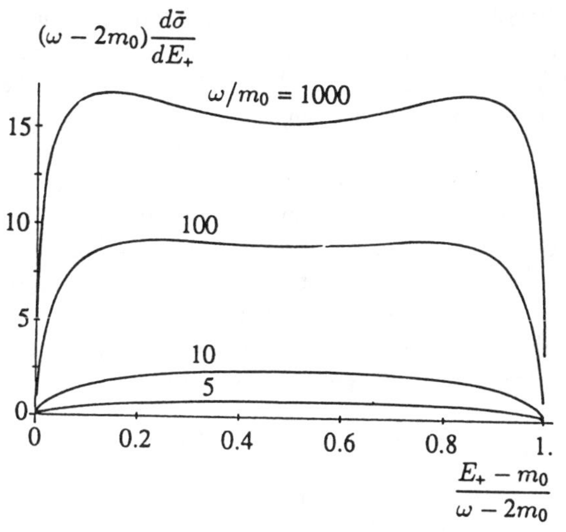

Fig. 3.53. The differential cross section for pair production by photons with energy ω according to the Bethe–Heitler formula (13)

$$\bar{\sigma}_{\text{ur}} \simeq \frac{Z^2\alpha^3}{m_0^2}\left(\frac{28}{9}\ln\frac{2\omega}{m_0} - \frac{218}{27}\right) \quad . \tag{15}$$

If the target consists of neutral atoms, electron screening will lead to a saturation of the pair production cross section: The rise of (15) at high photon energies is caused by the creation of pairs at increasingly larger distances from the nucleus. Electron screening acts to suppress these contributions.

EXAMPLE

3.17 The Method of Equivalent Photons

Throughout this chapter we have studied QED processes in which charged particles interact through the exchange of *virtual photons*. This terminology, however, has been somewhat artificial since we nowhere encountered true physical photons in these calculations. However, conditions can be found under which these *virtual* photons behave like ordinary *real* photons to a good approximation. Under these conditions the colliding charged particles can be replaced by an equivalent bunch of incoming photons with a certain energy distribution which can be calculated. This approximation serves two purposes: it can be used to simplify the description of various processes, and it also helps to visualize and understand qualitatively how high-energy scatterings proceed.

A very interesting application was already mentioned at the end of Exercise 3.15: By colliding charged particles at high energy the process of *photon-photon scattering* can be studied, even though no intense colliding beams of real photons are available to the experimentalist. In this example we will derive the method of equivalent photons and subsequently use it to calculate the cross section for the production of muon pairs through the reaction $e^+ + e^- \to e^+ + e^- + \mu^+ + \mu^-$. We will discover that at high collision energies this process dominates over the pair annihilation reaction $e^+ + e^- \to \mu^+ + \mu^-$ which we studied in Section 3.4.

Derivation of the Equivalent Photon Spectrum. For a start we will study reactions in which a single virtual photon is exchanged. A high-energy electron having initial momentum $p = (E, \boldsymbol{p})$ is scattered into the final state $p' = (E', \boldsymbol{p}')$ while emitting a virtual photon with momentum $k = p - p'$. This photon strikes a target where it gives rise to a reaction which produces a (possibly complex) many-particle final state X, see Fig. 3.54. The details of this reaction are not important for our study. As an example, the target might be a heavy nucleus and the final state an electron-positron pair produced in its Coulomb field.

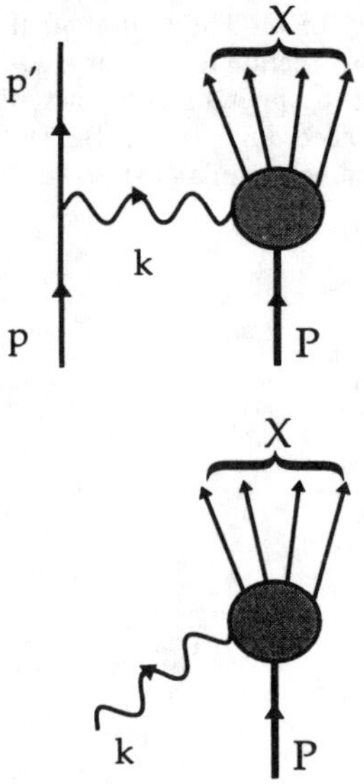

Fig. 3.54a,b. Feynman graphs describing a scattering process initiated by a virtual photon (**a**) or a real photon (**b**)

Our goal is to find a relation between the process shown in Fig. 3.54a and the analogous reaction which is triggered by an incoming *real* photon shown in Fig. 3.54b. Thus let us investigate the unpolarized cross section for the Feynman graph of Fig. 3.54a which is given by

$$\begin{aligned} d\bar{\sigma} &= \frac{mM}{\sqrt{(p \cdot P)^2 - m^2 M^2}} \overline{|M_{\text{fi}}|^2} \\ &\quad \times (2\pi)^4 \delta^4(P_X + p' - P - p) \frac{m d^3 p'}{(2\pi)^3 E'} \prod_{n=1}^{N} \frac{m_n dP_n'}{(2\pi)^3 E_n'} \quad , \end{aligned} \tag{1}$$

where the target is assumed to be a fermion of mass M and the final state X consists of N fermions (this assumption only affects the normalization factors). The invariant matrix element in (1) reads

$$\begin{aligned} M_{\text{fi}} &= j^\mu_{\text{fi}}(p', p) D_{\text{F}\mu\nu}(k) J^\nu_{\text{fi}}(P_n', P) \\ &= \frac{4\pi}{k^2} j^\mu_{\text{fi}}(p', p) J_{\text{fi}\mu}(P_n', P) \quad . \end{aligned} \tag{2}$$

Here $j^\mu_{\text{fi}} = e\, \bar{u} \gamma^\mu u$ is the familiar transition current of the electron and J^μ_{fi} is the transition current of the target which may have a complex structure and possibly is not known in detail. For the evaluation of (1) we need the squared and spin-averaged invariant matrix element (2). This will be written as in Sect. 3.2 in the form

$$\overline{|M_{\text{fi}}|^2} = (4\pi)^2 \frac{e^2}{(k^2)^2} L^{\mu\nu} H_{\mu\nu} \quad . \tag{3}$$

Here $L^{\mu\nu}$ is the *lepton tensor* which we encountered several times before:

$$\begin{aligned} L^{\mu\nu} &= \frac{1}{2} \text{Sp} \left(\frac{\not{p}' + m}{2m} \gamma^\mu \frac{\not{p} + m}{2m} \gamma^\nu \right) \\ &= \frac{1}{2} \frac{1}{m^2} \left(p'^\mu p^\nu + p'^\nu p^\mu - g^{\mu\nu}(p' \cdot p - m^2) \right) \quad . \end{aligned} \tag{4}$$

Out of habit we will call the object $H_{\mu\nu}$ the *hadron tensor* although this may be slightly misleading since the derivation also remains valid if no hadrons are involved in the reaction. The hadron tensor is obtained from

$$H^{\mu\nu} = \sum_{\text{Spin}} J^{\mu *}_{\text{fi}}(P_n', P) J^\nu_{\text{fi}}(P_n', P) \quad . \tag{5}$$

In general not much is known about the current J^μ_{fi}. However, we can rely on the principle of gauge invariance which implies electromagnetic current conservation. The hadron tensor therefore satisfies the four-dimensional transversality condition

Example 3.17.

$$k_\mu H^{\mu\nu} = H^{\mu\nu} k_\nu = 0 \,. \tag{6}$$

This helps to simplify the expression (4) for the lepton tensor a bit. Use of $p' = p - k$ and $k^2 = (p - p')^2 = 2m^2 - 2p \cdot p'$ leads to

$$L^{\mu\nu} = \frac{1}{2}\frac{1}{m^2}\left(2p^\mu p^\nu + \frac{1}{2}g^{\mu\nu}k^2\right) - \frac{1}{2}\frac{1}{m^2}\left(k^\mu p^\nu + k^\nu p^\mu\right) \quad , \tag{7}$$

where the second term can be discarded because of (6). The scattering cross section, integrated over the final state of the electron, then reads

$$d\bar{\sigma} = \int \frac{d^3p'}{|\boldsymbol{p}|E'} \frac{\alpha}{2\pi^2} \left(\frac{1}{k^2}\right)^2 \left(2p^\mu p^\nu + \frac{1}{2}g^{\mu\nu}k^2\right) H_{\mu\nu}\, 2\pi \, d\Gamma \quad , \tag{8}$$

where the target has been assumed to be at rest, $P = (M, \boldsymbol{0})$, leading to the flux factor $m/|\boldsymbol{p}|$. The symbol $d\Gamma$ designates the phase space volume of the target final state:

$$d\Gamma = (2\pi)^4 \delta^4(P_X - P - k) \prod_{n=1}^{N} \frac{m_n dP'_n}{(2\pi)^3 E'_n} \quad . \tag{9}$$

Now let us investigate the cross section for the analogous *photon induced* process according to Fig. 3.54b:

$$\begin{aligned} d\bar{\sigma}_\gamma &= \frac{2m \; 4\pi}{4\sqrt{(k \cdot P)}} \overline{|\epsilon \cdot J_{\mathrm{fi}}|^2} (2\pi)^4 \delta^4(P_X - P - k) \prod_{n=1}^{N} \frac{m_n dP'_n}{(2\pi)^3 E'_n} \\ &= \frac{1}{2\omega} 4\pi \sum_{\text{Spin}} \frac{1}{2} \sum_{\lambda} \epsilon^*_\mu(\boldsymbol{k}, \lambda) \epsilon_\nu(\boldsymbol{k}, \lambda) J^{\mu *}_{\mathrm{fi}} J^{\nu}_{\mathrm{fi}} \, d\Gamma \\ &= \frac{1}{2\omega} 2\pi \left(-H^{0\mu}_{\mu}\right) d\Gamma \quad , \end{aligned} \tag{10}$$

where we have used the completeness relation of the photon polarization vectors $\sum_\lambda \epsilon^*_\mu \epsilon_\nu = -g_{\mu\nu}$.

The expression (10) contains the *trace of the hadron tensor*. The index 0 is meant to imply that the involved photon is *real*, i.e. it satisfies the mass-shell condition $k^2 = \omega^2 - \boldsymbol{k}^2 = 0$. In contrast, in (5) there is no fixed relation between the frequency ω and the momentum $\boldsymbol{k}$ of the photon. We only know that the virtual photon has a space-like momentum $k^2 < 0$. The squared momentum k^2 is sometimes called the *virtuality* of the photon. Except for this difference the tensors $H_{\mu\nu}$ and $H^0_{\mu\nu}$ have exactly the same structure.

Now let us try to express the electron cross section (8) in terms of the photon cross section (10).[35] For this purpose it is helpful to choose a suitable gauge in which the transverse degrees of freedom, corresponding to physical photons, can be most easily singled out. This is achieved in the *Coulomb gauge* defined by the condition $\nabla \cdot \boldsymbol{A} = 0$. The choice of this gauge also affects the photon propagator, a topic which we will discuss in more detail in Sect. 4.2. The covariant Feynman propagator

[35] S.J. Brodsky, T. Kinoshita, H. Terazawa: Phys. Rev. **D4**, 1532 (1971).

Example 3.17.

$$D_{\mathrm{F}\mu\nu} = -\frac{4\pi g_{\mu\nu}}{k^2} \tag{11}$$

in this way will be replaced by the propagator in the Coulomb gauge which has the components

$$D^{\mathrm{C}}_{00} = \frac{4\pi}{\boldsymbol{k}^2} \quad , \quad D^{\mathrm{C}}_{ij} = \frac{4\pi}{k^2}\left(\delta_{ij} - \frac{k_i k_j}{\boldsymbol{k}^2}\right) \quad , \quad D^{\mathrm{C}}_{0i} = D^{\mathrm{C}}_{i0} = 0 \quad , \tag{12}$$

where the latin indices as usual run over space coordinates $i = 1,2,3$. Note: we will also use the summation convention for repeated spatial indices, but in this case no distinction between co- and contravariant indices will be made.

The photo-production cross section (10) in Coulomb gauge attains the following form

$$d\bar{\sigma}_\gamma = \frac{1}{2\omega} 2\pi H^{0\perp}_{ii}\, d\Gamma \quad . \tag{13}$$

The sign $\perp$ indicates that because of the Coulomb gauge condition the hadron tensor is transversal (in three dimensions).

The electron cross section (8) becomes

$$d\bar{\sigma} = \int \frac{d^3p'}{|\boldsymbol{p}|E'}\frac{\alpha}{2\pi^2}\left(2p^\mu p^\nu + \frac{1}{2}g^{\mu\nu}k^2\right)\frac{1}{4\pi}D^{\mathrm{C}}_{\mu\alpha}\frac{1}{4\pi}D^{\mathrm{C}}_{\nu\beta}H^{\alpha\beta}\, 2\pi\, d\Gamma \, . \tag{14}$$

Now we introduce a crucial approximation and *discard the contribution of the "scalar" propagator* D^{C}_{00}. As a motivation for this step we note that its contribution can not be related to the interaction of transverse photons. Thus we are left with the approximate cross section

$$d\bar{\sigma} = \int \frac{d^3p'}{|\boldsymbol{p}|E'}\frac{\alpha}{2\pi^2}\left(\frac{1}{k^2}\right)^2\left(2p_ip_j - \frac{1}{2}\delta_{ij}k^2\right)H^{\perp}_{ij}\, 2\pi\, d\Gamma \quad . \tag{15}$$

The transverse hadron tensor is obtained by applying two transverse projection operators which originate from the the photon propagator (12):

$$H^{\perp}_{ij} = \left(\delta_{ik} - \frac{k_i k_k}{\boldsymbol{k}^2}\right)\left(\delta_{jl} - \frac{k_j k_l}{\boldsymbol{k}^2}\right)H_{kl} \, . \tag{16}$$

The integrand of (15) contains a contraction which appears to be more involved than the simple trace of the tensor which had entered (13). However, closer inspection reveals that the additional term $p_ip_jH^{\perp}_{ij}$ in fact also can be reduced to the trace $H^{\perp}_{ii}$ when the integration over the azimuthal angle of the final electron momentum $\boldsymbol{p}'$ is performed. In Exercise 3.18 the following relation between these two expressions will be shown

$$\int_0^{2\pi} d\varphi\, p_ip_jH^{\perp}_{ij} \simeq \frac{E^2E'^2}{2\boldsymbol{k}^2}\sin^2\theta \int_0^{2\pi} d\varphi\, H^{\perp}_{ii} \quad . \tag{17}$$

Here θ is the angle between $\boldsymbol{p}$ and $\boldsymbol{p}'$, i.e. the scattering angle of the electron. Relation (17) is valid only for the case of *forward scattering* (small scattering angles) and all the following results will be restricted to this case. Furthermore we will make the *ultrarelativistic approximation* $E \gg m$, $|\boldsymbol{p}| \simeq E$ and thus will neglect the electron mass wherever possible. *Example 3.17.*

Using (17) the electron scattering cross section becomes

$$d\bar{\sigma} = \frac{\alpha}{2\pi^2} \int \frac{d^3p'}{EE'} \left(\frac{1}{k^2}\right)^2 \left(-\frac{k^2}{2} + \frac{E^2E'^2}{\boldsymbol{k}^2} \sin^2\theta\right) H_{ii}^{\perp}\, 2\pi\, d\Gamma \quad . \tag{18}$$

In the last approximation step we now identify

$$H_{ii}^{\perp} \simeq H_{ii}^{\perp 0} \quad , \tag{19}$$

i.e. we use $k^2 = 0$ for the photon momentum, thus neglecting the fact that the transition current $H_{ii}^{\perp}$ strictly speaking has to be evaluated for off-shell photons. Under this condition the integral in (18) is found to contain the photon cross section (13) as a factor:

$$d\bar{\sigma} = \frac{\alpha}{2\pi^2} \int \frac{d^3p'}{EE'} \left(\frac{1}{k^2}\right)^2 \left(-\frac{k^2}{2} + \frac{E^2E'^2}{\boldsymbol{k}^2} \sin^2\theta\right) 2\omega\, d\bar{\sigma}_\gamma \quad . \tag{20}$$

The integration over the final electron momentum can be rewritten as follows:

$$\begin{aligned} \int d^3p &= \int_0^{|\boldsymbol{p}|} d|\boldsymbol{p}'|\, |\boldsymbol{p}'|^2 \int_{-1}^{+1} d\cos\theta \int_0^{2\pi} d\varphi \simeq 2\pi \int_0^E dE'\, E'^2 \int_{-1}^{+1} d\cos\theta \\ &\simeq 2\pi \int_0^E dE'\, E'^2 \frac{1}{2EE'} \int_{k_+^2}^{k_-^2} dk^2 \quad , \end{aligned} \tag{21}$$

where we have used

$$k^2 = (p-p')^2 = 2m^2 - 2p\cdot p' = 2m^2 - 2EE' + 2|\boldsymbol{p}||\boldsymbol{p}'|\cos\theta \quad . \tag{22}$$

The maximum squared momentum transfer (at $\theta = \pi$) in the ultrarelativistic limit is given by

$$k_+^2 = 2m^2 - 2EE' - 2|\boldsymbol{p}||\boldsymbol{p}'| \simeq -4EE' \,. \tag{23}$$

For the minimum momentum transfer (at $\theta = 0$) the same approximation would give zero. This would make the integral diverge because of the pole originating from the photon propagator. Thus we must be more careful and take into account the finite electron mass. Inserting the Taylor expansion of $|\boldsymbol{p}| = \sqrt{E^2 - m^2} \simeq E(1 - m^2/(2E^2))$ into (22) gives

$$\begin{aligned} k^2 &= 2m^2 - 2EE' + 2|\boldsymbol{p}||\boldsymbol{p}'|\cos\theta \\ &\simeq 2m^2 - 2EE' + 2EE'\Big(1 - \frac{m^2}{2E^2}\Big)\Big(1 - \frac{m^2}{2E'^2}\Big)\cos\theta \\ &\simeq 2m^2 - 2EE'(1-\cos\theta) - m^2\frac{E^2+E'^2}{EE'}\cos\theta \quad . \end{aligned} \tag{24}$$

Considering forward scattering ($\theta = 0$) this leads to the minimum squared momentum transfer

$$k_-^2 \simeq -m^2 \frac{\omega^2}{EE'} \quad , \tag{25}$$

Example 3.17.

the scale of which is set by the squared electron mass. As a consequence the integral will be dominated by contributions from photons with *small virtuality* $k^2 \simeq k_-^2 = O(m^2)$ being close to the mass shell. These photons originate mostly from *peripheral collisions*. Equation (24) allows us to estimate that the largest contribution to the scattering cross section arises from an angular region around

$$\theta \sim \frac{m\omega}{EE'} \ll 1 \quad . \tag{26}$$

Using (21) the scattering cross section (20) can be written as follows:

$$\begin{aligned} d\bar{\sigma} &= \frac{\alpha}{\pi} \int_0^E dE' \, E'^2 \frac{2\omega}{2EE'} \int_{k_+^2}^{k_-^2} dk^2 \left(\frac{1}{k^2}\right)^2 \left(-\frac{k^2}{2} + \frac{E^2 E'^2}{k^2} \sin^2\theta \right) d\bar{\sigma}_\gamma(\omega) \\ &= \int_0^E \frac{d\omega}{\omega} N(\omega) \, d\bar{\sigma}_\gamma(\omega) \quad . \end{aligned} \tag{27}$$

The integration extends over the energy ω of the photon and the function $N(\omega)/\omega$ can be interpreted as the *energy spectrum of equivalent photons* emitted by the scattering electron. This equation is at the heart of the *method of equivalent photons* which also is known by the name *Weizsäcker-Williams approximation*[36]. Originally the equivalent photon spectrum $N(\omega)$ was obtained classically by Fourier transforming the Poynting vector of the electromagnetic field generated by a fast moving charge. Above we have deduced $N(\omega)$ from the transition current given by quantum electrodynamics. The result reads

$$\begin{aligned} N(\omega) &= \frac{\alpha}{\pi} \frac{\omega^2}{E^2} \int_{k_+^2}^{k_-^2} dk^2 \left(\frac{1}{k^2}\right)^2 \left(-\frac{k^2}{2} + \frac{E^2 E'^2}{k^2} \sin^2\theta \right) \\ &= \frac{\alpha}{\pi} \frac{\omega^2}{E^2} \int_{k_+^2}^{k_-^2} dk^2 \left(\frac{1}{k^2}\right)^2 \left(-\frac{k^2}{2} + \frac{-m^2\omega^2 - \frac{1}{4}k^4 - k^2 EE'}{\omega^2 - k^2} \right) \quad , \end{aligned} \tag{28}$$

where the relation (22) was used. The first term in the integrand leads to a logarithm

$$\begin{aligned} \int_{k_+^2}^{k_-^2} dk^2 \left(\frac{1}{k^2}\right)^2 \left(-\frac{k^2}{2}\right) &= -\frac{1}{2} \ln \frac{k_-^2}{k_+^2} = -\frac{1}{2} \ln \frac{-m^2\omega^2/EE'}{-4EE'} \\ &= \ln \frac{2EE'}{m\omega} \quad . \end{aligned} \tag{29}$$

The remainder of the integral can also be solved in closed form. The resulting *equivalent photon spectrum* can be written as

$$\begin{aligned} N(\omega) = \frac{\alpha}{\pi} \frac{1}{E^2} \Big[&(E^2 + E'^2) \ln \frac{2E}{m} + \frac{1}{2}(E - E')^2 \ln \frac{E'}{E - E'} \\ &+ \frac{1}{2}(E + E')^2 \ln \frac{E'}{E + E'} - EE' \Big] \quad . \end{aligned} \tag{30}$$

[36] The basic idea was formulated by Fermi already in 1924 to describe the energy loss of α particles in matter, E. Fermi: Z. Physik **29**, 315 (1924). Later the method was formulated in general terms by C.F. v. Weizsäcker: Z. Physik **88**, 612 (1934) and E.J. Williams: Phys. Rev. **45**, 729 (1934). A detailed derivation with special emphasis on relativistic heavy ion collisions can be found in M. Vidović, M. Greiner, C. Best, G. Soff: Phys. Rev. **C47**, 2308 (1993); A useful review article is C.A. Bertulani, G. Baur: Phys. Rep. **161**, 299 (1988) .

Example 3.17.

In the ultrarelativistic limit this result is dominated by the *logarithmic increase with energy* E. The remaining terms in (30) only depend on E'/E, i.e. on the energy loss of the electron.

Photon-Photon Collisions. The method of equivalent photons can also be applied if the target is not a charged particle but a second virtual photon which itself originates from a transition current. Here we will not repeat the derivation which is very similar to the case of a single virtual photon, and shall instead immediately present the plausible result. In complete analogy with (27) the cross section for the creation of a final state X in electron-electron scattering according to the graph of Fig. 3.55a reads

$$\bar{\sigma}_{ee\to eeX} = \int_0^{E_1} \frac{d\omega_1}{\omega_1} N(\omega_1) \int_0^{E_2} \frac{d\omega_2}{\omega_2} N(\omega_2)\, \bar{\sigma}_{\gamma\gamma\to X}(\omega_1,\omega_2) \quad . \tag{31}$$

Here $\bar{\sigma}_{\gamma\gamma\to X}$ is the cross section for the creation of X in a collision of two real photons with energies ω_1 and ω_2.

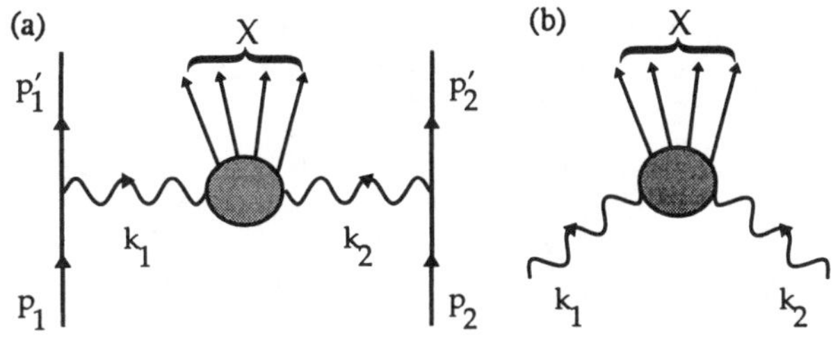

Fig. 3.55a,b. Feynman graphs for two-photon scattering involving two virtual photons (**a**) or two real photons (**b**)

Now we will apply (31) to calculate the cross section for *muon pair creation* through the process $e^+ + e^- \to e^+ + e^- + \mu^+ + \mu^-$. To achieve this we need the two-photon pair creation cross section for $\gamma + \gamma \to \mu^+ + \mu^-$, which has been calculated in Exercise 3.15. We will use the Breit-Wheeler formula derived in (13) of this exercise and express it in terms of the Mandelstam variable $s = (k_1 + k_2)^2$, i.e. the invariant mass of the created pair (which should not be confused with the Mandelstam variable $s_0 = (p_1 + p_2)^2$ of the whole collision). The variable v used in Exercise 3.15 is the velocity of the muons in the center of momentum system which is related to $s = (2E_\mu)^2$ through

$$v = \frac{|\boldsymbol{p}_\mu|}{E_\mu} = \sqrt{\frac{s/4 - M_\mu^2}{s/4}} = \sqrt{\frac{s - 4M_\mu^2}{s}} \quad . \tag{32}$$

In terms of the variable s the Breit-Wheeler formula takes the form

$$\bar{\sigma}_{\gamma\gamma\to\mu^+\mu^-} = \frac{4\pi\alpha^2}{s}\left[\left(2 + \frac{8M_\mu^2}{s} - \frac{16M_\mu^4}{s^2}\right)\ln\frac{\sqrt{s}+\sqrt{s-4M_\mu^2}}{2M_\mu} - \sqrt{1-\frac{4M_\mu^2}{s}}\left(1+4\frac{M_\mu^2}{s}\right)\right] \quad . \tag{33}$$

Example 3.17.

We are left with the task to solve the double integral in (31). To achieve this we transform to the set of variables

$$s = 4\,\omega_1\omega_2 \quad , \quad \omega = \omega_1 + \omega_2 \quad . \tag{34}$$

Actually the variable s defined in (34) coincides with the Mandelstam-s only under the condition that the momentum vector of the scattered electrons $\boldsymbol{p}_1'$ and $\boldsymbol{p}_2'$ are antiparallel. In the center of momentum frame of the electrons, $\boldsymbol{p}_1 + \boldsymbol{p}_2 = 0$, we have

$$\begin{aligned} s &= (k_1 + k_2)^2 = (\omega_1 + \omega_2)^2 - (\boldsymbol{p}_1 - \boldsymbol{p}_1' + \boldsymbol{p}_2 - \boldsymbol{p}_2')^2 \\ &= (\omega_1 + \omega_2)^2 - (\boldsymbol{p}_1' + \boldsymbol{p}_2')^2 \simeq (\omega_1 + \omega_2)^2 - (|\boldsymbol{p}_1'| - |\boldsymbol{p}_2'|)^2 \\ &\simeq (\omega_1 + \omega_2)^2 - (E_1' - E_2')^2 = (\omega_1 + \omega_2)^2 - (E - \omega_1 - E + \omega_2)^2 \\ &= (\omega_1 + \omega_2)^2 - (\omega_1 - \omega_2)^2 = 4\,\omega_1\omega_2 \quad . \end{aligned} \tag{35}$$

Since peripheral collisions with small scattering angles θ_1, θ_2 make the dominant contribution this approximation is well justified.

The Jacobian of the transformation (34) reads

$$\frac{\partial(s,\omega)}{\partial(\omega_1,\omega_2)} = \begin{vmatrix} 4\omega_2 & 4\omega_1 \\ 1 & 1 \end{vmatrix} = 4\,|\omega_2 - \omega_1| = 4\sqrt{\omega^2 - s} \quad . \tag{36}$$

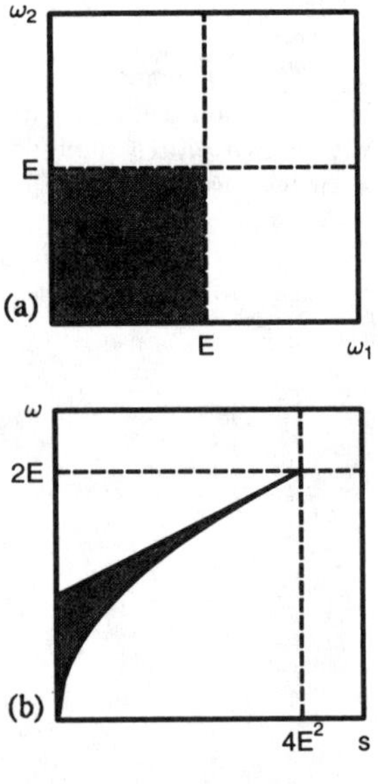

Fig. 3.56a,b. The kinematically allowed regions of integration in the $\omega_1 - \omega_2$ plane (**a**) and the $s - \omega$ plane (**b**)

Using this result the pair production cross section (31) takes the following form

$$\bar{\sigma}_{ee\to ee\mu\mu} = 2\int_{4M_\mu^2}^{4E^2} ds\,\frac{1}{s}\int_{\sqrt{s}}^{E+s/4E} d\omega\,\frac{1}{\sqrt{\omega^2 - s}}\,N(\omega_1)\,N(\omega_2)\,\bar{\sigma}_{\gamma\gamma\to\mu\mu}(s) \quad , \tag{37}$$

where $\omega_{1,2} = \frac{1}{2}\big(\omega \pm \sqrt{\omega^2 - s}\,\big)$. The limits of integration in (36) need some further consideration. The transformation (34) maps the quadratic range of integration in the $\omega_1 - \omega_2$ plane to the acutely shaped area in the $s - \omega$ plane shown in the bottom part of Fig. 3.56. This region is bounded from below by the curve $\omega(s) = \sqrt{s}$ which corresponds to equal photon energies $\omega_1 = \omega_2 = \omega/2$. The upper boundary is given by the straight line $\omega(s) = E + s/4E$ where one of the electrons is completely stopped, $\omega_1 = E, \omega_2 = s/4E$ or $\omega_2 = E, \omega_1 = s/4E$. Obviously each point in the $s - \omega$ plane can be reached twice, corresponding to either $\omega_1 > \omega_2$ or $\omega_2 > \omega_1$. This explains the extra factor of 2 in (37). Both boundary lines intersect at the point of total energy transfer $s = 4E^2, \omega = 2E$.

We now evalutate the ω integral in (37), restricting our attention to the leading logarithmic term in the photon spectrum (30):

$$\begin{aligned} &\int_{\sqrt{s}}^{E+s/4E} d\omega\,\frac{1}{\sqrt{\omega^2 - s}}\,N(\omega_1)N(\omega_2) \\ &\simeq \left(\ln\frac{2E}{m}\right)^2 \int_{\sqrt{s}}^{E+s/4E} d\omega\,\frac{1}{\sqrt{\omega^2 - s}} \\ &\quad\times \frac{E^2 + \left(E - \frac{1}{2}(\omega + \sqrt{\omega^2 - s})\right)^2}{E^2}\,\frac{E^2 + \left(E - \frac{1}{2}(\omega - \sqrt{\omega^2 - s})\right)^2}{E^2} \\ &\equiv f\left(\frac{\sqrt{s}}{2E}\right) \quad . \end{aligned} \tag{38}$$

Example 3.17.

With some effort this integral can be solved exactly. The result reads, expressed in terms of the parameter $\kappa = \sqrt{s}/2E$:

$$f(\kappa) = (2+\kappa^2)^2 \ln\frac{1}{\kappa} - (1-\kappa^2)(3+\kappa^2)\,. \tag{39}$$

This finally allows us to evaluate the total cross section for muon pair creation as a function of the invariant mass s of the pair:

$$\begin{aligned}\bar{\sigma}_{ee\to ee\mu\mu} &= \frac{\alpha^2}{\pi^2}\left(\ln\frac{2E}{m}\right)^2 \int_{4M_\mu^2}^{4E^2} ds\,\frac{1}{s}\,f\Big(\frac{\sqrt{s}}{2E}\Big)\,\bar{\sigma}_{\gamma\gamma\to\mu\mu}(s)\\ &= \frac{2\alpha^4}{\pi}\frac{1}{M_\mu^2}\left(\ln\frac{2E}{m}\right)^2 \int_1^{(E/M_\mu)^2} dx\, f\Big(\frac{M_\mu}{E}\sqrt{x}\Big)\\ &\quad\times\frac{1}{x}\left[\Big(2+\frac{2}{x}-\frac{1}{x^2}\Big)\ln\big(\sqrt{x}+\sqrt{x-1}\big) - \sqrt{1-\frac{1}{x}}\Big(1+\frac{1}{x}\Big)\right].\end{aligned} \tag{40}$$

In the last step we have transformed to the dimensionless variable $x = s/4M_\mu^2$ and used formula (33) for the photon-induced production process. While the integral in (40) in general has to be solved numerically, in the ultrarelativistic limit an analytical aproximation can be found. An inspection of the integrand reveals that for $E \gg M_\mu$ the function $f(\sqrt{x}M_\mu/E)$ falls off more slowly than the remaining factors. Therefore we replace this function by the value it takes at the lower boundary $f(M_\mu/E) \simeq 4\ln(E/M_\mu)$. If the upper boundary is extended to infinity the remaining integral can be solved in closed form, leading to the numerical factor 14/9.

Thus the *total cross section for the production of muon pairs* in electron collisions at high energies is given by

$$\begin{aligned}\bar{\sigma}_{ee\to ee\mu\mu} &\simeq \frac{2\alpha^4}{\pi}\frac{1}{M_\mu^2}\left(\ln\frac{2E}{m}\right)^2 4\ln\frac{E}{M_\mu}\,\frac{14}{9}\\ &= \frac{112\alpha^4}{9\pi}\frac{1}{M_\mu^2}\left(\ln\frac{2E}{m}\right)^2 \ln\frac{E}{M_\mu}\,.\end{aligned} \tag{41}$$

This result is valid in *logarithmic accuracy*, i.e. terms of the order unity are neglected compared to the logarithms in (41). Note that for the two-photon mechanism it does not matter whether an electron and a positron or two electrons are colliding.

The *rise of the cross section with incident energy* is very interesting. For comparison we refer to the result (3.145) for the annihilation cross section of an electron positron pair into muons:

$$\bar{\sigma}_{e^+e^-\to\mu^+\mu^-} = \frac{\pi}{3}\frac{\alpha^2}{E^2}\,, \tag{42}$$

which *falls off* with energy. This difference in behaviour has a simple qualitative explanation: The reaction (42) requires that electron and positron meet and annihilate at the same point in space and time, a process which becomes less probable with increasing energy. On the other hand the pair creation according to (41) mainly occurs in peripheral collisions; viewed classically, for increasing incident energy collisions at larger and larger distances (impact parameters) contribute so that the

Example 3.17.

cross section is enhanced. As an interesting consequence at sufficiently high energy (41) will be the dominant process for muon pair production, although it arises from a fourth order graph which should be suppressed by a factor α^2 compared to the annihilation graph.

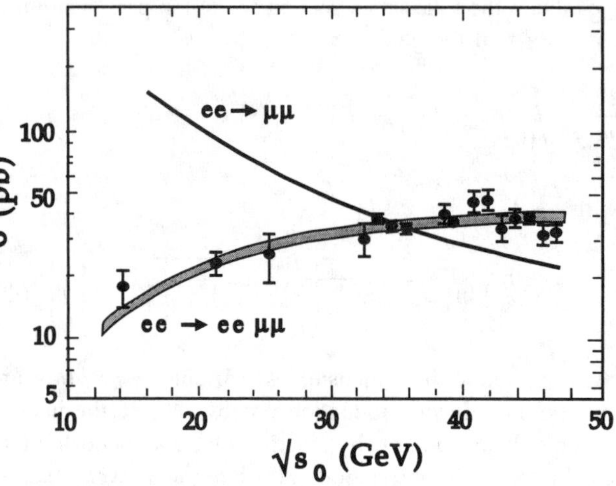

Fig. 3.57. Energy dependence of the total cross section for muon pair production by the process $e^+ + e^- \to e^+ + e^- + \mu^+ + \mu^-$. The data points agree with the QED prediction if the detector response is taken into account (shaded area). For comparison also the cross section for the process $e^+ + e^- \to \mu^+ + \mu^-$ is shown which falls off with energy

Figure 3.57 shows experimental results obtained at the PETRA collider at DESY for muon pair creation[37] which nicely show the increase of the cross section. The data points refer to an "untagged" experiment in which the scattered electrons (positrons) are not detected and mostly remain in the beam pipe, being deflected only by a very small angle. Experiments are also performed under single or double tagging conditions where the collisions partners are observed at finite deflection angles. In this case the equivalent photon spectrum gets modified[38].

We remark that the processes $e^+ + e^- \to e^+ + e^- + \mu^+ + \mu^-$ and $e^+ + e^- \to \mu^+ + \mu^-$ can be clearly separated in the experiment since the muon pairs in the former case are produced with rather low energy while in the latter case they carry the whole collision energy.

The method of equivalent photons provides fairly accurate values for cross sections in high-energy collisions. To make precise checks of QED predictions, however, the Feynman graphs should be evaluated exactly[39], also taking into account interference terms. When evaluating the results from high-energy experiments the numerically obtained predictions are used as an input for Monte-Carlo simulations which also take into account the response of the detector system. In this way the theoretical curve in Fig. 3.57 was generated.

Two-photon collisions of the type we discussed in this exercise have several uses. They can serve to check quantum electrodynamics and to study predictions on

[37] B. Adeva et. al. (Mark J Collaboration): Phys. Rev. **D38**, 2655 (1988)
[38] J.H. Field: Nucl. Phys. **B168**, 477 (1980)
[39] Pioneering works on the two-photon process are V.M. Budnev, I.F. Ginzburg, G.V. Meledin, V.G. Serbo: Phys. Rep. **15**, 181 (1975); G. Bonneau, M. Gourdin, F. Martin: Nucl. Phys. **B54**, 573 (1973)

Example 3.17.

the "photon structure function". Furthermore they provide a comparatively "clean" source for the creation of new particles, e.g. Higgs bosons, supersymmetric particles, glueballs, heavy mesons, etc.[40]

EXERCISE

3.18 Angular Integration of the Hadron Tensor

Problem. Show the validity of (17) in Example 3.17

$$\int_0^{2\pi} \frac{d\varphi}{2\pi}\, p_i p_j H_{ij}^{\perp} \simeq \frac{\boldsymbol{p}^2 \boldsymbol{p}'^2}{2\boldsymbol{k}^2} \sin^2\theta \int_0^{2\pi} \frac{d\varphi}{2\pi}\, H_{ii}^{\perp} \quad , \tag{1}$$

where the integration runs over the azimuthal angle of the scattered electron.

Solution. The object $H_{ij}^{\perp}$ is obtained from the hadron tensor by applying two transverse projection operators according to

$$H_{ij}^{\perp} = \left(\delta_{ik} - \frac{k_i k_k}{\boldsymbol{k}^2}\right)\left(\delta_{jl} - \frac{k_j k_l}{\boldsymbol{k}^2}\right) H_{kl} \quad . \tag{2}$$

H_{kl} is a three-dimensional tensor which can be constructed from the momentum vectors of the incoming and outgoing particles. The vectors available are the momentum $\boldsymbol{k}$ of the incoming photon and the initial and final target momenta. In the following we use only a single target momentum vector $\boldsymbol{\ell}$ but this can be generalized to more involved cases. We adapt the following general approach for the hadron tensor

$$H_{kl} = A\,\delta_{kl} + B\,k_k \ell_l + C \ell_k k_l + D\,\ell_k \ell_l + E\,k_k k_l \quad .$$

Here $A, \ldots, E$ are yet unspecified functions which can depend on the scalar quantities $\boldsymbol{k}^2, \boldsymbol{\ell}^2, \boldsymbol{k}\cdot\boldsymbol{\ell}$. After transverse projection according to (2) only the A and D terms survive:

$$H_{ij}^{\perp} = A\left(\delta_{ik} - \frac{k_i k_k}{\boldsymbol{k}^2}\right) + D\left(\ell_i - \frac{\boldsymbol{k}\cdot\boldsymbol{\ell}}{\boldsymbol{k}^2} k_i\right)\left(\ell_j - \frac{\boldsymbol{k}\cdot\boldsymbol{\ell}}{\boldsymbol{k}^2} k_j\right) \quad . \tag{4}$$

The functions under the integrals in (1) therefore are given by

$$H_{ii}^{\perp} = 2A + D\left(\boldsymbol{\ell}^2 - \frac{(\boldsymbol{k}\cdot\boldsymbol{\ell})^2}{\boldsymbol{k}^2}\right) \quad , \tag{5a}$$

$$p_i p_j H_{ij}^{\perp} = A\left(\boldsymbol{p}^2 - \frac{(\boldsymbol{p}\cdot\boldsymbol{k})^2}{\boldsymbol{k}^2}\right) + D\left(\boldsymbol{p}\cdot\boldsymbol{\ell} - \frac{\boldsymbol{k}\cdot\boldsymbol{\ell}}{\boldsymbol{k}^2}\boldsymbol{p}\cdot\boldsymbol{k}\right)^2 \quad . \tag{5b}$$

In order to perform the azimuthal integration we introduce the following coordinate system: The z-axis points in the direction of the incident electron, i.e. the vector $\boldsymbol{p}$, and the orientation is chosen such that $\boldsymbol{\ell}$ lies in the xz-plane. Expressed in terms of spherical coordinate the vectors of interest are given by

[40] see M. Greiner, M. Vidović, G. Soff: Phys. Rev. **C47**, 2288 (1993)

Exercise 3.18

$$p = p\,(0,0,1) \quad , \tag{6a}$$
$$\ell = \ell\,(\sin\beta,\, 0,\, \cos\beta) \quad , \tag{6b}$$
$$p' = p'(\sin\theta\cos\varphi,\, \sin\theta\sin\varphi,\, \cos\theta) \quad , \tag{6c}$$
$$k = p - p' = (-p'\sin\theta\cos\varphi,\, -p'\sin\theta\sin\varphi,\, p - p'\cos\theta) \quad , \tag{6d}$$

leading to the scalar products

$$p\cdot\ell = p\,\ell\cos\beta \quad , \tag{7a}$$
$$p\cdot k = p(p - p'\cos\theta) \quad , \tag{7b}$$
$$k^2 = p'^2\sin^2\theta + (p - p'\cos\theta)^2 \quad , \tag{7c}$$
$$k\cdot\ell = -\ell p'\sin\theta\cos\varphi\sin\beta + \ell(p - p'\cos\theta)\cos\beta \quad . \tag{7d}$$

We notice that the only quantity which depends on the angle φ is $k\cdot\ell$. As a consequence the coefficients of the A-terms in (5) are isotropic with respect to the azimuthal angle. Using (7a–d) the factors appearing in (5) are given as follows

$$p^2 - \frac{(p\cdot k)^2}{k^2} = \frac{p^2p'^2}{k^2}\sin^2\theta \quad , \tag{8a}$$

$$\begin{aligned}\ell^2 - \frac{(k\cdot\ell)^2}{k^2} = \frac{\ell^2}{k^2}\Big[&p'^2\sin^2\theta(1 - \cos^2\varphi\sin^2\beta) + (p - p'\cos\theta)^2\sin^2\beta \\ &- 2p'(p - p'\cos\theta)\sin\theta\cos\varphi\sin\beta\cos\beta\Big] \quad ,\end{aligned} \tag{8b}$$

$$\begin{aligned}\left(p\cdot\ell - \frac{k\cdot\ell}{k^2}p\cdot k\right)^2 = \frac{\ell^2}{k^2}\frac{p^2p'^2}{2k^2}\sin^2\theta\Big[&2p'^2\sin^2\theta\cos^2\beta \\ &+ 4p'(p - p'\cos\theta)\sin\theta\cos\varphi\sin\beta\cos\beta \\ &+ 2(p - p'\cos\theta)^2\cos^2\varphi\sin^2\beta\Big] \quad .\end{aligned} \tag{8c}$$

The angular averaging leads to the replacements $\langle\cos\varphi\rangle = 0$ and $\langle\cos^2\varphi\rangle = \frac{1}{2}$. Thus (5a) becomes

$$\int_0^{2\pi}\frac{d\varphi}{2\pi}\,H_{ii}^{\perp} = 2A + D\,\frac{\ell^2}{k^2}\left[p'^2\sin^2\theta\left(1 - \frac{1}{2}\sin^2\beta\right) + \left(p - p'\cos\theta\right)^2\sin^2\beta\right] \tag{9}$$

while (5b) leads to

$$\begin{aligned}\int_0^{2\pi}\frac{d\varphi}{2\pi}\,p_ip_jH_{ij}^{\perp} = \frac{p^2p'^2}{2k^2}\sin^2\theta\,2A + \frac{p^2p'^2}{2k^2}\sin^2\theta\,D\,\frac{\ell^2}{k^2}\Big[&2p'^2\sin^2\theta\cos^2\beta \\ &+ (p - p'\cos\theta)^2\sin^2\beta\Big] \quad .\end{aligned} \tag{10}$$

A comparison of these two expressions essentially confirms that they are proportional to each other, as claimed in (1). This is not exactly true, since the first terms in the square brackets do not agree. However, these terms are proportional to $\sin^2\theta$ and therefore are suppressed if the scattering angle is small. This argument also applies to the φ-dependence of the functions A and D: The term $\cos\varphi$ in the scalar product $k\cdot\ell$ which gives rise to a possible azimuthal dependence of these functions is suppressed by the small factor $\sin\theta$.

3.9 Biographical Notes

BETHE, Hans Albrecht, German-American physicist. *2.7.1906 in Straßburg. B. studied physics at Frankfurt and Munich. He was research assistant and Privatdozent at the Universities of Frankfurt (1928), Stuttgart, Munich (1930-32), Manchester and Bristol. Since 1935 he has been professor for physics at Cornell University (Ithaka, NY). In WW II he was leader of the theoretical physics division of the Manhattan project at Los Alamos. B. has made numerous essential contributions to various areas such as nuclear physics, atomic structure physics, solid state spectroscopy, Quantum Electrodynamics. For his explanation of the nuclear fusion processes in the interior of main sequence stars B. received the Nobel price for physics in 1967.

BHABHA, Homi Jehangir, Indian physicist. *30.10.1909 in Bombay, †24.1.1966 at Mont Blanc in an airplane accident. B. studied physics at Bombay and Cambridge (PhD 1932). Since 1945 he was professor for theoretical physics and director of the Tata Institute for Fundamental Research at Bombay. He was president of the Indian Atomic Energy Commission and of the IUPAP. B. gave the first relativistic description of electron-positron scattering. B. also worked in the fields of nuclear physics and cosmic radiation.

BREIT, Gregory, American physicist. *14.7.1899 in Nikolaev (Russia), † 13.9.1981 in Salem (Oregon). B. emigrated to the U.S. in 1915. He was educated at Johns Hopkins University (Baltimore) where he received his PhD in electrical engineering in 1921. He was professor of physics at the Universities of Minnesota, Wisconsin, New York University, Yale University (for 21 years) and finally the State University of New York at Buffalo. B. made important contributions in various areas of physics. With M. Tuve he demonstrated the existence of the Heaviside layer by reflecting radar pulses from the ionosphere. He worked on the theory of molecular beam interaction (the Breit-Rabi equation) and proposed the method of optical pumping (the basis for laser radiation). In the field of nuclear reactions B. developed the description of resonances (the Breit-Wigner formula), worked on the charge independence of nucleon-nucleon scattering and initiated the study of heavy-ion reactions. He also developed the basic principles of various particle accelerators.

DALITZ, Richard Henry, British Physicist. *28.2.1925 in Dimboola (Australia). D. studied physics at the Universities of Melbourne and Cambridge (PhD 1950). He worked as research assistant and lecturer at Bristol and Birmingham. He became professor at the University of Chicago (1956) and at Oxford (1963). His main research contributions have been in hadronic physics, where he studied the decay and reaction properties of mesons and baryons, the interaction of Λ-hyperons etc. He introduced the Dalitz plot for the analysis of the many-particle phase space.

MØLLER, Christian, Danish physicist. *22.12.1904 in Notmark (Denmark), †14.1.1980. M. studied physics in Copenhagen (with N. Bohr), Rome and Cambridge. He spent his scientific career at the University of Copenhagen (1933-75). M. worked on scattering theory (the Møller operators) and S-matrix theory. Later his main interest was directed to the theory of relativity. In particular he addressed the question of conservation and localization of energy and momentum in general relativity.

MOTT, Sir Nevill Francis, British physicist. *30.9.1905 in Leeds. M. received his education at Cambridge University where he graduated with a master degree in 1930. He was lecturer of mathematics at Cambridge and became professor of physics at the University of Bristol (1933). In 1954 he went back to Cambridge and became head of the Cavendish Laboratory. M. developed the quantum theory of atomic collisions on which he wrote an influential monograph in 1933 (with H.S.M. Massey). His subsequent work concentrated on solid state physics where he made important contributions to various subjects, e.g. the band structure model, dislocations, defects, the theory of plasticity, metal-insulator transitions (the Mott transition) etc. In 1977 he was awarded the Nobel prize for physics (with J.H. van Vleck and P.W. Anderson).

NISHINA, Yoshio, Japanese physicist. *6.12.1890 in Okayama, †10.1.1951 in Tokyo. N. studied physics at Tokyo University and did his postgraduate work in europe. He became a collaborator of N. Bohr in Copenhagen. After returning to Japan he was appointed one of the leaders of the Institute of Physico-Chemical Research in Tokyo. During his stay at Copenhagen N. derived the theory of the Compton effect (with O. Klein). Later he became the founder of experimental nuclear and cosmic ray research in Japan. He supervised the construction of several particle accelerators.

ROSENBLUTH, Marshal N., American physicist. *5.2.1927 in Albany (NY). R. studied at Harvard and at theUniversity of Chicago where he got his PhD in 1949 under the direction of E. Teller and E. Fermi. He worked at Los Alamos and at the General Atomics Laboratory and in 1960 went to the University of California at San Diego. In 1967 he went to Princeton University and to the Institute for Advanced Studies. Since 1980 he is director of the Institute for Fusion Studies at the University of Texas in Austin. R. worked on the analysis of the scattering of relativistic electrons to study the charge distibution within nuclei. His main area of research is plasma physics where he developed the theory of inhomogeneous plasmas. He made suggestions to avoid plasma instabilities which were essential for the development of the tokamak reactors for controlled thermonuclear fusion. In 1985 R. received the Fermi prize.

WHEELER, John Archibald, American physicist. *9.7.1911 in Jacksonville (Florida). W. studied physics at Johns Hopkins University (PhD 1933). He was a postdoctoral fellow at the Universities of Copenhagen and North Carolina. In 1938 he joined the faculty of Princeton University. After retirement he went to the University of Texas in Austin (1976). W. made important contributions in various fields of theoretical physics. His main interests have been in gravitation theory and cosmology and in the quantum theory of measurement.

4. Summary: The Feynman Rules of QED

In the last chapter we analysed a variety of scattering processes. In this way we have gained enough experience to extract a set of rules – the Feynman rules – which in principle will allow the calculation of any QED process no matter how complicated it is.

Let us consider the most general elastic or inelastic *scattering of two particles* of the type

$$1+2 \rightarrow 1'+2'+\ldots+n \quad ,$$

which includes the possibility of pair creation and photon emission. The matrix element S_{fi} can be written as

$$S_{fi} = \mathrm{i}(2\pi)^4\delta^4\left(p_1+p_2-\sum_{i=1}^{n} p_i'\right) M_{fi} \prod_{i=1}^{2}\sqrt{\frac{N_i}{2E_iV}} \prod_{i=1}^{n}\sqrt{\frac{N_i'}{2E_i'V}} \quad , \tag{4.1}$$

where the essential physics is contained in the Lorentz covariant amplitude M_{fi}. The square root factors are due to the normalization of the incoming and outgoing plane waves. Within the conventions we use they are different for fermions and photons; this is expressed by the normalization factor N_i:

$$N_i = \begin{cases} 4\pi & \text{for} \quad \text{photons} \\ 2m_0 & \text{for} \quad \text{spin-}\tfrac{1}{2}\text{ particles} \end{cases} \quad . \tag{4.2}$$

These factors become clear for spin-$\frac{1}{2}$ particles, e.g., from (3.2) in Section 3.1 and for photons from (3.170), (3.184) in Section 3.6.

From (4.1) one can derive the general expression for the differential cross section:

$$\begin{aligned} d\sigma = {} & \frac{1}{4\sqrt{(p_1\cdot p_2)^2 - m_1^2 m_2^2}} N_1 N_2 (2\pi)^4 \delta^4\left(p_1+p_2-\sum_{i=1}^{n} p_i'\right) S \\ & \times |M_{fi}|^2 \frac{N_1'\,d^3p_1'}{2E_1'(2\pi)^3} \cdots \frac{N_n'\,d^3p_n'}{2E_n'(2\pi)^3} \quad . \end{aligned} \tag{4.3}$$

The square root in the denominator originates from the incoming particle current written in a Lorentz invariant way, see (3.79) in Section 3.2.

The degeneracy factor S becomes important when the final state contains identical particles. Since configurations differing only by a permutation of the particles describe the same quantum-mechanical state, the phase-space is reduced in this case. This is taken into account by the statistical factor

$$S = \prod_k \frac{1}{g_k!} \quad , \tag{4.4}$$

if there are g_k particles of the kind k in the final state.

In the case of electron–electron scattering this means that the cross section is reduced by a factor of 1/2. Another example of practical importance is the process of multiple photon bremsstrahlung, where the factor $g_k!$ can become very large. This factor arises because for g_k identical particles in the final state there are exactly $g_k!$ possibilities of arranging (counting) these particles, but only one such arrangement is measured experimentally!

To compare (4.3) with experiment one has to integrate the differential cross section $d\sigma$ over the phase-space intervals which are not distinguished in the measurement. In addition one has to average over the initial polarization and to sum over the final polarizations, if these polarizations are not measured.

The invariant amplitude M_{fi} can be expanded into a perturbation series in powers of the coupling constant e using the propagator method. The following rules allow the calculation of the expansion coefficients. They are given below in the form which is most useful for practical calculations, namely in momentum space.

4.1 The Feynman Rules of QED in Momentum Space

1) In the nth order of perturbation theory one has to draw all possible topologically distinct Feynman diagrams with n vertices that have the prescribed number of particles in the initial and final states (external lines).

2) With each external line one has to associate the following factors:

   a) incoming electron: $u(p,s)$

   b) incoming positron: $\bar{v}(p,s)$
   (outgoing electron with negative energy);

   c) incoming photon: $\varepsilon_\mu(\boldsymbol{k},\lambda)$

   d) outgoing electron: $\bar{u}(p,s)$

   e) outgoing positron: $v(p,s)$
   (incoming electron of negative energy) ;

   f) outgoing photon: $\varepsilon^*_\mu(\boldsymbol{k},\lambda)$

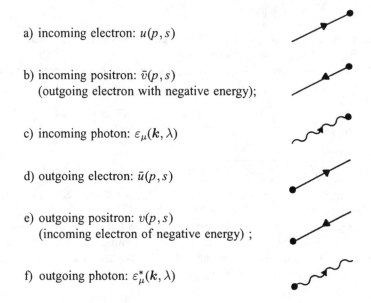

3) Each internal line connecting two vertices has to be associated with a propagator

   a) electron: $iS_F(p) = \dfrac{i(\not{p} + m_0)}{p^2 - m_0^2 + i\varepsilon}$

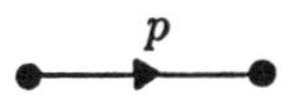

   b) photon: $iD_F^{\mu\nu}(k) = \dfrac{-i\,4\pi\, g^{\mu\nu}}{k^2 + i\varepsilon}$

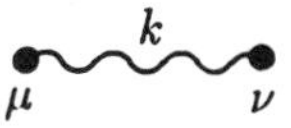

4) Each vertex is associated with a factor

   $-ie\gamma_\mu$

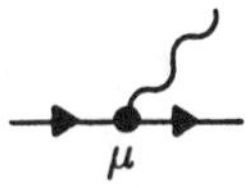

   The index μ has to be multiplied with that of the photon line and summed over.

5) The conservation of four-momentum holds at each vertex. One has to integrate over all momentum variables p that cannot be fixed (internal loops): $\displaystyle\int \frac{d^4p}{(2\pi)^4}$

6) The amplitudes of all graphs have to be added coherently, taking into account the following phase factors:

   a) a factor of -1 for each incoming positron (outgoing electron with negative energy);

   b) a factor of -1 in the case that two graphs differ only by the exchange of two fermion lines – this also holds for the exchange of an incoming (outgoing) particle line with an outgoing (incoming) antiparticle line, since the latter is an incoming (outgoing) particle line with negative energy;

   c) a factor of -1 for each closed fermion loop.

Here we add the following remarks:

To 1: For the construction of Feynman graphs, only the topological structure is important. Since the theory was formulated in a relativistically covariant way, all possible time orderings are automatically taken into account. As long as the ordering of the vertices along the fermion lines is kept, the graphs can be arbitrarily deformed without changing their meaning.

To 5: First one associates a four-momentum δ-function to each vertex and then integrates over the momenta of all internal lines. Since one δ-function is needed for the conservation of the total momentum, for a graph of order n with I internal lines there remain $I - (n-1)$ integrations over internal lines. We illustrate this for a graph of eighth order (Fig. 4.1).

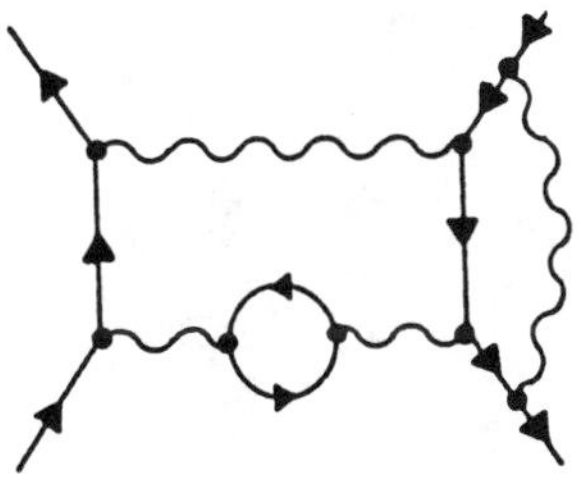

Fig. 4.1. A graph of eighth order with 10 internal lines. One must integrate over $10 - (8-1) = 3$ momenta of internal loops

To 6: The sign factor in a) was discussed for the S matrix of (2.42). There it is denoted by ε_f. The minus sign in b) was explained in the examples of electron–electron and electron–positron scatterings calculated in first order. It originates from the antisymmetry of the wave function required by Fermi–Dirac statistics. Rule 6c

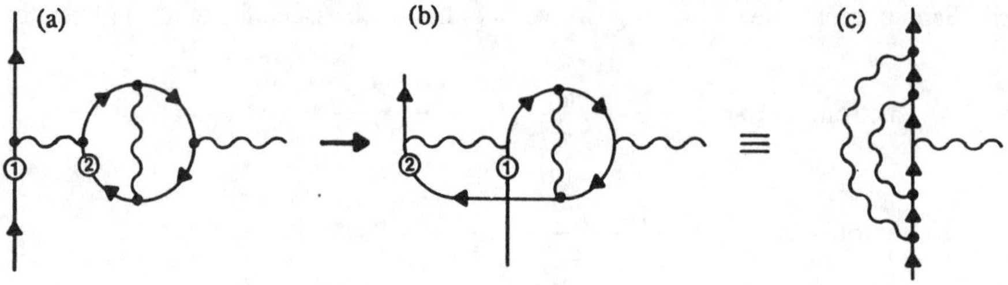

Fig. 4.2a–c. In graph **(b)** the lines (1) and (2) of graph **(a)** are exchanged explaining the minus-sign which accompanies closed fermion loops

is new. Nevertheless, it can be derived directly from 6b. Let us consider, e.g., the graph of Fig. 4.2a which contains an electron loop and may be part of a bigger graph. Exchanging the two electron lines one gets the diagram 4.2b, which can also be drawn in the way shown in 4.2c. Consequently the graph with the closed loop gets a phase factor of -1 compared to 4.2c.

A further very useful statement, which can be derived from the other rules, applies to graphs with closed electron lines.

Furry's *Theorem:* Graphs which contain electron loops with an odd number of photon vertices can be omitted in the expansion of M_{fi}. This will be shown in Exercise 4.1.

EXERCISE

4.1 Furry's Theorem

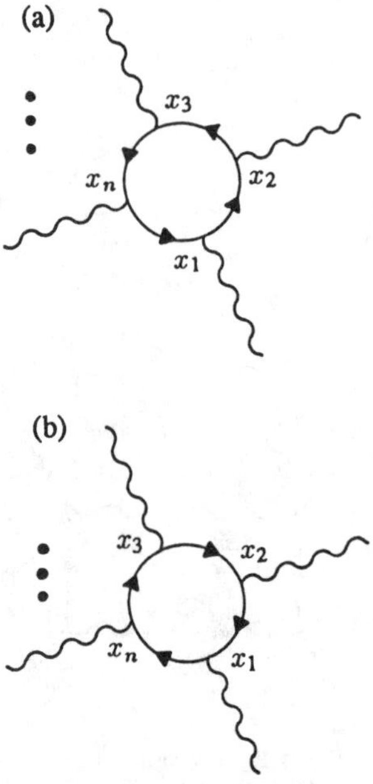

Fig. 4.3a,b. Two graphs with opposite directions of the internal fermion loop

Problem. Show that Feynman diagrams containing a closed electron loop with an *odd number of photon vertices* can be omitted in the calculation of physical processes.

Solution. Consider a process that can be described by a graph containing an electron loop with n vertices. According to Fig. 4.3a and 4.3b for each diagram there is another one where the direction of circulation within the loop is reversed. We shall show that the contribution from each cancels the other for odd n.

The relevant contribution to the S-matrix element describing the loop (a) has the form

$$\begin{aligned}M_{\mathrm{a}} &= (-\mathrm{i}e\gamma_{\mu_1})_{\alpha\beta}\,(\mathrm{i}S_{\mathrm{F}}(x_1-x_n))_{\beta\gamma}\,(-\mathrm{i}e\gamma_{\mu_n})_{\gamma\delta}\,\big(\mathrm{i}S_{\mathrm{F}}(x_n-x_{n-1})\big)_{\delta\varepsilon}\times\dots\\&\quad\times(\mathrm{i}S_{\mathrm{F}}(x_3-x_2))_{\rho\kappa}\,(-\mathrm{i}e\gamma_{\mu_2})_{\kappa\tau}\,(\mathrm{i}S_{\mathrm{F}}(x_2-x_1))_{\tau\alpha}\\&=\mathrm{Tr}\big[(-\mathrm{i}e\gamma_{\mu_1})\,(\mathrm{i}S_{\mathrm{F}}(x_1-x_n))(-\mathrm{i}e\gamma_{\mu_n})\,\big(\mathrm{i}S_{\mathrm{F}}(x_n-x_{n-1})\big)\times\dots\\&\quad\times(\mathrm{i}S_{\mathrm{F}}(x_3-x_2))\,(-\mathrm{i}e\gamma_{\mu_2})\,(\mathrm{i}S_{\mathrm{F}}(x_2-x_1))\big]\quad.\end{aligned}\tag{1}$$

The trace originates from the fact that the first γ matrix is multiplied with the last propagator S_{F} since the loop closes here. Owing to the Feynman rules each γ matrix is multiplied with the propagator following it. Since the starting point is arbitrary, the last propagator has to be multiplied with the first γ matrix, leading to a trace. In an analogous manner the contribution from the graph (b) reads

Exercise 4.1.

$$M_b = \mathrm{Tr}\big[(-\mathrm{i}e\gamma_{\mu_1})\,(\mathrm{i}S_F(x_1-x_2))(-\mathrm{i}e\gamma_{\mu_2})\,(\mathrm{i}S_F(x_2-x_3))\times\dots \\ \times\left(\mathrm{i}S_F(x_{n-1}-x_n)\right)\,(-\mathrm{i}e\gamma_{\mu_n})\,(\mathrm{i}S_F(x_n-x_1))\big] \quad . \tag{2}$$

The traces in (1) and (2) are closely related to each other. To see this we make use of the charge conjugation matrix $\hat{C} = \mathrm{i}\gamma^2\gamma^0$ with the property[1]

$$\hat{C}\gamma_\mu\hat{C}^{-1} = -\gamma_\mu^T \quad . \tag{3}$$

Applied to the Feynman propagator in position space this transformation yields

$$\begin{aligned}\hat{C}S_F(x)\hat{C}^{-1} &= \int\frac{d^4p}{(2\pi)^4}\,\mathrm{e}^{-\mathrm{i}p\cdot x}\,\frac{p^\mu\hat{C}\gamma_\mu\hat{C}^{-1}+m_0\mathbb{1}}{p^2-m_0^2+\mathrm{i}\varepsilon} \\ &= \int\frac{d^4p}{(2\pi)^4}\,\mathrm{e}^{-\mathrm{i}p\cdot x}\,\frac{-p^\mu\gamma_\mu^T+m_0\mathbb{1}}{p^2-m_0^2+\mathrm{i}\varepsilon} \\ &= S_F^T(-x) \quad .\end{aligned} \tag{4}$$

Note the index T at the propagator in the last step, which indicates that S_F has to be transposed! Now we insert factors of $\hat{C}^{-1}\hat{C} = \mathbb{1}$ in (2):

$$M_b = \mathrm{Tr}\big[\hat{C}^{-1}\hat{C}\,(-\mathrm{i}e\gamma_{\mu_1})\hat{C}^{-1}\hat{C}\,(\mathrm{i}S_F(x_1-x_2))\hat{C}^{-1}\hat{C}\,(-\mathrm{i}e\gamma_{\mu_2})\hat{C}^{-1}\hat{C} \\ \times(\mathrm{i}S_F(x_2-x_3))\times\dots\times\hat{C}^{-1}\hat{C}\,\left(\mathrm{i}S_F(x_{n-1}-x_n)\right)\,\hat{C}^{-1}\hat{C} \\ \times(-\mathrm{i}e\gamma_{\mu_n})\hat{C}^{-1}\hat{C}\,(\mathrm{i}S_F(x_n-x_1))\big] \quad . \tag{5}$$

The first factor $\hat{C}^{-1}$ under the trace is permuted to the right side (using $\mathrm{Tr}[AB] = \mathrm{Tr}[BA]$). Furthermore we use (3) and (4), yielding

$$\begin{aligned}M_b &= (-1)^n\,\mathrm{Tr}\big[\left(-\mathrm{i}e\gamma_{\mu_1}^T\right)\,\left(\mathrm{i}S_F^T(x_2-x_1)\right)\,\left(-\mathrm{i}e\gamma_{\mu_2}^T\right)\,\left(\mathrm{i}S_F^T(x_3-x_2)\right)\times\cdots \\ &\quad\times\left(\mathrm{i}S_F^T(x_n-x_{n-1})\right)\,\left(-\mathrm{i}e\gamma_{\mu_n}^T\right)\,\left(\mathrm{i}S_F^T(x_1-x_n)\right)\big] \\ &= (-1)^n\,\mathrm{Tr}\big[(\mathrm{i}S_F(x_1-x_n))\,(-\mathrm{i}e\gamma_{\mu_n})\,\left(\mathrm{i}S_F(x_n-x_{n-1})\right)\times\cdots \\ &\quad\times(-\mathrm{i}e\gamma_{\mu_2})\,(\mathrm{i}S_F(x_2-x_1))\,(-\mathrm{i}e\gamma_{\mu_1})\big]^T \\ &= (-1)^n M_a \quad .\end{aligned} \tag{6}$$

Therefore the sum $M_a + M_b$ vanishes if n is an odd number! There is a plausible explanation for this result. In a closed loop there can be an electron as well as a positron "circling around". These particles interact with the electromagnetic field with an opposite sign of the charge. Thus their contributions cancel each other for an odd number of vertices.

Additional Remark: The existence of two contributions M_a and M_b having equal absolute value has the consequence that for even values of n the contribution to the amplitude made by a loop graph is doubled. One has to be careful, however, in the case $n = 2$. Here one might also be tempted to add two contributions, cf. Fig. 4.4a and 4.4b. This idea is incorrect, however, since both drawings represent exactly the same Feynman graph which has merely been drawn in two different fashions! In the case $n > 2$, on the other hand, this argument does not apply since

[1] See W. Greiner: Theoretical Physics, Vol. 3, *Relativistic Quantum Mechanics – Wave Equations* (Springer, Berlin, Heidelberg 1990), Sect. 12.1.

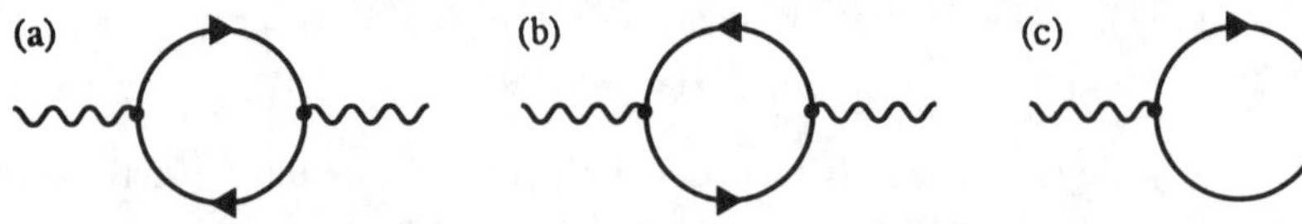

Fig. 4.4a–c. Two equivalent ways to draw a loop diagram with two external lines. (**c**): The tadpole diagram

the loops turning left and turning right lead to topologically distinct graphs which differ in the ordering of the vertices.

The case of a loop with a single vertex, $n = 1$, is somewhat pathological. Obviously there is no cancellation and Furry's theorem does not apply. The contribution of the diagram in Fig. 4.4c (which is known as the *tadpole graph*) does not vanish automatically. Clearly the photon line cannot refer to a free photon since this cannot simply disappear, violating energy and momentum conservation. The tadpole graph will emerge, however, in higher orders of perturbation theory where the loop is coupled via a virtual photon (with momentum $k = 0$) to an electron line in some more complicated graph. This leads to a contribution to the "self energy of the electron" which we will treat in the next chapter. It turns out that the tadpole contribution has no physically observable consequence since its size is independent of the momentum of the electron, in contrast to the self energy correction to be discussed in Sect. 5.3. It can be fully absorbed into a (divergent) renormalization constant which at the end drops out of any calculation. The same effect can be achieved more economically by simply leaving out any tadpole contributions from the outset.[2]

4.2 The Photon Propagator in Different Gauges

The form of the photon propagator D_{F} which was introduced in Section 3.2 cannot be determined uniquely. Until now we have chosen the form (in momentum space)

$$D_{\mathrm{F}\mu\nu} = -\frac{4\pi}{k^2 + \mathrm{i}\varepsilon}\, g_{\mu\nu} \quad , \tag{4.5}$$

which is only one of many ways of defining the photon propagator. The origin of this ambiguity becomes clear if one takes into account that the photon propagator is always sandwiched *between two transition-current elements* when one constructs S-matrix elements, e.g.

$$j^{\mu}_{43}(p_4, p_3)\, D_{\mathrm{F}\mu\nu}(k)\, j^{\nu}_{21}(p_2, p_1) \quad , \tag{4.6}$$

where $p_2 = p_1 - k$ and $p_4 = p_3 + k$. Now, the transition currents obey the equation of continuity. Their four-divergences vanish, i.e. in momentum space (in this context see also (3.210) in Sect. 3.6)

[2] The formal way to do this is quantum field theory consists in the prescription of "normal ordering" of the interaction Hamiltonian.

$$k_\nu j_{12}^\nu(p_1 - k, p_1) = 0 \quad , \tag{4.7a}$$

$$k_\mu j_{43}^\mu(p_3 + k, p_3) = 0 \tag{4.7b}$$

holds. Therefore one can add to $D_{F\mu\nu}$ the expression $k_\mu f_\nu(k) + k_\nu g_\mu(k)$ with arbitrary functions $f_\nu(k)$ and $g_\mu(k)$ without changing the result of the calculation. A somewhat restricted generalization of (4.5) keeping the symmetry between the two currents in (4.6) reads

$$D_{\mu\nu}(k) = -\frac{4\pi}{k^2 + \mathrm{i}\varepsilon} g_{\mu\nu} + k_\mu f_\nu(k) + k_\nu f_\mu(k) \quad , \tag{4.8}$$

with an arbitrary function $f_\mu(k)$ of dimension $[k^{-3}]$. The origin of this ambiguity of the photon propagator is the *gauge degree of freedom* of the electromagnetic field. Because of this there are no observable changes if the potential A_μ is subjected to the transformation

$$A_\mu(x) \to A_\mu(x) + \frac{\partial}{\partial x^\mu} \chi(x) \quad . \tag{4.9}$$

Therefore the propagator of the photon field becomes ambiguous too. The special choice we have made in (4.5) is called the *Feynman gauge*. This is the most common choice. We now introduce two other gauges which prove useful in some applications.

I) The *Landau Gauge*. If one chooses

$$f_\mu(k) = \frac{1}{2} \frac{4\pi k_\mu}{(k^2 + \mathrm{i}\varepsilon)^2} \tag{4.10}$$

for the function $f_\mu(k)$ in (4.4), then the propagator reads as follows:

$$D_{L\mu\nu}(k) = -\frac{4\pi}{k^2 + \mathrm{i}\varepsilon} \left(g_{\mu\nu} - \frac{k_\mu k_\nu}{k^2 + \mathrm{i}\varepsilon} \right) \quad . \tag{4.11}$$

In this form the propagator obeys the condition

$$k^\mu D_{L\mu\nu}(k) = 0 \tag{4.12}$$

in analogy to the Lorentz gauge

$$k^\mu A_\mu(k) = 0$$

of the potential in momentum space.

II) The *Coulomb Gauge*. For this we choose

$$f_0 = \frac{1}{2} \frac{4\pi}{k^2 + \mathrm{i}\varepsilon} \frac{k_0}{\boldsymbol{k}^2} \quad , \quad f_i = -\frac{1}{2} \frac{4\pi}{k^2 + \mathrm{i}\varepsilon} \frac{k_i}{\boldsymbol{k}^2} \quad , \tag{4.13}$$

where the latin indices i denote the space components ($i = 1, 2, 3$). Then the propagator takes the form

$$D_{Cij}(k) = \frac{4\pi}{k^2 + \mathrm{i}\varepsilon} \left(\delta_{ij} - \frac{k_i k_j}{\boldsymbol{k}^2} \right) \quad , \tag{4.14a}$$

$$D_{C0i}(k) = D_{Ci0}(k) = 0 \quad , \tag{4.14b}$$

$$D_{C00}(k) = \frac{4\pi}{\boldsymbol{k}^2} \quad . \tag{4.14c}$$

In (4.14c) we have used $k_0^2 - \boldsymbol{k}^2 = k^2$ so that one factor k^2 in the denominator is cancelled. This gauge obeys the condition

$$k^i D_{Ci\mu}(k) = 0 \quad , \quad (i = 1, 2, 3) \quad , \tag{4.15}$$

which corresponds to the Coulomb gauge of the potential, $\nabla \cdot \boldsymbol{A}(x) = 0$ or $k^i A_i(k) = 0$. The component D_{C00} is just the Fourier transform of the electrostatic potential $1/r$.

EXERCISE

4.2 Supplement: Systems of Units in Electrodynamics

In electrodynamics traditionally several different systems of units are used, which may lead to confusion if they are mixed up. In this volume we use *Gaussian units*, which we shall now compare with other systems of units. As a starting point we shall use the force laws for electrostatics and magnetostatics. Coulomb's law for the force between two charges e_1 and e_2 reads

$$\boldsymbol{F} = k_1 \frac{e_1 e_2}{r^3} \boldsymbol{r} \quad , \tag{1}$$

where k_1 is a constant of proportionality that is still arbitrary at the moment. The constant k_1 defines the unit of charge. One usually assigns an independent basic unit to the charge (e.g. 1 Coulomb = 1 A s). Then k_1 has the dimension [mass][length]^3[charge]$^{-2}$[time]$^{-2}$. In the MKSA *system*, which is part of the legally adopted SI system of units, this is achieved with the dielectric constant ε_0 by defining

$$k_1^{\text{MKSA}} = \frac{1}{4\pi\varepsilon_0} \quad . \tag{2}$$

In the *Gaussian system* one simply chooses

$$k_1^{\text{G}} = 1 \quad . \tag{3}$$

Charge is then a unit that can be derived from mechanical quantities. The system that is probably used most frequently in the literature of theoretical physics takes

$$k_1^{\text{HL}} = \frac{1}{4\pi} \tag{4}$$

and is called the "rationalized Gaussian system" or *Heaviside–Lorentz system.*

For the magnetostatic interaction Ampère's force law is valid

$$\boldsymbol{F} = k_2 \int\!\!\int d^3r_1 d^3r_2 \, \frac{\boldsymbol{j}_1 \times (\boldsymbol{j}_2 \times \boldsymbol{r}_{12})}{r_{12}^3} \quad , \tag{5}$$

where

$$k_2^{\text{MKSA}} = \frac{\mu_0}{4\pi} \tag{6}$$

Exercise 4.2.

in the MKSA system. The units of charge density and current density are always related to each other by the continuity equation

$$\nabla \cdot \boldsymbol{j} + \frac{\partial \rho}{\partial t} = 0 \quad , \tag{7}$$

which implies that the ratio of k_1 and k_2 has the dimension of a squared velocity. We identify this velocity with the velocity of light, which is characteristic for electrodynamics:

$$\frac{k_1}{k_2} = c^2 \quad . \tag{8}$$

Furthermore one can introduce a proportionality factor, k_3, in the definition of the magnetic field strength $\boldsymbol{B}$, so that the Lorentz force on a moving charge e reads

$$\boldsymbol{F} = e\left(\boldsymbol{E} + k_3 \frac{\boldsymbol{v}}{c} \times \boldsymbol{B}\right) \quad . \tag{9}$$

In the Gaussian system and in the Heaviside–Lorentz system one chooses

$$k_3^{\mathrm{G}} = k_3^{\mathrm{HL}} = 1 \quad , \tag{10}$$

whereas

$$k_3^{\mathrm{MKSA}} = c \quad . \tag{11}$$

Through (9) the definition of the field strengths $\boldsymbol{E}$ and $\boldsymbol{B}$ also depends on the choice of unit for charge.

Summarizing all this, Maxwell's equations read (in vacuum)

$$\nabla \cdot \boldsymbol{E} = 4\pi k_1 \rho \quad , \tag{12a}$$

$$\nabla \cdot \boldsymbol{B} = 0 \quad , \tag{12b}$$

$$\nabla \times \boldsymbol{E} = -\frac{k_3}{c}\frac{\partial \boldsymbol{B}}{\partial t} \quad , \tag{12c}$$

$$k_3 \nabla \times \boldsymbol{B} = \frac{4\pi k_1}{c}\boldsymbol{j} + \frac{1}{c}\frac{\partial \boldsymbol{E}}{\partial t} \quad . \tag{12d}$$

For a point charge (12a) leads to Coulomb's law (1) because $\boldsymbol{F} = e\boldsymbol{E}$. In the same way one recognizes that the inhomogeneous term in (12d) together with (9), namely $\boldsymbol{F} = k_3/c \int d^3r \boldsymbol{j} \times \boldsymbol{B}$, leads to Ampère's law (5). Maxwell's equations take the simplest form in the rationalized Heaviside–Lorentz system, since all factors 4π vanish in this case. On the other hand they reappear elsewhere, e.g. in Coulomb's law (1). This is quite reasonable because 4π is a "geometrical factor" that depends on the dimensionality of space. As such it should preferably not appear in the field equations but rather in their solutions.

Of course it is quite easy to see that physical observables do not depend on the choice of units. Especially in the calculation of S-matrix elements the combination $1/c(j_\mu A^\mu)$ always appears which is the interaction energy density. This combination is invariant since the current density transforms in the same way as the charge, i.e. because of (1) it is multiplied by a factor $\sqrt{k_1}$. Simultaneously the potential because of

Exercise 4.2.

$$E = -\nabla\phi - \frac{k_3}{c}\frac{\partial \boldsymbol{A}}{\partial t} \tag{13}$$

is multiplied by $1/\sqrt{k_1}$. Therefore, it follows that

$$e^{\mathrm{G}}A^{\mathrm{G}} = \left(\sqrt{4\pi}e^{\mathrm{HL}}\right)\left(\frac{1}{\sqrt{4\pi}}A^{\mathrm{HL}}\right) = e^{\mathrm{HL}}A^{\mathrm{HL}} \quad . \tag{14}$$

When Heaviside–Lorentz units are used, the Feynman rules change in the following way:

1) There is no factor $N_i = 4\pi$ for external photon lines.
2) For each vertex there is a factor $-\mathrm{i}e^{\mathrm{HL}}\gamma_\mu$.
3) The photon propagator does not contain the factor 4π, i.e. it reads $D_{\mathrm{F}}^{\mu\nu} = -g^{\mu\nu}/(k^2 + \mathrm{i}\varepsilon)$.

The final result of any calculation can always be expressed in terms of the dimensionless fine-structure constant $\alpha \approx 1/137.036$. This constant, however, is related in different ways to the elementary charge:

$$\alpha = \begin{cases} e^2/\hbar c & \text{Gauss} \\ e^2/4\pi\hbar c & \text{Heaviside–Lorentz} \\ e^2/4\pi\varepsilon_0\hbar c & \text{MKSA} \end{cases} \quad . \tag{15}$$

4.3 Biographical Note

FURRY, Wendell Hinkle, American physicist. *18.2.1907 in Prairieton (Indiana), †1984. F. received his Ph.D. at the University of Illinois in 1932. Subsequently he went to Caltech and in 1934 to Harvard where he became professor of physics. F. worked on positron theory, quantum field theory, the theory of molecular energies and the quantum theory of measurement.

5. The Scattering Matrix in Higher Orders

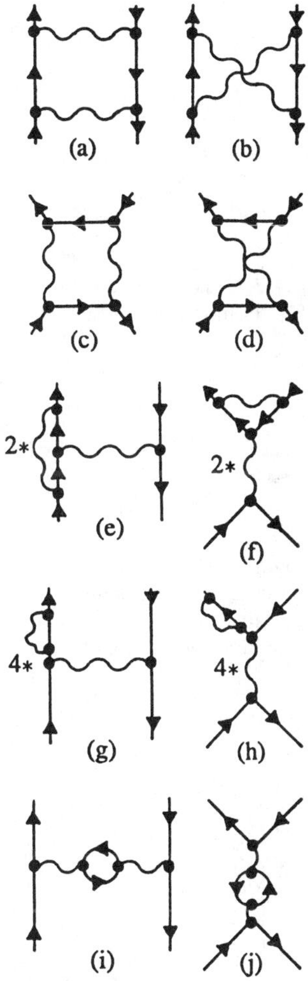

Fig. 5.1a–j. Survey of all Feynman graphs of fourth order for electron–positron scattering

In Chap. 3 many scattering processes were calculated to the lowest nonvanishing order of perturbation theory. Because of the small value of the coupling constant $\alpha \approx 1/137$ the first term alone frequently gives reliable results. Within a satisfactory theory, however, one should be able to calculate the contributions of higher orders too. As we shall see in QED – and the same is true in all quantum field theories – this leads to characteristic difficulties: some of the "small corrections" become infinitely large! Surmounting this problem has been an essential step in the development of the theory. In what follows we shall discuss "renormalization" in the lowest nontrivial order of perturbation theory. The finite theory obtained in this way can then be applied to calculate measurable effects of vacuum fluctuations, i.e. the interaction with virtually created particles.

5.1 Electron–Positron Scattering in Fourth Order

To survey the possible processes that occur in the higher orders of the perturbation expansion, we consider as an example the process of electron–positron scattering in fourth order. According to the general Feynman rules we must construct all topologically different graphs that have four vertices and the prescribed configuration of exterior lines. They are collected in Fig. 5.1. As one can see, a rather imposing list of 18 different diagrams results. The graphs (e–h) occur in different versions, being distinguished by the position of the photon loop, which for brevity have not all been drawn here.

As an exercise we write down the invariant matrix elements for some of the diagrams. We already encountered a diagram similar to Fig. 5.1a when we dealt with electron–proton scattering (Example 3.6). The invariant matrix element corresponding to Fig. 5.2 has the value

$$\begin{aligned} M_{fi}^{(a)} = -e^4 \int \frac{d^4q_1}{(2\pi)^4} \bar{u}(p_1',s_1')(-\mathrm{i}\gamma_\mu) \frac{\mathrm{i}}{\not p_1 - \not q_1 - m + \mathrm{i}\varepsilon} (-\mathrm{i}\gamma_\nu) u(p_1,s_1) \frac{-4\pi\mathrm{i}}{q_1^2 + \mathrm{i}\varepsilon} \\ \times \bar{v}(p_2,s_2)(-\mathrm{i}\gamma^\nu) \frac{\mathrm{i}}{-\not p_2 - \not q_1 - m + \mathrm{i}\varepsilon} (-\mathrm{i}\gamma^\mu) v(p_2',s_2') \frac{-4\pi\mathrm{i}}{(q+q_1)^2 + \mathrm{i}\varepsilon} \quad . \end{aligned} \tag{5.1}$$

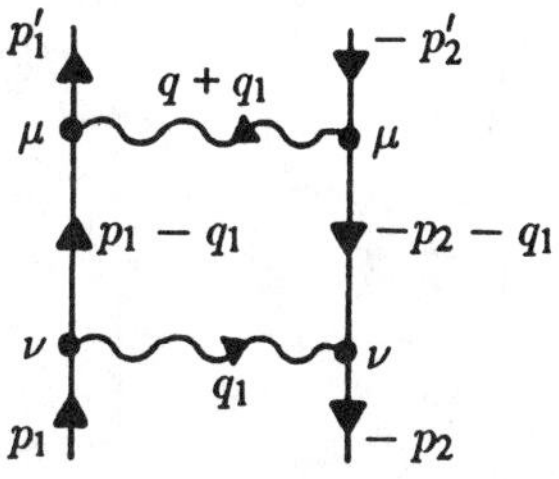

Fig. 5.2. The graph of Fig. 5.1a in momentum space

Here we have introduced the momentum transfer $q = p_1' - p_1 = p_2 - p_2'$; the factors $(\mathrm{i})^4(-\mathrm{i})^4$ cancel. The resulting minus sign arises because of Feynman rule 6a in Chap. 4 since there is an incoming positron.

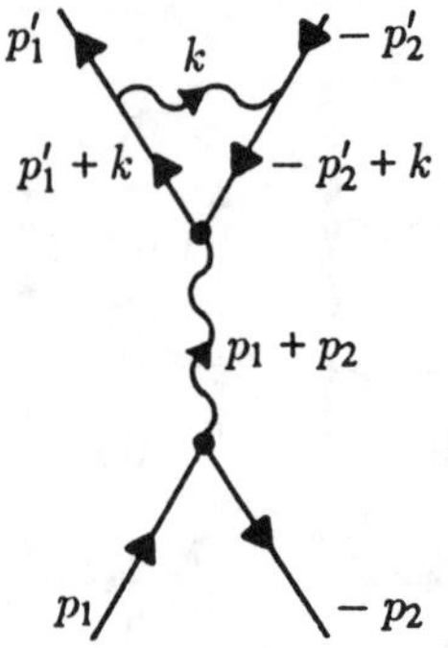

Fig. 5.3. The graph of Fig. 5.1f in momentum space

Next we study the new diagram (f) which is shown in Fig. 5.3. Its matrix element is

$$
\begin{aligned}
M_{fi}^{(f)} = &+e^4\bar{u}(p_1',s_1')\left[\int \frac{d^4k}{(2\pi)^4}(-\mathrm{i}\gamma_\nu)\frac{\mathrm{i}}{\not p_1' + \not k - m + \mathrm{i}\varepsilon}(-\mathrm{i}\gamma_\mu)\right.\\
&\times \left.\frac{\mathrm{i}}{-\not p_2' + \not k - m + \mathrm{i}\varepsilon}(-\mathrm{i}\gamma^\nu)\frac{-4\pi\mathrm{i}}{k^2+\mathrm{i}\varepsilon}\right] v(p_2',s_2')\\
&\times \bar{v}(p_2,s_2)(-\mathrm{i}\gamma^\mu)u(p_1,s_1)\frac{-4\pi\mathrm{i}}{(p_1+p_2)^2+\mathrm{i}\varepsilon} \quad .
\end{aligned} \tag{5.2}
$$

The brackets enclose that part of the graph, which the momentum integration ranges over. The sign of (5.2) relative to (5.1) is explained by Feynman rule 6b. Namely, graph (f) is obtainable from graph (a) by an exchange of two electron lines. To see this, one only needs to choose a convenient time ordering by deforming the diagram (we pass over to configuration space for this purpose), as indicated in Fig. 5.4. Then for a certain time interval one has a virtual electron which has to be antisymmetrized with the incoming real electron, to satisfy Fermi statistics. This explains the relative minus sign between (5.1) and (5.2).

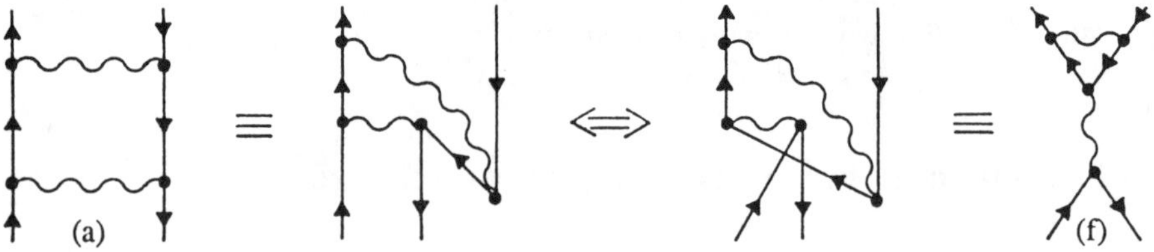

Fig. 5.4. The graphs (**a**) and (**f**) are exchange graphs. This explains the relative minus-sign between (5.1) and (5.2)

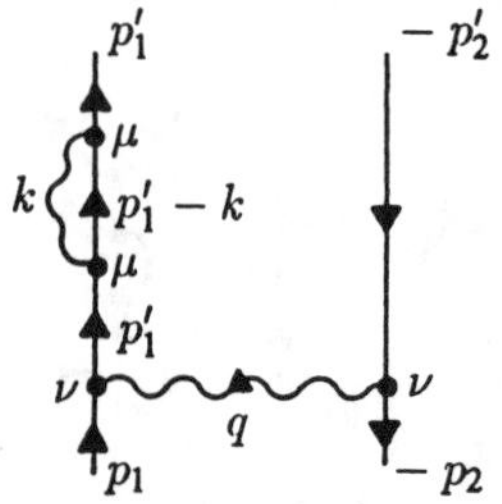

Fig. 5.5. The graph of Fig. 5.1g in momentum space

The matrix element for graph (g), Fig. 5.5, reads

$$
\begin{aligned}
M_{fi}^{(g)} = &-e^4\bar{u}(p_1',s_1')\left[\int \frac{d^4k}{(2\pi)^4}(-\mathrm{i}\gamma_\mu)\frac{\mathrm{i}}{\not p_1' - \not k - m + \mathrm{i}\varepsilon}(-\mathrm{i}\gamma^\mu)\frac{-4\pi\mathrm{i}}{k^2+\mathrm{i}\varepsilon}\right]\\
&\times \frac{\mathrm{i}}{\not p_1' - m + \mathrm{i}\varepsilon}(-\mathrm{i}\gamma_\nu)u(p_1,s_1)\,\bar{v}(p_2,s_2)(-\mathrm{i}\gamma^\nu)\,v(p_2',s_2')\frac{-4\pi\mathrm{i}}{q^2+\mathrm{i}\varepsilon} \ ,
\end{aligned} \tag{5.3}
$$

where again the loop integration extends only over some of the factors.

Let us finally consider graph (i) shown in Fig. 5.6. Because of the closed electron line it has an additional minus sign. The matrix element is

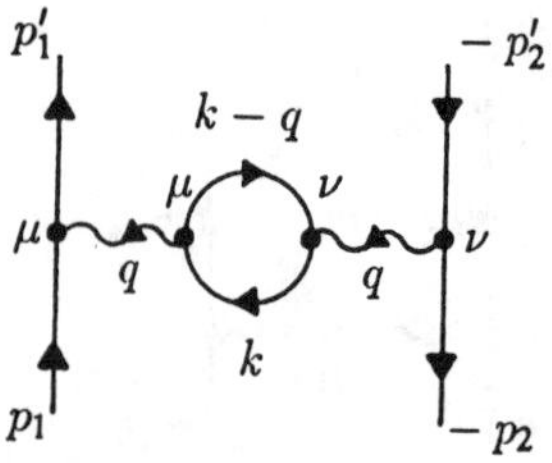

Fig. 5.6. The graph of Fig. 5.1i in momentum space

$$
\begin{aligned}
M_{fi}^{(i)} = &+e^4\bar{u}(p_1',s_1')(-\mathrm{i}\gamma^\mu)u(p_1,s_1)\frac{-4\pi\mathrm{i}}{q^2+\mathrm{i}\varepsilon}\\
&\times \left[\mathrm{Tr}\int \frac{d^4k}{(2\pi)^4}\frac{\mathrm{i}}{\not k - m + \mathrm{i}\varepsilon}(-\mathrm{i}\gamma_\nu)\frac{\mathrm{i}}{\not k - \not q - m + \mathrm{i}\varepsilon}(-\mathrm{i}\gamma_\mu)\right]\\
&\times \frac{-4\pi\mathrm{i}}{q^2+\mathrm{i}\varepsilon}\bar{v}(p_2,s_2)(-\mathrm{i}\gamma^\nu)v(p_2',s_2') \quad .
\end{aligned} \tag{5.4}
$$

The trace results from the fact that one multiplies the different matrices $-\mathrm{i}e\gamma_\mu$ and $\mathrm{i}S_{\mathrm{F}}$ in the electron loop cyclically when one follows the electron line (compare also

Exercise 4.1). This also ensures that the product of the 4 × 4 matrices γ_μ and S_F becomes a pure number.

An inspection of Fig. 5.1 shows that there are three characteristic subgraphs, which can occur at various places in a diagram. These subgraphs are sketched separately in Fig. 5.7.

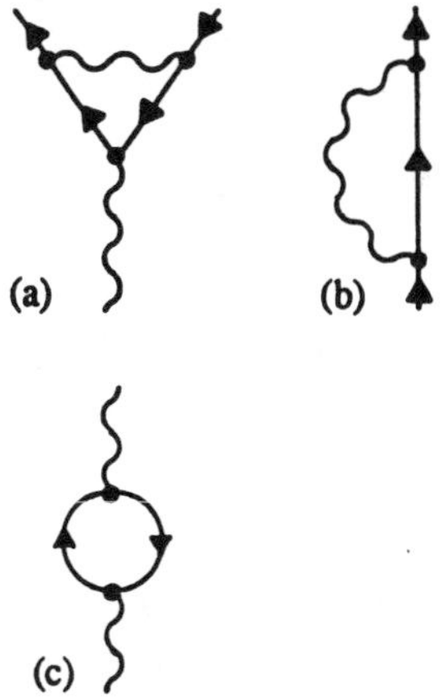

Fig. 5.7a–c. The three basic subgraphs involving a loop. (**a**) vertex correction, (**b**) self energy, (**c**) vacuum polarization

These three diagrams are termed *vertex correction*, *self-energy* of the electron, and *vacuum polarization*. Mathematically they show up in the occurrence of the four-dimensional momentum integrals (loop integrals) that were emphasized by the square brackets in (5.2–4).

Unfortunately all of these integrals are divergent as $k \to \infty$. This can be seen immediately by counting the powers of k in the integrand. While the four-dimensional volume element in the numerator grows like k^3, the denominators for the three processes of Fig. 5.7 are proportional to k^{-4}, k^{-3}, and k^{-2}, respectively, so that we expect a *logarithmic divergence* for the vertex correction, a *linear divergence* (for the time being) for the electron self-energy, and a *quadratic divergence* for the vacuum polarization. In the following sections we will consider these problems separately.

We finally remark that in the systematic construction of all Feynman graphs of fourth order we kept silent about some terms. Figure 5.8 shows two examples of graphs that separate into two disconnected parts.

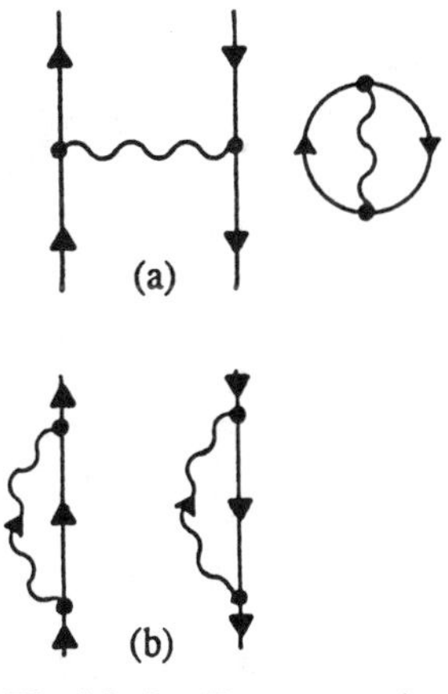

Fig. 5.8a,b. Two examples of disconnected graphs of fourth order

In Fig. 5.8a, for instance, a scattering in lowest order takes place, while independently in the vacuum a virtual pair is created and annihilated some time later. The S-matrix element for a process like that of Fig. 5.8a separates into a product of the matrix elements for the *"connected" part*, which contains the external lines, and the *"disconnected" vacuum bubble*. These "vacuum fluctuations", however, take place all the time, independent of whether there are real particles present or not. Thus every graph is multiplied by a factor of the kind

$S'_{fi} = S_{fi}\Big[\,1 +$  $+$ 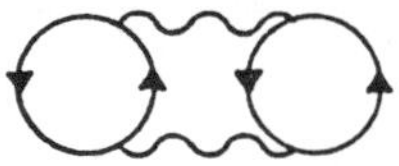 $+ \cdots \Big]$

$= S_{fi} C \quad .$

However, the constant C must have absolute value 1, because the "background noise" of the vacuum is always present, even in the absence of real particles. Thus the S-matrix element for vacuum–vacuum transitions is

$$S'_{0,0} = S_{0,0} C \quad .$$

Since the vacuum – regardless of whether we take into account the electromagnetic interaction – must be conserved, $|C|^2 = 1$ follows and C is only a phase factor without physical significance. *Disconnected graphs can therefore be neglected* in the expansion of the S matrix.

5.2 Vacuum Polarization

In this section we will calculate the influence of the creation of a virtual electron–positron pair (Fig. 5.7c) on the propagation of a photon. To do this we investigate how the *unperturbed* photon propagator

$$\mathrm{i}D_{\mathrm{F}\mu\nu}(q) = \frac{-4\pi\mathrm{i}}{q^2 + \mathrm{i}\varepsilon} g_{\mu\nu} \tag{5.5}$$

is modified by the correction of the order of e^2

$\mathrm{i}D'_{\mathrm{F}\mu\nu}(q) =$

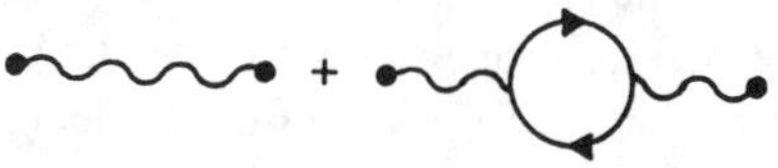

$$= \mathrm{i}D_{\mathrm{F}\mu\nu}(q) + \mathrm{i}D_{\mathrm{F}\mu\lambda}(q)\frac{\mathrm{i}\Pi^{\lambda\sigma}(q)}{4\pi}\mathrm{i}D_{\mathrm{F}\sigma\nu}(q) \quad . \tag{5.6}$$

Here the *polarization tensor*

$$\frac{\mathrm{i}\Pi_{\lambda\sigma}(q)}{4\pi} = -e^2 \int \frac{d^4k}{(2\pi)^4} \mathrm{Tr}\left[\gamma_\lambda \frac{1}{\not{k} - m + \mathrm{i}\varepsilon}\gamma_\sigma \frac{1}{\not{k} - \not{q} - m + \mathrm{i}\varepsilon}\right] \tag{5.7}$$

was introduced (cf. (5.4)). As already mentioned, counting of the powers of k in the integrand reveals that the integral is quadratically divergent. Before we embark on a calculation, inspite of this problem, we first examine some general features of $\Pi_{\mu\nu}(q)$. Since it is a Lorentz tensor, $\Pi_{\mu\nu}$ can be constructed out of the parts $g_{\mu\nu}$ and $q_\mu q_\nu$ and scalar functions of q^2. We write

$$\Pi_{\mu\nu}(q) = Dg_{\mu\nu} + g_{\mu\nu}q^2\Pi^{(1)}(q^2) + q_\mu q_\nu \Pi^{(2)}(q^2) \quad , \tag{5.8}$$

where the constant term $Dg_{\mu\nu}$ is the value of the polarization tensor for vanishing momentum transfer, $q \to 0$. The existence of D has unpleasant consequences. We now use the very generally valid identity for operators

$$\frac{1}{\hat{X} + \hat{Y}} = \frac{1}{\hat{X}} - \frac{1}{\hat{X}}\hat{Y}\frac{1}{\hat{X}} + \frac{1}{\hat{X}}\hat{Y}\frac{1}{\hat{X}}\hat{Y}\frac{1}{\hat{X}} \pm \dots \quad , \tag{5.9}$$

which can be easily verified by multiplication with $(\hat{X} + \hat{Y})$, and rewrite (5.6) by use of it. In the limit $q^2 \to 0$ and up to terms of higher order in $\alpha = e^2$ this yields

$$\begin{aligned}
\mathrm{i}D'_{\mathrm{F}\mu\nu}(q^2 \to 0) &= \frac{-4\pi\mathrm{i}}{q^2 + \mathrm{i}\varepsilon} g_{\mu\nu} + \frac{-4\pi\mathrm{i}}{q^2 + \mathrm{i}\varepsilon} g_{\mu\lambda}\frac{\mathrm{i}D}{4\pi} g^{\lambda\sigma}\frac{-4\pi\mathrm{i}}{q^2 + \mathrm{i}\varepsilon} g_{\sigma\nu} + \dots \\
&= \left(\frac{-4\pi\mathrm{i}}{q^2 + \mathrm{i}\varepsilon} + \frac{-4\pi\mathrm{i}}{q^2 + \mathrm{i}\varepsilon}\frac{\mathrm{i}D}{4\pi}\frac{-4\pi\mathrm{i}}{q^2 + \mathrm{i}\varepsilon} + \dots\right) g_{\mu\nu} \\
&= -\frac{4\pi\mathrm{i}g_{\mu\nu}}{q^2 + \mathrm{i}\varepsilon}\left(1 + \frac{D}{q^2} + \left(\frac{D}{q^2}\right)^2 + \dots\right) \\
&\simeq \frac{-4\pi\mathrm{i}g_{\mu\nu}}{q^2 - D + \mathrm{i}\varepsilon} \quad .
\end{aligned} \tag{5.10}$$

This is just the propagator of a boson with mass $\sqrt{D}$ (a "heavy photon"). Thus the value of $\Pi_{\mu\nu}(0)$ must equal zero, since we know the photon to be massless. The reason this is true has a profound root, namely the required *gauge invariance* of the theory.

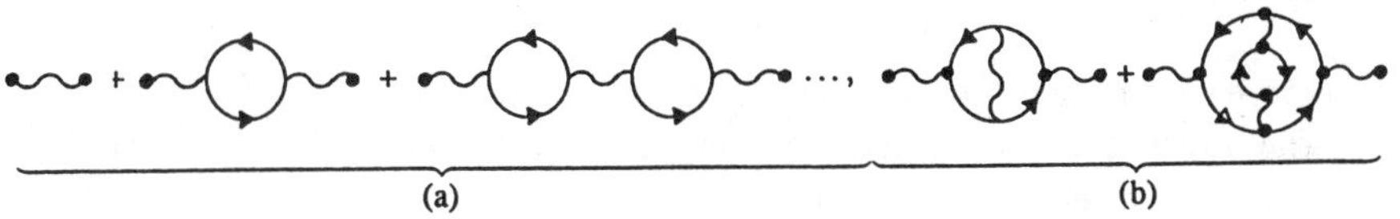

Fig. 5.9. Equation (5.10) sums up all graphs of part a . More complex graphs (part b)), however, are not included

Equation (5.10), by the way, is valid even more generally, because every term in the expansion (5.9) has its counterpart in a Feynman diagram with one more electron loop (see Fig. 5.9a). For obvious reasons summing up these graphs is called the chain approximation. However, the infinite series generated in this way sums up only a certain class of diagrams. There is still any number of more complex "bubbles", which also must be taken into account in the calculation of the exact photon propagator (see Fig. 5.9b).

To see how the constant D is connected with gauge invariance, we consider, for instance, the scattering of an electron at an external potential with the Fourier transform $A_\mu(q)$, see Fig. 5.10. The corresponding matrix element for the direct scattering is proportional to

$$M_{fi} \sim e\bar{u}_f\gamma^\mu u_i A_\mu(q) \quad . \tag{5.11}$$

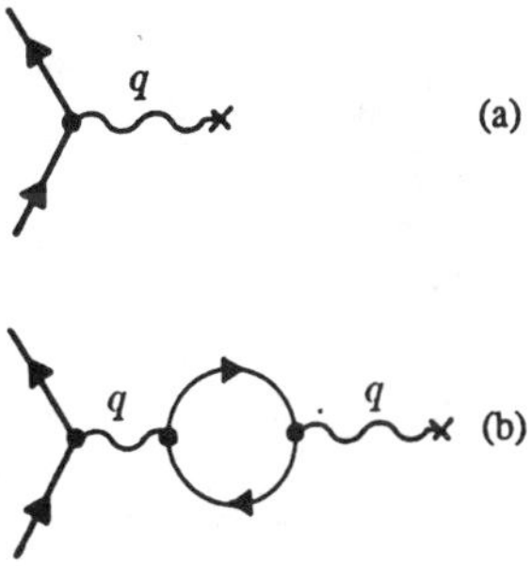

Fig. 5.10. The interaction of an electron with an external potential **(a)** is modified by vacuum polarization **(b)**

To account for the vacuum polarization of the exchanged photon the following correction must be added

$$M_{fi}^{\mathrm{VP}} \sim e\bar{u}_f\gamma^\mu u_i \frac{-4\pi\mathrm{i}}{q^2+\mathrm{i}\varepsilon}\frac{\mathrm{i}\Pi_{\mu\nu}(q)}{4\pi}A^\nu(q) \quad . \tag{5.12}$$

The principle of gauge invariance requires this expression to be invariant under a transformation

$$A_\mu(x) \longrightarrow A_\mu(x) - \frac{\partial}{\partial x^\mu}\chi(x) \quad \text{or} \quad A_\mu(q) \longrightarrow A_\mu(q) + \mathrm{i}q_\mu\chi(q) \quad . \tag{5.13}$$

Obviously this implies that

$$\Pi_{\mu\nu}(q)q^\nu = 0 \quad . \tag{5.14}$$

Because of the symmetry of $\Pi_{\mu\nu}$ (cf. (5.8)),

$$q^\mu \Pi_{\mu\nu}(q) = 0 \tag{5.15}$$

is valid too. This can also be interpreted as the condition of current conservation $q^\mu J_\mu^{\mathrm{VP}}(q)$ for a *polarization current*

$$J_\mu^{\mathrm{VP}}(q) = \frac{1}{4\pi}\mathrm{i}\Pi_{\mu\nu}(q)A^\nu(q) \tag{5.16}$$

induced by the external field. Equations (5.14) and (5.15) applied to (5.8) yield

$$D + q^2\big(\Pi^{(1)}(q^2) + \Pi^{(2)}(q^2)\big) = 0 \quad . \tag{5.17}$$

Since the equation has to be fulfilled for all q^2, the constant D must equal zero (assuming the functions $\Pi^{(i)}$ have no pole at $q^2 = 0$!) and the polarization tensor must assume the simple form

$$\Pi_{\mu\nu}(q^2) = (q^2 g_{\mu\nu} - q_\mu q_\nu)\Pi(q^2) \quad , \tag{5.18}$$

where we have introduced the polarization function $\Pi(q^2) \equiv \Pi^{(1)}(q^2) = -\Pi^{(2)}(q^2)$.

The direct calculation of D, however, seems to contradict this conclusion. According to (5.8) and (5.7) we get

$$\begin{aligned} D &= \frac{1}{4}\Pi^{\mu}{}_{\mu}(0) \\ &= \frac{1}{4}4\pi \mathrm{i}e^2 \int \frac{d^4k}{(2\pi)^4} \mathrm{Tr}\left[\gamma^\mu \frac{\not{k}+m}{k^2-m^2+\mathrm{i}\varepsilon}\gamma_\mu \frac{\not{k}+m}{k^2-m^2+\mathrm{i}\varepsilon}\right] \\ &= 8\pi \mathrm{i}e^2 \int \frac{d^4k}{(2\pi)^4} \frac{2m^2-k^2}{(k^2-m^2+\mathrm{i}\varepsilon)^2} \quad , \end{aligned} \tag{5.19}$$

where the trace was calculated in the usual way:

$$\mathrm{Tr}\big[\gamma^\mu(\not{k}+m)\gamma_\mu(\not{k}+m)\big] = \mathrm{Tr}\big[(-2\not{k}+4m)(\not{k}+m)\big] = 4(-2k^2+4m^2) \quad .$$

The expression (5.19) does not vanish at all, but represents a quadratically divergent integral. Since the original theory was gauge invariant, the result $D \neq 0$ is obviously due to the fact that the defining equation (5.7) for $\Pi_{\mu\nu}$ is an ambiguous mathematical expression. There are different ways of avoiding this problem and of forcing the convergence of these integrals by *"regularization"*. So, for instance, one can simply cut off the k integration at a large momentum value Λ, or one can introduce a damping factor, which for $k \gg \Lambda$ continuously approaches zero, for instance $\Lambda^2/(k^2+\Lambda^2)$.

We want to use the regularization prescription devised by Pauli and Villars.[1] It can be applied relatively simply and has the advantage that it conserves all invariances of the theory. The idea is to subtract from the integrand a function which has the same asymptotic behaviour, in order for the resulting integrand to fall off fast enough with increasing k. More precisely, a set of N (large) auxiliary masses M_i and constants C_i is introduced and the integrand is replaced as follows:

$$\begin{aligned} \Pi_{\mu\nu}(q) &= \int d^4k\, f_{\mu\nu}(q,k,m^2) \\ \longrightarrow \bar{\Pi}_{\mu\nu}(q) &= \int d^4k \Big(f_{\mu\nu}(q,k,m^2) + \sum_{i=1}^{N} C_i f_{\mu\nu}(q,k,M_i^2)\Big) \quad . \end{aligned} \tag{5.20}$$

The constants C_i, M_i are then determined such that the regularized polarization tensor $\bar{\Pi}_{\mu\nu}(q)$ is given by a convergent integral. Through this procedure it is possible to treat the integral with common methods (e.g. to interchange the order of integrations and summations etc.). At the end of the calculation the limit $M_i \to \infty$ must be performed. Since the cutoff procedure was chosen at will, the calculation makes sense only if physical observables do not depend on the parameters C_i, M_i. This can be achieved, as we shall see.

[1] W. Pauli and F. Villars: Rev. Mod. Phys. **21**, 434 (1949).

Now we explicitly calculate the *regularized polarization tensor*.

$$\bar{\Pi}_{\mu\nu}(q) = 4\pi \mathrm{i}e^2 \int \frac{d^4k}{(2\pi)^4} \left\{ \frac{\mathrm{Tr}[\gamma_\mu(\not{k}+m)\gamma_\nu(\not{k}-\not{p}+m)]}{(k^2-m^2+\mathrm{i}\varepsilon)\big[(k-q)^2-m^2+\mathrm{i}\varepsilon\big]} \right.$$
$$\left. + \sum_{i=1}^{N} C_i \frac{\mathrm{Tr}[\gamma_\mu(\not{k}+M_i)\gamma_\nu(\not{k}-\not{p}+M_i)]}{(k^2-M_i^2+\mathrm{i}\varepsilon)\big[(k-q)^2-M_i^2+\mathrm{i}\varepsilon\big]} \right\}$$
$$= 16\pi \mathrm{i}e^2 \int \frac{d^4k}{(2\pi)^4}$$
$$\times \left\{ \frac{k_\mu(k-q)_\nu + k_\nu(k-q)_\mu - g_{\mu\nu}(k^2-q\cdot k-m^2)}{(k^2-m^2+\mathrm{i}\varepsilon)\big[(k-q)^2-m^2+\mathrm{i}\varepsilon\big]} + \mathrm{reg} \right\} , \tag{5.21}$$

where the trace identities 2 and 3 of the Mathematical Supplement 3.3 have been used and "reg" stands for the regularization term. The calculation of momentum integrals of this type is rather difficult. Several techniques have been devised for this purpose which all depend on the introduction of new integrals over auxiliary variables and thus make the d^4k integration simple. We make use of the following integral representation of the causal propagator in momentum space:

$$\frac{\mathrm{i}}{k^2-m^2+\mathrm{i}\varepsilon} = \int_0^\infty d\alpha \, \exp\left[\mathrm{i}\alpha(k^2-m^2+\mathrm{i}\varepsilon)\right] \quad . \tag{5.22}$$

In this way the momentum integration becomes *Gaussian* and can be readily solved. For convenience we will drop the term $i\varepsilon$, which makes the integral convergent, in what follows and consider the mass as a complex number with a small negative imaginary part. Then we obtain

$$\bar{\Pi}_{\mu\nu} = -16\pi \mathrm{i}e^2 \int \frac{d^4k}{(2\pi)^4} \left(\Big[k_\mu(k-q)_\nu + k_\nu(k-q)_\mu - g_{\mu\nu}(k^2-q\cdot k-m^2)\Big] \right.$$
$$\left. \times \int_0^\infty d\alpha_1 \int_0^\infty d\alpha_2 \exp\Big\{\mathrm{i}\Big[\alpha_1(k^2-m^2) + \alpha_2\big((k-q)^2-m^2\big)\Big]\Big\} + \mathrm{reg} \right) . \tag{5.23}$$

To get rid of the polynomial in k and $k-q$, one can introduce two vectorial auxiliary variables z_1 and z_2 into the exponential factor and make use of the identity

$$\mathrm{i}k_\mu = \frac{\partial}{\partial z_1^\mu} \exp(\mathrm{i}k\cdot z_1)\bigg|_{z_1=0} \quad . \tag{5.24}$$

Thus (5.23) can be written as

$$\bar{\Pi}_{\mu\nu} = -16\pi \mathrm{i}e^2 \int \frac{d^4k}{(2\pi)^4} \left(\left[-\frac{\partial}{\partial z_1^\mu}\frac{\partial}{\partial z_2^\nu} - \frac{\partial}{\partial z_1^\nu}\frac{\partial}{\partial z_2^\mu} - g_{\mu\nu}\left(-\frac{\partial}{\partial z_1}\cdot\frac{\partial}{\partial z_2} - m^2\right) \right] \right.$$
$$\times \int_0^\infty d\alpha_1 \int_0^\infty d\alpha_2 \exp\Big\{\mathrm{i}\Big[\alpha_1(k^2-m^2) + \alpha_2\big((k-q)^2-m^2\big)$$
$$\left. + z_1\cdot k + z_2\cdot(k-q)\Big]\Big\} + \mathrm{reg} \right)\Bigg|_{z_1=z_2=0} \quad . \tag{5.25}$$

After the order of the integrations has been exchanged the k integral is easy to solve. According to the Mathematical Supplement 5.5, (6), the Gaussian momentum integral is solved by

$$\int \frac{d^4k}{(2\pi)^4} \exp\left[\mathrm{i}(ak^2 + b\cdot k)\right] = \frac{-\mathrm{i}}{(4\pi)^2 a^2} \exp\left(-\mathrm{i}b^2/4a\right) \quad , \tag{5.26}$$

and thus

$$\begin{aligned}
&\int \frac{d^4k}{(2\pi)^4} \exp\Big\{\mathrm{i}\big[k^2(\alpha_1+\alpha_2) + k\cdot(-2\alpha_2 q + z_1 + z_2) \\
&\qquad +(-m^2(\alpha_1+\alpha_2) + q^2\alpha_2 - q\cdot z_2)\big]\Big\} \\
&= \frac{-\mathrm{i}}{(4\pi)^2(\alpha_1+\alpha_2)^2} \exp\left\{\mathrm{i}\left[-m^2(\alpha_1+\alpha_2) + q^2\alpha_2 - \frac{\alpha_2^2 q^2}{\alpha_1+\alpha_2}\right]\right\} \\
&\quad \times \exp\left\{-\mathrm{i}\left[q\cdot z_2 + \frac{z_1^2 + 2z_1\cdot z_2 + z_2^2 - 4\alpha_2 q\cdot z_1 - 4\alpha_2 q\cdot z_2}{4(\alpha_1+\alpha_2)}\right]\right\} .
\end{aligned} \tag{5.27}$$

The required derivatives are easy to perform. One obtains

$$\begin{aligned}
\frac{\partial}{\partial z_2^\nu} \exp(\ldots) &= -\mathrm{i}\left[q_\nu + \frac{z_{1\nu} + z_{2\nu} - 2\alpha_2 q_\nu}{2(\alpha_1+\alpha_2)}\right] \exp(\ldots) \quad , \\
\frac{\partial}{\partial z_1^\mu} \exp(\ldots) &= -\mathrm{i}\left[\frac{z_{1\mu} + z_{2\mu} - 2\alpha_2 q_\mu}{2(\alpha_1+\alpha_2)}\right] \exp(\ldots) \quad , \\
\frac{\partial}{\partial z_1^\mu}\frac{\partial}{\partial z_2^\nu} \exp(\ldots) &= \left\{-\mathrm{i}\frac{g_{\mu\nu}}{2(\alpha_1+\alpha_2)}\right. \\
&\quad \left. - \left[q_\nu + \frac{z_{1\nu} + z_{2\nu} - 2\alpha_2 q_\nu}{2(\alpha_1+\alpha_2)}\right] \frac{z_{1\mu} + z_{2\mu} - 2\alpha_2 q_\mu}{2(\alpha_1+\alpha_2)}\right\} \exp(\ldots) \\
&\overset{z_1=z_2=0}{\longrightarrow} -\mathrm{i}\frac{g_{\mu\nu}}{2(\alpha_1+\alpha_2)} + \frac{\alpha_1\alpha_2}{(\alpha_1+\alpha_2)^2} q_\nu q_\mu \quad .
\end{aligned}$$

Contraction of the indices μ and ν leads to

$$\frac{\partial}{\partial z_1}\cdot\frac{\partial}{\partial z_2} \exp(\ldots) \longrightarrow -\frac{2\mathrm{i}}{\alpha_1+\alpha_2} + \frac{\alpha_1\alpha_2}{(\alpha_1+\alpha_2)^2} q^2 \quad .$$

Using this we get

$$\begin{aligned}
\bar{\Pi}_{\mu\nu}(q) &= +16\pi\mathrm{i}e^2 \int_0^\infty d\alpha_1 \int_0^\infty d\alpha_2 \\
&\quad \times \left(\frac{-\mathrm{i}}{(4\pi)^2(\alpha_1+\alpha_2)^2} \exp\left\{\mathrm{i}\left[-m^2(\alpha_1+\alpha_2) + \frac{\alpha_1\alpha_2}{\alpha_1+\alpha_2} q^2\right]\right\}\right. \\
&\quad \left. \times \left\{\frac{2\alpha_1\alpha_2}{(\alpha_1+\alpha_2)^2} q_\mu q_\nu - g_{\mu\nu}\left[\frac{\alpha_1\alpha_2}{(\alpha_1+\alpha_2)^2} q^2 - \frac{\mathrm{i}}{\alpha_1+\alpha_2} + m^2\right]\right\} + \text{reg}\right) .
\end{aligned} \tag{5.28}$$

The polynomial in the integrand is split into a gauge-invariant part of the form (5.18) and a remainder

$$\Big\{\dots\Big\} = \frac{2\alpha_1\alpha_2}{(\alpha_1+\alpha_2)^2}(q_\mu q_\nu - g_{\mu\nu}q^2) + g_{\mu\nu}\left[\frac{\alpha_1\alpha_2 q^2}{(\alpha_1+\alpha_2)^2} + \frac{\mathrm{i}}{\alpha_1+\alpha_2} - m^2\right] . \tag{5.29}$$

We can show that the offending non-gauge-invariant part – we will call it $\Delta\bar{\Pi}_{\mu\nu}$ – vanishes in the regularized theory. With the convention $C_0 = 1$, $M_0 = m$ the non-gauge-invariant part reads

$$\Delta\bar{\Pi}_{\mu\nu} = \frac{e^2}{\pi} g_{\mu\nu} \int_0^\infty\int_0^\infty \frac{d\alpha_1 d\alpha_2}{(\alpha_1+\alpha_2)^2} \sum_{i=0}^N C_i \left[\frac{\alpha_1\alpha_2 q^2}{(\alpha_1+\alpha_2)^2} + \frac{\mathrm{i}}{\alpha_1+\alpha_2} - M_i^2\right]$$
$$\times \exp\left\{\mathrm{i}\left[-M_i^2(\alpha_1+\alpha_2) + \frac{\alpha_1\alpha_2}{\alpha_1+\alpha_2}q^2\right]\right\} . \tag{5.30}$$

We note that the terms in the square brackets look quite similar. This can be exploited to simplify the integral. Again, an auxiliary variable ϱ is introduced and the expression is calculated at the point $\varrho = 1$:

$$\Delta\bar{\Pi}_{\mu\nu} = -\frac{e^2}{\pi} g_{\mu\nu} \mathrm{i}\varrho \frac{\partial}{\partial\varrho} \int_0^\infty\int_0^\infty \frac{d\alpha_1 d\alpha_2}{(\alpha_1+\alpha_2)^3}\frac{1}{\varrho}$$
$$\times \sum_{i=0}^N C_i \exp\left\{\mathrm{i}\varrho\left[-M_i^2(\alpha_1+\alpha_2) + \frac{\alpha_1\alpha_2}{\alpha_1+\alpha_2}q^2\right]\right\}\Bigg|_{\varrho=1} . \tag{5.31}$$

Moving the differentiation before the integrals is allowed here, because the constants C_i can be chosen such that the integral converges absolutely (at $\alpha_i \to 0$). Now we perform a scale-transformation $\alpha_i' = \varrho\alpha_i$ in the integral. As one can see, this has the effect of completely eliminating ϱ ! Thus the integral does not depend on ϱ at all, so the derivative $\partial/\partial\varrho$ acting on it vanishes. This *proves the gauge invariance of the regularized polarization tensor* $\bar{\Pi}_{\mu\nu}$.

Let us proceed with the calculation of the gauge-invariant polarization operator from (5.18). According to (5.28, 5.29) its regularized version is determined by

$$\bar{\Pi}(q^2) = -\frac{2e^2}{\pi}\int_0^\infty d\alpha_1 \int_0^\infty d\alpha_2 \frac{\alpha_1\alpha_2}{(\alpha_1+\alpha_2)^4}$$
$$\times \sum_{i=0}^N C_i \exp\left\{\mathrm{i}\left[-M_i^2(\alpha_1+\alpha_2) + \frac{\alpha_1\alpha_2}{\alpha_1+\alpha_2}q^2\right]\right\} . \tag{5.32}$$

Here we introduce a factor

$$1 = \int_0^\infty d\varrho\, \delta(\varrho - \alpha_1 - \alpha_2) \tag{5.33}$$

and then, again, perform a scale transformation $\alpha_i = \varrho\beta_i$:

$$
\begin{aligned}
\bar{\Pi}(q^2) &= -\frac{2e^2}{\pi}\int\limits_0^\infty d\alpha_1 \int\limits_0^\infty d\alpha_2 \int\limits_0^\infty d\varrho\, \delta(\varrho-\alpha_1-\alpha_2)\frac{\alpha_1\alpha_2}{\varrho^4} \\
&\quad \times \sum_{i=0}^{N} C_i \exp\left[\mathrm{i}\left(-M_i^2\varrho + \frac{\alpha_1\alpha_2}{\varrho}q^2\right)\right] \\
&= -\frac{2e^2}{\pi}\int\limits_0^\infty d\varrho \int\limits_0^\infty \varrho\, d\beta_1 \int\limits_0^\infty \varrho\, d\beta_2\, \delta(\varrho-\varrho\beta_1-\varrho\beta_2) \\
&\quad \times \frac{\varrho^2\beta_1\beta_2}{\varrho^4}\sum_{i=0}^{N} C_i \exp\left[\mathrm{i}(-M_i^2+\beta_1\beta_2 q^2)\varrho\right] \\
&= -\frac{2e^2}{\pi}\int\limits_0^1 d\beta_1 \int\limits_0^1 d\beta_2\, \beta_1\beta_2\delta(1-\beta_1-\beta_2) \\
&\quad \times \underbrace{\int\limits_0^\infty \frac{d\varrho}{\varrho}\sum_{i=0}^{N} C_i \exp\left[\mathrm{i}\varrho(-M_i^2+\beta_1\beta_2 q^2)\right]}_{\equiv I} \quad .
\end{aligned} \tag{5.34}
$$

The range of the β integrations can be restricted to the interval $0 \le \beta_i \le 1$ because of the δ function.

Without regularization the integral $I = \int_0^\infty d\varrho/\varrho \ldots$ would be *logarithmically divergent*. The gauge invariance of the polarization tensor which allowed factoring out two powers of momentum in (5.18) has thus reduced the degree of divergence by two. How do we have to choose now the values of C_i and M_i in order to obtain a finite result? Let us consider the ϱ integral and let us deform the contour of integration to the negative imaginary axis. Since the product $\beta_1\beta_2 = \beta_1(1-\beta_1)$ never exceeds the value $1/4$, this leads to an exponentially decreasing integrand as long as the condition $q^2 < 4m^2$ is fulfilled.[2] If we introduce a lower integration boundary η and consider the limit $\eta \to 0$, we get, using the substitution $t = \mathrm{i}\varrho(M_i^2 - \beta_1\beta_2 q^2)$, according to exercise 5.1

$$
\begin{aligned}
I &= \lim_{\eta\to 0}\int\limits_{-\mathrm{i}\eta}^{-\mathrm{i}\infty} \frac{d\varrho}{\varrho}\sum_{i=0}^{N} C_i \exp\left[-\mathrm{i}\varrho(M_i^2-\beta_1\beta_2 q^2)\right] \\
&= \lim_{\eta\to 0}\sum_{i=0}^{N} C_i \int\limits_{\eta(M_i^2-\beta_1\beta_2q^2)}^{\infty} \frac{dt}{t}\exp(-t)
\end{aligned}
$$

and with integration by parts

$$
I = -\lim_{\eta\to 0}\left(\sum_{i=0}^{N} C_i \ln\eta\right) - \sum_{i=0}^{N} C_i \ln(M_i^2-\beta_1\beta_2 q^2) + \sum_{i=0}^{N} C_i \int\limits_0^\infty dt \ln t\, \exp(-t) . \tag{5.35}
$$

[2] The value $q^2 = 4m^2$ is the threshold for the production of real electron–positron pairs as will be discussed in Exercise 5.2.

With this we have reached our goal: by the choice of the constants

$$\sum_{i=0}^{N} C_i = 0 \tag{5.36}$$

the first and the last term of (5.35) are made to vanish and we obtain a finite result:

$$\begin{aligned} I &= -\left[\ln(m^2 - \beta_1\beta_2 q^2) + \sum_{i=1}^{N} C_i \ln(M_i^2 - \beta_1\beta_2 q^2)\right] \\ &= -\left[\ln\left(1 - \beta_1\beta_2 \frac{q^2}{m^2}\right) + \sum_{i=1}^{N} C_i \ln \frac{M_i^2}{m^2} + \underbrace{\sum_{i=0}^{N} C_i \ln m^2}_{=0}\right] \\ &= -\left[\ln\left(1 - \beta_1\beta_2 \frac{q^2}{m^2}\right) - \ln \frac{\Lambda^2}{m^2}\right] \quad , \end{aligned} \tag{5.37}$$

where the term q^2 has been neglected compared to the large masses M_i, and the abbreviation

$$\sum_{i=1}^{N} C_i \ln \frac{M_i^2}{m^2} \equiv -\ln \frac{\Lambda^2}{m^2} \tag{5.38}$$

has been introduced defining an averaged *cutoff momentum* Λ.[3] With (5.37) the polarization function finally (5.34) reads

$$\begin{aligned} \bar{\Pi}(q^2) &= \frac{2e^2}{\pi} \int_0^1 d\beta\, \beta(1-\beta) \left\{ -\ln \frac{\Lambda^2}{m^2} + \ln\left[1 - \beta(1-\beta)\frac{q^2}{m^2}\right]\right\} \\ &= \frac{2e^2}{\pi} \left\{ -\frac{1}{6} \ln \frac{\Lambda^2}{m^2} + \int_0^1 d\beta\, \beta(1-\beta) \ln\left[1 - \beta(1-\beta)\frac{q^2}{m^2}\right]\right\} \\ &\equiv -\frac{e^2}{3\pi} \ln \frac{\Lambda^2}{m^2} + \Pi^{\mathrm{R}}(q^2) \quad . \end{aligned} \tag{5.39}$$

The remaining one-dimensional integral in $\Pi^{\mathrm{R}}(q^2)$ can be solved analytically. We shall not give the exact result here (this will be deferred to Exercise 5.2) and consider only the limit $q^2/m^2 \ll 1$. In this case one can expand the logarithm into a Taylor series,

$$\ln(1 - z) = -\left(z + \frac{z^2}{2} + \dots\right) \quad , \tag{5.40}$$

and obtains after an elementary integration

$$\Pi^{\mathrm{R}}(q^2) = -\frac{e^2}{\pi} \frac{q^2}{m^2} \left(\frac{1}{15} + \frac{1}{140} \frac{q^2}{m^2} + \dots\right) \quad . \tag{5.41}$$

[3] We remark that the regularization of $\bar{\Pi}(q^2)$ could have been achieved with a single subtraction term ($N = 1$), namely $C_1 = -1$, $M_1 = \Lambda$. This would not, however, have been enough to cancel the quadratically divergent, non-gauge-invariant term $\Delta\Pi_{\mu\nu}(q)$.

Thus the regularized vacuum-polarization tensor according to (5.39) consists of a constant term (up to a coefficient $g_{\mu\nu}q^2 - q_\mu q_\nu$) that *diverges logarithmically with the cutoff momentum* Λ, and a well-defined finite momentum-dependent part $\Pi^{\mathrm{R}}(q^2)$. What is the significance of these terms? In order to understand them we go back to the modified photon propagator (5.6) and consider its influence in a scattering process. For instance, the amplitude of Møller scattering reads

$M_{fi}^{(2)} =$ 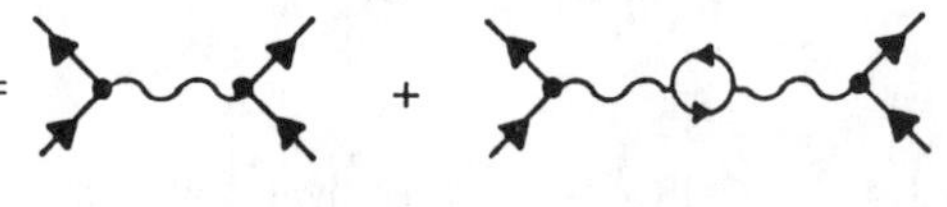

$$
\begin{aligned}
&= (-\mathrm{i}e\bar{u}_1'\gamma^\mu u_1)\frac{-4\pi\mathrm{i}}{q^2}\left[g_{\mu\nu} + \frac{\mathrm{i}}{4\pi}\bar{\Pi}_{\mu\tau}(q)\frac{-4\pi\mathrm{i}g^\tau{}_\nu}{q^2}\right](-\mathrm{i}e\bar{u}_2'\gamma^\nu u_2) \\
&= (-\mathrm{i}e\bar{u}_1'\gamma^\mu u_1)\mathrm{i}D_{\mathrm{F}}^{(0)}(q)\left[g_{\mu\nu} + (g_{\mu\tau}q^2 - q_\mu q_\tau)\bar{\Pi}(q^2)\frac{g^\tau{}_\nu}{q^2}\right](-\mathrm{i}e\bar{u}_2'\gamma^\nu u_2) \\
&= (-\mathrm{i}e\bar{u}_1'\gamma^\mu u_1)\mathrm{i}D_{\mathrm{F}}^{(0)}(q)\left[1 - \frac{e^2}{3\pi}\ln\frac{\Lambda^2}{m^2} + \Pi^{\mathrm{R}}(q^2)\right](-\mathrm{i}e\bar{u}_2'\gamma_\mu u_2) \quad , \qquad (5.42)
\end{aligned}
$$

where in the last transformation we used the gauge invariance of the transition currents, $q_\nu j_{fi}^\nu = 0$. In this way the term proportional to $q_\mu q_\nu$ drops out. One can prove this explicitly by considering that the u_1 are free Dirac spinors, namely

$$
\left.\begin{array}{l}(\not{p}_1 - m)u_1 = 0 \\ \bar{u}_1'(\not{p}_1' - m) = 0\end{array}\right\} \quad \text{and thus} \quad q_\nu\bar{u}_1\gamma^\nu u_1 = \bar{u}_1(\not{p}_1' - \not{p}_1)u_1 = 0 \quad , \qquad (5.43)
$$

because $q = (p_1' - p_1)$. Since (5.42) is to be valid only to the order of α^2, it can also be written as

$$
M_{fi}^{(2)} = (-\mathrm{i}e\bar{u}_1'\gamma^\mu u_1)\left[Z_3\left(1 + \Pi^{\mathrm{R}}(q^2)\right)\mathrm{i}D_{\mathrm{F}}^{(0)}(q)\right](-\mathrm{i}e\bar{u}_2'\gamma_\mu u_2) \quad , \qquad (5.44)
$$

with the constant factor

$$
Z_3 = 1 - \frac{e^2}{3\pi}\ln\frac{\Lambda^2}{m^2} \quad . \qquad (5.45)
$$

The expressions (5.44) and (5.42) differ by a term of the order of α^3 which is beyond the presently required accuracy. Compared to the matrix element in lowest order, the consideration of vacuum polarization thus yields a constant correction factor Z_3 and a momentum-dependent modification $\Pi^R(q^2)$. The latter approaches zero for small q^2 according to (5.41). Thus if one performs a scattering experiment with small q^2 (this corresponds classically to a "peripheral" scattering between two charges that are separated by a large distance), then the scattering amplitude in second order is simply given by

$$
M_{fi}^{(2)} = Z_3 M_{fi}^{(1)} \quad .
$$

The value of the electric charge, however, is empirically determined just by such experiments. *A particle with the "bare" charge e for a distant observer seems to carry the "renormalized" charge* e_{R} , with

$$
e_{\mathrm{R}} = \sqrt{Z_3}\, e \quad . \qquad (5.46)
$$

The bare charge e, however, in principle is *not observable*, since the interaction between electron and photon fields cannot be "switched off". In this manner we have circumvented the problem of the dependence on the cutoff Λ in a most elegant way: only the renormalized charge is relevant for physical observations; it is experimentally determined to be $e_\mathrm{R}^2 \simeq 1/137$. The magnitude of the renormalization constant Z_3 and the bare charge e do not matter at all. To calculate any process one simply uses the charge e_R and then has to deal only with well-defined finite quantities!

Of course this argument is nevertheless a little unsatisfactory. If divergencies occur in the calculation of the renormalization constant, this is a hint that the theory is fundamentally not entirely consistent. For very large momenta, or, which is the same, for very small distances it breaks down. We can, however, convince ourselves that the region of validity will be very large. Since the divergence in (5.45) is only logarithmic, Z_3 will be significantly different from unity only if

$$\Lambda \simeq \mathrm{e}^{3\pi/2\alpha}\, m = 10^{280}\, m \quad . \tag{5.47}$$

According to the uncertainty relation this corresponds to a length of $\Delta x = h/\Lambda \simeq 10^{-293}$cm! In practice this is completely irrelevant, because the existence of other quantum fields limits the validity of pure QED anyway. For instance, as soon as there occur momenta in the region of the pion mass $m_\pi \simeq 270\, m_\mathrm{e}$, the strong interaction must be taken into account. On a still higher scale the weak interaction becomes important. According to present-day understanding both phenomena are unified in a single framework, i.e. the theory of electroweak interaction. However, in this more general theory, also, divergent renormalization integrals do occur (see W. Greiner and B. Müller: Theoretical Physics, Vol. 5, *Gauge Theory of Weak Interaction*).

It might well be possible that space and time are not continuous but consist of tiny space–time cells. Then no momenta would occur that correspond to lengths smaller than the size of a cell. The loop integrals (which are divergent in the case of continuous space–time) in this case would have their natural "cutoff parameter" at the maximum momentum. Renormalization would still be necessary, but the need for an artificial regularization would not arise.

Up to now the influence of vacuum polarization on *internal* photon lines has been examined. This leads to the occurrence of a modified photon propagator $Z_3(1+\Pi^R)D_\mathrm{F}^{(0)}$ in (5.44). Corrections of *external* photon lines must certainly be considered as well. These lines represent a potential $A_\mu(q)$ with a momentum on "mass-shell" $q^2 = 0$. The potential satisfies the Laplace equation, which in momentum space reads

$$q^2 A_\mu(q) = 0 \quad , \tag{5.48}$$

and the Lorentz gauge condition,

$$q^\mu A_\mu(q) = 0 \quad . \tag{5.49}$$

According to Fig. 5.11 one could try to form a renormalized potential A_μ^R out of the free A_μ^0 by the prescription

$$A_\mu^\mathrm{R} = A_\mu^0 + \mathrm{i}D_{\mathrm{F}\mu\nu}(q)\frac{\mathrm{i}\Pi^{\nu\sigma}(q)}{4\pi}A_\sigma^0 = A_\mu^0 + \frac{-4\pi\mathrm{i}}{q^2}\frac{\mathrm{i}}{4\pi}q^2\bar{\Pi}(q^2)A_\mu^0 \quad . \tag{5.50}$$

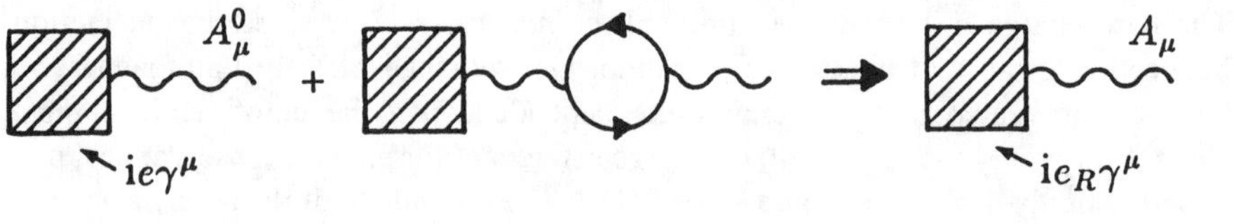

Fig. 5.11. Renormalization of an external photon–line

Unfortunately the second term is an undefined expression because it yields either $\bar{\Pi} A^0_\mu$ or the value zero (because of (5.48)), depending on how the factors are combined. A definite result is obtained from the following consideration. From (5.44) we know that the modified photon propagator $D_F^{(1)}(q^2)$ emerges from the free one by multiplication with the constant Z_3. The propagator of a field, however, is always a *quadratic* function of the wave function. Therefore it appears natural to renormalize the photon wave function and thus the potential by[4]

$$A^R_\mu = \sqrt{Z_3} A^0_\mu \quad . \tag{5.51}$$

This is also plausible because a photon never is really free, but at some time it has been emitted from a source at a large distance (or will be absorbed by an observer at a large distance). The charge of this source, however, must also be renormalized, according to $e_R = \sqrt{Z_3}\, e$. The argument again leads to (5.51). According to (5.51) it is thus sufficient just to *drop the contribution of the vacuum polarization bubbles in the calculation of a graph with external photon lines* and instead at the vertex use the charge e_R. By this procedure the bare charge e is completely eliminated from the matrix element: in the case of internal photon lines by using $D_F^R = Z_3 D_F$, which renormalizes two factors e at the two ends, and in the case of external photon lines by the replacement $A^R_\mu = \sqrt{Z_3} A^0_\mu$.

Thus we have learned that according to the renormalization procedure the value of the polarization function at zero momentum transfer $\bar{\Pi}(0) = -(e^2/3\pi)\ln(\Lambda^2/m^2)$ has no physical significance at all, because it is absorbed in the coupling constant. The *momentum-dependent contribution* $\Pi^R(q^2)$, however, leads to well-defined measurable effects. As has already been shown in (5.44), for instance, the scattering cross section of two electrons or other charged particles will be influenced. The binding energy of an electron in an atom is also affected.

One can understand this effect most clearly if one considers the Coulomb potential of an external static source of charge[5] $-Ze$, i.e. $A_0(\boldsymbol{x}) = -Ze/|\boldsymbol{x}|$, $\boldsymbol{A}(\boldsymbol{x}) = 0$. According to (5.50) the modified potential in momentum space reads

$$A_0'(\boldsymbol{q}) = A_0(\boldsymbol{q}) + \Pi^R(-\boldsymbol{q}^2) A_0(\boldsymbol{q}) \tag{5.52}$$

or in coordinate space

$$A_0'(\boldsymbol{x}) = \int \frac{d^3q}{(2\pi)^3} \exp(\mathrm{i}\boldsymbol{q}\cdot\boldsymbol{x}) \left(1 + \Pi^R(-\boldsymbol{q}^2)\right) A_0(\boldsymbol{q}) \quad . \tag{5.53}$$

[4] The finite part $\Pi^R(q^2)$ does not contribute to external lines because according to (5.41) it is proportional to q^2 and thus vanishes because of (5.48).

[5] In the following calculation we assume that charge renormalization has already been performed; however, we drop the index e_R for convenience.

Since a stationary source can absorb momentum but not energy we have set $q_0 = 0$ and thus $q^2 = -\boldsymbol{q}^2$. In the Exercise 5.3, (7) this is demonstrated explicitly. The original Coulomb potential in momentum space is

$$A_0(\boldsymbol{q}) = \int d^3x \exp(-\mathrm{i}\boldsymbol{q}\cdot\boldsymbol{x}) \frac{-Ze}{|\boldsymbol{x}|} = -4\pi \frac{Ze}{|\boldsymbol{q}|^2} \quad . \tag{5.54}$$

The Fourier integral in (5.53) can be approximately solved if one uses the Taylor series expansion (5.41) in lowest order in q^2:

$$\begin{aligned} A_0'(\boldsymbol{x}) &\simeq \int \frac{d^3q}{(2\pi)^3} \exp(\mathrm{i}\boldsymbol{q}\cdot\boldsymbol{x}) \left(1 + \frac{e^2}{15\pi m^2}\boldsymbol{q}^2\right) A_0(\boldsymbol{q}) \\ &= A_0(\boldsymbol{x}) + \frac{e^2}{15\pi m^2}(-\nabla^2) \int \frac{d^3q}{(2\pi)^3} \exp(\mathrm{i}\boldsymbol{q}\cdot\boldsymbol{x}) A_0(\boldsymbol{q}) \\ &= \left(1 - \frac{e^2}{15\pi m^2}\nabla^2\right) A_0(\boldsymbol{x}) \quad , \end{aligned} \tag{5.55}$$

where the fact has been taken into account that the momentum transfer is purely space-like, $q^2 = -\boldsymbol{q}^2$. Starting with a point-like charge $-Ze$, the potential energy reads

$$eA_0'(\boldsymbol{x}) = -\frac{Z\alpha}{|\boldsymbol{x}|} - \alpha Z\alpha \frac{4}{15m^2}\delta^3(\boldsymbol{x}) \equiv eA_0(\boldsymbol{x}) + e\Delta A_0(\boldsymbol{x}) \quad , \tag{5.56}$$

because of $\nabla^2(1/|\boldsymbol{x}|) = -4\pi\delta^3(\boldsymbol{x})$. In addition to the Coulomb potential a *short-range attractive additional potential* acts. This result was found very early on. It is only valid for low momentum transfer, i.e. in lowest approximation for $\Pi^{\mathrm{R}}(q^2)$ in (5.41). After Dirac and Heisenberg had discussed the effect of vacuum polarization a short time after the discovery of the positron, the resulting modification of the electromagnetic interaction was derived by **Uehling**[6] in 1935. Expression (5.53) is therefore often called the *Uehling potential.*

Since the motion of an electron in the field of a nucleus can be very accurately described by a static potential $A_0(\boldsymbol{x})$, one can immediately determine the *change of atomic binding energies* by means of (5.56). To calculate the expectation value of the additional potential ΔA_0, it is obviously sufficient to know the density of the electron wave function at the position of the nucleus. Nonrelativistic quantum mechanics yields for the hydrogen wave function with principal quantum number n and angular momentum l

$$|\Psi_{nl}(0)|^2 = \frac{m^3(Z\alpha)^3}{\pi n^3}\delta_{l0} \quad . \tag{5.57}$$

The energy shift due to vacuum polarization in first-order perturbation theory then simply reads

$$\begin{aligned} \Delta E_{nl}^{\mathrm{VP}} &= \langle\Psi_{nl}|e\Delta A_0|\Psi_{nl}\rangle = -\alpha(Z\alpha)\frac{4}{15m^2}|\Psi_{nl}(0)|^2 \\ &= -\frac{4m}{15\pi n^3}\alpha(Z\alpha)^4\delta_{l0} \quad . \end{aligned} \tag{5.58}$$

[6] E.A. Uehling: Phys. Rev. **48**, 55 (1935).

Because of the short range of $\Delta A_0(\boldsymbol{x})$ only s states ($l = 0$) are influenced, since all other wave functions have a node at the nucleus (owing to the angular-momentum barrier).

The historically most important example is the $2s$ state in hydrogen ($Z = 1$, $n = 2$, $l = 0$), which should be energetically degenerated with the $2p$ state for a pure Coulomb potential. Equation (5.58) on the other hand predicts an energy shift of

$$\Delta E_{2s}^{\mathrm{VP}} = -1.122 \times 10^{-7}\mathrm{eV} \quad . \tag{5.59}$$

In spite of its tiny magnitude a shift like this is very precisely measurable by investigation of the energy difference to the nonshifted $2p_{1/2}$ state. In the hydrogen atom transitions between these states can be stimulated by electromagnetic fields in the radio-frequency region. According to (5.59) we would expect a resonance effect at the frequency

$$\nu = \frac{E_{2s} - E_{2p}}{2\pi\hbar} = -27.1\,\mathrm{MHz} \quad . \tag{5.60}$$

Experimentally, however, it was found by **Lamb** and collaborators[7] using microwave techniques that the $2s$ state lies *above* the $2p_{1/2}$ state, namely by (we quote the modern value)

$$\nu_{\mathrm{exp}} = +1057.8\,\mathrm{MHz} \quad . \tag{5.61}$$

Later we shall see that the "Lamb shift" in hydrogen is mainly caused by the other two radiative corrections of Fig. 5.7. Experimental and theoretical precision are, however, by far sufficient to confirm the presence of the vacuum polarization energy shift according to (5.60).

To increase the effect according to (5.58) it is necessary to increase the density of the wave function at the nucleus. This can be achieved by increasing the nuclear charge number Z. In the nonrelativistic approximation the shift (5.58) increases like Z^4, while the binding energy increases only in proportion to Z^2. Actually the increase of ΔE^{VP} is even larger, because one must use the bound-state solutions of the Dirac equation, which are much more strongly localized at the nucleus than the Schrödinger wave functions (5.57). This is partly offset by the necessity of taking into account the finite extension of the nucleus.

For the extreme example of a hypothetical atom with $Z = 170$ the calculated vacuum polarization energy shift of the lowest bound state is $\Delta E_{1s}^{\mathrm{VP}} = -8$ keV. This amounts to about 1% of the binding energy, up from a fraction of 10^{-7} in hydrogen. Experimentally the effect of vacuum polarization is most clearly seen in muonic atoms. This is explained in Example 5.4.

The short-range delta-function polarization potential $\Delta A_0(\boldsymbol{x})$ in (5.56) rests on the approximation $\Pi^{\mathrm{R}}(q^2) \simeq -(e^2/15\pi)(q^2/m^2)$ for the polarization function (5.41). The task of Exercise 5.3 is to investigate the polarization potential induced by the charge of a nucleus more exactly.

[7] W.E. Lamb and R.C. Retherford: Phys. Rev. **72**, 241 (1947).

EXERCISE

5.1 Evaluation of an Integral

Problem. Perform the steps leading from the integral

$$I := \int_0^\infty \frac{d\varrho}{\varrho} \sum_{i=0}^N C_i \exp\left[\mathrm{i}\varrho(-M_i^2 + \beta_1\beta_2 q^2)\right] \quad (1)$$

in (5.34) to (5.35) in detail.

Solution. Since the sum in the integrand is finite, we write, with the abbreviation $M_i^2 - \beta_1\beta_2 q^2 =: B_i$

$$I = \sum_{i=0}^N C_i \int_0^\infty \frac{d\varrho}{\varrho} \mathrm{e}^{-\mathrm{i}\varrho B_i} =: \sum_{i=0}^N C_i I_i \quad (2)$$

The integral I_i is not well defined mathematically since the integrand diverges at the lower boundary. For this reason we replace I_i by

$$I_i = \lim_{\eta\to 0} \int_\eta^\infty \frac{d\varrho}{\varrho} \mathrm{e}^{-\mathrm{i}\varrho B_i} \quad (3)$$

with the understanding that the limit $\eta \to 0$ is to be taken only at the end of the calculation.

For further evaluation we now consider the following contour integral in the complex plane:

$$J_i := \int_C \frac{dz}{z} \mathrm{e}^{-\mathrm{i}zB_i} \quad , \quad (4)$$

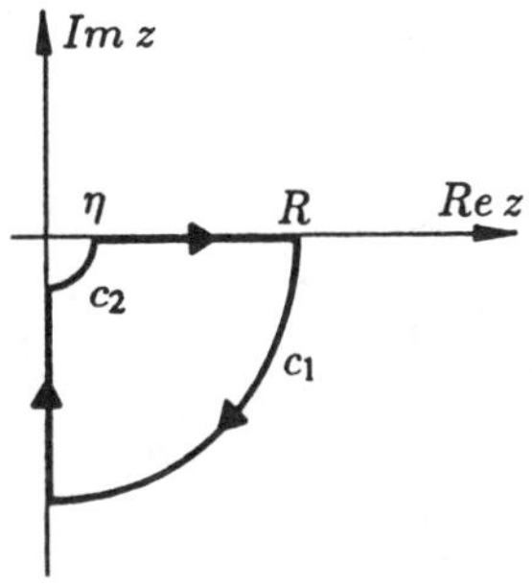

Fig. 5.12. The integration contour in the complex z-plane

which is extended along the curve C, as shown in Fig. 5.12. Obviously this integral can be split into four parts:

$$J_i = \int_\eta^R \frac{d\varrho}{\varrho} \mathrm{e}^{-\mathrm{i}\varrho B_i} + \int_{-\mathrm{i}R}^{-\mathrm{i}\eta} \frac{d\varrho}{\varrho} \mathrm{e}^{-\mathrm{i}\varrho B_i} + \int_{C_1} \frac{dz}{z} \mathrm{e}^{-\mathrm{i}zB_i} + \int_{C_2} \frac{dz}{z} \mathrm{e}^{-\mathrm{i}zB_i} \quad , \quad (5)$$

where the last two integrals are extended along the quarter circles C_1 and C_2 as indicated in Fig. 5.12. Now according to the theorem of residues the integral vanishes, $J_i = 0$, because in the region enclosed by the curve C the integrand is a regular function. This is valid for arbitrary positive real values of η and R. We may therefore perform the limits $R \to \infty$ and $\eta \to 0$ and obtain

$$0 = I_i + \lim_{\eta\to 0} \lim_{R\to\infty} \int_{-\mathrm{i}R}^{-\mathrm{i}\eta} \frac{d\varrho}{\varrho} \mathrm{e}^{-\mathrm{i}\varrho B_i} + \lim_{R\to\infty} \int_{C_1} \frac{dz}{z} \mathrm{e}^{-\mathrm{i}zB_i} + \lim_{\eta\to 0} \int_{C_2} \frac{dz}{z} \mathrm{e}^{-\mathrm{i}zB_i} \quad . \quad (6)$$

Exercise 5.1.

Using the parametrization $z = Re^{i\phi}$ we can rewrite the integral along C_1 in the following way:

$$\int_{C_1} \frac{dz}{z} \exp(-izB_i) = i \int_{2\pi}^{3\pi/2} d\phi \exp(-iR\cos\phi\, B_i) \exp(R\sin\phi\, B_i) \quad . \tag{7}$$

We now assume that the masses M_i^2 can be chosen larger than $\beta_1\beta_2 q^2$, i.e. $B_i > 0$.

While the absolute value of the oscillating factor $\exp(-iR\cos\phi\, B_i)$ has the value 1, in the limit $R \to \infty$ the second factor is exponentially suppressed (except for a narrow region close to $\phi = 2\pi$ which can be neglected since its extension is of the order of $1/R$). Consequently the whole integral along C_1 equals zero. Similar considerations for the integral along C_2 yield the value $i\pi/2$ (after the limit $\eta \to 0$!). Thus we finally get

$$\begin{aligned} I_i &= -\lim_{\eta\to 0}\lim_{R\to\infty} \int_{-iR}^{-i\eta} \frac{d\varrho}{\varrho} e^{-i\varrho B_i} - i\frac{\pi}{2} \\ &= \lim_{\eta\to 0} \int_{-i\eta}^{-i\infty} \frac{d\varrho}{\varrho} e^{-i\varrho B_i} - i\frac{\pi}{2} \quad , \end{aligned} \tag{8}$$

where convergence at the upper boundary is unproblematic. With the substitution $t = i\varrho B_i$ the integral reads

$$I_i = \lim_{\eta\to 0} \int_{\eta B_i}^{\infty} \frac{dt}{t} e^{-t} - i\frac{\pi}{2} \tag{9}$$

with a positive lower boundary because $B_i > 0$ was assumed. Integration by parts yields

$$\begin{aligned} I_i &= \lim_{\eta\to 0} \left(\ln t e^{-t} \Big|_{\eta B_i}^{\infty} + \int_{\eta B_i}^{\infty} dt \ln t e^{-t} \right) - i\frac{\pi}{2} \\ &= \lim_{\eta\to 0} \left[-e^{-\eta B_i} \left(\ln\eta + \ln B_i \right) + \int_{\eta B_i}^{\infty} dt \ln t e^{-t} \right] - i\frac{\pi}{2} \\ &= -\lim_{\eta\to 0} \ln\eta - \ln(M_i^2 - \beta_1\beta_2 q^2) + \int_0^{\infty} dt \ln t e^{-t} - i\frac{\pi}{2} \quad . \end{aligned} \tag{10}$$

Summation over all i from 0 to N yields (5.35) up to the constant term $-i\pi/2$, which can be neglected because of the choice $\sum_{i=0}^{N} C_i = 0$, cf. (5.36).

EXERCISE

5.2 The Photon Polarization Function

Problem. (a) Derive an explicit expression for the renormalized photon polarization function $\Pi^{\mathrm{R}}(q^2)$ given in (5.39) which is valid for all values of q^2. Show that $\Pi^{\mathrm{R}}(q^2)$ obtains an imaginary part in the region $q^2 > 4m^2$ and give an explanation of this observation.

(b) Prove that the photon polarization function can be expressed in terms of its imaginary part alone according to the "subtracted dispersion relation"

$$\Pi^{\mathrm{R}}(q^2) = \frac{1}{\pi} q^2 \int_0^\infty dq'^2 \frac{\mathrm{Im}\Pi^{\mathrm{R}}(q'^2)}{q'^2(q'^2 - q^2 - \mathrm{i}\varepsilon)} \quad . \tag{1}$$

Solution. (a) The polarization function is given by the integral representation

$$\begin{aligned} \Pi^{\mathrm{R}}(q^2) &= \frac{2\alpha}{\pi} \int_0^1 d\beta \beta(1-\beta) \ln\left[1 - \beta(1-\beta)\frac{q^2}{m^2}\right] \\ &= -\frac{2\alpha}{\pi} \int_0^1 d\beta \left(\frac{1}{2}\beta^2 - \frac{1}{3}\beta^3\right) \frac{1}{1 - \beta(1-\beta)\frac{q^2}{m^2}} \left[-\frac{q^2}{m^2}(1-2\beta)\right] \quad , \end{aligned} \tag{2}$$

where in the second line the logarithm has been eliminated through integration by parts. The denominator in (2) can be simplified by transforming to the new integration variable $v = 2\beta - 1$, i.e. $\beta(1-\beta) = \frac{1}{4}(1 - v^2)$:

$$\begin{aligned} \Pi^{\mathrm{R}}(q^2) &= \frac{2\alpha}{\pi} \frac{q^2}{m^2} \int_{-1}^{+1} dv \tfrac{1}{4}(1+v)^2 \left(\tfrac{1}{2} - \tfrac{1}{6} - \tfrac{1}{6}v\right) \frac{-v}{1 - \frac{q^2}{4m^2}(1 - v^2)} \\ &= -\frac{\alpha}{\pi} \frac{q^2}{4m^2} \int_{-1}^{+1} dv \frac{\frac{1}{3}v + \frac{1}{2}v^2 - \frac{1}{6}v^4}{1 - \frac{q^2}{4m^2}(1 - v^2)} \\ &= -\frac{\alpha}{\pi} \int_0^1 dv \frac{v^2(1 - \frac{1}{3}v^2)}{v^2 + \frac{4m^2}{q^2} - 1} \quad . \end{aligned} \tag{3}$$

In the last step use has been made of the symmetry of the integrand. The integral (2) can be solved by elementary means. We remind the reader of the basic indefinite integral

$$\int \frac{dv}{v^2 - c} = \begin{cases} \dfrac{1}{2}\dfrac{1}{\sqrt{c}} \ln \left|\dfrac{v - \sqrt{c}}{v + \sqrt{c}}\right| & \text{for} \quad c > 0 \\ \dfrac{1}{\sqrt{-c}} \arctan \dfrac{v}{\sqrt{-c}} & \text{for} \quad c < 0 \end{cases} \quad . \tag{4}$$

Identifying $c = 1 - 4m^2/q^2$ we have to distinguish three separate regions of the squared momentum

Exercise 5.2.

$$\begin{array}{llcl} \text{Region I} & -\infty < q^2 < 0 & : & 1 < c < \infty \quad , \\ \text{Region II} & 0 < q^2 \leq 4m^2 & : & -\infty < c \leq 0 \quad , \\ \text{Region III} & 4m^2 < q^2 < \infty & : & 0 < c < 1 \quad . \end{array}$$

Region III is of particular interest since here one of the two poles of the integrand enters the integration interval $v \in [0, 1]$. To get a well-defined result we have to remember that there is a general prescription for treating such poles, going back to the definition of the Feynman propagator. By giving a small negative imaginary part to the mass $m \to m - \mathrm{i}\varepsilon$, the pole in (2), (3) is shifted into the upper half of the complex plane. Equivalently the integration contour can be modified by inserting an infinitesimal half-circle extending into the negative half plane. Then the definite integral consists of the principal part integral plus half the residue of the pole at $v_0 = \sqrt{c}$:

$$\begin{aligned} I_0 &= \int_0^1 dv \frac{1}{v^2 - c - \mathrm{i}\varepsilon} = \left[\int_0^{v_0-\varepsilon} + \int_{v_0+\varepsilon}^1 \right] dv \frac{1}{v^2 - c} + \frac{1}{2} 2\pi \mathrm{i} \, \mathrm{Res} \left. \frac{1}{v^2 - c} \right|_{v_0} \\ &= \mathrm{P} \int_0^1 dv \frac{1}{v^2 - c} + \pi \mathrm{i} \frac{1}{2v_0} \quad . \end{aligned} \tag{5}$$

This corresponds to the well-known identity

$$\frac{1}{x - \mathrm{i}\varepsilon} = \mathrm{P}\frac{1}{x} + \mathrm{i}\pi\delta(x) \quad . \tag{6}$$

Thus we obtain in the three regions of interest

$$I_0 = \begin{cases} \frac{1}{\sqrt{c}} \frac{1}{2} \ln \frac{\sqrt{c}-1}{\sqrt{c}+1} & \text{for} \quad 1 < c < \infty \\ \frac{1}{\sqrt{-c}} \arctan \frac{1}{\sqrt{-c}} & \text{for} \quad -\infty < c \leq 0 \\ \frac{1}{\sqrt{c}} \frac{1}{2} \ln \frac{1-\sqrt{c}}{1+\sqrt{c}} + \mathrm{i} \frac{\pi}{2\sqrt{c}} & \text{for} \quad 0 < c < 1 \end{cases} \quad . \tag{7}$$

The integrals of the class

$$I_n = \int_0^1 dv \frac{v^n}{v^2 - c - \mathrm{i}\varepsilon} \tag{8}$$

needed in (3) can be traced back to I_0 using the following obvious recursion relation

$$I_n - cI_{n-2} = \int_0^1 dv v^{n-2} = \frac{1}{n-1} \quad . \tag{9}$$

Using (7) and (9) we obtain the following result for the polarization function

$$\Pi^{\mathrm{R}}(q^2) = \frac{\alpha}{\pi} \left[-\frac{5}{9} - \frac{4}{3} \frac{m^2}{q^2} + \frac{1}{3} \left(1 + \frac{2m^2}{q^2} \right) f(q^2) \right] , \tag{10}$$

with the abbreviation

Exercise 5.2.

$$f(q^2) = \sqrt{1-\frac{4m^2}{q^2}}\ln\frac{\sqrt{1-\frac{4m^2}{q^2}}+1}{\sqrt{1-\frac{4m^2}{q^2}}-1} \quad , \quad q^2<0$$

$$= 2\sqrt{\frac{4m^2}{q^2}-1}\arctan\frac{1}{\sqrt{\frac{4m^2}{q^2}-1}} \quad , \quad 0<q^2\leq 4m^2$$

$$= \sqrt{1-\frac{4m^2}{q^2}}\ln\frac{1+\sqrt{1-\frac{4m^2}{q^2}}}{1-\sqrt{1-\frac{4m^2}{q^2}}} - \mathrm{i}\pi\sqrt{1-\frac{4m^2}{q^2}} \quad , \quad 4m^2<q^2 \quad . \tag{11}$$

The Fig. 5.13 shows the function $\Pi^{\mathrm{R}}(q^2)$ multiplied by π/α. The function is smooth at $q^2=0$ but shows an *algebraic singularity* at the momentum $q^2=4m^2$. At this point of discontinuity an imaginary part $\mathrm{Im}\Pi^{\mathrm{R}}(q^2)$ emerges.

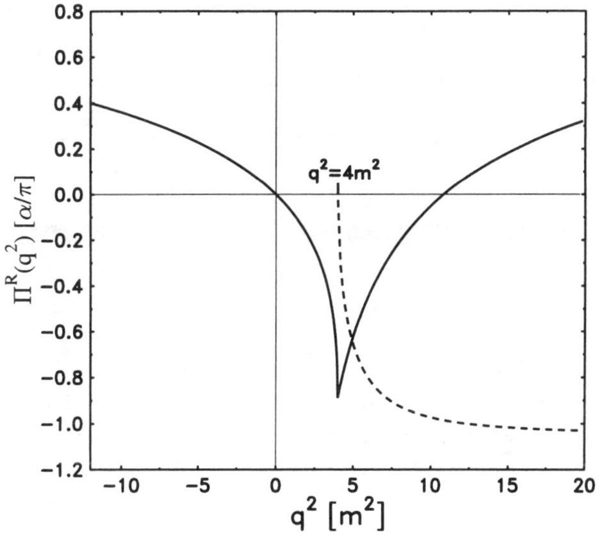

Fig. 5.13. The real part (full line) and imaginary part (dashed line) of the photon polarization function $\Pi^{\mathrm{R}}(q^2)$ in units of α/π drawn as a function of the squared momentum q^2

At large momentum the polarization function rises logarithmically:

$$\Pi^{\mathrm{R}}(q^2) \simeq \frac{\alpha}{\pi}\left(-\frac{5}{9}+\frac{1}{3}\ln\frac{|q^2|}{m^2}-\frac{1}{3}\mathrm{i}\pi\Theta(q^2-4m^2)\right) \quad . \tag{12}$$

The imaginary part of Π^{R} can be understood as follows: The value $q^2=4m^2$ is the threshold for the *production of real electron-positron pairs* by the electromagnetic field. The quantity $1+\Pi^{\mathrm{R}}$ plays the role of the *dielectric function of the vacuum.* As in the case of macroscopic polarizable media a negative imaginary part of the dielectric function signals the absorption of electromagnetic radiation. In our case the intensity of the A_μ-field is diminished while at the same time e^+e^--pairs are produced.

Exercise 5.2.

(b) The validity of (1) is easily verified. We insert

$$\mathrm{Im}\Pi^{\mathrm{R}}(q'^2) = -\frac{\alpha}{\pi}\frac{1}{3}\left(1+\frac{2m^2}{q'^2}\right)\sqrt{1-\frac{4m^2}{q'^2}}\pi\Theta(q'^2-4m^2) \tag{13}$$

from (10) and perform the variable transformation $q'^2 = 4m^2/(1-v^2)$ i.e. $v = \sqrt{1-m^2/q'^2}$. This immediately leads to

$$\frac{1}{\pi}q^2\int\limits_{4m^2}^{\infty} dq'^2 \frac{\mathrm{Im}\Pi^{\mathrm{R}}(q'^2)}{q'^2(q'^2-q^2-\mathrm{i}\varepsilon)} = -\frac{\alpha}{\pi}\int\limits_0^1 dv \frac{v^2(1-\frac{1}{3}v^2)}{v^2+\frac{4m^2}{q^2}-1-\mathrm{i}\varepsilon} = \Pi^{\mathrm{R}}(q^2) \tag{14}$$

according to (3). The name *subtracted* dispersion relation derives from the fact that (1) can be decomposed into partial fractions as follows

$$\Pi^{\mathrm{R}}(q^2) = \frac{1}{\pi}\int\limits_0^{\infty} dq'^2\,\mathrm{Im}\Pi^{\mathrm{R}}(q'^2)\left(\frac{1}{q'^2-q^2-\mathrm{i}\varepsilon} - \frac{1}{q'^2}\right) \quad . \tag{15}$$

Here a q^2-independent term is subtracted in order to make the integral finite. Without this subtraction the integral would diverge logarithmically. The dispersion relation (1) can be deduced from Cauchy's integral formula, taking proper account of the behaviour of $\Pi^{\mathrm{R}}(q^2)$ in the complex q^2 plane.

EXERCISE

5.3 The Uehling Potential

Problem. Calculate the potential generated by a given external point charge $-Ze$, taking into account the polarizability of the vacuum. What are the deviations from the Coulomb potential close to the charge centre and at a large distance? Find the polarization charge density induced in the vacuum. Hint: Use the identity

$$\left(\nabla^2-\mu^2\right)\frac{\mathrm{e}^{-\mu r}}{r} = -4\pi\delta^3(\boldsymbol{x}) \quad . \tag{1}$$

Solution. The potential is generated by a stationary external charge of density

$$j_\mu(x) = -Ze\,\delta^3(\boldsymbol{x})\,\delta_{\mu 0} \quad . \tag{2}$$

With the aid of the modified photon propagator D'_{F} from (5.6) this can be used to calculate the potential, taking into account the effect of vacuum polarization. It is again convenient to work in momentum space:

$$\begin{aligned} A'_\mu(x) &= \int d^4y\, D'_{\mathrm{F}\mu\nu}(x-y)\,j_\nu(y) \\ &= \int \frac{d^4q}{(2\pi)^4}\,\mathrm{e}^{-\mathrm{i}q\cdot x} D'_{\mathrm{F}\mu\nu}(q) j^\nu(q) \quad . \end{aligned} \tag{3}$$

The modified photon propagator can be expressed in terms of the renormalized vacuum polarization function. In momentum space this reads

Exercise 5.3.

$$
\begin{aligned}
D'_{\mathrm{F}\mu\nu}(q) &= D_{\mathrm{F}\mu\nu}(q) - D_{\mathrm{F}\mu\lambda}(q)\frac{\Pi^{\lambda\sigma}(q)}{4\pi}D_{\mathrm{F}\sigma\nu}(q) \\
&= \frac{-4\pi g_{\mu\nu}}{q^2} - \frac{-4\pi g_{\mu\lambda}}{q^2}\frac{q^2 g^{\lambda\sigma} - q^\lambda q^\sigma}{4\pi}\Pi^{\mathrm{R}}(q^2)\frac{-4\pi g_{\sigma\nu}}{q^2} \\
&= \frac{-4\pi g_{\mu\nu}}{q^2}\left(1 + \Pi^{\mathrm{R}}(q^2)\right) \quad . \qquad (4)
\end{aligned}
$$

Thus the modified potential (3) is

$$
\begin{aligned}
A'_\mu(x) &= \int \frac{d^4q}{(2\pi)^4}\, \mathrm{e}^{-\mathrm{i}q\cdot x}\left(1 + \Pi^{\mathrm{R}}(q^2)\right)D_{\mathrm{F}\mu\nu}(q)j^\nu(q) \\
&= \int \frac{d^4q}{(2\pi)^4}\, \mathrm{e}^{-\mathrm{i}q\cdot x}\left(1 + \Pi^{\mathrm{R}}(q^2)\right)A_\mu(q) \quad . \qquad (5)
\end{aligned}
$$

since the unmodified potential in momentum space is related to the current through $A_\mu(q) = D_{\mathrm{F}\mu\nu}(q)j^\nu(q)$. If the current source is stationary, $j^\nu(x) = j^\nu(\boldsymbol{x})$, the q_0-dependence in (5) drops out according to

$$
\begin{aligned}
j^\nu(q) &= \int d^4y\, \mathrm{e}^{\mathrm{i}q\cdot y} j^\nu(y) = \int dy_0\, \mathrm{e}^{\mathrm{i}q_0 y_0}\int d^3y\, \mathrm{e}^{-\mathrm{i}\boldsymbol{q}\cdot\boldsymbol{y}} j^\nu(\boldsymbol{y}) \\
&= 2\pi\delta(q_0) j^\nu(\boldsymbol{q}) \quad . \qquad (6)
\end{aligned}
$$

Thus (5) is reduced to

$$
\begin{aligned}
A'_\mu(\boldsymbol{x}) &= \int \frac{d^3q}{(2\pi)^3}\, \mathrm{e}^{\mathrm{i}\boldsymbol{q}\cdot\boldsymbol{x}}\left(1 + \Pi^{\mathrm{R}}(-\boldsymbol{q}^2)\right)D_{\mathrm{F}\mu\nu}(0,\boldsymbol{q})j^\nu(\boldsymbol{q}) \\
&= \int \frac{d^3q}{(2\pi)^3}\, \mathrm{e}^{\mathrm{i}\boldsymbol{q}\cdot\boldsymbol{x}}\left(1 + \Pi^{\mathrm{R}}(-\boldsymbol{q}^2)\right)A_\mu(\boldsymbol{q}) \quad . \qquad (7)
\end{aligned}
$$

In the case of the electrostatic point charge (2) we have

$$
j^\nu(\boldsymbol{q}) = \int d^3y\, \mathrm{e}^{-\mathrm{i}\boldsymbol{q}\cdot\boldsymbol{y}}\left(-Ze\,\delta_{\nu 0}\,\delta^3(\boldsymbol{y})\right) = -Ze\,\delta_{\nu 0} \quad , \qquad (8)
$$

so that (7) becomes

$$
A'_\mu(\boldsymbol{x}) = -Ze\int \frac{d^3q}{(2\pi)^3}\, \mathrm{e}^{\mathrm{i}\boldsymbol{q}\cdot\boldsymbol{x}}\left(1 + \Pi^{\mathrm{R}}(-\boldsymbol{q}^2)\right)D_{\mathrm{F}\mu 0}(0,\boldsymbol{q}) \qquad (9)
$$

or, with the insertion of the polarization function (5.39),

$$
\begin{aligned}
A'_0(\boldsymbol{x}) &= -Ze\int \frac{d^3q}{(2\pi)^3}\, \mathrm{e}^{\mathrm{i}\boldsymbol{q}\cdot\boldsymbol{x}}\,\frac{4\pi}{\boldsymbol{q}^2} \\
&\times\left[1 + \frac{2\alpha}{\pi}\int_0^1 d\beta\, \beta(1-\beta)\ln\left(1 + \frac{\boldsymbol{q}^2}{m^2}\beta(1-\beta)\right)\right] \quad . \qquad (10)
\end{aligned}
$$

Here we have tacitly assumed that charge renormalization has already been performed, i.e. e is the renormalized charge. From the identity (with $r \equiv |\boldsymbol{x}|$)

$$
\int \frac{d^3q}{(2\pi)^3}\frac{\exp(\mathrm{i}\boldsymbol{q}\cdot\boldsymbol{x})}{\boldsymbol{q}^2} = \frac{1}{4\pi r} \qquad (11)
$$

Exercise 5.3.

the Fourier transform of the first term of course yields just the ordinary Coulomb potential. When evaluating the correction term it is convenient to eliminate the logarithm by partial integration and to introduce the new variable of integration $v = 2\beta - 1$. According to (3) in Exercise 5.2 this leads to

$$\Pi^{\mathrm{R}}(-q^2) = \frac{\alpha}{\pi}\frac{q^2}{4m^2}\int\limits_0^1 dv \frac{v^2(1-\frac{1}{3}v^2)}{1+\frac{q^2}{4m^2}(1-v^2)} \quad . \tag{12}$$

After exchanging the order of integration in (10) the Fourier integral can be solved. To do this we make use of the formula

$$\int \frac{d^3q}{(2\pi)^3}\frac{\exp(\mathrm{i}\boldsymbol{q}\cdot\boldsymbol{x})}{q^2+a^2} = \frac{1}{4\pi}\frac{\exp(-ar)}{r} \quad , \tag{13}$$

which is easily proved by residue integration. The potential then reads

$$\begin{aligned} A_0'(\boldsymbol{x}) &= -Ze\left[\frac{1}{r} + \frac{2\alpha}{\pi}\frac{4\pi}{8m^2}\int\limits_0^1 dv\, v^2\left(1-\tfrac{1}{3}v^2\right)\int \frac{d^3q}{(2\pi)^3}\frac{\exp(\mathrm{i}\boldsymbol{q}\cdot\boldsymbol{x})}{1+\frac{q^2}{4m^2}(1-v^2)}\right] \\ &= -\frac{Ze}{r}\left[1+\frac{\alpha}{\pi}\int\limits_0^1 dv \frac{v^2(1-\frac{1}{3}v^2)}{1-v^2}\exp\left(-\frac{2m}{\sqrt{1-v^2}}r\right)\right] \quad . \end{aligned} \tag{14}$$

A further transformation $\zeta = (1-v^2)^{-1/2}$ or $v^2 = 1 - 1/\zeta^2$ with $v\,dv = \zeta^{-3}d\zeta$ simplifies the exponent:

$$\begin{aligned} A_0'(r) &= -\frac{Ze}{r}\left[1+\frac{\alpha}{\pi}\frac{1}{3}\int\limits_1^\infty \frac{d\zeta}{\zeta^3} v(3-v^2)\mathrm{e}^{-2m\zeta r}\right] \\ &= -\frac{Ze}{r}\left[1+\frac{2\alpha}{3\pi}\int\limits_1^\infty d\zeta\left(1+\frac{1}{2\zeta^2}\right)\frac{\sqrt{\zeta^2-1}}{\zeta^2}\mathrm{e}^{-2m\zeta r}\right] \quad . \end{aligned} \tag{15}$$

This is the commonly used integral representation for the *Uehling potential.* The integral cannot be evaluated in closed form but is easily solved numerically. To get an impression of the behaviour of the correction term one can obtain *asymptotic expressions* for the cases $mr \ll 1$ and $mr \gg 1$.

$mr \ll 1$. We first split the integral (15) at the point τ into two parts:

$$\int\limits_1^\infty d\zeta\,\mathrm{e}^{-2mr\zeta}\frac{\sqrt{\zeta^2-1}}{\zeta^2} = \int\limits_1^\tau \cdots + \int\limits_\tau^\infty \cdots \equiv I_1 + I_2 \quad , \tag{16}$$

where we choose $1/mr \gg \tau \gg 1$. Then in I_1 the exponential can be assumed constant ($\exp(-2mr\zeta) \approx 1$),

$$\begin{aligned} I_1 &\simeq \int\limits_1^\tau d\zeta \frac{\sqrt{\zeta^2-1}}{\zeta^2} = \left[-\frac{\sqrt{\zeta^2-1}}{\zeta} + \ln\left(\zeta+\sqrt{\zeta^2-1}\right)\right]_1^\tau \\ &\simeq \ln(2\tau) - 1 \quad . \end{aligned} \tag{17}$$

In the second integral one can approximate $\sqrt{\zeta^2 - 1} \simeq \zeta$, and thus

Exercise 5.3.

$$I_2 \simeq \int_\tau^\infty \frac{d\zeta}{\zeta}\, \mathrm{e}^{-2mr\zeta} \quad . \tag{18}$$

After partial integration the integral becomes convergent at the lower boundary:

$$\begin{aligned} I_2 &\simeq \mathrm{e}^{-2mr\zeta} \ln\zeta \big|_\tau^\infty + 2mr \int_\tau^\infty d\zeta \ln\zeta\, \mathrm{e}^{-2mr\zeta} \\ &= -\,\mathrm{e}^{-2mr\tau} \ln\tau + \int_{2mr\tau}^\infty du \ln\frac{u}{2mr}\, \mathrm{e}^{-u} \\ &\simeq -\ln\tau + \int_0^\infty du \ln u\, \mathrm{e}^{-u} + \ln\frac{1}{2mr} \int_0^\infty du\, \mathrm{e}^{-u} \\ &= -\ln\tau - C - \ln 2 + \ln\frac{1}{mr} \quad . \end{aligned} \tag{19}$$

Here the well-known definite integral $\int_0^\infty du \ln u \exp(-u)$ occurs, the value of which is equal to the negative of Euler's constant $C = 0.5772\ldots$, and thus

$$I_2 \simeq -\ln 2\tau + \ln\frac{1}{mr} - C \quad . \tag{20}$$

As intended, the dependence on the value of τ cancels when I_1 and I_2 are added. The second integral in (15), which has not yet been taken into account in (16), does not cause any trouble, because it converges for large ζ even without help from the exponential function. One can then approximate as follows:

$$\int_1^\infty d\zeta\, \mathrm{e}^{-2mr\zeta} \frac{\sqrt{\zeta^2-1}}{2\zeta^4} \simeq \int_1^\infty d\zeta \frac{\sqrt{\zeta^2-1}}{2\zeta^4} = \frac{1}{2} \frac{(\zeta^2-1)^{3/2}}{3\zeta^3}\bigg|_1^\infty = \frac{1}{6} \quad . \tag{21}$$

Then the asymptotic approximation for the potential reads

$$A_0'(r) \simeq -\frac{Ze}{r}\left[1 + \frac{2\alpha}{3\pi}\left(\ln\frac{1}{mr} - \frac{5}{6} - C\right)\right] \tag{22}$$

for $mr \ll 1$.

$mr \gg 1$. Here only the region $0 \le \zeta - 1 \ll 1/mr$ contributes to the integral, so that one can approximate $\zeta \simeq 1$ at various places:

$$\begin{aligned} \int_1^\infty d\zeta \left(1 + \frac{1}{2\zeta^2}\right) \frac{\sqrt{\zeta^2-1}}{\zeta^2} \mathrm{e}^{-2mr\zeta} &\simeq \int_1^\infty d\zeta \frac{3\sqrt{2}}{2} \sqrt{\zeta - 1}\, \mathrm{e}^{-2mr\zeta} \\ &= \frac{3\sqrt{2}}{2} \mathrm{e}^{-2mr} \int_0^\infty d\zeta' \sqrt{\zeta'}\, \mathrm{e}^{-2mr\zeta'} \quad . \end{aligned} \tag{23}$$

Exercise 5.3.

By use of the integral representation of the gamma function,

$$\int_0^\infty dt\, e^{-at} t^{z-1} = a^{-z}\Gamma(z) \quad , \tag{24}$$

and $\Gamma(3/2) = \sqrt{\pi}/2$ we get

$$A_0'(r) \simeq -\frac{Ze}{r}\left[1 + \frac{\alpha}{4\sqrt{\pi}}\frac{e^{-2mr}}{(mr)^{3/2}}\right] \tag{25}$$

for $mr \gg 1$.

We have thus found that an electron in the field of a point-like positive charge feels the interaction

$$eA_0'(r) = -\frac{Z\alpha}{r}Q(r) \quad , \tag{26}$$

where $Q(r)$ is a function which tends to 1 very quickly for large distances, according to (25). The deviations are perceptible only if r is smaller than about one Compton wavelength of the electron. The approximation (5.56) used earlier describes this effect only in the mean. The function $eQ(r)$ can be considered as an effective coupling "constant", which increases at small distances. It is also known by the name *running coupling constant.* Figure 5.14 shows the function $Q(r)$, obtained by numerical integration of (15). The logarithmic increase of the effective interaction strength at small r is clearly visible. However, only at extremely small distances does the function $Q(r)$ deviate considerably from the value 1.

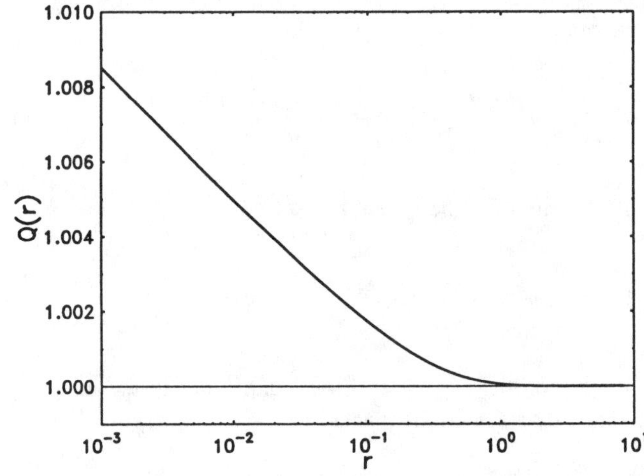

Fig. 5.14. The effective coupling strength $Q(r)$ as a function of distance r in units of the Compton wavelength

From the knowledge of the potential $A_0(\boldsymbol{x})$ the corresponding *polarization charge density* can be calculated using Poisson's equation. Using the integral representation (15) of the Uehling potential the vacuum polarization charge distribution is found to be

Exercise 5.3.

$$\begin{aligned}\rho_{\mathrm{VP}}(\boldsymbol{x}) &= -\frac{1}{4\pi}\nabla^2 A_0^{\mathrm{VP}} \\ &= -\frac{1}{4\pi}(-Ze)\frac{2\alpha}{3\pi}\int\limits_1^\infty d\zeta\left(1+\frac{1}{2\zeta^2}\right)\frac{\sqrt{\zeta^2-1}}{\zeta^2}\nabla^2\left(\frac{\mathrm{e}^{-2m\zeta r}}{r}\right) \quad . \qquad (27)\end{aligned}$$

Employing the identity (1) we find

$$\rho_{\mathrm{VP}}(\boldsymbol{x}) = -Ze\frac{2\alpha}{3\pi}\int\limits_1^\infty d\zeta\left(1+\frac{1}{2\zeta^2}\right)\frac{\sqrt{\zeta^2-1}}{\zeta^2}\left[\delta^3(\boldsymbol{x}) - \frac{m^2}{\pi}\zeta^2\frac{\mathrm{e}^{-2m\zeta r}}{r}\right] \quad . \qquad (28)$$

The induced vacuum charge thus consists of two components: There is a *positive* charge localized at $\boldsymbol{x} = 0$ and a *negative* charge cloud which extends over a region of the size of the Compton wavelength $1/m$. This is shown in Fig. 5.15a. Fig. 5.15b shows a schematic representation of the distribution of e^+e^- pairs induced in the vacuum. The behaviour of the induced charge cloud thus is just opposite to what one is used from the case of ordinary polarizable media! One would expect that a dipole layer is formed in which the opposite (i.e. negative) charge is located at the inside and the like (i.e. positive) charge at the outside. We will come back to the apparently paradoxical behaviour of the vacuum polarization charge at the end of Sect. 7.1.

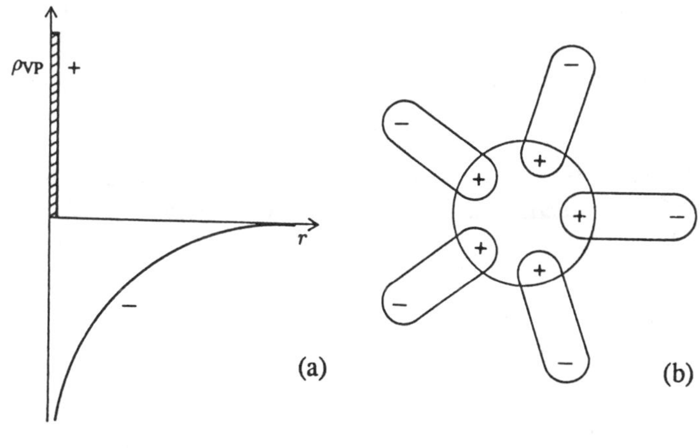

Fig. 5.15. (a) The vacuum polarization charge density $\rho_{\mathrm{VP}}(r)$ induced by a positively charged external point source. It consists of a positive part localized at the position of the source and a negatively charged extended polarization cloud. **(b)** Schematic drawing of the electron-positron pairs induced in the vacuum around an extended positive source.

The total induced vacuum charge is found to vanish. This is immediately seen using the Poisson equation and Gauss' theorem

$$\begin{aligned}Q_{\mathrm{VP}} &= \int d^3x\, \rho_{\mathrm{VP}}(\boldsymbol{x}) = -\frac{1}{4\pi}\int d^3x\, \nabla\cdot\left(\nabla A_0^{\mathrm{VP}}\right) \\ &= -\frac{1}{4\pi}\int d\boldsymbol{o}\cdot\nabla A_0^{\mathrm{VP}} = 0 \quad , \qquad (29)\end{aligned}$$

since according to (25) A_0^{VP} falls off faster than $1/r$ at large distances.

The result (28) has to be interpreted with care since the two contributions which cancel each other when calculating the total charge are *divergent* when taken separately. The $\delta^3(\boldsymbol{x})$ term is multiplied by an integral which diverges logarithmically at

Exercise 5.3.

the upper boundary $\zeta \to \infty$. The negative charge cloud at small distances behaves as

$$\begin{aligned}\rho_{\mathrm{VP}}^{(-)}(r) &= Ze\frac{2\alpha}{3\pi}\frac{m^2}{\pi}\frac{1}{r}\int_1^\infty d\zeta\left(1+\frac{1}{2\zeta^2}\right)\sqrt{\zeta^2-1}\,\mathrm{e}^{-2m\zeta r}\\ &\simeq Ze\frac{2\alpha}{3\pi}\frac{m^2}{\pi}\frac{1}{r}\int_1^\infty d\zeta\,\zeta\,\mathrm{e}^{-2m\zeta r}\\ &= Ze\frac{2\alpha}{3\pi}\frac{1}{4\pi}\frac{1}{r^3}\quad,\end{aligned} \tag{30}$$

which also can be deduced by applying the radial Laplace operator to (22). The integral of $\rho_{\mathrm{VP}}^{(-)}(r)$ thus diverges logarithmically.

Additional Remarks. The Uehling potential and the corresponding vacuum charge distribution can also be calculated for an *extended source*. In this case ρ_{VP} is a smooth function and no divergent quantities are found. We consider an electrostatic spherically symmetric source

$$j_0(\boldsymbol{x}) \equiv \rho(r) = -Zeh(r)\quad, \tag{31}$$

where $h(r)$ is a smooth distribution function which is normalized according to

$$\int d^3x\ h(r) = 4\pi\int_0^\infty dr\ r^2 h(r) = 1\quad. \tag{32}$$

The vacuum polarization potential at $\boldsymbol{x}$ then is obtained by folding the point-source Uehling potential with the extended source distribution. In momentum space this folding corresponds to a simple multiplication. According to (7) we have

$$A_0^{\mathrm{VP}}(\boldsymbol{x}) = \int\frac{d^3q}{(2\pi)^3}\,\mathrm{e}^{\mathrm{i}\boldsymbol{q}\cdot\boldsymbol{x}}\,\Pi^{\mathrm{R}}(-\boldsymbol{q}^2)\,\frac{4\pi}{\boldsymbol{q}^2}\,\tilde{\rho}(\boldsymbol{q})\quad, \tag{33}$$

where

$$\tilde{\rho}(\boldsymbol{q}) = \int d^3x\ \mathrm{e}^{-\mathrm{i}\boldsymbol{q}\cdot\boldsymbol{x}}\rho(\boldsymbol{x}) \tag{34}$$

is the Fourier transform of the charge distribution of the source. If $\rho(\boldsymbol{x})$ is spherically symmetric $\tilde{\rho}(\boldsymbol{q})$ will depend only on the absolute value of the momentum and the angular integration in (33) can be carried out:

$$A_0^{\mathrm{VP}}(r) = \frac{2}{\pi}\int d|\boldsymbol{q}|\,\frac{\sin|\boldsymbol{q}|r}{|\boldsymbol{q}|r}\,\Pi^{\mathrm{R}}(-\boldsymbol{q}^2)\,\tilde{\rho}(|\boldsymbol{q}|)\quad. \tag{35}$$

The renormalized polarization function $\Pi^{\mathrm{R}}(-\boldsymbol{q}^2)$ is known analytically (see (10) in Exercise 5.2), the remaining one-dimensional integral (35) in general has to be integrated numerically.

Exercise 5.3.

As an example let us consider the charge distribution of an extended atomic nucleus which can be approximated by the Fermi distribution function[8]

$$\rho(r) = \frac{\rho_0}{1 + e^{(r-R)/c}} \quad . \tag{36}$$

The parameter R characterizes the radius of the nucleus while c describes the "smearing" of its surface (the limit $c \to 0$ leads to a sharp surface). The function (36) can not be Fourier transformed in closed form but $\tilde{\rho}(q)$ can be expressed in terms of a rapidly converging series expansion.[9]

Figure 5.16 shows the Uehling potential induced by the extended source (36) for the example of the nucleus ^{208}Pb. The following values for the parameters were used: $Z = 82$, $R = 6.62$ fm, $c = 0.549$ fm. The dashed line shows the

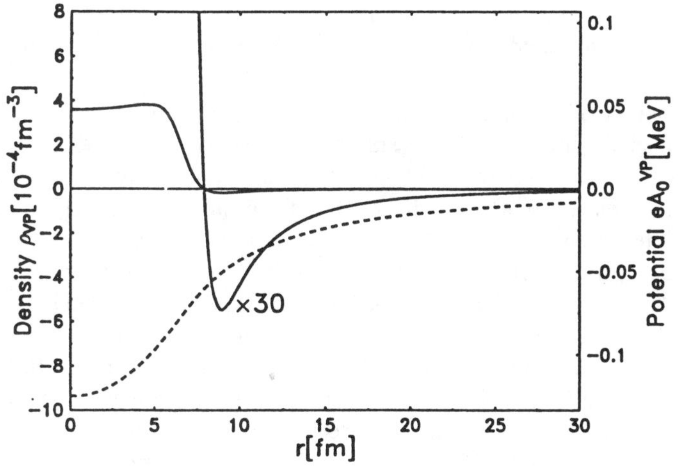

Fig. 5.16. The Uehling potential $eA_0^{VP}(r)$ induced by an extended ^{208}Pb nucleus (dashed line). The full lines show the corresponding vacuum polarization charge density $\rho_{VP}(r)$. It is positive inside the nucleus and negative outside

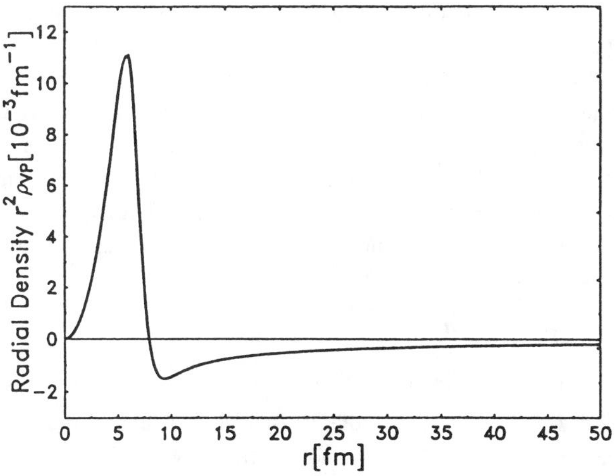

Fig. 5.17. The vacuum polarization charge density shown in Fig. 5.16 here is multiplied by the radial volume element, $r^2\rho_{VP}$. The total area under the curve has to vanish

[8] If a box distribution $\rho(r) = \rho_0\Theta(R - r)$ is used instead of (36) the induced vacuum charge turns out to be divergent at the nuclear surface.

[9] V. Hnizdo: J. Phys. **A21**, 3629 (1988).

Exercise 5.3.

Uehling potential eA_0^{VP} in units of MeV, the full line is the induced vacuum charge deduced with the help of Poisson's equation. The result is consistent with the earlier discussed case of a point-like source (see Fig. 5.14): In the interior of the source the charge density is enhanced by vacuum polarization. In the exterior there is a negative charge cloud (to make this better visible the curve has been multiplied by a factor 30 in the figure). The Fig. 5.17 shows the radial vacuum charge density $r^2\rho^{\text{VP}}$. The area under this curve vanishes when one integrates up to a distance of a few Compton wavelengths. Owing to the small value of α the total positive and negative induced charges are very small ($Q_{\text{VP}}^- = -Q_{\text{VP}}^+ = 0.037e$ compared to the charge of the source $Q = -82e$).

EXAMPLE

5.4 Muonic Atoms

A very effective method to increase the strength of vacuum polarization is to use negative muons (μ^-) instead of electrons as test charges. These particles obey – as far as we know – QED as strictly as electrons, differing from them, however, by their mass, which is larger by a factor $m_\mu/m_e = 206.77$. They decay by weak interaction[10] ($\mu^- \to e^- + \nu_\mu + \bar{\nu}_e$), but can be regarded as nearly stable, if one deals with electromagnetic processes, because of their long lifetime of 2.2×10^{-6}s.

Today muon beams can be produced in particle accelerators without great effort. Bombarding a target with high-energy protons produces π mesons, which decay into muons and neutrinos ($\pi^- \to \mu^- + \nu_\mu$) after about 2×10^{-8}s. Afterwards the muons are decelerated by collisions in matter. Finally a muon that has become sufficiently slow can be "captured" by an atom. Usually it is first captured in one of the outer shells (the main quantum number n is typically 14), from where it then goes down to the 1s level by a cascade of radiative transitions. In contrast to the free decay mentioned above it is then mostly captured by a proton of the nucleus ($\mu^- + p^+ \to n + \nu_\mu$). The cascade only needs a time of about $10^{-9} \dots 10^{-12}$s, which is very fast compared to the lifetime of the muon. In the transitions between the outer shells predominantly electrons are emitted (Auger transitions) while the transitions between inner shell orbits occur via photon emission. The X-rays that are emitted by these processes can be measured with high precision.

Muonic atoms are of particular interest for QED because the bound states are highly localized owing to the large muon mass. Indeed, the Bohr radius (with $\hbar = c = 1$),

$$a_{\text{B}}^{(\mu)} = \frac{n}{m_\mu(Z\alpha)} \quad , \tag{1}$$

is by a factor m_e/m_μ smaller than that of electronic atoms. It can therefore easily become smaller than the Compton wavelength of the electron $\lambda\!\!\!^{-}_e = 1/m_e$, which is the typical scale of vacuum fluctuations. Figure 5.18 gives an impression of the

[10] This is fully discussed in W. Greiner and B. Müller: Theoretical Physics, Vol. 5, *Gauge Theory of Weak Interactions* (Springer, Berlin, Heidelberg, New York 1993).

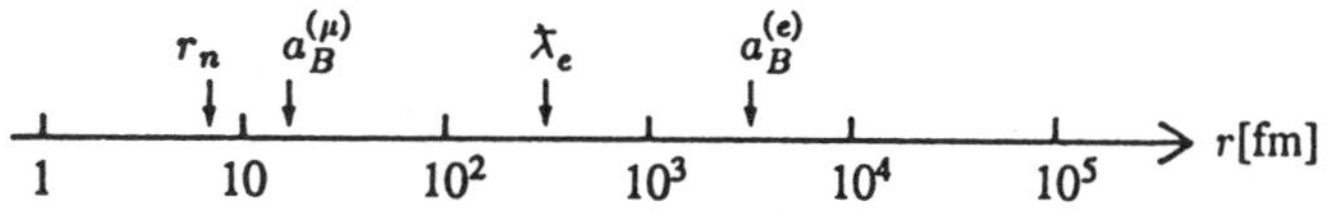

Fig. 5.18. The relevant length scales in a muonic atom

relevant scales in a heavy muonic atom. Here r_n stands for the nuclear radius; $a_{\mathrm{B}}^{(\mu)}$ and $a_{\mathrm{B}}^{(e)}$ are the Bohr radii of muons and electrons, respectively.

Muonic atoms therefore are the ideal tool for examining vacuum polarization. The simple estimate of the energy shift (5.58) is modified to

$$\Delta E_{nl}^{\mathrm{VP}} = -\alpha(Z\alpha)^4 \frac{4m_{\mathrm{e}}}{15\pi n^3} \delta_{l0} \left(\frac{m_\mu}{m_{\mathrm{e}}}\right)^3 \quad . \tag{2}$$

This follows immediately from (5.57). This simple approximation is not, however, sufficient to obtain adequate precision; one must use the exact form of the Uehling potential of Exercise 5.3, which requires a numerical integration.

Interestingly it turns out that the energy shift in muonic atoms is nearly exclusively due to vacuum polarization, in contrast to the Lamb shift in (electronic) hydrogen. The high precision of measurement even makes it possible to see contributions of higher order in $Z\alpha$. This is demonstrated in Table 5.1 for the example of a transition in muonic lead.

Table 5.1. Contributions to the $5g_{9/2} - 4f_{7/2}$ energy difference in eV in muonic lead

Dirac energy	$429\,339 \pm 2$
VP $\alpha Z\alpha$ Uehling	$2\,105 \pm 1$
VP $\alpha^2 Z\alpha$	15
VP $\alpha(Z\alpha)^{n\geq 3}$	-43
VP $\alpha^2(Z\alpha)^2$	1
other corrections	0 ± 2
electronic shielding	-82 ± 4
sum theory	$431\,336 \pm 6$
experiment*	$431\,331 \pm 8$

* T. Dubler et al.: Nucl. Phys. **A294**, 397 (1978).

The excellent agreement between theory and experiment is remarkable. This does not only confirm the (dominating) contribution of the Feynman graph of Fig. 5.10b, which is of the order of magnitude of $\alpha Z\alpha$: additionally, higher-order corrections to the photon propagator are confirmed, the graphs of which are displayed in Fig. 5.19. Note that the particles generated in the loops are electron–positron pairs. The contribution of $\mu^+\mu^-$ pairs is much smaller. (If there were only muons, we would of course have the same situation – except for the finite nuclear radius r_n! – as in electronic atoms, only the scale of energy would be enlarged by a factor of m_μ/m_{e}.)

The dominant higher-order contributions are due to the graphs that describe the multiple interaction of the e^+e^- loop with the field of the nucleus, i.e. $\alpha(Z\alpha)^3$,

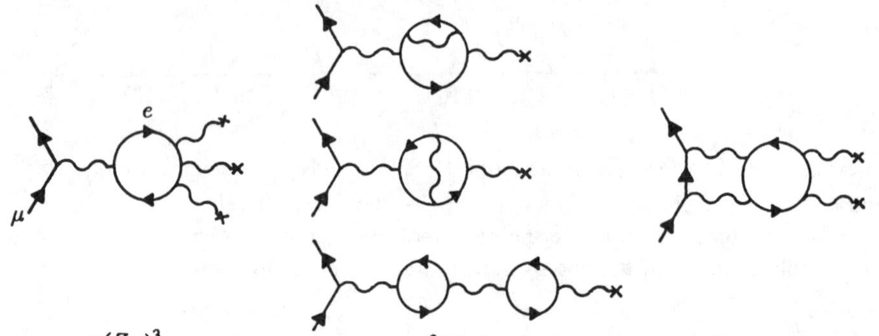

$\alpha(Z\alpha)^3$ $\alpha^2(Z\alpha)$ $\alpha^2(Z\alpha)^2$

Fig. 5.19. Typical higher-order vacuum polarization graphs contributing to the binding energy in muonic atoms

$\alpha(Z\alpha)^5$, etc. (because of Furry's theorem only odd powers contribute). Very extensive calculations are required to determine these higher-order contributions.

We remark finally that in heavy muonic atoms it is essential to take into account the finite extension of the nucleus. In particular, the lower orbitals do indeed pass through the interior of the nucleus to a large degree. Muons are therefore a sensitive probe for determining nuclear charge distributions and deformations.

MATHEMATICAL SUPPLEMENT

5.5 Gaussian Integrals

In the following we prove an integral formula needed in the calculation of radiative corrections:

$$\int d^4k \exp\left[\mathrm{i}(ak^2 + 2b\cdot k)\right] = \frac{\pi^2}{\mathrm{i}a^2}\exp\left(-\mathrm{i}b^2/a\right) \tag{1}$$

for $a > 0$. Equation (1) is split into four cartesian single integrals. The one-dimensional integral

$$\int_{-\infty}^{\infty} dx \exp\left[\mathrm{i}(ax^2 + 2bx)\right] \qquad (a > 0) \tag{2}$$

is subject to the transformation

$$x = \frac{1+\mathrm{i}}{\sqrt{2}}u - \frac{b}{a} \quad , \quad \mathrm{i}(ax^2 + 2bx) = -au^2 - \frac{\mathrm{i}b^2}{a} \quad , \tag{3}$$

i.e. a rotation by the angle $-\pi/4$ and a shift of the origin, which transforms (2) into a standard Gaussian integral:

$$\int_{-\infty}^{\infty} dx \exp\left[\mathrm{i}(ax^2 + 2bx)\right] = \frac{1+\mathrm{i}}{\sqrt{2}}\exp\left(-\mathrm{i}b^2/a\right)\int_{-\infty}^{\infty} du \exp\left(-au^2\right)$$

$$= \frac{1+\mathrm{i}}{\sqrt{2}}\sqrt{\frac{\pi}{a}}\exp\left(-\mathrm{i}b^2/a\right) \quad . \tag{4}$$

Mathematical Supplement 5.5.

Owing to the Minkowski metric in (1) we also need the complex conjugate relation

$$\int_{-\infty}^{\infty} \exp\left(-\mathrm{i}(ax^2 + 2bx)\right) dx = \frac{1-\mathrm{i}}{\sqrt{2}}\sqrt{\frac{\pi}{a}} \exp\left(+\mathrm{i}b^2/a\right) \quad . \tag{5}$$

The four-dimensional integral (1) then reads

$$\begin{aligned} &\int d^4k \exp\left[\mathrm{i}(ak^2 + 2b \cdot k)\right] \\ &= \int_{-\infty}^{\infty} dk_0 \exp\left[\mathrm{i}(ak_0^2 + 2b_0k_0)\right] \prod_{i=1}^{3} \int_{-\infty}^{\infty} dk_i \exp\left[-\mathrm{i}(ak_i^2 + 2b_ik_i)\right] \\ &= \frac{1+\mathrm{i}}{\sqrt{2}} \left(\frac{1-\mathrm{i}}{\sqrt{2}}\right)^3 \frac{\pi^2}{a^2} \exp\left[-\mathrm{i}(b_0^2 - \boldsymbol{b}^2)/a\right] \\ &= \frac{\pi^2}{\mathrm{i}a^2} \exp\left(-\mathrm{i}b^2/a\right) \quad . \end{aligned} \tag{6}$$

5.3 Self-Energy of the Electron

Just as the propagator of the photon was modified by the creation and subsequent annihilation of a pair, so also will the electron propagator be modified by the process where by a virtual photon is emitted and reabsorbed. We want to calculate how the undisturbed electron propagator

$$\mathrm{i}S_\mathrm{F}(p) = \frac{\mathrm{i}}{\not p - m + \mathrm{i}\varepsilon} \tag{5.62}$$

is modified by the Feynman graph of Fig. 5.7b. In order e^2 we obtain

$\mathrm{i}S'_\mathrm{F}(p) =$ 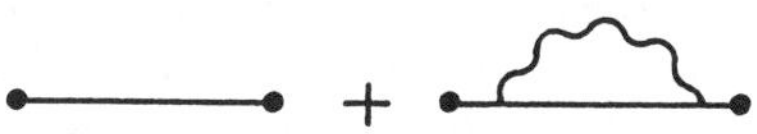

$$= \mathrm{i}S_\mathrm{F}(p) + \mathrm{i}S_\mathrm{F}(p)\,(-\mathrm{i}\Sigma(p))\,\mathrm{i}S_\mathrm{F}(p) \quad . \tag{5.63}$$

Here the *self-energy function*

$$-\mathrm{i}\Sigma(p) = (-\mathrm{i}e)^2 \int \frac{d^4k}{(2\pi)^4} \frac{-4\pi\mathrm{i}}{k^2 + \mathrm{i}\varepsilon} \gamma^\mu \frac{\mathrm{i}}{\not p - \not k - m + \mathrm{i}\varepsilon} \gamma_\mu \tag{5.64}$$

has been introduced (compare (5.3)). $\Sigma(p)$ is a 4×4 matrix in spinor space, in contrast to the polarization Lorentz tensor $\Pi_{\mu\nu}(q)$. One can decompose it into terms proportional to the unit matrix $\mathbb{1}$ and to $\not p$. However, the following decomposition is more convenient:

$$\Sigma(p) = A + B(\not{p} - m) + \Sigma^{\mathrm{R}}(p)(\not{p} - m)^2 \quad , \tag{5.65}$$

which in a sense corresponds to a Taylor expansion about the point $\not{p} = m$. The constants A, B and the residual term $\Sigma^{\mathrm{R}}(p)$ should not contain any γ matrices. The last term in (5.65) contains terms proportional to $\not{p}\not{p} = p^2$, proportional to $\not{p}$, and proportional to m^2. Such terms do of course already occur in the first two terms of (5.65). What is new about the third term is that these factors are multiplied with a momentum-dependent function (residual function) $\Sigma^{\mathrm{R}}(p)$. Before we perform the explicit calculation of (5.64) we examine the meaning of the first two contributions in (5.65). Using the identity (5.9), the "chain approximation" of (5.63) can be immediately summed up as a geometrical series. Similar to the photon propagator (5.10) this yields

$$\mathrm{i}S_{\mathrm{F}}'(p) = \frac{\mathrm{i}}{\not{p} - m - \Sigma(p) + \mathrm{i}\varepsilon} \quad . \tag{5.66}$$

Not only is (5.66) approximately valid up to order e^2, but it exactly sums up all graphs with arbitrarily many single photon lines put one after the other (the "chain approximation").

Let us now have a look at the modified propagator in the vicinity of $\not{p} = m$. This loose manner of speaking means that actually we are looking at matrix elements of S_{F}' between spinor wave functions that are on mass shell $p^2 = m^2$. Since the free spinors obey the Dirac equation $(\not{p} - m)u(p) = 0$, in the transition operator one can set $\not{p} = m$, if $\not{p}$ acts directly on a free spinor. This replacement would not be allowed for bound electrons, for which $\not{p}u(p) \neq mu(p)$.

From (5.66), together with (5.65), we get

$$\begin{aligned}
\mathrm{i}S_{\mathrm{F}}'(p) &= \frac{\mathrm{i}}{\not{p} - m - A - B(\not{p} - m) - (\not{p} - m)^2\Sigma^{\mathrm{R}}(p) + \mathrm{i}\varepsilon} \\
&\simeq \frac{\mathrm{i}}{(\not{p} - m - A)(1 - B)(1 - (\not{p} - m)\Sigma^{\mathrm{R}}(p)) + \mathrm{i}\varepsilon} \\
&\simeq \frac{(1 + B)\mathrm{i}}{(\not{p} - m - A)(1 - (\not{p} - m)\Sigma^{\mathrm{R}}(p)) + \mathrm{i}\varepsilon} \quad .
\end{aligned} \tag{5.67}$$

Here A, B, Σ^{R} were treated as small quantities of order e^2, and terms of higher order were neglected in several instances. In the first transformation, for instance, products like AB etc. have been dropped, and in the last line we have set $(1-B)^{-1} \simeq 1 + B$. Near the mass shell we can obviously neglect the momentum-dependent correction $(\not{p} - m)\Sigma^{\mathrm{R}}(p)$. Then the modified electron propagator reads simply

$$\mathrm{i}S_{\mathrm{F}}'(p) \simeq \frac{\mathrm{i}(1 + B)}{\not{p} - m - A + \mathrm{i}\varepsilon} \tag{5.68}$$

or

$$\mathrm{i}S_{\mathrm{F}}'(p) \simeq Z_2\, \mathrm{i}S_{\mathrm{F}}(p, m \to m + \delta m) \quad , \tag{5.69}$$

where the *electron renormalization constant*

$$Z_2 \equiv 1 + B \tag{5.70}$$

and the *self-energy*

$$\delta m \equiv A \tag{5.71}$$

have been introduced. The latter shifts the pole of the electron propagator from the value m to $m + \delta m$. This means that the electron, which originally had the *bare mass m*, now moves with the *physical mass*

$$m_{\mathrm{R}} = m + \delta m \quad , \tag{5.72}$$

if one takes into account the interaction with the self-generated electromagnetic field. Since it is completely impossible to switch off this interaction, the quantities m and δm separately do not have any physical significance, just as the bare charge had no physical significance in the case of charge renormalization (5.46). All observables, i.e. S-matrix elements, contain the renormalized mass $m_{\mathrm{R}} \simeq 0.511$ MeV. Thus we need not worry too much about the fact that δm will again prove to be a divergent expression. An analogous role is played by the factor Z_2 given by (5.70). It multiplies all electron propagators that occur in a diagram. Since internal electron lines are always located between two vertices with the factor $(-\mathrm{i}e\gamma_\mu)$ one can perform a *charge renormalization* to absorb the factors Z_2, namely

$$e'_{\mathrm{R}} = Z_2 e \quad . \tag{5.73}$$

In contrast to (5.46) there is no square root involved here, because there are always two electron lines that share a factor of e. The renormalization of the external lines again requires special consideration. Application of Feynman's rules to an external electron line of an arbitrary graph causes the self-energy replacement

$$u'(p) = u(p) + \mathrm{i}S_{\mathrm{F}}(p)\,(-\mathrm{i}\Sigma(p))\,u(p) \tag{5.74}$$

Fig. 5.20. Renormalization of an external fermion line

according to the graphs of Fig. 5.20. The direct application of this formula results in an undefined expression of the form $(\not{p} - m)^{-1}(\not{p} - m)u(p)$. However, since we know how to renormalize the propagator $S'_{\mathrm{F}}(p)$, and since according to (2.24) the latter is a quadratic function of the field amplitudes, we must perform a *wave-function renormalization*,

$$u'(p) = \sqrt{Z_2}\,u(p) \quad . \tag{5.75}$$

In the calculation of the matrix element of any Feynman diagram of arbitrary complexity the renormalization constant cancels out. Let us consider a graph of order n containing n_{e} external electron (or positron) lines. It is easy to see that there are $n - n_{\mathrm{e}}/2$ internal electron lines, each of which must be renormalized according to (5.69), taking into account the electron self-energy. This yields the renormalized matrix element

$$M'_{fi} \sim (Z_2)^{n-n_{\mathrm{e}}/2}\sqrt{Z_2}^{\,n_{\mathrm{e}}} M_{fi} = Z_2^n M_{fi} \quad , \tag{5.76}$$

which coincides exactly with the result one obtains by the renormalization of the electron charge according to (5.73) at each of the n vertices.

After these general considerations we are left with the task of explicitly calculating the self-energy function $\Sigma(p)$ and of identifying the terms of the expansion (5.65). As already mentioned, by counting the powers of k the integral (5.64) is linearly divergent; it must thus be regularized. However, (6.54) has the further problem of diverging at small values of k, too. For the time being this is not obvious; however, it will become clear in the following. To avoid this so-called *infrared divergence* we (just formally) attach a small mass μ to the photon. With this modification the self-energy function reads

$$\Sigma(p,\mu) = -4\pi \mathrm{i} e^2 \int \frac{d^4k}{(2\pi)^4} \frac{1}{k^2-\mu^2+\mathrm{i}\varepsilon} \gamma^\mu \frac{\not p - \not k + m}{(p-k)^2-m^2+\mathrm{i}\varepsilon} \gamma_\mu \quad . \tag{5.77}$$

(In the end, physical observables have to be calculated in the limit $\mu \to 0$. A detailed discussion of this problem will follow later.) In this way we will be prepared to face possible problems with the integrand of (5.77) for $k \to 0$.

To calculate (5.77) we again introduce the parameter representation (5.22) of the propagators. The numerator can be simplified according to the rules of the Mathematical Supplement 3.3, $\gamma^\mu(\not p - \not k + m)\gamma_\mu = -2\not p + 2\not k + 4m$. We remind the reader again of the idea of that short calculation: $\gamma^\mu \not p \gamma_\mu = p^\nu \gamma^\mu \gamma_\nu \gamma_\mu = p^\nu \gamma^\mu(-\gamma_\mu\gamma_\nu + 2g_{\mu\nu}) = -\gamma^\mu\gamma_\mu \not p + 2\not p = -4\not p + 2\not p = -2\not p$. If we again replace the factor proportional to k by a differentiation according to (5.24), we finally get the self-energy function, regularized by the Pauli–Villars method:

$$\begin{aligned}\bar\Sigma(p,\mu) = 4\pi\mathrm{i}e^2 \int_0^\infty d\alpha_1 \int_0^\infty d\alpha_2 \left(-2\not p - 2\mathrm{i}\gamma^\mu \frac{\partial}{\partial z^\mu} + 4m\right) \\ \times \int \frac{d^4k}{(2\pi)^4} \Big(\exp\big\{\mathrm{i}\big[\alpha_1(k^2-\mu^2) + \alpha_2((p-k)^2-m^2) \\ + z\cdot k\big]\big\} + \mathrm{reg}\Big)\Big|_{z=0}\end{aligned} \tag{5.78}$$

Evaluation of the k integration is possible with Gauss' formula (5.26):

$$\begin{aligned}\int \frac{d^4k}{(2\pi)^4} \cdots = \exp\left[\mathrm{i}(-\alpha_1\mu^2 + \alpha_2 p^2 - \alpha_2 m^2)\right] \frac{-\mathrm{i}}{(4\pi)^2(\alpha_1+\alpha_2)^2} \\ \times \exp\left[-\mathrm{i}(-2\alpha_2 p + z)^2/4(\alpha_1+\alpha_2)\right] \quad .\end{aligned}$$

After having performed the z differentiation and a trivial transformation we get for (5.78)

$$\begin{aligned}\bar\Sigma(p,\mu) = \frac{e^2}{2\pi} \int_0^\infty d\alpha_1 \int_0^\infty d\alpha_2 \frac{1}{(\alpha_1+\alpha_2)^2} \left\{\left(2m - \frac{\alpha_1}{\alpha_1+\alpha_2}\not p\right)\right. \\ \left. \times \exp\left[\mathrm{i}\left(\frac{\alpha_1\alpha_2}{\alpha_1+\alpha_2}p^2 - \alpha_1\mu^2 - \alpha_2 m^2\right)\right] + \mathrm{reg}\right\} \quad .\end{aligned} \tag{5.79}$$

If we introduce again the factor $1 = \int_0^\infty d\varrho\, \delta(\varrho - \alpha_1 - \alpha_2)$ of (5.33) and perform a scale transformation according to $\alpha_i = \varrho\beta_i$, we get

$$
\begin{aligned}
\bar{\Sigma}(p,\mu) &= \frac{e^2}{2\pi}\int_0^\infty d\varrho \int_0^\infty d\alpha_1 \int_0^\infty d\alpha_2\, \delta(\varrho-\alpha_1-\alpha_2)\frac{1}{\varrho^2} \\
&\quad\times\left\{\left(2m-\frac{\alpha_1}{\varrho}\not{p}\right)\exp\left[\mathrm{i}\left(\frac{\alpha_1\alpha_2}{\varrho}p^2-\alpha_1\mu^2-\alpha_2 m^2\right)\right]+\mathrm{reg}\right\} \\
&= \frac{e^2}{2\pi}\int_0^1 d\beta_1\int_0^1 d\beta_2\,\delta(1-\beta_1-\beta_2) \\
&\quad\times\left\{(2m-\beta_1\not{p})\int_0^\infty \frac{d\varrho}{\varrho}\exp\left[\mathrm{i}\varrho(\beta_1\beta_2p^2-\beta_1\mu^2-\beta_2m^2)\right]+\mathrm{reg}\right\} \quad .
\end{aligned}
\tag{5.80}
$$

Obviously the ϱ integral is *logarithmically divergent* at the lower boundary. To regularize it, it is sufficient to subtract a single term in the integrand ($C_1 = -1$), in which the photon mass μ is replaced by the large cutoff momentum $\mu_1 = \Lambda$, i.e.

$$
I = \int_0^\infty \frac{d\varrho}{\varrho}\left\{\exp\left[\mathrm{i}\varrho(\beta_1\beta_2p^2-\beta_1\mu^2-\beta_2m^2)\right]-\exp\left(-\mathrm{i}\varrho\beta_1\Lambda^2\right)\right\} \quad , \tag{5.81}
$$

where in the second exponent all other terms have been neglected compared to the large Λ^2. The integral can be solved by use of the formula

$$
\int_0^\infty \frac{d\varrho}{\varrho}\left(\mathrm{e}^{\mathrm{i}\varrho(z_1+\mathrm{i}\varepsilon)}-\mathrm{e}^{\mathrm{i}\varrho(z_2+\mathrm{i}\varepsilon)}\right)=\ln\frac{z_2}{z_1} \quad . \tag{5.82}
$$

In connection with (5.35) we showed how this result can be obtained by performing the limit $\lim_{\eta\to 0}\int_\eta^\infty \cdots$. Then the *regularized self-energy function* reads

$$
\bar{\Sigma}(p,\mu,\Lambda)=\frac{e^2}{2\pi}\int_0^1 d\beta(2m-\beta\not{p})\ln\left[\frac{\beta\Lambda^2}{(1-\beta)m^2+\beta\mu^2-\beta(1-\beta)p^2}\right] . \tag{5.83}
$$

If the momentum vector is *time-like* an effect that already showed up in the polarization function $\Pi^{\mathrm{R}}(q)$ of (5.39) occurs here again: If p^2 is sufficiently large, the argument of the logarithm in (5.83) obtains a zero, which moves into the region of integration $\beta \in [0,1]$. An examination of the quadratic form $(1-\beta)m^2+\beta\mu^2-\beta(1-\beta)p^2$ shows that this happens if $p^2 > (m+\mu)^2$. If the squared momentum passes over this threshold, $\bar{\Sigma}(p,\mu,\Lambda)$ gets an *imaginary part*. Physically this must be understood as the possibility of a virtual electron decaying into a real electron and a real photon, just like in Sect. 3.3 (Bhabha scattering), where a virtual photon turned into a real electron–positron pair. At $p^2 = (m+\mu)^2$, therefore, $\bar{\Sigma}(p,\mu,\Lambda)$ is a nonregular function. Thus the expansion intended in (5.65) is only possible if the fictitious photon mass μ is not set equal to zero! The reason for the difficulties in the case $\mu = 0$ is obviously due to the fact that an electron that is arbitrarily close to the mass shell is still able to emit real photons (with accordingly large wavelengths) and that for this reason *it is not at all possible to consider an isolated electron without a radiation cloud.* We will therefore keep μ finite and only at the very end will we perform the limit to the physical value $\mu = 0$.

The complete solution of the integral (5.83) is rather difficult.[11] We want only to identify the constants A and B of (5.65). The mass correction δm results from the calculation of the self-energy function on the mass shell, $\not p = m, p^2 = m^2$ (in the sense discussed above: sandwiching between free spinors $u(p)$):

$$\begin{aligned}\delta m \equiv A &= \bar{\Sigma}(p,\mu,\Lambda)\Big|_{\not p=m,p^2=m^2}\\ &= \frac{e^2 m}{2\pi}\int_0^1 d\beta(2-\beta)\ln\frac{\beta\Lambda^2}{(1-\beta)m^2+\beta\mu^2-\beta(1-\beta)m^2}\\ &= \frac{e^2 m}{2\pi}\int_0^1 d\beta(2-\beta)\ln\frac{\beta\Lambda^2}{m^2(1-\beta)^2+\beta\mu^2} \quad . \end{aligned} \tag{5.84}$$

In the limit $\mu \to 0$ this integral converges to a finite value, so that we can set $\mu = 0$. Splitting the logarithm yields

$$\delta m = \frac{e^2 m}{2\pi}\int_0^1 d\beta(2-\beta)\left[\ln\beta - 2\ln(1-\beta) + \ln\frac{\Lambda^2}{m^2}\right] \quad . \tag{5.85}$$

With the aid of the elementary integral

$$\int_0^1 dx\, x^n \, \ln x = -\frac{1}{(n+1)^2} \tag{5.86}$$

it follows immediately that

$$\delta m = m\frac{3\alpha}{4\pi}\ln\left(\frac{\Lambda^2}{m^2}+\frac{1}{2}\right) \quad . \tag{5.87}$$

The renormalization constant B results as the first term in the Taylor expansion of $\bar{\Sigma}$. With (5.83) and $p^2 = \not p^2$

$$\begin{aligned}B \equiv Z_2 - 1 &= \frac{\partial\bar{\Sigma}}{\partial\not p}\Big|_{\not p=m,p^2=m^2}\\ &= \frac{e^2}{2\pi}\int_0^1 d\beta\Bigg[-\beta\ln\frac{\beta\Lambda^2}{(1-\beta)m^2+\beta\mu^2-\beta(1-\beta)p^2}\\ &\quad -(2m-\beta\not p)(-)\frac{\beta(1-\beta)2\not p}{m^2(1-\beta)+\beta\mu^2-\beta(1-\beta)p^2}\Bigg]_{\not p=m,p^2=m^2}\\ &= \frac{-e^2}{2\pi}\int_0^1 d\beta\beta\left[\ln\frac{\beta\Lambda^2}{(1-\beta)^2m^2+\beta\mu^2} - \frac{2m^2(2-\beta)(1-\beta)}{m^2(1-\beta)^2+\beta\mu^2}\right] \quad . \end{aligned} \tag{5.88}$$

---

[11] It can be found in R. Karplus and N.M. Kroll: Phys. Rev. **77**, 536 (1950).

The first integral is solved just like (5.85); here we can set $\mu = 0$ without penalty:

$$I_1 = \int_0^1 d\beta\beta \left[\ln\beta - 2\ln(1-\beta) + \ln\Lambda^2/m^2\right] = +\frac{5}{4} + \frac{1}{2}\ln\frac{\Lambda^2}{m^2} \quad .$$

The second part can be approximated in logarithmic accuracy by

$$I_2 = -2\int_0^1 d\beta\beta \frac{(2-\beta)(1-\beta)}{(1-\beta)^2 + \beta\mu^2/m^2} \approx -2\int_0^{1-\mu/m} d\beta\,\beta \frac{(2-\beta)(1-\beta)}{(1-\beta)^2} \quad ,$$

where the cutoff term in the denominator has been replaced by a reduction of the upper bound of the integral. With the substitution $t = 1-\beta$ we get

$$I_2 \approx -2\int_{\mu/m}^1 dt\left(\frac{1}{t} - t\right) \simeq \ln\frac{\mu^2}{m^2} + 1$$

and thus

$$Z_2 = 1 + B = 1 - \frac{e^2}{2\pi}\left(\frac{1}{2}\ln\frac{\Lambda^2}{m^2} + \ln\frac{\mu^2}{m^2} + \frac{9}{4}\right) \quad . \tag{5.89}$$

We finally remark that the mass renormalization δm (5.87) and the renormalization Z_2 of the electron propagator (5.89) are only weakly, that is, logarithmically, divergent. The counting of the powers of k had led to the overly pessimistic prediction of a linear divergence.

Except for that, the renormalization constant Z_2 has very unpleasant features. It is infrared divergent for zero photon mass and in addition depends on the gauge of the photon field. If we had used instead of the "Feynman gauge" another form of the photon propagator (cf. Chap. 4), we would have got a result for Z_2 different from (5.89). This is not true for the mass renormalization δm, which is gauge invariant. In the next section we shall see that this twofold ambiguity of Z_2 has no harmful consequences.

5.4 The Vertex Correction

As the last of the radiative corrections fundamental for renormalization we examine the change of the vertex due to a virtual photon as depicted in Fig. 7a. According to the Feynman rules the factor $-ie\gamma_\mu$ is replaced by

$-ie\Lambda_\mu(p',p) =$ 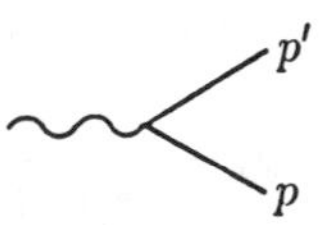

$+$ 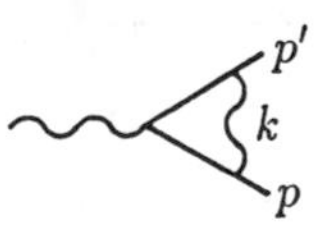

$$= -ie\gamma_\mu - ie\Gamma_\mu(p',p) \tag{5.90}$$

with the vertex function (cf. (5.2))

$$\Gamma_\mu(p',p) = -4\pi \mathrm{i} e^2 \int \frac{d^4k}{(2\pi)^4} \frac{1}{k^2-\mu^2+\mathrm{i}\varepsilon} \times \left(\gamma^\nu \frac{1}{\not p' - \not k - m + \mathrm{i}\varepsilon}\gamma_\mu \frac{1}{\not p - \not k - m + \mathrm{i}\varepsilon}\gamma_\nu\right) \quad . \tag{5.91}$$

This momentum loop again leads to a (logarithmically) divergent integral and must be renormalized. Additionally we have providently introduced once more the fictitious photon mass μ, in order to be able to handle a possible infrared divergence.

Although a complete evaluation of the vertex function $\Gamma_\mu(p',p)$ is feasible, it is quite demanding. We therefore restrict ourselves to the important special case that the electron lines of the vertex "lie on the mass shell". By this we mean that at the end of the calculations a matrix element of the form $\bar{u}'(p')\Gamma_\mu(p',p)u(p)$ between *free spinors* is to be formed. The following calculation is valid only under this condition,[12] although we will not always write down these spinors in what follows. This condition permits us to replace $\not p' \to m$ when acting to the left and $\not p \to m$ when acting to the right.

Now we decompose $\Gamma_\mu(p',p)$ into a sum of the limit for zero momentum transfer $q = p' - p = 0$ ("forward scattering") and the remainder

$$\begin{aligned}\Gamma_\mu(p',p) =& \Gamma_\mu(p,p) + \big(\Gamma_\mu(p',p) - \Gamma_\mu(p,p)\big), \\ \equiv& \Gamma_\mu(p,p) + \Gamma_\mu^{\mathrm{R}}(p',p) \quad .\end{aligned} \tag{5.92}$$

When constructing the forward-scattering part we do not have the vector q_μ at our disposal (it is equal to zero); therefore $\Gamma_\mu(p,p)$ can only be proportional to γ_μ or p_μ. The matrix elements between free spinors of both of these operators are, however, simply proportional to each other, if the momentum transfer vanishes, and can thus be transformed into each other. This follows from the *Gordon decomposition* of the Dirac current

$$\bar{u}(p')\gamma_\mu u(p) = \frac{1}{2m}\bar{u}(p')\big[(p+p')_\mu + \mathrm{i}\sigma_{\mu\nu}(p'-p)^\nu\big]u(p) \quad . \tag{5.93}$$

It is thus sufficient to use only γ_μ. This leads us to the ansatz

$$\Gamma_\mu(p,p) = L\gamma_\mu \quad . \tag{5.94}$$

L will soon prove to be a constant that diverges logarithmically in the cutoff momentum Λ, while the remainder of the vertex function, namely $\Gamma_\mu^{\mathrm{R}}(p',p)$ is a *well-defined* finite expression. One can easily understand this assertion if one expands the first electron propagator in the integral (5.91) at fixed momentum p according to the operator identity (5.9), that is,

$$\begin{aligned}\frac{1}{\not p' - \not k - m + \mathrm{i}\varepsilon} &= \frac{1}{\not p - \not k - m + (\not p' - \not p) + \mathrm{i}\varepsilon} \\ &= \frac{1}{\not p - \not k - m + \mathrm{i}\varepsilon} \\ &\quad - \frac{1}{\not p - \not k - m + \mathrm{i}\varepsilon}(\not p' - \not p)\frac{1}{\not p - \not k - m + \mathrm{i}\varepsilon} + \cdots \quad .\end{aligned} \tag{5.95}$$

[12] Explicit expressions for Γ_μ assuming that one or both electrons are on the mass shell are given in A.I. Akhiezer, V.B. Berestetskii, Quantum Electrodynamics, Wiley-Interscience, New York, 1965.

The first term, which is independent of p', is proportional to $|k|^{-1}$ for large values of k and is the reason for the logarithmic divergence of the integral (5.91). On the other hand, the following terms of the series, which vanish as $p' \to p$ and therefore lead to $\Gamma^{\mathrm{R}}_\mu(p',p)$, have higher powers of k in the denominator and render the momentum integration convergent. Since the electric charge is measured by scattering with low momentum transfer, the replacement

$$-\mathrm{i}e\gamma_\mu \to -\mathrm{i}e\gamma_\mu - \mathrm{i}eL\gamma_\mu + O(q) \tag{5.96}$$

makes us expect that one further *charge renormalization* will be necessary, namely

$$e''_{\mathrm{R}} = Z_1^{-1} e \quad , \tag{5.97}$$

where by convention the renormalization constant Z_1 has been introduced;

$$Z_1 = (1+L)^{-1} \simeq 1 - L \quad . \tag{5.98}$$

The divergent part $\Gamma_\mu(p,p)$, however, can be traced back to an already familiar result without any calculation. To do this, we differentiate the electron propagator $S_{\mathrm{F}} = (\not p - m + \mathrm{i}\varepsilon)^{-1}$ with respect to momentum. Because of $S_{\mathrm{F}}(p) S_{\mathrm{F}}^{-1}(p) = 1$ we obtain, according to the product rule,

$$\frac{\partial}{\partial p^\mu} S_{\mathrm{F}}(p) \cdot S_{\mathrm{F}}^{-1}(p) + S_{\mathrm{F}}(p) \frac{\partial}{\partial p^\mu}(\not p - m) = 0 \quad ,$$

or

$$\frac{\partial}{\partial p^\mu} S_{\mathrm{F}}(p) = -S_{\mathrm{F}}(p)\gamma_\mu S_{\mathrm{F}}(p) \quad . \tag{5.99}$$

The differentiation of the electron propagator with respect to momentum thus corresponds to the introduction of a vertex with zero momentum transfer, as indicated in Fig. 5.21a. Obviously it is then also possible, according to Fig. 5.21b, to *relate the vertex correction with* $q = 0$ *to the diagram of self-energy* through a simple differentiation:

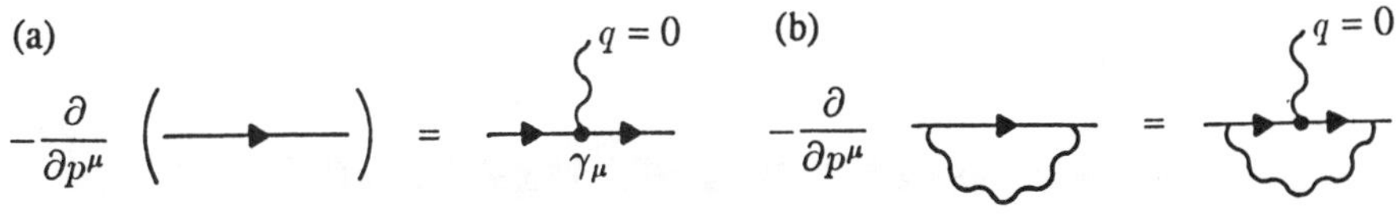

Fig. 5.21. Illustration of the Ward identity

$$\Gamma_\mu(p,p) = -\frac{\partial}{\partial p^\mu}\Sigma(p) \quad . \tag{5.100}$$

This follows immediately by application of (5.99) to the self-energy function (5.77). Equation (5.100), which is called the **Ward** *identity*,[13] has far-reaching consequences.

If one differentiates the expansion of the self-energy function (5.65) with respect to the momentum vector, one gets

[13] J.C. Ward, Phys. Rev. **78**, 182 (1950).

$$L\gamma_\mu = \Gamma_\mu(p,p) = -\frac{\partial}{\partial p^\mu}\Sigma(p) = -B\gamma_\mu + O(\not p - m) \quad . \tag{5.101}$$

On the mass shell

$$L = -B \quad , \quad \text{and thus} \quad Z_1 = Z_2 \quad . \tag{5.102}$$

The renormalization constants of self-energy and vertex correction are thus exactly equal and *simply cancel each other*! According to (5.46), (5.73) and (5.97) the final charge renormalization is

$$e_\mathrm{R} = Z_1^{-1} Z_2 \sqrt{Z_3}\, e = \sqrt{Z_3}\, e \quad , \tag{5.103}$$

which includes the effects of vacuum polarization, self-energy, and vertex correction. Since we consistently include only corrections of order e^2, the multiplicative treatment of the individual corrections is justified, because in lowest order $(1+\varepsilon_1+\varepsilon_2+\dots+\varepsilon_n) \simeq (1+\varepsilon_1)(1+\varepsilon_2)\cdots(1+\varepsilon_n)$.

The result (5.103), which states that charge renormalization is solely due to vacuum polarization, is most satisfactory. This is because the resulting renormalization of charge (5.103) in contrast to (5.73) and also to (5.97) does not depend on the fictitious photon mass μ and the gauge chosen, which indeed were both chosen at will. Yet the Ward identity has a much more fundamental significance: it ensures the *universality of the electromagnetic interaction.* The reason for this is that self-energy (Z_2) and vertex correction (Z_1) will look different for each charged particle (e, μ, p, $\cdots$). Equation (5.103), however, ensures that the renormalization of charge does not depend on what species of particle one is dealing with, but that it is only a consequence of the photon propagator being modified by virtual pair creation. Thus if the bare charges e of two elementary particles are equal, the Ward identity (5.103) ensures that the physically observable charges e_R are equal too. If the Ward identity did not hold the bare charges would have to differ by exactly the amount which ensures that the difference due to renormalization is cancelled. This is absurd! – The measured elementary charge e is a universal constant. A remarkable result of quantum electrodynamics is the fact that this property remains true *in all orders* of perturbation theory (we have restricted ourselves to the lowest nontrivial order e^2).

EXAMPLE

5.6 The Form Factor of the Electron

We want to study those physically observable consequences arising from the replacement of the vertex factor $-\mathrm{i}e\gamma_\mu$ by the complex expression (5.90), which has the form $-\mathrm{i}e(\gamma_\mu + \Gamma_\mu(p',p))$. As we did in our general considerations of the previous section here the electron will be approximately regarded to be free (i.e. on the mass shell).

To evaluate the momentum integral (5.91) we first simplify the numerator which will be called X. Since the vertex function is to be located between free spinors, we have $\bar u(p')(\not p' - m) = 0, (\not p - m)u(p) = 0$ and can replace $\not p' \to m, \not p \to m$. To

Example 5.6.

be able to do this, however, one must place the matrix $\not{p}'$ totally at the left and the matrix $\not{p}$ totally to the right, so that each stands adjacent to its eigenspinor. Using the commutation rules of the Dirac matrices we get

$$\begin{aligned} X &= \gamma^\nu(\not{p}' - \not{k} + m)\gamma_\mu(\not{p} - \not{k} + m)\gamma_\nu \\ &= \left[(-\not{p}' + \not{k} + m)\gamma^\nu + 2(p' - k)^\nu\right]\gamma_\mu\left[\gamma_\nu(-\not{p} + \not{k} + m) + 2(p - k)_\nu\right] \\ &= \left[\not{k}\gamma^\nu + 2(p' - k)^\nu\right]\gamma_\mu\left[\gamma_\nu\not{k} + 2(p - k)_\nu\right] \quad . \end{aligned} \tag{1}$$

This expression can be transformed into

$$X = 4\left\{\gamma_\mu\left[(p' - k)\cdot(p - k) - k^2/2\right] + (p' + p - k)_\mu\not{k} - mk_\mu\right\} \quad . \tag{2}$$

The vertex function thus reads

$$\begin{aligned} \Gamma_\mu(p', p) &= -4 \times 4\pi \mathrm{i} e^2 \int \frac{d^4k}{(2\pi)^4} \\ &\times \frac{\gamma_\mu[(p' - k)\cdot(p - k) - k^2/2] + (p' + p - k)_\mu\not{k} - mk_\mu}{(k^2 - \mu^2 + \mathrm{i}\varepsilon)(k^2 - 2p'\cdot k + \mathrm{i}\varepsilon)(k^2 - 2p\cdot k + \mathrm{i}\varepsilon)} \quad , \end{aligned} \tag{3}$$

where in the denominator again the mass-shell condition $p^2 = p'^2 = m^2$ and $k^2 = \mu^2$ was used. Obviously the evaluation of this integral is rather laborious. However, one can employ the methods already used in the calculation of vacuum polarization and self-energy. For this reason we shall pass more quickly over the intermediate steps.

Again one can introduce the integral representation of the propagator (5.22), where now because of the three propagators in (3) three parameter integrations are required. Using the Gaussian integral (5.26) we get the identity

$$\begin{aligned} &\int \frac{d^4k}{(2\pi)^4}\exp(\mathrm{i}k\cdot z)\frac{1}{(k^2 - \mu^2 + \mathrm{i}\varepsilon)(k^2 - 2p'\cdot k + \mathrm{i}\varepsilon)(k^2 - 2p\cdot k + \mathrm{i}\varepsilon)} \\ &= \frac{1}{(4\pi)^2}\int_0^\infty \frac{d\alpha_1 d\alpha_2 d\alpha_3}{(\alpha_1 + \alpha_2 + \alpha_3)^2}\exp\left\{-\mathrm{i}\left[\frac{(z/2 - p'\alpha_2 - p\alpha_3)^2}{\alpha_1 + \alpha_2 + \alpha_3} + \mu^2\alpha_1\right]\right\} \quad . \end{aligned} \tag{4}$$

The momentum factors k in the numerator (3) can be converted into a differentiation of the exponent with respect to the auxiliary variable z, as in (5.24). Thus one may replace $k\exp(\mathrm{i}kz) = -\mathrm{i}\partial/\partial z\exp(\mathrm{i}k\cdot z)$, which leads to

$$k \to \frac{-z/2 + p'\alpha_2 + p\alpha_3}{\alpha_1 + \alpha_2 + \alpha_3} \quad . \tag{5}$$

Finally the integral (3) assumes the form

Example 5.6.

$$\Gamma_\mu(p',p) = -\mathrm{i}\frac{e^2}{\pi}\int\limits_0^\infty \frac{d\alpha_1 d\alpha_2 d\alpha_3}{(\alpha_1+\alpha_2+\alpha_3)^3}\left\{\gamma_\mu\left[(\alpha_1+\alpha_2+\alpha_3)p\cdot p'\right.\right.$$
$$+\frac{1}{2}(\alpha_2+\alpha_3)(p+p')^2+\frac{1}{2}\frac{m^2(\alpha_2+\alpha_3)^2-\alpha_2\alpha_3 q^2}{\alpha_1+\alpha_2+\alpha_3}+\left.\frac{\mathrm{i}}{2}\right]$$
$$\left.+\frac{m}{2}(p+p')_\mu\frac{\alpha_1(\alpha_2+\alpha_3)}{\alpha_1+\alpha_2+\alpha_3}\right\}$$
$$\times\exp\left\{-\mathrm{i}\left[\mu^2\alpha_1+\frac{m^2(\alpha_2+\alpha_3)^2-\alpha_2\alpha_3 q^2}{\alpha_1+\alpha_2+\alpha_3}\right]\right\} \quad . \tag{6}$$

Here $q = p' - p$ denotes the momentum transfer. In the derivation of (6) we used $p^2 = p'^2 = m^2, p'\cdot p = m^2 - q^2/2$ several times, for instance in order to obtain

$$(p'\alpha_2 + p\alpha_3)^2 = m^2(\alpha_2+\alpha_3)^2 - q^2\alpha_2\alpha_3 \quad .$$

In addition we replaced $\not p \to m, \not p' \to m$. A further trick was to make use of the symmetry of the integrand under the exchange of α_2 and α_3. Terms like

$$\int\limits_0^\infty d\alpha_2 d\alpha_3(\alpha_2-\alpha_3)f(\alpha_2,\alpha_3) \quad ,$$

with a symmetric function $f(\alpha_2,\alpha_3) = f(\alpha_3,\alpha_2)$, are equal to zero. The term $\mathrm{i}/2$ in the fourth line of (6) is due to the factor k^2, because in the twofold differentiation $(-\mathrm{i}\partial/\partial z)$ of (4) the product rule must be applied. The argument of the exponential can be made real and negative by a rotation about $-\pi/2$ in the complex α_i plane. To this end we take into account that the momentum transfer to a free particle is space-like, $q^2 < 0$. One can thus substitute

$$\alpha_i \to -\mathrm{i}\alpha_i \tag{7}$$

and afterwards have the integration extend to $+\infty$ again on the real axis. Additionally, as in (5.33), we introduce an auxiliary integration with respect to a variable ϱ according to

$$\int\limits_0^\infty d\varrho\,\delta(\varrho-\alpha_1-\alpha_2-\alpha_3) = 1 \quad . \tag{8}$$

The vertex function (6) then reads

$$\Gamma_\mu(p',p) = -\frac{e^2}{\pi}\int\limits_0^\infty d\alpha_1 d\alpha_2 d\alpha_3\int\limits_0^\infty d\varrho\,\delta(\varrho-\alpha_1-\alpha_2-\alpha_3)\frac{1}{\varrho^3}$$
$$\times\left(\gamma_\mu\left\{\varrho p\cdot p' - \frac{1}{2}(\alpha_2+\alpha_3)(p+p')^2\right.\right.$$
$$\left.+\frac{1}{2\varrho}\left[m^2(\alpha_2+\alpha_3)^2-\alpha_2\alpha_3 q^2\right]-\frac{1}{2}\right\}$$
$$\left.+\frac{m}{2}(p+p')_\mu\frac{\alpha_1(\alpha_2+\alpha_3)}{\varrho}\right)$$
$$\times\exp\left\{-\frac{1}{\varrho}\left[-\alpha_2\alpha_3 q^2+(\alpha_2+\alpha_3)^2 m^2+\alpha_1\varrho\mu^2\right]\right\} \quad . \tag{9}$$

Example 5.6.

Now an exchange of the order of integrations follows, and afterwards a scale transformation $\alpha_i = \varrho\beta_i$, so that we finally get

$$\Gamma_\mu(p',p) = -\frac{e^2}{\pi}\int_0^\infty d\beta_1 d\beta_2 d\beta_3 \delta(1-\beta_1-\beta_2-\beta_3)$$
$$\times \int_0^\infty d\varrho\Bigg(\gamma_\mu\Bigg\{p\cdot p' - \frac{1}{2}(\beta_2+\beta_3)(p+p')^2$$
$$+\frac{1}{2}\left[m^2(\beta_2+\beta_3)^2 - \beta_2\beta_3 q^2\right] - \frac{1}{2\varrho}\Bigg\} + \frac{m}{2}(p+p')_\mu\beta_1(\beta_2+\beta_3)\Bigg)$$
$$\times \exp\{-\varrho[-\beta_2\beta_3 q^2 + (\beta_2+\beta_3)^2 m^2 + \beta_1\mu^2]\} \quad . \tag{10}$$

The ϱ integral has a part that is logarithmically divergent because of the term $\gamma_\mu(-1/(2\varrho))$. We know from (5.102) that such a divergence occurs. Regularization is performed according to (5.92) by subtracting the term for forward scattering ($q = 0$):

$$\Gamma^{\mathrm{R}}_\mu(p',p) = \Gamma_\mu(p',p) - \Gamma_\mu(p,p) \quad . \tag{11}$$

Strictly speaking we would have already been obliged to take this into account in the derivation of (10), in order to justify the formal manipulations.

We want to represent the regularized vertex function in a more clearly arranged form. Therefore we write

$$\Gamma^{\mathrm{R}}_\mu(p',p) = \gamma_\mu F_1(q^2) + \frac{\mathrm{i}}{2m}\sigma_{\mu\nu}q^\nu F_2(q^2) \quad . \tag{12}$$

This is exactly the ansatz we had found in Exercise 3.5 by general considerations on the interaction of a photon with a spin-1/2 particle. The functions $F_1(q^2)$ and $F_2(q^2)$ are called *form factors*. Obviously the electron gets an apparent internal structure by the interaction with the virtual radiation field and differs in its behaviour from a pure Dirac particle. The form factors in (12) can be re-expressed by integrals over the β variables. To obtain the form given in (12) we use the Gordon decomposition

$$\bar{u}(p')\gamma_\mu u(p) = \frac{1}{2m}\bar{u}(p')\left[(p+p')_\mu + \mathrm{i}\sigma_{\mu\nu}q^\nu\right]u(p) \quad . \tag{13}$$

The integration over ϱ is easily done. For the first form factor we obtain

$$F_1(q^2) = -\frac{e^2}{\pi}\int_0^\infty d\beta_1 d\beta_2 d\beta_3\, \delta(1-\beta_1-\beta_2-\beta_3)$$
$$\times\Bigg\{\Bigg[m^2 - \frac{q^2}{2} - (\beta_2+\beta_3)\left(2m^2 - \frac{q^2}{2}\right)$$
$$+\frac{1}{2}m^2(\beta_2+\beta_3)^2 - \beta_2\beta_3 q^2 + m^2\beta_1(\beta_2+\beta_3)\Bigg]$$
$$\times\left[m^2(\beta_2+\beta_3)^2 + \beta_1\mu^2 - \beta_2\beta_3 q^2\right]^{-1}$$
$$+\frac{1}{2}\ln\left[m^2(\beta_2+\beta_3)^2 + \beta_1\mu^2 - \beta_2\beta_3 q^2\right] - \underbrace{\cdots}_{(q=0)}\Bigg\} \quad . \tag{14}$$

Example 5.6.

The logarithmic integral originates from the $1/\varrho$ term in the integrand. We used

$$\begin{aligned}\int\limits_0^\infty \frac{d\varrho}{\varrho}\left(e^{-\varrho A}-e^{-\varrho B}\right) &:= \lim_{\eta\to 0}\int\limits_\eta^\infty \frac{d\varrho}{\varrho}\left(e^{-\varrho A}-e^{-\varrho B}\right)\\ &= \lim_{\eta\to 0}\left(\int\limits_{\eta A}^\infty - \int\limits_{\eta B}^\infty\right)\frac{d\varrho}{\varrho}e^{-\varrho} = \lim_{\eta\to 0}\int\limits_{\eta A}^{\eta B}\frac{d\varrho}{\varrho}e^{-\varrho}\\ &= \lim_{\eta\to 0}\int\limits_{\eta A}^{\eta B}\frac{d\varrho}{\varrho} = \ln B - \ln A \quad .\end{aligned} \tag{15}$$

In a similar form this integral occurred in (5.35) and (5.82). The product $p'\cdot p$ has been replaced by $m^2-q^2/2$, because $q^2=(p'-p)^2=p'^2+p^2-2p'\cdot p = 2m^2-2p'\cdot p$. The term $m^2\beta_1(\beta_2+\beta_3)$ in the numerator of the first term of (14) follows from $(m/2)(p+p')_\mu\beta_1(\beta_2+\beta_3)$ by replacing $(p+p')_\mu = 2m\gamma_\mu - i\sigma_{\mu\nu}q^\nu$.

The result for the second form factor is less involved. There is no need to regularize and the result is

$$\begin{aligned}F_2(q^2) &= \frac{e^2}{\pi}\int\limits_0^\infty d\beta_1 d\beta_2 d\beta_3 \delta(1-\beta_1-\beta_2-\beta_3)\\ &\quad\times\frac{m^2\beta_1(\beta_2+\beta_3)}{m^2(\beta_2+\beta_3)^2+\beta_1\mu^2-\beta_2\beta_3 q^2} \quad .\end{aligned} \tag{16}$$

We do not want to go into the details of the general evaluation of the integrals (14) and (16), which is rather difficult. A simple result is obtained in the limit of *low momentum transfer*, $q^2\to 0$, which is also of special physical significance. The form factor $F_2(q^2)$ can be calculated immediately. Since no infrared divergence occurs, we set $\mu^2=0$. Equation (16) then reads

$$\begin{aligned}F_2(0) &= \frac{e^2}{\pi}\int\limits_0^1 d\beta_2\int\limits_0^{1-\beta_2} d\beta_3\frac{1-\beta_2-\beta_3}{\beta_2+\beta_3}\\ &= \frac{e^2}{\pi}\int\limits_0^1 d\beta_2\left[\ln(\beta_2+\beta_3)-\beta_3\right]_0^{1-\beta_2}\\ &= \frac{e^2}{\pi}\left[(-\beta_2\ln\beta_2+\beta_2)-\beta_2+\frac{1}{2}\beta_2^2\right]_0^1\\ &= \frac{\alpha}{2\pi} \quad .\end{aligned} \tag{17}$$

In the limit $q^2\to 0$ the function $F_1(q^2)$ contains a divergent part, which leads to charge renormalization and which is eliminated by the regularization in (14). We obtain in the lowest nonvanishing order

$$F_1(q^2) \simeq \frac{\alpha}{3\pi}\frac{q^2}{m^2}\left(\ln\frac{m}{\mu}-\frac{3}{8}\right) \quad , \tag{18}$$

Example 5.6.

which can be checked by the elementary but rather lengthy integration of (14). Thus a logarithmic divergence occurs, similar to the case of the self-energy graph (5.89). The regularized vertex function $\Gamma^{\mathrm{R}}_{\mu}(p',p)$ for free spinors and low momentum transfer is now completely determined by (12), (17), and (18).

To understand the physical consequences of this result, we examine the *interaction energy of an electron with a static external electromagnetic field* A^{ext}_{μ}, just as in Exercise 3.5:

$$\begin{aligned} W &= \int d^3x\, j_\mu A^\mu_{\mathrm{ext}} \\ &= e\int d^3x \bar{\psi}_{p'} \left(\gamma_\mu + \Gamma^{\mathrm{R}}_{\mu}(p',p) + \frac{\mathrm{i}\Pi^{\mathrm{R}}_{\mu\nu}}{4\pi} \mathrm{i}D^{\nu\sigma}_{\mathrm{F}}\gamma_\sigma \right) \psi_p A^\mu_{\mathrm{ext}} \quad . \end{aligned} \tag{19}$$

Here the graphs (a), (b) and (c) of Fig. 5.22 were added. They all contribute to the interaction of a fermion with an external field.

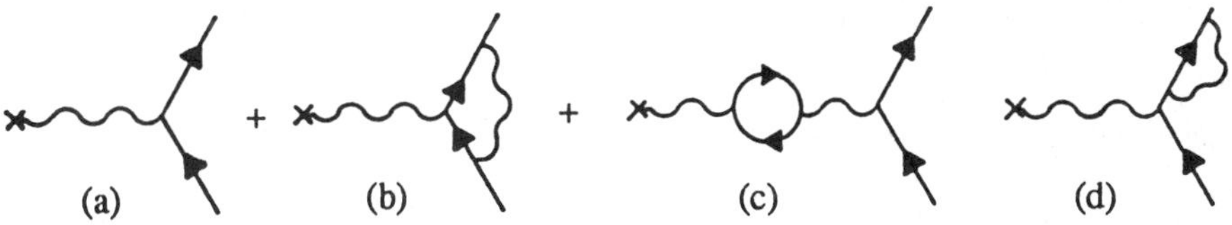

Fig. 5.22. The graphs (**a–c**) contribute to the interaction of a fermion with an external field

The self-energy correction can be left out, because for free particles it only contributes to charge and mass renormalization. This is so because free particles are on the mass shell, so that the term $\Sigma^{\mathrm{R}}(p)(\not{p}-m)^2$ in (5.65) does not contribute because of $\not{p} = m$. Insertion of (17) and (18) as well as (5.18), (5.41), and $D^{\nu\sigma}_{\mathrm{F}} = -4\pi g^{\nu\sigma}/q^2$ yields for small values of q^2

$$\begin{aligned} W \simeq e\int d^3x \bar{\psi}_{p'} \Bigg\{ \gamma_\mu \left[1 + \frac{\alpha}{3\pi}\frac{q^2}{m^2}\left(\ln\frac{m}{\mu} - \frac{3}{8} - \frac{1}{5} \right)\right] \\ + \frac{\alpha}{2\pi}\frac{\mathrm{i}}{2m}\sigma_{\mu\nu}q^\nu \Bigg\} \psi_p A^\mu_{\mathrm{ext}} \quad , \end{aligned} \tag{20}$$

where the term $-1/5$ results from vacuum polarization (5.41). Using the Gordon decomposition this can be written as

$$\begin{aligned} W \simeq e\int d^3x \bar{\psi}_{p'} \Bigg\{ \frac{1}{2m}(p+p')_\mu \left[1 + \frac{\alpha}{3\pi}\frac{q^2}{m^2}\left(\ln\frac{m}{\mu} - \frac{3}{8} - \frac{1}{5} \right)\right] \\ + \left(1+\frac{\alpha}{2\pi}\right) \frac{\mathrm{i}}{2m}\sigma_{\mu\nu}q^\nu \Bigg\} \psi_p A^\mu_{\mathrm{ext}} \quad . \end{aligned} \tag{21}$$

Since we are interested only in slowly varying ('quasistationary') fields, the correction proportional to q^3 in the last term was dropped. The momentum factors can be transformed to gradients in configuration space: $q_\mu \to \mathrm{i}\partial_\mu$ acts on the photon field and $p'_\mu = -\mathrm{i}\overleftarrow{\partial}_\mu$, $p_\mu = \mathrm{i}\partial_\mu$ act on the spinor field to the left and the right, respectively. This leads to

Example 5.6.

$$W \simeq e \int d^3x \left\{ \frac{\mathrm{i}}{2m} \bar{\psi}_{p'}(x) \overleftrightarrow{\partial}_\mu \psi_p(x) \left[1 - \frac{\alpha}{3\pi} \frac{1}{m^2} \left(\ln \frac{m}{\mu} - \frac{3}{8} - \frac{1}{5} \right) \Box \right] A^\mu_{\mathrm{ext}} - \left(1 + \frac{\alpha}{2\pi} \right) \frac{1}{2m} \bar{\psi}_p(x) \sigma_{\mu\nu} \psi_p(x) \partial^\nu A^\mu_{\mathrm{ext}} \right\} \quad . \tag{22}$$

The first term contains the 'convection current' of the electron, which interacts with the potential. In the special case of a purely magnetic field the second part can be identified as the magnetic dipole energy. To see this, we introduce the electromagnetic field strength tensor $F^{\mu\nu} = \partial^\mu A^\nu - \partial^\nu A^\mu$ and use the antisymmetry of $\sigma_{\mu\nu} = \mathrm{i}/2\ [\gamma_\mu, \gamma_\nu]$; thus

$$W_{\mathrm{mag}} \simeq e\big(1 + \frac{\alpha}{2\pi}\big) \frac{1}{4m} \int d^3x\, \bar{\psi}(x) \sigma_{\mu\nu} \psi(x) F^{\mu\nu} \quad . \tag{23}$$

In the case of a purely magnetic field $F^{12} = -B^3$, $\sigma_{12} = \Sigma_3$ and cyclic permutations. Thus the interaction energy becomes

$$\begin{aligned} W_{\mathrm{mag}} &\simeq -\frac{e}{4m}\big(1 + \frac{\alpha}{2\pi}\big)\, 2 \int d^3x\, \bar{\psi}(x) \boldsymbol{\Sigma} \psi(x) \cdot \boldsymbol{B} \\ &= - < \boldsymbol{\mu} > \cdot \boldsymbol{B} \quad , \end{aligned} \tag{24}$$

with the *magnetic moment* ($< \boldsymbol{s} > = < \boldsymbol{\Sigma} > /2$)

$$\begin{aligned} < \boldsymbol{\mu} > &= \frac{e\hbar}{2mc}\big(1 + \frac{\alpha}{2\pi}\big) 2 < \boldsymbol{S} > \\ &= 2\big(1 + \frac{\alpha}{2\pi}\big) \mu_{\mathrm{B}} < \boldsymbol{S} > = g \mu_{\mathrm{B}} < \boldsymbol{S} > \quad . \end{aligned} \tag{25}$$

Here we have for once also written down the natural constants $\hbar$ and c. The factor $\hbar$ is due to $q \to \mathrm{i}\hbar\partial_x$ and the c is due to the Gordon decomposition, which strictly speaking contains the factor $1/2mc$. As is to be expected, the magnetic moment is thus proportional to the spin expectation value of the electron. In units of the Bohr magneton $\mu_{\mathrm{B}} = e\hbar/2mc$ the proportionality factor (Landé's g factor) is

$$g = 2\left(1 + \frac{1}{2}\frac{\alpha}{\pi}\right) \simeq 2(1 + 0.001\,161\,41) \quad . \tag{26}$$

The difference of the g-factor from 2 is called the *anomaly* of the electron. It is one of the most important predictions of quantum electrodynamics and since its first calculation by Schwinger[14] it has been measured with impressive accuracy. A modern experimental value is[15]

$$g_{\mathrm{exp}} = 2(1 + 0.001\,159\,652\,193) \quad , \tag{27}$$

where only the last digit is uncertain.

Obviously Schwinger's prediction is perfectly confirmed within the range of validity of second order. To understand the result (27) completely, higher-order terms must be taken into account. In Fig. 5.23 we show the graphs of fourth order (α^2). One finds that only the five diagrams (a–e) contribute to the magnetic moment. In sixth order 72 Feynman diagrams contribute. With increasing accuracy

[14] J. Schwinger: Phys. Rev. **73**, 416 (1948) and **76**, 790 (1949).

[15] R.S. Van Dyck, Jr., P.B. Schwinberg, H.G. Dehmelt: Phys. Rev. Lett. **59**, 26 (1987).

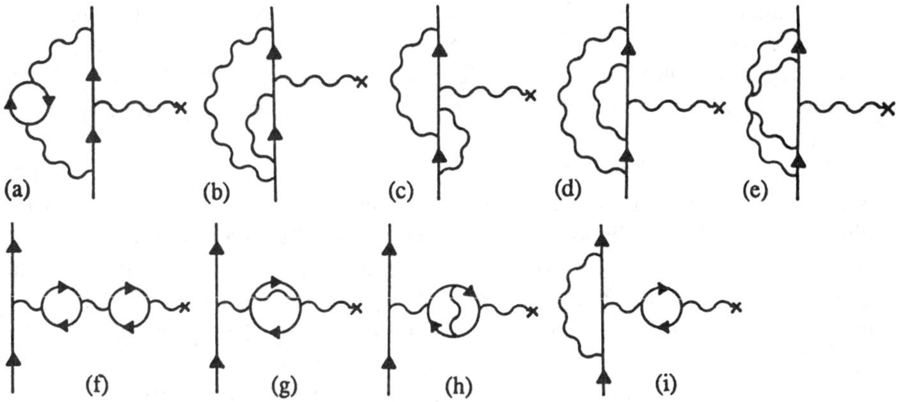

Fig. 5.23a–i. The radiation corrections of fourth order to photon–electron interaction

of measurement 891 Feynman diagrams of order α^4 have to be calculated and a number of further corrections (virtual hadron creation!) must be taken into account.

The pure-QED contributions usually are represented by the coefficients C_i of a power series in α/π which turns out to be the natural expansion parameter for this problem:

$$g_{\text{theor}} = 2\left[1 + C_1\left(\frac{\alpha}{\pi}\right) + C_2\left(\frac{\alpha}{\pi}\right)^2 + C_3\left(\frac{\alpha}{\pi}\right)^3 + \dots\right] \quad . \tag{28}$$

Above we have evaluated Schwinger's second-order term $C_1 = 1/2$. It took nearly a decade to obtain the correct analytical expression for the fourth-order coefficient which reads[16]

$$\begin{aligned} C_2 &= \tfrac{197}{144} + (\tfrac{1}{2} - 3\ \ln 2)\zeta(2) + \tfrac{3}{4}\zeta(3) \\ &= -0.328\,478\,965\dots \quad . \end{aligned} \tag{29}$$

Here $\zeta(n) = 1 + 2^{-n} + 3^{-n} + \dots$ is Riemann's zeta function. At the time of writing many, but not all, of the sixth-order graphs have been calculated analytically. For the remaining graphs (involving overlapping photon loops) one has to solve multidimensional integrals numerically. This is also true for the eighth-order contributions. The following values for the coefficients C_3 and C_4 have been found[17]

$$\begin{aligned} C_3 &= 1.176\,11 \pm 0.000\,42 \quad , \\ C_4 &= -1.434 \pm 0.138 \quad , \end{aligned} \tag{30}$$

where the errors result from the statistical uncertainty of the Monte-Carlo integrations. Using these coefficients the theoretical g-factor is

$$g_{\text{theor}} = 2(1 + 0.001\,159\,652\,140 \pm 0.000\,000\,000\,028) \quad . \tag{31}$$

The larget part of the error is caused by the uncertain knowledge of the fine-structure constant α. Equation (31) is in remarkably good agreement with the

[16] A. Peterman: Helv. Phys. Act. **30**, 407 (1957); C.M. Sommerfield: Ann. Phys. (N.Y.) **5**, 26 (1958).

[17] Many of these very sophisticated and laborious calculations have been worked out by T. Kinoshita and collaborators. The subject is extensively reviewed in the book T. Kinoshita: *Quantum Electrodynamics* (World Scientific, Singapore, 1990).

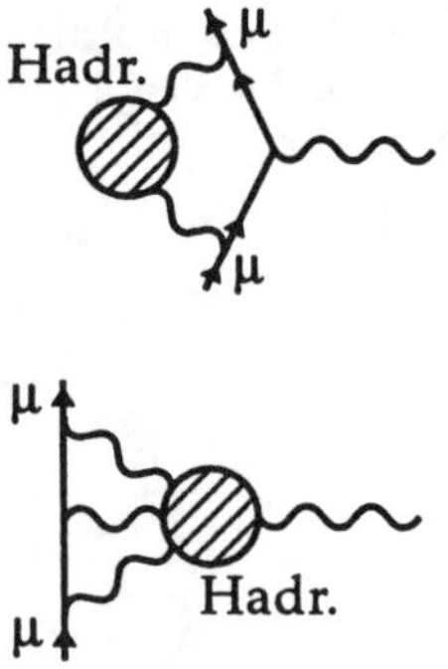

Fig. 5.24. Hadronic corrections to the anomalous magnetic moment of the muon

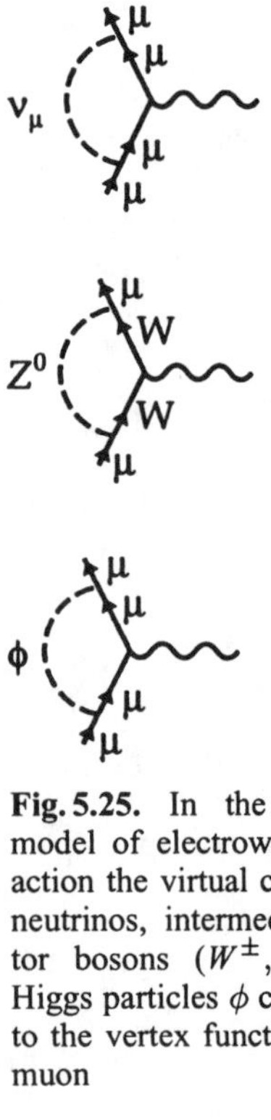

Fig. 5.25. In the standard model of electroweak interaction the virtual creation of neutrinos, intermediate vector bosons ($W^\pm, Z^0$), and Higgs particles ϕ contributes to the vertex function of the muon

experimental result (27). This proves most clearly that the "bare" electron is an *ideal point-like Dirac particle*. There are indeed deviations, but they are solely due to the interaction with the radiation field, which can be calculated with arbitrary accuracy. The observable extended structure is thus no "intrinsic" property of the electron.

The same conclusion is true for the muon. Its $(g-2)$ anomaly is slightly larger than that of (27). This is, however, well understood, because in graphs with virtual fermion loops, for instance (a) in Fig. 5.23, e^+e^- pairs are preferentially created by the muon. Therefore the result of the sum is no longer independent of the lepton mass. The theoretical prediction is somewhat less precise since the hadronic corrections make a larger contribution, owing to the higher energy scale, i.e. smaller length scale, of the vacuum fluctuations probed by the muon.

The presently available theoretical and experimental[18] values for the anomalous magnetic moment of the muon

$$g^\mu_{\text{th}} = 2(1 + .001\,165\,919 \pm .000\,000\,002) \quad , \tag{32}$$

$$g^\mu_{\text{exp}} = 2(1 + .001\,165\,923 \pm .000\,000\,009) \quad , \tag{33}$$

are in perfect agreement. The theoretical value, in addition to the QED radiative correction, also contains a contribution caused by the *virtual creation of hadrons*

$$\Delta a_\mu(\text{hadr.}) = (70 \pm 2) \cdot 10^{-9} \quad . \tag{34}$$

The two graphs responsible for this contribution are shown in Fig. 5.24. Here the hatched blobs stand for a multitude of complicated hadronic states coupling to the virtual photons. While it is impossible to calculate these processes from first principles, their contribution nevertheless can be deduced quite reliably using experimental data on the electromagnetic production of hadrons. A further correction to the g factor of the muon is caused by the *weak interaction*. In the Glashow-Salam-Weinberg standard model which unifies electromagnetic and weak interaction there are three (lowest order) additional graphs involving vertex corrections, see Fig. 5.25. Their combined contribution has been calculated as

$$\Delta a_\mu(\text{weak}) = (7 \pm 0.1) \cdot 10^{-9} \quad . \tag{35}$$

Thus a rather modest increase in the accuracy of (33) and (34) will allow a check of the prediction made by the standard model. Finally we mention that the comparison of g^μ_{th} and g^μ_{exp} also provides information on the existence of new particles,[19] on the mass of the Higgs boson, etc.

[18] J. Bailey et al., Nucl. Phys. **B150**, 1 (1979).

[19] J. Reinhardt, A. Schäfer, B. Müller, W. Greiner, Phys. Rev. **C33**, 194 (1986).

EXAMPLE

5.7 The Infrared Catastrophe

The result (18) of Example 5.6 for the form factor $F_1(q^2)$ of the electron is obviously incomplete. The expression is infrared divergent, i.e. it increases infinitely if the fictitious photon mass μ tends to zero. In order to understand this apparently absurd behaviour we recall another process, in which an infrared divergence occurred: The process of bremsstrahlung, discussed in Sect. 3.6. The scattering cross section for the emission of photons calculated there diverges, too, if the emission of arbitrarily soft photons is taken into account. We shall now show that these two divergences cancel each other and that the result is independent of the photon mass μ, which had been introduced, after all, only as a means of computation.

Indeed, it is not sufficient to sum up only the graphs of Fig. 5.22 in order to describe the scattering process of the electron. In addition to the purely elastic scattering there is also the possibility of *inelastic scattering*, in which a photon of frequency ω is emitted, and the outgoing electron correspondingly has a somewhat reduced energy. The crucial point is now that every experimental apparatus has only a *finite energy resolution* ΔE. The emission of a photon with $\omega < \Delta E$ thus cannot be detected. Purely elastic collisions and collisions in which photons with a very long wavelength are emitted are nondistinguishable, and consequently their cross sections must be added. This is sketched in Fig. 5.26. The scattering amplitudes of the graphs corresponding to elastic scattering as well as to the bremsstrahlung of a photon are summed up coherently, while both contributions must then be added incoherently (because the quantum-mechanical final states are different). The sum must then be taken over all photon energies from the lower bound μ up to the value ΔE.

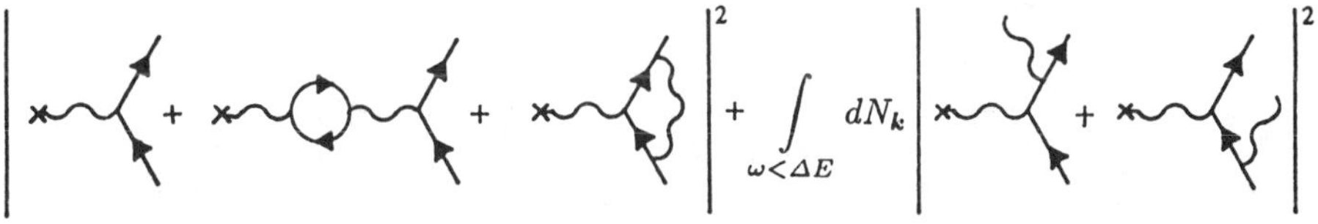

Fig. 5.26. Processes in which photons with long wavelengths are emitted which cannot be distinguished from elastic scattering

To describe the correction quantitatively we compare the S-matrix element for elastic scattering at a Coulomb potential (3.8) to the corresponding result for the emission of bremsstrahlung, (3.193) and (3.201) in Sect. 3.6. In the limit $\omega \to 0$ we find

$$S_{fi}^{\mathrm{Br}} = e\sqrt{\frac{4\pi}{2\omega V}}\left(\frac{\varepsilon\cdot p'}{k\cdot p'} - \frac{\varepsilon\cdot p}{k\cdot p}\right) S_{fi}^{\mathrm{elast}} \quad . \tag{1}$$

Thus the S-matrix element for elastic scattering at the Coulomb potential for photons of long wavelength separates into an expression for the elastic scattering and a factor that depends on the photon energy. We can include the latter as a correction to the form factor of the electron. To do this we must first sum over the photon polarizations λ that are not observed and integrate over the energy:

Example 5.7.

$$\int\limits_{\omega<\Delta E} dN_k \sum_\lambda |F_1^{\mathrm{Br}}|^2 = \int\limits_{\omega<\Delta E} \frac{V|\boldsymbol{k}|^2 d|\boldsymbol{k}| d\Omega_k}{(2\pi)^3} \frac{e^2 4\pi}{2\omega V} \times \sum_\lambda \left| \frac{\varepsilon^{(\lambda)}\cdot p'}{k\cdot p'} - \frac{\varepsilon^{(\lambda)}\cdot p}{k\cdot p}\right|^2 \quad . \tag{2}$$

The polarization sum can be evaluated with the aid of (3.220) from Sect. 3.6:

$$\sum_\lambda \left(\varepsilon^{(\lambda)}\cdot a\right)\left(\varepsilon^{(\lambda)}\cdot b\right) = -a\cdot b \quad , \tag{3}$$

and thus

$$\sum_\lambda |F_1^{\mathrm{Br}}|^2 = \frac{4\pi}{2\omega V} e^2 \left[\frac{-m^2}{(k\cdot p')^2} + \frac{-m^2}{(k\cdot p)^2} + \frac{2p\cdot p'}{(k\cdot p)(k\cdot p')}\right] \quad . \tag{4}$$

Now we restrict ourselves to the *nonrelativistic limit* $|\boldsymbol{p}|, |\boldsymbol{p}'| \ll m$. Noting that

$$\begin{aligned} p\cdot k &= E\omega - \boldsymbol{p}\cdot\boldsymbol{k} = m\sqrt{1+|\boldsymbol{p}|^2/m^2}\,\omega - \boldsymbol{p}\cdot\boldsymbol{k} \\ &\simeq m\omega + \frac{|\boldsymbol{p}|^2\omega}{2m} - \boldsymbol{p}\cdot\boldsymbol{k} \end{aligned}$$

and analogously for $p'\cdot k$, we neglect terms of higher than second order in $|\boldsymbol{p}|$. With the momentum transfer $\boldsymbol{q} = \boldsymbol{p}' - \boldsymbol{p}$ (4) reduces to

$$\sum_\lambda |F_1^{\mathrm{Br}}|^2 = \frac{4\pi}{2\omega V} \frac{e^2|\boldsymbol{q}|^2}{m^2\omega^2} \left[1 - \frac{(\boldsymbol{q}\cdot\boldsymbol{k})^2}{|\boldsymbol{q}|^2\omega^2}\right] \quad . \tag{5}$$

Now the integration over the photon energy in (2) is performed, taking into account $|\boldsymbol{k}|^2 = \omega^2 - \mu^2$ and $|\boldsymbol{k}|d|\boldsymbol{k}| = \omega d\omega$:

$$\begin{aligned} &\int\limits_{\omega<\Delta E} dN_k \sum_\lambda |F_1^{\mathrm{Br}}|^2 \\ &= \frac{4\pi}{(2\pi)^3} \frac{e^2|\boldsymbol{q}|^2}{2m^2} \int\limits_\mu^{\Delta E} \frac{d\omega}{\omega^2} \sqrt{\omega^2-\mu^2} \int d\Omega_k \left[1 - \frac{(\boldsymbol{q}\cdot\boldsymbol{k})^2}{|\boldsymbol{q}|^2\omega^2}\right] \\ &= \frac{e^2|\boldsymbol{q}|^2}{\pi m^2} \int\limits_\mu^{\Delta E} \frac{d\omega}{\omega^2} \sqrt{\omega^2-\mu^2} \left(1 - \frac{1}{3}\frac{\omega^2-\mu^2}{\omega^2}\right) \quad . \end{aligned} \tag{6}$$

This integral can be evaluated with the aid of

$$\begin{aligned} \int dx \frac{\sqrt{x^2-a^2}}{x^2} &= -\frac{\sqrt{x^2-a^2}}{x^2} + \ln\left(x + \sqrt{x^2-a^2}\right) \quad , \\ \int dx \frac{\sqrt{x^2-a^2}}{x^4} &= \frac{1}{3a^2}\frac{(x^2-a^2)^{3/2}}{x^3} \quad . \end{aligned}$$

The result is ($\mu \ll \Delta E$)

$$\int dN_k \sum_\lambda |F_1^{\mathrm{Br}}|^2 = \frac{2e^2|\boldsymbol{q}|^2}{3\pi m^2} \left(\ln\frac{2\Delta E}{\mu} - \frac{5}{6}\right) \quad . \tag{7}$$

Example 5.7.

This result has to be combined with the form factor due to the virtual radiation corrections according to Fig. 5.26. Denoting by $\tilde{F}_1(q^2)$ the square-bracket factor from Example 5.6, (20), we get

$$\left|F_1'(q^2)\right|^2 := \left|\tilde{F}_1(q^2)\right|^2 + \int dN_k |F_1^{\mathrm{Br}}(q^2,k)|^2$$
$$= \left|1 + \frac{\alpha q^2}{3\pi m^2}\left(\ln\frac{m}{\mu} - \frac{3}{8} - \frac{1}{5}\right)\right|^2 + \frac{2\alpha(-q^2)}{3\pi m^2}\left(\ln\frac{2\Delta E}{\mu} - \frac{5}{6}\right)$$

or

$$F_1'(q^2) \simeq 1 + \frac{\alpha}{3\pi m^2}q^2\left(\ln\frac{m}{2\Delta E} + \frac{5}{6} - \frac{3}{8} - \frac{1}{5}\right) \quad . \tag{8}$$

In the last step only the contribution of lowest order in α has been taken into account, to generate the *'infrared-corrected' form factor* $F_1'(q^2)$. The momentum dependent part of this function modifies the strength of the interaction at the vertex according to $e\gamma_\mu \to F_1'(q^2)\gamma_\mu$.

As we have already said, this expression for the form factor of the electron is independent of the photon mass μ. In a careful analysis the problem of the infrared catastrophe has turned out to be fictitious. It arises if one does not account for the fact that the electron is always surrounded by a "cloud" of photons of long wave-lengths, which can be virtual (vertex correction) as well as real (soft bremsstrahlung). Inconsistent results are obtained if one tries to separate the electron from its radiation cloud, for instance by insisting on a final state without photons. It can be shown, in fact, that in every scattering process of charged particles an arbitrary number of photons with long wavelengths is emitted. Purely elastic scattering does not exist in a theory with massless particles. This origin of the infrared catastrophe was recognized very early on.[20] In higher orders of perturbation theory the catastrophe can also be removed by combining internal and external radiation corrections.

The energy ΔE in (8) is determined by the experimental arrangement and the result depends on the resolution one has achieved. In particular, in the study of high-energy collisions in elementary-particle physics, radiation corrections must be carefully evaluated and taken into account in the interpretation of experimental results.

EXAMPLE

5.8 The Energy Shift of Atomic Levels

Besides the $(g-2)$ anomaly of the magnetic moment, the "vacuum fluctuations" predicted by QED show up most clearly in their influence on the energy of bound

[20] F. Bloch and A. Nordsieck: Phys. Rev. **52**, 54 (1937); D.R. Yennie, S.C. Frautschi, H. Sunra: Ann. Phys. (NY) **13**, 379 (1961).

Example 5.8.

states in atoms. Since atomic transition energies can be measured by spectroscopic methods with extreme accuracy, the calculation of level shifts provides a sensitive touchstone for the theory. The exact treatment of this problem is, however, very difficult. After all, we are dealing with bound states, which can interact with the charge *Ze* of the atomic nucleus any number of times. Therefore one must not use free plane waves, as was done in the preceding scattering problems, but one has to use the solutions of the Dirac equation in the Coulomb field. This is depicted in terms of graphs in Fig. 5.27, where the double lines stand for the exact electron propagator in the Coulomb field. (a) shows the self-energy (which now automatically contains the vertex correction). (b) denotes the vacuum polarization. Both sketched graphs are of first order in α with respect to the emission of the virtual photon, but they take into account the interaction with the nuclear charge in any order. The corresponding calculation can be performed with the aid of the exact Feynman propagator of the Dirac equation in the Coulomb field. This is a rather difficult task and can be handled only with the help of numerical calculations.

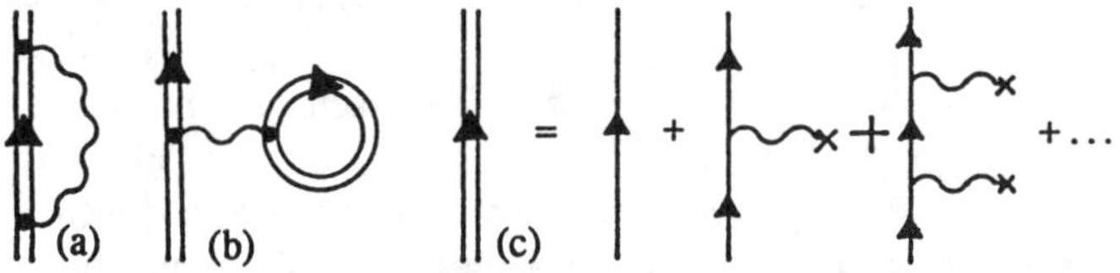

Fig. 5.27a–c. Graphs for self-energy and vacuum polarization in an atom

Fortunately the circumstances in light atoms ($Z\alpha \ll 1$) allow for an approximative method of high accuracy. Its basis is the fact that the atomic binding energies are of the order of magnitude of $(Z\alpha)^2 m$, and thus electron states in light atoms are highly nonrelativistic. One can thus split the evaluation of the graph (a) into two steps.

(i) If virtual photons of high energy $\omega \geq K \gg (Z\alpha)^2 m$ are emitted, the effect of the Coulomb potential can be neglected and one may use free states for the calculation.

(ii) If the virtually emitted photon has low energy $\omega \leq K \ll m$ on the other hand, then the initial, intermediate and final states are all nonrelativistic. Thus one can simply use ordinary quantum-mechanical perturbation theory for the Schrödinger equation.

This procedure will work successfully if one can choose the separating energy K such that

$$(Z\alpha)^2 m \ll K \ll m \tag{1}$$

is valid. In light atoms this poses no problem. Adding the energy shifts obtained in both regions must yield a result independent of K, as long as (1) can be fulfilled.

The Contribution of High Frequencies. We shall calculate the energy shift of an atomic state ψ_ν due to the emission of virtual photons of high frequency $\omega \gg K$. ν denotes the quantum numbers $(njlm)$ of the state. We use the form factors $F_1(q^2)$, $F_2(q^2)$ *for free states*, calculated in Example 5.6. In first-order perturbation theory the energy shift then assumes the form (cf. Example 5.6, (20))

Example 5.8.

$$\delta E_\nu^{>} = e\int d^3x\,\psi_\nu^\dagger(\boldsymbol{x})$$
$$\times\left[\frac{\alpha}{3\pi m^2}\left(\ln\frac{m}{\mu}-\frac{3}{8}-\frac{1}{5}\right)\triangle A^0(\boldsymbol{x})+\frac{\alpha}{2\pi}\frac{\mathrm{i}}{2m}\boldsymbol{\gamma}\cdot\boldsymbol{E}\right]\psi_\nu(\boldsymbol{x}) \quad . \tag{2}$$

Here we have used $q^2=q_0^2-\boldsymbol{q}^2\to(\mathrm{i}\nabla_\mu)\,(\mathrm{i}\nabla^\mu)=-(\partial^2/\partial t^2-\Delta)$. Since we are dealing with a static potential $A^0(\boldsymbol{x})$, only the space-like part, i.e. $+\Delta$, contributes.

$$A^0(\boldsymbol{x})=-\frac{Ze}{|\boldsymbol{x}|} \tag{3}$$

is the Coulomb potential of the nucleus. Only this component of the four-potential is different from zero in our case. $\boldsymbol{E}=-\nabla A^0$ is the corresponding field strength. We have used $\gamma^0\sigma_{0k}\partial A^0/\partial x_k=-\gamma^0\sigma_{0k}\partial A^0/\partial x^k=\gamma^0(\mathrm{i}/2)(\gamma_0\gamma_k-\gamma_k\gamma_0)E^k=\mathrm{i}\gamma_kE^k=-\mathrm{i}\boldsymbol{\gamma}\cdot\boldsymbol{E}$. The term $-1/5$ is due to vacuum polarization; the term proportional to $\boldsymbol{\gamma}\cdot\boldsymbol{E}$ describes the effect of the anomalous magnetic moment.

In order to ensure that only high-frequency photons are considered, we would have to perform the derivation of the form factor $F_1(q^2)$ once more, with the restriction that the region of integration extends over frequencies $\omega>K$, but without the photon mass μ. The result of this calculation, which we do not want to go through here, leads to an elimination of the infrared divergence, just as in Example 5.7, (8). The cutoff parameter K now plays the role of ΔE, which is the energy below which no photons are observed. In (2) one must then replace

$$\ln\frac{m}{\mu}\to\ln\frac{m}{2K}+\frac{5}{6} \quad . \tag{4}$$

With this the energy shift reads

$$\delta E_\nu^{>}=\delta E_\nu^{(1)}+\delta E_\nu^{(2)} \quad , \tag{5}$$

where

$$\delta E_\nu^{(1)}=\frac{e\alpha}{3\pi m^2}\left(\ln\frac{m}{2K}+\frac{5}{6}-\frac{3}{8}-\frac{1}{5}\right)\langle\nu|\nabla^2A^0|\nu\rangle \tag{6a}$$

and

$$\delta E_\nu^{(2)}=\frac{-\mathrm{i}e\alpha}{4\pi m}\langle\nu|\boldsymbol{\gamma}\cdot\nabla A^0|\nu\rangle \quad , \tag{6b}$$

and where we have used the bracket notation for the expectation values. The contribution (6b) of the anomalous magnetic moment can be further transformed. To do this we notice that $\boldsymbol{\gamma}$ is an odd operator, which connects the large and small components of the wave function ψ_ν. To get a nonzero result, one must use the nonrelativistic approximation (for brevity we drop the index ν)

$$\psi=\begin{pmatrix}\phi\\ \chi\end{pmatrix} \tag{7}$$

with

$$\chi=-\frac{\mathrm{i}}{2m}\boldsymbol{\sigma}\cdot\nabla\phi \quad .$$

Example 5.8.

With (7) and $\gamma = \begin{pmatrix} 0 & \sigma \\ -\sigma & 0 \end{pmatrix}$ the matrix element in (6b) reads

$$\begin{aligned} \langle\nu|\gamma\cdot\nabla A^0|\nu\rangle &= \int d^3x\,[\phi^*(\sigma\cdot\nabla A^0)\chi - \chi^*(\sigma\cdot\nabla A^0)\phi] \\ &= -\frac{\mathrm{i}}{2m}\int d^3x\,[\phi^*(\sigma\cdot\nabla A^0)(\sigma\cdot\nabla\phi) + (\sigma\cdot\nabla\phi^*)(\sigma\cdot\nabla A^0)\phi]\,. \end{aligned} \tag{8}$$

Now we can simply use the Schrödinger wave function for ϕ. The second term in (8) we transform by partial integration

$$-\frac{\mathrm{i}}{2m}\int d^3x(\sigma\cdot\nabla\phi^*)(\sigma\cdot\nabla A^0)\phi = +\frac{\mathrm{i}}{2m}\int d^3x\,\phi^*\sigma\cdot\nabla\left[(\sigma\cdot\nabla A^0)\phi\right] \quad . \tag{9}$$

Then the identity

$$\sigma\cdot A\,\sigma\cdot B = A\cdot B + \mathrm{i}\sigma\cdot(A\times B) \tag{10}$$

is used twice:

$$\begin{aligned} &\phi^*(\sigma\cdot\nabla A^0)(\sigma\cdot\nabla\phi) - \phi^*\sigma\cdot\nabla\left[(\sigma\cdot\nabla A^0\phi)\right] \\ &\quad= \phi^*\left[(\nabla A^0)\cdot(\nabla\phi) + \mathrm{i}\sigma\cdot(\nabla A^0)\times(\nabla\phi)\right] \\ &\qquad - \phi^*\left\{\left[(\nabla^2 + \mathrm{i}\sigma\cdot\nabla\times\nabla)A^0\right]\phi + \left[(\nabla\phi)\cdot(\nabla A^0) + \mathrm{i}\sigma\cdot(\nabla\phi)\times(\nabla A^0)\right]\right\} \\ &\quad= -\phi^*(\nabla^2 A^0)\phi + 2\mathrm{i}\phi^*\sigma\cdot(\nabla A^0)\times(\nabla\phi) \quad . \end{aligned} \tag{11}$$

In the last term the angular-momentum operator $L = -\mathrm{i}r\times\nabla$ can be introduced if one takes into account the spherical symmetry of the potential and thus $\nabla A^0 = (r/r)(dA^0/dr)$:

$$2\mathrm{i}\phi^*\sigma\cdot(\nabla A^0)\times(\nabla\phi) = -2\phi^*\frac{1}{r}\frac{dA^0}{dr}\sigma\cdot L\phi \quad . \tag{12}$$

With (8), (11), and (12) the energy shift thus reads

$$\delta E_\nu^{(2)} = \frac{e\alpha}{8\pi m^2}\left(\langle\nu\,|\nabla^2 A^0|\,\nu\rangle + 4\left\langle\nu\left|\frac{1}{r}\frac{dA^0}{dr}S\cdot L\right|\nu\right\rangle\right) \quad . \tag{13}$$

The explicit calculation of the expectation values occurring in (6a) and (13) provides no difficulties.

The Contribution of Low Frequencies. We want to evaluate the energy shift $\delta E^<$ due to emission and reabsorption of soft photons with frequency $\omega < K$ in nonrelativistic perturbation theory in the sense of the approximation scheme discussed at the beginning. The unperturbed problem is described by the Schrödinger equation

$$\hat{H}_0\psi_\nu = \left(-\frac{\nabla^2}{2m} + eA^0(x)\right)\psi_\nu = E\psi_\nu \quad . \tag{14}$$

In addition, there is the perturbation operator

Example 5.8.

$$\hat{H}' = 2\frac{\mathrm{i}e}{2m}\boldsymbol{A}\cdot\nabla \quad , \tag{15}$$

where $\boldsymbol{A}$ is the potential of the radiation field (in transverse gauge). Since $\hat{H}'$ creates and annihilates photons, the energy shift in second-order perturbation theory results as

$$\delta E_\nu^< = \sum_{\nu',\boldsymbol{k},\lambda} \frac{|\langle\nu',\boldsymbol{k},\lambda|\hat{H}'|\nu\rangle|^2}{E_\nu - (E_{\nu'}+\omega)} \quad . \tag{16}$$

Here one must sum or integrate over all electron states ν' and over the momenta $\boldsymbol{k}$ and the polarizations λ of the virtual photon. With the normalization and phase-space factors of the photon field (16) becomes

$$\delta E_\nu^< = \sum_{\nu'}\sum_{\lambda}\int_{\omega<K} \frac{d^3k}{2\omega(2\pi)^3}\frac{4\pi\alpha}{(2m)^2}\frac{\left|\langle\nu'|2\mathrm{i}\,\mathrm{e}^{\mathrm{i}\boldsymbol{k}\cdot\boldsymbol{x}}\boldsymbol{\varepsilon}(\boldsymbol{k},\lambda)\cdot\nabla|\nu\rangle\right|^2}{E_\nu - E_{\nu'}-\omega} \quad . \tag{17}$$

The photon field is introduced according to Sect. 3.6, (3.170). The normalization factor of this plane wave is just $\sqrt{4\pi/2\omega V}$. For a further simplification we now make the "long-wavelength approximation" and replace $\exp(\mathrm{i}\boldsymbol{k}\cdot\boldsymbol{x})$ by 1. This is justified because the typical length scale of the state ϕ_ν is given by the Bohr radius $a_\mathrm{B} = 1/(Z\alpha m)$. This leads to $\boldsymbol{k}\cdot\boldsymbol{x} \leq K a_\mathrm{B} = K/(Z\alpha m)$ and one can choose K such that the exponent remains small in the region allowed by (1).

Generally for the transverse polarization vectors, the following "completeness relation" holds:

$$\int d\Omega_k \sum_{\lambda=1}^{2} \varepsilon_i^*(\boldsymbol{k},\lambda)\varepsilon_j(\boldsymbol{k},\lambda) = \frac{2}{3}4\pi\delta_{ij} \quad . \tag{18}$$

This can be checked by using the polarization vectors

$$\begin{aligned} \boldsymbol{\varepsilon}(\boldsymbol{k},1) &= (\cos\theta\cos\varphi, \cos\theta\sin\varphi, -\sin\theta) \quad , \\ \boldsymbol{\varepsilon}(\boldsymbol{k},2) &= (-\sin\varphi, \cos\varphi, 0) \end{aligned} \tag{19}$$

and integrating over $\int d(\cos\theta)d\varphi$.

The summation over polarizations and the integration over the photon solid angle are easily performed:

$$\int d\Omega_k \sum_{\lambda} \left|\langle\nu'|\mathrm{i}\boldsymbol{\varepsilon}^{(\lambda)}\cdot\nabla|\nu\rangle\right|^2 = \frac{2}{3}4\pi\left|\langle\nu'|\mathrm{i}\nabla|\nu\rangle\right|^2 \quad . \tag{20}$$

Then (17) reads

$$\delta E_\nu^< = \frac{2\alpha}{3\pi}\int_0^K \omega d\omega \sum_{\nu'} \frac{\left|\langle\nu'|\hat{\boldsymbol{v}}|\nu\rangle\right|^2}{E_\nu - E_{\nu'}-\omega} \quad , \tag{21}$$

with the velocity operator $\hat{\boldsymbol{v}} = \hat{\boldsymbol{p}}/m = -\mathrm{i}\nabla/m$. Equation (21), however, does not yet correctly describe the physically observable energy shift, because it contains the contribution of the low-energy photons to the *mass renormalization*. On can

Example 5.8.

realize this by recognizing that there is also a shift for free states. For free electrons the velocity operator is diagonal and the shift simply reads

$$\delta E^{<}_{\mathrm{free}} = \frac{2\alpha}{3\pi}\int_0^K \omega d\omega \frac{\boldsymbol{v}^2}{-\omega} \quad . \tag{22}$$

One gets the physical energy shift by subtracting expression (22) from (21). With $\langle \nu|\boldsymbol{v}^2|\nu\rangle = \Sigma_{\nu'}|\langle \nu'|\boldsymbol{v}|\nu\rangle|^2$ the expression renormalized in this way reads

$$\begin{aligned}(\delta E^{<}_{\nu})_{\mathrm{ren}} &= \delta E^{<}_{\nu} - \delta E^{<}_{\mathrm{free}} \\ &= \frac{2\alpha}{3\pi}\int_0^K \omega d\omega \sum_{\nu'} |\langle \nu'|\hat{\boldsymbol{v}}|\nu\rangle|^2 \left(\frac{1}{E_\nu - E_{\nu'} - \omega} + \frac{1}{\omega}\right) \\ &= \frac{2\alpha}{3\pi}\int_0^K d\omega \sum_{\nu'} |\langle \nu'|\hat{\boldsymbol{v}}|\nu\rangle|^2 \frac{E_\nu - E_{\nu'}}{E_\nu - E_{\nu'} - \omega} \quad . \end{aligned} \tag{23}$$

If we bear in mind that K is to be large compared to all energy differences $E_{\nu'} - E_\nu$, the integration over ω yields for (a) $E_{\nu'} - E_\nu > 0$

$$\begin{aligned}-\int_0^K \frac{d\omega}{\omega + (E_{\nu'} - E_\nu)} &= -\ln \frac{K + (E_{\nu'} - E_\nu)}{E_{\nu'} - E_\nu} \\ &\simeq -\ln \frac{K}{E_{\nu'} - E_\nu}\end{aligned} \tag{24}$$

and for (b) $E_\nu - E_{\nu'} > 0$

$$\begin{aligned}-\int_0^K \frac{d\omega}{\omega - (E_\nu - E_{\nu'})} &= -\left[\int_0^{E_\nu - E_{\nu'} - \varepsilon} \cdots + \int_{E_\nu - E_{\nu'} + \varepsilon}^K \cdots\right] \\ &= -\left[\ln \frac{\varepsilon}{E_\nu - E_{\nu'}} + \ln \frac{K - (E_\nu - E_{\nu'})}{\varepsilon}\right] \\ &\simeq -\ln \frac{K}{E_\nu - E_{\nu'}} \quad .\end{aligned} \tag{25}$$

This leads to

$$(\delta E^{<}_{\nu})_{\mathrm{ren}} = \frac{2\alpha}{3\pi}\sum_{\nu'} |\langle \nu'|\hat{\boldsymbol{v}}|\nu\rangle|^2 (E_{\nu'} - E_\nu) \ln \frac{K}{|E_{\nu'} - E_\nu|} \quad . \tag{26}$$

The evaluation of this sum can not be done analytically. It is useful, however, to rescale the logarithm according to

$$\ln \frac{K}{|E_{\nu'} - E_\nu|} = \ln \frac{2K}{m} - 2\ln(Z\alpha) + \ln \frac{(Z\alpha)^2 m/2}{|E_{\nu'} - E_\nu|} \quad , \tag{27}$$

where in the last term the energy differences are referred to the unit of binding energy (the Rydberg). The first two terms are independent of ν'. The corresponding sum over intermediate states reads

Example 5.8.

$$S = \sum_{\nu'} |\langle\nu'|\hat{\boldsymbol{v}}|\nu\rangle|^2 (E_{\nu'} - E_\nu)$$

$$= \sum_{\nu'} \langle\nu|\hat{\boldsymbol{v}}|\nu'\rangle\langle\nu'|\hat{\boldsymbol{v}}|\nu\rangle(E_{\nu'} - E_\nu) \quad . \tag{28}$$

This expression can be evaluated by a commutator trick frequently used for "sum rules" of this kind. Because of (14) we can replace the energy eigenvalues by the Hamiltonian $\hat{H}_0$

$$S = \frac{1}{2}\sum_{\nu'}\Big(\langle\nu|\hat{\boldsymbol{v}}\hat{H}_0|\nu'\rangle\langle\nu'|\hat{\boldsymbol{v}}|\nu\rangle + \langle\nu|\hat{\boldsymbol{v}}|\nu'\rangle\langle\nu'|\hat{H}_0\hat{\boldsymbol{v}}|\nu\rangle$$

$$- \langle\nu|\hat{H}_0\hat{\boldsymbol{v}}|\nu'\rangle\langle\nu'|\hat{\boldsymbol{v}}|\nu\rangle - \langle\nu|\hat{\boldsymbol{v}}|\nu'\rangle\langle\nu'|\hat{\boldsymbol{v}}\hat{H}_0|\nu\rangle\Big)$$

$$= \frac{1}{2m^2}\langle\nu|2\hat{\boldsymbol{p}}\hat{H}_0\hat{\boldsymbol{p}} - \hat{H}_0\hat{\boldsymbol{p}}\hat{\boldsymbol{p}} - \hat{\boldsymbol{p}}\hat{\boldsymbol{p}}\hat{H}_0|\nu\rangle$$

$$= \frac{1}{2m^2}\langle\nu|[\hat{\boldsymbol{p}},\hat{H}_0]\hat{\boldsymbol{p}} - \hat{\boldsymbol{p}}[\hat{\boldsymbol{p}},\hat{H}_0]|\nu\rangle$$

$$= -\frac{1}{2m^2}\langle\nu|[\hat{\boldsymbol{p}},[\hat{\boldsymbol{p}},\hat{H}_0]]|\nu\rangle \quad , \tag{29}$$

where the closure relation has been used. Because of

$$[\hat{\boldsymbol{p}},\hat{H}_0] = -\mathrm{i}e[\nabla, A^0] = -\mathrm{i}e(\nabla A^0)$$

we get

$$S = \frac{e}{2m^2}\langle\nu|(\nabla^2 A^0)|\nu\rangle \quad . \tag{30}$$

The energy shift due to long-wavelength photons is thus

$$(\delta E_\nu^<)_{\mathrm{ren}} = \frac{e\alpha}{3\pi m^2}\langle\nu|(\nabla^2 A^0)|\nu\rangle\left[\ln\frac{2K}{m} - 2\ln(Z\alpha)\right]$$

$$+ \frac{2\alpha}{3\pi}\sum_{\nu'}|\langle\nu'|\hat{\boldsymbol{v}}|\nu\rangle|^2(E_{\nu'} - E_\nu)\ln\frac{(Z\alpha)^2 m/2}{|E_{\nu'} - E_\nu|} \quad . \tag{31}$$

The Total Energy Shift. Now the contributions of (6a), (13), and (31) must be summed up. As one can see, the logarithmic terms with $\ln(2K/m)$ just cancel each other. As intended the result is therefore independent of the choice of the separating energy K ! The total energy shift is thus given by the following well-defined result:

$$\delta E_\nu = \frac{e\alpha}{3\pi m^2}\left(\frac{5}{6} - \frac{3}{8} - \frac{1}{5} + \frac{3}{8} - 2\ln(Z\alpha)\right)\left\langle\nu\left|\nabla^2 A^0\right|\nu\right\rangle$$

$$+ \frac{e\alpha}{2\pi m^2}\left\langle\nu\left|\frac{1}{r}\frac{dA^0}{dr}\boldsymbol{S}\cdot\boldsymbol{L}\right|\nu\right\rangle$$

$$+ \frac{2\alpha}{3\pi}\sum_{\nu'}|\langle\nu'|\hat{\boldsymbol{v}}|\nu\rangle|^2(E_{\nu'} - E_\nu)\ln\frac{(Z\alpha)^2 m/2}{|E_{\nu'} - E_\nu|} \quad . \tag{32}$$

Now we want to evaluate the matrix elements of (32) as far as possible. The potential (3) satisfies

Example 5.8.

$$\nabla^2 A^0 = 4\pi Ze\delta^3(\boldsymbol{x}) \tag{33}$$

and thus

$$\langle\nu|\nabla^2 A^0|\nu\rangle = 4\pi Ze|\psi_\nu(\mathbf{0})|^2 \quad . \tag{34}$$

The density of the nonrelativistic wave function in a hydrogen-like atom at the origin has the value

$$|\psi_n(\mathbf{0})|^2 \quad = \quad \frac{(Z\alpha)^3 m^3}{\pi n^3}\delta_{l0} \quad . \tag{35}$$

This means that the first term in (32) contributes only for s states ($l = 0$).

For the second matrix element we use the squared expression for the total angular momentum, $\boldsymbol{J} = \boldsymbol{L} + \boldsymbol{S}$, and thus

$$\langle \boldsymbol{L}\cdot\boldsymbol{S}\rangle = \frac{1}{2}\left[j(j+1) - l(l+1) - \frac{3}{4}\right] = \frac{1}{2}\begin{pmatrix} l \\ -l-1 \end{pmatrix} \quad , \tag{36}$$

where the upper (lower) expression is valid for the case $j = l + 1/2$ ($j = l - 1/2$). For s waves (36) vanishes. Because of $(1/r)(dA^0/dr) = -Ze/r^3$ the expectation value of the operator $1/r^3$ is required. Nonrelativistic quantum mechanics yields for this problem, which is in principle elementary,

$$\left\langle \nu \left| \frac{1}{r^3} \right| \nu \right\rangle = \frac{2(Z\alpha)^3 m^3}{l(l+1)(2l+1)n^3} \quad . \tag{37}$$

Finally, we define the quantity

$$L_{nl} = \frac{n^3}{2m(Z\alpha)^4}\sum_{\nu'} |\langle\nu'|\hat{\boldsymbol{v}}|\nu\rangle|^2 (E_{\nu'} - E_\nu)\ln\frac{(Z\alpha)^2 m/2}{(E_{\nu'} - E_\nu)} \quad , \tag{38}$$

which is also known as the *Bethe logarithm* and must be evaluated numerically.

Inserting (34–38) into (32) we finally end up with the expression for the *energy shift of an atomic level,*

$$\begin{aligned} \delta E_{njl} = {} & \frac{4m}{3\pi n^3}\alpha(Z\alpha)^4\left\{ L_{nl} + \left[\frac{19}{30} - 2\ln(Z\alpha)\right]\delta_{l0} \right. \\ \pm {} & \left. \frac{3}{4}\frac{1}{(2j+1)(2l+1)}(1-\delta_{l0})\right\} \end{aligned} \tag{39}$$

for states with $j = l \pm 1/2$. Compared to the unperturbed binding energies

$$E_n \simeq -\frac{(Z\alpha)^2 m}{2n^2} \tag{40}$$

(without spin–orbit splitting) the energy shift is suppressed by a factor $\alpha(Z\alpha)^2$ and thus is very small. Nevertheless the influence of δE_{njl} can be experimentally measured with very high accuracy. In particular (39) predicts that the degeneracy of states with equal total angular momentum j, which is still valid in the Dirac theory for point-like nuclei, is broken. The classical example of this is the energy splitting between the states $2s_{1/2}$ and $2p_{1/2}$ in the hydrogen atom. It was measured

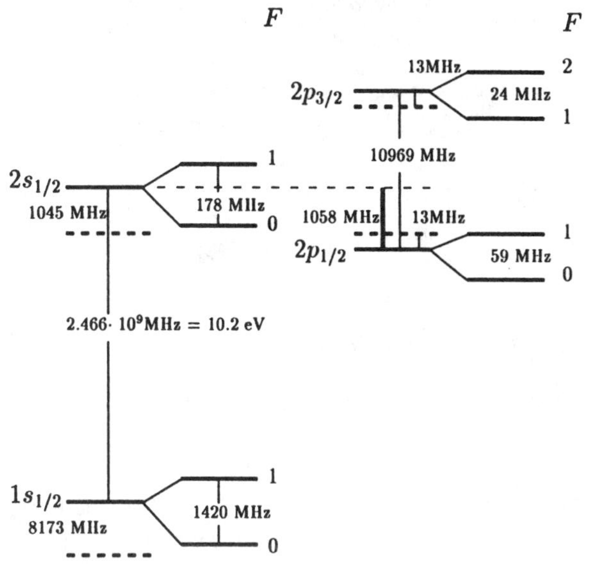

Fig. 5.28. The QED corrections to the energy levels of the K and L shell in the hydrogen atom. The dashed lines indicate the position the unshifted Dirac levels. The heavy line marks the Lamb shift

for the first time by Lamb and collaborators[21] with newly developed methods of microwave spectroscopy and is called the *Lamb shift*.

Figure 5.28 shows a drawing (not a scale) of the innermost energy levels of the hydrogen atom. For each of the states $1s_{1/2}, 2s_{1/2}, 2p_{1/2}, 2p_{3/2}$ the QED shift δE_{njl} with respect to the Dirac energies (dashed lines) is shown. One should note that in reality there is additional structure arising from the interaction of the electron with the magnetic moment of the proton (spin $S = 1/2$). As a result the *hyperfine splitting* transforms each energy level into a doublet of states with different values of the total spin F. When quoting the Lamb shift one always refers to the central energies of the doublets, tacitly removing the hyperfine structure.[22]

To calculate the value of the Lamb shift, the sum L_{nl} of (38) must be evaluated numerically. The result is

$$\begin{aligned} L_{20} &\simeq -2.811\,77 \quad , \\ L_{21} &\simeq +0.030\,02 \quad . \end{aligned} \tag{41}$$

Thus we get in hydrogen ($Z = 1$)

$$\begin{aligned} \delta E_{2s_{1/2}} &= \frac{4m}{3\pi 2^3}\,\alpha(Z\alpha)^4 \left[-2.81177 - 2\ln(Z\alpha) + \frac{19}{30}\right] \\ &= \frac{m}{6\pi}\,\alpha(Z\alpha)^4(+7.662\,05) \end{aligned}$$

---

[21] W.E. Lamb and R.C. Retherford: Phys. Rev. **72**, 241 (1947).

[22] The hyperfine splitting of the hydrogenic ground state ($\delta E_{1s}^{\rm hfs} = 1.420\,405\,751\,766\,7\,\mathrm{GHz}$ which gives rise to the famous 21 cm hydrogen line) is one of the best studied physical observables (known to an accuracy of 10^{-12}). Unfortunately theory is not on a par with this experimental achievement since predictions depend on the detailed internal structure of the proton which is not known well enough. Therefore the last 6 digits in the quoted number have to remain unexplained.

Example 5.8

$$= +4.298\,28 \times 10^{-6}\text{eV}$$
$$= +1039.3\,\text{MHz} \quad , \tag{42a}$$

where in the last step the energy was converted into a frequency using $\delta E = 2\pi\hbar\nu$ ($\hbar = 6.58212 \times 10^{-16}$ eV sec). Analogously it follows that

$$\begin{aligned}\delta E_{2p_{1/2}} &= \frac{4m}{3\pi 2^3}\,\alpha(Z\alpha)^4\left(+0.03002 - \frac{1}{8}\right)\\ &= \frac{m}{6\pi}\,\alpha(Z\alpha)^4(-0.09498)\\ &= -5.328 \times 10^{-8}\text{eV}\\ &= -12.9\,\text{MHz} \quad .\end{aligned} \tag{42b}$$

The resulting hydrogenic *Lamb shift* is thus

$$L^{\text{theor}} = \delta E_{2s_{1/2}} - \delta E_{2p_{1/2}} = +1052.2\,\text{MHz} \quad . \tag{43}$$

This contains a contribution of -27.1 MHz due to vacuum polarization (the term $-1/5$ in (2)). The prediction in (43) is in quite good agreement with todays measured value[23]

$$L^{\text{exp}} = +1057.845 \pm 0.009\,\text{MHz} \quad . \tag{44}$$

To understand the order of magnitude of L the nonrelativistic calculation (31) is quite sufficient. The latter was performed by Bethe[24] immediately after the discovery of the effect by Lamb and Retherford. With the quite arbitrary choice of the cutoff energy $K = m$ he obtained the value $L = 1040$ MHz. To understand the experimental value (44) fully, however, the contribution of the high-energy photons must be taken into account. To achieve a still better quantitative agreement, several contributions of higher order must be included,[25] in particular $\alpha(Z\alpha)^5$ and $\alpha^2(Z\alpha)^4$, as well as recoil corrections due to the finite nuclear mass; see Table 5.2. The result is today's theoretical value for the Lamb shift[26]

$$L^{\text{theor}} = +1057.855 \pm 0.014\text{ MHz} \quad , \tag{45}$$

which is in excellent agreement with the experiment. The binding energies in hydrogen in the framework of quantum electrodynamics are thus *understood with a relative accuracy of* 10^{-11}, a remarkable success! The achievable agreement is limited by the uncertainty in the knowledge of the proton radius.

Using the method of recoilless two-photon laser spectroscopy it has recently become possible to directly measure the energy difference $E_{2s_{1/2}} - E_{1s_{1/2}}$ to a high precision.[27] The results are in full agreement with QED predictions. Note that

---

[23] S.R. Lundeen and F.M. Pipkin: Phys. Rev. Lett **46**, 232 (1981).
[24] H.A. Bethe: Phys. Rev. **72**, 339 (1947).
[25] For example B.E. Lautrup, A. Peterman, E. de Rafael: Phys. Rep. **3**, 193 (1972).
[26] See the review J.R. Sapirstein and D.R. Yennie in T. Kinoshita: *Quantum Electrodynamics*, (World Scientific, Singapore, 1990).
[27] R.G. Beausoleil, D.H. MacIntyre, C.J. Fort, E.A. Hildum, C. Couillard, T.W. Hänsch, Phys. Rev. **A35**, 4878 (1987).

Table 5.2. Contributions to the energy splitting $E_{2s_{1/2}} - E_{2p_{1/2}}$ in the hydrogen atom.

Example 5.8

Contribution	ΔE [MHz]
$\alpha(Z\alpha)^4$	+1 050.559
$\alpha(Z\alpha)^5$	+7.129
$\alpha(Z\alpha)^6$	−0.419
$\alpha^2(Z\alpha)^4$	+0.101
recoil	+0.358
finite nuclear radius	+0.127
sum	1 057.855

the accuracy of these measurements in principle can be increased nearly without bound. In contrast, the original Lamb shift measurements encounter a natural limit since the $2p_{1/2}$ state has a short life time leading to an energy broadening of the order 100 MHz. Thus modern measurements of the Lamb shift already determine the position of the resonance signal to an accuracy of 10^{-4} of the line width. The direct measurement of the $1s - 2s$ energy difference, on the other hand, is not hampered by this problem: the ground state of course is completely stable and the $2s_{1/2}$ level is metastable (life time $\sim 1/7$ s) so that line broadening of this transition is negligible.

Today's efforts aim at the determination of level shifts in hydrogen-like ions with high atomic numbers Z. Such highly charged ions up to U^{91+} can be generated in heavy-ion accelerators. The relative strength of the radiative corrections should be much larger in such systems than in hydrogen. For an adequate theoretical understanding it is necessary, however, to avoid the approximations we made in calculating the Lamb shift but to use instead the full propagator of the Dirac equation with Coulomb potential.[28] The result then is accurate to all orders of the expansion parameter $Z\alpha$.

Values for the energy shift of the $1s$ state in various one-electron atoms are depicted in Fig. 5.29. This figure shows the energy shift divided by the fourth power of the nuclear charge Z or, to be more specific, the function $F(Z)$ which is defined by

$$\delta E_{1s} = \frac{\alpha}{\pi}\frac{(Z\alpha)^4}{n^3} m\, F(Z) \quad . \tag{46}$$

The measured values are in good agreement with the curve representing the theoretical prediction. In the heaviest ion studied, U^{91+}, a group working at the ESR heavy ion storage ring at GSI (Darmstadt)[29] has measured a value of $\delta E_{1s} = 413 \pm 65$ eV which has to be compared with the theoretical value of $\delta E_{1s} = 454$ eV.

[28] P.J. Mohr: Ann. Phys. (NY) **88**, 26 and 52 (1974); P.J. Mohr: in *Physics of Strong Fields*, W. Greiner (ed.) (Plenum, London and New York 1987).

[29] Th. Stöhlker, P.H. Mokler, K. Beckert, F. Bosch, H. Eickhoff, B. Franzke, M. Jung, T. Kandler, C. Kozhuharov, R. Mooshammer, F. Nolden, H. Reich, R. Rymuza, P. Spädtke, M. Steck: Phys. Rev. Lett. **71**, 2184 (1993).

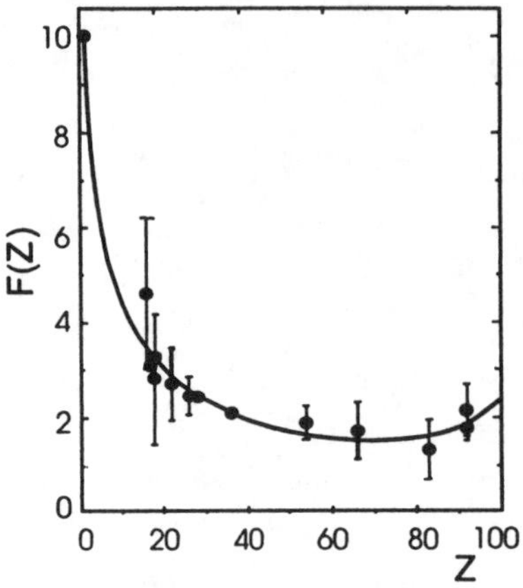

Fig. 5.29. Comparison of the experimental and theoretical results for the energy shift of the 1*s* state in single-electron atoms drawn as a function of the of nuclear charge *Z*

5.5 Biographical Notes

BLOCH, Felix, Swiss-American physicist. *13.10.1905 in Zurich, †10.9.1983 in Zurich. B. studied physics at the Universities of Zurich and Leipzig where he obtained his PhD in 1928 under the direction of W. Heisenberg. After stays at Copenhagen (with N. Bohr) and various other places he became Privatdozent in Leipzig (1932). In 1934 he emigrated to the US and joined the department of physics at Stanford University. In 1954 B. became the first director of CERN (Geneva). B. worked on the quantum theory of metals (his PhD thesis) and ferromagnetism where he established the nature of the boundaries between magnetic domains (Bloch walls). He also studied the stopping of energetic particles in matter. In Stanford he worked on neutron polarization, the magnetic scattering of neutrons, and the magnetic moment of nuclei. For his discovery of nuclear magnetic induction B. was awarded the 1952 Nobel Prize for physics (with E. Purcell).

LAMB, Willis Eugene, American physicist. *12.7.1913 in Los Angeles. L. studied physics at the University of California (PhD in 1938). He became research assistant and later professor at Columbia University (1938-1952), Stanford, Oxford and Yale University (1962-72) and the University of Arizona. L. worked in atomic and solid state physics and microwave spectroscopy. In 1955 he received the Nobel prize for physics (with P. Kusch) for the measurement of the fine structure splitting of the atomic levels in hydrogen (the Lamb shift).

UEHLING, Edwin Albrecht, American physicist. *27.01.1901 in Lowell (Wisc.), † 18.5.1985. Ue. studied physics at the University of Wisconsin and worked as a radio engineer for Bell Telephone Labs and other companies. In 1932 he obtained his PhD at the University of Michigan under the direction of G. Uhlenbeck. Subsequently he was a postdoctoral fellow at the institutes for theoretical physics at Copenhagen and Leipzig (with W. Heisenberg), Berkeley, and Pasadena (with J.R. Oppenheimer). Since 1936 Ue. was professor of physics at the University of Washington. Ue. worked on the quantum theory of transport processes and on the effect of vacuum polarization. He also made contributions to condensed matter physics, in particular the theory of ferroelectricity.

WARD, John Clive, British physicist. *1924 in London. W. studied mathematics and physics at Oxford (DPhil in 1949). He taught at the Universities of Maryland, Miami, and Johns Hopkins University. Since 1966 he has been professor of physics at Macquarie University (Australia). W. has worked on renormalization theory and the effect of gauge invariance in a field theory (the Ward identities). With A. Salam he has investigated the connection between the weak, electromagnetic and strong interactions, preparing the way for unified field theories. W. also contributed to statistical mechanics, working on the two-dimensional Ising model.

6. Two-Particle Systems

6.1 The Bethe–Salpeter Equation

For many purposes it is necessary to have a better description for the interaction of two particles than the perturbation theory we have discussed so far. This is particularly so if we deal with *bound states*: the particles stay together infinitely long and they can therefore interact arbitrarily often as depicted in Fig. 6.1. It is clear that this situation cannot be described by the summation of a few Feynman diagrams. If one of the particles is much heavier than the other the problem can be considerably simplified: one solves the Dirac equation for the light particle with an additional external potential, which is produced by the heavier particle. Because of the large mass ratio the influence of the lighter particle on the heavier one can be neglected.

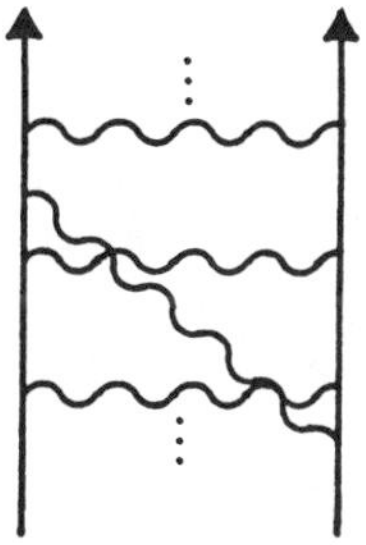

Fig. 6.1. The particles in a bound state can interact infinitely often

The interaction of two particles with equal or comparable masses is much more difficult to describe since the "recoil effects" cannot be neglected. The most famous example of this is positronium, i.e. the bound system of an electron and a positron. Here one cannot distinguish between the source of the field and the test particle. One must rather treat both particles on an equal footing. There is a further complication due to the fact that the interaction propagates with finite velocity, leading to "retardation effects". In a correct relativistic theory there is no preferred common time coordinate. The wave function of the two-particle system depends on two time coordinates and therefore its interpretation is difficult.

In the following we want to write down an exact equation for the two-particle system, at least in principle, even if it turns out that it is too difficult to find an exact solution for this equation. Then we shall study an approximation that again leads to a kind of Dirac equation with an interaction potential (the Breit interaction, see Exercise 6.4).

In order to study the behaviour of two equal particles a and b, we generalize the definition of the propagator. Looking only at one particle, the wave function of that particle is described by a spinor $\psi(x)$ with four components. As shown in Exercise 6.1 the corresponding initial value problem is solved by the Feynman propagator $S_F(x_2, x_1)$:

$$\psi(x_2) = -\mathrm{i} \int d\sigma(x_1)\, S_F(x_2, x_1)\gamma^\mu n_\mu(x_1)\psi(x_1) \quad , \tag{6.1}$$

where we have to integrate over a closed three-dimensional hypersurface that includes the space-time point x_2; $n_\mu(x_1)$ is the exterior normal vector of this hypersurface.

This relation can be extended to the two-particle case. Obviously such a system must be described by a wave function $\psi_{ab}(x_1, x_2)$ that depends on eight coordinates: six space coordinates $(\boldsymbol{x}_1, \boldsymbol{x}_2)$ and two time coordinates (t_1, t_2). Furthermore $\psi_{ab}(x_1, x_2)$ has two spinor indices, i.e. $4 \times 4 = 16$ components! The indices a and b refer to the spinor spaces of the two particles. The appearance of two time coordinates, in contrast to the nonrelativistic many-particle theory, makes the interpretation of this wave function more difficult. Alas, this cannot be avoided if one wants to treat space and time coordinates in the same way.

In analogy to (6.1) we can define a propagator for the two-particle wave function

$$\psi_{ab}(x_3, x_4) = \int d\sigma(x_1)\, d\sigma(x_2)\, S^{ab}(x_3, x_4; x_1, x_2)\, \not n(x_1) \not n(x_2) \psi_{ab}(x_1, x_2) \ , \qquad (6.2)$$

where one now has to integrate over two three-dimensional hypersurfaces which include the points x_3 and x_4. As in Exercise 6.1, (6.2) can be specialized to the case of two space-like hypersurfaces, i.e. to an integration over d^3x_n at a fixed time coordinate t_n.

If the two particles do not interact, the wave function simply is given by a product

$$\psi^0_{ab}(x_1, x_2) = \psi_a(x_1)\psi_b(x_2) \qquad (6.3)$$

so that (6.2) factorizes into two integrals of the form (6.1) if one sets

$$S^{0\,ab}(x_3, x_4; x_1, x_2) = \mathrm{i}S^a_\mathrm{F}(x_3, x_1)\, \mathrm{i}S^b_\mathrm{F}(x_4, x_2) \quad . \qquad (6.4)$$

Going beyond this trivial case we have to consider what happens if the two particles interact electromagnetically. In perturbation theory (expanding in powers of the coupling constant e) the first correction to (6.4) can be found easily,[1] it is given by the one photon exchange diagram

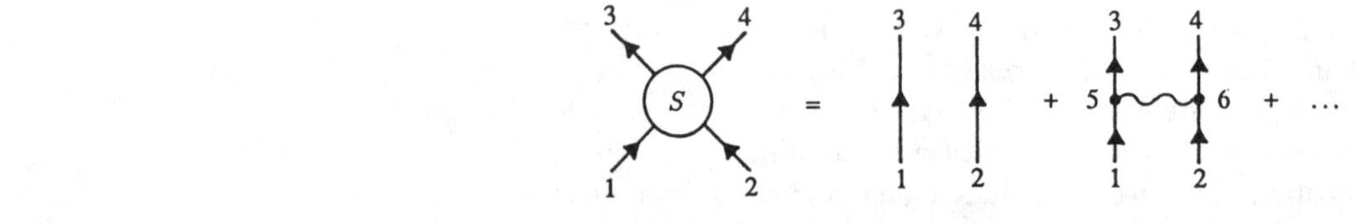

(6.5)

The bubble denoted by S represents the full two-particle propagator whereas the unconnected fermion lines represent the free two-particle propagator S^0 given in (6.4). Using the Feynman rules we can translate (6.5) into the formula

$$\begin{aligned} S^{ab}(x_3, x_4; x_1, x_2) &= \mathrm{i}S^a_\mathrm{F}(x_3, x_1)\, \mathrm{i}S^b_\mathrm{F}(x_4, x_2) \\ &\quad + \int d^4x_5\, d^4x_6\, \mathrm{i}S^a_\mathrm{F}(x_3, x_5)\, \mathrm{i}S^b_\mathrm{F}(x_4, x_6) \\ &\quad \times \left[(-\mathrm{i}e_a)\gamma^a_\mu\, \mathrm{i}D^{\mu\nu}_\mathrm{F}(x_5, x_6)(-\mathrm{i}e_b)\gamma^b_\nu\right]\, \mathrm{i}S^a_\mathrm{F}(x_5, x_1)\, \mathrm{i}S^b_\mathrm{F}(x_6, x_2) \\ &\quad + \cdots \quad . \end{aligned} \qquad (6.6)$$

[1] For simplicity we assume that particles a and b are distinguishable, since otherwise exchange graphs would appear. The case of a particle interacting with its antiparticle will be considered in Example 6.2.

Note that in this and the following equations there are two independent spinor indices. For example the 4×4 matrix γ^a_μ gets multiplied by the matrices $S^a_\mathrm{F}(x_3,x_5)$ and $S^a_\mathrm{F}(x_5,x_1)$, whereas γ^b_ν is multiplied by $S^b_\mathrm{F}(x_4,x_6)$ and $S^b_\mathrm{F}(x_6,x_2)$. *The two-particle propagator therefore has the character of a* 16×16 *matrix.* In (6.5) and (6.6) we have written down only the first term of an infinite series. The result of the infinite series can be represented in terms of a "black box" by a function $K(x_3,x_4;x_1,x_2)$ which is called the *interaction kernel*. The exact form of (6.5) then reads

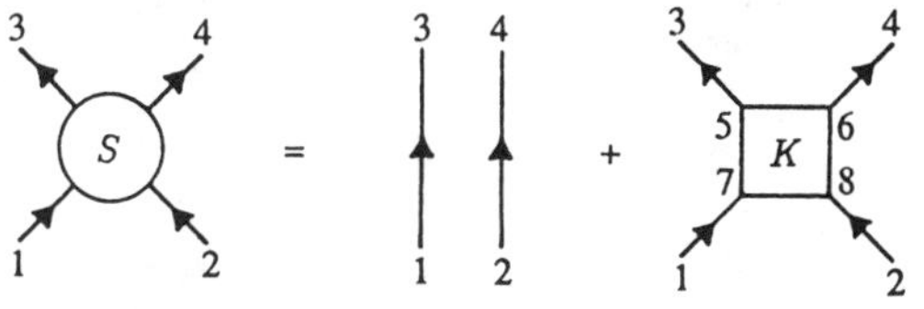

(6.7)

or, written out explicitly,

$$
\begin{aligned}
S^{ab}(x_3,x_4;x_1,x_2) &= \mathrm{i}S^a_\mathrm{F}(x_3,x_1)\,\mathrm{i}S^b_\mathrm{F}(x_4,x_2) \\
&\quad + \int d^4x_5\,d^4x_6\,d^4x_7\,d^4x_8\;\mathrm{i}S^a_\mathrm{F}(x_3,x_5)\,\mathrm{i}S^b_\mathrm{F}(x_4,x_6) \\
&\quad \times K^{ab}(x_5,x_6;x_7,x_8)\;\mathrm{i}S^a_\mathrm{F}(x_7,x_1)\,\mathrm{i}S^b_\mathrm{F}(x_8,x_2) \quad . \qquad (6.8)
\end{aligned}
$$

We have not gained much by doing this since K is an extremely complicated function. Only in first-order perturbation theory does the kernel become very simple, that is, according to (6.6)

$$K^{ab}_0(x_5,x_6;x_7,x_8) = (-\mathrm{i}e_a)\gamma^a_\mu\,\mathrm{i}D^{\mu\nu}_\mathrm{F}(x_5,x_6)(-\mathrm{i}e_b)\gamma^b_\nu\,\delta^4(x_5-x_7)\,\delta^4(x_6-x_8)\ . \qquad (6.9)$$

The complete function K is a sum over infinitely many graphs of arbitrarily high order. Some examples are shown in Fig. 6.2. To go beyond perturbation theory one can apply some cunning and combine at least a certain subset of the terms contributing to the infinite sum K. To that end we define the notion of a *reducible interaction kernel*. It is characterized by the fact that it can be split into two unconnected parts by cutting two fermion lines. Correspondingly a kernel is called *irreducible* if it is so densely interwoven that such a dissection is not possible.[2]

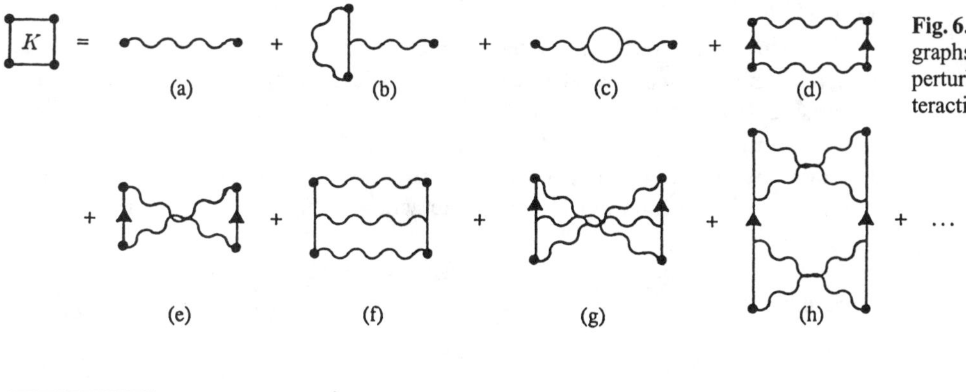

Fig. 6.2a–h. A few typical graphs contributing to the perturbation series for the interaction kernel K

[2] The cut has to be applied in the "horizontal" direction. The graph (a) to (c) in Fig. 6.2 thus are not reducible.

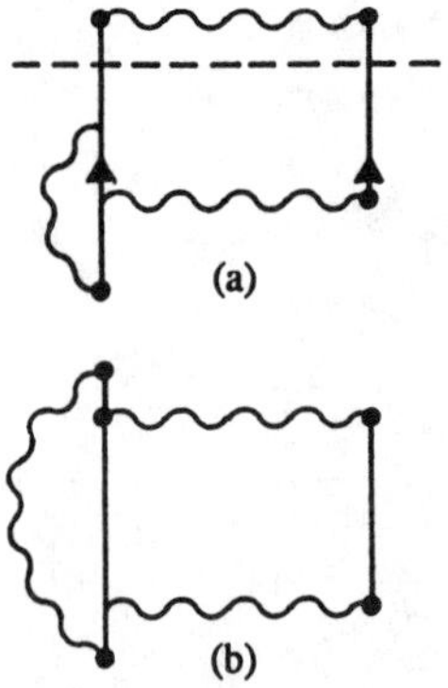

Fig. 6.3. Examples for a reducible (**a**) and an irreducible (**b**) third-order interaction

Examples of reducible and irreducible graphs are shown in Fig. 6.3. In Fig. 6.2 only graphs d, f, and h are reducible. We shall denote the sum of all irreducible contributions to K as the *irreducible interaction kernel* $\overline{K}$.

It is obvious that each reducible graph can be described by joining together several irreducible graphs from the set which contributes to the kernel $\overline{K}$. But this can be obtained by a simple modification of (6.7):

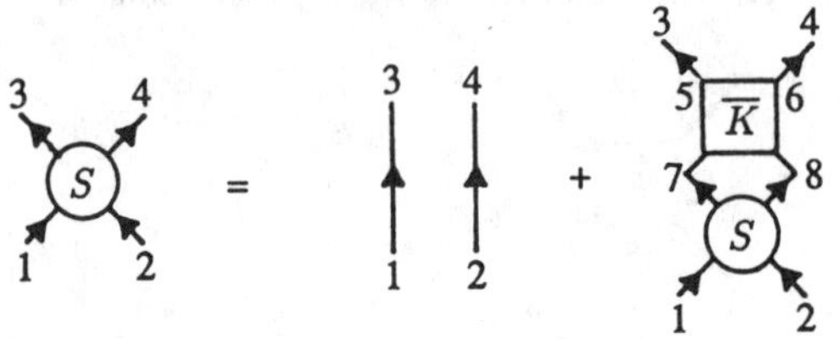

(6.10)

or, written out explicitly

$$S^{ab}(x_3,x_4;x_1,x_2) = \mathrm{i}S_{\mathrm{F}}^{a}(x_3,x_1)\,\mathrm{i}S_{\mathrm{F}}^{b}(x_4,x_2) + \int d^4x_5\,d^4x_6\,d^4x_7\,d^4x_8\;\mathrm{i}S_{\mathrm{F}}^{a}(x_3,x_5)\,\mathrm{i}S_{\mathrm{F}}^{b}(x_4,x_6) \times \overline{K}^{ab}(x_5,x_6;x_7,x_8)S^{ab}(x_7,x_8;x_1,x_2) \quad . \tag{6.11}$$

In this equation the complete kernel K has been replaced by the irreducible kernel $\overline{K}$, but in return for that we have also replaced two independent one-particle propagators by the complete two-particle propagator. The fact that (6.10) is equivalent to (6.7) becomes immediately clear if one solves (6.10) iteratively, i.e. if one repeatedly inserts the left-hand side of (6.10) into the right-hand side:

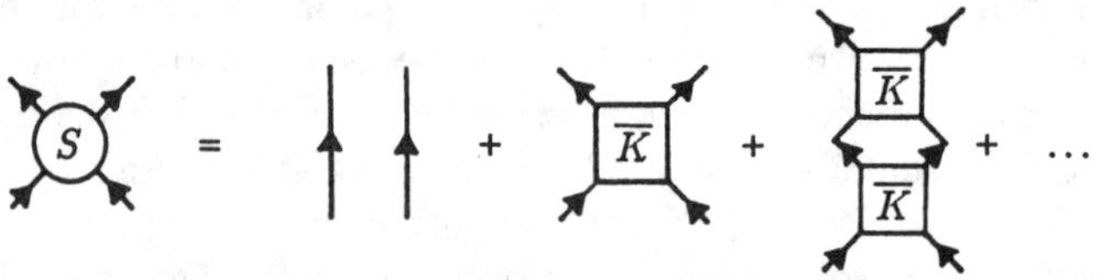

(6.12)

Therefore the iteration of (6.10) ensures that all possible combinations of irreducible Feynman graphs are combined and in this way one gets the complete sum K from $\overline{K}$.

The irreducible kernel $\overline{K}$ contains an infinite number of Feynman graphs as well and cannot be calculated exactly. Nevertheless, compared to (6.7), (6.10) has a decisive advantage: looking at (6.12) one sees that the solution automatically contains an *infinite series of interaction*, even if $\overline{K}$ itself is calculated within perturbation theory at low order! As we have discussed at the beginning of this section the inclusion of an infinite number of interactions is necessary if one is interested in bound systems.

For many practical purposes one restricts oneself to the lowest order of the irreducible kernel $\overline{K}$ (Fig. 6.2a), i.e. to the one-photon exchange

$$\overline{K}^{ab}(x_5,x_6;x_7,x_8) \approx K_0^{ab}(x_5,x_6;x_7,x_8) \tag{6.13}$$

given in (6.9). This prescription is called the *ladder approximation*. This name suggests itself if we look at the iterated equation (6.12):

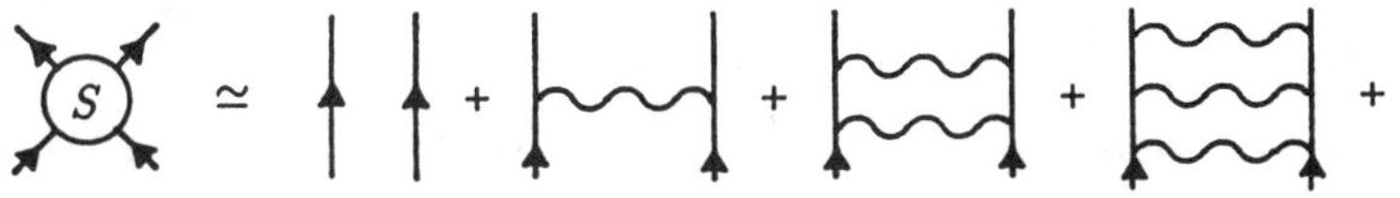

... (6.14)

In this approximation a Lorentz frame can be found in which only one photon is exchanged at a given time, but this process can be repeated an arbitrary number of times. Nevertheless, one has to be aware that there is a multitude of possible graphs. Within the ladder approximation one considers only a very special class of them owing to the restriction of $\overline{K}$, even though this class contains an infinite number of graphs. The quality of this approximation can only be proved by its success.

We want to remark that in principle one should not use the free Feynman propagators S_{F} when calculating (6.11). Instead one should use "dressed" propagators that contain the interaction with their own photon field to all orders (cf. Chap. 5). Then one has taken into account all self-interaction graphs in (6.10). In the same way one should use the exact photon propagators and vertex functions when calculating $\overline{K}$. The renormalization problems related to that will not be discussed here.

In order to get an equation for the two-particle wave function we insert (6.11) into (6.2):

$$\begin{aligned}\psi_{ab}(x_3,x_4) =& \int d\sigma(x_1)\,d\sigma(x_2)\,\mathrm{i}S_{\mathrm{F}}^{a}(x_3,x_1)\,\mathrm{i}S_{\mathrm{F}}^{b}(x_4,x_2)\,\not{n}(x_1)\,\not{n}(x_2)\psi_{ab}(x_1,x_2)\\ &+\int d\sigma(x_1)\,d\sigma(x_2)\int d^4x_5\,d^4x_6\,d^4x_7\,d^4x_8\,\mathrm{i}S_{\mathrm{F}}^{a}(x_3,x_5)\,\mathrm{i}S_{\mathrm{F}}^{b}(x_4,x_6)\\ &\times\overline{K}^{ab}(x_5,x_6;x_7,x_8)\,S^{ab}(x_7,x_8;x_1,x_2)\,\not{n}(x_1)\,\not{n}(x_2)\,\psi_{ab}(x_1,x_2)\quad,\end{aligned} \tag{6.15}$$

or, if we abbreviate the first term by $\phi_{ab}(x_3,x_4)$, insert (6.2) in the second term and rename some indices,

$$\begin{aligned}\psi_{ab}(x_1,x_2) =& \,\phi_{ab}(x_1,x_2)+\int d^4x_3\,d^4x_4\,d^4x_5\,d^4x_6\,\mathrm{i}S_{\mathrm{F}}^{a}(x_1,x_5)\,\mathrm{i}S_{\mathrm{F}}^{b}(x_2,x_6)\\ &\times\overline{K}^{ab}(x_5,x_6;x_3,x_4)\psi_{ab}(x_3,x_4)\quad.\end{aligned} \tag{6.16}$$

This is the *Bethe–***Salpeter** *equation.*[3] It is a complicated inhomogenous integral equation of the Fredholm type. Its mathematical structure is the price we have to pay in order to go beyond perturbation theory.

$\phi_{ab}(x_1,x_2)$ is the *free* two-particle wave function. If one is interested in bound states, i.e. localized states, $\phi_{ab}(x_1,x_2)$ drops out of (6.16) and the integral equation becomes homogeneous (see the supplementary remarks at the end of this section). The Bethe–Salpeter equation (6.16) can also be written in another form if one multiplies with the free Dirac operators $(\mathrm{i}\not{\nabla}_1-m_a)$ and $(\mathrm{i}\not{\nabla}_2-m_b)$ on the left-hand side. Since the one-particle propagators obey the relations

$$\begin{aligned}(\mathrm{i}\not{\nabla}_1-m_a)S_{\mathrm{F}}^{a}(x_1,x_5) &= \delta^4(x_1-x_5)\quad,\\ (\mathrm{i}\not{\nabla}_2-m_b)S_{\mathrm{F}}^{b}(x_2,x_6) &= \delta^4(x_2-x_6)\quad,\end{aligned} \tag{6.17}$$

[3] H.A. Bethe and E.E. Salpeter: Phys. Rev. **82**, 309 (1951) and **84**, 1232 (1951).

it follows that

$$(\mathrm{i}\nabla\!\!\!/_1 - m_a)(\mathrm{i}\nabla\!\!\!/_2 - m_b)\psi_{ab}(x_1,x_2) = -\int d^4x_3\, d^4x_4\, \overline{K}^{ab}(x_1,x_2;x_3,x_4)\psi_{ab}(x_3,x_4) \ . \tag{6.18}$$

In this form the Bethe–Salpeter equation is an integro-differential equation (in eight variables). For practical purposes there is another useful form of this equation, obtained by transforming it into momentum space.

If we define the wave function in momentum space as

$$\chi_{ab}(p_1,p_2) = \frac{1}{(2\pi)^4}\int d^4x_1 d^4x_2\, \mathrm{e}^{\mathrm{i}(p_1\cdot x_1+p_2\cdot x_2)}\psi_{ab}(x_1,x_2) \quad , \tag{6.19}$$

then the Fourier transform of (6.18) reads

$$\begin{aligned}&\frac{1}{(2\pi)^4}\int d^4x_1 d^4x_2\, \mathrm{e}^{\mathrm{i}(p_1\cdot x_1+p_2\cdot x_2)}(\mathrm{i}\nabla\!\!\!/_1 - m_a)(\mathrm{i}\nabla\!\!\!/ - m_b)\psi_{ab}(x_1,x_2)\\ &= -\frac{1}{(2\pi)^4}\int d^4x_1\, d^4x_2\, d^4x_3\, d^4x_4\, \mathrm{e}^{\mathrm{i}(p_1\cdot x_1+p_2\cdot x_2)}\overline{K}^{ab}(x_1,x_2;x_3,x_4)\psi_{ab}(x_3,x_4) \ .\end{aligned} \tag{6.20}$$

On the left-hand side we integrate by parts so that the gradient operators act on the exponential function, and on the right-hand side we insert two delta functions $\delta^4(x_3'-x_3)$, $\delta^4(x_4'-x_4)$:

$$\begin{aligned}&\frac{1}{(2\pi)^4}\int d^4x_1\, d^4x_2\, \left[(-\mathrm{i}\nabla\!\!\!/_1 - m_a)(-\mathrm{i}\nabla\!\!\!/_2 - m_b)\,\mathrm{e}^{\mathrm{i}(p_1\cdot x_1+p_2\cdot x_2)}\right]\psi_{ab}(x_1,x_2)\\ &= -\frac{1}{(2\pi)^4}\int d^4x_1\, d^4x_2\, d^4x_3\, d^4x_4\, d^4x_3'\, d^4x_4'\, \delta^4(x_3'-x_3)\,\delta^4(x_4'-x_4)\\ &\quad\times \mathrm{e}^{\mathrm{i}(p_1\cdot x_1+p_2\cdot x_2)}\overline{K}^{ab}(x_1,x_2;x_3,x_4)\,\psi_{ab}(x_3',x_4') \quad .\end{aligned} \tag{6.21}$$

Using the integral representation of the delta function,

$$\delta^4(x_3'-x_3) = \frac{1}{(2\pi)^4}\int d^4p_1'\, \mathrm{e}^{\mathrm{i}p_1'\cdot(x_3'-x_3)} \quad , \tag{6.22}$$

one can express the right-hand side of (6.21) as a product of momentum-space wave functions and the interaction kernel in momentum space:

$$\begin{aligned}&\overline{K}^{ab}(p_1,p_2;p_3,p_4)\\ &= \frac{1}{(2\pi)^8}\int d^4x_1\, d^4x_2\, d^4x_3\, d^4x_4\, \mathrm{e}^{\mathrm{i}(p_1\cdot x_1+p_2\cdot x_2-p_3\cdot x_3-p_4\cdot x_4)}\overline{K}^{ab}(x_1,x_2;x_3,x_4) \quad .\end{aligned} \tag{6.23}$$

The *Bethe–Salpeter equation in momentum space* then reads

$$\begin{aligned}&(p\!\!\!/_1 - m_a)(p\!\!\!/_2 - m_b)\chi_{ab}(p_1,p_2)\\ &= -\int d^4p_1'\, d^4p_2'\, \overline{K}^{ab}(p_1,p_2;p_1',p_2')\,\chi(p_1',p_2') \quad .\end{aligned} \tag{6.24}$$

When treating a two-particle system it is always advantageous to transform to absolute and relative coordinates. For simplicity we assume that both particles have the same mass $m = m_a = m_b$, and we define

$$P = p_1 + p_2 \quad , \quad p = \frac{1}{2}(p_1 - p_2) \quad , \tag{6.25}$$

or

$$p_1 = \frac{1}{2}P + p \quad , \quad p_2 = \frac{1}{2}P - p \quad . \tag{6.26}$$

Since the interaction described by the kernel $\overline{K}^{ab}$ must conserve momentum, i.e. $p_1 + p_2 = p_1' + p_2'$, one can make the ansatz

$$\overline{K}^{ab}(p_1, p_2; p_1', p_2') = \delta^4(P - P')\overline{K}^{ab}(p, p'; P) \quad . \tag{6.27}$$

Using (6.26) and (6.27), (6.24) then reads

$$\begin{aligned}\left(\frac{1}{2}\not{P}^a + \not{p}^a - m\right)\left(\frac{1}{2}\not{P}^b - \not{p}^b - m\right)\chi_{ab}(p, P) \\ = -\int d^4p'\, d^4P' \left|\frac{\partial(p_1', p_2')}{\partial(p', P')}\right| \overline{K}^{ab}(p, p'; P')\,\delta^4(P - P')\,\chi_{ab}(p', P')\end{aligned} \tag{6.28}$$

or

$$\begin{aligned}\left(\frac{1}{2}\not{P}^a + \not{p}^a - m\right)\left(\frac{1}{2}\not{P}^b - \not{p}^b - m\right)\chi_{ab}(p, P) \\ = -\int d^4p'\, \overline{K}^{ab}(p, p'; P)\,\chi_{ab}(p', P) \quad ,\end{aligned} \tag{6.29}$$

since the Jacobian determinant for the transformation of the volume element in (6.27) is equal to one. In (6.29) P plays only the role of a parameter. One can look at a wave function with a given value K of the "centre-of-mass momentum"

$$\chi_{ab}(p, P) = \delta^4(P - K)\,\chi_{ab}(p) \quad . \tag{6.30}$$

Integrating (6.29) over P and defining $p' = p + k$, one gets the final result

$$\begin{aligned}\left(\frac{1}{2}\not{K}^a + \not{p}^a - m\right)\left(\frac{1}{2}\not{K}^b - \not{p}^b - m\right)\chi_{ab}(p) \\ = -\int d^4k\, \overline{K}^{ab}(p, p + k; K)\,\chi_{ab}(p + k) \quad .\end{aligned} \tag{6.31}$$

This integral equation has discrete eigensolutions for K and – in principle – it allows us to determine the spectrum of a bound system of two fermions. The binding energy E_B can be read off from the eigenvalue K in the "centre-of-mass system" defined by $\boldsymbol{p}_1 + \boldsymbol{p}_2 = \boldsymbol{0}$, where it takes the form

$$K = (2m - E_B, \boldsymbol{0}) \quad . \tag{6.32}$$

Unfortunately the interaction kernel $\overline{K}^{ab}$ is very complicated and cannot be written down in a closed form. But even if one restricts oneself to the simplest case of the ladder approximation (6.13), the structure of (6.31) is still so complicated that one will not succeed in finding exact solutions. Only a simplified problem, the binding of two spin-0 particles with a scalar interaction,[4] can be solved completely. It

[4] G.C. Wick: Phys. Rev. **96**, 1124 (1954); R.E. Cutkosky, ibid. p. 1135.

turns out that the Bethe–Salpeter equation is beset with a number of serious difficulties (unphysical states, wrong limit $m_b/m_a \to \infty$, etc.). For problems within the framework of QED, especially those concerning the spectrum of positronium, perturbative approximation techniques are adequate (however, the ladder approximation does not suffice, one must take into account also the graph from Fig. 6.2e). Furthermore the numerical solution of the Bethe–Salpeter equation is an important tool for calculating bound states in the realm of elementary particle physics (quark–antiquark systems, i.e. mesons).[5]

Supplement. In the following we want to justify our assertion that the free solution $\phi_{ab}(x_1, x_2)$ of the Bethe–Salpeter equation drops out if one studies bound states. The argument is based on the energy and momentum balance.

For the integration limits $\sigma(x_1')$ and $\sigma(x_2')$ of the initial value integral (6.15) we choose space-like hypersurfaces with $t_1' = t_2' = t$ in the distant past:

$$\phi_{ab}(x_1, x_2) = \int d^3x_1' \int d^3x_2' \, \mathrm{i}S_F^a(x_1; \boldsymbol{x}_1', t)\, \mathrm{i}S_F^b(x_2; \boldsymbol{x}_2', t)\, \gamma_0^a \gamma_0^b \, \psi_{ab}(\boldsymbol{x}_1', t; \boldsymbol{x}_2', t) \, . \tag{6.33}$$

Because of $t_1 > t$ and $t_2 > t$ only that part of the Feynman propagator contributes which propagates forward in time, namely (cf. Problem 2.1)

$$\mathrm{i}S_F(x', x) \longrightarrow \int d^3p \, \frac{m}{E} \Lambda_+(p) \, \mathrm{e}^{-\mathrm{i}p\cdot(x'-x)} \quad . \tag{6.34}$$

The two-particle wave function ψ_{ab} with equal time argument t should have the following form:

$$\psi_{ab}(\boldsymbol{x}_1', t; \boldsymbol{x}_2', t) = \mathrm{e}^{-\mathrm{i}K_0 t} \, \mathrm{e}^{\mathrm{i}\boldsymbol{K}\cdot(\boldsymbol{x}_1' + \boldsymbol{x}_2')/2} \, \chi_{ab}(\boldsymbol{x}_1' - \boldsymbol{x}_2') \quad . \tag{6.35}$$

Here the first factor describes the time development with the total energy K_0, whereas the second factor describes the motion of the centre of mass with momentum $\boldsymbol{K}$. $\chi_{ab}(\boldsymbol{x}_1' - \boldsymbol{x}_2')$ is the wave fuction of the relative motion of the bound state.

K_0 and $\boldsymbol{K}$ have to satisfy the usual dispersion relation

$$K_0^2 - \boldsymbol{K}^2 = M^2 \quad , \tag{6.36}$$

i.e. seen from the outside, with respect to the centre-of-mass motion, the bound system behaves like a single particle. However, its mass M is reduced by the binding energy E_B (the mass defect):

$$M = 2m - E_B \quad . \tag{6.37}$$

Now we insert (6.34) and (6.35) into (6.33)

$$\begin{aligned} \phi_{ab}(x_1, x_2) = & \int d^3x_1' \, d^3x_2' \int d^3p' \, \frac{m}{E'} \Lambda_+(p') \, \mathrm{e}^{-\mathrm{i}p_0'(t_1 - t)} \, \mathrm{e}^{+\mathrm{i}\boldsymbol{p}'\cdot(\boldsymbol{x}_1 - \boldsymbol{x}_1')} \\ & \times \int d^3p \, \frac{m}{E} \Lambda_+(p) \, \mathrm{e}^{-\mathrm{i}p_0(t_2 - t)} \, \mathrm{e}^{+\mathrm{i}\boldsymbol{p}\cdot(\boldsymbol{x}_2 - \boldsymbol{x}_2')} \\ & \times \mathrm{e}^{-\mathrm{i}K_0 t} \, \mathrm{e}^{\mathrm{i}\boldsymbol{K}\cdot(\boldsymbol{x}_1' + \boldsymbol{x}_2')/2} \, \gamma_0^a \, \gamma_0^b \, \chi_{ab}(\boldsymbol{x}_1' - \boldsymbol{x}_2') \quad . \end{aligned} \tag{6.38}$$

[5] An extensive bibliography on the Bethe–Salpeter equation and its applications can be found in: N. Nakanishi: Prog. Theor. Phys. Suppl. **95**, 78 (1988).

Next it is useful to transform this to centre-of-mass coordinates and relative coordinates:

$$\boldsymbol{x} = \boldsymbol{x}_1' - \boldsymbol{x}_2' \quad , \quad \boldsymbol{X} = \frac{1}{2}(\boldsymbol{x}_1' + \boldsymbol{x}_2') \quad , \tag{6.39a}$$

or

$$\boldsymbol{x}_1' = \boldsymbol{X} + \frac{1}{2}\boldsymbol{x} \quad , \quad \boldsymbol{x}_2' = \boldsymbol{X} - \frac{1}{2}\boldsymbol{x} \quad . \tag{6.39b}$$

This yields

$$\begin{aligned}\phi_{ab}(x_1,x_2) = &\int d^3p' \, \frac{m}{E'} \, \Lambda_+(p') \, \mathrm{e}^{-\mathrm{i}p'\cdot x_1} \int d^3p \, \frac{m}{E} \, \Lambda_+(p) \, \mathrm{e}^{-\mathrm{i}p\cdot x_2} \\ &\times \mathrm{e}^{\mathrm{i}(p_0'+p_0-K_0)t} (2\pi)^3 \, \delta^3(\boldsymbol{K}-\boldsymbol{p}-\boldsymbol{p}') \int d^3x \, \mathrm{e}^{\mathrm{i}\frac{1}{2}\boldsymbol{x}\cdot(\boldsymbol{p}'-\boldsymbol{p})} \, \gamma_0^a \, \gamma_0^b \, \chi_{ab}(\boldsymbol{x}) \, .\end{aligned} \tag{6.40}$$

Therefore, the relation

$$\boldsymbol{K} = \boldsymbol{p} + \boldsymbol{p}' \tag{6.41a}$$

must hold. On the other hand (6.40) leads to an analogous condition for K_0. This is because ϕ_{ab} is only independent of the arbitrary choice of the starting time t if the oscillating factor vanishes,

$$K_0 = p_0 + p_0' \quad . \tag{6.41b}$$

For a somewhat more rigorous derivation of this argument one could take the average over some time interval, in order to eliminate strongly oscillating contributions, e.g. by the prescription

$$\phi_{ab}(x_1, x_2) = \lim_{T\to-\infty} \int_{2T}^{T} dt \, \ldots \quad .$$

Using (6.41) the dispersion relation (6.36) takes the form

$$(p_0 + p_0')^2 - (\boldsymbol{p} + \boldsymbol{p}')^2 = M^2 \quad ,$$

or

$$\begin{aligned} 2m^2 + 2p_0p_0' - 2\boldsymbol{p}\cdot\boldsymbol{p}' &= M^2 \quad , \\ \sqrt{m^2+\boldsymbol{p}^2}\,\sqrt{m^2+\boldsymbol{p}'^2} - \boldsymbol{p}\cdot\boldsymbol{p}' &= m^2 - \frac{1}{2}(4m^2 - M^2) \quad . \end{aligned} \tag{6.42}$$

This equation cannot be satisfied if there is a mass defect due to binding, $M^2 < (2m)^2$, for the following reasons. For $\boldsymbol{p} = \boldsymbol{p}' = 0$ the left-hand side is larger than the right-hand side, and this also remains true for finite momenta (where $\boldsymbol{p} \parallel \boldsymbol{p}'$ is the most favourable case) , since $\sqrt{(1+x^2)(1+y^2)} - xy \geq 1$, which can be verified immediately by taking the square of this expression. Thus we have proved that the inhomogeneous term of the Bethe–Salpeter equation, ϕ_{ab} from (6.33), vanishes for bound states.

EXERCISE

6.1 Solution of the Initial-Value Problem for Fermions

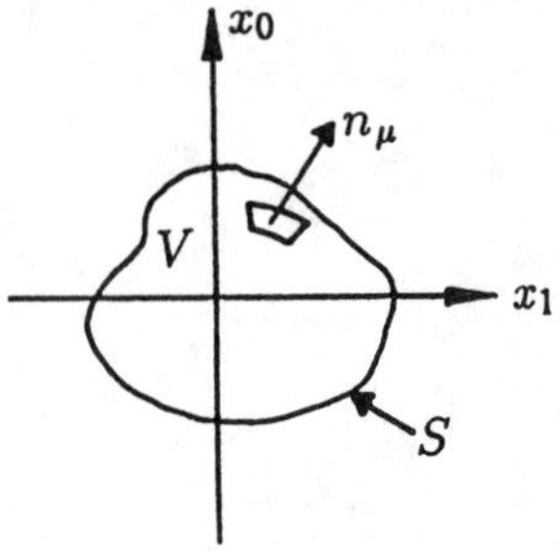

Fig. 6.4. A four-dimensional space–time volume V with surface S

Problem. Using Gauss' theorem in four dimensions determine the wave function $\psi(x)$ at points x within a space–time volume V provided that its value is known on a closed surface S, see Fig. 6.4. Derive an analogous relation also for the adjoint spinor $\bar{\psi}(x)$.

Solution. Gauss' theorem in four dimensions takes the form

$$\int_V d^4x \, \frac{\partial F_\mu}{\partial x_\mu}(x) = \int_S d\sigma(x) \, F_\mu(x) n^\mu(x) \quad . \tag{1}$$

Here F_μ is a vector field and $n^\mu(x)$ is the exterior normal on the surface element $d\sigma(x)$. We choose the function $F_\mu(x') \equiv \mathrm{i}S_\mathrm{F}\,(x-x')\gamma_\mu\psi(x')$. Then the divergence entering the volume integral is given by

$$\begin{aligned} &\frac{\partial}{\partial x'_\mu}\left(\mathrm{i}S_\mathrm{F}\,(x-x')\gamma_\mu\psi(x')\right) \\ &= \left(\mathrm{i}\frac{\partial}{\partial x'_\mu}S_\mathrm{F}\,(x-x')\right)\gamma_\mu\psi(x') + S_\mathrm{F}\,(x-x')\gamma_\mu\left(\mathrm{i}\frac{\partial}{\partial x'_\mu}\psi(x')\right) \\ &= \left(\mathrm{i}\frac{\partial}{\partial x'_\mu}S_\mathrm{F}\,(x-x')\right)\gamma_\mu\psi(x') + mS_\mathrm{F}\,(x-x')\,\psi(x') \\ &= \left[S_\mathrm{F}\,(x-x')\left(\mathrm{i}\gamma_\mu\frac{\overleftarrow{\partial}}{\partial x'_\mu} + m\right)\right]\psi(x') \quad . \end{aligned} \tag{2}$$

Here we have made use of the fact that $\psi(x)$ fulfills the free Dirac equation. The expression in square brackets reduces to a delta function, since the equation

$$\left(i\gamma_\mu\frac{\partial}{\partial x_\mu} - m\right)S_\mathrm{F}\,(x-x') = \delta^4(x-x') \tag{3}$$

defining the Feynman propagator is equivalent to

$$S_\mathrm{F}\,(x-x')\left(\mathrm{i}\gamma_\mu\frac{\overleftarrow{\partial}}{\partial x'_\mu} + m\right) = -\delta^4(x-x') \quad , \tag{4}$$

which can be shown using the momentum space representation (2.19). Therefore the left-hand side of (1) becomes

$$\begin{aligned} \int_V d^4x' \, \frac{\partial F_\mu}{\partial x'_\mu}(x') &= \int_V d^4x' \, \left(-\delta^4(x-x')\right)\psi(x') \\ &= -\psi(x) \end{aligned} \tag{5}$$

if $x \in V$. The right-hand side of (1) becomes

$$\int_S d\sigma(x') \, F_\mu(x') n^\mu(x') = \mathrm{i}\int_S d\sigma(x') \, S_\mathrm{F}\,(x-x')\gamma_\mu n^\mu(x')\psi(x') \quad , \tag{6}$$

Exercise 6.1.

with the result

$$\psi(x) = -\mathrm{i} \int_S d\sigma(x')\, S_\mathrm{F}(x - x') \not{n}(x')\psi(x') \quad . \tag{7}$$

This result can be specialized by choosing a four-dimensional volume V delimited by two flat hypersurfaces at constant times, e.g. t_1 and t_2, see Fig. 6.5. The side faces are assumed to lie infinitely far away, so that the volume V comprises the whole three-dimensional space but only a restricted time interval. The normal vectors are $n_1 = (-1, \mathbf{0})$ and $n_2 = (1, \mathbf{0})$. Then (7) takes the form

$$\psi(x) = \mathrm{i} \int_{t_1} d^3x'\, S_\mathrm{F}(x - x')\gamma_0\psi(x') - \mathrm{i} \int_{t_2} d^3x'\, S_\mathrm{F}(x - x')\gamma_0\psi(x') \quad , \tag{8}$$

with $t_1 < t < t_2$.

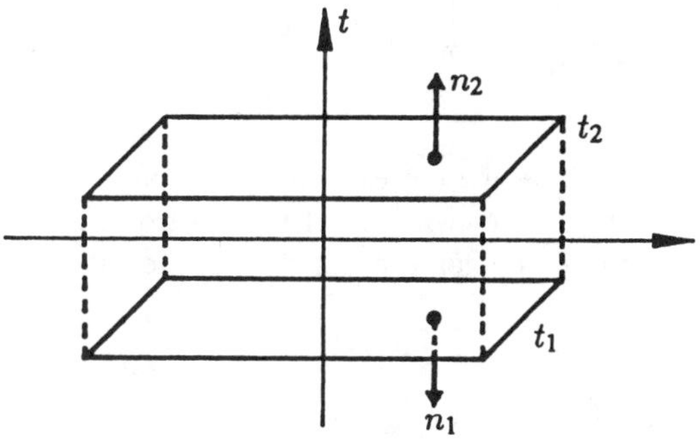

Fig. 6.5. A special choice for the space–time volume bounded by flat hypersurfaces at constant times t_1 and t_2

If the wave function has only components with positive (negative) frequencies, then it follows from the properties of the propagator (see (2.25, 2.26)) that only the integral at the time t_1 (t_2) contributes so (8) is in agreement with the result of Exercise 2.4. Therefore (7), which solves the boundary-value problem for a spinor field, is the covariant generalization of the former result of Chap. 2.

The calculation for the adjoint spinor $\bar{\psi}(x)$ proceeds in an analogous fashion. We now choose the vector field as $F_\mu(x') = -\mathrm{i}\bar{\psi}(x')\gamma_\mu S_\mathrm{F}(x' - x)$. Using the Dirac equation for the adjoint spinor

$$\bar{\psi}(x)\left(\mathrm{i}\gamma_\mu \frac{\partial}{\partial x_\mu} + m\right) = 0 \tag{9}$$

and (3) the left hand side of (1) can be reduced to $-\psi(x)$. Thus in analogy to (7) the adjoint spinor satisfies

$$\bar{\psi}(x) = \mathrm{i} \int d\sigma(x')\, \bar{\psi}(x') \not{n}(x')\, S_\mathrm{F}(x' - x) \quad . \tag{10}$$

EXERCISE

6.2 The Bethe–Salpeter Equation for Positronium

Problem. Until now we have assumed that the interacting particles a and b are not related to each other. What will change in the case of a particle–antiparticle system such as positronium (e^+e^-)? Derive the Bethe–Salpeter equation in momentum space (6.31) considering the interaction kernel $\overline{K}$ in lowest order.

Solution. The electron positron system can be treated in very much the same way as the electron electron problem. In the expansion in terms of Feynman graphs the direction of the arrow of one fermion is inverted. As we have already discussed in connection with Bhabha scattering (Section 3.4) the exchange graph describes the process of virtual pair annihilation.

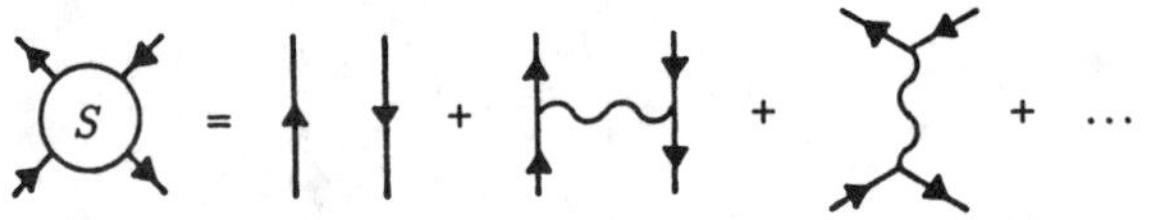 (1)

Constructing the corresponding two-particle wave function we take into account the antiparticle character of the positron by choosing an adjoint spinor (incoming electron ↔ outgoing positron). For the case of no interaction this means

$$\psi^0_{\mu\sigma}(x_1,x_2) = \psi_\mu(x_1)\,\overline{\psi}_\sigma(x_2) \quad . \tag{2}$$

Without interaction the explicit form of the electron spinor would be a plane wave $\psi(x_1) = u(p,s)\,\mathrm{e}^{-\mathrm{i}p\cdot x_1}$. In the same way the positron spinor would be given by $\bar{\psi}(x_2) = \bar{v}(p',s')\,\mathrm{e}^{+\mathrm{i}p'\cdot x_2}$. In accordance with the Feynman rules the incoming electron is thus described by $u(p,s)$ and the incoming positron by $\bar{v}(p',s')$. The propagation of an adjoint spinor was derived in Exercise 6.1, (12):

$$\bar{\psi}(x) = \mathrm{i}\int d\sigma(x')\,\bar{\psi}(x')\,\not{n}(x')\,S_\mathrm{F}(x',x) \quad . \tag{3}$$

For the two-particle wave function this means

$$\begin{aligned}\psi_{\mu\sigma}(x_3,x_4) = \int & d\sigma(x_1)\,d\sigma(x_2)\,S_{\mu\sigma\nu\tau}(x_3,x_4;x_1,x_2)\\ &\times \not{n}_{\nu\nu'}(x_1)\not{n}_{\tau'\tau}(x_2)\,\psi_{\nu'\tau'}(x_1,x_2) \quad .\end{aligned} \tag{4}$$

Note the reversed order of the primed and unprimed indices of the matrices $\not{n}$. This becomes clear by writing $\not{n}_{\nu\nu'}(x_1)\,\psi_{\nu'\tau'}(x_1,x_2)\,\not{n}_{\tau'\tau}(x_2)$ and is related to the reversed order of $\not{n}\psi$ in (6.1) and $\bar{\psi}\not{n}$ in (3), respectively. The two-particle propagator reads as follows:

$$\begin{aligned}&S_{\mu\sigma\nu\tau}(x_3,x_4;x_1,x_2)\\ &\quad= -\mathrm{i}S_{\mathrm{F}\mu\nu}(x_3,x_1)\,\mathrm{i}S_{\mathrm{F}\tau\sigma}(x_2,x_4) - \int d^4x_5\,d^4x_6\,d^4x_7\,d^4x_8\;\mathrm{i}S_{\mathrm{F}\mu\mu'}(x_3,x_5)\\ &\quad\times \mathrm{i}S_{\mathrm{F}\sigma'\sigma}(x_6,x_4)K_{\mu'\sigma'\nu'\tau'}(x_5,x_6;x_7,x_8)\mathrm{i}S_{\mathrm{F}\nu'\nu}(x_7,x_1)\mathrm{i}S_{\mathrm{F}\tau\tau'}(x_2,x_8) \;.\end{aligned} \tag{5}$$

Exercise 6.2.

Equation (5) differs from (6.8) only by the "reversed" Feynman propagators for the positron. This shows up especially in the exchanged primed and umprimed indices of the matrices $S_{\mathrm{F}\nu'\nu}(x_7,x_1)$ and $S_{\mathrm{F}\tau\tau'}(x_2,x_8)$ for electrons and positrons, respectively. This has the same origin as the change of the positions of the indices in both matrices $\not{\eta}$ in (4). Here one must observe carefully the ordering of S, $\not{\eta}$, and ψ in (6.1) and (3). The positron propagators always have "exchanged" indices.

Now one can again introduce an irreducible interaction kernel $\overline{K}$ and iterate (5).

$$\begin{aligned}
&S_{\mu\sigma\nu\tau}(x_3,x_4;x_1,x_2)\\
&\quad= -\mathrm{i}S_{\mathrm{F}\mu\nu}(x_3,x_1)\,\mathrm{i}S_{\mathrm{F}\tau\sigma}(x_2,x_4) + \int d^4x_5\,d^4x_6\,d^4x_7\,d^4x_8\,\mathrm{i}S_{\mathrm{F}\mu\mu'}(x_3,x_5)\\
&\qquad\times \mathrm{i}S_{\mathrm{F}\sigma'\sigma}(x_6,x_4)\overline{K}_{\mu'\sigma'\nu'\tau'}(x_5,x_6;x_7,x_8)S_{\nu'\tau'\nu\tau}(x_7,x_8;x_1,x_2) \quad . \qquad (6)
\end{aligned}$$

Inserting this into (4) yields the integral form of the Bethe–Salpeter equation

$$\begin{aligned}
&\psi_{\mu\sigma}(x_1,x_2)\\
&\quad= \phi_{\mu\sigma}(x_1,x_2) + \int d^4x_3\,d^4x_4\,d^4x_5\,d^4x_6\\
&\qquad\times \mathrm{i}S_{\mathrm{F}\mu\mu'}(x_1,x_5)\mathrm{i}S_{\mathrm{F}\sigma'\sigma}(x_6,x_2)\overline{K}_{\mu'\sigma'\nu'\tau'}(x_5,x_6\,;x_3,x_4)\psi_{\nu'\tau'}(x_3,x_4) \quad . \qquad (7)
\end{aligned}$$

For bound states (i.e. localized states) the free solution $\phi_{\mu\sigma}(x_1,x_2)$ can again be omitted.

In order to get an integro-differential equation like (6.18) we make use of

$$(\mathrm{i}\not{\nabla}_1 - m)_{\mu\nu}\,S_{\mathrm{F}\nu\mu'}(x_1,x_5) = \delta^4(x_1-x_5)\,\delta_{\mu\mu'} \qquad (8a)$$

and

$$(\mathrm{i}\not{\nabla}_2 + m)_{\tau\sigma}\,S_{\mathrm{F}\sigma'\tau}(x_6,x_2) = -\delta^4(x_2-x_6)\,\delta_{\sigma\sigma'} \quad . \qquad (8b)$$

Equation (8b) can be derived immediately from the integral representation of the Feynman propagator in momentum space. Applying the Dirac operators from (8) to (7) and renaming some indices yields

$$\begin{aligned}
&(\mathrm{i}\not{\nabla}_1 - m)_{\mu\mu'}\,(\mathrm{i}\not{\nabla}_2 + m)_{\sigma'\sigma}\,\psi_{\mu'\sigma'}(x_1,x_2)\\
&\quad= \int d^4x_3\,d^4x_4\,\overline{K}_{\mu\sigma\nu\tau}(x_1,x_2\,;x_3,x_4)\,\psi_{\nu\tau}(x_3,x_4) \qquad (9)
\end{aligned}$$

in analogy to (6.18).

This result can be transformed into momentum space, too. Again defining the interacting kernel as

$$\begin{aligned}
\overline{K}_{\mu\sigma\nu\tau}(p_1,p_2\,;p_3,p_4) = {}& \frac{1}{(2\pi)^8}\int d^4x_1\,d^4x_2\,d^4x_3\,d^4x_4\\
&\times \mathrm{e}^{\mathrm{i}(p_1\cdot x_1+p_2\cdot x_2-p_3\cdot x_3-p_4\cdot x_4)}\overline{K}_{\mu\sigma\nu\tau}(x_1,x_2\,;x_3,x_4) \qquad (10)
\end{aligned}$$

we can directly use the previous calculation with the result

$$\begin{aligned}
&\left(\frac{1}{2}\not{P} + \not{p} - m\right)_{\mu\mu'}\left(\frac{1}{2}\not{P} - \not{p} + m\right)_{\sigma'\sigma}\chi_{\mu'\sigma'}(p,P)\\
&\quad= \int d^4p'\,\overline{K}_{\mu\sigma\nu\tau}(p,p';P)\,\chi_{\nu\tau}(p',P) \quad . \qquad (11)
\end{aligned}$$

Exercise 6.2.

This equation differs from (6.29) only by the second Dirac operator and the plus sign in the interaction term.

In order to write down the kernel $\overline{K}$ in lowest order of the perturbation series in e the contributions from direct scattering and virtual annihilation must be taken into account. The corresponding Feynman graphs of Fig. 6.6 can be translated into formulae:

$$\begin{aligned}
&\overline{K}^0_{\mu\sigma\nu\tau}(p_1,p_2;p_1',p_2')\\
&= \frac{1}{(2\pi)^8}\int d^4x_1\,d^4x_2\,d^4x_1'\,d^4x_2'\,\mathrm{e}^{\mathrm{i}(p_1\cdot x_1+p_2\cdot x_2-p_1'\cdot x_1'-p_2'\cdot x_2')}(-\mathrm{i}e)^2\\
&\times\Big(\gamma^\alpha_{\mu\nu}\mathrm{i}D_{\mathrm{F}\alpha\beta}(x_1,x_2)\,\gamma^\beta_{\tau\sigma}\,\delta^4(x_1-x_1')\,\delta^4(x_2-x_2')\\
&-\gamma^\alpha_{\mu\sigma}\mathrm{i}D_{\mathrm{F}\alpha\beta}(x_1,x_1')\,\gamma^\beta_{\tau\nu}\,\delta^4(x_1-x_2)\,\delta^4(x_1'-x_2')\Big)\\
&= \frac{1}{(2\pi)^8}\int d^4x_1\,d^4x_2\,\mathrm{e}^{\mathrm{i}[(p_1-p_1')\cdot x_1+(p_2-p_2')\cdot x_2]}(-\mathrm{i}e)^2\gamma^\alpha_{\mu\nu}\gamma^\beta_{\tau\sigma}\mathrm{i}D_{\mathrm{F}\alpha\beta}(x_1,x_2)\\
&-\frac{1}{(2\pi)^8}\int d^4x_1\,d^4x_1'\,\mathrm{e}^{\mathrm{i}[(p_1+p_2)\cdot x_1-(p_1'+p_2')\cdot x_1']}(-\mathrm{i}e)^2\gamma^\alpha_{\mu\sigma}\gamma^\beta_{\tau\nu}\mathrm{i}D_{\mathrm{F}\alpha\beta}(x_1,x_1')\quad.
\end{aligned}\tag{12}$$

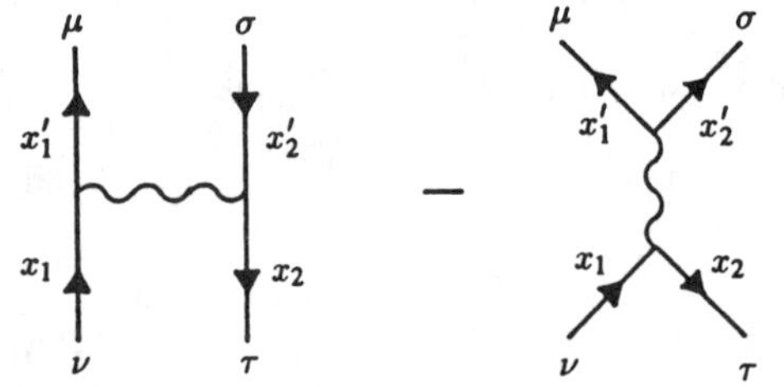

Fig. 6.6. The lowest-order direct and exchange graphs. The fermion lines are "amputated", i.e. they do not enter the expression for the integral kernel $\overline{K}$

In the first integral we now substitute $u = x_1 - x_2$, $v = x_1 + x_2$ and in the second $u = x_1 - x_1'$, $v = x_1 + x_1'$. After ordering the terms in the exponent we find that the integration over v breaks down and yields a delta function of 4-momentum conservation. The integration over u yields the Fourier transform of the photon propagator. The result is

$$\begin{aligned}
\overline{K}^0_{\mu\sigma\nu\tau}(p_1,p_2;p_1',p_2') &= \delta^4(P-P')\frac{(-\mathrm{i}e)^2}{(2\pi)^4}\gamma^\alpha_{\mu\nu}\gamma^\beta_{\tau\sigma}\mathrm{i}D_{\mathrm{F}\alpha\beta}(p-p')\\
&\quad-\delta^4(P-P')\frac{(-\mathrm{i}e)^2}{(2\pi)^4}\gamma^\alpha_{\mu\sigma}\gamma^\beta_{\tau\nu}\mathrm{i}D_{\mathrm{F}\alpha\beta}(P)\quad,
\end{aligned}\tag{13}$$

with

$$P = p_1 + p_2\quad,\qquad P' = p_1' + p_2'\quad,$$
$$p = \frac{1}{2}(p_1 - p_2)\quad,\qquad p' = \frac{1}{2}(p_1' - p_2')$$

or

$$\overline{K}^0_{\mu\sigma\nu\tau}(p,p';P) = \frac{-\mathrm{i}e^2}{(2\pi)^4}\gamma^\alpha_{\mu\nu}\gamma^\beta_{\tau\sigma}D_{\mathrm{F}\alpha\beta}(p-p') + \frac{\mathrm{i}e^2}{(2\pi)^4}\gamma^\alpha_{\mu\sigma}\gamma^\beta_{\tau\nu}D_{\mathrm{F}\alpha\beta}(P)\ .\tag{14}$$

Exercise 6.2.

As in (6.30) we now separate off the centre-of-mass momentum P. After integrating over P in analogy to (6.31) one gets the *Bethe–Salpeter equation for the electron positron system in the ladder approximation:*

$$\left(\frac{1}{2}\not{K}+\not{p}-m\right)_{\mu\mu'}\left(\frac{1}{2}\not{K}-\not{p}+m\right)_{\sigma'\sigma}\chi_{\mu'\sigma'}(p)$$
$$=\left[(V_{\text{dir}}+V_{\text{anni}})\,\chi(p)\right]_{\mu\sigma}\quad, \tag{15}$$

with the direct interaction

$$\left[V_{\text{dir}}\,\chi(p)\right]_{\mu\sigma}=\frac{-\mathrm{i}e^2}{(2\pi)^4}\gamma^{\alpha}_{\mu\nu}\gamma^{\beta}_{\tau\sigma}\int d^4p'\,D_{\mathrm{F}\alpha\beta}(p-p')\,\chi_{\nu\tau}(p')\quad, \tag{16a}$$

and the annihilation interaction

$$\left[V_{\text{anni}}\,\chi(p)\right]_{\mu\sigma}=\frac{\mathrm{i}e^2}{(2\pi)^4}\gamma^{\alpha}_{\mu\sigma}\gamma^{\beta}_{\tau\nu}D_{\mathrm{F}\alpha\beta}(K)\int d^4p'\,\chi_{\nu\tau}(p')\quad. \tag{16b}$$

In the Feynman gauge the photon propagator is

$$D_{\mathrm{F}\alpha\beta}(q)=-\frac{4\pi g_{\alpha\beta}}{q^2} \tag{17}$$

and the interaction terms take the form

$$V_{\text{dir}}\,\chi(p)=\frac{\mathrm{i}e^2}{4\pi^3}\int\frac{d^4p'}{(p-p')^2}\gamma^{\alpha}\chi(p')\gamma_{\alpha}\quad, \tag{18a}$$

$$V_{\text{anni}}\,\chi(p)=-\frac{\mathrm{i}e^2}{4\pi^3}\frac{1}{K^2}\gamma^{\alpha}\int d^4p'\,\mathrm{Tr}\left[\gamma_{\alpha}\chi(p')\right]\quad, \tag{18b}$$

where we have not written out the Dirac indices. The direct term (18a) has exactly the same form as for a system of two particles of different kinds with opposite charges, $e_b=-e_a$, (cf. (5) in Example 6.3). The annihilation interaction enters with the same power of e, but nevertheless it is much weaker since the denominator is very large, namely $K^2\approx(2m)^2$ for weakly bound systems. In practice one therefore first solves (15) without the annihilation interaction and then treats (18b) as a perturbation.

Supplement. One can bring the Bethe–Salpeter equation (15) into a form that is symmetric with respect to particles and antiparticles by using the charge conjugation transformation. The charge conjugated wave function reads[6]

$$\psi_c=\hat{C}\bar{\psi}^T=\hat{C}\gamma_0\psi^*\quad. \tag{19}$$

The transformation matrix $\hat{C}$ satisfies the condition

$$\hat{C}\gamma^{\mu}\hat{C}^{-1}=-\gamma^{\mu T} \tag{20}$$

and furthermore

$$\hat{C}^T=\hat{C}^{\dagger}=\hat{C}^{-1}=-\hat{C}\quad. \tag{21}$$

[6] See W. Greiner: Theoretical Physics, Vol. 3, *Relativistic Quantum Mechanics – Wave Equations* (Springer, Berlin, Heidelberg 1990), Chap. 12.

Exercise 6.2.

We apply the charge conjugation matrix $\hat{C}$ on the right index of the two-particle wave function χ and introduce a transformed wave function ψ:

$$\psi_{\mu\sigma}(p) = C_{\sigma\sigma'}\chi_{\mu\sigma'}(p) \tag{22}$$

or in compact notation

$$\psi(p) = \chi(p)\,\hat{C}^T \quad . \tag{23}$$

In turn we can use (21) and get

$$\chi(p) = \psi(p)\,(\hat{C}^T)^{-1} = \psi(p)\,\hat{C} \quad . \tag{24}$$

This transformation is inserted into (15) and the equation is multiplied by $\hat{C}^{-1}$ from the right-hand side. Because of (20) only the relative sign between momentum and mass terms in the Dirac operator is changed. After some manipulation one gets

$$\begin{aligned}
&\left(\frac{K\!\!\!/}{2} + p\!\!\!/ - m\right)_1 \left(\frac{K\!\!\!/}{2} - p\!\!\!/ - m\right)_2 \psi(p) \\
&= \frac{\mathrm{i}e^2}{4\pi^3} \int \frac{d^4p'}{(p-p')^2}\,(\gamma^\alpha)_1\,(\gamma_\alpha)_2\,\psi(p') \\
&\quad - \frac{\mathrm{i}e^2}{4\pi^3}\,\frac{1}{K^2}\,\gamma^\alpha\hat{C}\int d^4p'\,\mathrm{Tr}\left[\gamma_\alpha\psi(p')\,\hat{C}\right] \quad .
\end{aligned} \tag{25}$$

EXAMPLE

6.3 The Nonretarded Limit of the Bethe–Salpeter Equation

We continue to study the Bethe–Salpeter equation in the form (6.31) within the *ladder approximation* and introduce an approximation in which it takes the form of a simple Dirac equation with an interaction potential.

To this end we first have to determine the interaction kernel $\overline{K}^{ab}(p,p';P)$ in momentum space. In the ladder approximation (6.9) according to (6.23) one essentially has to take the Fourier transform of the photon propagator $D_\mathrm{F}(x-x')$:

$$\begin{aligned}
&\overline{K}_0^{ab}(p_1,p_2;p_1',p_2') \\
&= \frac{1}{(2\pi)^8}\int d^4x_1\,d^4x_2\,d^4x_1'\,d^4x_2'\,\mathrm{e}^{\mathrm{i}(p_1\cdot x_1 + p_2\cdot x_2 - p_1'\cdot x_1' - p_2'\cdot x_2')} \\
&\quad \times (-\mathrm{i}e_a)(-\mathrm{i}e_b)\,\gamma_a^\mu\gamma_\mu^b\,\mathrm{i}D_\mathrm{F}(x_1-x_2)\delta^4(x_1-x_1')\,\delta^4(x_2-x_2') \quad .
\end{aligned} \tag{1}$$

Transforming to the new variables $u = x_1 - x_2$ and $v = x_1 + x_2$ yields

$$\begin{aligned}
\overline{K}_0^{ab}(p_1,p_2;p_1',p_2') &= \frac{(-\mathrm{i}e_a)(-\mathrm{i}e_b)}{(2\pi)^4}\,\gamma_\mu^a\gamma_\nu^b\,\delta^4\left[(p_1+p_2)-(p_1'+p_2')\right] \\
&\quad \times \mathrm{i}D_\mathrm{F}^{\mu\nu}\left(\frac{1}{2}(p_1-p_2) - \frac{1}{2}(p_1'-p_2')\right)
\end{aligned} \tag{2}$$

or, according to (6.27), *Example 6.3.*

$$\overline{K}_0^{ab}(p,p';P) = \frac{-\mathrm{i}e_a e_b}{(2\pi)^4}\,\gamma_\mu^a\, D_{\mathrm{F}}^{\mu\nu}(p-p')\,\gamma_\nu^b \quad . \tag{3}$$

In Feynman gauge, which we have used so far, the momentum-space representation of the photon propagator reads

$$D_{\mathrm{F}}^{\mu\nu}(q) = \int d^4x\, \mathrm{e}^{\mathrm{i}q\cdot x}\, D_{\mathrm{F}}^{\mu\nu}(x) = -\frac{4\pi g^{\mu\nu}}{q^2+\mathrm{i}\varepsilon} \quad . \tag{4}$$

Thus in momentum space the Bethe–Salpeter equation (6.31) in the ladder approximation takes the form

$$\begin{aligned}&\left(\frac{1}{2}\,\not{K}^a + \not{p}^a - m\right)\left(\frac{1}{2}\,\not{K}^b - \not{p}^b - m\right)\chi_{ab}(p)\\ &\quad = \mathrm{i}e_a e_b \int \frac{d^4k}{(2\pi)^4}\,\gamma_\mu^a\, D_{\mathrm{F}}^{\mu\nu}(k)\,\gamma_\nu^b\,\chi_{ab}(p+k) \quad .\end{aligned} \tag{5}$$

The integration kernel in (5) is still too complicated for practical purposes. The problem can be simplified if one *neglects the frequency dependence*, i.e. if one replaces

$$D_{\mathrm{F}}^{\mu\nu}(k_0,\boldsymbol{k}) \quad \to \quad D_{\mathrm{F}}^{\mu\nu}(0,\boldsymbol{k}) \quad . \tag{6}$$

As we shall examine further in Example 6.4 this means that one *neglects the retardation* of the interaction. If we multiply (5) by $\gamma_0^a\gamma_0^b$, we get

$$\hat{F}\chi_{ab}(p) = \frac{\mathrm{i}e_a e_b}{(2\pi)^4}\int d^4k\,\gamma_0^a\gamma_\mu^a\gamma_0^b\gamma_\nu^b\, D_{\mathrm{F}}^{\mu\nu}(0,\boldsymbol{k})\,\chi_{ab}(p+k) \tag{7}$$

with the abbreviation

$$\hat{F} = \gamma_0^a\left(\frac{1}{2}\gamma_\mu^a K^\mu + \gamma_\mu^a p^\mu - m\right)\gamma_0^b\left(\frac{1}{2}\gamma_\nu^b K^\nu - \gamma_\nu^b p^\nu - m\right) \quad . \tag{8}$$

In the centre-of-mass system characterized by $\boldsymbol{p}_1+\boldsymbol{p}_2 = \boldsymbol{0}$ the vector K representing the total momentum is purely time-like, that is

$$K = (2m - E_{\mathrm{B}}, \boldsymbol{0}) = (E, \boldsymbol{0}) \quad , \tag{9}$$

so that we get

$$\begin{aligned}\hat{F}(p) &= \left(\frac{1}{2}E - \boldsymbol{\alpha}_a\cdot\boldsymbol{p} - \beta_a m + p_0\right)\left(\frac{1}{2}E + \boldsymbol{\alpha}_b\cdot\boldsymbol{p} - \beta_b m - p_0\right)\\ &= \left(\frac{1}{2}E - \hat{H}_a(\boldsymbol{p}) + p_0\right)\left(\frac{1}{2}E - \hat{H}_b(\boldsymbol{p}) - p_0\right) \quad ,\end{aligned} \tag{10}$$

with the free Dirac Hamiltonians

$$\hat{H}_a(\boldsymbol{p}) = \boldsymbol{\alpha}_a\cdot\boldsymbol{p} + \beta_a m \quad , \tag{11a}$$

$$\hat{H}_b(\boldsymbol{p}) = -\boldsymbol{\alpha}_b\cdot\boldsymbol{p} + \beta_b m \quad . \tag{11b}$$

Example 6.3.

Using the approximation (6) it is possible to separate the frequency integration. We define a new wave function that depends only on the spatial components of the momentum vector:

$$\phi_{ab}(\boldsymbol{p}) = \int dp_0\, \chi_{ab}(p_0, \boldsymbol{p}) \quad . \tag{12}$$

After performing the k_0 integration on the right-hand side of (7) we thus get the approximate result

$$\hat{F}(p)\chi_{ab}(p) = \frac{\mathrm{i}e_a e_b}{2\pi} \int \frac{d^3k}{(2\pi)^3}\, \gamma_0^a \gamma_\mu^a \gamma_0^b \gamma_\nu^b\, D_{\mathrm{F}}^{\mu\nu}(0, \boldsymbol{k})\, \phi_{ab}(\boldsymbol{p} + \boldsymbol{k}) \quad , \tag{13}$$

or, written in a shorthand notation,

$$\hat{F}(p)\chi_{ab}(p) = \Gamma(\boldsymbol{p}) \quad . \tag{14}$$

To make use of this equation one must integrate over the variable p_0 on the left-hand side of (13) according to (12). To this end we invert the operator $\hat{F}(p)$, i.e. we bring it to the right-hand side. This can be done by the following trick: introduce *projection operators* for the components of the wave function with positive and negative frequency, namely

$$\hat{\Lambda}_\pm^n(\boldsymbol{p}) = \frac{\omega(\boldsymbol{p}) \pm \hat{H}_n(\boldsymbol{p})}{2\omega(\boldsymbol{p})} \qquad \text{with } n = a,b \quad , \tag{15}$$

where $\omega(\boldsymbol{p}) = +\sqrt{m^2 + \boldsymbol{p}^2}$. Because of $\hat{H}_n^2(\boldsymbol{p}) = \omega^2(\boldsymbol{p})$ one can verify right away that the operators (15) fulfill the usual rules for orthogonal projection operators:

$$(\hat{\Lambda}_\pm^n)^2 = \hat{\Lambda}_\pm^n \quad , \quad \hat{\Lambda}_+^n \hat{\Lambda}_-^n = 0 \quad , \quad \hat{\Lambda}_+^n + \hat{\Lambda}_-^n = \mathbb{1}^n \quad . \tag{16}$$

Furthermore there is the important relation

$$\hat{H}_n(\boldsymbol{p})\hat{\Lambda}_\pm^n(\boldsymbol{p}) = \hat{\Lambda}_\pm^n(\boldsymbol{p})\hat{H}_n(\boldsymbol{p}) = \pm\omega(\boldsymbol{p})\hat{\Lambda}_\pm^n(\boldsymbol{p}) \quad . \tag{17}$$

After applying the projection operator (15) the Hamiltonians in (10) can thus be replaced by the frequency ω (times the unit matrix):

$$\begin{aligned} &\hat{\Lambda}_\pm^a(\boldsymbol{p})\hat{\Lambda}_\pm^b(\boldsymbol{p})\hat{F}(p)\chi_{ab}(p) \\ &\quad = \left(\frac{1}{2}E \mp \omega(\boldsymbol{p}) + p_0\right)\left(\frac{1}{2}E \mp \omega(\boldsymbol{p}) - p_0\right) \hat{\Lambda}_\pm^a(\boldsymbol{p})\hat{\Lambda}_\pm^b(\boldsymbol{p})\chi_{ab}(p) \quad , \end{aligned} \tag{18}$$

where all of the four combinations of signs are admitted. The Bethe–Salpeter equation (14) then becomes a *system of four projected equations*:

$$\left(\frac{1}{2}E \mp \omega(\boldsymbol{p}) + p_0\right)\left(\frac{1}{2}E \mp \omega(\boldsymbol{p}) - p_0\right) \chi_{\pm\pm}(p) = \hat{\Lambda}_\pm^a(\boldsymbol{p})\hat{\Lambda}_\pm^b(\boldsymbol{p})\Gamma(\boldsymbol{p}) \, , \tag{19}$$

with the *projected wave functions*

$$\chi_{\pm\pm}(p) = \hat{\Lambda}_\pm^a(\boldsymbol{p})\hat{\Lambda}_\pm^b(\boldsymbol{p})\chi_{ab}(p) \quad . \tag{20}$$

Example 6.3.

In a completely analogous way one defines the projected wave functions $\phi_{\pm\pm}(\boldsymbol{p}) = \Lambda^a_\pm(\boldsymbol{p})\Lambda^b_\pm(\boldsymbol{p})\phi_{ab}(\boldsymbol{p})$ by integrating over p_0 (cf. (12)). Making use of (17) the operator on the left-hand side of (13) has been replaced by a simple number, which can be brought to the right-hand side by division. Thus it is also possible to eliminate the frequency variable p_0 on the left-hand side of (19) by integrating according to (12), i.e.

$$\begin{aligned}\phi_{\pm\pm}(\boldsymbol{p}) &= \int dp_0 \chi_{\pm\pm}(p)\\ &= \int dp_0 \frac{1}{\frac{1}{2}E \mp \omega(\boldsymbol{p}) + p_0} \frac{1}{\frac{1}{2}E \mp \omega(\boldsymbol{p}) - p_0} \hat{\Lambda}^a_\pm(\boldsymbol{p})\hat{\Lambda}^b_\pm(\boldsymbol{p})\Gamma(\boldsymbol{p}) \quad . \qquad (21)\end{aligned}$$

To make (21) unique we have to determine how to treat the poles when integrating over p_0. In this context we remember the rule that we have already used frequently, which states that the condition of causality can be fulfilled by giving the particle mass a small negative imaginary part, i.e. $m \to m - \mathrm{i}\varepsilon$, and consequently also $\omega(\boldsymbol{p}) \to \omega(\boldsymbol{p}) - \mathrm{i}\delta$. This means that

$$\phi_{\pm\pm}(\boldsymbol{p}) = \left(\int dp_0 \frac{1}{\frac{1}{2}E \mp \omega(\boldsymbol{p}) + p_0 \pm \mathrm{i}\delta} \frac{1}{\frac{1}{2}E \mp \omega(\boldsymbol{p}) - p_0 \pm \mathrm{i}\delta}\right) \Gamma_{\pm\pm}(\boldsymbol{p}) \quad , \qquad (22)$$

where

$$\Gamma_{\pm\pm}(\boldsymbol{p}) = \hat{\Lambda}^a_\pm(\boldsymbol{p})\hat{\Lambda}^b_\pm(\boldsymbol{p})\Gamma(\boldsymbol{p}) \quad .$$

The p_0 integration can be performed using the theorem of residues, where the integration path is closed by a half circle in the upper or lower half plane. This is possible because the integrand falls off like $1/|p_0|^2$ for large p_0 and therefore does not contribute to the integral. The integrand has two poles at

$$p_0 = \frac{1}{2}E \mp \omega(\boldsymbol{p}) \pm \mathrm{i}\delta \quad , \qquad (23a)$$

$$p_0 = -\frac{1}{2}E \pm \omega(\boldsymbol{p}) \mp \mathrm{i}\delta \quad . \qquad (23b)$$

As we have already emphasized, both signs in (22) can be chosen independently. If both signs of δ in (23a) and (23b) are equal, then both poles are in the same half plane. Since one can close the integration path in the other half plane, where the integrand is regular, the theorem of residues yields

$$\phi_{+-}(\boldsymbol{p}) = \phi_{-+}(\boldsymbol{p}) = 0 \quad . \qquad (24)$$

In the opposite case one gets

$$\int_{-\infty}^{\infty} dz \frac{1}{z - a \mp \mathrm{i}\delta} \frac{1}{z + b \pm \mathrm{i}\delta} = \pm \frac{2\pi\mathrm{i}}{a + b} \qquad (25)$$

and therefore

Example 6.3.

$$\phi_{++}(\boldsymbol{p}) = -\frac{2\pi\mathrm{i}}{E - 2\omega(\boldsymbol{p})}\Gamma_{++}(\boldsymbol{p}) \quad , \tag{26a}$$

$$\phi_{--}(\boldsymbol{p}) = \frac{2\pi\mathrm{i}}{E + 2\omega(\boldsymbol{p})}\Gamma_{--}(\boldsymbol{p}) \quad . \tag{26b}$$

Owing to (24) and (26) the complete wave function $\phi(\boldsymbol{p}) = \phi_{++} + \phi_{+-} + \phi_{-+} + \phi_{--}$ then obeys the integral equation

$$\phi(\boldsymbol{p}) = 2\pi\mathrm{i}\left(\frac{\hat{\Lambda}_{++}(\boldsymbol{p})}{2\omega - E} + \frac{\hat{\Lambda}_{--}(\boldsymbol{p})}{2\omega + E}\right)\Gamma(\boldsymbol{p}) \quad . \tag{27}$$

We eliminate the denominators in (26) and subtract the resulting equations:

$$\begin{aligned}&\big(E - 2\omega(\boldsymbol{p})\big)\phi_{++}(\boldsymbol{p}) + \big(E + 2\omega(\boldsymbol{p})\big)\phi_{--}(\boldsymbol{p})\\ &\quad = -2\pi\mathrm{i}\big(\Gamma_{++}(\boldsymbol{p}) - \Gamma_{--}(\boldsymbol{p})\big) \quad .\end{aligned} \tag{28}$$

According to (17) the frequency $\omega(p)$ can be replaced by the Hamiltonians $\hat{H}_n$. To do this in a symmetric fashion we identify

$$\begin{aligned}&-2\omega\phi_{++} + 2\omega\phi_{--}\\ &\quad = \left(-\omega\hat{\Lambda}^a_+\hat{\Lambda}^b_+ - \hat{\Lambda}^a_+\omega\hat{\Lambda}^b_+ + \omega\hat{\Lambda}^a_-\hat{\Lambda}^b_- + \hat{\Lambda}^a_-\omega\hat{\Lambda}^b_-\right)\phi\\ &\quad = \left(-\hat{\Lambda}^a_+\hat{\Lambda}^b_+\hat{H}_a - \hat{\Lambda}^a_+\hat{\Lambda}^b_+\hat{H}_b - \hat{\Lambda}^a_-\hat{\Lambda}^b_-\hat{H}_a - \hat{\Lambda}^a_-\hat{\Lambda}^b_-\hat{H}_b\right)\phi\\ &\quad = \left(\hat{\Lambda}^a_+\hat{\Lambda}^b_+ + \hat{\Lambda}^a_-\hat{\Lambda}^b_-\right)\left(-\hat{H}_a - \hat{H}_b\right)\phi\\ &\quad = \left(\mathbb{1} - \hat{\Lambda}^a_+\hat{\Lambda}^b_- - \hat{\Lambda}^a_-\hat{\Lambda}^b_+\right)\left(-\hat{H}_a - \hat{H}_b\right)\phi\\ &\quad = \left(-\hat{H}_a - \hat{H}_b\right)\phi \quad .\end{aligned}$$

In the last step the contributions from the mixed projection operators $\hat{\Lambda}^a_\pm\hat{\Lambda}^b_\mp$ have vanished since, because of (16), one can replace $\hat{H}_n$ by $\pm\omega$ and then use (24). If we now define a mixed projection operator

$$\hat{\Lambda}(\boldsymbol{p}) = \hat{\Lambda}_{++}(\boldsymbol{p}) - \hat{\Lambda}_{--}(\boldsymbol{p}) = \hat{\Lambda}^a_+\hat{\Lambda}^b_+(\boldsymbol{p}) - \hat{\Lambda}^a_-\hat{\Lambda}^b_-(\boldsymbol{p}) \quad , \tag{29}$$

then the right-hand side of (28) simply becomes $-2\pi\mathrm{i}\hat{\Lambda}(\boldsymbol{p})\,\Gamma(\boldsymbol{p})$. Equation (28) then becomes the *Bethe–Salpeter equation in the nonretarded approximation*:[7]

$$\big(E - \hat{H}_a(\boldsymbol{p}) - \hat{H}_b(\boldsymbol{p})\big)\phi(\boldsymbol{p}) = -2\pi\mathrm{i}\hat{\Lambda}(\boldsymbol{p})\,\Gamma(\boldsymbol{p}) \quad , \tag{30}$$

where in the ladder approximation $\Gamma(\boldsymbol{p})$ is given by the right-hand side of (13).

Equation (30) can also be brought into a more familiar form if one assumes that the *solutions with negative energy can be neglected*, i.e. if one makes the approximation

$$\hat{\Lambda}(\boldsymbol{p}) \quad \to \quad \mathbb{1} \tag{31}$$

in (30). Equation (30) will now be transformed into coordinate space according to

$$\phi(\boldsymbol{r}) = \int\frac{d^3p}{(2\pi)^3}\,\mathrm{e}^{-\mathrm{i}\boldsymbol{p}\cdot\boldsymbol{r}}\,\phi(\boldsymbol{p}) \quad , \tag{32}$$

[7] E.E. Salpeter: Phys. Rev. **87**, 328 (1952).

i.e. using $\Gamma(\boldsymbol{p})$ defined in (13), (14) *Example 6.3.*

$$\int \frac{d^3p}{(2\pi)^3}\, \mathrm{e}^{-\mathrm{i}\boldsymbol{p}\cdot\boldsymbol{r}} \left(E - \hat{H}_a(\boldsymbol{p}) - \hat{H}_b(\boldsymbol{p})\right)\phi(\boldsymbol{p})$$
$$= e_a e_b \int \frac{d^3p}{(2\pi)^3}\, \mathrm{e}^{-\mathrm{i}\boldsymbol{p}\cdot\boldsymbol{r}} \int \frac{d^3k}{(2\pi)^3}\, \gamma_0^a\gamma_\mu^a\gamma_0^b\gamma_\nu^b\, D_{\mathrm{F}}^{\mu\nu}(0,\boldsymbol{k})\,\phi(\boldsymbol{p}+\boldsymbol{k}) \quad , \tag{33}$$

or, with (32),

$$\left(E - \hat{H}_a(\mathrm{i}\nabla) - \hat{H}_b(\mathrm{i}\nabla)\right)\phi(\boldsymbol{r})$$
$$= \left(e_a e_b \int \frac{d^3k}{(2\pi)^3}\, \mathrm{e}^{\mathrm{i}\boldsymbol{k}\cdot\boldsymbol{r}}\, \gamma_0^a\gamma_\mu^a\gamma_0^b\gamma_\nu^b\, D_{\mathrm{F}}^{\mu\nu}(0,\boldsymbol{k})\right)\phi(\boldsymbol{r})$$
$$= U(\boldsymbol{r})\,\phi(\boldsymbol{r}) \quad . \tag{34}$$

To arrive at this result the integration variable $\boldsymbol{p}$ was shifted to $\boldsymbol{p}+\boldsymbol{k}$ in (33). Equation (34) has the same form as the ordinary Dirac equation with an *effective interaction potential* $U(\boldsymbol{r})$. In order to calculate this potential it is advantageous to use the photon propagator in *Coulomb gauge* (cf. Chap. 4, (13-15)). In this gauge D_{F}^{00} depends only on the spatial components $\boldsymbol{k}$, so that here the approximation (6) (neglecting the frequency dependence) has a less drastic effect than in the commonly used Feynman gauge of (4). Thus we insert

$$D_{\mathrm{F}}^{ij}(\boldsymbol{k}) = \frac{-4\pi}{k^2}\left(\frac{k^i k^j}{|\boldsymbol{k}|^2} - \delta_{ij}\right) \to \frac{4\pi}{|\boldsymbol{k}|^2}\left(\frac{k^i k^j}{|\boldsymbol{k}|^2} - \delta_{ij}\right) \quad , \tag{35a}$$

$$D_{\mathrm{F}}^{00}(\boldsymbol{k}) = +\frac{4\pi}{|\boldsymbol{k}|^2} \quad , \tag{35b}$$

$$D_{\mathrm{F}}^{0i}(\boldsymbol{k}) = D_{\mathrm{F}}^{i0}(\boldsymbol{k}) = 0 \quad . \tag{35c}$$

The interaction potential then reads

$$U(\boldsymbol{r}) = e_a e_b 4\pi \int \frac{d^3k}{(2\pi)^3}\, \mathrm{e}^{\mathrm{i}\boldsymbol{k}\cdot\boldsymbol{r}} \left[\frac{1}{|\boldsymbol{k}|^2}\left(\frac{\boldsymbol{\alpha}_a\cdot\boldsymbol{k}\,\boldsymbol{\alpha}_b\cdot\boldsymbol{k}}{|\boldsymbol{k}|^2} - \boldsymbol{\alpha}_a\cdot\boldsymbol{\alpha}_b\right) + \frac{1}{|\boldsymbol{k}|^2}\right] . \tag{36}$$

One of the Fourier integrals we need,

$$\int \frac{d^3k}{(2\pi)^3}\,\frac{\mathrm{e}^{i\boldsymbol{k}\cdot\boldsymbol{r}}}{|\boldsymbol{k}|^2} = \frac{1}{4\pi r} \quad , \tag{37}$$

is the familiar Coulomb potential. The other one can be derived from this with a few tricks. Using $\nabla_{\boldsymbol{k}}(1/|\boldsymbol{k}|^2) = -2\boldsymbol{k}/|\boldsymbol{k}|^4$ we write

$$I \equiv \int \frac{d^3k}{(2\pi)^3}\, \mathrm{e}^{i\boldsymbol{k}\cdot\boldsymbol{r}}\, \frac{(\boldsymbol{a}\cdot\boldsymbol{k})(\boldsymbol{b}\cdot\boldsymbol{k})}{|\boldsymbol{k}|^4} = -\mathrm{i}(\boldsymbol{a}\cdot\nabla_{\boldsymbol{r}}) \int \frac{d^3k}{(2\pi)^3}\, \mathrm{e}^{\mathrm{i}\boldsymbol{k}\cdot\boldsymbol{r}} \left(\boldsymbol{b}\cdot\frac{\boldsymbol{k}}{|\boldsymbol{k}|^4}\right)$$
$$= \frac{\mathrm{i}}{2}(\boldsymbol{a}\cdot\nabla_{\boldsymbol{r}}) \int \frac{d^3k}{(2\pi)^3}\, \mathrm{e}^{\mathrm{i}\boldsymbol{k}\cdot\boldsymbol{r}}\, (\boldsymbol{b}\cdot\nabla_{\boldsymbol{k}})\,\frac{1}{|\boldsymbol{k}|^2} \quad , \tag{38}$$

Example 6.3.

or, after partial integration,

$$I = -\frac{\mathrm{i}}{2}(\boldsymbol{a}\cdot\nabla_r)\int\frac{d^3k}{(2\pi)^3}\left(\boldsymbol{b}\cdot\nabla_k\,\mathrm{e}^{\mathrm{i}\boldsymbol{k}\cdot\boldsymbol{r}}\right)\frac{1}{|\boldsymbol{k}|^2}$$
$$= \frac{1}{2}(\boldsymbol{a}\cdot\nabla_r)(\boldsymbol{b}\cdot\boldsymbol{r})\int\frac{d^3k}{(2\pi)^3}\frac{\mathrm{e}^{\mathrm{i}\boldsymbol{k}\cdot\boldsymbol{r}}}{|\boldsymbol{k}|^2} = \frac{1}{2}(\boldsymbol{a}\cdot\nabla_r)(\boldsymbol{b}\cdot\boldsymbol{r})\frac{1}{4\pi r} \quad . \tag{39}$$

On evaluating the gradients we have

$$I = \frac{1}{4\pi}\frac{1}{2r}\left(\boldsymbol{a}\cdot\boldsymbol{b} - \frac{(\boldsymbol{a}\cdot\boldsymbol{r})(\boldsymbol{b}\cdot\boldsymbol{r})}{r^2}\right) \quad . \tag{40}$$

Thus the effective potential reads

$$U(\boldsymbol{r}) = e_a e_b\left[\frac{1}{2r}\left(\boldsymbol{\alpha}_a\cdot\boldsymbol{\alpha}_b - \frac{(\boldsymbol{\alpha}_a\cdot\boldsymbol{r})(\boldsymbol{\alpha}_b\cdot\boldsymbol{r})}{r^2}\right) - \frac{\boldsymbol{\alpha}_a\cdot\boldsymbol{\alpha}_b}{r} + \frac{1}{r}\right]$$
$$= e_a e_b\left[\frac{1}{r} - \frac{1}{2r}\left(\boldsymbol{\alpha}_a\cdot\boldsymbol{\alpha}_b + \frac{(\boldsymbol{\alpha}_a\cdot\boldsymbol{r})(\boldsymbol{\alpha}_b\cdot\boldsymbol{r})}{r^2}\right)\right] \quad . \tag{41}$$

This potential is known as the *Breit interaction*. We shall derive it in an alternative way in Example 6.4.

EXAMPLE

6.4 The Breit Interaction

The interaction between two Dirac particles is described covariantly by the exchange of virtual photons. We will attempt to describe this interaction in a nonrelativistic approximation by a potential $U(\boldsymbol{r})$ (which does not depend on the time coordinate). Of course this should lead to the Coulomb potential in the static limit. In addition we want to calculate the relativistic corrections up to the order $(v/c)^2$.

We shall proceed as follows. First we shall calculate the S-matrix element for the one-photon exchange and then we shall examine which potential $U(\boldsymbol{r})$ yields the same result in the desired approximation. As we have discussed in Example 3.2 the S-matrix element for the graph

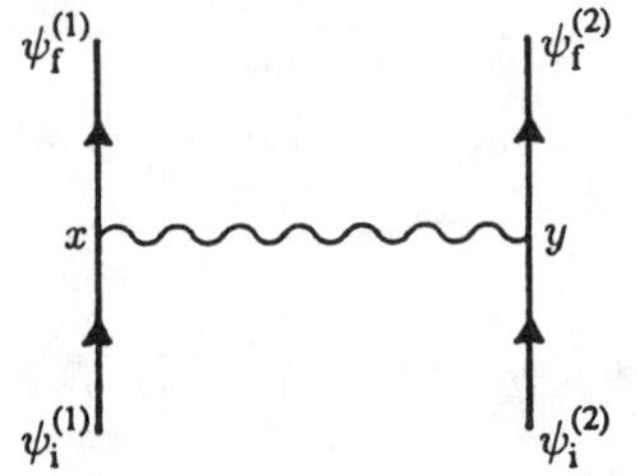

has the form

Example 6.4.

$$S_{fi} = -\mathrm{i}\int d^4x \int d^4y\, j_{fi}^{(2)}(y)\, D_\mathrm{F}(y-x)\, j_{fi}^{(1)}(x) \tag{1}$$

with the photon propagator D_F and the two transition currents

$$\begin{aligned} j_{fi}^{(n)}(x) &= e_n \bar{\psi}_f^{(n)}(x)\gamma\psi_i^{(n)}(x) \\ &= j_{fi}^{(n)}(\boldsymbol{x})\,\mathrm{e}^{\mathrm{i}\omega_{fi}^{(n)}t_x} \quad , \quad n = 1,2 \quad . \end{aligned} \tag{2}$$

In the second line the time dependence has been separated off, thus defining the transition frequency $\omega_{fi}^{(n)} = E_f^{(n)} - E_i^{(n)}$. The 4-vector index has been omitted for brevity. If the particles are not distinguishable one must subtract the exchange term in (1).

Inserting (2) and the Fourier representation of the propagator (in Feynman gauge) into (1) yields

$$S_{fi} = -\mathrm{i}\int d^4x \int d^4y \int \frac{d^4k}{(2\pi)^4}\, j_{fi}^{(2)}(\boldsymbol{y})\,\mathrm{e}^{\mathrm{i}\omega_{fi}^{(2)}t_y}\, \frac{-4\pi\mathrm{e}^{-\mathrm{i}k\cdot(y-x)}}{k^2+\mathrm{i}\varepsilon}\, j_{fi}^{(1)}(\boldsymbol{x})\,\mathrm{e}^{\mathrm{i}\omega_{fi}^{(1)}t_x} \quad . \tag{3}$$

The t_y integration yields a δ function. With $k = (\omega, \boldsymbol{k})$ we get

$$\begin{aligned} S_{fi} = 4\pi\mathrm{i}\int d^3x \int d^3y \int dt_x \int \frac{d\omega}{2\pi}\, 2\pi\delta(\omega - \omega_{fi}^{(2)})\,\mathrm{e}^{\mathrm{i}(\omega_{fi}^{(1)}+\omega)t_x} \\ \times j_{fi}^{(2)}(\boldsymbol{y})\, j_{fi}^{(1)}(\boldsymbol{x}) \int \frac{d^3k}{(2\pi)^3}\, \frac{\mathrm{e}^{+\mathrm{i}\boldsymbol{k}\cdot(\boldsymbol{y}-\boldsymbol{x})}}{\omega^2 - \boldsymbol{k}^2 + \mathrm{i}\varepsilon} \quad . \end{aligned} \tag{4}$$

After introducing spherical coordinates we can carry out the integration over the momentum with residue integration:

$$\begin{aligned} \int \frac{d^3k}{(2\pi)^3}\, \frac{\mathrm{e}^{+\mathrm{i}\boldsymbol{k}\cdot\boldsymbol{r}}}{\omega^2 - \boldsymbol{k}^2 + \mathrm{i}\varepsilon} &= \frac{1}{(2\pi)^3}\int_0^\infty k^2 dk \int_{-1}^{1} d\cos\theta_k \int_0^{2\pi} d\phi_k\, \frac{\mathrm{e}^{+\mathrm{i}kr\cos\theta_k}}{\omega^2 - \boldsymbol{k}^2 + \mathrm{i}\varepsilon} \\ &= \frac{\mathrm{i}}{(2\pi)^2}\frac{1}{r}\int_0^\infty dk\, \frac{k}{k^2 - \omega^2 - \mathrm{i}\varepsilon}\left(\mathrm{e}^{\mathrm{i}kr} - \mathrm{e}^{-\mathrm{i}kr}\right) \\ &= \frac{\mathrm{i}}{(2\pi)^2}\frac{1}{r}\int_{-\infty}^\infty dk\, \frac{k}{k^2 - \omega^2 - \mathrm{i}\varepsilon}\,\mathrm{e}^{\mathrm{i}kr} \quad . \end{aligned} \tag{5}$$

The poles are located at $k = \pm(\omega + \mathrm{i}\varepsilon'\mathrm{sgn}(\omega))$. The residue integral (which must be closed in the upper half plane) thus encircles only one pole yielding

$$\int \frac{d^3k}{(2\pi)^3}\, \frac{\mathrm{e}^{-\mathrm{i}\boldsymbol{k}\cdot\boldsymbol{r}}}{\omega^2 - \boldsymbol{k}^2 + \mathrm{i}\varepsilon} = -\frac{\mathrm{e}^{+\mathrm{i}|\omega|r}}{4\pi r} \quad . \tag{6}$$

If one also carries out the integrations over ω and t_x, then the S-matrix element reads

Example 6.4.

$$S_{fi} = -2\pi \mathrm{i}\delta(\omega_{fi}^{(1)} + \omega_{fi}^{(2)}) \int d^3x \int d^3y j_{fi}^{(2)}(\boldsymbol{y}) \frac{\mathrm{e}^{\mathrm{i}|\omega_{fi}|\cdot|\boldsymbol{y}-\boldsymbol{x}|}}{|\boldsymbol{y}-\boldsymbol{x}|} j_{fi}^{(1)}(\boldsymbol{x}) \quad . \tag{7}$$

As usual the δ function ensures the conservation of energy. Apart from that, the magnitude of S_{fi} is determined by a coupling between the transition currents $j_{fi}^{(1)}$ and $j_{fi}^{(2)}$ with a *frequency-dependent interaction.*

In order to understand this more deeply we go back one step and keep the time integration. Using $|\omega_{fi}^{(2)}| = |\omega_{fi}^{(1)}|$, which follows from (7), we get

$$S_{fi} = -\mathrm{i} \int d^3y \int dt\, j_{fi}^{(2)}(\boldsymbol{y})\, \mathrm{e}^{\mathrm{i}\omega_{fi}^{(2)}t} \int d^3x \, \frac{\mathrm{e}^{\mathrm{i}(\omega_{fi}^{(1)}t + |\omega_{fi}^{(1)}|\cdot|\boldsymbol{y}-\boldsymbol{x}|)}}{|\boldsymbol{y}-\boldsymbol{x}|} j_{fi}^{(1)}(\boldsymbol{x}) \quad . \tag{8}$$

If we consider a scattering process in which the energy of particle 1 is transferred to particle 2, then $\omega_{fi}^{(1)} < 0$, and $|\omega_{fi}^{(1)}| = -\omega_{fi}^{(1)}$, so that, with (2b), (8) can be cast in the form

$$S_{fi} = -\mathrm{i} \int dt \int d^3y \, j_{fi}^{(2)}(\boldsymbol{y},t) A_{fi}^{(1)}(\boldsymbol{y},t) \quad , \tag{9}$$

where

$$\begin{aligned} A_{fi}^{(1)}(\boldsymbol{y},t) &= \int d^3x \, j_{fi}^{(1)}(\boldsymbol{x}) \, \frac{\mathrm{e}^{\mathrm{i}\omega_{fi}^{(1)}(t-|\boldsymbol{y}-\boldsymbol{x}|)}}{|\boldsymbol{y}-\boldsymbol{x}|} \\ &= \int d^3x \, \frac{1}{|\boldsymbol{x}-\boldsymbol{y}|} j_{fi}^{(1)}(\boldsymbol{x}, t - |\boldsymbol{y}-\boldsymbol{x}|) \quad . \end{aligned} \tag{10}$$

Thus (9) implies that the transition current $j_{fi}^{(2)}(\boldsymbol{y})$ interacts with the electromagnetic field that was emitted by the current $j_{fi}^{(2)}(\boldsymbol{x})$ of the other particle at an earlier time, where the time difference is given by $|\boldsymbol{y}-\boldsymbol{x}| \equiv |\boldsymbol{y}-\boldsymbol{x}|/c$. Therefore (10) is just the *retarded potential.* The frequency-dependent factor in (7) is thus responsible for the retardation due to the finite propagation velocity of the interaction. For small particle velocities ($v/c \ll 1$) we are therefore justified in replacing the exponential by the lowest-order terms of its Taylor expansion:

$$\frac{\mathrm{e}^{\mathrm{i}|\omega_{fi}|\cdot|\boldsymbol{y}-\boldsymbol{x}|/c}}{|\boldsymbol{y}-\boldsymbol{x}|} \approx \frac{1}{|\boldsymbol{y}-\boldsymbol{x}|} + \frac{\mathrm{i}}{c}|\omega_{fi}| - \frac{1}{2c^2}|\omega_{fi}|^2|\boldsymbol{y}-\boldsymbol{x}| \quad . \tag{11}$$

In the following we will write out the powers of the velocity of light c explicitly. The approximated S-matrix element reads

$$\begin{aligned} S_{fi} &= -2\pi\mathrm{i}\delta\left(\omega_{fi}^{(1)} + \omega_{fi}^{(2)}\right) e_1 e_2 \int d^3x \int d^3y \\ &\quad \times \overline{\psi}_f^{(2)}(\boldsymbol{y})\overline{\psi}_f^{(1)}(\boldsymbol{x})\gamma_\mu^{(1)}\gamma^{(2)\mu} \frac{\mathrm{e}^{\mathrm{i}|\omega_{fi}|\cdot|\boldsymbol{y}-\boldsymbol{x}|}}{|\boldsymbol{y}-\boldsymbol{x}|} \psi_i^{(2)}(\boldsymbol{y})\psi_i^{(1)}(\boldsymbol{x}) \\ &\simeq -2\pi\mathrm{i}\delta\left(\omega_{fi}^{(1)} + \omega_{fi}^{(2)}\right) e_1 e_2 \int d^3x \int d^3y \\ &\quad \times \psi_f^{(2)\dagger}(\boldsymbol{y})\psi_f^{(1)\dagger}(\boldsymbol{x}) \left(\mathbb{1}^{(1)}\mathbb{1}^{(2)} - \boldsymbol{\alpha}^{(1)}\cdot\boldsymbol{\alpha}^{(2)}\right) \\ &\quad \times \left(\frac{1}{|\boldsymbol{y}-\boldsymbol{x}|} + \mathrm{i}\frac{|\omega_{fi}|}{c} - \frac{1}{2c^2}\omega_{fi}^2|\boldsymbol{y}-\boldsymbol{x}|\right) \psi_i^{(2)}(\boldsymbol{y})\psi_i^{(1)}(\boldsymbol{x}) \quad . \end{aligned} \tag{12}$$

Example 6.4.

In (12) only the terms of lowest order in v/c are to be taken into account. Since $c\boldsymbol{\alpha}$ is the velocity operator in the Dirac theory the consistent approximation is

$$\begin{aligned}
&\left(\mathbb{1}^{(1)}\mathbb{1}^{(2)} - \boldsymbol{\alpha}^{(1)}\cdot\boldsymbol{\alpha}^{(2)}\right)\left(\frac{1}{|\boldsymbol{y}-\boldsymbol{x}|} + \mathrm{i}\frac{|\omega_{fi}|}{c} - \frac{1}{2c^2}\omega_{fi}^2|\boldsymbol{y}-\boldsymbol{x}|\right)\\
&\approx \frac{\mathbb{1}^{(1)}\mathbb{1}^{(2)}}{|\boldsymbol{y}-\boldsymbol{x}|} + \frac{\mathrm{i}}{c}\,\mathbb{1}^{(1)}\mathbb{1}^{(2)}|\omega_{fi}| - \frac{\boldsymbol{\alpha}^{(1)}\cdot\boldsymbol{\alpha}^{(2)}}{|\boldsymbol{y}-\boldsymbol{x}|} - \frac{1}{2c^2}\,\omega_{fi}^2|\boldsymbol{y}-\boldsymbol{x}|\,\mathbb{1}^{(1)}\mathbb{1}^{(2)} \quad ,
\end{aligned} \tag{13}$$

up to the order $1/c^2$. If this is inserted into (12) the contribution of the term of order $1/c$ obviously vanishes owing to the orthogonality of the wave functions $\psi_f^{(n)}$ and $\psi_i^{(n)}$. The last quadratic correction term in (13) looks somewhat unpleasant, since it contains ω_{fi} and thus depends on the initial and final states. Here one can take advantage of a commutator trick. The wave functions are assumed to be stationary eigenstates of a Hamiltonian, so that

$$\begin{aligned}
\hat{H}^{(1)}(\boldsymbol{x})\,\psi_i^{(1)}(\boldsymbol{x}) &= E_{\mathrm{i}}^{(1)}\,\psi_i^{(1)}(\boldsymbol{x}) \quad ,\\
\hat{H}^{(1)}(\boldsymbol{x})\,\psi_f^{(1)}(\boldsymbol{x}) &= E_{\mathrm{f}}^{(1)}\,\psi_f^{(1)}(\boldsymbol{x}) \quad ,
\end{aligned} \tag{14}$$

and analogously the equation involving $\hat{H}^{(2)}(\boldsymbol{x})$. Therefore, one can replace in (13) each energy by the corresponding Hamiltonian, if one makes sure that it acts immediately on the wave function. With $\omega_{fi}^{(1)} = -\omega_{fi}^{(2)}$ one gets

$$\begin{aligned}
-\omega_{fi}^2|\boldsymbol{x}-\boldsymbol{y}| &= \left(E_{\mathrm{f}}^{(1)} - E_{\mathrm{i}}^{(1)}\right)\left(E_{\mathrm{f}}^{(2)} - E_{\mathrm{i}}^{(2)}\right)|\boldsymbol{x}-\boldsymbol{y}|\\
&= |\boldsymbol{x}-\boldsymbol{y}|\hat{H}^{(1)}\hat{H}^{(2)} - \hat{H}^{(1)}|\boldsymbol{x}-\boldsymbol{y}|\hat{H}^{(2)}\\
&\quad - \hat{H}^{(2)}|\boldsymbol{x}-\boldsymbol{y}|\hat{H}^{(1)} + \hat{H}^{(1)}\hat{H}^{(2)}|\boldsymbol{x}-\boldsymbol{y}|\\
&= \left[\hat{H}^{(1)},[\hat{H}^{(2)},|\boldsymbol{x}-\boldsymbol{y}|]\right] \quad .
\end{aligned} \tag{15}$$

Thus one has to evaluate the double commutator of the distance function $|\boldsymbol{x}-\boldsymbol{y}|$ with the Hamiltonians of the two particles. These Hamiltonians are of the form

$$\hat{H}^{(1)} = c\boldsymbol{\alpha}^{(1)}\cdot\hat{\boldsymbol{p}}_x + \beta^{(1)}m_1c^2 + \hat{H}_{\mathrm{ext}}^{(1)}(\boldsymbol{x}) \quad , \tag{16}$$

where only the momentum operator does not commute with $|\boldsymbol{x}-\boldsymbol{y}|$. $\hat{H}_{\mathrm{ext}}$ here denotes the interaction with a possibly present (stationary) external potential, which does not change the following considerations. Using $[\hat{\boldsymbol{p}},f(\boldsymbol{x})] = -\mathrm{i}\nabla f(\boldsymbol{x})$ we get

$$\left[\hat{H}^{(1)},[\hat{H}^{(2)},|\boldsymbol{x}-\boldsymbol{y}|]\right] = c^2\boldsymbol{\alpha}^{(1)}\cdot(-\mathrm{i}\nabla_{\boldsymbol{x}})\left[\boldsymbol{\alpha}^{(2)}\cdot(-\mathrm{i}\nabla_{\boldsymbol{y}})|\boldsymbol{x}-\boldsymbol{y}|\right] \quad . \tag{17}$$

The differentiation of $|\boldsymbol{x}-\boldsymbol{y}| = \sqrt{\sum_k (x_k-y_k)^2}$ does not present any difficulties and leads to

$$\left[\hat{H}^{(1)},[\hat{H}^{(2)},|\boldsymbol{x}-\boldsymbol{y}|]\right] = c^2\left[\frac{\boldsymbol{\alpha}^{(1)}\cdot\boldsymbol{\alpha}^{(2)}}{|\boldsymbol{x}-\boldsymbol{y}|} - \frac{\boldsymbol{\alpha}^{(1)}\cdot(\boldsymbol{x}-\boldsymbol{y})\,\boldsymbol{\alpha}^{(2)}\cdot(\boldsymbol{x}-\boldsymbol{y})}{|\boldsymbol{x}-\boldsymbol{y}|^3}\right]. \tag{18}$$

Obviously the third term in (13) and the first term in (18) can be combined. The final result reads

$$\begin{aligned}
S_{fi} = &-2\pi\mathrm{i}\delta\left(\omega_{fi}^{(1)} + \omega_{fi}^{(2)}\right)e_1e_2\int d^3x\,d^3y\,\psi_f^{(2)\dagger}(\boldsymbol{y})\psi_f^{(1)\dagger}(\boldsymbol{x})\\
&\times\left[\frac{\mathbb{1}^{(1)}\mathbb{1}^{(2)}}{|\boldsymbol{x}-\boldsymbol{y}|} - \frac{\boldsymbol{\alpha}^{(1)}\cdot\boldsymbol{\alpha}^{(2)} + (\boldsymbol{\alpha}^{(1)}\cdot\boldsymbol{n})(\boldsymbol{\alpha}^{(2)}\cdot\boldsymbol{n})}{2|\boldsymbol{x}-\boldsymbol{y}|}\right]\psi_i^{(2)}(\boldsymbol{y})\psi_i^{(1)}(\boldsymbol{x}) \quad ,
\end{aligned} \tag{19}$$

Example 6.4.

with the direction vector $\boldsymbol{n} = (\boldsymbol{x} - \boldsymbol{y})/|\boldsymbol{x} - \boldsymbol{y}|$. Equation (19) can be interpreted in such a way that the particles scatter at each other via a (nonretarded) *effective interaction* $U(\boldsymbol{x} - \boldsymbol{y})$, namely

$$S_{fi} = -2\pi \mathrm{i}\delta\left(\omega_{fi}^{(1)} + \omega_{fi}^{(2)}\right) \int d^3x \int d^3y \times \psi_f^{(2)\dagger}(\boldsymbol{y})\psi_f^{(1)\dagger}(\boldsymbol{x})U(\boldsymbol{x} - \boldsymbol{y})\psi_i^{(2)}(\boldsymbol{y})\psi_i^{(1)}(\boldsymbol{x}) \quad . \tag{20}$$

The effective interaction operator has the form

$$\begin{aligned} U(\boldsymbol{x} - \boldsymbol{y}) &= \frac{e_1 e_2}{|\boldsymbol{x} - \boldsymbol{y}|} - \frac{e_1 e_2}{2} \frac{\boldsymbol{\alpha}^{(1)}\cdot\boldsymbol{\alpha}^{(2)} + (\boldsymbol{\alpha}^{(1)}\cdot\boldsymbol{n})(\boldsymbol{\alpha}^{(2)}\cdot\boldsymbol{n})}{|\boldsymbol{x} - \boldsymbol{y}|} \\ &\equiv U_{\mathrm{C}}(\boldsymbol{x} - \boldsymbol{y}) + U_{\mathrm{B}}(\boldsymbol{x} - \boldsymbol{y}) \quad . \end{aligned} \tag{21}$$

As expected the first term is the Coulomb potential between the particles. Additionally one gets a correction term U_{B}, quadratic in the velocity, which is known as the *Breit interaction*. It is interesting to note that (21) agrees with the retarded interaction of two *classical* particles if one replaces the Dirac matrices by the classical velocity $\boldsymbol{\alpha}^{(i)} \to \boldsymbol{v}_i/c$.

If one makes the nonrelativistic approximation also for the wave functions $\psi_{i,f}^{(n)}$, (21) can be reduced to a sum of contributions that one recognizes as spin-orbit and spin-spin interactions. For further information see e.g. H.A. Bethe and E.E. Salpeter: Quantum Mechanics of One- and Two-Electron Atoms (Springer, Berlin, 1957), and Example 6.5.

The Breit interaction is very useful for the calculation of energy shifts in many-particle systems and in positronium (where one also has to take into account the Feynman graph for virtual pair annihilation). However, one must keep in mind that (21) is only an approximation. Its use is strictly justified only in perturbation theory. If one simply includes the potential U in the Dirac equation as an interaction potential, this can lead to wrong results. In particular one has to avoid the mixing of solutions with positive and negative energies, which can be done by the introduction of projection operators.[8]

EXAMPLE

6.5 Nonrelativistic Reduction of the Two-Body Equation Applied to Positronium

We have learned earlier in this chapter that a two-fermion system within QED has a formally exact description in terms of the Bethe–Salpeter equation. One has to construct a $4 \times 4 = 16$ – component two-body wave function and a two-body propagator S_{F}^{ab}. For an exact description one would need to know the interaction kernel K^{ab}, which is formally represented by an infinite sum over all possible Feynman graphs. In practice, however, this is too demanding and the Bethe–Salpeter equation can be applied only in a simplified form, i.e. the ladder approximation. In this

[8] J. Sucher: Phys. Rev. **A22**, 348 (1980).

Example 6.5.

example we will use the ladder approximation to sketch the derivation of a nonrelativistic Hamiltonian for the positronium system, which is one of the cornerstones of QED.

Although we shall start from the full two-body problem, it is reasonable to reduce the corresponding 16-component equation to those components which are largest in the nonrelativistic limit. This both helps in practical calculations and leads to an equation that has a clear physical interpretation. The justification for this procedure, of course, rests on the fact that positronium is a very weakly bound system. The electron and positron are localized in orbitals having twice (owing to the reduced mass) the hydrogen Bohr radius, which is large compared to the Compton wavelength

$$r_0 \approx \frac{2}{m\alpha} \gg \frac{1}{m} \quad .$$

The binding energy is of the order $\alpha^2 m$, which is very small compared to the rest mass $2m$. In the ordinary single-particle Dirac theory the nonrelativistic approximation leads from the 4-component Dirac equation to the 2-component Pauli equation. In an elegant and systematic way this reduction is achieved using the Foldy Wouthuysen technique.[9]

The idea of this method is to apply a unitary transformation $\hat{U}_\mathrm{F}$ which eliminates the "odd" operators $\mathcal{O}$ contained in the Hamiltonian. The name odd is given to those operators which couple the large and small components of the Dirac spinor; typical examples are the matrices $\boldsymbol{\alpha}$ and $\boldsymbol{\gamma}$. Correspondingly those operators which are diagonal with respect to the large and small components are called "even", designated by the letter $\mathcal{E}$. Examples of this class of operators are $\mathbb{1}$, β, $\boldsymbol{\Sigma}$. The task then is to construct a unitary operator $\hat{U}_\mathrm{F}$ such that the odd parts of the Hamiltonian are eliminated. If the original Hamiltonian is written in the form

$$H = \beta m + \mathcal{E} + \mathcal{O} \quad , \tag{1}$$

according to Foldy and Wouthuysen the transformed Hamiltonian reads

$$\begin{aligned} \tilde{H} &= \hat{U}_\mathrm{F}^{-1} H \hat{U}_\mathrm{F} \\ &= \beta m + \mathcal{E} + \frac{\beta}{2m}\mathcal{O}^2 + \frac{1}{8m^2}\left[[\mathcal{O},\mathcal{E}],\mathcal{O}\right] - \frac{\beta}{8m^3}\mathcal{O}^4 + \ldots \quad . \end{aligned} \tag{2}$$

This procedure has been generalized by Chraplyvy[10] in such a way that it can be applied to the two-body problem.

The principal idea of the method remains unchanged. However, the Hamiltonian H now consists of 16×16 matrices, which are constructed from direct products of the 4×4 one-body Dirac matrices referring to particles 1 and 2. Generalizing the notion of even and odd operators we now have to distinguish between four cases:

even even operators ($\mathcal{E}\mathcal{E}$); e.g. $\mathbb{1}$, $\beta^{(1)}$, $\beta^{(2)}$, $\boldsymbol{\Sigma}^{(1)}$, $\boldsymbol{\Sigma}^{(1)} \cdot \boldsymbol{\Sigma}^{(2)}$;
even odd operators ($\mathcal{E}\mathcal{O}$), e.g. $\boldsymbol{\alpha}^{(2)}$, $\boldsymbol{\gamma}^{(2)}$;

[9] See W. Greiner: Theoretical Physics, Vol. 3, *Relativistic Quantum Mechanics – Wave Equations* (Springer, Berlin, Heidelberg 1990), Chap. 11.
[10] Z.V. Chraplyvy: Phys. Rev. **91**, 388 (1953) and **92**, 1310 (1953).

Example 6.5.

odd even operators $(\mathcal{OE})$, e.g. $\boldsymbol{\alpha}^{(1)}$, $\gamma^{(1)}$;
odd odd operators $(\mathcal{OO})$, e.g. $\boldsymbol{\alpha}^{(1)}\cdot\boldsymbol{\alpha}^{(2)}$, $\gamma^{(1)}\cdot\gamma^{(2)}$.

Here we have used the shorthand notation

$$\boldsymbol{\alpha}^{(1)} \equiv \boldsymbol{\alpha}\otimes\mathbb{1} \quad\text{and}\quad \boldsymbol{\alpha}^{(2)} \equiv \mathbb{1}\otimes\boldsymbol{\alpha} \quad . \tag{3}$$

The operators of type $(\mathcal{EO})$, $(\mathcal{OE})$, $(\mathcal{OO})$ couple between the 16 components of the two-body equation, while $(\mathcal{EE})$ operators are diagonal with respect to large and small components. Thus if we are able to find a unitary transformation $\hat{U}_{12}$ that eliminates the operators $(\mathcal{OE})$, $(\mathcal{EO})$, $(\mathcal{OO})$ then the 16-component equation will decouple into a set of four independent 4-component equations. One of them will contain the Pauli approximation to the two-body equation while the remaining three equations describe the admixture of negative-energy states and can be shown to vanish in the nonrelativistic limit. In analogy to the Foldy Wouthuysen transformation operator $\hat{U}_{\mathrm{F}}$, a unitary operator $\hat{U}_{12}$ can also be constructed for the 16-component two-body equation that decouples the large components to any chosen order in $1/m$.

We start from the general two-body Hamiltonian

$$H_{12} = \beta^{(1)}m_1 + \beta^{(2)}m_2 + (\mathcal{EE}) + (\mathcal{EO}) + (\mathcal{OE}) + (\mathcal{OO}) \quad . \tag{4}$$

As shown by Chraplyvy in a lengthy calculation, which we will not reproduce here, the transformed Hamiltonian has the form

$$\begin{aligned}
\tilde{H}_{12} &\equiv \hat{U}_{12}^{-1}H_{12}\hat{U}_{12} \\
&= \beta^{(1)}m_1 + \beta^{(2)}m_2 + (\mathcal{EE}) + \frac{\beta^{(1)}}{2m_1}(\mathcal{OE})^2 + \frac{\beta^{(2)}}{2m_2}(\mathcal{EO})^2 \\
&\quad - \frac{\beta^{(1)}}{8m_1^3}(\mathcal{OE})^4 - \frac{\beta^{(2)}}{8m_2^3}(\mathcal{EO})^4 + \frac{1}{8m_1^2}\left[[(\mathcal{OE}),(\mathcal{EE})]_-,(\mathcal{OE})\right]_- \\
&\quad + \frac{1}{8m_2^2}\left[[(\mathcal{EO}),(\mathcal{EE})]_-,(\mathcal{EO})\right]_- + \frac{\beta^{(1)}\beta^{(2)}}{4m_1m_2}\left[[(\mathcal{OE}),(\mathcal{OO})]_+,(\mathcal{EO})\right]_+ \\
&\quad + \frac{\beta^{(1)}+\beta^{(2)}}{4(m_1+m_2)}(\mathcal{OO})^2 + \dots \quad .
\end{aligned} \tag{5}$$

In principle $\tilde{H}_{12}$ still acts on a 16-component wave function. However, now it is possible to separate off a 4-component equation that contains "large components" only.

In the case of the ordinary single-particle Foldy Wouthuysen transformation this is achieved by replacing each 4×4 Dirac matrix

$$M = \begin{pmatrix} a & b \\ c & d \end{pmatrix}_{4\times 4}$$

simply by a, which is a 2×2 matrix, e.g.

$$\beta = \begin{pmatrix} \mathbb{1} & 0 \\ 0 & -\mathbb{1} \end{pmatrix}_{4\times 4} \rightarrow \mathbb{1}_{2\times 2} \quad .$$

The analogous replacement in the two-body case reads, e.g.

Example 6.5.

$$\beta^{(1)} = \begin{pmatrix} 1\!\!1 & & & \\ & 1\!\!1 & & \\ & & -1\!\!1 & \\ & & & -1\!\!1 \end{pmatrix}_{16\times 16} \rightarrow 1\!\!1_{4\times 4} \quad ,$$

$$\beta^{(2)} = \begin{pmatrix} 1\!\!1 & & & \\ & -1\!\!1 & & \\ & & 1\!\!1 & \\ & & & -1\!\!1 \end{pmatrix}_{16\times 16} \rightarrow 1\!\!1_{4\times 4} \quad .$$

Application to the Positronium System. We start from (34), Example 6.3, which was derived from the Bethe–Salpeter equation

$$\left(E - H^{(1)}(\mathrm{i}\nabla) - H^{(2)}(\mathrm{i}\nabla)\right) \Phi(r) = U(r)\Phi(r) \quad , \tag{6}$$

where $U(r)$ is the Breit interaction defined in (41), Example 6.3. Here

$$H^{(1)} = \boldsymbol{\alpha}^{(1)} \cdot \boldsymbol{p} + \beta^{(1)} m_1 \quad ,$$
$$H^{(2)} = -\boldsymbol{\alpha}^{(2)} \cdot \boldsymbol{p} + \beta^{(2)} m_2 \quad .$$

Written out explicitly (6) reads

$$H\Phi(\boldsymbol{r}) = \left\{ \boldsymbol{\alpha}^{(1)} \cdot \boldsymbol{p} - \boldsymbol{\alpha}^{(2)} \cdot \boldsymbol{p} + \beta^{(1)} m_1 + \beta^{(2)} m_2 + \frac{e_1 e_2}{r} \left[1\!\!1 - \frac{\boldsymbol{\alpha}^{(1)} \cdot \boldsymbol{\alpha}^{(2)}}{2} - \frac{(\boldsymbol{\alpha}^{(1)} \cdot \boldsymbol{r})(\boldsymbol{\alpha}^{(2)} \cdot \boldsymbol{r})}{2r^2} \right] \right\} \Phi(\boldsymbol{r}) = E\Phi(\boldsymbol{r}) \quad . \tag{7}$$

The positronium system is described if we set $m_1 = m_2 \equiv m$, $e_1 = -e_2 \equiv e$. To apply the reduction method discussed above we have to identify the various types of even and odd operators in the Hamiltonian:

$$\begin{aligned} (\mathcal{E}\mathcal{E}) &= -\frac{e^2}{r} 1\!\!1 \quad , \\ (\mathcal{E}\mathcal{O}) &= -\boldsymbol{\alpha}^{(2)} \cdot \boldsymbol{p} \quad , \\ (\mathcal{O}\mathcal{E}) &= +\boldsymbol{\alpha}^{(1)} \cdot \boldsymbol{p} \quad , \\ (\mathcal{O}\mathcal{O}) &= +\frac{e^2}{2r} \left[\boldsymbol{\alpha}^{(1)} \cdot \boldsymbol{\alpha}^{(2)} + \frac{(\boldsymbol{\alpha}^{(1)} \cdot \boldsymbol{r})(\boldsymbol{\alpha}^{(2)} \cdot \boldsymbol{r})}{r^2} \right] \quad . \end{aligned} \tag{8}$$

From this the following ingredients entering (5) can be deduced:

$$(\mathcal{O}\mathcal{E})^2 = (\mathcal{E}\mathcal{O})^2 = \boldsymbol{p}^2 \quad , \tag{9a}$$

$$(\mathcal{O}\mathcal{E})^4 = (\mathcal{E}\mathcal{O})^4 = \boldsymbol{p}^4 \quad , \tag{9b}$$

$$\begin{aligned} &\left\{ \left[[(\mathcal{O}\mathcal{E}), (\mathcal{E}\mathcal{E})]_-, (\mathcal{O}\mathcal{E}) \right]_- + \left[[(\mathcal{E}\mathcal{O}), (\mathcal{E}\mathcal{E})]_-, (\mathcal{E}\mathcal{O}) \right]_- \right\} \\ &\quad = 2e^2 \left[(\boldsymbol{\alpha}^{(1)} + \boldsymbol{\alpha}^{(2)}) \cdot \left(\frac{\boldsymbol{r}}{r^3} \times \boldsymbol{p} \right) + 4\pi\delta^3(\boldsymbol{r}) \right] \quad , \end{aligned} \tag{9c}$$

Example 6.5.

$$[[(\mathcal{O}\mathcal{E}),(\mathcal{O}\mathcal{O})]_+,(\mathcal{E}\mathcal{O})]_+$$
$$= -2e^2\left[\frac{\boldsymbol{p}^2}{r} + \boldsymbol{r}\left(\frac{\boldsymbol{r}}{r^3}\cdot\boldsymbol{p}\right)\boldsymbol{p} - \left(\boldsymbol{\alpha}^{(1)}+\boldsymbol{\alpha}^{(2)}\right)\left(\frac{\boldsymbol{r}}{r^3}\times\boldsymbol{p}\right)\right]$$
$$- e^2\left[\frac{\boldsymbol{\alpha}^{(1)}\cdot\boldsymbol{\alpha}^{(2)}}{r^3} - \frac{3\left(\boldsymbol{\alpha}^{(1)}\cdot\boldsymbol{r}\right)\left(\boldsymbol{\alpha}^{(2)}\cdot\boldsymbol{r}\right)}{r^5} - \frac{8\pi}{3}\boldsymbol{\alpha}^{(1)}\cdot\boldsymbol{\alpha}^{(2)}\delta^3(\boldsymbol{r})\right] \quad , \tag{9d}$$

and

$$(\mathcal{O}\mathcal{O})^2 = \frac{e^4}{2r^2}\left[3 - 2\boldsymbol{\alpha}^{(1)}\cdot\boldsymbol{\alpha}^{(2)} + \frac{\left(\boldsymbol{\alpha}^{(1)}\cdot\boldsymbol{r}\right)\left(\boldsymbol{\alpha}^{(2)}\cdot\boldsymbol{r}\right)}{r^2}\right] \quad . \tag{9e}$$

We insert these expressions into (5) and perform the replacement $\beta^{(1)} \to \mathbb{1}$, $\beta^{(2)} \to \mathbb{1}$, etc., as discussed above. The wave function consists of a direct product of two Pauli spinors. This corresponds to the approximation that both particles are described by the two upper components of their original bispinor. Thus the generalized Pauli equation reads

$$H_{12}^{\mathrm{Pauli}}\psi_{12}(\boldsymbol{r}) = E\psi_{12}(\boldsymbol{r}) \quad , \tag{10}$$

where

$$H_{12}^{\mathrm{Pauli}} \equiv H_0 + H_1 + H_2 + H_3 + H_4$$

and

$$\begin{aligned} H_0 &= -\frac{e^2}{r} + \frac{\boldsymbol{p}^2}{m} \quad , \\ H_1 &= -\frac{1}{4m^3}\boldsymbol{p}^4 \quad , \\ H_2 &= -\frac{e^2}{2m^2}\left[\frac{\boldsymbol{p}^2}{r} + \boldsymbol{r}\left(\frac{\boldsymbol{r}}{r^3}\cdot\boldsymbol{p}\right)\boldsymbol{p}\right] + \frac{e^2}{m^2}\pi\delta^3(\boldsymbol{r}) \quad , \\ H_3 &= \frac{3e^2}{4m^2}\left(\frac{\boldsymbol{r}}{r^3}\times\boldsymbol{p}\right)\left(\boldsymbol{\sigma}^{(1)}+\boldsymbol{\sigma}^{(2)}\right) \quad , \\ H_4 &= -\frac{e^2}{4m^2}\left[\frac{\boldsymbol{\sigma}^{(1)}\cdot\boldsymbol{\sigma}^{(2)}}{r^3} - \frac{3(\boldsymbol{\sigma}^{(1)}\cdot\boldsymbol{r})(\boldsymbol{\sigma}^{(2)}\cdot\boldsymbol{r})}{r^5} - \frac{8\pi}{3}(\boldsymbol{\sigma}^{(1)}\cdot\boldsymbol{\sigma}^{(2)})\delta^3(\boldsymbol{r})\right] \quad . \end{aligned} \tag{11}$$

Terms of the order e^4 have been neglected. The various parts of the two-body Pauli Hamiltonian have an intuitive interpretation. H_0 is the nonrelativistic Schrödinger operator (the rest mass has been subtracted from (10)). H_1 describes the first-order corrections due to the velocity dependence of the mass. H_2 contains the classical relativistic correction to the interaction of two charged particles due to the effect of retardation and the Darwin term due to Zitterbewegung. H_3 is the interaction between the total spin $\boldsymbol{s} = \boldsymbol{s}^{(1)}+\boldsymbol{s}^{(2)} = \frac{1}{2}(\boldsymbol{\sigma}^{(1)}+\boldsymbol{\sigma}^{(2)})$ and the relative orbital angular momentum of the particle, $\boldsymbol{L} = \boldsymbol{r}\times\boldsymbol{p}$. H_4 is the magnetic dipole interaction between the magnetic moments of the electron and positron (spin spin interaction).

With some more thought we come to the conclusion that the Hamiltonian (11) does not contain the whole truth about the positronium system. We have treated the two particles as independent objects having opposite charge. Because we are dealing with a particle antiparticle system, however, the process of virtual pair annihilation also has to be taken into account. Thus the annihilation interaction introduced in Exercise 6.2 has to be included. We will not go through the steps leading to the nonrelativistic limit of this interaction but only quote the resulting additional contribution to the Hamiltonian:

Example 6.5.

$$H_5 = \frac{e^2\pi}{2m^2}\left(3 + \boldsymbol{\sigma}^{(1)} \cdot \boldsymbol{\sigma}^{(2)}\right)\delta^3(\boldsymbol{r}) \quad .$$

Note that this interaction only contributes for states with angular orbital momentum $L = 0$ (because of the delta function) and total spin $S = 1$ (since $\boldsymbol{\sigma}^{(1)} \cdot \boldsymbol{\sigma}^{(2)} = -3$ for singlet states).

Including the Hamiltonians H_1 to H_5 as perturbations to the H_0 problem will lead to the energy levels of positronium[11] correct to order α^4.

6.2 Biographical Note

SALPETER, Edwin Ernest, American physicist. *3.12.1924 in Vienna. S. studied at Sydney University (Australia) and at the University of Birmingham where he got his PhD in 1948. He became research associate and later professor at Cornell University and Director of Cornell's Center of Radiophysics and Space Research. S. worked with H. Bethe on few-electron systems and the quantum mechanical two-body problem (the Bethe–Salpeter equation). S.'s main field of research has been astrophysics where he worked on nuclear fusion mechanisms in stars, the structure of collapsed stars, interstellar matter etc.

[11] See for example M.A. Stroscio: Phys. Rep. **22**, 215 (1975).

7. Quantum Electrodynamics of Strong Fields

Up to now, our considerations in this book have mainly treated the behaviour of electrons (or positrons) under the influence of weak perturbations. In this chapter we want to deal with phenomena that occur in the presence of strong electromagnetic fields.[1]

We shall see that a novel "effect of zeroth order" occurs in this case which cannot be described by perturbation theory as usual: the ground state (the "vacuum") of the theory becomes unstable and changes at a certain strength of the potential. At first, we shall discuss the process qualitatively.

Normally, i.e. in weak fields, the energetically lowest stable state is characterized by the fact that no (real) particles are present; in the case of QED this means neither electrons nor positrons. In the absence of an external field the Dirac equation possesses only continuum solutions with energies $E_p = \pm\sqrt{m_0^2c^4 + p^2c^2}$. The vacuum state is determined by the requirement that all positive energy states are empty and all negative energy states are occupied (see Fig. 7.1).

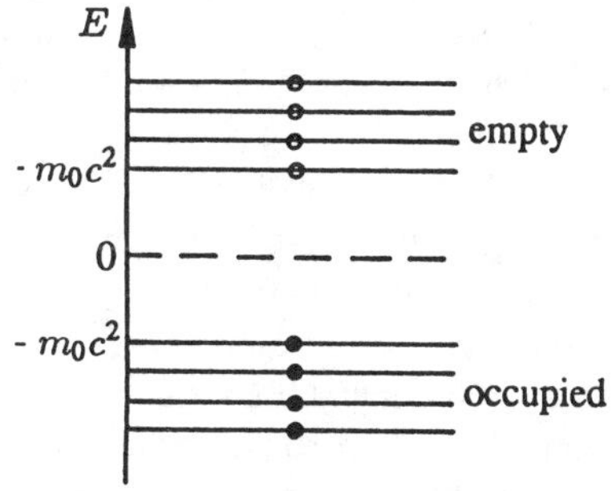

Fig. 7.1. The vacuum of the Dirac theory in the absence of an external field

In Dirac's hole picture this means that neither free electrons nor free positrons (i.e. holes in the lower continuum) are present. The formally infinitely large energy and charge of the "Dirac sea" are unobservable in principle and are "renormalized away". Therefore, the physically observable vacuum (without electromagnetic field) is free of particles and is electrically neutral. We now switch on an external electromagnetic field $A_\mu(x)$. The approximation of an external field means that $A_\mu(x)$ is assumed to be given classically and that it is not be influenced by the electrons. In reality such a field is very well represented by that of an atomic nu-

[1] Detailed presentations may be found in W. Greiner, B. Müller, J. Rafelski: *Quantum Electrodynamics of Strong Fields* (Springer, Berlin, Heidelberg 1985).

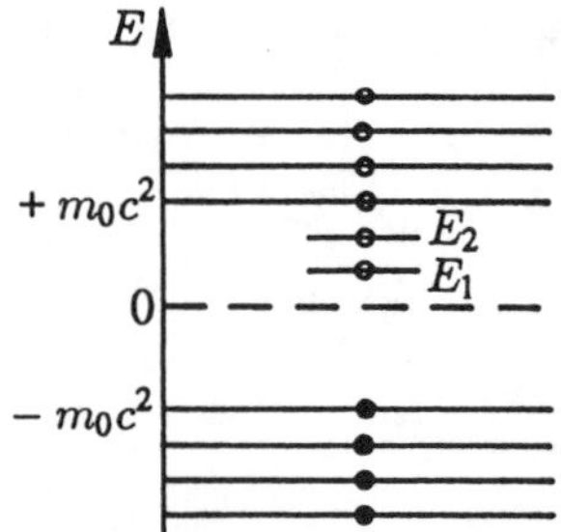

Fig. 7.2. The spectrum of the Dirac equation in the presence of a weak external binding potential

cleus with charge $-Ze$ since its mass is extremely large compared to the electron mass ($m_p/m_e \sim 2000$). Macroscopic fields and coherent electromagnetic radiation can also be described in terms of an external field.

In order to realize the essential effect we consider an electrostatic potential well $A_0(x)$ of finite extent. If the potential is positive, electrons can be bound in the well. This means that one or more electronic levels $E_1, E_2, \ldots$ are lowered into the energy gap between $E = +m_0c^2$ and $E = -m_0c^2$. The corresponding wave functions are spatially localized in the region of the potential well. The basic structure of the vacuum will not be changed, however, because it is still energetically favourable to have the lowered states unoccupied (Fig. 7.2).

This seems to change if the potential is so strong that E_1 becomes negative, i.e. if the binding energy $E_B = m_0c^2 - E_1$ surpasses the value m_0c^2. By adding one electron to the system, more energy can be emitted than is contained in the rest mass of the electron. This leads to the curious phenomenon that according to mass–energy equivalence the system becomes lighter than it would be without the electron! However, since charge conservation must be fulfilled rigorously there is no way to produce a single electron; only electron–positron pairs can be created. The energy threshold for this is $2m_0c^2$ and hence a system with empty energy levels remains stable also in the range $-m_0c^2 < E_1 < 0$ (Fig.7.3a). Things change, however, as soon as a state falls below the threshold $E = -m_0c^2$. An electron–positron pair can then be created without the need to expend additional energy (e.g. from an incoming photon). There is even a surplus which is available as kinetic energy, namely $E_{kin} = E_B - 2m_0c^2 = |E_1| - m_0c^2$. When *passing beyond a certain critical potential strength* the following will happen: *An e^+e^- pair is created spontaneously*. The electron is attracted and stays inside the potential well, whereas the positron is repelled and thus escapes with kinetic energy E_{kin} (Fig. 7.3b).

This process may also be interpreted in the following way: as the potential strength is increased *the state E_1 "dives into the lower continuum"* and merges with it. If it has been empty previously then the Dirac sea now contains an additional hole which physically corresponds to a positron. After a certain length of time (we shall calculate this time more precisely) a new stable ground state is formed: the previously empty (in the subcritical case) potential well is now filled with one electron or, in other words, the previously *neutral vacuum* has decayed into a *charged vacuum*. Of course, global charge conservation is not violated by this. The emitted positron carries a positive unit of charge to infinity, i.e. it escapes from any arbitrarily large but finite volume surrounding the potential well.

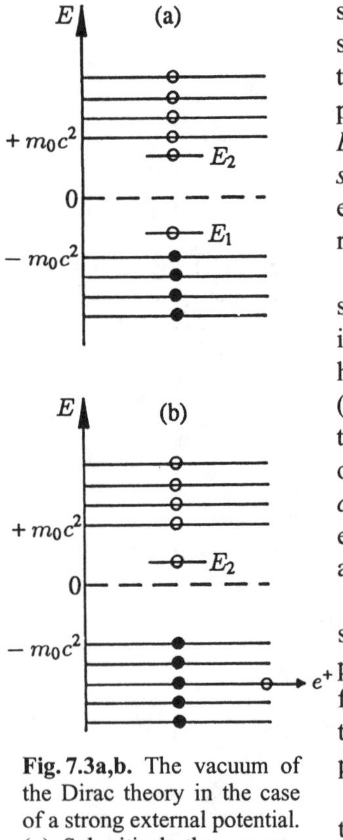

Fig. 7.3a,b. The vacuum of the Dirac theory in the case of a strong external potential. (**a**) Subcritical: the vacuum is neutral. (**b**) Supercritical: a positron is emitted spontaneously and the vacuum gets charged

The charge Q_{vac} which the supercritical vacuum assumes depends on the strength of the potential, namely on the number of levels $E_1, E_2, \ldots$ that have passed below the threshold $-m_0c^2$ and are thus submerged. If there is no magnetic field present, all levels are twofold degenerate (equal energy for both spin orientations $s_z = \pm 1/2$), so that Q_{vac} runs over the values $2e, 4e, \ldots$ with increasing potential strength.

Before we give a quantitative description it is useful to consider qualitatively the structure of the quantum–mechanical wave functions in the various cases. We are looking for eigensolutions ψ of the stationary Dirac equation with a scalar electromagnetic potential $A_0(x)$ such that

$$\left(c\boldsymbol{\alpha} \cdot \boldsymbol{p} + \beta m_0c^2 + eA_0(x)\right) \psi(x) = E\psi(x) \quad . \tag{7.1}$$

Here it is essential that the energy E and potential $V(x) = eA_0(x)$ occur in the combination $E - V(x)$. This means that the boundaries of the energy gap (which is in the range between $+m_0c^2$ and $-m_0c^2$ in the field-free case) are shifted locally by the potential. This is shown in Fig. 7.4 for the case of an attractive potential well.

Inside the energy gap is the classically forbidden region, where only exponentially decaying wave functions are possible. Outside the gap, i.e. for $|E - V| > m_0c^2$, there are oscillating solutions. Since $V(x)$ is assumed to vanish asymptotically, we obtain for $E > m_0c^2$ and $E < -m_0c^2$ oscillating continuum wave functions that extend to infinity. The effect of the potential well is to deform these wave functions. In ordinary weak potentials this distortion does not lead to a qualitative change of the character of the wave function. In Fig. 7.5 two typical wave functions of the upper (a) and lower (b) continuum are sketched. (The picture is somewhat schematic since strictly we have to deal with four-component spinor functions.) It is the attractive potential which makes it possible that there are spatially localized bound wave functions at discrete energies in the region $-m_0c^2 < E < +m_0c^2$ (case (c) in Fig. 7.5). However, if the potential well is deep enough one (or more) of the bound states can fall below the threshold $-m_0c^2$. The wave function then necessarily has to change its character, because it can no longer decay exponentially but extends to infinity. Its shape is displayed as case (d) in Fig. 7.5.

We can now distinguish three regions: in the interior of the potential well the wave function resembles that of an ordinary bound state; outside it oscillates as a continuum wave. Both regions are connected by a zone in which the energy is in the "forbidden" energy gap.

This situation strongly reminds us of a process familiar from nonrelativistic quantum mechanics, namely the *tunnel effect* that occurs, for example in α decay or in solid-state physics. Tunneling is possible whenever a particle must pass through a region of space without having the classically required energy as, for example shown in Fig. 7.6 for a potential barrier.

A particle sitting in the potential well is in a "quasi-bound" state. It can tunnel through the energetically forbidden region of the potential barrier with a certain decay rate (probability per time). We can now say in full analogy that a hole in the overcritical bound state tunnels through the energy gap between the upper and the lower continuum in order to escape as a positron.

It is well known that the decay rate in a tunnel process decreases exponentially with the width and height of the barrier to be passed (the "Gamow factor"). This is also valid for the decay of the neutral vacuum. This explains why spontaneous pair production is not observed in macroscopic electrostatic fields. Potential differences of several megavolts can be easily produced but typically they extend over a range of meters which makes the pair production rate extremely small. A rough estimate tells us that pair production becomes considerable if the potential $\Delta V = e\Delta A_0$ changes by a value of two rest-mass units $2m_0c^2$ over a characteristic length scale which is set by the Compton wavelength of the electron $\lambda\!\!\!^{-} = \hbar/m_0c$. Because of $\boldsymbol{E} = -\nabla A_0$ this leads to a *critical field strength* of the order of magnitude of

$$E_{\text{cr}} \sim \frac{\Delta A_0}{\Delta x} = \frac{2m_0c^2}{e\hbar/m_0c} = \frac{2m_0^2c^3}{e\hbar} = \frac{1}{e}\frac{2 \times 511\,\text{keV}}{386\,\text{fm}} = 2.6 \times 10^{16}\,\frac{\text{V}}{\text{cm}} \quad . \tag{7.2}$$

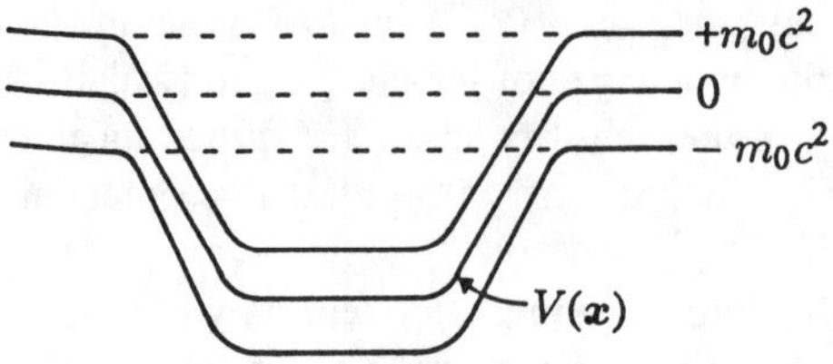

Fig. 7.4. The energy gap of the Dirac equation in the presence of an electrostatic potential well

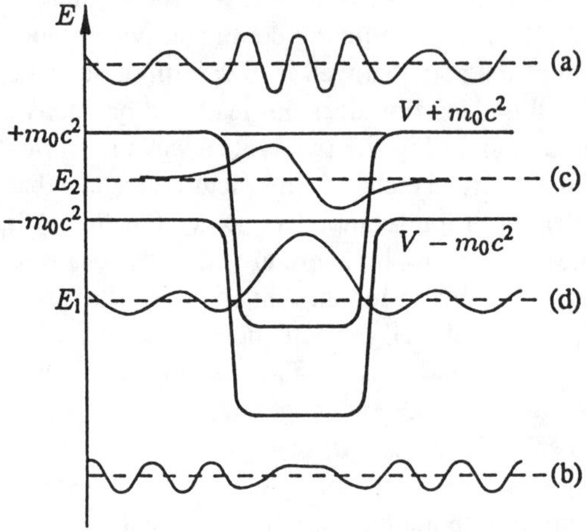

Fig. 7.5. Schematic representation of the Dirac wave functions of a deep potential well: **(a)** free electron state, **(b)** free positron state, **(c)** bound electron state, **(d)** resonance in the lower continuum

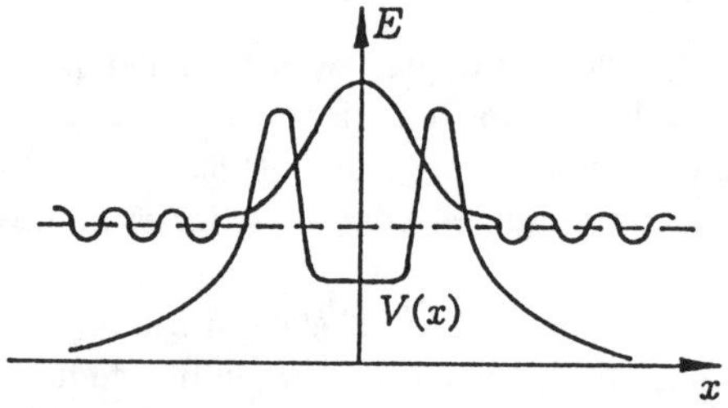

Fig. 7.6. Illustration of the tunnel effect in nonrelativistic quantum mechanics

We shall encounter this critical field strength once again in Sect. 7.3 where a homogeneous electric field will be studied. Field strengths of magnitude E_{cr} occur only in microscopic systems.[2] On the other hand the size of the potential region has to be large enough that (by the uncertainly relation) localization of the wave function is possible. There is only one experimentally accessible system in which the decay of the vacuum may be examined: a heavy atom with very high nuclear charge number Z.

[2] We shall not take up here the matter of possible laser fields of very high intensity.

7.1 Strong Fields in Atoms

Nature provides atomic nuclei as an almost ideal source of strong external electric fields. A nucleus of charge Z and mass number A produces a spherically symmetric electric potential of the form

$$V(r) = ZU(r) = \begin{cases} -Ze^2/r & \text{for} \quad r > R \\ -(Ze^2/R)f(r) & \text{for} \quad r < R \end{cases} , \tag{7.3}$$

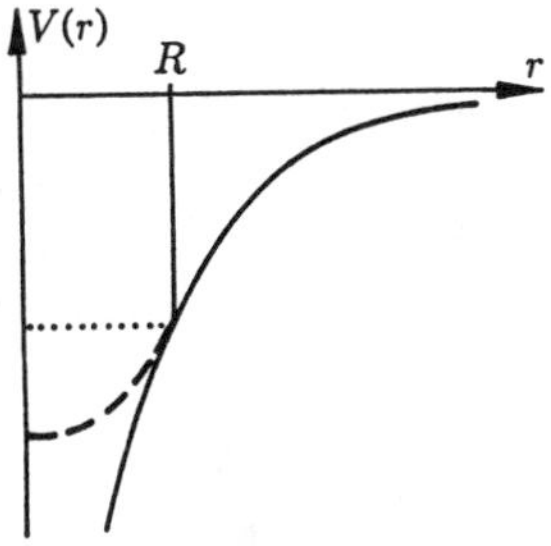

Fig. 7.7. The Coulomb potential of a point nucleus (solid line), of a homogeneously charged sphere (dashed line) and of a charged spherical shell (dotted line)

where the charge radius R of the nucleus is given approximately by

$$R = 1.2A^{1/3}\text{fm} \quad . \tag{7.4}$$

If one models the nucleus as a homogeneously charged sphere with a sharp edge, which is sufficient for most purposes, then the bottom of the potential well is parabolic, i.e.

$$f(r) = \frac{1}{2}\left(3 - \frac{r^2}{R^2}\right) \quad . \tag{7.5}$$

As a crude but simple approximation one sometimes uses $f(r) = 1$, corresponding to a charged spherical shell. The three cases are shown in Fig. 7.7.

The potential in the interior and the maximum electric field strength (at the nuclear surface), e.g. for a uranium nucleus ($Z = 92, A = 238$), are very large, namely

$$|V_{\text{max}}| = \frac{3}{2}\frac{Ze^2}{R} \simeq 26.7 \text{ MeV} \simeq 52 \;\; m_0c^2$$

and

$$|\boldsymbol{E}_{\text{max}}| = \frac{Ze}{R^2} \simeq 2.4 \; 10^{19}\frac{V}{\text{cm}} \simeq 900 \; E_{\text{cr}} \quad .$$

While these numbers by far exceed the values of $2m_0c^2$ and E_{cr} one cannot call the electric field of a uranium nucleus supercritical. In addition the region of space over which the field extends has to be large enough to make the localization of a quantum-mechanical wave function possible. This condition is not met in ordinary atoms. In contrast to the situation shown in Fig. 7.5 the electronic levels here are not bound very deeply.

In order to investigate the behaviour of the states the Dirac equation (7.1) has to be solved for the central potential (7.3). The procedure is described extensively in Chap. 9 of the volume *RQM* and we shall only briefly repeat it here.

The angular-momentum operator and the parity operator commute with the Dirac Hamiltonian $H_{\text{D}} = \boldsymbol{\alpha} \cdot \boldsymbol{p} + \beta m_0 + V(r)$ for a potential of the form (7.3). Hence the wave functions can be classified according to their angular momentum j and parity π. The ansatz for the bispinor wave function is

$$\psi_{jm}(\boldsymbol{x}) = \frac{1}{r}\begin{pmatrix} u_1(r)\chi_{\kappa m}(\Omega) \\ \mathrm{i}u_2(r)\chi_{-\kappa m}(\Omega) \end{pmatrix} , \tag{7.6}$$

with the two-component spherical spinors

$$\chi_{\kappa m}(\Omega) = \sum_{\mu=\pm 1/2} \left(l \tfrac{1}{2} j \middle| m-\mu \, \mu m\right) Y_{lm-\mu}(\Omega)\chi_\mu \quad , \tag{7.7}$$

where $\chi_{1/2} = \binom{1}{0}$, $\chi_{-1/2} = \binom{0}{1}$. κ is the eigenvalue of the operator $\hat{K} = \beta(\boldsymbol{\sigma}\cdot\boldsymbol{L}+1)$ and has the value

$$\kappa = \begin{cases} -(l+1) & = -(j+\frac{1}{2}) \quad \text{for} \quad j = l+\frac{1}{2} \\ l & = +(j+\frac{1}{2}) \quad \text{for} \quad j = l-\frac{1}{2} \end{cases} \quad .$$

The Dirac equation (7.1) reduces to a system of two coupled ordinary differential equations of first order ($\hbar = c = 1$)

$$\begin{aligned} \frac{d}{dr}u_1 &= -\frac{\kappa}{r}u_1 + (E + m_0 - V(r))u_2 \quad , \\ \frac{d}{dr}u_2 &= -(E - m_0 - V(r))u_1 + \frac{\kappa}{r}u_2 \quad . \end{aligned} \tag{7.8}$$

For all energies $E > m_0$, $E < -m_0$, (7.8) possesses continuum solutions which are regular at the origin ($r = 0$) and oscillate asymptotically ($r \to \infty$). In the energy gap $-m_0 < E < m_0$ only at certain discrete energy eigenvalues E_{nj} are solutions found that fulfill the regularity requirement at $r \to 0$ and $r \to \infty$ simultaneously and are thus normalizable:

$$\int_0^\infty dr(u_1^2 + u_2^2) = 1 \quad . \tag{7.9}$$

The system (7.8) can be solved numerically without difficulty in order to determine the solutions for arbitrary potentials. For some potentials the solution may also be found analytically. For this purpose it is useful to rewrite the system (7.8) into a single differential equation of second order.

A closed solution is obtained in the case of a *pure Coulomb potential*, $V(r) = -Z\alpha/r$ for all r. It is composed essentially of confluent hypergeometric functions. The energy eigenvalues satisfy Sommerfeld's well-known fine-structure formula,

$$E_{nj} = m_0 \left[1 + \left(\frac{Z\alpha}{n - |\kappa| + \sqrt{\kappa^2 - Z^2\alpha^2}}\right)^2\right]^{-\frac{1}{2}} \quad , \tag{7.10}$$

with the principal quantum number $n = 1, 2, \ldots$. Obviously this formula is no longer valid for charges $Z\alpha > |\kappa|$, because then the root $\gamma = \sqrt{\kappa^2 - Z^2\alpha^2}$ becomes imaginary. This is seen especially clearly for the most deeply bound $1s$-state ($\kappa = -1, n = 1$) whose energy is $E_{1s} = m_0\sqrt{1 - (Z\alpha)^2}$. The function $E_{1s}(z)$ breaks off with vertical tangent $dE_{1s}/dZ \to -\infty$ as $Z\alpha \to 1$.

We recognize the reason for this behaviour in the shape of the wave function near the origin. For $r \to 0$ the Coulomb potential dominates: $|\, V \,| \gg E, m_0$. By elimination of u_2 in the system (7.8) (cf. Exercise 7.1), we get the differential equation for u_1 ($r \to 0$) :

$$u_1'' + \frac{1}{r}u_1' + \frac{(Z\alpha)^2 - \kappa^2}{r^2}u_1 = 0 \quad . \tag{7.11}$$

The regular solution of (7.11) is

$$u_1(r) \sim r^{\sqrt{\kappa^2-(Z\alpha)^2}} \quad \text{for} \quad r \to 0 \quad . \tag{7.12}$$

The exponent becomes imaginary in the case $Z\alpha > |\kappa|$. If we construct a real solution it will oscillate with infinite frequency like $\sin(\sqrt{(Z\alpha)^2 - \kappa^2}\ln r + \delta)$. Since such a singular wave function is not acceptable, there is no solution of the problem of an electron in a pure Coulomb potential for $Z\alpha > |\kappa|$.[3]

This is because we did not formulate the problem correctly. In reality the source of the Coulomb field has finite extension so that the potential V remains finite as given in (7.3). In this case the singular behaviour (7.12) of the wave function at the origin does not arise. The energy levels can then be traced continuously beyond the point $Z\alpha = |\kappa|$.

Solving the Dirac equation with the truncated Coulomb potential (7.3) is more involved compared to the case of the pure Coulomb potential. A way to do this is to construct at first two solutions $u_1^{(i)}, u_2^{(i)}$ in the inner region and $u_1^{(a)}, u_2^{(a)}$ in the outer region. In each region there are two linearly independent solutions of the system (7.8) of differential equations of first order. The inner solution is required to be regular at the origin, $r \to 0$, whereas the outer solution should decay exponentially at infinity, $r \to \infty$. Both solutions have to be matched at a point R_0 that can in principle be chosen arbitrarily. We can do this by making the ratios of the large to the small component of the wave function equal,

$$\frac{u_1^{(i)}(R_0)}{u_2^{(i)}(R_0)} = \frac{u_1^{(a)}(R_0)}{u_2^{(a)}(R_0)} \quad . \tag{7.13}$$

This condition is not fulfilled in general. The two solutions can be joined continuously to a total wave function regular at both boundaries only for certain discrete energy eigenvalues (which can be determined by iteration). In contrast to (7.10) this procedure normally cannot been done analytically.

EXERCISE

7.1 The Wave Function at the Diving Point

Problem. The Dirac wave function at the diving point (where the energy eigenvalue is $E = -m_0$) can be determined analytically with relative ease.

a) Write the radial Dirac equation (7.8) as a differential equation of second order for u_1 by elimination of u_2.

b) Show that, in the case of the Coulomb potential $V(r) = -Z\alpha/r$, at the peculiar energy $E = -m_0$ the wave function u_1 satisfies Bessel's differential equation, and find the solution regular at infinity.

Hint: use the substitution $\varrho = \sqrt{8m_0 Z\alpha r}$.

c) Find an equation which determines the critical charge Z_{cr} of the $1s$-state for the truncated Coulomb potential ($V(r) = -Z\alpha/R$ for $r \leq R, V(r) = -Z\alpha/r$

[3] For a resolution of this problem see P. Gärtner et al.: Z. Phys. **A300**, 143 (1981).

Exercise 7.1.

for $r \geq R$) by matching the solutions in the inner and outer regions according to (7.13).

Solution. a) We differentiate the differential equation (7.8a) with respect to r, which gives

$$u_1'' = \frac{\kappa}{r^2}u_1 - \frac{\kappa}{r}u_1' - V'u_2 + (E + m_0 - V)u_2' \quad . \tag{1}$$

Applying (7.8a) once again we can eliminate u_2 from this equation,

$$u_2 = \frac{u_1' + \frac{\kappa}{r}u_1}{E + m_0 - V} \quad . \tag{2}$$

By use of the second differential equation (7.8b) we eliminate u_2',

$$u_2' = -(E - m_0 - V)u_1 + \frac{\kappa}{r}u_2 \quad . \tag{3}$$

The result from (1), (2), and (3) is

$$u_1'' + \frac{V'}{E + m_0 - V}u_1' + \left[(E - V)^2 - m_0^2 - \frac{\kappa(\kappa+1)}{r^2} + \frac{\kappa}{r}\frac{V'}{E + m_0 - V}\right]u_1 = 0 \quad . \tag{4}$$

b) At the special energy $E = -m_0$ (4) reduces by use of the Coulomb potential $V = -Z\alpha/r$ to

$$u_1'' + \frac{1}{r}u_1' - \left(\frac{2m_0 Z\alpha}{r} - \frac{(Z\alpha)^2 - \kappa^2}{r^2}\right)u_1 = 0 \quad . \tag{5}$$

The suggested substitution

$$\varrho^2 = 8m_0 Z\alpha r \tag{6}$$

leads to

$$\frac{d}{dr} = \frac{4m_0 Z\alpha}{\varrho}\frac{d}{d\varrho} \quad , \tag{7a}$$

$$\frac{d^2}{dr^2} = \frac{(4m_0 Z\alpha)^2}{\varrho^2}\left(-\frac{1}{\varrho}\frac{d}{d\varrho} + \frac{d^2}{d\varrho^2}\right) \quad . \tag{7b}$$

Insertion into (5) and multiplication by $\varrho^2/(4Z\alpha)^2$ leads to

$$\frac{d^2}{d\varrho^2}u_1 + \frac{1}{\varrho}\frac{d}{d\varrho}u_1 - \left\{1 - \frac{4[(Z\alpha)^2 - \kappa^2]}{\varrho^2}\right\}u_1 = 0 \quad . \tag{8}$$

This is just the differential equation obeyed by the modified Bessel functions

$$f'' + \frac{1}{\varrho}f' - \left(1 + \frac{\mu^2}{\varrho^2}\right)f = 0 \quad . \tag{9}$$

This differential equation is solved by the linearly independent solutions $K_\mu(\varrho)$ and $I_\mu(\varrho)$. The function $I_\mu(\varrho)$ is to be rejected since it increases exponentially at

Exercise 7.1.

$$u_1(r) = cK_{i\nu}\left(\sqrt{8m_0Z\alpha r}\right) \quad , \tag{10}$$

with a normalising constant c that is of no interest here. The index of the Bessel function is purely imaginary in the case $Z\alpha >| \kappa |$, being given by

$$\nu = 2\sqrt{(Z\alpha)^2 - \kappa^2} \quad . \tag{11}$$

The lower component of the wave function follows from (2)

$$u_2 = \frac{c}{Z\alpha}\left[\frac{1}{2}\sqrt{8m_0Z\alpha r}\; K'_{i\nu}\left(\sqrt{8m_0Z\alpha r}\right) + \kappa K_{i\nu}\left(\sqrt{8m_0Z\alpha r}\right)\right] \quad , \tag{12}$$

where $K'_{i\nu}$ denotes the derivative of the Bessel function with respect to its argument and (7a) has been used. Remarkably, the "critical wave function" decays exponentially as a function of $\sqrt{r}$ for large values of r, since[4]

$$K_{i\nu}(z) \to \sqrt{\frac{\pi}{2z}}e^{-z} \quad \text{for} \quad z \to \infty \quad . \tag{13}$$

The limit $r \to 0$ is more involved. By use of the (9.6.7), and (6.1.31) of Abramowitz we get

$$K_{i\nu}(z) \to \sqrt{\frac{\pi}{\nu \sinh \pi\nu}} \sin\left(\nu \ln\frac{2}{z} + \arg\Gamma(1 + i\nu)\right) \quad . \tag{14}$$

As we have already discussed in connection with (7) this means that the wave function oscillates like $\sin\left(\sqrt{(Z\alpha)^2 - \kappa^2}\; \ln r + \delta\right)$.

c) In order to obtain a wave function regular also at the origin the solutions (10), (12) have to be matched to an inner solution $u^{(i)}$ at the nuclear radius R. Since the potential in this region is constant, $V_0 = -Z\alpha/R$ the free spherical solutions of the Dirac equation can be taken (*RQM*, Chap. 9). They are

$$\begin{aligned} u_1^{(i)} &= c'\, \beta r\, j_l(\beta r) \quad , \\ u_2^{(i)} &= c'\, \mathrm{sgn}\kappa \frac{\beta}{E - V_0 + m_0} \beta r\, j_{\bar{l}}(\beta r) \quad , \end{aligned} \tag{15}$$

where j_l denotes the spherical Bessel function and $\beta = \sqrt{(E - V_0)^2 - m_0^2}$. The order of the Bessel functions is $l = -\kappa - 1, \bar{l} = -\kappa$ for $\kappa < 0$ and $l = \kappa, \bar{l} = \kappa - 1$ for $\kappa > 0$.

In the special case $\kappa = -1$ ($s_{1/2}$ states) (15) reads

$$\begin{aligned} u_1^{(i)} &= c' \sin\beta r \quad , \\ u_2^{(i)} &= -c' \frac{\beta}{E - V_0 + m_0}\left(\frac{\sin\beta r}{\beta r} - \cos\beta r\right) \quad . \end{aligned} \tag{16}$$

The matching condition (13) for the inner and outer solutions then has the form

$$-\frac{\sin\beta R}{\frac{\sin\beta R}{\beta R} - \cos\beta R}\frac{E - V_0 + m_0}{\beta} = \frac{Z\alpha K_{i\nu}}{\frac{1}{2}\sqrt{8m_0Z\alpha R}\;\; K'_{i\nu} - K_{i\nu}} \tag{17}$$

[4] M. Abramowitz, I.A. Stegun: *Handbook of Mathematical Functions* (Dover), Chap. 9.7.

Exercise 7.1.

to be evaluated at the energy $E = -m_0c^2$. The second factor on the left-hand side can be approximated by 1 because of $| V_0 | \gg m_0$. By the same reason $\beta \simeq | V_0 | = Z\alpha/R$. Then (17) may be rewritten

$$\frac{\varrho K'_{\mathrm{i}\nu}(\varrho)}{K_{\mathrm{i}\nu}(\varrho)} = 2(Z\alpha)\cot(Z\alpha) \quad . \tag{18}$$

where $\varrho = \sqrt{8m_0Z\alpha R}$. This is a transcendental equation for Z. Its solutions are the critical charges for the states $1s_{1/2}, 2s_{1/2}$ etc. The Figure 7.8 displays the graphical solution of (18). To obtain reasonably realistic values of Z_{cr} the following assumption was made for the nuclear radius R as a function of Z:

$$R = (2.5Z)^{1/3} \times 1.2\,\mathrm{fm} \quad .$$

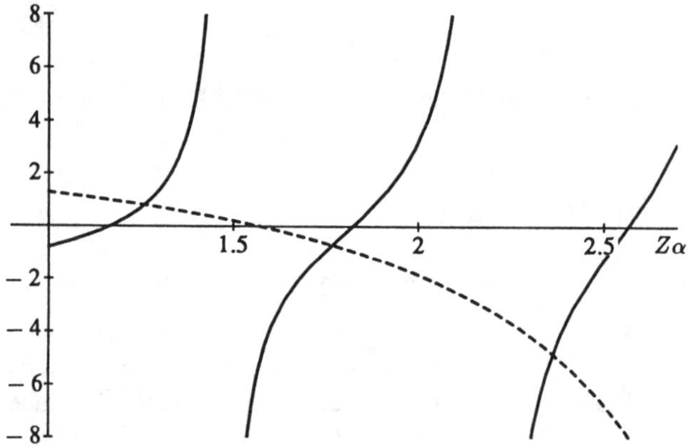

Fig. 7.8. Graphical determination of the critical nuclear charge for s-states. The full (dashed) curve is the left- (right-) hand side of (18). The curves intersect at $Z_{\mathrm{cr}} = 172, 242, 324, \ldots$

Figure 7.9 displays the result of numerical calculations for the energies of several bound states as a function of nuclear charge Z. The energy levels descend progressively with increasing Z. The energy of the lowest state ($1s, \kappa = -1$) becomes negative when $Z > 150$. The $1s$ level finally reaches the value $E_{1s} = -m_0c^2$ at a *critical charge* $Z_{\mathrm{cr}}^{1s} \simeq 173$. The same happens for the $2p_{1/2}$ state with $\kappa = +1$ at $Z_{\mathrm{cr}}^{2p_{1/2}} \simeq 185$, whereas higher states reach the lower continuum at much larger nuclear charges.

The large gain in binding energy of the levels having $\kappa = \pm 1$ is accompanied by drastic changes in the wave functions. The radial density $r^2\psi^\dagger(r)\psi(r)$ of the $1s$ wave function is plotted in Fig. 7.10 for the three nuclear charges $Z = 100, 135,$ and 170. A scaled representation has been chosen that does already account for the "trivial" shrinking of the atomic radius like $1/Z$. As is well known the non-relativistic $1s$ hydrogen wave function is of the form

$$\psi_{1s} = 2\left(\frac{Z}{a_{\mathrm{B}}}\right)^{3/2} e^{-rZ/a_{\mathrm{B}}} \quad ,$$

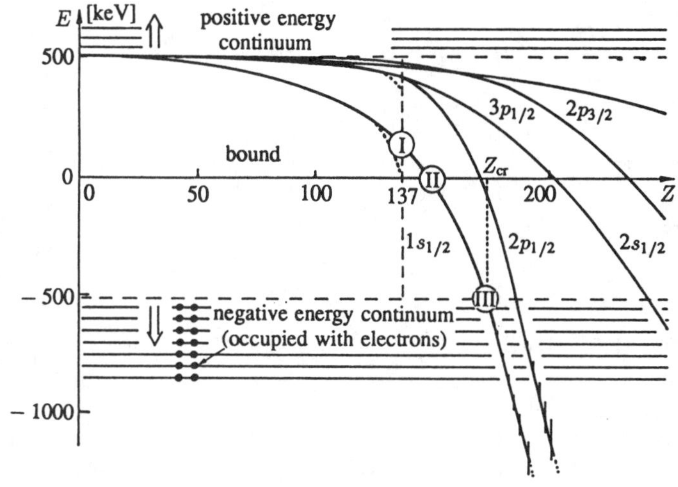

Fig. 7.9. The lowest bound states of the Dirac equation for atoms with nuclear charge Z. The Sommerfeld energies of the states with $\kappa = -1$ $(ns_{1/2})$ and $\kappa = +1$ $(np_{1/2})$ break off with a vertical tangent at $Z\alpha = 1$ (dotted curve). With the finite nuclear radius taken into account all levels reach the edge of the lower continuum $E = -m_0c^2$ at a corresponding critical charge Z_{cr}. The bound states can be followed into the lower continuum as resonances (the energy width is magnified by a factor of 10)

with Bohr's radius $a_B = \lambdabar_e/\alpha = 52918\,\text{fm}$. The corresponding density is independent of Z in the representation used in Fig. 7.10. We see that the influence of relativistic effects is still rather small in normal atoms ($Z \leq 100$). If one proceeds into the range $Z\alpha > 1$, the wave function entirely changes its shape and shrinks to a fraction of its normal extent a_B/Z. A total *"collapse" of the wave function* is avoided only by the finite nuclear radius R. The strong increase of the electron density at the origin $\psi(0)^\dagger\psi(0)$ as a function of Z can be seen even more distinctly in the inset of Fig. 7.9. The nonrelativistic Z-dependence again has been factored out by scaling with Z^{-3}. The Dirac $1s$ wave function increases in density by a factor of about 1000 at large Z compared to the nonrelativistic value. This effect is even more drastic in the case of the $2p_{1/2}$ state which nonrelativistically has a node at $r = 0$ but whose density progessively resembles that of the $1s$ state at large Z. This increase is effected by the lower component u_2 of the wave function which normally is small.

The physically most interesting effect occurs if the critical nuclear charge Z_{cr} is surpassed. As we have discussed in the previous section, the bound state "dives" into the lower continuum of the Dirac equation and two (because of the spin degeneracy, $m = \pm 1/2$) positrons can be emitted spontaneously. This shows up mathematically by the fact that the continuum contains a resonance. The previously bound state does no longer exist as a discrete eigensolution of the Dirac equation but is mixed with the continuum. This admixture is concentrated around a mean resonance energy E_r in a narrow region with width Γ. This is shown schematically in Fig. 7.11.

The figure illustrates how a state that has been discrete in the subcritical potential (a) is spread over a large number of neighbouring continuum states in the

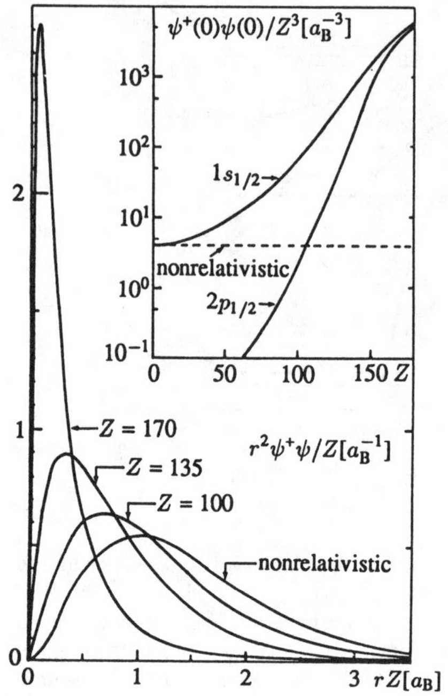

Fig. 7.10. The radial electron density $r^2\psi^\dagger\psi$ of the 1s state divided by Z as a function of the radial distance times Z. The nonrelativistic density is independent of Z in this scaled representation. In contrast, we see that the Dirac wave functions shrink strongly in the range $Z\alpha \geq 1$. The inset displays the electron density at the origin scaled by Z^{-3} as a function of Z

supercritical case (b). The continuum has been discretised for better illustration in Fig. 7.11. We achieve this by a trick: we enclose the system into a box and impose a boundary condition at the surface that is fulfilled only for discrete energy values (which are, however, very dense in the limit $V \to \infty$).

For a quantitative description of the resonance the Dirac equation (7.8) with the supercritical potential has to be solved for various energy values E in the lower continuum. An inner solution $u_1^{(i)}, u_2^{(i)}$ regular at the origin can again be given similar to the case of bound states. The outer solution $u_1^{(a)}, u_2^{(a)}$ is not determined uniquely, however, because the two linearly independent solutions are both bounded at $r \to \infty$ and thus both are admissible. Therefore the matching condition (7.13) does not give a restriction for the energy E, i.e. we obtain a continuum. The condition (7.13) now determines the asymptotic behaviour of the solution. We will not carry out explicitly the construction of the continuum solutions here but refer to the literature.[5]

One can construct real wave functions whose radial part displays the asymptotic behaviour ($r \to \infty$)

$$\begin{pmatrix} u_1 \\ u_2 \end{pmatrix} \sim \frac{1}{\sqrt{\pi p}} \begin{pmatrix} \sqrt{E + m_0} & \cos(pr + \Delta) \\ -\sqrt{E - m_0} & \sin(pr + \Delta) \end{pmatrix}$$

for $E > m_0$ and

$$\begin{pmatrix} u_1 \\ u_2 \end{pmatrix} \sim \frac{1}{\sqrt{\pi p}} \begin{pmatrix} \sqrt{-E - m_0} & \cos(pr + \Delta) \\ \sqrt{-E + m_0} & \sin(pr + \Delta) \end{pmatrix} \tag{7.14}$$

[5] B. Müller, J. Rafelski, W. Greiner: Nuovo Cim. **18**, 551 (1973).

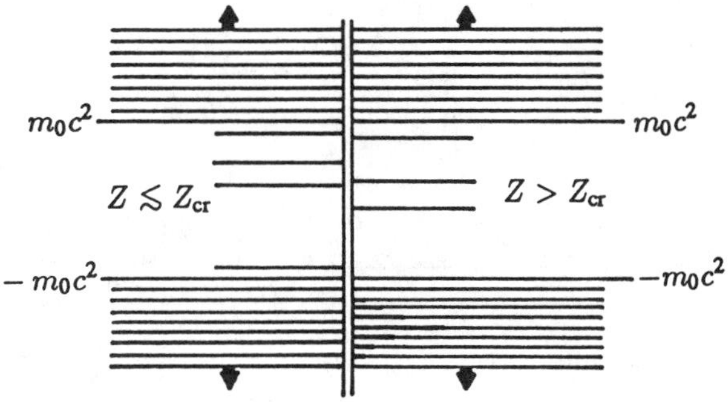

Fig. 7.11. A state that is discrete in the subcritical case $Z < Z_{cr}$ becomes distributed with a certain width over many neighbouring continuum states in a supercritical potential $Z > Z_{cr}$

for $E < m_0$. Here $p = \sqrt{E^2 - m_0^2}$, $\quad \Delta = \delta_E + \delta_{\log}$ with

$$\delta_{\log} = y\left(\ln\frac{2pr}{|y|} + 1\right) - \frac{\pi}{4} \quad , \tag{7.15}$$

where $y = Z\alpha E/p$. $\delta_{\log}$ is a phase shift growing logarithmically with r. It occurs also in nonrelativistic quantum mechanics and is due to the long range of the Coulomb potential. δ_E is the physical phase shift which is determined by the shape of the potential in the inner region. The wave function (7.14) is normalised "on the energy scale", that means

$$\int d^3x\, \Psi^\dagger_{E'}(\boldsymbol{x})\Psi_E(\boldsymbol{x}) = \delta(E - E') \quad .$$

Normally the phase shift δ_E is a function only slowly varying with energy (its value is zero at the edge of the continuum, $E = \pm m_0$). When calculating the wave function of the lower continuum in the case $Z > Z_{cr}$, however, we find that there is an energy region where the phase shift δ_E suddenly varies strongly. δ_E rises about a value of π in a narrow energy range $E_r - \Gamma \leq E \leq E_r + \Gamma$. This is the characteristic signature of a resonance. Inspection of the space dependence of the corresponding wave function shows that the probability density at small distances is strongly enhanced compared to the off-resonance case ($|E - E_r| \gg \Gamma$). Quantitative values are shown in Fig. 7.12 for nuclear charge number $Z = 184$. In the displayed case the 1s resonance has a width of $\Gamma \simeq 0.004\, m_0c^2 \simeq 2$ keV. The wave function u_1 in the inner region is enhanced by a value of 50 corresponding to a factor of 2500 in density.

A phase analysis of the continuum enables us to follow the 1s state having turned into a resonance as a function of Z also in the supercritical case. We already encountered this result in Fig. 7.9: the state moves even deeper into the the continuum and its width increases rapidly starting from a value $\Gamma = 0$ at $Z = Z_{cr}$. However, Γ always remains very small compared to m_0 in the physically accessible range of Z values.

Since the exact continuum solutions of the Dirac equation are rather inconvenient it is useful for our physical understanding to have an analytic model of the resonance. In Example 7.2 we shall present a formalism that yields an easily interpretable closed expression for the wave function Ψ_E.

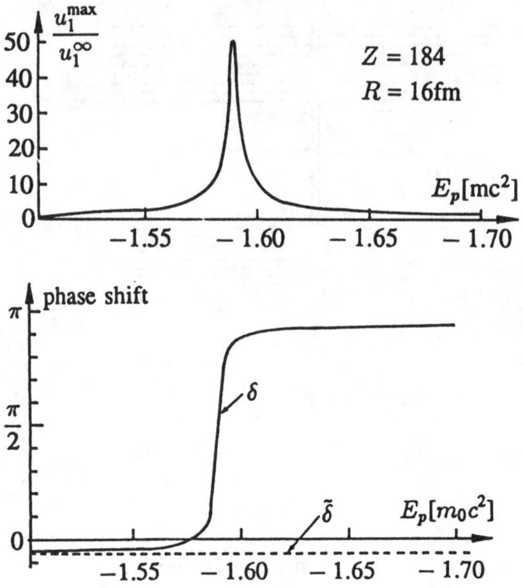

Fig. 7.12. Resonance enhancement of the $s_{1/2}$ lower continuum wave function (u_1 component) and rise of phase shift δ displayed for a supercritical nucleus with $Z = 184$. $\tilde{\delta}$ is the scattering phase without resonance

EXAMPLE

7.2 Fano's Formalism for the Description of Resonances

It is a common problem in applications of quantum mechanics that an initially stable system becomes unstable and can decay if a small perturbation is switched on. Mathematically speaking one starts with a system characterized by a Hamiltonian H_0 which possesses (at least) one discrete and normalisable eigenstate ϕ_0 with energy E_0, i.e.

$$H_0\phi_0 = E_0\phi_0 \quad . \tag{1}$$

Furthermore, H_0 is assumed to have a continuous spectrum, i.e.

$$H_0\psi_E = E\psi_E \tag{2}$$

for a certain range of energy values. (In the case of spontaneous positron production, which is of interest here, this range is $-\infty < E < -m_0c^2$). We require that the wave functions ψ_E asymptotically ($r \to \infty$) are stationary standing waves. The following orthonormality conditions hold:

$$\langle\phi_0|\phi_0\rangle = 1 \quad , \tag{3a}$$
$$\langle\psi_E|\phi_0\rangle = 0 \quad , \tag{3b}$$
$$\langle\psi_{E'}|\psi_E\rangle = \delta(E' - E) \quad . \tag{3c}$$

The continuum thus is normalised "to a delta function". An additional interaction is added now, in the form of a perturbation potential V':

$$H = H_0 + V' \quad . \tag{4}$$

Example 7.2.

Here it can happen that the discrete state ϕ_0 is "lost" under the influence of the interaction. It amalgamates with the continuum (it is "embedded" into it) which is now described by the equation

$$H\Psi_E = E\Psi_E \tag{5}$$

again with the normalisation

$$\langle\Psi_{E'}|\Psi_E\rangle = \delta(E'-E) \quad . \tag{6}$$

If this occurs it will be no longer possible to keep the system in a localised stationary state. Physically, the following happens: The previously stable system *decays* by emission of one of its constituents (since a continuous spectrum is always related to a motion to infinity). In a time-dependent description it can actually be shown that the probability to encounter the state ϕ_0 decreases to zero exponentially in time after the perturbation V' has been switched on (cf. Exercise 7.4).

From the point of view of quantum mechanical scattering theory the presence of a previously bound state ϕ_0 manifests as a *resonance* with its well known signatures: The scattering cross section grows in the vicinity of the resonance energy, the scattering phase varies rapidly as a function of E, the density of the wave function $|\Psi_E(\boldsymbol{x})|^2$ is strongly enhanced at small distances. We now want to examine the properties of the solution of (5) making use of the solutions of the unperturbed problem (1), (2). To achieve this we employ a method that was developed by U. Fano for the case of autoionisation of excited states in atomic physics.[6]

The new continuum wave function is expanded as

$$\Psi_E(\boldsymbol{x}) = a(E)\phi_0(\boldsymbol{x}) + \int dE' h_{E'}(E)\psi_{E'}(\boldsymbol{x}) \quad , \tag{7}$$

with unknown functions $a(E)$ and $h_{E'}(E)$ that have to be determined. The integral extends over the entire range of the continuum. Note: In (7) it was assumed that Ψ_E is in that Hilbert space that is spanned by the set $\{\phi_0, \psi_{E'}\}$. If the spectrum of H_0 does contain additional (discrete or continuous) parts, these states could also be admixed by the action of V'. Equation (7) will still be a useful ansatz, however, as long as the other states are "sufficiently far away" and couple only weakly.

The expansion coefficients are given by

$$a(E) = \langle\phi_0|\Psi_E\rangle \quad , \tag{8a}$$

$$h_{E'}(E) = \langle\psi_{E'}|\Psi_E\rangle \quad , \tag{8b}$$

but this does not help us much as long as Ψ_E is unknown. We now use (5) and project onto $\langle\phi_0|$ and $\langle\psi_{E'}|$. By use of (3) and (7) the system of equations

$$(E - E_0 - \Delta E)a(E) = \int dE'\ V^*_{E'} h_{E'}(E) \tag{9a}$$

and

[6] U. Fano: Phys. Rev. **124**, 1866 (1961). The cases of several bound states or several continua are also treated here. The application to QED resonances was developed in B. Müller, J. Rafelski, W. Greiner: Z. Physik **257**, 62 and 183 (1972).

Example 7.2.

$$(E - E')h_{E'}(E) = V_{E'}a(E) + \int dE'' U_{E'E''}h_{E''}(E) \tag{9b}$$

is obtained. The abbreviations used are:

$$E_0 + \Delta E = \langle\phi_0|H|\phi_0\rangle$$
$$= \langle\phi_0|H_0|\phi_0\rangle + \langle\phi_0|V'|\phi_0\rangle \quad , \tag{10a}$$
$$V_E = \langle\psi_E|H|\phi_0\rangle = \langle\psi_E|V'|\phi_0\rangle \quad , \tag{10b}$$
$$U_{E'E''} = \langle\psi_{E'}|V'|\phi_{E''}\rangle \quad . \tag{10c}$$

E_0 is the original energy of the bound state of H_0. The expectation value ΔE of $H - H_0 = V'$ in this state is equal to the energy shift in first-order perturbation theory. The matrix element V_E specifies how strongly the bound state couples to the continuum, and $U_{E'E''}$ describes the mixing among the continuum states. Additionally, there is the normalization condition (6), namely

$$a^*(E)a(E') + \int dE'' h^*_{E''}(E)h_{E''}(E') = \delta(E - E') \quad . \tag{11}$$

In total the coupled system of integral equations (9a,b) and (11) has to be solved. An analytic solution may be found if the coupling term $U_{E'E''}$ in (9b) is neglected,

$$\int dE'' U_{E'E''}h_{E''}(E) \simeq 0 \quad . \tag{12}$$

This could in principle be achieved by “prediagonalizing” the continuum ψ_E, which means by a unitary transformation

$$\chi_E = \int dE' \, M_{EE'} \, \psi_{E'} \quad , \tag{13}$$

with suitably chosen coefficients $M_{EE'}$, such that $\langle\chi_{E'}|H|\chi_E\rangle$ becomes diagonal. This is, however, difficult to perform practically. On the other hand the distortion of the continuum by $U_{E'E''}$ does not change the solution of the problem qualitatively so that we shall use the approximation (12) in the following. In order to solve the system of (9) and (11) we split the expansion coefficient $h_{E'}(E)$ into

$$h_{E'}(E) = C_{E'}(E)a(E) = 1 - P_0(T) \quad . \tag{14}$$

In this way $a(E)$ is eliminated from (9a) and (9b), i.e.

$$E - E_0 - \Delta E = \int dE' \, V^*_{E'}C_{E'}(E) \tag{15a}$$

and

$$(E - E')C_{E'}(E) = V_{E'} \quad . \tag{15b}$$

Equation (15b) may be immediately solved formally,

$$C_{E'}(E) = \frac{V_{E'}}{E - E'} \quad . \tag{16}$$

Here the question arises how to treat the pole at $E = E'$. In the treatment of scattering problems usually such poles are shifted by a small imaginary part into the

Example 7.2.

complex plane and the requirement of causality is used. Following Fano we want to go a different way, however, that is adapted to the chosen boundary conditions (standing waves). Therefore we replace (16) by

$$C_{E'}(E) = P\frac{V_{E'}}{E-E'} + g(E)V_E\delta(E-E') \quad , \tag{17}$$

where P means that Cauchy's principal value is to be taken in the energy integration. $C_{E'}(E)$ (17) solves (15b) for any arbitrary function $g(E)$ because of $x\delta(x) = 0$. The unknown function $g(E)$ may now be determined by insertion into (15a),

$$E - E_0 - \Delta E = P\int dE'\frac{|V_{E'}|^2}{E-E'} + g(E)|V_E|^2 \tag{18}$$

or

$$g(E) = \frac{1}{|V_E|^2}\left[E - E_0 - \Delta E - F(E)\right] \quad , \tag{19}$$

where $F(E)$ is the abbreviation of

$$F(E) = P\int dE'\frac{|V_{E'}|^2}{E-E'} \quad . \tag{20}$$

$a(E)$ has to be determined from the normalization condition (11). Insertion of (17) yields

$$\begin{aligned} a^*(E)a(E')\Bigg[1 + \int dE''\, V^*_{E''}\left(P\frac{1}{E-E''} + g(E)\delta(E-E'')\right) \\ \times\left(P\frac{1}{E'-E''} + g(E')\delta(E'-E'')\right)V_{E''}\Bigg] = \delta(E-E') \quad , \end{aligned} \tag{21}$$

or

$$\begin{aligned} a^*(E)a(E')\Big(1 + \int dE''|V_{E''}|^2 P\frac{1}{E-E''}P\frac{1}{E'-E''} \\ + g(E')|V_{E'}|^2 P\frac{1}{E-E'} + g(E)|V_E|^2 P\frac{1}{E'-E} \\ + g^2(E)|V_E|^2\delta(E-E')\Big) = \delta(E-E') \quad , \end{aligned} \tag{22}$$

where we have used the following properties of the delta function:

$$\delta(E-E'')\delta(E'-E'') = \delta(E-E')\delta(E-E'') \quad , \tag{23}$$

$$\int dE' f(E')\delta(E-E') = f(E) \quad . \tag{24}$$

The product of the two principle-value factors in (22) is troublesome if the poles coincide, i.e. $E = E'$. In Exercise 7.3 we shall show that

$$\begin{aligned} P\frac{1}{E-E''}P\frac{1}{E'-E''} \\ = -P\frac{1}{E-E'}\left(P\frac{1}{E-E''} - P\frac{1}{E'-E''}\right) + \pi^2\delta(E-E'')\delta(E'-E'') \end{aligned} \tag{25}$$

Example 7.2.

holds for such an expression. Inserting this into (22) together with definition (20) yields

$$\begin{aligned} a^*(E)a(E')\Big[1 - P\frac{1}{E-E'}(F(E) - F(E')) \\ + g(E')|V_{E}'|^2 P\frac{1}{E-E'} - g(E)|V_E|^2 P\frac{1}{E-E'} \\ + \pi^2|V_E|^2\delta(E-E') + g^2(E)|V_E|^2\delta(E-E')\Big] = \delta(E-E') \quad . \end{aligned} \tag{26}$$

We now collect the principal-value terms

$$\begin{aligned} -P\frac{1}{E-E'}\left[(F(E) + g(E)|V_E|^2) - (F(E') + g(E')|V_{E'}|^2)\right] \\ = -P\frac{1}{E-E'}(E-E') = -1 \quad , \end{aligned}$$

where (19) has been used for $g(E)$. Obviously all terms without the factor $\delta(E-E')$ cancel out in (26) such that a consistent equation remains:

$$a^*(E)a(E')\left(\pi^2|V_E|^2 + g^2(E)|V_E|^2\right)\delta(E-E') = \delta(E-E') \quad , \tag{27}$$

or

$$\begin{aligned} |a(E)|^2 &= \frac{1}{|V_E|^2(g^2(E)+\pi^2)} = \frac{|V_E|^2}{(E - E_0 - \Delta E - F(E))^2 + \pi^2|V_E|^4} \\ &= \frac{\Gamma_E/2\pi}{(E-E_\mathrm{r})^2 + \Gamma_E^2/4} \quad , \end{aligned} \tag{28}$$

with the abbreviations

$$\Gamma_E = 2\pi|V_E|^2 \quad , \tag{29a}$$

$$E_\mathrm{r} = E_0 + \Delta E + F(E) \quad . \tag{29b}$$

The expansion coefficient $a(E)$ can be written as

$$a(E) = \frac{1}{V_E\sqrt{g^2(E)+\pi^2}} = \frac{V_E^*}{\sqrt{(E-E_\mathrm{r})^2 + \Gamma_E^2/4}} \quad . \tag{30a}$$

If the matrix element V_E is real we obtain

$$a(E) = \sqrt{\frac{\Gamma_E/2\pi}{(E-E_\mathrm{r})^2 + \Gamma_E^2/4}} \quad . \tag{30b}$$

Because of (14) and (15)

Example 7.2.

$$h_{E'}(E) = \frac{V_E^*}{\sqrt{(E-E_r)^2+\Gamma_E^2/4}}\left(P\frac{V_{E'}}{E-E'}+\frac{E-E_r}{V_E^*}\delta(E-E')\right) \tag{31}$$

holds. This completes the solution of the system of (9a,b) and (11). Let us now consider the structure of the newly constructed continuum wave function $\Psi_E(\boldsymbol{x})$ from (7). Ψ_E is a superposition of unperturbed continuum waves $\Psi_{E'}$ with maximum weight at $E = E'$ as well as of the bound state ϕ_0. The latter is characterized by a wave function which is localized at a finite distance, e.g. Bohr's radius, and decays rapidly in the asymptotic region. The strength of this localized contribution to $\Psi_E(\boldsymbol{x})$,

$$|\langle\phi_0|\Psi_E\rangle|^2 = |a(E)|^2 = \frac{\Gamma_E/2\pi}{(E-E_r)^2+\Gamma_E^2/4} \quad , \tag{32}$$

has a maximum at the *resonance energy* $E = E_r$. (Rigorously this is an implicit equation, because E_r itself depends (weakly) on E through the principal value integral $F(E)$! $F(E)$ describes the additional level shift due to the coupling to the continuum.) If one moves away from E_r in energy the admixture (32) falls off on a scale determined by the *resonance width* Γ_E. Equation (32) has the shape of a *Breit–Wigner* curve characteristic of resonances. The enhancement of the wave function Ψ_E near the resonance is displayed schematically in Fig. 7.13 for the case of an attractive potential well.

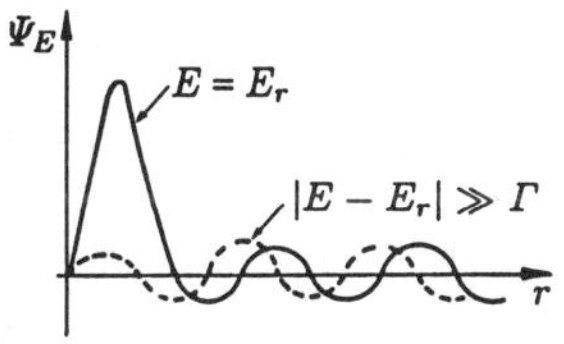

Fig. 7.13. The wave function $\Psi_E(r)$ in the vicinity of the resonance and far away from it

The *asymptotic behaviour* of the wave function Ψ_E for $r \to \infty$ may be examined using (7), (30), and (31). For simplicity, we restrict ourselves to the nonrelativistic case (the Schrödinger equation) with a rapidly decreasing central potential. The unperturbed continuum wave ψ_E then behaves like

$$\psi_E(r) \to N_E \sin(pr + \delta_E) \quad , \tag{33}$$

with momentum $p = \sqrt{2mE}$ and a normalization constant N_E, which is of no interest here. The information about the effective scattering potential is contained in the phase shift δ_E. The new continuum wave function Ψ_E depends on r like

$$\begin{aligned}\Psi_E(r) &= a(E)\phi_0 + \int dE' h_E'(E)\psi_{E'}(r)\\ &\to a(E)\int dE'\left(P\frac{V_{E'}}{E-E'}+\frac{E-E_r}{V_E^*}\delta(E-E')\right)N_{E'}\sin(p'r+\delta_{E'}) \quad ,\end{aligned} \tag{34}$$

where we have used the fact that $\phi_0(r)$ asymptotically approaches zero. By use of the identity

$$P\frac{1}{x} = \frac{1}{2}\left(\frac{1}{x+\mathrm{i}\varepsilon}+\frac{1}{x-\mathrm{i}\varepsilon}\right) \tag{35}$$

the principal-value integral in (34) reads

$$\begin{aligned}&\int dE' P\frac{V_{E'}}{E-E'}N_{E'}\sin(p'r+\delta_{E'})\\ &= \frac{\mathrm{i}}{4}\int dE' V_{E'}N_{E'}\left(\frac{1}{E'-E-\mathrm{i}\varepsilon}+\frac{1}{E'-E+\mathrm{i}\varepsilon}\right)\\ &\times\left(\mathrm{e}^{\mathrm{i}(p'r+\delta_{E'})}-\mathrm{e}^{-\mathrm{i}(p'r+\delta_{E'})}\right) \quad .\end{aligned} \tag{36}$$

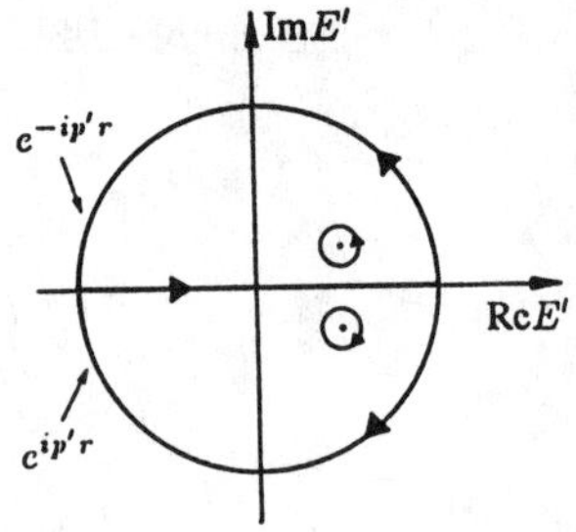

Fig. 7.14. Integration contours in (36)

This integral has simple poles at $E' = E \pm \mathrm{i}\varepsilon$. It can easily be solved by use of the theorem of residues if the limits of integration are extended to infinity, see Fig. 7.14. (This is an approximation since in reality the continuous spectrum is bounded on one side, e.g. $0 < E' < \infty$ in the nonrelativistic case or $mc^2 < |E'| < \infty$ for the Dirac equation.) The integration contour then can be closed by a semicircle at infinity. If the principal value of the square root in the complex plane is chosen for $p' = \sqrt{2mE'}$,

$$\mathrm{sgn}(\mathrm{Im}\{p'\}) = \mathrm{sgn}(\mathrm{Im}\{E'\}) \quad , \tag{37}$$

then the contour has to be closed in the upper (lower) half plane for the case $\mathrm{e}^{\mathrm{i}p'r}(\mathrm{e}^{-\mathrm{i}p'r})$. Two of the four integrals in (36) vanish provided that $V_{E'}N_{E'}$ is holomorphic; the remaining two integrals yield

$$\pm\frac{\mathrm{i}}{4}\int\limits_{-\infty}^{\infty} dE'\, V_{E'}N_{E'}\frac{\mathrm{e}^{\mathrm{i}\pm(p'r+\delta_{E'})}}{E'-E\mp\mathrm{i}\varepsilon} = -\frac{\pi}{2}V_E N_E\,\mathrm{e}^{\pm\mathrm{i}(pr+\delta_E)} \quad . \tag{38}$$

This leads to

$$\int dE' P\frac{V_{E'}}{E-E'}N_{E'}\sin(p'r+\delta_{E'}) = -\pi V_E N_E\cos(pr+\delta_E) \quad . \tag{39}$$

Hence (34) combined with (30a) reads

$$\begin{aligned}\Psi_E(r) \to a(E)\Big[-\pi V_E N_E\cos(pr+\delta_E) + \big((E-E_\mathrm{r})/V_E^*\big)N_E\sin(pr+\delta_E)\Big]\\ = \frac{N_E}{\sqrt{(E-E_\mathrm{r})^2+\Gamma_E^2/4}}\big[-(\Gamma_E/2)\cos(pr+\delta_E)+(E-E_\mathrm{r})\sin(pr+\delta_E)\big]\\ = N_E\sin(pr+\delta_E+\Delta_E) \quad ,\end{aligned} \tag{40}$$

with the *phase shift*

$$\Delta_E = -\arctan\frac{\Gamma_E/2}{E-E_\mathrm{r}} \quad . \tag{41}$$

The modified continuum wave $\Psi_E(r)$ thus displays exactly the same asymptotic behaviour as $\psi_E(r)$ but it is shifted by an angle Δ_E. As Fig. 7.15 shows the phase shift is nearly constant for $|E-E_\mathrm{r}| \gg \Gamma_E \sim \Gamma_{E_\mathrm{r}}$ and has no effect in (40). In the vicinity of E_r, however, it "jumps" very rapidly by an angle of π. This behaviour is well known from the quantum-mechanical theory of resonant scattering.

Finally, we express the expansion coefficients $a(E)$, $h_{E'}(E)$ in terms of the phase shift. Equation (30b) yields

$$\begin{aligned}a(E) &= \frac{1}{\pi V_E}\frac{\Gamma_E/2}{\sqrt{(E-E_\mathrm{r})^2+\Gamma_E^2/4}}\\ &= \frac{1}{\pi V_E}\sqrt{\frac{\tan^2\Delta_E}{1+\tan^2\Delta_E}} = \frac{\sin\Delta_E}{\pi V_E}\end{aligned} \tag{42}$$

and (31) becomes

$$\begin{aligned}h_{E'}(E) &= \frac{\sin\Delta_E}{\pi V_E}P\frac{V_{E'}}{E-E'} + \sin\Delta_E\frac{E-E_\mathrm{r}}{\pi|V_E|^2}\delta(E-E')\\ &= \frac{\sin\Delta_E}{\pi V_E}P\frac{V_{E'}}{E-E'} - \cos\Delta_E\,\delta(E-E') \quad .\end{aligned} \tag{43}$$

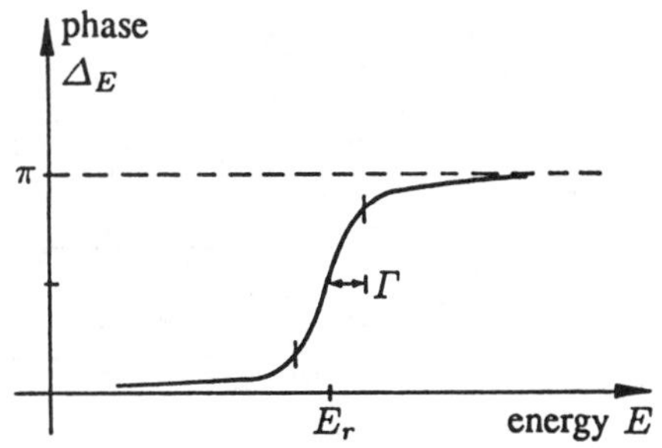

Fig. 7.15. The additional phase shift Δ_E in the vicinity of the resonance

EXERCISE

7.3 The Product of Two Principal-Value Poles

Problem. Prove the identity

$$P\frac{1}{E-E''}\,P\frac{1}{E'-E''} = P\frac{1}{E-E'}\left(P\frac{1}{E'-E''}-P\frac{1}{E-E''}\right)+\pi^2\delta(E-E'')\delta(E'-E'') \quad , \qquad (1)$$

which displays the behaviour of the product of two principal-value singularities as a function of $E'-E$.

Solution. A formal proof can be easily given by use of the identity

$$\frac{1}{x\pm \mathrm{i}\varepsilon} = P\frac{1}{x}\mp \mathrm{i}\pi\delta(x) \quad , \qquad (2)$$

which holds in the limit $\varepsilon\to +0$. The left-hand side can then be written as

$$\left(\frac{1}{E-E''+\mathrm{i}\varepsilon}+\mathrm{i}\pi\delta(E-E'')\right)\left(\frac{1}{E'-E''-\mathrm{i}\varepsilon'}-\mathrm{i}\pi\delta(E'-E'')\right)$$
$$= \frac{1}{(E-E''+\mathrm{i}\varepsilon)(E'-E''-\mathrm{i}\varepsilon')}+\mathrm{i}\pi\frac{\delta(E-E'')}{(E'-E''-\mathrm{i}\varepsilon')}-\mathrm{i}\pi\frac{\delta(E'-E'')}{(E-E''+\mathrm{i}\varepsilon)}$$
$$+\pi^2\delta(E-E'')\delta(E'-E'') \quad . \qquad (3)$$

The relation

$$P\frac{1}{E-E'}\left(P\frac{1}{E'-E''}-P\frac{1}{E-E''}\right)$$
$$=\left(\frac{1}{E-E'+\mathrm{i}\varepsilon}+\mathrm{i}\pi\delta(E-E')\right)$$
$$\times\left(\frac{1}{E'-E''-\mathrm{i}\varepsilon'}-\mathrm{i}\pi\delta(E'-E'')-\frac{1}{E-E''+\mathrm{i}\varepsilon''}-\mathrm{i}\pi\delta(E-E'')\right) \qquad (4)$$

holds for the first term on the right-hand side of the conjecture. The delta function in the first factor can be omitted, because the second factor vanishes for $E=E'$,

Exercise 7.3.

$$
\begin{aligned}
&\frac{1}{E-E'+\mathrm{i}\varepsilon}\left(\frac{1}{E'-E''-\mathrm{i}\varepsilon'}-\frac{1}{E-E''+\mathrm{i}\varepsilon''}\right)\\
&\quad+\mathrm{i}\pi\frac{\delta(E-E'')}{E'-E''-\mathrm{i}\varepsilon}-\mathrm{i}\pi\frac{\delta(E'-E'')}{E-E''+\mathrm{i}\varepsilon}\\
&=\frac{(E-E''+\mathrm{i}\varepsilon'')-(E'-E''-\mathrm{i}\varepsilon')}{(E-E'+\mathrm{i}\varepsilon)(E'-E''-\mathrm{i}\varepsilon')(E-E''+\mathrm{i}\varepsilon'')}\\
&\quad+\mathrm{i}\pi\frac{\delta(E-E'')}{E'-E''-\mathrm{i}\varepsilon}-\mathrm{i}\pi\frac{\delta(E'-E'')}{E-E''+\mathrm{i}\varepsilon}\quad .
\end{aligned}
\tag{5}
$$

Keeping in mind that $\varepsilon, \varepsilon', \varepsilon''$ are infinitesimal quantities whose value does not matter, the first factor can be cancelled in the denominator. Comparison of (5) and (3) proves the conjecture (1).

Remark: A less formal proof can be given by use of the integral representation of the principal-value distribution

$$
P\frac{1}{x}=\frac{\mathrm{i}}{2}\int\limits_{-\infty}^{\infty} du\,\mathrm{sgn}\,u\,\mathrm{e}^{-\mathrm{i}xu}\quad . \tag{6}
$$

The validity of (6) can be shown by performing the limit

$$
\begin{aligned}
&\lim_{\varepsilon\to 0}\frac{\mathrm{i}}{2}\int\limits_{-\infty}^{\infty} du\,\mathrm{sgn}\,u\,\mathrm{e}^{-\mathrm{i}xu-\varepsilon|u|}\\
&=\lim_{\varepsilon\to 0}\frac{\mathrm{i}}{2}\left(-\int\limits_{-\infty}^{0} du\,u\,\mathrm{e}^{-(\mathrm{i}x-\varepsilon)u}+\int\limits_{0}^{+\infty} du\,u\,\mathrm{e}^{-(\mathrm{i}x+\varepsilon)u}\right)\\
&=\lim_{\varepsilon\to 0}\frac{\mathrm{i}}{2}\left(\frac{1}{\mathrm{i}x-\varepsilon}+\frac{1}{\mathrm{i}x+\varepsilon}\right)=\lim_{\varepsilon\to 0}\frac{x}{x^2+\varepsilon^2}=P\frac{1}{x}\quad .
\end{aligned}
\tag{7}
$$

Further calculations using this representation can be found in the work by U. Fano quoted at the beginning of Example 7.2.

EXERCISE

7.4 Time-Dependent Decay of the Vacuum

Problem. In order to describe the decay of a supercritical system as a function of time, we make the following gedankenexperiment. The potential strength is increased suddenly at the time $t = 0$ so that the state which previously was bound slightly subcritically becomes a resonance in the lower continuum. The subcritical system is restored at $t = T$ by reducing the potential strength. By use of Fano's formalism (Exercise 7.2), calculate the final hole probability in the bound state and the spectrum of the emitted positrons depending on the "diving duration".

Exercise 7.4.

Solution. We use the notation of Example 7.2. The Hamiltonian is now a function of time, namely

$$H(t)=\begin{cases} H_0 \quad , & \text{for} \quad t<0 \\ H=H_0+V' \quad , & \text{for} \quad 0\le t\le T \\ H_0 \quad , & \text{for} \quad t>T \end{cases} \quad . \tag{1}$$

A time-dependent wave function $\Psi(t)$ has to be constructed in such a way that it describes a bound hole state ϕ_0 for $t<0$. A piecewise ansatz can be made for $\Psi(t)$ in the three regions:

$$\Psi(t)=\begin{cases} \phi_0 \mathrm{e}^{-\mathrm{i}E_0 t} & \text{for } t<0 \\ \int dE\, \tilde{c}(E)\Psi_E\, \mathrm{e}^{-\mathrm{i}Et} & \text{for } 0\le t\le T \\ c_0\phi_0 \mathrm{e}^{-\mathrm{i}E_0(t-T)}+\int dE\; c(E)\psi_E\, \mathrm{e}^{-\mathrm{i}E(t-T)} & \text{for } t>T \end{cases} \quad . \tag{2}$$

The expansion coefficients $\tilde{c}(E)$ or c_0 and $c(E)$ are independent of time since by assumption the wave functions Ψ_E or ψ_E and ϕ_0 are eigenstates of the Hamiltonians H or H_0. Hence they develop freely with a time dependence $\exp(-\mathrm{i}Et)$. $\Psi(t)$ is required to be continuous at $t=0$ and $t=T$. This means that

$$\phi_0=\int dE\; \tilde{c}(E)\Psi_E \tag{3}$$

and

$$\int dE\; \tilde{c}(E)\Psi_E \mathrm{e}^{-\mathrm{i}ET}=c_0\phi_0+\int dE\; c(E)\psi_E \quad . \tag{4}$$

By projection onto $\langle\Psi_E|$ or $\langle\phi_0|$ and $\langle\psi_E|$, equations determining the coefficients $\tilde{c}(E)$ and $c(E)$ are obtained using (3a,b,c) and (8a,b) of Example 7.2:

$$\tilde{c}(E)=a^*(E) \tag{5}$$

and

$$c_0=\int dE\; \tilde{c}(E)a(E)\,\mathrm{e}^{-\mathrm{i}ET} \quad , \tag{6a}$$

$$c_E=\int dE'\; \tilde{c}(E')h_E(E')\,\mathrm{e}^{-\mathrm{i}E'T} \quad . \tag{6b}$$

Here two Fourier integrals over Fano's expansion coefficients $|a(E)|^2$ or $a^*(E)h_E(E')$ have to be calculated. The integration interval extends from $E=-\infty$ to $E=-m_0c^2$. In order to be able to proceed with an analytical calculation we make an approximation[7] and replace the upper limit by $E=+\infty$. The integrals can be solved by residue integration with the assumption that the matrix element $V_E=\langle\Psi_E|V'|\phi_0\rangle$ is an analytic function of energy.

[7] A finite integration interval leads to a nonexponential decay law. In relation to this problem see L. Fonda et al.: Rep. Progr. Phys. **41**, 587 (1978).

Exercise 7.4.

Using Exercise 7.2, (32) the discrete expansion coefficient follows as

$$c_0 = \int_{-\infty}^{+\infty} dE \, \frac{\Gamma_E/2\pi}{(E-E_\mathrm{r})^2 + \Gamma_E^2/4} \mathrm{e}^{-\mathrm{i}ET} = \mathrm{e}^{-\mathrm{i}E_\mathrm{r}T} \mathrm{e}^{-\Gamma_0 T/2} \quad . \tag{7}$$

Here the integration path has been closed in the lower half-plane. The relevant quantity is the squared absolute value of c_0,

$$P_0(T) = |c_0|^2 = \mathrm{e}^{-\Gamma_0 T} \quad . \tag{8}$$

The probability of finding a hole in state ϕ_0 after a time T thus decreases exponentially as determined by the decay width $\Gamma_0 = \Gamma_{E_\mathrm{r}}$. In order to see how the positron probability correspondingly builds up with time, the integral

$$c(E) \simeq \int_{-\infty}^{+\infty} dE' \frac{|V_{E'}|^2}{(E'-E_\mathrm{r})^2 + \Gamma_{E'}^2/4} \left(P \frac{V_E}{E'-E} + \frac{E-E_\mathrm{r}}{V_E^*} \delta(E-E') \right) \mathrm{e}^{-\mathrm{i}E'T} \tag{9}$$

has to be solved using Exercise 7.2, (30a) and (31). We reformulate the principal-value term:

$$P\frac{V_E}{E'-E} = \frac{V_E}{E'-E-\mathrm{i}\varepsilon} - \mathrm{i}\pi V_E \delta(E-E') \quad , \tag{10}$$

which is useful because a pole in the upper half-plane does not contribute. Equation (9) then reads

$$\begin{aligned} c(E) \simeq & \int_{-\infty}^{+\infty} dE' \frac{|V_{E'}|^2}{(E'-E_\mathrm{r})^2 + \Gamma_{E'}^2/4} \frac{V_E}{E'-E-\mathrm{i}\varepsilon} \mathrm{e}^{-\mathrm{i}E'T} \\ & + \frac{V_E}{(E-E_\mathrm{r})^2 + \Gamma_E^2/4} (E - E_\mathrm{r} - \mathrm{i}\pi |V_E|^2) \mathrm{e}^{-\mathrm{i}ET} \quad . \end{aligned} \tag{11}$$

Only the pole at the point $E' = E_\mathrm{r} - \mathrm{i}\Gamma_0/2$ contributes in the residue integration. The second term in (11) can be simplified since $|V_E|^2 = \Gamma_E/2\pi$. We then obtain

$$c(E) \simeq \frac{V_E}{E - E_\mathrm{r} + \mathrm{i}\Gamma_E/2} \mathrm{e}^{-\mathrm{i}ET} - \frac{V_{E_\mathrm{r}}}{E - E_\mathrm{r} + \mathrm{i}\Gamma_0/2} \mathrm{e}^{-\mathrm{i}E_\mathrm{r}T} \mathrm{e}^{-\Gamma_0 T/2} \quad . \tag{12}$$

The energy spectrum of the produced positrons is given by the absolute squared of the expansion coefficient $c(E)$. With the assumption that Γ_E depends only weakly on E and can be replaced by Γ_0 we obtain

$$\frac{dP}{dE} = |c(E)|^2 = \frac{\Gamma_0/2\pi}{(E-E_\mathrm{r})^2 + \Gamma_0^2/4} |1 - \mathrm{e}^{-\mathrm{i}(E-E_\mathrm{r})T} \mathrm{e}^{-\Gamma_0 T/2}|^2 \quad . \tag{13}$$

It can be easily checked that the norm of the wave function $\Psi(t)$ is conserved in spite of the diverse approximations, i.e.

Exercise 7.4.

$$\int_{-\infty}^{+\infty} dE \, \frac{dP}{dE} = 1 - e^{-\Gamma_0 T} = 1 - P_0(T) \quad . \tag{14}$$

Two limiting cases of the spectrum (13) can be discussed, see Fig. 7.16.

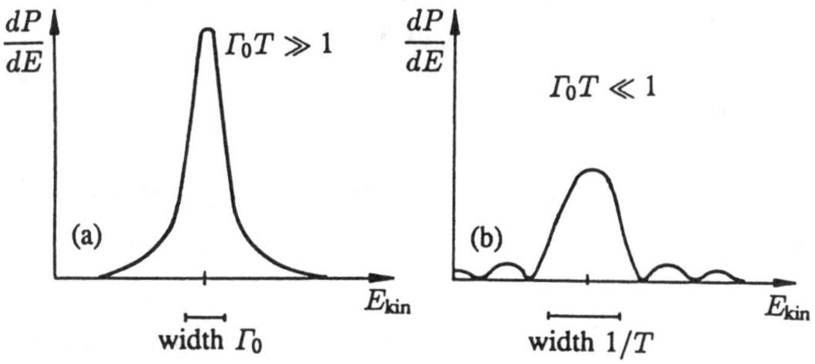

Fig. 7.16. The spectra of emitted positrons in the limits of a large (**a**) and small (**b**) time of supercriticality T

a) $\Gamma_0 T \gg 1$: If the 'diving time' is large compared to the natural decay time $1/\Gamma_0$, a Breit–Wigner spectrum with line width Γ_0 results:

$$\frac{dP}{dE} = \frac{\Gamma_0/2\pi}{(E - E_r)^2 + \Gamma_0^2/4} \quad . \tag{15}$$

b) $\Gamma_0 T \ll 1$: If the potential is supercritical only during a short time interval T, then

$$\frac{dP}{dE} = \frac{\Gamma_0}{2\pi} \frac{|1 - e^{i(E-E_r)T}|^2}{(E - E_r)^2} = \frac{\Gamma_0 T^2}{2\pi} \left[\frac{\sin(E - E_r)T/2}{(E - E_r)T/2} \right]^2 \quad . \tag{16}$$

This is an oscillating function with maximum at $E = E_r$ having a width decreasing with the inverse of T. The peak height increases quadratically and its contents linearly in T until saturation is reached at $T > 1/\Gamma_0$.

The method presented in Example 7.2 can be immediately applied to the case of a supercritical atom. The eigenstates of the subcritical Hamiltonian ($Z < Z_{cr}$) can be used as a basis for the expansion of the supercritical wave functions

$$\Psi_E = a(E)\phi_0 + \int dE' h_{E'}(E')\psi_{E'} \quad . \tag{7.16}$$

The perturbative potential V' then is of the form

$$V' = V(Z) - V(Z_{cr}) \approx (Z - Z_{cr})U(r) \quad , \tag{7.17}$$

where $U(r)$ depends on the shape of the truncated Coulomb potential according to (7.3). The bound state "dives" into the continuum with a linear dependence on the charge excess $Z' = Z - Z_{cr}$ because of (29b) and (10a) of Example 7.2. A value of about $E_{1s} \simeq -m_0c^2 - Z' \times 30\text{keV}$ is found numerically. The width Γ grows about quadratically with Z' according to (29a),(10b). At small diving energies (near the threshold), however, it is strongly suppressed owing to the Coulomb repulsion of the wave function ψ_E in (10b).

In Exercise 7.4 we showed with the help of Fano's formalism that a hole in the 1s state decays by positron emission if the potential is made supercritical. An exponential decay law holds; (inserting $\hbar$)

$$P_{1s}(t) = P_{1s}(0)\mathrm{e}^{-\Gamma t/\hbar} \quad . \tag{7.18}$$

The lifetime τ of the hole is determined by the inverse of the width

$$\tau = \hbar/\Gamma \quad . \tag{7.19}$$

Typical decay times are of the order $\tau = 10^{-19}$ s. A stable state is reached after the positron emission (or if the 1s state had been occupied right from the beginning). The K shell of the supercritical atom is now filled with two electrons; hence, the new vacuum state has a double negative charge. The resonance property of the K shell of a stable supercritical atom could be noticed only in the following situation: in *positron scattering* on such an atom a sudden strong increase of the cross section should occur at the energy $E^{\text{kin}}_{e^+} = |E_{1s}| - m_0c^2$ since the positron can penetrate into the potential well and stay there for some time.

We still have to demonstrate that the supercritical continuum Ψ_E does in fact contain the charge distribution of the K shell. To this end we calculate the total charge density $\rho(r, Z)$ of the lower continuum assuming that all of its states are occupied by electrons with charge e. This is of course a mathematically undefined and badly divergent quantity. However, we obtain a meaningful expression if we subtract the charge density of the continuum in the subcritical case from it:

$$\Delta\rho(r) = \rho(r,Z) - \rho(r,Z_{\text{cr}}) = 2e\left(\int dE\ \Psi^\dagger_E(r)\Psi_E(r) - \int dE\ \psi^\dagger_E(r)\psi(r)\right) \quad , \tag{7.20}$$

where the integrals extend from $-\infty$ to $-m_0c^2$. With the wave function (7.16) we get

$$\begin{aligned}\Delta\rho(r) = 2e\Bigg[&\left(\int dE|a(E)|^2\right)\phi^\dagger_0\phi_0 + \int dE'\left(\int dE\ a^*(E)h_{E'}(E)\right)\phi^\dagger_0\psi_{E'}\\ &+\int dE'\left(\int dE\ a(E)h^*_{E'}(E)\right)\psi^\dagger_{E'}\phi_0\\ &+\int dE'\int dE''\left(\int dE\ h^*_{E''}(E)h_{E'}(E)\right)\psi^\dagger_{E''}\psi_{E'} - \int dE\ \psi^\dagger_E\psi_E\Bigg] \quad .\end{aligned} \tag{7.21}$$

The factor 2 takes account of the spin degeneracy. The integrals over the expansion coefficients (in brackets) can be evaluated approximately with residue integration. As is calculated in Exercise 7.4 (set $T = 0$ there!) the following relations hold:

$$\int dE\ |a(E)|^2 = 1 \quad , \qquad \int dE\ a^*(E)h_{E'}(E) = 0 \quad .$$

Similarly it can be shown that

$$\int dE\ h^*_{E''}(E)h_{E'}(E) = \delta(E' - E'') \quad .$$

Using these relations the incremental charge density is:

$$\begin{aligned}\Delta\rho(r) &= 2e\left(|\phi_0(r)|^2 + \int dE|\psi_E(r)|^2 - \int dE|\psi_E(r)|^2\right)\\ &= 2e|\phi_0(r)|^2 \quad . \end{aligned} \tag{7.22}$$

The 'excess charge' of the supercritical vacuum thus corresponds exactly to a doubly occupied 1s state!

This result has a further very profound interpretation. The virtual particles which occupy the 'empty' states of the Dirac equation are responsible for the effect of vacuum polarization. Using tools from quantum field theory it can be argued that the corresponding charge distribution has the form

$$\rho_{\mathrm{VP}}(r) = \frac{e}{2}\left(\sum_{n<F}|\phi_n(r)|^2 - \sum_{p>F}|\phi_p(r)|^2\right) \quad . \tag{7.23}$$

Here we sum over all states of the spectrum. The "Fermi surface" F denotes the border between occupied and empty levels. The first sum extends over all states of the 'Dirac sea' with the energy eigenvalues $E < -m_0c^2$. A charge $e = -|e|$ is attributed to them ('occupied by electrons'). The second sum runs over all levels of the upper continuum and the bound states that evolved from it. These levels are assigned a charge of $-e = |e|$ ('occupied by positrons'). Equation (7.23) takes an average value of both pictures, thus taking account of the equivalence of electrons and positrons which is required by charge conjugation symmetry. In the field-free case the two sums cancel. For weak potentials a displacement charge distribution $\rho_{\mathrm{VP}}(r)$ remains but the integral over it still vanishes. Since the wave functions ϕ_p are pulled closer to the nucleus by the Coulomb field and the ϕ_n are pushed out (7.23) explains the apparently paradoxical sign of the vacuum-polarization charge cloud encountered in Sect. 5.3, see e.g. Fig. 5.15.

The case is entirely different for $Z > Z_{\mathrm{cr}}$. The 1s state then moves from the second sum in (7.23) into the first sum. Because of (7.22) ρ_{VP} in this case equals

$$\rho_{\mathrm{VP}}(r) = 2e|\phi_0(r)|^2 \quad , \tag{7.24}$$

which means that a *real vacuum polarization* does occur. This is another way of describing the charged vacuum.

7.2 Strong Fields in Heavy Ion Collisions

As we have seen in the previous section the decay of the neutral vacuum in an atom is expected to occur only if the nuclear charge number exceeds the value of $Z_{\mathrm{cr}} \approx 173$. Since the stable nuclei found in nature, as well as the superheavy elements produced synthetically, have much smaller charge numbers the quantum electrodynamics of strong fields might appear to be a purely academic problem. However, the following considerations open the possibility of an experimental access to the field. To examine the effect of a strong field it is not absolutely necessary to maintain the field for an infinitely long time. The supercritical source of charge only has to be concentrated for a certain time in a sufficiently small region of space to allow the electron–positron field to adapt itself to this situation.

A possible experimental way of producing strong fields is given by collisions of heavy ions. A beam of heavy ions is accelerated to an energy of several MeV per nucleon and is shot onto a target of an element that also has a high charge number. The high kinetic energy of the projectile makes it possible to overcome the Coulomb repulsion and to put both nuclei close together. In the supercritical systems which are of interest here, the Coulomb barrier which has to be surmounted to put the nuclei into contact amounts to an energy of about 6 MeV/nucleon, corresponding to a projectile velocity of about $v_{\mathrm{ion}} = 0.1c$. In comparison to this the bound electrons in the inner shells move with relativistic velocity, $v_{\mathrm{el}} \approx c$. We can conclude from the ratio

$$v_{\mathrm{el}}/v_{\mathrm{ion}} \gg 1 \tag{7.25}$$

that the electrons in the inner shells will be able to adapt themselves at any time to the Coulomb field of two nuclei of charge number Z_1, Z_2 at distance $R(t)$ acting at that moment. If R were constant with time we would have to deal with a stationary molecule. But since the corresponding electron orbitals can be formed only temporarily and imperfectly in the collision the system is called a *quasi-molecule*.

It is useful for the understanding of this term to consider first the 'adiabatic approximation'. For arbitrarily slow nuclear motion R (the extreme limit of (7.25)) the electron wave function ϕ_n would be described by the solution of the stationary two-centre Dirac equation,

$$H_{\mathrm{TCD}}(\boldsymbol{R})\phi_n = \big(\boldsymbol{\alpha}\cdot\boldsymbol{p} + \beta m_0 + V_{\mathrm{TC}}(\boldsymbol{r},\boldsymbol{R})\big)\phi_n = E_n(R)\phi_n \quad . \tag{7.26}$$

Here V_{TC} is the combined Coulomb potential of both nuclei. If we approximate them to be pointlike then

$$V_{\mathrm{TC}}(\boldsymbol{r},\boldsymbol{R}) = -\frac{Z_1\alpha}{|\boldsymbol{r}-\boldsymbol{R}_1|} - \frac{Z_2\alpha}{|\boldsymbol{r}-\boldsymbol{R}_2|} \quad , \tag{7.27}$$

with the distance vectors $\boldsymbol{R}_1, \boldsymbol{R}_2$ of the nuclei with respect to the origin and $\boldsymbol{R} = \boldsymbol{R}_2 - \boldsymbol{R}_1$. The potential (7.27) is not spherically symmetric. Therefore the angular momentum-operator $\hat{J}$ does not commute with H_{TCD} and the solutions cannot be classified by their angular momenta. The only valid quantum numbers are the projection m of the angular momentum onto the internuclear axis $\boldsymbol{R}$ (since (7.27) has cylindrical symmetry) and parity in the special case $Z_1 = Z_2$. In contrast to the spherical case (7.6) the partial differential equation (7.26) cannot be separated into a "radial part" and an "angular part", i.e. split into three one-dimensional ordinary differential equations. Therefore the solution of the two-centre Dirac equation is possible only numerically and with great effort.

We do not want to treat the problem in more detail here but cite only the essential result. It is contained in the *correlation diagram*, which depicts the eigenenergies $E_n(R)$ as functions of nuclear distance R. The curves vary continuously between the limiting cases $R \to 0$ (unified atom) and $R \to \infty$ (two separate atoms). As an example, Fig. 7.17 shows the correlation diagram of the heaviest experimentally accessible system, uranium + curium with total charge $Z = Z_1 + Z_2 = 92 + 96 = 188$.

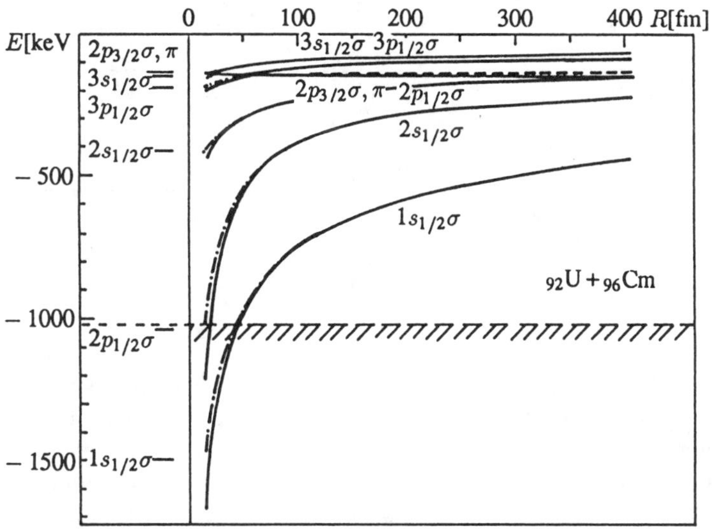

Fig. 7.17. The binding energies of the most deeply bound states of the quasi-molecule $U + Cm$ plotted as function of the distance between the nuclei R. The finite radius of both nuclei is taken into account with the dot-dashed curves

As expected (we already know the limiting case of the unified system) some of the energy levels drop down steeply when the nuclei are put closely together. For the lowest state $1s\sigma$ (the greek letters $\sigma, \pi, \delta, \ldots$ denote the magnetic quantum numbers $|m| = 1/2, 3/2, 5/2, \ldots$ in the molecular nomenclature) there is a *critical distance* R_{cr} at which the binding energy exceeds the threshold $2m_0c^2$. The vacuum becomes unstable if the two nuclei approach one another any further, and spontaneous emission of two positrons ($m = \pm 1/2$) is possible. The critical distance has the value $R_{\mathrm{cr}} = 43$ fm in the considered case ($U + Cm$). The region in which such a quasi-molecule can be supercritical is thus rather small, which will turn out to be a serious problem. The motion $R(t)$ along the trajectory is strongly accelerated by the mutual Coulomb repulsion of the nuclei. Hence the colliding nuclei can only approach close enough to cross the critical distance in a short time interval Δt. The correlation diagram Fig. 7.17 can immediately be translated into a representation of the energies with respect to the time axis since the function $R(t)$ is known (it is a Rutherford hyperbola in the force field $F(R) = Z_1 Z_2 e^2/R^2$). Figure 7.18 shows the result. We can easily estimate that the "dive" of the 1s level into the Dirac sea lasts only for about

$$\Delta t \sim \frac{2R_{cr}}{v_{ion}} \sim 2 \times 10^{-21}\ \mathrm{s} \quad . \tag{7.28}$$

This value is smaller by about two orders of magnitude than the decay time of the resonance. Hence we can only expect that at most a small fraction of the existing holes in the $1s\sigma$ state leads to spontaneous positron emission. This process is labeled by arrow c in Fig. 7.18.

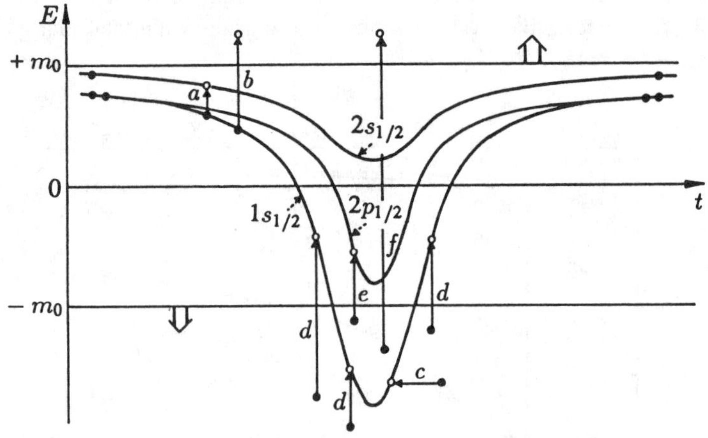

Fig. 7.18. The time dependence of the quasi-molecular energy levels in a supercritical heavy ion collision. The arrows denote various excitation processes which lead to the production of holes and positrons

The fact that the orbital motion $R(t)$ is rather fast compared to the "typical length scale" of the superheavy quasi molecule also has another significant consequence: the electron wave function *does not develop fully adiabatically* in the sense described above. An electron (or hole) which was in a certain molecular eigenstate ϕ_i at the beginning of the collision can be transferred with a certain probability into different states ϕ_k by the "collision dynamics". This can lead to the following physical processes as indicated by the arrows in Fig. 7.18:
(a) hole production in an inner shell by *excitation* of an electron to a higher state; (b) hole production by *ionization* of an electron to the continuum; (d) and (e) *induced positron production* by excitation of an electron from the lower continuum to an empty bound level; (f) *direct pair production.*

In all these processes kinetic energy can be transferred from the nuclear motion to the electrons.

The dynamical production of positrons (d, e, f) superposes with the *spontaneous decay* of the 1s resonance (c) and cannot be separated from this contribution. The observation of positron emission in heavy ion collisions therefore does not automatically mean the confirmation of the predicted vacuum decay. On the other hand, excitations of the kind (a,b) provide a good opportunity to create holes in the $1s\sigma$ level in the incoming phase of the collision which are then available for positron production. Otherwise one would have to deal with bare, i.e. fully ionized, atoms which is much more difficult experimentally.

We now want to move (very briefly) to the field of atomic scattering theory and discuss how the various excitation processes of electrons and positrons in the heavy ion collision may be calculated.[8] Because of the very large nuclear masses it is allowed to treat the scattering in the *semiclassical approximation* (which we already did implicitly in the qualitative discussion). Here, the nuclear motion is given classically as a function of time. The *time-dependent two-centre Dirac equation* is then solved,

$$\mathrm{i}\frac{\partial}{\partial t}\psi_i = H_{\mathrm{TCD}}(\boldsymbol{R}(t))\psi_i \quad . \tag{7.29}$$

In contrast to (7.26) the Hamiltonian here depends parametrically on time via the given trajectory $\boldsymbol{R}(t)$. The index of the wave function ψ_i indicates that it evolves from a well-defined initial state ϕ_i in the limit $t \to -\infty$. Direct solution of the system of partial differential equations (7.30) is a very difficult task. The problem becomes more tractable if the time-dependent wave function is expanded with respect to a complete set of basis wave functions ϕ_k chosen as suitably as possible:

$$\psi_i(t) = \sum_k a_{ik}(t)\phi_k(R)\mathrm{e}^{-\mathrm{i}\chi_k(t)} \quad . \tag{7.30}$$

The sum extends over the entire discrete and continuous spectrum here. $\chi_k(t)$ is an arbitrary phase factor which can be freely choosen. We assume for simplification that the basis wave functions ϕ_k form an orthonormal system,

$$\langle\phi_k|\phi_l\rangle = \delta_{kl} \quad . \tag{7.31}$$

Inserting the expansion (7.30) into (7.29) leads to

$$\mathrm{i}\sum_l \left(\dot{a}_{il}\phi_l + a_{il}\dot{\phi}_l - \mathrm{i}\dot{\chi}_l a_{il}\phi_l\right)\mathrm{e}^{-\mathrm{i}\chi_l} = \sum_l a_{il}H_{\mathrm{TCD}}\,\phi_l\,\mathrm{e}^{-\mathrm{i}\chi_l} \quad . \tag{7.32}$$

Because of the orthogonality (7.31), projection on $\langle\phi_k|$ and multiplication by $-\mathrm{i}$ yields

$$\dot{a}_{ik} = -\sum_l a_{il}\left(\langle\phi_k|\frac{\partial}{\partial t}|\phi_l\rangle - \mathrm{i}\langle\phi_k|H_{\mathrm{TCD}}|\phi_l\rangle + \mathrm{i}\dot{\chi}_l\delta_{kl}\right)\mathrm{e}^{-\mathrm{i}(\chi_l-\chi_k)} \quad . \tag{7.33}$$

The contribution of $\langle\phi_k|\dot{\phi}_k\rangle$ vanishes because of the normalization $\langle\phi_k|\phi_k\rangle = 1$ and $\langle\phi_k|\dot{\phi}_k\rangle + \langle\dot{\phi}_k|\phi_k\rangle = 0$.

The diagonal term $l = k$ in (7.33) can be eliminated by the choice of phase

$$\chi_k(t) = \int_{t_0}^{t} dt'\langle\phi_k|H_{\mathrm{TCD}}|\phi_k\rangle \quad , \tag{7.34}$$

with arbitrary t_0; hence

$$\dot{a}_{ik} = -\sum_{l\neq k} a_{ik}\left(\langle\phi_k|\frac{\partial}{\partial t}|\phi_l\rangle - \mathrm{i}\langle\phi_k|H_{\mathrm{TCD}}|\phi_l\rangle\right)\mathrm{e}^{-\mathrm{i}(\chi_l-\chi_k)} \quad . \tag{7.35}$$

[8] See J. Reinhardt, W. Greiner: Heavy ion atomic physics, in *Heavy Ion Science*, Vol. 5, edited by D.A. Bromley (Plenum, New York, 1984).

The partial differential equation (7.29) now has been reduced to a set of coupled ordinary differential equations. The dimension of this system is infinite, however! With a suitable choice of the basis ϕ_k it may be sufficient to include only a few terms in the expansion to achieve convergence. If the coupling matrix elements in (7.35) are small, then the a_{ik} vary only a little with time and we can set them to be approximately constant on the right-hand side. Equation (7.35) then leads to a simple integral over time and we obtain the well-known formula of time-dependent perturbation theory.

If the condition of adiabaticity (7.25) is fulfilled approximately, it is most favourable to use quasi-molecular states as the basis, i.e. the solutions of (7.26). The system of coupled-channel equations then reads

$$\dot{a}_{ik} = -\sum_{l \neq k} a_{il} \langle \phi_k | \frac{\partial}{\partial t} | \phi_l \rangle \exp\left[-\mathrm{i} \int_{t_0}^{t} dt' (E_l - E_k) \right] \quad , \tag{7.36}$$

since the matrix element of the Hamiltonian,

$$\langle \phi_k | H_{\mathrm{TCD}} | \phi_l \rangle = E_k \delta_{kl} \quad , \tag{7.37}$$

vanishes between two orthogonal eigenstates. In order to calculate excitation probabilities, the differential equations of system (7.36) are truncated to a finite number of channels and are integrated starting from the boundary condition $a_{ik}(t \to -\infty) = \delta_{ik}$. The quantity $|a_{ik}(t \to +\infty)|^2$ specifies the probability for an electron to be transferred from state i to state k in the course of the collision. We now have to take into account that, in general, several electrons are present. If we neglect the (comparatively weak) mutual interactions among the electrons, their contributions can be summed up independently. The *number of electrons* N_k in a level k which was empty before the collision is then

$$N_k = \sum_{i<F} |a_{ik}(\infty)|^2 \quad , \tag{7.38}$$

where the sum extends over all levels occupied before the collision. This is indicated by $i < F$ with F denoting the "Fermi surface" up to which the electronic levels are filled. The entire lower continuum is also included in this set of states.

The *number of holes* $\bar{N}_l$ in an originally occupied state l can be calculated in a similar way leading to the result

$$\bar{N}_l = \sum_{i>F} |a_{il}|^2 \quad . \tag{7.39}$$

This expression is also valid for the production of positrons (i.e. holes in the lower continuum).

All atomic excitation processes in heavy ion collisions at incident energies around the Coulomb barrier can be calculated using the formalism of (7.26), (7.36), (7.38) and (7.39). It is found that the transition rates are very large. The probability of knocking an electron out of the K shell in a close collision of two very heavy atoms exceeds 10%. Higher levels become almost fully ionized. Measurements of K-hole production and of the energy distribution of the ionized electrons (which are called δ electrons) give important insights about the action of the strong field. The predictions of the theory (and thus the concept of the formation of quasi-molecules according to Fig. 7.18, which is basic to it) have been confirmed to a large degree. In particular we can clearly conclude from the measurements that in a Coulomb potential with $Z\alpha \geq 1$, (1) electronic binding energies of magnitude m_0c^2 and beyond can be reached, and (2) the wave functions shrink very much, as was shown in Fig. 7.10.

It proves to be very difficult, however, to give evidence about the process of spontaneous positron production. To begin with, the formalism has to be extended in order to describe overcritical collisions, i.e. $Z_1 + Z_2 > Z_{\mathrm{cr}}$ and $R_{\mathrm{min}} < R_{\mathrm{cr}}$. As we have seen, the $1s$ state vanishes out of the bound spectrum beyond the critical point. Instead it is hidden as a narrow time-dependent resonance structure in the continuum states Ψ_E. A straightforward solution of (7.36) is thus made almost impossible. This difficulty can be overcome if we analyse the wave function Ψ_E at the resonance energy E_{r} in more detail, see Fig. 7.5d. Besides a localized part at small distances an oscillating tail with small amplitude extends to infinity. If we cut off this tail, i.e. artificially forbid tunnelling through the particle–antiparticle gap, it should be possible to construct a "$1s$-like" bound wave function $\tilde{\phi}_{1s}$. This new state $\tilde{\phi}_{1s}$ is of course not linearly independent from the states Ψ_E of the lower continuum. Hence, it must not be inserted into the basis expansion (7.30) of the time-dependent wave function. Rather, the bound-state contribution first has to be projected out of the continuum. This is done using the projection operator

$$\hat{P} = 1 - |\tilde{\phi}_{1s}\rangle\langle\tilde{\phi}_{1s}| \quad . \tag{7.40}$$

A modified continuum $\tilde{\phi}_E$ can then be defined which does not possess the resonance property any more. $\tilde{\phi}_E$ is an eigenstate of the projected Hamiltonian

$$(E - \hat{P}H\hat{P})|\tilde{\phi}_E\rangle = 0 \quad . \tag{7.41}$$

If we assume orthogonality to the higher electron states ϕ_n (with $E_n > -m_0c^2$), (7.41) can be written as

$$(E - H)|\tilde{\phi}_E\rangle = -\langle\tilde{\phi}_{1s}|H|\tilde{\phi}_E\rangle|\tilde{\phi}_{1s}\rangle \quad . \tag{7.42}$$

Instead of Ψ_E the states $\tilde{\phi}_{1s}$ and $\tilde{\phi}_E$ can now be inserted in the expansion (7.30). This does not change the formalism in principle, but it has an important consequence: Since the modified states $\tilde{\phi}_{1s}$ and $\tilde{\phi}_E$ are not exact eigenstates of the Hamiltonian H (7.37) is no longer valid. Therefore we have to perform the following replacement in the differential equations (7.36):

$$\langle\phi_E|\partial/\partial t|\phi_{1s}\rangle \to \langle\tilde{\phi}_E|\partial/\partial t|\tilde{\phi}_{1s}\rangle + \mathrm{i}\langle\tilde{\phi}_E|H|\tilde{\phi}_{1s}\rangle \quad . \tag{7.43}$$

The second matrix element plays exactly the role of V_E in Fano's formalism, Example 7.2. It is responsible for the decay of a 1s hole by *spontaneous positron production* with a width

$$\Gamma_{1s} = 2\pi|\langle\tilde{\phi}_E|H|\tilde{\phi}_{1s}\rangle|^2 \quad . \tag{7.44}$$

The first matrix element in (7.43) is effective only if the colliding nuclei are in motion. It describes the "induced" positron emission.

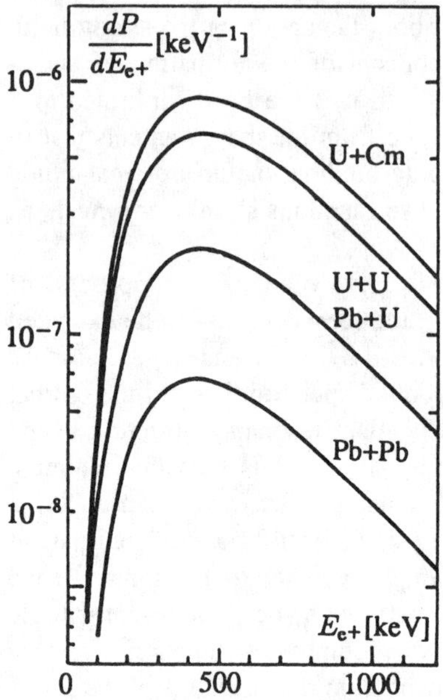

Fig. 7.19. The calculated energy distribution of the positrons emitted in various heavy ion collisions. The intensity increases strongly with the charge of the participating nuclei. The predictions shown in this figure have been fully confirmed by experiment

If the positron emission is calculated using the formalism developed above, the result, however, is somewhat disappointing. Figure 7.19 shows the predicted distribution of emitted positrons dP_{e^+}/dE_{e^+} as a function of their kinetic energy for central collisions at 5.9 MeV/nucleon projectile energy.[9]

Different collision systems ($Z = Z_1 + Z_2 = 164, 174, 184, 188$) are compared. A striking increase in the positron yield with charge Z is seen. One finds, approximately

$$P_{e^+} \propto Z^{20} = (Z_1 + Z_2)^{20} \quad . \tag{7.45}$$

This is clear evidence that the strong combined Coulomb field of the two nuclei is active here. It must be kept in mind that from a treatment by perturbation theory only an increase with $Z_1^2 Z_2^2$ is to be expected (compare the Feynman graphs of lowest order for pair production in the collision of two charged particles, Fig. 7.20a). The result (7.45) stresses the importance of the exchange of very many photons as depicted symbolically in Fig. 7.20b. This nonperturbative effect in pair production has been fully confirmed by the experiments of the groups around P. Kienle, H. Backe, E. Kankeleit, and J.S. Greenberg working at the GSI heavy ion accelerator (Darmstadt).[10] Besides the strong increase with Z the positron spectra shown in Fig. 7.19 give no clear evidence of spontaneous pair production. This should not come as a surprise since due to the short collision time $\Delta t \simeq 10^{-21}$ sec the occurence of narrow structures in the emission spectrum would be in conflict with Heisenberg's uncertainty relation.

[9] J. Reinhardt, B. Müller, W. Greiner: Phys. Rev. **A24**, 103 (1981).

[10] H. Backe, L. Handschug, F. Hessberger, E. Kankeleit, L. Richter, F. Weik, R. Willwater, H. Bokemeyer, P. Vincent, J. Nakayama, J.S. Greenberg: Phys. Rev. Lett. **40**, 1443 (1978); C. Kozhuharov, P. Kienle, E. Berdermann, H. Bokemeyer, J.S. Greenberg, Y. Nakayama, P. Vincent, H. Backe, L. Handschug, E. Kankeleit: Phys. Rev. Lett. **42**, 376 (1979).

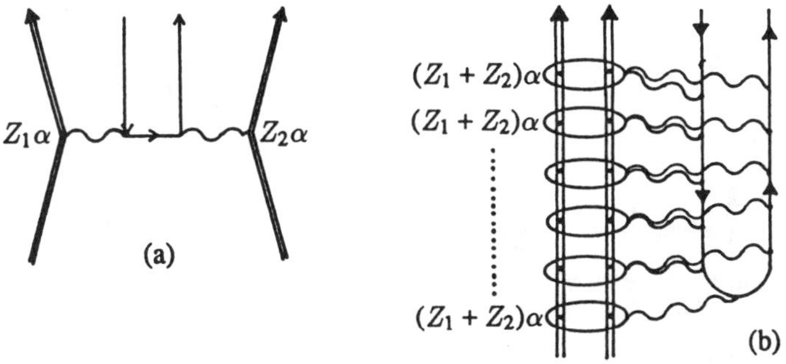

Fig. 7.20. Pair production in the collision of two nuclei: **(a)** Feynman graph of lowest order; **(b)** exchange of many photons at large Z. This non-perturbative way of pair creation can be interpreted as a shake-off of the vacuum polarization cloud in the collision of two heavy nuclei

The situation would change if there were a mechanism leading to a *time delay* in the collision. Such an effect can happen if the nuclei have enough energy to surmount the Coulomb repulsion and touch each other so that the attractive nuclear force sets in, see Fig. 7.21. Let us suppose that the nuclei "stick together" for a time T and separate again subsequently. Spontaneous positron production should then be enhanced in supercritical systems. The holes present in the 1s level are emitted as positrons owing to the decay coupling of (7.43) with a lifetime $\tau = \hbar/\Gamma$. If the time delay T is comparable to τ, a sharp line will build up in the positron spectrum whose position is determined by the diving depth of the submerged 1s state. We already discussed this in Exercise 7.4. Figure 7.22 shows this effect very clearly for various assumed times up to 10^{-20} s in $U + U$ collisions.

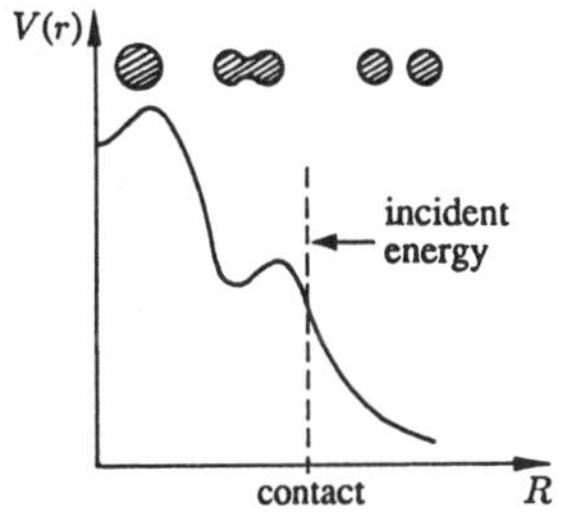

Fig. 7.21. The attractive nuclear force may produce a "pocket" in the internuclear potential $V(r)$

Positron production in heavy ion collision has been examined experimentally in recent years, at the Gesellschaft für Schwerionenforschung (GSI) at Darmstadt. While the predictions of Fig. 7.19 have been confirmed convincingly, in addition narrow line structures in the spectrum have been discovered.[11] As an example Fig. 7.23 shows the positron emission in uranium–curium collisions ($Z = 188$) at a projectile energy of 5.8 MeV/nucleon at which the nuclei just touch each other.

At first it appeared that the lines originate from spontaneous positron production in a small fraction of scattering events with enhanced collision time in which the nuclei stick together owing to the strong interaction. However, lines at about the same energy position were also observed in collision systems with subcritical charge $Z_1 + Z_2 < Z_{cr}$ where no spontaneous pair production should occur.[12] Subsequently it was discovered[13] that the line positrons seem to be accompanied by electrons of equally sharp energy emitted in the opposite direction! The only natural explanation of such a phenomenon seemed to be a scenario where neutral "objects" X with masses in the range $m_X = 1.5 \cdots 2$ MeV are produced which

[11] J. Schweppe, A. Gruppe, K. Bethge, H. Bokemeyer, T. Cowan, H. Folger, J.S. Greenberg, H. Grein, S. Ito, R. Schule, D. Schwalm, K.E. Stiebing, N. Trautmann, P. Vincent, and M. Waldschmidt: Phys. Rev. Lett. **51**, 2261 (1983); M. Clemente, E. Berdermann, P. Kienle, H. Tsertos, W. Wagner, C. Kozhuharov, F. Bosch, and W. Koenig: Phys. Lett. **137B**, 41 (1984).

[12] Detailed information is found in W. Greiner (ed.): *Physics of Strong Fields* (Plenum, New York 1987).

[13] T. Cowan, H. Backe, K. Bethge, H. Bokemeyer, H. Folger, J.S. Greenberg, K. Sakaguchi, D. Schwalm, J. Schweppe, K.E. Stiebing, P. Vincent: Phys. Rev. Lett. **56**, 444 (1986); W. Koenig, E. Berdermann, F. Bosch, S. Huchler, P. Kienle, C. Kozhuharov, A. Schröter, S. Schuhbeck, H. Tsertos: Phys. Lett. **B218**, 12 (1989).

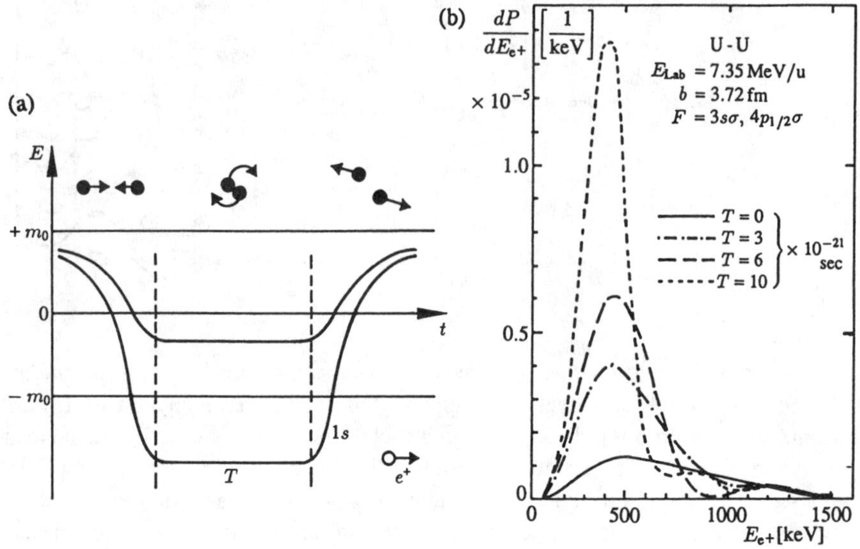

Fig. 7.22. (**a**) Two heavy ions are assumed to stick together for some time T. During this time the Coulomb potential is supercritical and positrons can be emitted spontaneously from the 1s level. (**b**) A line builds up in the spectrum of the emitted positrons with increasing time T

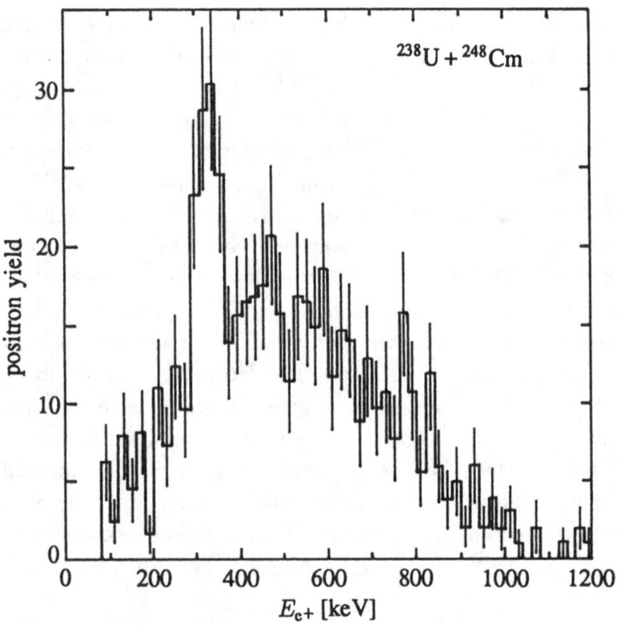

Fig. 7.23. Experimental spectrum of positrons emitted in $U + Cm$ collisions. A narrow line is visible at kinetic positron energy $E_{e^+} =$ 320 keV

decay according to $X \to e^+ + e^-$. There has been widespread theoretical speculation on the possible nature of such objects.[14] Meanwhile a considerable number of positron line structures with varying properties have been observed. Despite considerable effort no convincing and consistent explanation of the phenomenon has been found. The question of the source of the intriguing positron peaks will probably not be settled until more and better experimental data are available.

[14] See, e.g., A. Schäfer: J. Phys. **G15**, 373 (1989).

7.3 The Effective Lagrangian of the Electromagnetic Field

If we consider the electromagnetic field in isolation, it satisfies the linear Maxwell equations, and the superposition principle holds. There are no charges in empty space in the classical theory, and since the photons also do not bear charge, and thus do not interact among themselves, their field is described by the free noninteracting Lagrange function (more precisely, the Lagrange density)

$$\mathcal{L}_0 = \frac{1}{8\pi}\left(\boldsymbol{E}^2 - \boldsymbol{H}^2\right) \quad , \tag{7.46}$$

where $\boldsymbol{E}$ and $\boldsymbol{H}$ denote the electric and magnetic field strengths. Since $\mathcal{L}_0$ depends quadratically on $\boldsymbol{E}$ and $\boldsymbol{H}$, it is ensured that *the ensuing field equations are linear*.

The situation changes, however, when we move to quantum theory. Now the possibility exists of creating virtual particles, in particular electron–positron pairs, by a photon. Since they are charged they can interact with further photons (before they annihilate). In particular two photons can scatter off one another. The corresponding Feynman graph of lowest order is displayed in Fig. 7.24a. In the same way it is possible to scatter photons off an external electromagnetic field, (cf. Fig. 7.24b). The crosses denote the external field, which may be provided e.g. by a heavy nucleus of charge $-Ze$.

In the construction of these diagrams it has been taken into account that according to Furry's theorem (see Exercise 4.1) electron loops with an odd number of photon vertices do not contribute. Hence the process of the scattering of light on light, Fig. 7.24a, is of fourth order; its cross section has to be proportional to α^4 and thus is so small that it could not be verified experimentally yet.

The situation is more favourable for the scattering of photons off the electromagnetic field of a nucleus, Fig. 7.24b. This process, which is known as **Delbrück** *scattering*,[15] scales with $(Z\alpha)^4\alpha^2$ and has been found experimentally[16] using high-energy photons (several MeV). We also mention the "splitting" of a photon into two owing to the scattering at a nucleus[17] (cf. Fig. 7.24c), as another interesting process that has been observed experimentally.

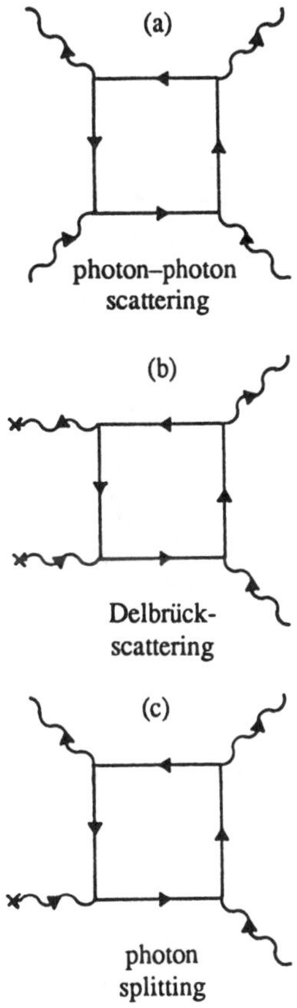

Fig. 7.24. (**a**) Photon–photon scattering. (**b**) Delbrück scattering. (**c**) Photon splitting

The vacuum of QED is a polarizable medium owing to virtual processes and obtains novel physical properties. One may try to describe this effect by replacing the Lagrangian $\mathcal{L}_0$ of the electromagnetic field by an *effective Lagrangian* $\mathcal{L}_{\text{eff}}$. This will contain corrections in higher orders in $\boldsymbol{E}$ and $\boldsymbol{H}$ and lead to nonlinear field equations. In the limiting case of a stationary and homogeneous electromagnetic field an "exact" closed expression can be given for $\mathcal{L}_{\text{eff}}$. This result was found in a

[15] M. Delbrück: Z. Physik **84**, 144 (1933); P. Papatzacos and K. Mork: Phys. Rep. **21**, 81 (1975).

[16] See e.g. S. Kahane and R. Moreh: Phys. Lett. **47B**, 351 (1973); P. Rullhusen et al.: Phys. Rev. **C27**, 559 (1983).

[17] G. Jarlskog et al.: Phys. Rev. **D8**, 3813 (1973).

pioneering work by **Heisenberg** and **Euler**.[18] We shall follow in part a derivation given by **Weisskopf**.[19]

To begin with we recall that there are two Lorentz-invariant quantities that characterize the electromagnetic field, namely

$$\begin{aligned} I_1 &= \boldsymbol{H}^2 - \boldsymbol{E}^2 \quad , \\ I_2 &= \boldsymbol{H} \cdot \boldsymbol{E} \quad . \end{aligned} \tag{7.47}$$

The effective Lagrangian can thus be expressed as a function of these invariants

$$\begin{aligned} \mathcal{L}_{\text{eff}}(\boldsymbol{H}, \boldsymbol{E}) &= \mathcal{L}_{\text{eff}}(I_1, I_2) \\ &= \mathcal{L}_{\text{eff}}(\boldsymbol{H}^2 - \boldsymbol{E}^2, \boldsymbol{H} \cdot \boldsymbol{E}) \quad . \end{aligned} \tag{7.48}$$

We remind the reader that the scalars I_1 and I_2 can be obtained by contraction of the electromagnetic field tensor $F_{\mu\nu}$, which is defined by

$$\begin{aligned} F^{\mu\nu} &= \partial^\mu F^\nu - \partial^\nu F^\mu \\ &= \begin{pmatrix} 0 & -E_1 & -E_2 & -E_3 \\ E_1 & 0 & -H_3 & H_2 \\ E_2 & H_3 & 0 & -H_1 \\ E_3 & -H_2 & H_1 & 0 \end{pmatrix} \quad . \end{aligned} \tag{7.49}$$

We also introduce the dual-field tensor obtained by contraction of $F_{\mu\nu}$ with the completely antisymetric unit tensor (the Levi–Civita tensor)

$$\begin{aligned} {}^*F^{\mu\nu} &= \frac{1}{2}\varepsilon^{\mu\nu\lambda\sigma} F_{\lambda\sigma} = F^{\mu\nu}(\boldsymbol{E} \leftrightarrow \boldsymbol{H}) \\ &= \begin{pmatrix} 0 & -H_1 & -H_2 & -H_3 \\ H_1 & 0 & -E_3 & E_2 \\ H_2 & E_3 & 0 & -E_1 \\ H_3 & -E_2 & E_1 & 0 \end{pmatrix} \quad . \end{aligned} \tag{7.50}$$

We may construct two scalars by contraction of these tensors, namely

$$F^{\mu\nu} F_{\mu\nu} = 2(\boldsymbol{H}^2 - \boldsymbol{E}^2) = 2I_1 \quad , \tag{7.51a}$$

$$F^{\mu\nu}\, {}^*F_{\mu\nu} = -4\boldsymbol{H} \cdot \boldsymbol{E} = -4I_2 \quad . \tag{7.51b}$$

The Lagrange function is gauge invariant because it depends only on the field strengths. Let us calculate the *energy* W_0 *of the vacuum* per unit volume as a function of the field strength. Proceeding quite naively we sum up the energy eigenvalues $\varepsilon_{p\sigma} < -m$ of all the electrons in the "Dirac sea" to obtain the total energy E_0. From this value the potential energy U_0 in the electric field has to be subtracted. The energy E_0 contains the potential energy U_0 of the electrons of the Dirac sea in the external field in addition to the pure energy W_0 of the vacuum. Since we are interested only in the pure energy of the vacuum the contribution U_0 has to be subtracted from E_0:

[18] W. Heisenberg and H. Euler: Z. Physik **38**, 314 (1936).

[19] V. Weisskopf: Kgl. Dankse Vid. Selskab., Math.-fys. Medd. **XIV**, 166 (1936); for a modern treatment see e.g. W. Dittrich, M. Reuter: *Effective Lagrangians in Quantum Electrodynamics* (Springer, Berlin, Heidelberg 1985).

$$W_0 = E_0 - U_0 \quad , \quad E_0 = \sum_{p\sigma} \varepsilon_{p\sigma}^{(-)} \quad , \tag{7.52}$$

$$U_0 = \sum_{p\sigma} \int d^3x \, \psi_{p\sigma}^{(-)\dagger} e A_0(\boldsymbol{x}) \psi_{p\sigma}^{(-)} \quad , \tag{7.53}$$

where $A_0(\boldsymbol{x})$ is the electrostatic potential. Here the sum extends over all momenta $\boldsymbol{p}$ and all spin directions; only the states with negative energy $(-)$ are taken into account. U_0 may be expressed in terms of E_0 by a trick. To do this, we make the following general consideration valid in quantum mechanics.

Let $\hat{H}(\lambda)$ be a self-adjoint Hamiltonian that depends analytically on a parameter λ and $\psi_n(\lambda)$ a normalized eigenfunction

$$\hat{H}(\lambda)\psi_n(\lambda) = \varepsilon_n(\lambda)\psi_n(\lambda) \quad . \tag{7.54}$$

The derivative of the energy eigenvalue with respect to the parameter λ then obeys

$$\frac{\partial \varepsilon_n}{\partial \lambda} = \left\langle \psi_n \left| \frac{\partial \hat{H}}{\partial \lambda} \right| \psi_n \right\rangle \quad , \tag{7.55}$$

since by differentiation of (7.54) and projection onto $\langle \psi_n |$ we get

$$\frac{\partial \varepsilon_n}{\partial \lambda} = \left\langle \psi_n \left| \frac{\partial \hat{H}}{\partial \lambda} \right| \psi_n \right\rangle + \left\langle \psi_n \left| (\hat{H} - \varepsilon_n) \frac{\partial}{\partial \lambda} \right| \psi_n \right\rangle \quad .$$

The last term is zero because of $\langle \psi_n | \hat{H} = \langle \psi_n | \varepsilon_n$.

Now we use this general statement by writing for the potential of a stationary, homogeneous $\boldsymbol{E}$ field

$$A_0(\boldsymbol{x}) = -\boldsymbol{E} \cdot \boldsymbol{x}$$

and use the field strength as the parameter λ. Thus

$$U_0 = \boldsymbol{E} \cdot \sum_{p\sigma} \int d^3x \, \psi_{p\sigma}^{(-)\dagger} \frac{\partial \hat{H}}{\partial \boldsymbol{E}} \psi_{p\sigma}^{(-)} = \boldsymbol{E} \cdot \frac{\partial E_0}{\partial \boldsymbol{E}} \quad , \tag{7.56}$$

and hence

$$W_0 = E_0 - U_0 = E_0 - \boldsymbol{E} \cdot \frac{\partial E_0}{\partial \boldsymbol{E}} \quad . \tag{7.57}$$

This relation serves to switch from the energy to the Lagrange function. The relationship between the energy (Hamiltonian) and the Lagrangian for a system having the generalized coordinates q_i in general reads

$$W = \sum_i \dot{q}_i \frac{\partial \mathcal{L}}{\partial \dot{q}_i} - \mathcal{L} \quad . \tag{7.58}$$

In electrodynamics the potentials A_0 and $\boldsymbol{A}$ play the role of the generalized coordinates q_i. Because of the relation $\boldsymbol{E} = -\dot{\boldsymbol{A}} - \nabla A_0$ and $\boldsymbol{H} = \nabla \times \boldsymbol{A}$, there is a dependence on a generalized velocity $(\dot{q}_i)$ in the Lagrangian only in the time derivative of the vector potential. But differentiation with respect to $\dot{\boldsymbol{A}}$ is equivalent to differentiation with respect to $\boldsymbol{E}$. Hence (7.58) can also be written as

$$W = \boldsymbol{E} \cdot \frac{\partial \mathcal{L}}{\partial \boldsymbol{E}} - \mathcal{L} \quad . \tag{7.59}$$

Thus we find that the change of the Lagrangian density of the electromagnetic field is given, up to a sign, by the additional energy density E_0:

$$\mathcal{L}_{\text{eff}} = \mathcal{L}_0 + \mathcal{L}' \quad , \tag{7.60a}$$

with

$$\mathcal{L}' = -E_0^{(\text{ren})} \quad . \tag{7.60b}$$

In (7.60b) we have indicated that the expression of (7.52) still has to be renormalized. In particular the energy of the vacuum in the absence of the electromagnetic field has to be subtracted, because it cannot be observed.

In order to calculate E_0 we restrict ourselves for the beginning to the case of a *pure magnetic field*, $\boldsymbol{E} = 0$. The energy eigenvalues can be given exactly according to Exercise 7.6:

$$\varepsilon_{p\sigma}^{(-)} = -\sqrt{m^2 + p_z^2 + |e|H(2n + 1 + \sigma)} \quad , \tag{7.61}$$

where $n = 0, 1, 2, \ldots$ and $\sigma = \pm 1$. The density of states per momentum interval is $|e|H/2\pi dp_z/2\pi$; cf. Exercise 7.6. Hence

$$\begin{aligned}
\mathcal{L}' &= -E_0 \\
&= \int\limits_{-\infty}^{+\infty} \frac{dp_z}{2\pi} \frac{|e|H}{2\pi} \sum_{n\sigma} \sqrt{m^2 + p_z^2 + |e|H(2n + 1 + \sigma)} \\
&= \frac{|e|H}{(2\pi)^2} \int\limits_{-\infty}^{+\infty} dp_z \left(\sqrt{m^2 + p_z^2} + 2 \sum_{n=1}^{\infty} \sqrt{m^2 + p_z^2 + 2|e|Hn} \right) \quad .
\end{aligned} \tag{7.62}$$

Here we have taken account of the fact that all states are doubly degenerate except for the level with $n = 0, \sigma = -1$. The states with quantum numbers $n, \sigma = +1$ and with $n - 1$, $\sigma = -1$ have the same energy. Only for the state $n = 0, \sigma = -1$ can such a partner not be found.

Obviously (7.62) is highly divergent. As we shall see we can nevertheless split off a physically meaningful finite expression. To this end we first *regularize* (7.62) by introducing a suitably chosen cutoff factor. With the abbreviation

$$F(n, \lambda) = \int\limits_0^{\infty} dp_z \sqrt{m^2 + p_z^2 + 2|e|Hn}\; \mathrm{e}^{-\lambda\sqrt{m^2+p_z^2+2|e|Hn}} \tag{7.63}$$

the regularized equation (7.62) reads

$$\mathcal{L}'(\lambda) = \frac{|e|H}{\pi^2} \left(\frac{1}{2} F(0, \lambda) + \sum_{n=1}^{\infty} F(n, \lambda) \right) \quad . \tag{7.64}$$

λ is a cutoff parameter with dimension one over energy. The limit $\lambda \to 0$ should be taken at the end of the calculation. Physically meaningful quantities must no longer depend on λ then. Hence they have to approach a finite limiting value. Equation (7.64) may be rewritten using the summation formula of Euler and MacLaurin[20]

$$\sum_{n=0}^{N} F(n,\lambda) = \frac{1}{2}F(0,\lambda) + \frac{1}{2}F(N,\lambda) + \int_0^N dn\, F(n,\lambda) + \sum_{k=1}^{\infty} \frac{B_{2k}}{(2k)!} \left(F^{(2k-1)}(N,\lambda) - F^{(2k-1)}(0,\lambda) \right) \quad . \tag{7.65}$$

Here $F^{(k)}(x,\lambda)$ denotes the kth derivative of the function $F(x,\lambda)$ with respect to x. The B_{2n} are Bernoulli's numbers $B_2 = 1/6, B_4 = -1/30, B_6 = 1/42, \ldots$. Because of (7.63), $F(n,\lambda)$ and all its derivatives decay exponentially at large n (for $\lambda \neq 0$) so that the limit $N \to \infty$ can be taken in (7.65), leading to the result

$$\sum_{n=0}^{\infty} F(n,\lambda) = \frac{1}{2}F(0,\lambda) + \int_0^\infty dn\, F(n,\lambda) - \sum_{k=1}^{\infty} \frac{B_{2k}}{(2k)!} F^{(2k-1)}(0,\lambda) \quad . \tag{7.66}$$

Hence, (7.64) can be written as

$$\mathcal{L}'(\lambda) = \frac{|e|H}{\pi^2} \left(\int_0^\infty dn\, F(n,\lambda) - \sum_{k=1}^{\infty} \frac{B_{2k}}{(2k)!} F^{(2k-1)}(0,\lambda) \right) \quad . \tag{7.67}$$

The integral in (7.63) defining the function $F(n,\lambda)$ can be evaluated explicitly. Substituting $x = \sqrt{p_z^2 + a^2}/a$ and $a^2 = m^2 + 2|e|Hn$ we obtain

$$\begin{aligned} F(n,\lambda) &= \int_0^\infty dp_z\, \sqrt{p_z^2 + a^2}\, \mathrm{e}^{-\lambda\sqrt{p_z^2+a^2}} = a^2 \int_1^\infty dx\, \frac{x^2}{\sqrt{x^2-1}}\, \mathrm{e}^{-\lambda a x} \\ &= a^2 \left(\frac{1}{-a} \right)^2 \frac{d^2}{d\lambda^2} \int_1^\infty dx\, \frac{\mathrm{e}^{-\lambda a x}}{\sqrt{x^2-1}} \\ &= \frac{d^2}{d\lambda^2} K_0(\lambda a) = a^2 \frac{d^2}{dz^2} K_0(z) \quad . \end{aligned} \tag{7.68}$$

Here $K_0(z)$ is the modified Bessel function of the second kind (the McDonald function) and we have substituted $z = \lambda a$.

The derivative $K_0''(z)$ is evaluated by use of recursion relations for the Bessel functions (see I.S. Gradshteyn and I.M. Ryzhik: Table of Integrals, Series, and Products, Academic Press, 1965, no. 8.486). In particular, $K_0' = -K_1$, $K_1' = K_1/z - K_2$, and thus

[20] See for example G. Arfken: *Mathematical Methods for Physicists* (Academic Press, New York 1970) Chapter 5.9.

$$F(n,\lambda) = -a^2\left(\frac{1}{\lambda a}K_1(\lambda a) - K_2(\lambda a)\right)$$
$$= -\frac{1}{\lambda^2}\left(zK_1(z) - z^2K_2(z)\right) \quad . \tag{7.69}$$

In (7.67) we need the derivatives of this function with respect to n. Because of $z = \lambda a = \lambda\sqrt{m^2 + 2|e|Hn}$, $zdz = \lambda^2|e|Hdn$, the mth derivative may be written as

$$\left(\frac{d}{dn}\right)^m = \left(\lambda^2|e|H\right)^m\left(\frac{1}{z}\frac{d}{dz}\right)^m \quad . \tag{7.70}$$

These derivatives lead just to simple modified Bessel functions (see Gradshteyn/Ryzhik no. 8.486.14):

$$\left(\frac{1}{z}\frac{d}{dz}\right)^m\left(z^\nu K_\nu(z)\right) = (-1)^m z^{\nu-m}K_{\nu-m}(z) \quad . \tag{7.71}$$

With $z(n=0) = \lambda m$ the regularized Lagrangian (7.67) now reads

$$\begin{aligned}\mathcal{L}'(\lambda) = &-\frac{1}{\pi^2}\frac{1}{\lambda^4}\int\limits_{\lambda m}^{\infty} dz\; z^2\left(K_1(z) - zK_2(z)\right)\\ &-\frac{|e|H}{\pi^2}\sum_{k=1}^{\infty}\frac{B_{2k}}{(2k)!}\left(-\frac{1}{\lambda^2}\right)\left(\lambda^2|e|H\right)^{2k-1}\\ &\times(-1)^{2k-1}\left[(\lambda m)^{2-2k}K_{2-2k}(\lambda m) - (\lambda m)^{3-2k}K_{3-2k}(\lambda m)\right] \quad .\end{aligned} \tag{7.72}$$

Let us now consider the structure of this expression. It is a power series in even powers of the field strength H multiplied by the elementary charge e,

$$\mathcal{L}'(\lambda) = C_0(\lambda) + C_2(\lambda)(eH)^2 + \sum_{k=2}^{\infty} C_{2k}(\lambda)(eH)^{2k} \quad . \tag{7.73}$$

It turns out that the first two coefficients, $C_0(\lambda)$ and $C_2(\lambda)$, diverge if the cutoff parameter λ approaches 0. The higher coefficients $C_4, C_6, \ldots$, however, are finite (see below): The divergence of $C_0(\lambda)$ follows from

$$C_0(\lambda) = -\frac{1}{\pi^2}\frac{1}{\lambda^4}\int\limits_{\lambda m}^{\infty} dz\; z^2\left(K_1(z) - zK_2(z)\right) = O\left(\frac{1}{\lambda^4}\right) \to \infty \quad , \tag{7.74}$$

because the integral converges at the lower bound. The term which is quadratic in the field strength results from (7.72):

$$C_2(\lambda) = -\frac{1}{\pi^2}\frac{B_2}{2}\left(K_0(\lambda m) - \lambda m K_1(\lambda m)\right) \quad . \tag{7.75}$$

The asymptotic behaviour of the Bessel function for $z \to 0$ is

$$\begin{aligned}&K_m(z) \to (m-1)!\; 2^{m-1}z^{-m} \quad \text{for} \quad m > 0 \quad ,\\ &K_0(z) \to -\ln(z) \quad .\end{aligned} \tag{7.76}$$

Hence $C_2(\lambda)$ diverges logarithmically as a function of the cutoff parameter λ

$$C_2(\lambda) \rightarrow \frac{1}{\pi^2}\frac{1}{12}\ln(\lambda m) \quad . \tag{7.77}$$

It has been clear from the beginning that divergence problems of this kind had to occur. The energies of all states of the lower continuum were summed up in the ansatz (7.72, 7.53). The constant C_0 is just the "total energy of the Dirac sea" and as such is not observable. This identification may be verified formally by converting the expression for C_0 into a three-dimensional momentum integral. The substitutions $p_\perp^2 = 2|e|Hn$ and $d^2p_\perp = p_\perp dp_\perp d\varphi$ in cylindrical coordinates lead to

$$\begin{aligned} C_0(\lambda) &= \frac{|e|H}{\pi^2}\int_0^\infty dnF(n,\lambda) \\ &= \frac{|e|H}{\pi^2}\int_0^\infty dn\int_0^\infty dp_z\sqrt{m^2+p_z^2+2|e|Hn}\;\mathrm{e}^{-\lambda\sqrt{m^2+p_z^2+2|e|Hn}} \\ &= \frac{1}{2\pi^2}\int_{-\infty}^\infty dp_z\int_0^\infty dp_\perp p_\perp\int_0^{2\pi}\frac{d\varphi}{2\pi}\sqrt{m^2+p_z^2+p_\perp^2}\,\mathrm{e}^{-\lambda\sqrt{m^2+p_z^2+p_\perp^2}} \\ &= \int\frac{d^3p}{(2\pi)^3}\,2\sqrt{m^2+p^2}\;\mathrm{e}^{-\lambda\sqrt{m^2+p^2}} \quad . \end{aligned} \tag{7.78}$$

This is just the regularized expression for the negative of the energy of the lower continuum in the absence of an external field. Thus we have to subtract C_0 in (7.73) in order to obtain a meaningful expression. Furthermore the term $C_2e^2H^2$ has exactly the form of the free Lagrangian (7.46) so that we can group the terms in the following way:

$$\begin{aligned} \mathcal{L}_{\mathrm{eff}} &= \mathcal{L}_0 + \mathcal{L}' - C_0 \\ &= \mathcal{L}_0 + C_2(eH)^2 + \left[\mathcal{L}' - C_0 - C_2(eH)^2\right] \\ &= (1+8\pi C_2e^2)\frac{H^2}{8\pi} + \sum_{n=2}^\infty C_{2n}(eH)^{2n} \quad . \end{aligned} \tag{7.79}$$

Hence the free Lagrangian is multiplied by a constant $(1+8\pi e^2C_2)$. Once again, the presence of such a factor cannot be observed physically. Since it is effective in all experiments, the constant factor only leads to a redefinition of the field strength and the charge. We can formally define a *renormalized elementary charge* by

$$e_{\mathrm{R}} = \frac{e}{\sqrt{1+8\pi e^2C_2}} \tag{7.80}$$

and the corresponding *renormalized field strength* by

$$H_{\mathrm{R}} = \frac{e}{e_{\mathrm{R}}}\,H = \sqrt{1+8\pi e^2C_2}\;H \quad . \tag{7.81}$$

The Lagrangian of the magnetic field expressed in terms of these quantities reads

$$\mathcal{L}_{\text{eff}} = \frac{1}{8\pi} H_{\mathrm{R}}^2 + \sum_{n=2}^{\infty} C_{2n}(e_{\mathrm{R}} H_{\mathrm{R}})^{2n} = \mathcal{L}_{0\mathrm{R}} + \mathcal{L}'_{\mathrm{R}} \quad . \tag{7.82}$$

This expression has the "correct" limit at small field strengths. Thus we may consider H_{R} and e_{R} to be the physically observable quantities. In the following we shall omit the index R for brevity.

The renormalized correction $\mathcal{L}'$ to the Lagrangian then reads, because of (7.76) (using $K_{-n}(z) = K_n(z)$),

$$\begin{aligned}
\mathcal{L}' &= -\frac{1}{\pi^2} \lim_{\lambda\to 0} \sum_{k=2}^{\infty} \frac{B_{2k}}{(2k)!} \frac{1}{\lambda^4} (\lambda^2 |e| H)^{2k} \\
&\quad \times \left[(\lambda m)^{2-2k} K_{2k-2}(\lambda m) - (\lambda m)^{3-2k} K_{2k-3}(\lambda m) \right] \\
&= -\frac{1}{\pi^2} \lim_{\lambda\to 0} \sum_{k=2}^{\infty} \frac{B_{2k}}{(2k)!} \lambda^{4k-4} (eH)^{2k} (\lambda m)^{2-2k} (2k-3)! 2^{2k-3} (\lambda m)^{-2k+2} \\
&= -\frac{1}{\pi^2} \sum_{k=2}^{\infty} \frac{B_{2k}}{(2k)!} \frac{1}{8} (2eH)^{2k} m^{4-4k} (2k-3)! \\
&= -\frac{1}{8\pi^2} \sum_{k=2}^{\infty} (2eH)^{2k} B_{2k} m^{4-4k} \frac{\Gamma(2k-2)}{(2k)!} \quad .
\end{aligned} \tag{7.83}$$

It is usual to express this result in terms of an integral representation. To this end we write the gamma function as

$$\Gamma(z) = \int_0^{\infty} d\eta \, \mathrm{e}^{-\eta} \eta^{z-1} \tag{7.84}$$

and obtain

$$\begin{aligned}
\mathcal{L}' &= -\frac{1}{8\pi^2} \sum_{k=2}^{\infty} (2|e|H)^{2k} B_{2k} m^{4-4k} \frac{1}{(2k)!} \int_0^{\infty} d\eta \, \mathrm{e}^{-\eta} \eta^{2k-3} \\
&= -\frac{1}{8\pi^2} |e| H m^2 \int_0^{\infty} d\eta \, \frac{\mathrm{e}^{-\eta}}{\eta^2} \sum_{k=2}^{\infty} \frac{2^{2k}}{(2k)!} B_{2k} \left(\frac{|e| H \eta}{m^2} \right)^{2k-1} \quad .
\end{aligned} \tag{7.85}$$

Careful inspection of the series in this expression reveals that it is identical to the Taylor expansion of the hyperbolic cotangent function,

$$\coth(x) = \frac{1}{x} + \frac{x}{3} + \sum_{k=2}^{\infty} \frac{2^{2k}}{(2k)!} B_{2k} x^{2k-1} \quad . \tag{7.86}$$

Introducing the dimensionless field strength

$$\tilde{H} = \frac{H}{H_{\mathrm{cr}}} = \frac{|e|}{m^2} H \quad , \tag{7.87}$$

we get as the final result in compact form

$$\mathcal{L}'(\boldsymbol{E} = 0, \boldsymbol{H}) = \frac{m^4}{8\pi^2} \int_0^{\infty} d\eta \frac{\mathrm{e}^{-\eta}}{\eta^3} \left(-\tilde{H}\eta \coth(\tilde{H}\eta) + 1 + \frac{1}{3} (\tilde{H}\eta)^2 \right) \quad . \tag{7.88}$$

The investigation is more difficult for general electromagnetic fields, because we cannot find an expression analogous to (7.61). However, the case of a constant *pure electric field* ($\boldsymbol{H} = 0$) can be reduced to (7.88) by a trick. To do this we note that the result may be expressed as a function of the two invariants $\boldsymbol{H}^2 - \boldsymbol{E}^2$ and $\boldsymbol{H} \cdot \boldsymbol{E}$ according to (7.48). Thus one immediately sees that

$$\begin{aligned} \mathcal{L}'(\boldsymbol{E}, \boldsymbol{H} = 0) &= \mathcal{L}'(I_1 = 0 - \boldsymbol{E}^2, I_2 = 0) \\ &= \mathcal{L}'(I_1 = (\mathrm{i}\boldsymbol{E})^2, I_2 = 0) \\ &= \mathcal{L}'(\boldsymbol{E} = 0, \boldsymbol{H} = \mathrm{i}E) \quad . \end{aligned} \tag{7.89}$$

Hence one can use the solution for the magnetic field and replace H by $\mathrm{i}E$! Because $\coth(\mathrm{i}x) = -\mathrm{i}\cot(x)$, this leads to

$$\mathcal{L}'(\boldsymbol{E}, \boldsymbol{H} = 0) = \frac{m^4}{8\pi^2} \int_0^\infty d\eta \frac{\mathrm{e}^{-\eta}}{\eta^3} \left[-\tilde{E}\eta \cot(\tilde{E}\eta) + 1 - \frac{1}{3}(\tilde{E}\eta)^2 \right] \quad , \tag{7.90}$$

with the reduced electric field strength

$$\tilde{E} = \frac{E}{E_{\mathrm{cr}}} = \frac{|e|}{m^2} E \quad . \tag{7.91}$$

For the sake of completeness we also quote without proof the extension to the case of constant parallel electric and magnetic fields. This is sufficient for a unique expansion of $\mathcal{L}'$ in the invariants I_1 and I_2 (7.47). The required calculation is quite lengthy; we refer to the cited original publications. The result reads

$$\begin{aligned} \mathcal{L}'(\boldsymbol{H} || \boldsymbol{E}) &= \frac{m^4}{8\pi^2} \int_0^\infty d\eta \, \frac{\mathrm{e}^{-\eta}}{\eta^3} \\ &\times \left[-\tilde{E}\eta \, \cot(\tilde{E}\eta) \tilde{H}\eta \coth(\tilde{H}\eta) + 1 - \frac{1}{3}(\tilde{E}^2 - \tilde{H}^2)\eta^2 \right] \quad , \end{aligned} \tag{7.92}$$

which obviously contains (7.88) and (7.90) as limiting cases. $\mathcal{L}_0 + \mathcal{L}'$ is the *effective Lagrangian of the electromagnetic field* which dates back to Heisenberg and Euler (1936). A formally more satisfying derivation based on the "proper time method" was given later by Schwinger.[21]

We now examine some consequences of (7.92). First, let us consider the limiting case of *weak fields*, i.e. $\tilde{E} \ll 1, \tilde{H} \ll 1$. A Taylor expansion corresponding to (7.86) up to the third term yields

$$\begin{aligned} \mathcal{L}' &= \frac{m^4}{8\pi^2} \frac{1}{45} \int_0^\infty d\eta \, \eta \mathrm{e}^{-\eta} \, (\tilde{E}^4 + \tilde{H}^4 + 5\tilde{E}^2\tilde{H}^2) \\ &= \frac{1}{8\pi} \frac{\mathrm{e}^4}{45\pi m^4} \left[(\boldsymbol{H}^2 - \boldsymbol{E}^2)^2 + 7(\boldsymbol{E} \cdot \boldsymbol{H})^2 \right] \quad . \end{aligned} \tag{7.93}$$

By the way, this result is valid in every frame of reference, because it has been expressed in terms of the invariants I_1 and I_2. Amongst other things, we conclude

[21] J. Schwinger: Phys. Rev. **82**, 664 (1951).

from (7.93) that there are no nonlinear corrections for plane wave since both invariants then vanish.

For the limiting case of *strong magnetic fields*, i.e. $\tilde{H} \gg 1$, we will be satisfied with a rough estimate in logarithmic approximation. With the substitution $\tau = \eta\tilde{H}$, (7.88) can be written as

$$\mathcal{L}' = \frac{m^4\tilde{H}^2}{8\pi^2}\int\limits_0^\infty d\tau \frac{e^{-\tau/\tilde{H}}}{\tau}\left(\frac{1}{3} + \frac{1-\tau\coth(\tau)}{\tau^2}\right) \quad . \tag{7.94}$$

For $\tau \leq 1$ the integrand is attenuated by the expression in parentheses (because $\coth\tau = 1/\tau + \tau/3 - \tau^3/45 \pm \cdots$) and for large $\tau \geq \tilde{H}$ by the exponential factor. Thus it is a reasonable approximation to replace the integration bounds by these values and further neglect the variation of the second term in parentheses and of $\exp\left(-\tau/H\right)$ in this range. Then we obtain

$$\mathcal{L}' = \frac{m^4\tilde{H}^2}{8\pi^2}\int\limits_1^{\tilde{H}} \frac{d\tau}{\tau}\frac{1}{3} = \frac{m^4\tilde{H}^2}{24\pi^2}\ln(\tilde{H}) \quad . \tag{7.95}$$

If we compare this to the free Lagrangian $\mathcal{L}_0$, we see that the nonlinear effects always stay small in QED,

$$\frac{\mathcal{L}'}{\mathcal{L}_0} = \frac{e^2}{3\pi}\ln\frac{|e|H}{m^2} \quad . \tag{7.96}$$

In order to have $\mathcal{L}' = \mathcal{L}_0$ one would have to reach entirely unrealistic field strengths with the order of magnitude

$$H = H_{\mathrm{cr}}\, e^{3\pi/\alpha} = 10^{560}\, H_{\mathrm{cr}} \quad . \tag{7.97}$$

Of course, this is due to the small electromagnetic compling constant.

Finally we are led to the most interesting result by considering the Lagrangian of *strong electric fields*. At first the result (7.90) is not well defined, because the cotangent has poles on the real axis. The integral for the energy density,

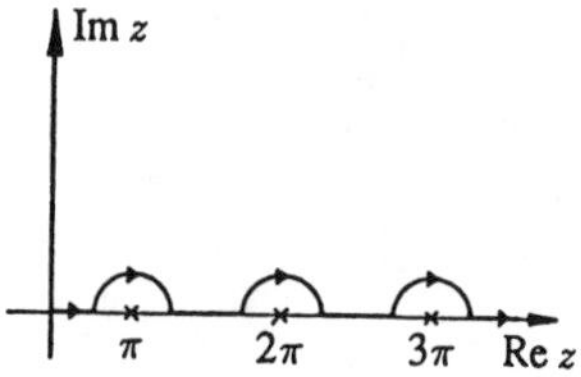

Fig. 7.25. The deformed integration contour to be chosen in (7.98)

$$E_0 = \frac{m^4\tilde{E}^2}{8\pi^2}\int\limits_0^\infty d\tau \frac{e^{\tau/\tilde{E}}}{\tau^3}\left(\tau\cot(\tau) - 1 + \frac{1}{3}\tau^2\right) \quad , \tag{7.98}$$

can be given a value by choosing a contour in the complex τ plane. If the poles are circumvented in the upper half plane, see Fig. 7.25, the energy obtains a negative imaginary part (the sign will become clear later). We calculate its magnitude by taking half of the negative residuum at each pole:

$$\begin{aligned}\mathrm{Im}\{E_0\} &= -\frac{1}{2}2\pi i\sum_{n=1}^\infty \mathrm{Res}|_{\tau=n\pi} \\ &= -\frac{1}{2}2\pi i\frac{m^4\tilde{E}^2}{8\pi^2}\sum_{n=1}^\infty \frac{e^{-n\pi/\tilde{E}}}{(n\pi)^3}(n\pi) \\ &= -i\frac{e^2E^2}{8\pi^3}\sum_{n=1}^\infty \frac{1}{n^2}e^{-n\pi m^2/|e|E} \quad . \end{aligned} \tag{7.99}$$

In order to understand this result we recall that complex energies characterize the decay of a quantum-mechanical state. In fact the probability of a time-dependent state $|\Phi(t)\rangle = \mathrm{e}^{-\mathrm{i}Et}|\phi\rangle$ is

$$P(t) = \langle\Phi(t)|\Phi(t)\rangle = \mathrm{e}^{-\mathrm{i}(E-E^*)t}\langle\phi|\phi\rangle = \mathrm{e}^{2\mathrm{Im}(Et)} \quad . \tag{7.100}$$

In our context this means that the vacuum state, which originally is free of particles, *decays spontaneously in a strong electric field by creation of electron–positron pairs*. The particle creation rate per unit volume and time is

$$\begin{aligned} w &= 2\,\mathrm{Im}(\mathcal{L}') \\ &= \frac{1}{4\pi^3}\left(\frac{eE\hbar}{m^2c^3}\right)^2 \underbrace{\frac{mc^2}{\hbar}}_{\frac{1}{\text{time}}} \underbrace{\left(\frac{mc}{\hbar}\right)^3}_{\frac{1}{\text{volume}}} \sum_{n=1}^{\infty}\frac{1}{n^2}\exp\left(-n\frac{\pi m^2c^3}{|e|E\hbar}\right) \quad , \end{aligned} \tag{7.101}$$

where the constants $\hbar$ and c have been written out explicitly this time.

It is very remarkable that the result (7.101) has an essential singularity in the limit $e \longrightarrow 0$. Thus *pair creation in a strong field is a nonperturbative effect* which cannot be calculated by a series expansion in the coupling constant.

Finally one has to keep in mind that (7.101) was calculated with the assumption that no real particles are present. When the field strength E becomes large, this is no longer justified and the reaction of the created pairs on the electric field would have to be taken into account by a self-consistent ansatz.[22]

At the beginning of this chapter we gave an intuitive interpretation of pair creation in an electric field: in the vacuum continuously short-living "virtual" e^+e^- pairs are created and annihilated again by quantum fluctuations. These pairs can be separated spatially by the external electric field and converted into real particles by expenditure of energy. For this to become possible the potential energy has to vary by an amount $\Delta V = |e|E\,\Delta l > 2mc^2$ in the range of about one Compton wavelength, $\Delta l \simeq \hbar/mc$ which leads to the value of the critical field strength.

A more quantitative interpretation results from Dirac's hole picture. Figure 7.26 shows the space dependence of the potential energy $V(x) = eEx$ as well as the corresponding energy gap of the Dirac equation between $mc^2 + V(x)$ and $-mc^2 + V(x)$. In full analogy to Fig. 7.5 pair creation results from the tunnelling of an electron from the "Dirac sea" through this classically forbidden zone. The probability for such a tunnel process is described by a penetration factor (the "Gamow factor") given by

$$P \simeq \exp\left(-\frac{2}{\hbar}\int_{x_-}^{x_+} q(x)dx\right) \quad , \tag{7.102}$$

with the imaginary momentum

$$q(x) = \sqrt{m^2c^2 - (W - eEx)^2/c^2} \quad . \tag{7.103}$$

[22] A complete solution of this problem is still lacking, see Y. Kluger, J.M. Eisenberg, B. Svetitsky, F. Cooper, E. Mottola, Phys. Rev. **D45**, 4659, (1992).

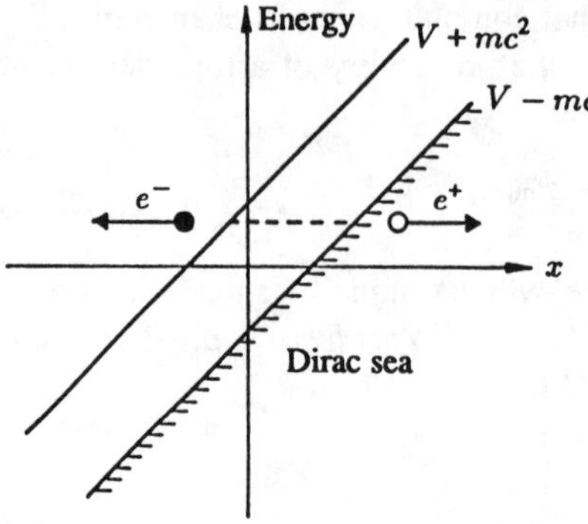

Fig. 7.26. e^+e^- pair creation in a constant electric field can be visualized as a tunneling process through the mass gap of the Dirac equation

$x_\pm$ denotes the classical turning points where $q(x_\pm) = 0$, W is the energy. We find

$$P \simeq \exp\left(-\frac{2}{\hbar c} \int\limits_{x_-}^{x_+} \sqrt{m^2c^4 - (W - eEx)^2}\, dx \right)$$

$$= \exp\left(-\frac{2}{\hbar c} \frac{1}{|e|E} m^2c^4 \int\limits_{-1}^{+1} \sqrt{1-u^2}\, du \right)$$

$$= \exp\left(-\frac{\pi m^2 c^3}{|e|E\hbar} \right) \quad . \tag{7.104}$$

This is exactly the exponential factor in (7.99) (for the one-pair term, $n = 1$)!

EXERCISE

7.5 An Alternative Derivation of the Effective Lagrangian

Problem. The effective Lagrangian of the electromagnetic field in the form (7.88) of Sect. 7.3 may also be derived with less mathematical effort. In a first step differentiate (7.62) twice with respect to the parameter m^2 and sum up the series using the integral representation

$$\frac{1}{m^2} = \int\limits_0^\infty d\eta\, \mathrm{e}^{-m^2\eta} \tag{1}$$

and integrate the resulting function twice over m^2. Finally a renormalization has to be performed.

Solution. The following expression has to be calculated:

$$\mathcal{L}'(\boldsymbol{E} = 0, \boldsymbol{H})$$
$$= \frac{|e|H}{(2\pi)^2}\, 2 \int\limits_0^\infty dp_z \left(\sqrt{m^2 + p_z^2} + 2 \sum_{n=1}^\infty \sqrt{m^2 + p_z^2 + 2|e|Hn} \right) \quad . \tag{2}$$

Exercise 7.5.

We formally treat this expression as if it were convergent, ignoring the need for a regularization prescription. Double differentiation by m^2 leads to

$$\begin{aligned} \Lambda(H) &:= \frac{d^2\mathcal{L}'}{(dm^2)^2} \\ &= -\frac{|e|H}{(2\pi)^2}\frac{1}{2}\int_0^\infty dp_z \left[\frac{1}{(m^2+p_z^2)^{\frac{3}{2}}} + 2\sum_{n=1}^{\infty}\frac{1}{(m^2+2|e|Hn+p_z^2)^{\frac{3}{2}}}\right] \quad . \end{aligned} \tag{3}$$

The p_z integral now is convergent and simply yields

$$\int_0^\infty dp_z \frac{1}{(p_z^2+a^2)^{3/2}} = \left.\frac{p_z}{a^2\sqrt{p_z^2+a^2}}\right|_0^\infty = \frac{1}{a^2} \quad , \tag{4}$$

and hence

$$\Lambda(H) = -\frac{|e|H}{8\pi^2}\left(\frac{1}{m^2} + 2\sum_{n=1}^{\infty}\frac{1}{m^2+2|e|Hn}\right) \quad . \tag{5}$$

The integral representation (1) leads to

$$\begin{aligned} \Lambda(H) &= -\frac{|e|H}{8\pi^2}\int_0^\infty d\eta \left(\mathrm{e}^{-m^2\eta} + 2\sum_{n=1}^{\infty}\mathrm{e}^{-(m^2+2|e|Hn)\eta}\right) \\ &= -\frac{|e|H}{8\pi^2}\int_0^\infty d\eta\, \mathrm{e}^{-m^2\eta}\left(2\sum_{n=0}^{\infty}\mathrm{e}^{-2|e|Hn\eta} - 1\right) \quad . \end{aligned} \tag{6}$$

The sum over n here is just a geometrical series! It can be summed in closed form to yield

$$\begin{aligned} \Lambda(H) &= -\frac{|e|H}{8\pi^2}\int_0^\infty d\eta\; \mathrm{e}^{-m^2\eta}\left(\frac{2}{1-\mathrm{e}^{-2|e|H\eta}} - 1\right) \\ &= -\frac{|e|H}{8\pi^2}\int_0^\infty d\eta\; \mathrm{e}^{-m^2\eta}\frac{1+\mathrm{e}^{-2|e|H\eta}}{1-\mathrm{e}^{-2|e|H\eta}} \\ &= -\frac{1}{8\pi^2}\int_0^\infty d\eta\; \frac{\mathrm{e}^{-m^2\eta}}{\eta}|e|H\eta\; \coth(|e|H\eta) \quad . \end{aligned} \tag{7}$$

Twofold integration over the variable m^2 yields

$$\mathcal{L}' = -\frac{1}{8\pi^2}\int_0^\infty d\eta\; \mathrm{e}^{-m^2\eta}\frac{1}{\eta^3}|e|H\eta\; \coth(|e|H\eta) + C_2 + C_1 m^2 \quad . \tag{8}$$

The integration constants C_1, C_2 can in principle be arbitrary functions of H but they must not depend on m^2. This is a very strong restriction. From dimensional

Exercise 7.5.

considerations the Lagrange density $\mathcal{L}'$ may depend only on the dimensionless ratio H/H_{cr}. In more precise terms the functional dependence on H and m should read

$$\begin{aligned}\mathcal{L}'(H,m) &= m^4 f(H/H_{\mathrm{cr}}) = m^4 f\left(\frac{|e|H}{m^2}\right)\\ &= \left(\frac{mc}{\hbar}\right)^3 mc^2 f\left(\frac{|e|H}{m^2}\frac{\hbar}{c^3}\right) \quad .\end{aligned} \tag{9}$$

The constants $\hbar$ and c have been inserted explicitly in the last step. This demonstrates that the factor m^4 indeed bears the correct dimension of the Lagrangian density, namely energy per volume. Taking into account (9) the integration constants in (7) can only have the form

$$C_1 = C_1'\, m^2 \frac{|e|H}{m^2} \quad , \quad C_2 = C_2'\, e^2 H^2 \quad . \tag{10}$$

The effective Lagrangian is connected to $\mathcal{L}'$ as follows (cf. (7.79)):

$$\mathcal{L}_{\mathrm{eff}} = \mathcal{L}_0 + \mathcal{L}' - C_0 = \mathcal{L}_{0\mathrm{R}} + \mathcal{L}'_{\mathrm{R}} \quad . \tag{11}$$

$\mathcal{L}'$ contains ill-defined contributions proportional to $(eH)^0$ and $(eH)^2$. The finite renormalized Lagrangian $\mathcal{L}'_{\mathrm{R}}$ is obtained from $\mathcal{L}'$ by

$$\mathcal{L}'_{\mathrm{R}} = \mathcal{L}' - C_0 \quad . \tag{12}$$

Here the constants C_0, C_1, C_2 have to be chosen such that the power series expansion of $\mathcal{L}'_{\mathrm{R}}$ with respect to the field strength H does not contain terms of the order ≤ 2. To achieve this we have to choose the integration constant $C_1 = 0$. Furthermore the renormalization condition leads to

$$C_0 = -\frac{1}{8\pi^2} m^4 \int\limits_0^\infty d\eta' \, \frac{\mathrm{e}^{-\eta'}}{\eta'^3} \tag{13a}$$

and

$$C_2 = \frac{1}{8\pi^2} \mathrm{e}^2 H^2 \frac{1}{3} \int\limits_0^\infty d\eta' \, \frac{\mathrm{e}^{-\eta'}}{\eta'} \quad , \tag{13b}$$

because $x\coth(x) = 1 + 1/3x^2 + \cdots$, with the substitution of variables $\eta' = m^2\eta$. Since we did not introduce a regularization prescription, these are divergent integrals.

Hence the renormalized effective Lagrangian of the electromagnetic field reads

$$\mathcal{L}'_{\mathrm{R}} = -\frac{m^4}{8\pi^2} \int\limits_0^\infty d\eta' \, \frac{\mathrm{e}^{-\eta'}}{\eta'^3} \left(\tilde{H}\eta' \coth(\tilde{H}\eta') - 1 - \frac{1}{3}(\tilde{H}\eta')^2\right) \quad , \tag{14}$$

with $\tilde{H} = H/H_{\mathrm{cr}} = |e|H/m^2$. This result is identical to (7.88).

EXERCISE

7.6 The Solution of the Dirac Equation in a Homogeneous Magnetic Field

Problem. Show that the energy of a Dirac particle in a homogenous magnetic field $\boldsymbol{H} = H\boldsymbol{e}_z$ is given by the expression

$$\varepsilon_{p\sigma} = \pm\sqrt{m^2 + p_z^2 + |e|H(2n+1-\sigma)} \quad .$$

p_z is the momentum in the z direction and $\sigma = \pm 1$ is the projection of the spin. Use the Dirac equation for the bispinor $\psi = \binom{\phi}{\chi}$ and reduce the problem to the differential equation of the harmonic oscillator by elimination of χ. Show that the density of states per momentum interval in the volume $V = 1$ is given by

$$\frac{dN}{dp_z} = \frac{|e|H}{(2\pi)^2} \quad .$$

Solution. According to in *RQM*, Chap. 2, the Dirac equation in two-component notation reads ($\hbar = 1$)

$$\begin{aligned} \mathrm{i}\frac{\partial}{\partial t}\phi &= \boldsymbol{\sigma}\cdot(\hat{\boldsymbol{p}} - e\boldsymbol{A})\chi + eA_0\phi + m\phi \quad , \\ \mathrm{i}\frac{\partial}{\partial t}\chi &= \boldsymbol{\sigma}\cdot(\hat{\boldsymbol{p}} - e\boldsymbol{A})\phi + eA_0\chi - m\chi \quad . \end{aligned} \tag{1}$$

The stationary solutions for a constant purely magnetic field ($A_0 = 0$, $\boldsymbol{A}$ independent of time) are obtained from

$$\begin{aligned} (\varepsilon - m)\phi &= \boldsymbol{\sigma}\cdot(\hat{\boldsymbol{p}} - e\boldsymbol{A})\chi \quad , \\ (\varepsilon + m)\chi &= \boldsymbol{\sigma}\cdot(\hat{\boldsymbol{p}} - e\boldsymbol{A})\phi \quad . \end{aligned} \tag{2}$$

We multiply the first equation by $(\varepsilon + m)$ and eliminate χ,

$$(\varepsilon^2 - m^2)\phi = \boldsymbol{\sigma}\cdot(\hat{\boldsymbol{p}} - e\boldsymbol{A})\,\boldsymbol{\sigma}\cdot(\hat{\boldsymbol{p}} - e\boldsymbol{A})\phi \quad . \tag{3}$$

Next we use the identity

$$(\boldsymbol{\sigma}\cdot\boldsymbol{a})(\boldsymbol{\sigma}\cdot\boldsymbol{b}) = \boldsymbol{a}\cdot\boldsymbol{b} + \mathrm{i}\boldsymbol{\sigma}\cdot\boldsymbol{a}\times\boldsymbol{b} \tag{4}$$

and the gradient property of the momentum operator $\hat{\boldsymbol{p}} = -\mathrm{i}\nabla$:

$$\begin{aligned} (\varepsilon^2 - m^2)\phi &= \left[(\hat{\boldsymbol{p}} - e\boldsymbol{A})^2 + \mathrm{i}\boldsymbol{\sigma}\cdot(\hat{\boldsymbol{p}} - e\boldsymbol{A})\times(\hat{\boldsymbol{p}} - e\boldsymbol{A})\right]\phi \\ &= \left[(\hat{\boldsymbol{p}} - e\boldsymbol{A})^2 - e\boldsymbol{\sigma}\cdot\boldsymbol{H}\right]\phi \\ &= \left[\hat{\boldsymbol{p}}^2 - 2e\boldsymbol{A}\cdot\hat{\boldsymbol{p}} + e^2\boldsymbol{A}^2 - e\,\boldsymbol{\sigma}\cdot\boldsymbol{H}\right]\phi \\ &= \left[\hat{\boldsymbol{p}}^2 + e^2H^2x^2 - eH(\sigma_z + 2x\hat{p}_y)\right]\phi \quad . \end{aligned} \tag{5}$$

The vector potential was chosen to be $\boldsymbol{A} = (0, Hx, 0)$ in the last transformation, and $\nabla\cdot\boldsymbol{A} = 0$ and $\boldsymbol{H} = \nabla\times\boldsymbol{A}$ have been used. We notice that the right-hand side of (5) obviously commutes with the components of the momentum operator $\hat{p}_y$ and $\hat{p}_z$. Consequently the ansatz

Exercise 7.6.

$$\phi_\sigma(\boldsymbol{x}) = \mathrm{e}^{\mathrm{i}(p_y y + p_z z)} f(x)\chi_\sigma \tag{6}$$

presents itself where χ_σ is the unit spinor. Insertion into (5) immediately yields

$$(\varepsilon^2 - m^2)f(x) = \left(-\frac{d^2}{dx^2} + p_y^2 + p_z^2 + e^2H^2x^2 - 2eHxp_y - eH\sigma\right)f(x) \quad , \tag{7}$$

which can be written as

$$\left[-\frac{d^2}{dx^2} + e^2H^2\left(x - \frac{p_y}{eH}\right)^2\right]f(x) = (\varepsilon^2 - m^2 - p_z^2 + eH\sigma)f(x) \quad . \tag{8}$$

This is just the Schrödinger equation of the harmonic oscillator in the variable $\xi = x - p_y/eH$. The "oscillator energy" is $\hbar\omega = 2|e|H$. The eigenvalues thus are $\lambda_n = (n + 1/2)h\omega = (2n+1)|e|H$. Hence

$$\varepsilon^2 - m^2 - p_z^2 + eH\sigma = (2n+1)|e|H \quad ,$$

or

$$\varepsilon_{p\sigma} = \pm\sqrt{m^2 + p_z^2 + |e|H(2n + 1 + \sigma)} \quad . \tag{9}$$

This is the relativistic generalization of the *Landau levels* of a particle in a magnetic field.

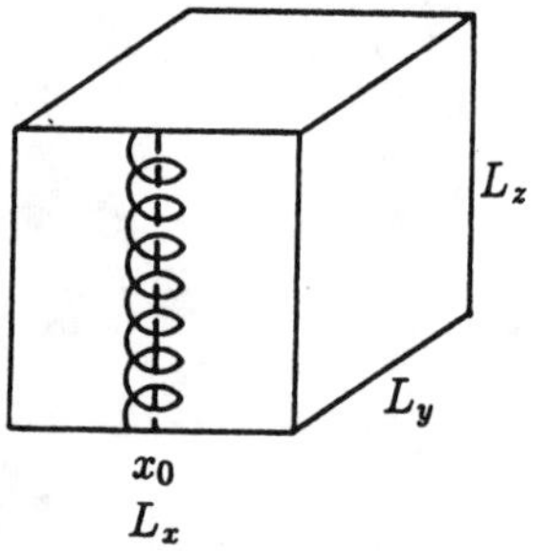

Fig. 7.27. Normalization box

In order to determine the density of states we note that the energy levels (9) are infinitely degenerate since the momentum p_y does not appear in the formula. In the classical framework our solution describes helical motion of the electron with freely chosen momentum components in y and z direction but orbiting around a fixed centre,

$$x_0 = \frac{p_y}{eH} \quad . \tag{10}$$

If we put the particle into a box with dimensions L_x, L_y, L_z (Fig. 7.27), the y and z motions are quantized by the boundary conditions and the number of states reads ($\hbar = 1$)

$$\Delta N = \frac{L_y}{2\pi}\Delta p_y \frac{L_z}{2\pi}\Delta p_z \quad . \tag{11}$$

$\Delta p_y = eH\Delta x_0$ holds because of (10). We sum over the allowed values $0 < x_0 < L_x$ and obtain, as conjectured above,

$$\Delta N = \frac{L_y}{2\pi}|e|HL_x\frac{L_z}{2\pi}\Delta p_z = \frac{|e|H}{(2\pi)^2}\Delta p_z V \quad . \tag{12}$$

7.4 Biographical Notes

DELBRÜCK, Max, German-American physicist and biologist. *4. 9. 1906 in Berlin, †9. 3. 1981 in Pasadena. D. studied physics in Tübingen, Berlin, and Göttingen (1926–29) and was a postdoctoral fellow at Copenhagen with Niels Bohr. Beginning in 1932 he was at the Kaiser Wilhelm Institute for Chemistry in Berlin, working as a theoretician with L. Meitner and O. Hahn. At this time he published a short addendum to an experimental paper on the coherent scattering of hard γ rays where he pointed out the possible interaction of photons with the vacuum polarization charge induced by a nucleus. The existence of this effect was demonstrated much later and called Delbrück scattering by H. Bethe. In 1938 D. left theoretical physics and turned to biology. He worked on genetics and bacterial viruses at Caltech where he became professor of biology in 1947, after having been on the faculty of Vanderbilt University (Nashville) for several years. D. made fundamental contributions to the the mechanisms of genetic recombination and to fields of bacterial and virus genetics in general. In 1969 he received the Nobel prize for medicine (with S. Luria and A. Hershey).

EULER, Hans, German physicist. *6. 10. 1909 in Meran, †23. 6. 1941. E. studied physics at the Universities of Göttingen, Frankfurt and Leipzig. In 1935 he obtained his PhD under the direction of W. Heisenberg. E. stayed a close collaborator of Heisenberg until 1940 when he joined the german air force. The airplane in which he served as a meteorological observer was shot down in 1941. In his PhD thesis E. calculated the QED process of scattering light by light. Subsequently he worked on the theory of high-energy cosmic ray collisions.

WEISSKOPF, Viktor Frederick, American physicist. *19. 9. 1908 in Vienna. W. studied physics at the Universities of Vienna and Göttingen where he received his PhD in (1931). W. worked at the Universities of Berlin, Copenhagen, Cambridge, Zurich (1933–36) and at the University at Rochester. Since 1945 he was professor of physics at the Massachusetts Institute of Technology (Cambridge). From 1961 to 1965 he served as director general of CERN. In 1934 W. formulated a relativistically covariant quantum field theory of bosons (with W. Pauli) and in 1936 introduced concepts which led to renormalization theory. He also contributed to nuclear physics where he introduced the concept of temperature to describe excited states (1937) and introduced the optical model for nuclear reactions (1950, with H. Feshbach). In 1956 W. received the Max Planck Medal of the German Physical Society.

8. Quantum Electrodynamics of Spinless Bosons

All considerations in this book so far have dealt with electrons (or muons) and their antiparticles, i.e. particles with spin $1/2$, which are described by the Dirac equation. The theory of quantum electrodynamics describes the interaction of these particles with each other and with the photon field in a very satisfying way. In principle, nothing prevents us from trying to construct a similar theory also for other kinds of particles. The simplest case is scalar (or pseudoscalar) bosons, which have spin 0. In the following we want to develop the theory of the electromagnetic interaction of these particles, which is also known as "scalar electrodynamics". We can take over many ingredients of the theory from Chapters 2 and 3 either directly or with only marginal modifications.

In contrast to the case of spinor electrodynamics, however, the importance of this theory is limited, because there are no elementary charged scalar particles in nature. The best candidates for this role are the pseudoscalar mesons, especially π and K. They are unstable and decay by virtue of weak interaction. For example, most of the charged pions decay through the channel

$$\pi^{\pm} \to \mu^{\pm} + \nu_{\mu} \quad ,$$

with a lifetime of about 2.6×10^{-8} s. Since this lifetime is very long on a natural time-scale, the pion can be considered stable with good approximation. The following problem is more basic: Unlike elementary leptons (e, μ, τ), which are point-like particles, pions have an internal structure. As is well known, they are now regarded as being composed of two quarks with spin $1/2$. In *Quantum Mechanics – Symmetries* by W. Greiner and B. Müller we discussed this fact extensively. Since these are subject to the strong interaction, there are massive effects of vacuum polarization, and the physically observed pion is a complex "cloud" of virtual particles. Even if one considers only the interaction with the electromagnetic field, we must not neglect this inner structure. Obviously scalar electrodynamics is completely inadequate for describing the coupling of mesons with each other, because this is dominated by the strong interaction. Nevertheless, in spite of the limited applicability it is very instructive to transfer the theory of quantum electrodynamics from the spinor to the scalar case.

8.1 The Klein–Gordon Equation

First, let us recall some properties of the relativistic wave equation for spin-0 particles.[1] The wave function $\phi(x)$ for free scalar particles obeys the Klein–Gordon equation

$$(\partial^\mu \partial_\mu + m_0^2)\phi(x) = 0 \quad . \tag{8.1}$$

The related current-density vector is[2]

$$\begin{aligned} j_\mu &= \mathrm{i}e(\phi^* \partial_\mu \phi - \phi \partial_\mu \phi^*) \\ &\equiv \mathrm{i}e\phi^* \overleftrightarrow{\partial}_\mu \phi \quad , \end{aligned} \tag{8.2}$$

which satisfies the continuity equation. In contrast to the Dirac case, here the charge-density j_0 can adopt both signs. The complete system of solutions of (8.1) consists of two classes of plane waves with positive and negative eigenfrequencies and momenta,

$$\varphi_{\boldsymbol{p}}^{(+)}(x) = N_p \, \mathrm{e}^{-\mathrm{i}(E_p t - \boldsymbol{p}\cdot\boldsymbol{x})} \tag{8.3a}$$

and

$$\varphi_{\boldsymbol{p}}^{(-)}(x) = N_p \, \mathrm{e}^{+\mathrm{i}(E_p t - \boldsymbol{p}\cdot\boldsymbol{x})} \quad , \tag{8.3b}$$

where

$$E_p = +\sqrt{m_0^2 + \boldsymbol{p}^2} \quad . \tag{8.4}$$

The normalization can be determined via the total charge $Q = \int d^3x \, j_0(\boldsymbol{x}, t)$, which is a conserved quantity. The charge of the solutions with positive frequency is fixed to the value $Q = +e$, whereas we require $Q = -e$ for the solutions with negative frequency. Thus, in general, a Klein–Gordon equation describes a superposition of particle and antiparticle contributions with positive and negative charges, e.g. π^- and π^+ mesons. (Neutral scalar particles will not be considered here.) Later on we shall return to the physical interpretation of these contributions.

In the case of the continuum normalization "to a δ function" one has to choose

$$N_p = \sqrt{\frac{1}{2E_p(2\pi)^3}} \quad , \tag{8.5}$$

and the normalization condition for the plane waves reads

$$\int d^3x \, \varphi_{\boldsymbol{p}'}^{(\pm)*}(x) \, \mathrm{i} \, \overleftrightarrow{\partial}_0 \, \varphi_{\boldsymbol{p}}^{(\pm)}(x) = \pm\delta^3(\boldsymbol{p}' - \boldsymbol{p}) \quad , \tag{8.6a}$$

$$\int d^3x \, \varphi_{\boldsymbol{p}'}^{(\pm)*}(x) \, \mathrm{i} \, \overleftrightarrow{\partial}_0 \, \varphi_{\boldsymbol{p}}^{(\mp)}(x) = 0 \quad . \tag{8.6b}$$

[1] (See W. Greiner: Theoretical Physics, Vol. 3, *Relativistic Quantum Mechanics – Wave Equations* (Springer, Berlin, Heidelberg 1990), Chap. 1.

[2] Here we use a notation which is a bit different from that used in *Relativistic Quantum Mechanics*. There the analog of (8.2) contains an additional factor $1/2m_0$, which is compensated by a different normalization of the wave function.

In general, (8.2) defines a modified "scalar product" $(\phi_2|\phi_1)$ in the following way:

$$(\phi_2|\phi_1) := \int d^3x\, \phi_2^*(x)\, \mathrm{i}\, \overleftrightarrow{\partial}_0\, \phi_1(x) \quad , \tag{8.7}$$

for which the positivity condition $(\phi|\phi) \geq 0$ is no longer valid.

The interaction with the electromagnetic field is introduced as usual by the *minimal coupling* prescription

$$\hat{p}_\mu \to \hat{p}_\mu - eA_\mu \quad .$$

The Klein–Gordon equation in the presence of an electromagnetic field then reads

$$\left[(\hat{p}^\mu - eA^\mu)(\hat{p}_\mu - eA_\mu) - m_0^2\right] \phi(x) = 0 \tag{8.8}$$

or

$$\left[\partial^\mu \partial_\mu + m_0^2\right] \phi(x) = -\hat{V}\, \phi(x) \quad . \tag{8.9}$$

Here we have formally introduced the potential operator $\hat{V}$. Explicitly it is given by

$$\hat{V}\phi = \mathrm{i}e(\partial_\mu A^\mu + A^\mu \partial_\mu)\phi - e^2 A^\mu A_\mu \phi \quad . \tag{8.10}$$

As the Klein–Gordon equation is of second order in the coordinates, the coupling term in (8.9) has a quite complicated structure: It contains gradients ∂_μ and moreover is *nonlinear* in A_μ because of the quadratic last term.

In addition the current density of the scalar field is modified by the presence of an electromagnetic potential, namely

$$j_\mu = \mathrm{i}e\phi^*\, \overleftrightarrow{\partial}_\mu\, \phi - 2e^2 A_\mu \phi^* \phi \quad . \tag{8.11}$$

Equation (8.9) will be used to calculate the scattering of a scalar particle at a given potential A_μ. Our experience with spinor QED suggests the development of a perturbation theory based on the use of the propagator of the free Klein–Gordon equation.

8.2 The Feynman Propagator for Scalar Particles

In complete analogy to our considerations in Chap. 2 for the Dirac equation we define a propagator $\Delta_\mathrm{F}(x'-x)$ that solves the Klein–Gordon equation with a point-like unit source:

$$(\partial'^\mu \partial'_\mu + m_0^2)\Delta_\mathrm{F}(x'-x) = -\delta^4(x'-x) \quad . \tag{8.12}$$

The choice of the minus sign of the source term on the right-hand side is natural since the interaction term in (8.9) also has a negative sign. A solution of this differential equation can be obtained, as usual, by Fourier transformation

$$\Delta_\mathrm{F}(x'-x) = \int \frac{d^4p}{(2\pi)^4}\, \mathrm{e}^{-\mathrm{i}p\cdot(x'-x)} \frac{1}{p^2 - m_0^2 + \mathrm{i}\varepsilon} \quad , \tag{8.13a}$$

i.e., in momentum space,

$$\Delta_{\rm F}(p) = \frac{1}{p^2 - m_0^2 + {\rm i}\varepsilon} \quad . \tag{8.13b}$$

In this we have again used the Stückelberg–Feynman prescription for circumventing the poles at $p^2 = m_0^2$. We shall soon convince ourselves that this means that the wave functions with positive frequency $+E_p$ propagate *forward in time*, while those with negative frequency $-E_p$ propagate *backward in time*. The relation to the spin-1/2 Feynman propagator $S_{\rm F}$ is very simple, namely

$$S_{\rm F}(x' - x) = ({\rm i}\gamma_\mu \partial^\mu - m_0)\Delta_{\rm F}(x' - x) \quad . \tag{8.14}$$

In order to investigate the action of the propagator on a wave function we first express $\Delta_{\rm F}(x' - x)$ as a sum over the complete set of solutions (8.3). For this purpose the p_0-integration in (8.13) is performed with the help of the theorem of residues, in the course of which the contour for $t' > t \quad (t' < t)$ is to be closed in the lower (upper) half-plane. This calculation proceeds exactly like that at the beginning of the second chapter for the Feynman propagator of Dirac particles, and one obtains

$$\begin{aligned}\Delta_{\rm F}(x' - x) &= \int \frac{d^3p}{(2\pi)^3}\, {\rm e}^{{\rm i}\boldsymbol{p}\cdot(\boldsymbol{x}' - \boldsymbol{x})} \int \frac{dp_0}{2\pi} \frac{{\rm e}^{-{\rm i}p_0(t'-t)}}{p_0^2 - E_p^2 + {\rm i}\epsilon} \\ &= -{\rm i} \int \frac{d^3p}{(2\pi)^3} \frac{{\rm e}^{{\rm i}\boldsymbol{p}\cdot(\boldsymbol{x}' - \boldsymbol{x})}}{2E_p} \left[\Theta(t' - t){\rm e}^{-{\rm i}E_p(t'-t)} + \Theta(t - t'){\rm e}^{+{\rm i}E_p(t'-t)}\right] .\end{aligned} \tag{8.15}$$

Using (8.3) this can be written as

$$\begin{aligned}\Delta_{\rm F}(x' - x) = &-{\rm i}\Theta(t' - t) \int d^3p\, \varphi_{\boldsymbol{p}}^{(+)}(x')\varphi_{\boldsymbol{p}}^{(+)*}(x) \\ &- {\rm i}\Theta(t - t') \int d^3p\, \varphi_{\boldsymbol{p}}^{(-)}(x')\varphi_{\boldsymbol{p}}^{(-)*}(x) \quad ,\end{aligned} \tag{8.16}$$

where we have substituted $\boldsymbol{p} \to -\boldsymbol{p}$ in the second integral.

Now we consider a general wave function $\phi(x)$ composed of contributions with positive and negative frequency,

$$\phi(x) = \int d^3p\, a_{\boldsymbol{p}}\varphi_{\boldsymbol{p}}^{(+)}(x) + \int d^3p\, b_{\boldsymbol{p}}\varphi_{\boldsymbol{p}}^{(-)}(x) \equiv \phi^{(+)}(x) + \phi^{(-)}(x) \quad , \tag{8.17}$$

and apply the propagator $\Delta_{\rm F}(x' - x)$. To make use of the modified scalar product (8.7) the operator ${\rm i}\,\overleftrightarrow{\partial_0}$ is sandwiched between the propagator and the wave function

$$\begin{aligned}&\int d^3x\, \Delta_{\rm F}(x' - x)\, {\rm i}\, \overleftrightarrow{\partial_0}\, \phi(x) \\ &= -{\rm i} \int d^3p \left(\Theta(t' - t)\varphi_{\boldsymbol{p}}^{(+)}(x') \int d^3x\, \varphi_{\boldsymbol{p}}^{(+)*}(x)\, {\rm i}\, \overleftrightarrow{\partial_0}\, \phi(x)\right. \\ &\quad \left. + \,\Theta(t - t')\, \varphi_{\boldsymbol{p}}^{(-)}(x') \int d^3x\, \varphi_{\boldsymbol{p}}^{(-)*}(x)\, {\rm i}\, \overleftrightarrow{\partial_0}\, \phi(x)\right) \quad .\end{aligned} \tag{8.18}$$

If now $\phi(x)$ is decomposed into components according to (8.17), then the orthonormalization relations (8.6a,b) can be applied. As we had conjectured, the result is

$$\int d^3x\, \Delta_F(x'-x)\, \mathrm{i}\, \overleftrightarrow{\partial_0}\, \phi(x) = -\mathrm{i}\Theta(t'-t)\phi^{(+)}(x') + \mathrm{i}\Theta(t-t')\phi^{(-)}(x') \,. \quad (8.19)$$

This result is completely analogous to (2.25) and (2.26) of Chap. 2, which were derived for Dirac spinors.

8.3 The Scattering of Spin-0 Bosons

Along the lines of Chap. 3 we can also treat scattering processes of spin-0 bosons with the help of the propagator. Let us represent $\Delta_F(x'-x)$ by a dashed line between x and x'. In Fig. 8.1a, for example, a free boson runs from the point $(\boldsymbol{x}, t)$ to the point $(\boldsymbol{x}', t')$ with $t' > t$. It is a particle with positive energy and thus charge $Q = +e$. For the sake of simplicity we shall speak of a pion (that is to say π^- because of the convention $e = -|e|$). Under the influence of the perturbation potential $\hat{V}$ the particle can be scattered at $(\boldsymbol{x}_1, t_1)$ (Fig. 8.1b). It can also be scattered several times Fig. 8.1c. However, in the latter case the propagator $\Delta_F(x_2 - x_1)$ also permits the time ordering $t_2 < t_1$, which means that the π^- propagates *backward in time* from x_1 to x_2, as in Fig. 8.2a. But physically a particle π^-, that is emitted backward in time at the point x_1 renders the same effect as if an antiparticle π^+ were absorbed. Hence one can also draw the Feynman graph in the form of Fig. 8.2b so that particles with both signs of charge that all propagate forward in time occur. In this language the process in Fig. 8.2 then means that a π^- comes in and then at the point $(\boldsymbol{x}_2, t_2)$ a $\pi^-\pi^+$ pair is created; later on, the π^+ annihilates with the incoming π^- at $(\boldsymbol{x}_1, t_1)$, whereas the other π^- keeps propagating into the future.

This stands in complete analogy to the case of spinor electrodynamics; one just has to change the names $\pi^\pm$ and $e^\pm$. Nevertheless, there is one difference: As the bosons do not obey Pauli's exclusion principle, *one cannot use Dirac's hole picture*; it does not make sense to interpret the π^+ as a vacancy in a sea of π^- particles whose other states are completely occupied.

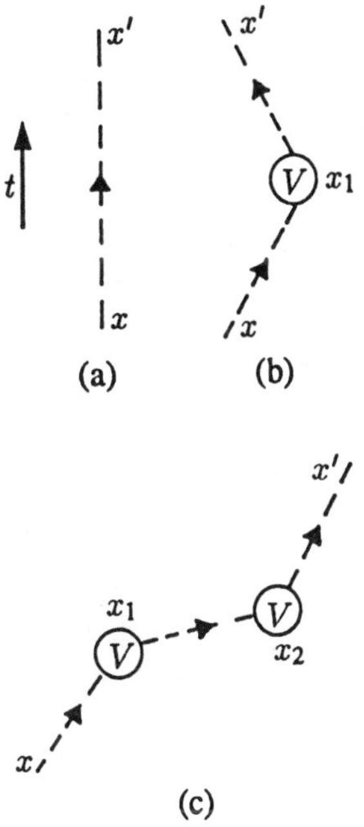

Fig. 8.1a–c. Free propagation and scattering of a boson

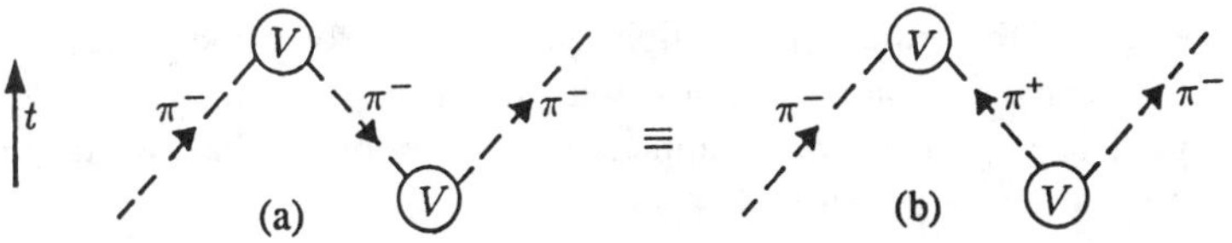

Fig. 8.2a,b. A π^- running backward in time is a π^+

Therefore, without having recourse to this auxiliary concept, we formulate Feynman's interpretation as a postulate for all particles (both bosons and fermions): *The emission (absorption) of a particle with 4-momentum p^μ is physically equivalent to the absorption (emission) of an antiparticle with 4-momentum $-p^\mu$.*

Now we give a quantitative formulation of these considerations and calculate the S matrix for scattering processes. For this purpose the Klein–Gordon equation

(8.9) for a wave function $\phi(x)$ with the perturbation "potential" $\hat{V}(x)$ given by (8.10) is to be solved. Using the Feynman propagator from (8.12) we get

$$\phi(x) = \varphi(x) + \int d^4y\, \Delta_F(x-y)\hat{V}(y)\phi(y) \quad , \tag{8.20}$$

where $\varphi(x)$ stands for a free wave satisfying $(\partial_\mu\partial^\mu + m_0^2)\varphi(x) = 0$. In the case of *particle scattering* $\boldsymbol{p}_i \to \boldsymbol{p}_f$ the wave function ϕ must satisfy the boundary condition $\phi(\boldsymbol{x},t) \to \varphi_{p_i}^{(+)}(\boldsymbol{x},t)$ for $t \to -\infty$. The S matrix results from a projection on the final state $\varphi_{p_f}^{(+)}(\boldsymbol{x},t)$ for $t \to +\infty$:

$$\begin{aligned} S_{fi} &= \lim_{t\to+\infty} (\varphi_{p_f}^{(+)}(x)|\phi_{p_i}(x)) \\ &= \lim_{t\to+\infty} \int d^3x\, \varphi_{p_f}^{(+)*}(\boldsymbol{x},t)\,\mathrm{i}\,\overleftrightarrow{\partial_0}\,\phi_{p_i}(\boldsymbol{x},t) \quad . \end{aligned} \tag{8.21}$$

With the representation (8.16) of the Feynman propagator and (8.20) we get for this expression

$$\begin{aligned} S_{fi} &= \lim_{t\to\infty} \int d^3x\, \varphi_{p_f}^{(+)*}(\boldsymbol{x},t)\,\mathrm{i}\,\overleftrightarrow{\partial_0} \left(\varphi_{p_i}(\boldsymbol{x},t) + \int d^4y\, \Delta_F(x-y)\hat{V}(y)\phi_{p_i}(y)\right) \\ &= \lim_{t\to\infty} \int d^3x\, \varphi_{p_f}^{(+)*}(\boldsymbol{x},t)\,\mathrm{i}\,\overleftrightarrow{\partial_0}\,\varphi_{p_i}(\boldsymbol{x},t) \\ &\quad + \lim_{t\to\infty} \int d^3x\, \varphi_{p_f}^{(+)*}(\boldsymbol{x},t)\,\mathrm{i}\,\overleftrightarrow{\partial_0} \\ &\quad \times \int d^4y\, d^3p\, (-\mathrm{i})\Theta(t-t_y)\varphi_{\boldsymbol{p}}^{(+)}(\boldsymbol{x},t)\varphi_{\boldsymbol{p}}^{(+)*}(y)\hat{V}(y)\phi_{p_i}(y) \\ &= \lim_{t\to\infty} \int d^3x\, \varphi_{p_f}^{(+)*}(\boldsymbol{x},t)\,\mathrm{i}\,\overleftrightarrow{\partial_0}\,\varphi_{p_i}^{(+)}(\boldsymbol{x},t) \\ &\quad - \mathrm{i} \lim_{t\to\infty} \int d^3p\, \underbrace{\left(\int d^3x\, \varphi_{p_f}^{(+)*}(x)\,\mathrm{i}\,\overleftrightarrow{\partial_0}\,\varphi_{\boldsymbol{p}}^{(+)}(x)\right)}_{=+\delta^3(p_f-p)} \int d^4y\, \varphi_{\boldsymbol{p}}^{(+)*}(y)\hat{V}(y)\phi_{p_i}(y)\,, \end{aligned} \tag{8.22}$$

which leads to

$$S_{fi} = \delta^3(\boldsymbol{p}_f - \boldsymbol{p}_i) - \mathrm{i}\int d^4y\, \varphi_{p_f}^{(+)*}(y)\hat{V}(y)\phi_{p_i}(y) \tag{8.23}$$

because of the orthonormalization relation (8.6) for plane waves.

Whenever we are concerned with *scattering of antiparticles* $\boldsymbol{p}_i \to \boldsymbol{p}_f$, we let a particle with negative 4-momentum $-p_f$ propagate backward into the past and project at $t \to -\infty$ on $\varphi_{p_i}^{(-)}(\boldsymbol{x},t)$:

$$\begin{aligned} S_{fi} &= -\lim_{t\to-\infty} \left(\varphi_{p_i}^{(-)}(x)|\phi_{p_f}(x)\right) \\ &= -\lim_{t\to-\infty} \int d^3x\, \varphi_{p_i}^{(-)*}(\boldsymbol{x},t)\,\mathrm{i}\,\overleftrightarrow{\partial_0}\,\phi_{p_f}(\boldsymbol{x},t) \quad , \end{aligned} \tag{8.24}$$

where the minus sign is only a convention. Here it becomes clear that – from a purely calculational point of view – the antiparticle enters the interaction region with momentum p_f and leaves with momentum p_i. So we obtain

$$S_{fi} = \delta^3(\boldsymbol{p}_f - \boldsymbol{p}_i) - \mathrm{i} \int d^4y \, \varphi^{(-)*}_{p_i}(y) \hat{V}(y) \phi_{p_f}(y) \quad . \tag{8.25}$$

Pair annihilation and pair creation can be calculated in a completely analogous way.

The expressions (8.23) and (8.25) together with the integral equation (8.20) for $\phi(y)$ again suggest a *perturbation expansion*. For this one proceeds in an iterative way: in the nth step of the iteration one inserts the expression for the wave function $\phi^{(n)}$ into the right-hand side of (8.20) in order to calculate the $(n+1)$th approximation $\phi^{(n+1)}$. For particle scattering the (formal) perturbation series

$$S_{fi} = \sum_{n=1}^{\infty} S^{(n)}_{fi} \tag{8.26a}$$

$$= \sum_{n=1}^{\infty} \int d^4x_n \ldots \int d^4x_1 \varphi^{(+)*}_{p_f}(x_n)$$
$$\times (-\mathrm{i})\hat{V}(x_n)\mathrm{i}\Delta_{\mathrm{F}}(x_n - x_{n-1})(-\mathrm{i})\hat{V}(x_{n-1}) \ldots \varphi^{(+)}_{p_i}(x_1) \quad . \tag{8.26b}$$

is obtained. However, here an important difference to the analogous formula of the spin-1/2 theory arises. Whereas there the expansion (8.26a) had the form of a power series in the charge e, here this is no longer the case: every term $S^{(n)}_{fi}$ contains a *mixture of powers* between e^n and e^{2n}. If one wants to keep an expansion in the coupling strength e, which is the reasonable way to do perturbation theory, the various contributions have to be rearranged and grouped together. The reason for this can be found in the form of the interaction potential $\hat{V}$ from (8.10), which contains both a linear term eA and a quadratic term e^2A^2.

This feature of $\hat{V}$ also has an important consequence whenever the terms of the perturbation series are represented by Feynman graphs. In scalar electrodynamics there are *two kinds of photon vertex*, as represented in Fig. 8.3. According to the four-leg vertex (Fig. 8.3b) (also known as the "seagull vertex" because of its shape) the boson can absorb and/or emit two photons simultaneously. Hence in contrast to spinor QED there is now a richer "zoology" of Feynman graphs contributing to a given process. However, the two-photon vertex contains the factor e^2; so if a calculation in lowest order suffices, its contribution can, circumstances permitting, be left out (for an example where this is not true see Exercise 8.2!).

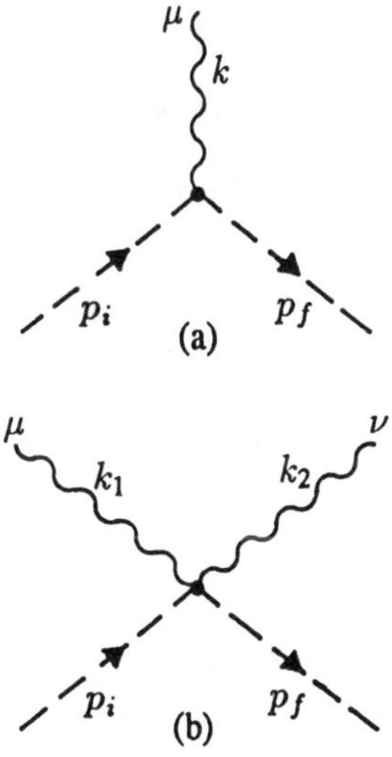

Fig. 8.3a,b. The two vertices of the coupling between photon and spin-0 boson

Let us now investigate which mathematical factors are assigned to the vertices when evaluating Feynman graphs. For this purpose we construct an S matrix element of first order (according to the expansion (8.26)):

$$S^{(1)}_{fi} = \int d^4x \, \varphi^{(+)*}_f(x)(-\mathrm{i})\hat{V}(x)\varphi^{(+)}_i(x)$$
$$= \int \frac{d^4x}{(2\pi)^3} \frac{1}{\sqrt{2E_f 2E_i}} \, \mathrm{e}^{+\mathrm{i}p_f \cdot x} \left[e(\partial^\mu A_\mu + A_\mu \partial^\mu) + \mathrm{i}e^2 g_{\mu\nu} A^\mu A^\nu \right] \mathrm{e}^{-\mathrm{i}p_i \cdot x}$$
$$\equiv S^{(1a)}_{fi} + S^{(1b)}_{fi} \quad . \tag{8.27}$$

The contribution that is linear in e can be simplified by a partial integration:

$$
\begin{aligned}
S_{fi}^{(1a)} &= \frac{1}{\sqrt{2E_f 2E_i (2\pi)^3}} \int d^4x \, e^{ip_f\cdot x} e(\partial^\mu A_\mu + A_\mu \partial^\mu) e^{-ip_i\cdot x} \\
&= \frac{1}{\sqrt{2E_f 2E_i (2\pi)^3}} \int d^4x \, e \left[-(ip_f^\mu) + (-ip_i^\mu)\right] A_\mu(x) e^{i(p_f - p_i)\cdot x} \\
&= \left[(-ie)(p_f^\mu + p_i^\mu)\right] \frac{1}{\sqrt{2E_f 2E_i (2\pi)^3}} A_\mu(p_f - p_i) \quad . \qquad (8.28)
\end{aligned}
$$

In the second line a surface integral at infinity, arising from the divergence $\partial^\mu \left(e^{i(p_f - p_i)x} A_\mu(x)\right)$, was dropped. Therefore the *one-photon vertex* in Fig. 8.3a is assigned a factor $-ie(p_f^\mu + p_i^\mu)$. This expression plays the same role as the factor $-ie\gamma^\mu$ at the electron-photon-vertex. In addition the S-matrix element contains a factor originating from the normalization of the plane waves and the Fourier-transformed electromagnetic potential $A_\mu(q)$ at a momentum transfer $q = p_f - p_i$. The quadratic coupling in (8.27) leads to

$$
S_{fi}^{(1b)} = (ie^2 g_{\mu\nu}) \frac{1}{\sqrt{2E_f 2E_i (2\pi)^3}} \int d^4x \, e^{i(p_f - p_i)\cdot x} A^\mu(x) A^\nu(x) \quad . \qquad (8.29)
$$

The Fourier transform of the product of vector potentials can be changed into a convolution integral in momentum space:

$$
S_{fi}^{(1b)} = (ie^2 g_{\mu\nu}) \frac{1}{\sqrt{2E_f 2E_i (2\pi)^3}} \int \frac{d^4k}{(2\pi)^4} A^\mu(q-k) A^\nu(k) \quad . \qquad (8.30)
$$

One has to be careful with the interpretation of (8.29) and (8.30) because at this vertex there are *two separate photons* to be emitted/absorbed. Thus the A^μ field contains two parts, let us call them $A^\mu(1)$ and $A^\mu(2)$, and if one takes the square, only the mixed terms are to be considered:

$$
\begin{aligned}
A^\mu A_\mu &= (A^\mu(1) + A^\mu(2))(A_\mu(1) + A_\mu(2)) \\
&\to A^\mu(1) A_\mu(2) + A^\mu(2) A_\mu(1) = 2A^\mu(1) A_\mu(2) \quad , \qquad (8.31)
\end{aligned}
$$

because a single photon cannot interact twice. This reasoning leads to an additional factor of 2. So one has to assign a factor $2ie^2 g_{\mu\nu}$ to the *two-photon vertex* from Fig. 8.3b. It is multiplied with the familiar normalization factor and with the product of the two photon fields.

The necessity for the additional factor 2 becomes even more obvious if one draws all possible Feynman graphs for a process of second order in e, e.g. in the case of pair annihilation. As one can see in Fig. 8.4, two photons are emitted. Since these are indistinguishable Bose particles, the final state is to be symmetrized. In the case of Fig. 8.4a with two separated emissions this leads to the exchange graph Fig 8.4c. Also in the case of a two-photon vertex the two photons can be related in different ways. But as the exchange graph Fig. 8.4d has the same structure as Fig. 8.4b, it is simply accounted for by doubling the vertex factor to a value of $2ie^2 g_{\mu\nu}$.

Nevertheless, this rule, which can be generalized to arbitrary processes, has one exception: graphs that contain a *closed photon loop* as in Fig. 8.5. Here a symmetrization has to be performed, too, which yields a factor 2. But if one uses $2ie^2 g_{\mu\nu}$ for both vertices, one will have double counted. Thus, to correct this, every closed photon loop has to be multiplied by the factor $1/2$.

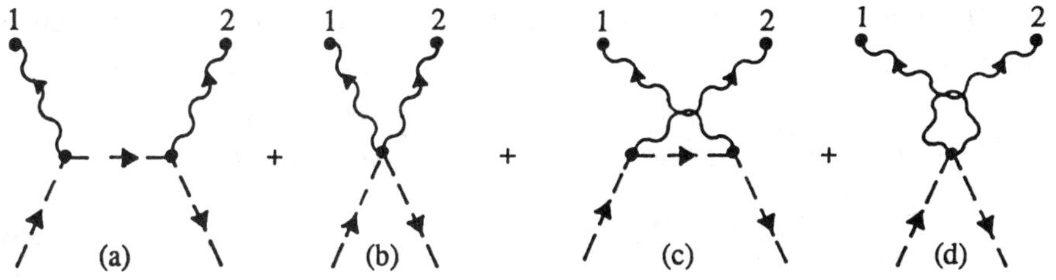

Fig. 8.4a–d. The graphs of second order for pair annihilation **(b)** and **(d)** have the same value

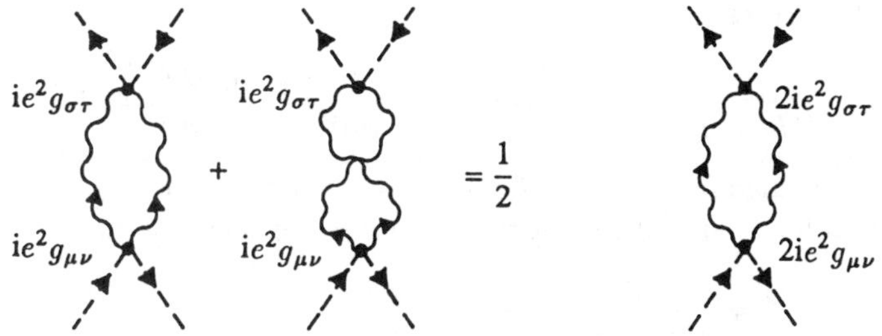

Fig. 8.5. Symmetrization for a graph with a closed photon loop

The electromagnetic field A_μ in (8.27) can also have the transition current j_{fi} of a spin-0 boson as a source:

$$A^\mu(x) = \int d^4y\, D_\mathrm{F}(x-y) j^\mu_{fi}(y) \quad . \tag{8.32}$$

Again here two terms of order e and e^2 occur in the transition current according to (8.11), which corresponds to the graphs in Fig. 8.3. This must be so, because the result has to be symmetric under the exchange of absorption and emission of virtual photons.

Except for the modified vertex factors and propagators, all further steps in the calculation of cross sections proceed as in Chap. 3. Compared with the spin-1/2 theory, the calculations are considerably simplified, because we no longer have to sum over particle polarizations, and we do not need the algebra of γ matrices.

Let us now consider the case of the scattering of a (structureless) pion at a time-independent *external Coulomb potential.* According to Section 3.1 the cross section is given by

$$d\sigma = \frac{1}{j_\mathrm{in}} \frac{|S_{fi}|^2}{T} \frac{Vd^3p_f}{(2\pi)^3} \quad . \tag{8.33}$$

The S-matrix element in *first order* was already given in (8.27). For the Fourier-transformed Coulomb potential we have

$$A_\mu(q) = \int d^4x\, \mathrm{e}^{\mathrm{i}q\cdot x} \frac{-Ze}{|\boldsymbol{x}|} g_{\mu 0} = 2\pi\delta(q_0) \frac{-Ze4\pi}{|\boldsymbol{q}|^2} g_{\mu 0} \quad . \tag{8.34}$$

If one uses a normalization of one particle per box with a finite volume V instead of (8.5), i.e. $N_p = 1/\sqrt{2E_pV}$, the incoming (particle) current is given by

$$j_\mathrm{in} = -\varphi^{(+)}_{p_i} \mathrm{i} \overleftrightarrow{\nabla} \varphi^{(+)}_{p_i} = -\mathrm{i}\frac{1}{2E_iV}(2\mathrm{i}p) = \frac{1}{V}\frac{\boldsymbol{p}_i}{E_i} = \frac{\boldsymbol{v}_i}{V} \quad . \tag{8.35}$$

With the usual replacement $\left(2\pi\delta(E_f - E_i)\right)^2 \to T\, 2\pi\delta(E_f - E_i)$ the cross section reads

$$d\sigma = \frac{V}{|\boldsymbol{v}_i|}\frac{1}{T}\left|(-\mathrm{i}e)\frac{(E_f + E_i)}{\sqrt{2E_f 2E_i V^2}} 2\pi\delta(E_f - E_i)\frac{-Ze4\pi}{|\boldsymbol{q}|^2}\right|^2 \frac{Vd^3p_f}{(2\pi)^3}$$

$$= \frac{4Z^2e^4}{|\boldsymbol{q}|^4}\frac{d^3p_f}{|\boldsymbol{v}_i|}\delta(E_f - E_i) \quad . \tag{8.36}$$

Thus we are led to the differential cross section for scattering into the solid angle element $d\Omega_f$:

$$\frac{d\sigma}{d\Omega_f} = \int \frac{|\boldsymbol{p}_f|^2 d|\boldsymbol{p}_f|}{|\boldsymbol{v}_i|}\delta(E_f - E_i)\frac{4Z^2\alpha^2}{|\boldsymbol{q}|^4} = \frac{4Z^2\alpha^2E^2}{|\boldsymbol{q}|^4} \quad . \tag{8.37}$$

With the momentum transfer $|\boldsymbol{q}|^2 = 4|\boldsymbol{p}|^2\sin^2\theta/2$, cf. (3.38), we can rewrite (8.37) as a function of the scattering angle θ in the centre-of-mass system:

$$\frac{d\sigma}{d\Omega_f} = \frac{Z^2\alpha^2}{4|\boldsymbol{p}|^2v^2\sin^4\theta/2} \quad . \tag{8.38}$$

Comparing this result with that of Section 3.1 one notices that the cross section for electron scattering contains an additional factor $(1 - \beta^2\sin^2\theta/2)$. This discrepancy is caused by the magnetic moment of the spin-1/2 particle; in the limit of low velocities both results agree since then the magnetic interaction is negligible.

In Exercises 8.1 and 8.2 two more processes, Compton scattering and pair production, will be calculated.

8.4 The Feynman Rules of Scalar Electrodynamics

The rules for calculating cross sections, which were compiled in Chap. 4, are to be extended in the following way, if spin-0 bosons are involved.

The cross section is calculated according to (4.3). The normalization factor for in- or outcoming bosons is

$$N_i = 1 \quad .$$

This corresponds to the normalization $\sqrt{1/(2E_iV)}$ of the external boson lines (normalization to one particle in a box with volume V). To calculate the invariant amplitude M_{fi} of a process the corresponding Feynman diagrams have to be drawn and translated into algebraic expressions with the help of the following set of rules.

The Feynman Rules for Spin-0 Particles.

1. The external lines

   a) incoming boson 

   b) outcoming boson 

   are assigned a factor 1.

2. Every internal boson line is assigned a factor

$$\mathrm{i}\Delta_{\mathrm{F}}(p) = \frac{\mathrm{i}}{p^2 - m_0^2 + \mathrm{i}\epsilon} \quad .$$

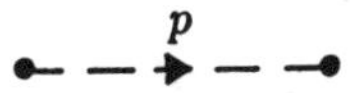

3. The vertices are described by the factors

   a) one-photon vertex: $-\mathrm{i}e(p'_\mu + p_\mu)$

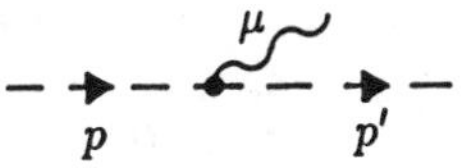

   b) two-photon vertex: $2\mathrm{i}e^2 g_{\mu\nu}$.

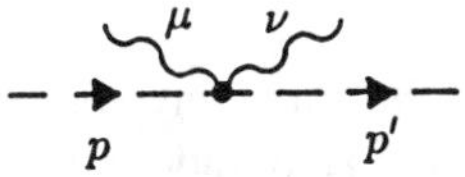

4. Every closed photon loop is assigned a factor 1/2.

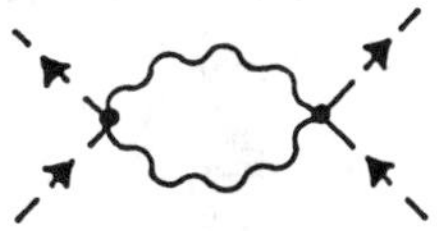

5. There are no extra factors −1.

Rule 5 is evident because the changing of boson lines always yields a factor +1 according to Bose statistics.

EXERCISE

8.1 Compton Scattering at Bosons

Problem. Calculate the cross section for photon scattering at spin-0 bosons in lowest order, in analogy to Section 3.7 Check the gauge invariance of the scattering amplitude.

Solution. In lowest order (e^2) there are three different graphs, which have to be added coherently, see Fig. 8.6. The invariant amplitude can be easily constructed according to the Feynman rules:

$$\begin{aligned} M_{fi}^{(a)} &= (-\mathrm{i}e)(p_f + p_f + k')^\mu \, \varepsilon_\mu^*(\boldsymbol{k}', \lambda')\mathrm{i}\Delta_{\mathrm{F}}(p_i + k) \\ &\quad \times (-\mathrm{i}e)(p_i + p_i + k)^\nu \varepsilon_\nu(\boldsymbol{k}, \lambda) \quad , \end{aligned} \tag{1a}$$

$$\begin{aligned} M_{fi}^{(b)} &= (-\mathrm{i}e)(p_f + p_f - k)^\nu \, \varepsilon_\nu(\boldsymbol{k}, \lambda)\mathrm{i}\Delta_{\mathrm{F}}(p_f - k) \\ &\quad \times (-\mathrm{i}e)(p_i + p_i - k')^\mu \varepsilon_\mu^*(\boldsymbol{k}', \lambda') \quad , \end{aligned} \tag{1b}$$

$$M_{fi}^{(c)} = 2\mathrm{i}e^2 g^{\mu\nu} \varepsilon_\mu^*(\boldsymbol{k}', \lambda')\varepsilon_\nu(\boldsymbol{k}, \lambda) \quad . \tag{1c}$$

The complete invariant amplitude can be written as

$$M_{fi} = \varepsilon_\mu^*(\boldsymbol{k}', \lambda') \, T^{\mu\nu} \, \varepsilon_\nu(\boldsymbol{k}, \lambda) \quad . \tag{2}$$

Here the *Compton tensor* $T^{\mu\nu}(p_i, p_f, k, k')$ has been introduced. It has the form

$$T^{\mu\nu} = -\mathrm{i}e^2 \left[\frac{(2p_f + k')^\mu (2p_i + k)^\nu}{(p_i + k)^2 - m_0^2} + \frac{(2p_i - k')^\mu (2p_f - k)^\nu}{(p_f - k)^2 - m_0^2} - 2g^{\mu\nu} \right] . \tag{3}$$

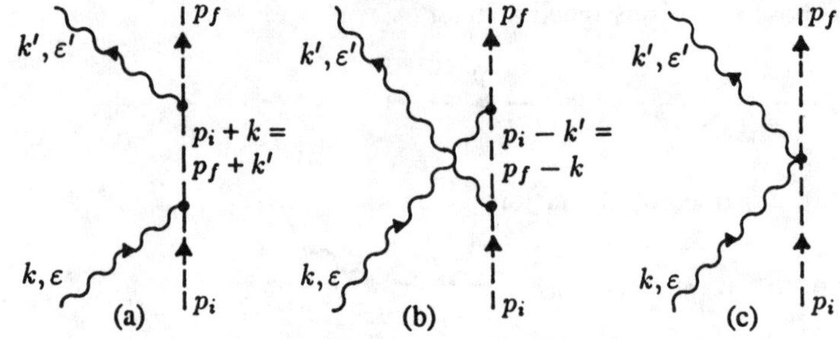

Fig. 8.6a–c. The three second-order Feynman diagrams for Compton scattering on a spinless boson

Let us first check the gauge invariance of this expression. As explained in Section 3.6, the value of the amplitude must not change when the potential is re-gauged, $\epsilon_\nu(k) \to \epsilon_\nu(k) + k_\nu \Lambda(k)$, i.e.

$$T^{\mu\nu}k_\nu = 0 \quad , \quad k_\mu T^{\mu\nu} = 0 \quad . \tag{4}$$

Now we rewrite the denominators in (3):

$$(p_i + k)^2 - m_0^2 = p_i^2 + 2k \cdot p_i + k^2 - m_0^2 = 2k \cdot p_i + k^2 \quad , \tag{5a}$$

$$(p_f - k)^2 - m_0^2 = p_f^2 - 2k \cdot p_f + k^2 - m_0^2 = -2k \cdot p_f + k^2 \quad . \tag{5b}$$

Hence we find that

$$\begin{aligned} T^{\mu\nu}k_\nu &= -\mathrm{i}e^2 \left[\frac{(2p_f + k')^\mu (2p_i \cdot k + k^2)}{(p_i + k)^2 - m_0^2} + \frac{(2p_i - k')^\mu (2p_f \cdot k - k^2)}{(p_f - k)^2 - m_0^2} - 2k^\mu \right] \\ &= -\mathrm{i}e^2[(2p_f + k')^\mu - (2p_i - k')^\mu - 2k^\mu] \\ &= -\mathrm{i}e^2 2(p_f + k' - p_i - k)^\mu = 0 \quad . \end{aligned} \tag{6}$$

In the same way one can prove (4b). Obviously the two-photon vertex is indispensible for satisfying the condition of gauge invariance in a given order e^n of the perturbation series.

According to the rules of Chap. 4 the cross section now reads

$$d\sigma = \frac{(4\pi)^2}{4\sqrt{(p_i \cdot k)^2 - 0}} (2\pi)^4 \delta^4(p_f + k' - p_i - k) |M_{fi}|^2 \frac{d^3p_f}{2E_f(2\pi)^3} \frac{d^3k'}{2\omega'(2\pi)^3} \quad , \tag{7}$$

where $\omega' = |\boldsymbol{k}'|$ is the photon energy. The somewhat tedious integration over the final momenta p_f and k' was performed in Section 3.7, (3.249), with the result

$$\int \frac{d^3p_f}{2E_f} \frac{d^3k'}{2\omega'} \delta^4(p_f + k' - p_i - k) = \frac{1}{2m_0} \int d\Omega_{k'} \frac{\omega'^2}{2\omega} \quad . \tag{8}$$

Thus the differential cross section is

$$\frac{d\sigma}{d\Omega_{k'}} = \frac{1}{|p_i \cdot k|} \frac{1}{2m_0} \frac{\omega'^2}{2\omega} |M_{fi}|^2 \quad . \tag{9}$$

Exercise 8.1.

In the *laboratory system*, where the particle is initially at rest, we have $p_i = (m_0, \mathbf{0})$ and therefore $p_i \cdot k = m_0\omega$. In order to calculate $|M_{fi}|^2$ we now choose, as in Section 3.7, the following *special gauge*:

$$\varepsilon^\mu(\boldsymbol{k}, \lambda) = \big(0, \varepsilon_\mu(\boldsymbol{k}, \lambda)\big) \quad \text{with} \quad \varepsilon(\boldsymbol{k}, \lambda) \cdot \boldsymbol{k} = 0 \quad , \tag{10a}$$

$$\varepsilon^\mu(\boldsymbol{k}', \lambda') = \big(0, \varepsilon_\mu(\boldsymbol{k}', \lambda')\big) \quad \text{with} \quad \varepsilon(\boldsymbol{k}', \lambda') \cdot \boldsymbol{k}' = 0 \quad . \tag{10b}$$

Because of the transversality of this gauge all terms with $\epsilon \cdot k$ and $\epsilon' \cdot k'$ vanish, and (2) is simplified to

$$\begin{aligned} M_{fi} &= -\mathrm{i}e^2 \left(\frac{2\epsilon'^* \cdot p_f \, 2\epsilon \cdot p_i}{2m_0\omega} + \frac{2\epsilon'^* \cdot p_i \, 2\epsilon \cdot p_f}{-2m_0\omega'} - 2\epsilon'^* \cdot \epsilon \right) \\ &= -2\mathrm{i}e^2 \varepsilon^*(\boldsymbol{k}', \lambda') \cdot \varepsilon(\boldsymbol{k}, \lambda) \quad . \end{aligned} \tag{11}$$

With the help of the gauge (10) the first two terms could be dropped, because $\epsilon \cdot p_i = (0, \boldsymbol{\epsilon}) \cdot (m_0, \mathbf{0}) = 0$ and $\epsilon' \cdot p_i = (0, \boldsymbol{\epsilon}\,') \cdot (m_0, \mathbf{0}) = 0$. Thus the Compton scattering of spin-0 bosons is completely described just by the graph (c)! The cross section from (9) reads

$$\frac{d\sigma}{d\Omega_{k'}} = \frac{e^4}{m_0^2} \frac{\omega'^2}{\omega^2} \left| \varepsilon^*(\boldsymbol{k}', \lambda') \cdot \varepsilon(\boldsymbol{k}, \lambda) \right|^2 \quad . \tag{12}$$

If one does not observe the polarization of the photons, one has to average over λ and to sum over λ'. According to Section 3.7, (3.286) we have

$$\frac{1}{2} \sum_{\lambda, \lambda'} \left| \varepsilon^*(\boldsymbol{k}', \lambda') \cdot \varepsilon(\boldsymbol{k}, \lambda) \right|^2 = \frac{1}{2}(1 + \cos^2\theta) \quad , \tag{13}$$

where θ is the angle between $\boldsymbol{k}$ and $\boldsymbol{k}'$. The unpolarized Compton cross section then becomes

$$\frac{d\sigma^{\text{unpol}}}{d\Omega_{k'}} = \frac{1}{2} r_0^2 \frac{\omega'^2}{\omega^2} \left(1 + \cos^2\theta\right) \quad , \tag{14}$$

with the classical electromagnetic radius of the particle $r_0 = e^2/m_0c^2$. This result is a bit simpler than that obtained for Compton scattering at spin-1/2 particles. The Klein-Nishina formula of Section 3.7 is recovered by replacing

$$1 + \cos^2\theta \longrightarrow \left(\frac{\omega'}{\omega} + \frac{\omega}{\omega'} - 1 \right) + \cos^2\theta$$

in (14). In the limit of low photon energies where $\omega' \simeq \omega$ the cross section thus is independent of the spin of the particle.

EXERCISE

8.2 The Electro-Production of Pion Pairs

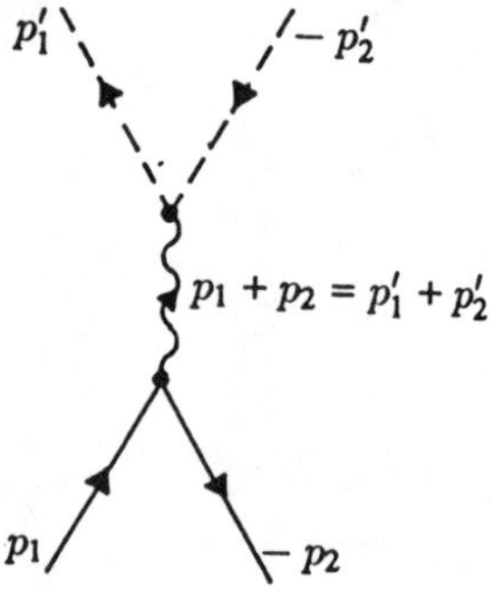

Fig. 8.7. The lowest-order graph for the electromagnetic annihilation of an electron–positron pair into a pair of spinless charged bosons

Problem. Calculate the cross section for the process $e^+ + e^- \to \pi^+ + \pi^-$, in the course of which an electron-positron pair annihilates into a virtual photon, which then decays into a pair of charged pions. The cross section was measured in accelerator experiments and has a maximum of $\sigma \simeq 1.4 \times 10^{-30}\mathrm{cm}^2$ at the energy $E_{\mathrm{tot}} = 770$ MeV (i.e. $E = 385$ MeV for particles and antiparticles if they collide with equal energy). Compare this with the theoretical result.

Solution. In lowest order the graph of Fig. 8.7 has to be calculated, where p_1 and p_2 denote the 4-momenta of the incoming electron and positron and p_1' (p_2') those of the π^- (π^+). According to Chap. 4 the cross section is given by

$$d\sigma = \frac{1}{4\sqrt{(p_1\cdot p_2)^2 - m_0^4}}(2m_0)^2(2\pi)^4\,\delta^4(p_1+p_2-p_1'-p_2')|M_{fi}|^2 \times \frac{d^3p_1'}{2E_1'(2\pi)^3}\,\frac{d^3p_2'}{2E_2'(2\pi)^3} \quad . \tag{1}$$

Now we proceed with the calculation in the centre-of-mass system (which is identical with the laboratory system in experiments with storage-ring colliders) so that we have

$$p_1 = (E, \boldsymbol{p}) \quad , \quad p_2 = (E, -\boldsymbol{p}) \quad , \tag{2a}$$
$$p_1' = (E, \boldsymbol{p}') \quad , \quad p_2' = (E, -\boldsymbol{p}') \quad . \tag{2b}$$

The flux factor is

$$\sqrt{(p_1\cdot p_2)^2 - m_0^4} = \sqrt{(E^2+\boldsymbol{p}^2) - m_0^4} = 2E|\boldsymbol{p}| \quad . \tag{3}$$

For the calculation of the invariant matrix element M_{fi} we employ the Feynman rules for scalar and spinor particles:

$$\begin{aligned} M_{fi} &= -(-\mathrm{i}e)(p_1'-p_2')_\mu\,\mathrm{i}D_{\mathrm{F}}^{\mu\nu}(p_1+p_2)\overline{v}(p_2,s_2)(-\mathrm{i}e\gamma_\nu)u(p_1,s_1) \\ &= \mathrm{i}e^2\frac{4\pi}{(p_1+p_2)^2}\,\overline{v}(p_2,s_2)(\not{p}_1'-\not{p}_2')u(p_1,s_1) \quad . \end{aligned} \tag{4}$$

With the total momentum

$$q = p_1 + p_2 = (2E, \boldsymbol{0}) \tag{5}$$

the differential cross section, averaged over the initial spins s_1 and s_2, now reads

$$\begin{aligned} d\overline{\sigma} &= \frac{1}{4}\sum_{s_1 s_2}\frac{m_0^2}{2E^3|\boldsymbol{p}|}\delta^4(p_1+p_2-p_1'-p_2')\frac{e^4}{q^4} \\ &\quad\times\left|\overline{v}(p_2,s_2)(\not{p}_1'-\not{p}_2')u(p_1,s_1)\right|^2 d^3p_1'\,d^3p_2' \\ &= \frac{e^4}{q^4}\frac{m_0^2}{2E^3|\boldsymbol{p}|}\delta^4(p_1+p_2-p_1'-p_2')\,S\,d^3p_1'\,d^3p_2' \quad . \end{aligned} \tag{6}$$

Exercise 8.2.

The calculation of the spin sum

$$S = \frac{1}{4}\sum_{s_1 s_2} \left|\overline{v}(p_2,s_2)(\not{p}'_1 - \not{p}'_2)u(p_1,s_1)\right|^2 \tag{7}$$

is simplified if one rewrites

$$\begin{aligned} &\overline{v}(p_2,s_2)(\not{p}'_1 - \not{p}'_2)u(p_1,s_1) \\ &\quad = \overline{v}(p_2,s_2)\left[(\not{p}'_1 + \not{p}'_2) - 2\not{p}'_2\right]u(p_1,s_1) \\ &\quad = \overline{v}(p_2,s_2)(\not{p}_1 + \not{p}_2)u(p_1,s_1) - 2\,\overline{v}(p_2,s_2)\not{p}'_2 u(p_1,s_1) \quad . \end{aligned} \tag{8}$$

The first term of the sum vanishes because of

$$\begin{aligned} &\not{p}_1 u(p_1,s_1) = m_0 u(p_1,s_1) \quad , \\ &\overline{v}(p_2,s_2)\not{p}_2 = -m_0 \overline{v}(p_2,s_2) \quad . \end{aligned}$$

According to the rules from Chap. 3 the spin sum can be rewritten as a trace:

$$\begin{aligned} S &= \sum_{s_1 s_2} \left(\overline{u}(p_1,s_1)\not{p}'_2 v(p_2,s_2)\right)\left(\overline{v}(p_2,s_2)\not{p}'_2 u(p_1,s_1)\right) \\ &= -\mathrm{Tr}\left(\not{p}'_2 \frac{-\not{p}_2 + m_0}{2m_0}\not{p}'_2 \frac{\not{p}_1 + m_0}{2m_0}\right) \quad . \end{aligned} \tag{9}$$

The evaluation of this trace according to the rules from Section 3 yields

$$\begin{aligned} S &= \frac{1}{4m_0^2}\left(\mathrm{Tr}\,\not{p}'_2\not{p}_2\not{p}'_2\not{p}_1 - m_0^2\,\mathrm{Tr}\,\not{p}'_2\not{p}'_2\right) \\ &= \frac{1}{m_0^2}\left(2p_1\cdot p'_2\, p_2\cdot p'_2 - M_0^2 p_1\cdot p_2 - m_0^2 M_0^2\right) \quad , \end{aligned} \tag{10}$$

where m_0 is the rest mass of the electron and M_0 that of the pion. So in the centre-of-mass system we obtain with (2) and the mass-shell constraints $E^2 = |\boldsymbol{p}|^2 + m_0^2, E^2 = |\boldsymbol{p}'|^2 + M_0^2$

$$\begin{aligned} S &= \frac{1}{m_0^2}\left[2(E^2 + \boldsymbol{p}\cdot\boldsymbol{p}')(E^2 - \boldsymbol{p}\cdot\boldsymbol{p}') - M_0^2(E^2 + |\boldsymbol{p}|^2) - m_0^2 M_0^2\right] \\ &= \frac{2}{m_0^2}\left[E^2|\boldsymbol{p}'|^2 - (\boldsymbol{p}\cdot\boldsymbol{p}')^2\right] \quad . \end{aligned} \tag{11}$$

With that the total cross section (6) reads

$$\begin{aligned} \overline{\sigma} &= \frac{e^4}{q^4}\frac{1}{E^3|\boldsymbol{p}|}\int d^3p'_1 d^3p'_2\, \delta^4(p_1 + p_2 - p'_1 - p'_2)\left[E^2|\boldsymbol{p}'|^2 - (\boldsymbol{p}\cdot\boldsymbol{p}')^2\right] \\ &= \frac{e^4}{q^4}\frac{1}{E^3|\boldsymbol{p}|}\int |\boldsymbol{p}'|^2 d|\boldsymbol{p}'|d\phi' d\cos\theta' \delta(2E - 2E') \\ &\quad \times \left(E^2|\boldsymbol{p}'|^2 - |\boldsymbol{p}|^2|\boldsymbol{p}'|^2\cos^2\theta'\right) \quad . \end{aligned} \tag{12}$$

The δ function describing energy conservation can be rewritten as

$$\delta(2E - 2E') = \frac{1}{2}\frac{E}{|\boldsymbol{p}'|}\delta\left(|\boldsymbol{p}'| - \sqrt{E^2 - M_0^2}\right) \quad . \tag{13}$$

Exercise 8.2.

Providing the threshold condition $L \geq M_0^2$ is met (12) becomes

$$\overline{\sigma} = \frac{e^4}{q^4}\frac{1}{E^3|\boldsymbol{p}|}2\pi\frac{1}{2}\frac{E}{|\boldsymbol{p}'|}|\boldsymbol{p}'|^4\int_{-1}^{+1} d\cos\theta'(E^2 - |\boldsymbol{p}|^2\cos^2\theta')$$

$$= \frac{\pi e^4}{q^4}\frac{|\boldsymbol{p}'|^3}{E^2|\boldsymbol{p}|}\left(2E^2 - \frac{2}{3}|\boldsymbol{p}|^2\right)$$

$$= \frac{\pi\alpha^2}{3q^2}\frac{|\boldsymbol{p}'|^3}{E^3}\frac{E}{|\boldsymbol{p}|}\left(1 + \frac{1}{2}\frac{m_0^2}{E^2}\right) \quad . \tag{14}$$

As the electron mass m_0 is negligible compared with the incident energy E, the exact result (14) can be simplified to

$$\overline{\sigma} \simeq \frac{\pi\alpha^2}{3q^2}\beta'^3 = \frac{\pi\alpha^2\beta'^3}{12E^2} \quad , \tag{15}$$

where $\beta' = |\boldsymbol{p}'|/E$ is the velocity of the pions.

In order to calculate the value of $\overline{\sigma}$ at the given energy $E = 385$ MeV, we have to insert the appropriate powers of $\hbar$ and c into (15). Obviously one has to multiply by $\hbar^2c^2 \simeq (197\ \mathrm{MeV}\cdot\mathrm{fm})^2$ so that $\overline{\sigma}$ gets the dimension of an area:

$$\overline{\sigma} = \frac{\pi\alpha^2\beta'^3}{12E^2}\hbar^2c^2$$

$$\simeq \frac{\pi\cdot 0.932^3\cdot 197^2}{12\cdot 137^2\cdot 385^2}\mathrm{fm}^2 = 3\times 10^{-32}\mathrm{cm}^2 \quad , \tag{16}$$

with the pion velocity $\beta' = \sqrt{E^2 - M_0^2}/E \simeq 0.932$. So the measured cross section for pion pairs is much greater than what the result (14) predicts, almost by a factor 50.

It is evident that the assumption of structureless particles that interact only by virtue of the electromagnetic field is not justified. The reason for the large cross section at the given energy is that the virtual photon is first converted into a ρ^0 meson with the same quantum numbers (Spin 1, negative parity), which then can decay into pions or take part in other processes involving the strong interaction. This property of virtual photons is described by the "vector dominance" model. The "resonance energy" is equal to the mass of this particle, which amounts to $M_\rho = 770$ MeV for the vector meson ρ^0. In fact, a whole family of shortliving mesons was observed as resonances in the cross section for e^+e^- annihilation. The purely electromagnetic production mechanism for pions only plays the role of a background process.

Appendix

In this appendix we collect a number of bibliographic references for the interested reader who either wants to learn more about Quantum Electrodynamics or is interested in the imbedding of QED in the more general framework of Quantum Field Theory.

1. Books which contain details on the formulation of QED and the calculation of various processes:

- A. Akhiezer and V.B. Berestetskii: *Quantum Electrodynamics*, Interscience, New York (1965)
- J.M. Jauch and F. Rohrlich: *The Theory of Photons and Electrons*, Springer-Verlag, New York, Heidelberg, Berlin (1976)
- G. Källen: *Quantum Electrodynamics*, Springer-Verlag, Berlin (1972)
- I. Bialynicki-Birula and Z. Bialynicka-Birula: *Quantum Electrodynamics*, Pergamon, Oxford (1975)
- V.B. Berestetzkii, E.M. Lifshitz, and L.P. Pitaevskii: *Relativistic Quantum Theory*, Pergamon Press, Oxford (1971)

2. Two books covering topics related to QED with strong external fields:

- W. Greiner, B. Müller, J. Rafelski: *Quantum Electrodynamics of Strong Fields*, Springer-Verlag, Berlin (1985)
- V.L. Ginzburg (ed.): *Issues in Intense-Field Quantum Electrodynamics*, Nova Science Publ., Commack, N.Y. (1987)

3. The most recent information on the status of QED experiments contrasted with theory has to be extracted from original research papers and from review articles published in conference proceedings. In addition the following book provides a good overview:

- T. Kinoshita (ed.): *Quantum Electrodynamics*, World Scientific, Singapore (1990)

4. An old but still useful collection of reprints of many of the basic original papers related to QED:

- J. Schwinger (ed.): *Quantum Electrodynamics*, Dover, New York (1958)

5. Classical textbooks on Quantum Field Theory, in chronological order:

- N.N. Bogoliubov and D.V. Shirkov: *Introduction to the Theory of Quantized Fields*, Interscience, New York (1959)
- S.S. Schweber: *An Introduction to Relativistic Quantum Field Theory*, Harper & Row, New York (1962)
- J.D. Bjorken and S.D. Drell: *Relativistic Quantum Mechanics*, and *Relativistic Quantum Fields*, Mc.Graw-Hill, New York (1964)

• D. Lurié: *Particles and Fields*, Interscience, New York (1968)
• C. Itzykson, J.-B. Zuber: *Quantum Field Theory*, McGraw-Hill, New York (1980)

6. Some further references on quantum fields and gauge theories, which emphasize the path integral formulation:
• L. H. Ryder: *Quantum Field Theory*, Cambridge University Press, Cambridge (1985)
• D. Bailin and A. Love: *Introduction to Gauge Field Theory*, Adam Hilger, Bristol, Boston (1986)
• R.J. Rivers: *Path Integral Methods in Quantum Field Theory*, Cambridge University Press, Cambridge (1987)
• S. Pokorski: *Gauge Field Theories*, Cambridge University Press, Cambridge (1987)

Subject Index